STUDENT'S SOLUTIONS MANUAL

VIKTOR MAYMESKUL

University of South Florida

FUNDAMENTALS OF DIFFERENTIAL EQUATIONS

EIGHTH EDITION

and

FUNDAMENTALS OF DIFFERENTIAL EQUATIONS AND BOUNDARY VALUE PROBLEMS

SIXTH EDITION

R. Kent Nagle

Edward B. Saff

Vanderbilt University

Arthur David Snider

University of South Florida

PEARSON

Boston   San Francisco   New York
London   Toronto   Sydney   Tokyo   Singapore   Madrid
Mexico City   Munich   Paris   Cape Town   Hong Kong   Montreal

The author and publisher of this book have used their best efforts in preparing this book. These efforts include the development, research, and testing of the theories and programs to determine their effectiveness. The author and publisher make no warranty of any kind, expressed or implied, with regard to these programs or the documentation contained in this book. The author and publisher shall not be liable in any event for incidental or consequential damages in connection with, or arising out of, the furnishing, performance, or use of these programs.

Reproduced by Pearson from electronic files supplied by the author.

Copyright © 2012, 2008, 2004 Pearson Education, Inc.
Publishing as Pearson, 75 Arlington Street, Boston, MA 02116.

All rights reserved. No part of this publication may be reproduced, stored in a retrieval system, or transmitted, in any form or by any means, electronic, mechanical, photocopying, recording, or otherwise, without the prior written permission of the publisher. Printed in the United States of America.

ISBN-13: 978-0-321-74834-8
ISBN-10: 0-321-74834-4

8 V036 16 15 14

www.pearsonhighered.com

PEARSON

Contents

Copyright © 2012 Pearson Education, Inc. Publishing as Addison-Wesley.

Copyright © 2012 Pearson Education, Inc. Publishing as Addison-Wesley.

CHAPTER 8: Series Solutions of Differential Equations   457

CHAPTER 9: Matrix Methods for Linear Systems   523

CHAPTER 10: Partial Differential Equations   577

CHAPTER 11: Eigenvalue Problems and Sturm-Liouville Eguations   621

Copyright © 2012 Pearson Education, Inc. Publishing as Addison-Wesley.

CHAPTER 1: Introduction

EXERCISES 1.1:   Background

1. This equation is an ODE because it contains no partial derivatives. Since the highest order derivative is d^2y/dx^2, the equation is a second order equation. This same term also shows us that the independent variable is x and the dependent variable is y. This equation is linear.

3. This equation is a PDE of the second order because it contains second partial derivatives. x and y are independent variables, and u is the dependent variable.

5. This equation is an ODE of the first order with the independent variable t and the dependent variable x. It is nonlinear.

7. ODE of the second order with the independent variable x and the dependent variable y, nonlinear.

9. ODE of the fourth order with the independent variable x and the dependent variable y, linear.

11. This equation contains partial derivatives, thus it is a PDE. Since the highest partial derivative is of the second order, the given equation is a second order equation. The terms $\partial N/\partial t$ and $\partial N/\partial r$ show that the independent variables are t and r and the dependent variable is N.

13. The rate of change of a quantity means its derivative. Denoting the proportionality coefficient between dp/dt and $p(t)$ by k ($k > 0$), we get

$$\frac{dp}{dt} = kp.$$

15. In this problem, $T \geq M$ (coffee is hotter than the air), and T is a decreasing function of t, that is $dT/dt \leq 0$. Thus,

$$\frac{dT}{dt} = k(M - T),$$

Copyright © 2012 Pearson Education, Inc. Publishing as Addison-Wesley.

Chapter 1

where $k > 0$ is the proportionality constant.

17. In classical physics, the instantaneous acceleration a of an object moving in a straight line is given by the second derivative of distance x with respect to time t; that is,

$$\frac{d^2x}{dt^2} = a.$$

Integrating both sides with respect to t and using the given fact that a is constant, we obtain

$$\frac{dx}{dt} = at + C. \tag{1.1}$$

The instantaneous velocity v of an object is given by the first derivative of distance x with respect to time t. At the beginning of the race ($t = 0$) both racers have zero velocity. Therefore, we have $C = 0$. Integrating equation (1.1) with respect to t yields

$$x = \frac{1}{2}at^2 + C_1.$$

We let the starting position for both competitors to be $x = 0$ at $t = 0$. Therefore, $C_1 = 0$. This gives us a general equation used for both racers as

$$x = \frac{1}{2}at^2 \qquad \text{or} \qquad t = \sqrt{\frac{2x}{a}},$$

where the acceleration constant a has different values for Kevin and for Alison. Alison covers the last $1/4$ of the full distance L in 3 seconds. This means Alison's acceleration a_A is determined by

$$t_A - t_{3/4} = 3 = \sqrt{\frac{2L}{a_A}} - \sqrt{\frac{2(3L/4)}{a_A}},$$

where t_A is the time it takes for Alison to finish the race. Solving this equation for a_A gives,

$$a_A = \frac{\left(\sqrt{2} - \sqrt{3/2}\right)^2}{9}L.$$

Therefore, the time required for Alison to finish the race is given by

$$t_A = \sqrt{\frac{2L}{\left(\sqrt{2} - \sqrt{3/2}\right)^2 (L/9)}} = \frac{3}{\sqrt{2} - \sqrt{3/2}}\sqrt{2} = 12 + 6\sqrt{3} \approx 22.39 \,(\text{sec}).$$

Kevin covers the last $1/3$ of the distance L in 4 seconds. This means Kevin's acceleration a_K is given by

$$t_K - t_{2/3} = 4 = \sqrt{\frac{2L}{a_K}} - \sqrt{\frac{2(2L/3)}{a_K}},$$

     Copyright © 2012 Pearson Education, Inc. Publishing as Addison-Wesley.

where t_K is the time required for Kevin to finish the race. Solving this equation for a_K gives

$$a_K = \frac{\left(\sqrt{2} - \sqrt{4/3}\right)^2}{16} L.$$

Therefore, the time required for Kevin to finish the race is given by

$$t_K = \sqrt{\frac{2L}{\left(\sqrt{2} - \sqrt{4/3}\right)^2 (L/16)}} = \frac{4}{\sqrt{2} - \sqrt{4/3}} \sqrt{2} = 12 + 4\sqrt{6} \approx 21.80 \ (\text{sec}).$$

The time required for Kevin to finish the race is less than that for Alison; therefore, Kevin wins the race by $6\sqrt{3} - 4\sqrt{6} \approx 0.594$ seconds.

EXERCISES 1.2: Solutions and Initial Value Problems

1. (a) Writing the given equation in the form $y^2 = 3 - x$, we see that it defines two functions of x on $x \le 3$, $y = \pm\sqrt{3 - x}$. Differentiation yields

$$\begin{aligned}
\frac{dy}{dx} &= \frac{d}{dx}\left(\pm\sqrt{3 - x}\right) = \pm\frac{d}{dx}\left[(3 - x)^{1/2}\right] \\
&= \pm\frac{1}{2}(3 - x)^{-1/2}(-1) = -\frac{1}{\pm 2\sqrt{3 - x}} = -\frac{1}{2y}.
\end{aligned}$$

(b) Solving for y yields

$$y^3(x - x\sin x) = 1 \quad \Rightarrow \quad y^3 = \frac{1}{x(1 - \sin x)}$$

$$\Rightarrow \quad y = \frac{1}{\sqrt[3]{x(1 - \sin x)}} = [x(1 - \sin x)]^{-1/3}.$$

The domain of this function is $x \ne 0$ and

$$\sin x \ne 1 \quad \Rightarrow \quad x \ne \frac{\pi}{2} + 2k\pi, \quad k = 0, \pm 1, \pm 2, \dots.$$

For $0 < x < \pi/2$, one has

$$\begin{aligned}
\frac{dy}{dx} &= \frac{d}{dx}\left\{[x(1 - \sin x)]^{-1/3}\right\} = -\frac{1}{3}[x(1 - \sin x)]^{-1/3 - 1}\frac{d}{dx}[x(1 - \sin x)] \\
&= -\frac{1}{3}[x(1 - \sin x)]^{-1}[x(1 - \sin x)]^{-1/3}[(1 - \sin x) + x(-\cos x)] \\
&= \frac{(x\cos x + \sin x - 1)y}{3x(1 - \sin x)}.
\end{aligned}$$

We also remark that the given relation is an implicit solution on any interval not containing points $x = 0, \pi/2 + 2k\pi, \ k = 0, \pm 1, \pm 2, \dots.$

Copyright © 2012 Pearson Education, Inc. Publishing as Addison-Wesley.

3. Differentiating the function $x = 2\cos t - 3\sin t$ twice, we obtain

$$x' = -2\sin t - 3\cos t, \quad x'' = -2\cos t + 3\sin t.$$

Thus,

$$x'' + x = (-2\cos t + 3\sin t) + (2\cos t - 3\sin t) = 0$$

for any t on $(-\infty, \infty)$. Thus, the answer is "Yes".

5. Substituting $x = \cos 2t$ and $x' = -2\sin 2t$ into the given equation yields

$$(-2\sin 2t) + t\cos 2t = \sin 2t \qquad \Leftrightarrow \qquad t\cos 2t = 3\sin 2t\,.$$

Clearly, this is not an identity and, therefore, the function $x = \cos 2t$ is not a solution.

7. Using the chain rule, we have

$$y = 3\sin 2x + e^{-x},$$

$$y' = 3(\cos 2x)(2x)' + e^{-x}(-x)' = 6\cos 2x - e^{-x},$$

$$y'' = 6(-\sin 2x)(2x)' - e^{-x}(-x)' = -12\sin 2x + e^{-x}.$$

Therefore,

$$y'' + 4y = \left(-12\sin 2x + e^{-x}\right) + 4\left(3\sin 2x + e^{-x}\right) = 5e^{-x},$$

which is the right-hand side of the given equation. So, $y = 3\sin 2x + e^{-x}$ is a solution.

9. Taking derivatives of both sides of the given relation with respect to x yields

$$\frac{d}{dx}\left(y - \ln y\right) = \frac{d}{dx}\left(x^2 + 1\right) \qquad \Rightarrow \qquad \frac{dy}{dx} - \frac{1}{y}\frac{dy}{dx} = 2x$$

$$\Rightarrow \qquad \frac{dy}{dx}\left(1 - \frac{1}{y}\right) = 2x \qquad \Rightarrow \qquad \frac{dy}{dx}\frac{y-1}{y} = 2x \qquad \Rightarrow \qquad \frac{dy}{dx} = \frac{2xy}{y-1}\,.$$

Thus, the relation $y - \ln y = x^2 + 1$ is an implicit solution to the equation $y' = 2xy/(y-1)$.

11. Differentiating the equation $e^{xy} + y = x - 1$ implicitly with respect to x yields

$$\frac{d}{dx}\left(e^{xy} + y\right) = \frac{d}{dx}\left(x - 1\right) \qquad \Rightarrow \qquad e^{xy}\frac{d}{dx}\left(xy\right) + \frac{dy}{dx} = 1$$

$$\Rightarrow \qquad e^{xy}\left(y + x\frac{dy}{dx}\right) + \frac{dy}{dx} = 1 \qquad \Rightarrow \qquad ye^{xy} + \frac{dy}{dx}\left(xe^{xy} + 1\right) = 1$$

$$\Rightarrow \qquad \frac{dy}{dx} = \frac{1 - ye^{xy}}{1 + xe^{xy}} = \frac{e^{xy}\left(e^{-xy} - y\right)}{e^{xy}\left(e^{-xy} + x\right)} = \frac{e^{-xy} - y}{e^{-xy} + x}\,.$$

Therefore, the function $y(x)$ defined by $e^{xy} + y = x - 1$ is an implicit solution to the given differential equation.

 Copyright © 2012 Pearson Education, Inc. Publishing as Addison-Wesley.

13. Differentiating the equation $\sin y + xy - x^3 = 2$ implicitly with respect to x, we obtain

$$y' \cos y + xy' + y - 3x^2 = 0$$

$$\Rightarrow \quad (\cos y + x)y' = 3x^2 - y \qquad \rightarrow \qquad y' = \frac{3x^2 - y}{\cos y + x}.$$

Differentiating the second equation above again, we obtain

$$(-y' \sin y + 1)y' + (\cos y + x)y'' = 6x - y'$$

$$\Rightarrow \quad (\cos y + x)y'' = 6x - y' + (y')^2 \sin y - y' = 6x - 2y' + (y')^2 \sin y$$

$$\Rightarrow \quad y'' = \frac{6x - 2y' + (y')^2 \sin y}{\cos y + x}.$$

Multiplying the right-hand side of this last equation by $y'/y' = 1$ and using the fact that

$$y' = \frac{3x^2 - y}{\cos y + x},$$

we get

$$\begin{aligned} y'' &= \frac{6x - 2y' + (y')^2 \sin y}{\cos y + x} \cdot \frac{y'}{(3x^2 - y)/(\cos y + x)} \\ &= \frac{6xy' - 2(y')^2 + (y')^3 \sin y}{3x^2 - y}. \end{aligned}$$

Thus, y is an implicit solution to the differential equation.

15. Differentiating $\phi(x)$, we find that

$$\begin{aligned} \phi'(x) &= \left(\frac{2}{1 - ce^x} \right)' = \left[2 \left(1 - ce^x \right)^{-1} \right]' \\ &= 2(-1) \left(1 - ce^x \right)^{-2} \left(1 - ce^x \right)' = 2ce^x \left(1 - ce^x \right)^{-2}. \end{aligned} \qquad (1.2)$$

On the other hand, substitution of $\phi(x)$ for y into the right-hand side of the given equation yields

$$\begin{aligned} \frac{\phi(x)(\phi(x) - 2)}{2} &= \frac{1}{2} \frac{2}{1 - ce^x} \left(\frac{2}{1 - ce^x} - 2 \right) \\ &= \frac{2}{1 - ce^x} \left(\frac{1}{1 - ce^x} - 1 \right) = \frac{2}{1 - ce^x} \frac{1 - (1 - ce^x)}{1 - ce^x} = \frac{2ce^x}{(1 - ce^x)^2}, \end{aligned}$$

which is identical to $\phi'(x)$ found in (1.2).

17. We differentiate $\phi(x)$ and substitute ϕ and ϕ' into the differential equation for y and y'. This yields

$$\phi(x) = Ce^{3x} + 1 \qquad \Rightarrow \qquad \frac{d\phi(x)}{dx} = \left(Ce^{3x} + 1 \right)' = 3Ce^{3x}$$

Copyright © 2012 Pearson Education, Inc. Publishing as Addison-Wesley.

and

$$\frac{d\phi}{dx} - 3\phi = (3Ce^{3x}) - 3\left(Ce^{3x} + 1\right) = (3C - 3C)e^{3x} - 3 = -3,$$

which holds for any constant C and any x on $(-\infty, \infty)$. Therefore, $\phi(x) = Ce^{3x} + 1$ is a one-parameter family of solutions to $y' - 3y = -3$ on $(-\infty, \infty)$. Graphs of these functions for $C = 0, \pm 0.5, \pm 1$, and ± 2 are sketched in Figure **1–A**.

19. Since, for any real-valued differentiable function $y(x)$, $(dy/dx)^2 + y^2 \geq 0$, we have $(dy/dx)^2 + y^2 + 4 \geq 4$. Hence, it can never be zero.

21. For $\phi(x) = x^m$, we have

$$\phi'(x) = mx^{m-1}, \qquad \phi''(x) = m(m-1)x^{m-2}.$$

(a) Substituting these expressions into the differential equation

$$3x^2 y'' + 11xy' - 3y = 0$$

gives

$$3x^2 \left[m(m-1)x^{m-2}\right] + 11x \left[mx^{m-1}\right] - 3x^m = 0$$

$$\Rightarrow \quad 3m(m-1)x^m + 11mx^m - 3x^m = 0$$

$$\Rightarrow \quad [3m(m-1) + 11m - 3] x^m = 0 \qquad \Rightarrow \qquad \left[3m^2 + 8m - 3\right] x^m = 0.$$

For the last equation to hold on an interval for x, we must have

$$3m^2 + 8m - 3 = (3m - 1)(m + 3) = 0.$$

Thus, either $(3m - 1) = 0$ or $(m + 3) = 0$, which yields $m = 1/3, -3$.

(b) Substituting the expressions for $\phi(x)$, $\phi'(x)$, and $\phi''(x)$ into the differential equation, $x^2 y'' - xy' - 5y = 0$, gives

$$x^2 \left[m(m-1)x^{m-2}\right] - x \left[mx^{m-1}\right] - 5x^m = 0 \qquad \Rightarrow \qquad \left[m^2 - 2m - 5\right] x^m = 0.$$

For the last equation to hold on an interval for x, we must have $m^2 - 2m - 5 = 0$. To solve for m we use the quadratic formula:

$$m = \frac{2 \pm \sqrt{4 + 20}}{2} = 1 \pm \sqrt{6}.$$

Copyright © 2012 Pearson Education, Inc. Publishing as Addison-Wesley.

23. In this problem, $f(x,y) = y^4 - x^4$ and so

$$\frac{\partial f}{\partial y} = \frac{\partial\left(y^4 - x^4\right)}{\partial y} = 4y^3.$$

Clearly, f and $\partial f/\partial y$ (being polynomials) are continuous on the whole xy-plane. Thus the hypotheses of Theorem 1 are satisfied, and the initial value problem has a unique solution for *any* initial data, in particular, for $y(0) = 7$.

25. Writing

$$\frac{dx}{dt} = -\frac{4t}{3x} = -(4/3)tx^{-1},$$

we see that $f(t,x) = -(4/3)tx^{-1}$ and $\partial f(t,x)/\partial x = \partial[-(4/3)tx^{-1}]/\partial x = (4/3)tx^{-2}$. The functions $f(t,x)$ and $\partial f(t,x)/\partial x$ are not continuous only when $x = 0$. Therefore, they are continuous in any rectangle R that contains the point $(2, -\pi)$, but does not intersect the t-axis; for instance, $R = \{(t,x): 1 < t < 3, \; -2\pi < x < 0\}$. Thus, Theorem 1 applies, and the given initial problem has a unique solution.

27. Rewriting the differential equation as $dy/dx = x/y$, we conclude that $f(x,y) = x/y$. Since f is not continuous when $y = 0$, there is no rectangle containing the point $(1,0)$ in which f is continuous. Therefore, Theorem 1 cannot be applied.

29. **(a)** Clearly, both functions $\phi_1(x) \equiv 0$ and $\phi_2(x) = (x-2)^3$ satisfy the initial condition, $y(2) = 0$. We check that they also satisfy the differential equation $dy/dx = 3y^{2/3}$.

$$\frac{d\phi_1}{dx} = \frac{d}{dx}\,(0) = 0 = 3\phi_1(x)^{2/3}\,;$$

$$\frac{d\phi_2}{dx} = \frac{d}{dx}\,\left[(x-2)^3\right] = 3(x-2)^2 = 3\left[(x-2)^3\right]^{2/3} = 3\phi_2(x)^{2/3}\,.$$

Hence, both functions, $\phi_1(x)$ and $\phi_2(x)$, are solutions to the initial value problem of Example 9.

(b) In this initial value problem,

$$f(x,y) = 3y^{2/3} \qquad \Rightarrow \qquad \frac{\partial f(x,y)}{\partial y} = 3\,\frac{2}{3}\,y^{2/3-1} = \frac{2}{y^{1/3}}\,,$$

$x_0 = 0$ and $y_0 = 10^{-7}$. The function $f(x,y)$ is continuous everywhere; $\partial f(x,y)/\partial y$ is continuous in any region which does not intersect the x-axis (where $y = 0$). In particular, both functions, $f(x,y)$ and $\partial f(x,y)/\partial y$, are continuous in the rectangle

$$R = \left\{(x,y): -1 < x < 1, \; (1/2)10^{-7} < y < (2)10^{-7}\right\}$$

Copyright © 2012 Pearson Education, Inc. Publishing as Addison-Wesley.

containing the initial point $(0, 10^{-7})$. Thus, it follows from Theorem 1 that the given initial value problem has a unique solution in an interval about $x_0 = 0$.

31. (a) To apply Theorem 1, we first must write the equation in the form $y' = f(x, y)$. Here, $f(x, y) = 4xy^{-1}$ and $\partial f(x, y)/\partial y = -4xy^{-2}$. Neither f nor $\partial f/\partial y$ are continuous or defined when $y = 0$. Therefore, there is no rectangle, containing $(x_0, 0)$, in which both f and $\partial f/\partial y$ are continuous. So, Theorem 1 cannot be applied.

(b) Suppose for the moment that there is such a solution $y(x)$ with $y(x_0) = 0$ and $x_0 \neq 0$. Substituting into the differential equation, we get

$$y(x_0)y'(x_0) - 4x_0 = 0,$$

which implies that

$$0 \cdot y'(x_0) - 4x_0 = 0 \qquad \Rightarrow \qquad 4x_0 = 0.$$

Thus, $x_0 = 0$ – a contradiction.

(c) Taking $C = 0$ in the implicit solution $4x^2 - y^2 = C$ given in Example 5 yields $4x^2 - y^2 = 0$ or $y = \pm 2x$. Both solutions, $y = 2x$ and $y = -2x$, satisfy $y(0) = 0$.

EXERCISES 1.3: Direction Fields

1. (a) Starting from the initial point $(0, -2)$ and following the direction markers we get the curve shown in Fig. **1–B** on page 21.

Thus, the solution curve to the initial value problem $dy/dx = 2x + y$, $y(0) = -2$, is the line with slope

$$\frac{dy}{dx}(0) = (2x + y)|_{x=0} = y(0) = -2$$

and y-intercept $y(0) = -2$. Using the slope-intercept form of an equation of a line, we get $y = -2x - 2$.

(b) This time, we start from the point $(-1, 3)$ and obtain the curve shown in Fig. **1–C** on page 21.

(c) From Fig. **1–C**, we conclude that

$$\lim_{x \to \infty} y(x) = \infty, \qquad \lim_{x \to -\infty} y(x) = \infty.$$

Copyright © 2012 Pearson Education, Inc. Publishing as Addison-Wesley.

3. From Figure B.3 in the answers section of the text, we conclude that, regardless of the initial velocity, $v(0)$, the corresponding solution curve $v = v(t)$ has the line $v = 8$ as a horizontal asymptote, that is, $\lim_{t\to\infty} v(t) = 8$. This explains the name "terminal velocity" for the value $v = 8$.

5. (a) The graph of the directional field is shown in Figure B.4 in the answers of the text.

   (b), (c) The direction field indicates that all solution curves (other than $p(t) \equiv 0$) will approach the horizontal asymptote $p = 1.5$ as $t \to +\infty$. Thus, $\lim_{t\to+\infty} p(t) = 1.5$.

   (d) No. The direction field shows that populations greater than 1500 will steadily decrease, but can never reach 1500 or any smaller value, i.e., the solution curves cannot cross the line $p = 1.5$. Indeed, the constant function $p(t) \equiv 1.5$ is a solution to the given logistic equation, and the uniqueness part of Theorem 1 prevents intersections of solution curves.

7. (a) The graph of the directional field is shown in Figure B.5 in the answers of the text.

   (b) The direction field indicates that all solution curves with $p(0) > 1$ will approach the horizontal line (asymptote) $p = 2$ as $t \to +\infty$. Thus, $\lim_{t\to+\infty} p(t) = 2$ if $p(0) = 4$.

   (c) The direction field shows that an initial population between 1000 and 2000 (that is, $1 < p(0) < 2$) will approach the horizontal line $p = 2$ as $t \to +\infty$. Thus, $\lim_{t\to+\infty} p(t) = 2$ if $p(0) = 1.7$.

   (d) The direction field shows that an initial population less than 1000 $(0 \leq p(0) < 1)$ will approach zero as $t \to +\infty$. Thus, $\lim_{t\to+\infty} p(t) = 0$ if $p(0) = 0.8$.

   (e) As noted in part (d), the line $p = 1$ is an asymptote. The direction field indicates that a population of 900 $(p(0) = 0.9)$ steadily decreases with time and, therefore, cannot increase to 1100.

9. (a) The function $\phi(x)$, being a solution to the given initial value problem, satisfies

$$\frac{d\phi}{dx} = x - \phi(x), \qquad \phi(0) = 1. \tag{1.3}$$

   Thus
$$\frac{d^2\phi}{dx^2} = \frac{d}{dx}\left(\frac{d\phi}{dx}\right) = \frac{d}{dx}(x - \phi(x)) = 1 - \frac{d\phi}{dx} = 1 - x + \phi(x),$$

   where we have used (1.3) substituting (twice) $x - \phi(x)$ for $d\phi/dx$.

Copyright © 2012 Pearson Education, Inc. Publishing as Addison-Wesley.

(b) First we note that any solution to the given differential equation on an interval I is continuously differentiable on I. Indeed, if $y(x)$ is a solution on I, then $y'(x)$ does exist on I, and so $y(x)$ is continuous on I because it is differentiable. This immediately implies that $y'(x)$ is continuous as it is the difference of two continuous functions, x and $y(x)$.

From (1.3) we conclude that

$$\frac{d\phi}{dx}\Big|_{x=0} = [x - \phi(x)]\big|_{x=0} = 0 - \phi(0) = -1 < 0$$

and so the continuity of $\phi'(x)$ implies that, for $|x|$ small enough, $\phi'(x) < 0$. Negative derivative of a function results that the function itself is decreasing.

When x increases from zero, as far as $\phi(x) > x$, one has $\phi'(x) < 0$ and so $\phi(x)$ decreases. On the other hand, the function $y = x$ increases unboundedly as $x \to \infty$. Thus, by intermediate value theorem, there is a point, say, $x^* > 0$, where the curve $y = \phi(x)$ crosses the line $y = x$. At this point, $\phi(x^*) = x^*$ and, hence, $\phi'(x^*) = x^* - \phi(x^*) = 0$.

(c) From (b) we conclude that x^* is a critical point for $\phi(x)$ (its derivative vanishes at this point). Also, from part (a), we see that

$$\phi''(x^*) = 1 - \phi'(x^*) = 1 > 0.$$

Hence, by second derivative test, $\phi(x)$ has a relative minimum at x^*.

(d) Remark that the arguments, used in part (c), can be applied to *any* point $\widetilde{x}$, where $\phi'(\widetilde{x}) = 0$, to conclude that $\phi(x)$ has a relative minimum at $\widetilde{x}$. Since a continuously differentiable function on an interval cannot have two relative minima on an interval without having a point of relative maximum, we conclude that x^* is the only point where $\phi'(x) = 0$. Continuity of $\phi'(x)$ implies that it has the same sign for all $x > x^*$ and, therefore, it is positive there since it is positive for $x > x^*$ and close to x^* ($\phi'(x^*) = 0$ and $\phi''(x^*) > 0$). Positive derivative makes $\phi(x)$ increasing for $x > x^*$. It asymptotically approaches the line $y = x - 1$. (See the part (e) below.)

(e) For $y = x - 1$, $dy/dx = 1$ and $x - y = x - (x - 1) = 1$. Thus the given differential equation is satisfied, and $y = x - 1$ is indeed a solution.

To show that the curve $y = \phi(x)$ always stays above the line $y = x - 1$, we note

 Copyright © 2012 Pearson Education, Inc. Publishing as Addison-Wesley.

that the initial value problem

$$\frac{dy}{dx} = x - y, \qquad y(x_0) = y_0 \tag{1.4}$$

has a unique solution for any x_0 and y_0. Indeed, functions $f(x, y) = x - y$ and $\partial f/\partial y \equiv -1$ are continuous on the whole xy-plane, and Theorem 1 applies. This implies that the curve $y = \phi(x)$ always stays above the line $y = x - 1$. Indeed,

$$\phi(0) = 1 > -1 = (x - 1)\big|_{x=0},$$

and the existence of a point $\widetilde{x}$ with $\phi(\widetilde{x}) \leq (\widetilde{x} - 1)$ would imply, by intermediate value theorem, the existence of a point x_0, $0 < x_0 \leq \widetilde{x}$, satisfying the equation $y_0 := \phi(x_0) = x_0 - 1$ and, therefore, there would be two solutions to the initial value problem (1.4).

Since, from part (a), $\phi''(x) = 1 - \phi'(x) = 1 - x + \phi(x) = \phi(x) - (x - 1) > 0$, we also conclude that $\phi'(x)$ is an increasing function and $\phi'(x) < 1$. Thus there exists $\lim_{x \to \infty} \phi'(x) \leq 1$. The strict inequality would imply that the values of the function $y = \phi(x)$, for x large enough, become smaller than those of $y = x - 1$. Therefore,

$$\lim_{x \to \infty} \phi'(x) = 1 \qquad \Leftrightarrow \qquad \lim_{x \to \infty} [x - \phi(x)] = 1,$$

and so the line $y = x - 1$ is a slant asymptote for $\phi(x)$.

(f), (g) The direction field for given differential equation and the curve $y = \phi(x)$ are shown in Figure B.6 in the answers of the text.

11. For the equation $\partial y/\partial x = -x/y$, the isoclines are the curves $-x/y = c$. These are lines that pass through the origin and have equations of the form $y = mx$, where $m = -1/c$, $c \neq 0$. If we let $c = 0$ in $-x/y = c$, we see that the y-axis ($x = 0$) is also an isocline. Each element of the direction field associated with a point on an isocline has slope c and is, therefore, perpendicular to that isocline. Since circles have the property that at any point on the circle the tangent at that point is perpendicular to a line from that point to the center of the circle, we see that the solution curves will be circles with their centers at the origin. But since we cannot have $y = 0$ (since $-x/y$ would then have a zero in the denominator) the solutions will not be defined on the x-axis. (Note, however, that a related form of this differential equation is $yy' + x = 0$. This equation has implicit solutions given by the equations $y^2 + x^2 = C$. These solutions will be circles.) The

Copyright © 2012 Pearson Education, Inc. Publishing as Addison-Wesley.

graph of $\phi(x)$, the solution to the equation satisfying the initial condition $y(0) = 4$, is the upper semicircle with center at the origin and passing through the point $(0, 4)$ (see Figure B.7 in the answers of the text).

13. For this equation, the isoclines are given by $2x = c$. These are vertical lines $x = c/2$. Each element of the direction field associated with a point on $x = c/2$ has slope c. (See Figure B.8 in the answers of the text.)

15. For the equation $dy/dx = 2x^2 - y$, the isoclines are the curves $2x^2 - y = c$, or $y = 2x^2 - c$. The curve $y = 2x^2 - c$ is a parabola which is open upward and has the vertex at $(0, -c)$. Three of them, for $c = -1$, 0, and 2 (dotted curves), as well as the solution curve, satisfying the initial condition $y(0) = 0$, are depicted in Figure B.9.

17. The isoclines for the equation

$$\frac{dy}{dx} = 3 - y + \frac{1}{x}$$

are given by

$$3 - y + \frac{1}{x} = c \qquad \Leftrightarrow \qquad y = \frac{1}{x} + 3 - c,$$

which are hyperbolas having $x = 0$ as a vertical asymptote and $y = 3 - c$ as a horizontal asymptote. Each element of the direction field associated with a point on such a hyperbola has slope c. For $x > 0$ large enough: if an isocline is located *above* the line $y = 3$, then $c \le 0$, and so the elements of the direction field have *negative* or *zero slope*; if an isocline is located *below* the line $y = 3$, then $c > 0$, and so the elements of the direction field have *positive slope*. In other words, for $x > 0$ large enough, at any point above the line $y = 3$ a solution curve decreases passing through this point, and any solution curve increases passing through a point below $y = 3$. The direction field for this differential equation is depicted in Figure **1–D** on page 21. From this picture we conclude that any solution to the differential equation $dy/dx = 3 - y + 1/x$ has the line $y = 3$ as a horizontal asymptote.

19. Integrating both sides of the equation $dy/y = -dx/x$ yields

$$\int \frac{1}{y}\, dy = -\int \frac{1}{x}\, dx$$
$$\Rightarrow \qquad \ln |y| = -\ln |x| + C_1$$
$$\Rightarrow \qquad |y| = e^{-\ln |x| + C_1} = e^{C_1} e^{-\ln |x|} = \frac{C_2}{|x|},$$

 Copyright © 2012 Pearson Education, Inc. Publishing as Addison-Wesley.

where C_1 is an arbitrary constant and so $C_2 := e^{C_1}$ is an arbitrary *positive* constant. The last equality can be written as $y = \pm C_2 x^{-1} = C x^{-1}$, where $C = \pm C_2$ is *any nonzero* constant. The value $C = 0$ gives $y \equiv 0$ (for $x \neq 0$), which is also a solution.

EXERCISES 1.4:   The Approximation Method of Euler

1. In this problem, $x_0 = 0$, $y_0 = 4$, $h = 0.1$, and $f(x, y) = -x/y$. Thus, the recursive formulas given in equations (2) and (3) of the text become

$$x_{n+1} = x_n + h = x_n + 0.1\,,$$

$$y_{n+1} = y_n + hf(x_n, y_n) = y_n + 0.1\left(-\frac{x_n}{y_n}\right), \qquad n = 0, 1, 2, \ldots .$$

To find an approximation for the solution at the point $x_1 = x_0 + 0.1 = 0.1$, we let $n = 0$ in the last recursive formula to find

$$y_1 = y_0 + 0.1\left(-\frac{x_0}{y_0}\right) = 4 + 0.1(0) = 4.$$

To approximate the value of the solution at the point $x_2 = x_1 + 0.1 = 0.2$, we let $n = 1$ in the last recursive formula to obtain

$$y_2 = y_1 + 0.1\left(-\frac{x_1}{y_1}\right) = 4 + 0.1\left(-\frac{0.1}{4}\right) = 4 - \frac{1}{400} = 3.9975 \approx 3.998\,.$$

Continuing in this way we find

$$x_3 = x_2 + 0.1 = 0.3, \quad y_3 = y_2 + 0.1\left(-\frac{x_2}{y_2}\right) = 3.9975 + 0.1\left(-\frac{0.2}{3.9975}\right) \approx 3.992\,,$$

$$x_4 = 0.4\,, \qquad\qquad y_4 \approx 3.985\,,$$

$$x_5 = 0.5\,, \qquad\qquad y_5 \approx 3.975\,,$$

where all of the answers have been rounded off to three decimal places.

3. Here $x_0 = 0$, $y_0 = 1$, and $f(x, y) = x + y$. So,

$$x_{n+1} = x_n + h = x_n + 0.1\,,$$

$$y_{n+1} = y_n + hf(x_n, y_n) = y_n + 0.1(x_n + y_n)\,, \qquad n = 0, 1, 2, \ldots .$$

Letting $n = 0, 1, 2, 3$, and 4, we recursively find

$$x_1 = x_0 + h = 0.1\,, \quad y_1 = y_0 + 0.1(x_0 + y_0) = 1 + 0.1(0 + 1) = 1.1\,,$$

Copyright © 2012 Pearson Education, Inc. Publishing as Addison-Wesley.

$$x_2 = x_1 + h = 0.2, \quad y_2 = y_1 + 0.1\,(x_1 + y_1) = 1.1 + 0.1(0.1 + 1.1) = 1.22,$$
$$x_3 = x_2 + h = 0.3, \quad y_3 = y_2 + 0.1\,(x_2 + y_2) = 1.22 + 0.1(0.2 + 1.22) = 1.362,$$
$$x_4 = x_3 + h = 0.4, \quad y_4 = y_3 + 0.1\,(x_3 + y_3) = 1.362 + 0.1(0.3 + 1.362) = 1.528,$$
$$x_5 = x_4 + h = 0.5, \quad y_5 = y_4 + 0.1\,(x_4 + y_4) = 1.5282 + 0.1(0.4 + 1.5282) = 1.721,$$

where all of the answers have been rounded off to three decimal places.

5. In this problem, $x_0 = 1$, $y_0 = 0$, and $f(x, y) = x - y^2$. So, we let $n = 0, 1, 2, 3,$ and 4, in the recursive formulas and find

$$x_1 = x_0 + h = 1.1, \quad y_1 = y_0 + 0.1\,(x_0 - y_0^2) = 0 + 0.1(1 - 0^2) = 0.1,$$
$$x_2 = x_1 + h = 1.2, \quad y_2 = y_1 + 0.1\,(x_1 - y_1^2) = 0.1 + 0.1(1.1 - 0.1^2) = 0.209,$$
$$x_3 = x_2 + h = 1.3, \quad y_3 = y_2 + 0.1\,(x_2 - y_2^2) = 0.209 + 0.1(1.2 - 0.209^2) = 0.325,$$
$$x_4 = x_3 + h = 1.4, \quad y_4 = y_3 + 0.1\,(x_3 - y_3^2) = 0.325 + 0.1(1.3 - 0.325^2) = 0.444,$$
$$x_5 = x_4 + h = 1.5, \quad y_5 = y_4 + 0.1\,(x_4 - y_4^2) = 0.444 + 0.1(1.4 - 0.444^2) = 0.564,$$

where all of the answers have been rounded off to three decimal places.

7. The initial values are $x_0 = y_0 = 0$, $f(x, y) = 1 - \sin y$. If number of steps is N, then the step $h = (\pi - x_0)/N = \pi/N$.

For $N = 1$, $h = \pi$,

$$x_1 = x_0 + h = \pi, \quad y_1 = y_0 + h(1 - \sin y_0) = \pi \approx 3.142.$$

For $N = 2$, $h = \pi/2$,

$$x_1 = x_0 + \pi/2 = \pi/2, \quad y_1 = y_0 + h(1 - \sin y_0) = \pi/2 \approx 1.571,$$
$$x_2 = x_1 + \pi/2 = \pi, \quad y_2 = y_1 + h(1 - \sin y_1) = \pi/2 \approx 1.571.$$

We continue with $N = 4$ and 8, and fill in Table **1–A** on page 19, where the approximations to $\phi(\pi)$ are rounded to three decimal places.

9. To approximate the solution on the whole interval $[1, 2]$ by Euler's method with the step $h = 0.1$, we first approximate the solution at the points

$$x_n = 1 + 0.1n, \quad n = 1, \ldots, 10.$$

 Copyright © 2012 Pearson Education, Inc. Publishing as Addison-Wesley.

Then, on each subinterval $[x_n, x_{n+1}]$, we approximate the solution by the linear interval, connecting (x_n, y_n) with (x_{n+1}, y_{n+1}), $n = 0, 1, \ldots, 9$. Since

$$f(x, y) = x^{-2} - yx^{-1} - y^2,$$

the recursive formulas have the form

$$x_{n+1} = x_n + 0.1, \quad y_{n+1} = y_n + 0.1 \left(\frac{1}{x_n^2} - \frac{y_n}{x_n} - y_n^2 \right), \quad n = 0, 1, \ldots, 9,$$

$x_0 = 1$, $y_0 = -1$. Therefore,

$$x_1 = 1 + 0.1 = 1.1, \quad y_1 = -1 + 0.1 \left(\frac{1}{1^2} - \frac{-1}{1} - (-1)^2 \right) = -0.9;$$

$$x_2 = 1.1 + 0.1 = 1.2, \quad y_2 = -0.9 + 0.1 \left(\frac{1}{1.1^2} - \frac{-0.9}{1.1} - (-0.9)^2 \right) \approx -0.81654;$$

$$x_3 = 1.2 + 0.1 = 1.3, \quad y_3 = -0.81654 + 0.1 \left(\frac{1}{1.2^2} - \frac{-0.81654}{1.2} - (-0.81654)^2 \right)$$
$$\approx -0.74572;$$

$$x_4 = 1.3 + 0.1 = 1.4, \quad y_4 = -0.74572 + 0.1 \left(\frac{1}{1.3^2} - \frac{-0.74572}{1.3} - (-0.74572)^2 \right)$$
$$\approx -0.68480;$$

$\vdots$

Further computations (rounded to five decimal places) are shown in Table **1–B** on page 19.

The function $y(x) = -1/x = x^{-1}$, obviously, satisfies the initial condition, $y(1) = -1$. Further we compute both sides of the given differential equation:

$$y'(x) = \left(-x^{-1} \right)' = x^{-2},$$
$$f(x, y(x)) = x^{-2} - \left(-x^{-1} \right) x^{-1} - \left(-x^{-1} \right)^2 = x^{-2} + x^{-2} - x^{-2} = x^{-2}.$$

Thus, the function $y(x) = -1/x$ is, indeed, the solution to the given initial value problem.

The graphs of the obtained polygonal line approximation and the actual solution are sketched in Figure **1–E**.

11. In this problem, the independent variable is t and the dependent variable is x. We have

$$f(t, x) = 1 + x^2, \quad t_0 = 0, \quad \text{and} \quad x_0 = 0.$$

Copyright © 2012 Pearson Education, Inc. Publishing as Addison-Wesley.

The function $\phi(t) = \tan t$ satisfies the initial condition: $\phi(0) = \tan 0 = 0$. The differential equation is also satisfied. Indeed,

$$\frac{d\phi}{dt} = \sec^2 t = 1 + \tan^2 t = 1 + \phi(t)^2.$$

Therefore, $\phi(t)$ is the solution to the given initial value problem.

For approximation of $\phi(t)$ at the point $t = 1$ with the step size $h = 2^{-1}$ ($N = 2$ steps), the recursive formulas for Euler's method are

$$t_{n+1} = t_n + 0.5\,,$$

$$x_{n+1} = x_n + 0.5 \left(1 + x_n^2\right).$$

Applying these formulas with $n = 0, 1$, we obtain

$$x_1 = x_0 + 0.5 \left(1 + x_0^2\right) = 0.5\,,$$

$$x_2 = x_1 + 0.5 \left(1 + x_1^2\right) = 0.5 + 0.5 \left(1 + 0.5^2\right) = 1.125\,.$$

We now take $h = 2^{-2}$ to get

$$x_1 = x_0 + 0.25 \left(1 + x_0^2\right) = 0.25\,,$$

$$x_2 = x_1 + 0.25 \left(1 + x_1^2\right) = 0.25 + 0.25 \left(1 + 0.25^2\right) = 0.5156\,,$$

$$x_3 = x_2 + 0.25 \left(1 + x_2^2\right) = 0.515625 + 0.25 \left(1 + 0.515625^2\right) \approx 0.8321\,,$$

$$x_4 = x_3 + 0.25 \left(1 + x_3^2\right) = 0.832092285 + 0.25 \left(1 + 0.832092285^2\right) \approx 1.2552\,.$$

We now take $h = 2^{-3}, 2^{-4}, \ldots$ (halving the step size each time), repeat computations, and stop when the difference between two successive approximations of $x(1)$ is within ± 0.01. The results of computations are shown in Table **1–C** on page 19. From this table we conclude that $h = 2^{-10}$ (or less) guarantees the required accuracy, and $\phi(1) \approx 1.56$ accurate to two decimal places.

Since $x'(t) > 0$ so that $x(t)$ is an increasing function, $x(0) < 1$, and $x(1) > 1$, we look for the unique solution to the equation $x(t_0) = 1$ on the interval $[0, 1]$. Choosing $h = 0.04$ in the Euler's method, we find that

$$x(0.76) \approx 0.928 < 1 \quad \text{and} \quad x(0.80) \approx 1.002 > 1\,.$$

Thus $x(t_0) = 1$ for some t_0 in $(0.76, 0.80)$. We can take $t_0 \approx (0.76 + 0.8)/2 = 0.78 \pm 0.02$.

The actual solution, $\phi(t) = \tan t$, gives $\phi(1) = \tan 1 \approx 1.557$ and $t_0 = \tan^{-1} 1 \approx 0.785$.

 Copyright © 2012 Pearson Education, Inc. Publishing as Addison-Wesley.

13. From Problem 12,

$$y_n = (1 + 1/n)^n$$

and so

$$\lim_{n \to \infty} \frac{e - y_n}{1/n}$$

is a 0/0 indeterminate form. If we let $h = 1/n$ in y_n and use L'Hospital's rule, we get

$$\lim_{n \to \infty} \frac{e - y_n}{1/n} = \lim_{h \to 0} \frac{e - (1 + h)^{1/h}}{h} = \lim_{h \to 0} \frac{g(h)}{h} = \lim_{h \to 0} g'(h),$$

where

$$g(h) = e - (1 + h)^{1/h}.$$

Writing $(1 + h)^{1/h}$ as $e^{\ln(1+h)/h}$ the function $g(h)$ becomes

$$g(h) = e - e^{\ln(1+h)/h}.$$

The first derivative is given by

$$g'(h) = 0 - \frac{d}{dh} \left[e^{\ln(1+h)/h} \right] = -e^{\ln(1+h)/h} \frac{d}{dh} \left[h^{-1} \ln(1 + h) \right].$$

Substituting Maclaurin's series for $\ln(1 + h)$, we obtain

$$g'(h) = -(1 + h)^{1/h} \frac{d}{dh} \left[h^{-1} \left(h - \frac{1}{2} h^2 + \frac{1}{3} h^3 - \frac{1}{4} h^4 + \cdots \right) \right]$$

$$= -(1 + h)^{1/h} \frac{d}{dh} \left[1 - \frac{1}{2} h + \frac{1}{3} h^2 - \frac{1}{4} h^3 + \cdots \right]$$

$$= -(1 + h)^{1/h} \left[-\frac{1}{2} + \frac{2}{3} h - \frac{3}{4} h^2 + \cdots \right]$$

Hence,

$$\lim_{h \to 0} g'(h) = \lim_{h \to 0} \left[-(1 + h)^{1/h} \left(-\frac{1}{2} + \frac{2}{3} h - \frac{3}{4} h^2 + \cdots \right) \right]$$

$$= \left[- \lim_{h \to 0} (1 + h)^{1/h} \right] \cdot \left[\lim_{h \to 0} \left(-\frac{1}{2} + \frac{2}{3} h - \frac{3}{4} h^2 + \cdots \right) \right].$$

From calculus we know that

$$e = \lim_{h \to 0} (1 + h)^{1/h},$$

which gives

$$\lim_{h \to 0} g'(h) = -e \left(-\frac{1}{2} \right) = \frac{e}{2}.$$

So, we have

$$\lim_{n \to \infty} \frac{e - y_n}{1/n} = \frac{e}{2}.$$

15. The independent variable in this problem is the time t and the dependent variable is the temperature $T(t)$ of a body. Thus, we will use the recursive formulas (2) and (3) with x replaced by t and y replaced by T. In the differential equation describing the Newton's Law of Cooling,

$$f(t, T) = K(M(t) - T).$$

With the suggested values of $K = 1\,(\text{min})^{-1}$, $M(t) \equiv 70°$, $h = 0.1$, and the initial condition $T(0) = 100°$, the initial value problem becomes

$$T' = 70 - T, \qquad T(0) = 100,$$

and so the recursive formulas are

$$t_{n+1} = t_n + 0.1,$$

$$T_{n+1} = T_n + 0.1\,(70 - T_n).$$

For $n = 0$, we get

$$t_1 = t_0 + 0.1 = 0.1,$$

$$T_1 = T_0 + 0.1(70 - T_0) = 100 + 0.1(70 - 100) = 97;$$

for $n = 1$,

$$t_2 = t_1 + 0.1 = 0.2,$$

$$T_2 = T_1 + 0.1(70 - T_1) = 97 + 0.1(70 - 97) = 94.3;$$

for $n = 2$,

$$t_3 = t_2 + 0.1 = 0.3,$$

$$T_3 = T_2 + 0.1(70 - T_2) = 94.3 + 0.1(70 - 94.3) = 91.87;$$

for $n = 3$,

$$t_4 = t_3 + 0.1 = 0.4,$$

$$T_4 = T_3 + 0.1(70 - T_3) = 91.87 + 0.1(70 - 91.87) = 89.683.$$

By continuing this way and rounding results to three decimal places, we get Table **1–D** on page 20.

From this table, we conclude that

     Copyright © 2012 Pearson Education, Inc. Publishing as Addison-Wesley.

(a) the temperature of the body after 1 minute is $T(1) \approx 80.460°$ and

(b) its temperature after 2 minutes is $T(2) \approx 73.647°$.

TABLES

Table 1–A: Euler's approximations to $y' = 1 - \sin y$, $y(0) = 0$, with N steps.

N	h	$\phi(\pi)$
1	π	3.142
2	$\pi/2$	1.571
4	$\pi/4$	1.207
8	$\pi/8$	1.148

Table 1–B: Euler's method approximations for the solutions of $y' = x^{-2} - yx^{-1} - y^2$, $y(1) = -1$, on $[1, 2]$ with $h = 0.1$.

n	x_n	y_n	n	x_n	y_n
1	1.1	−0.90000	6	1.6	−0.58511
2	1.2	−0.81654	7	1.7	−0.54371
3	1.3	−0.74572	8	1.8	−0.50669
4	1.4	0.68180	9	1.9	−0.47335
5	1.5	−0.63176	10	2.0	−0.44314

Table 1–C: Euler's approximations to the solution of $x'(t) = 1 + x^2$, $x(0) = 0$.

h	$x(1)$	h	$x(1)$
2^{-3}	1.36694	2^{-7}	1.54134
2^{-4}	1.44724	2^{-8}	1.54927
2^{-5}	1.49747	2^{-9}	1.55331
2^{-6}	1.52603	2^{-10}	1.55535

Copyright © 2012 Pearson Education, Inc. Publishing as Addison-Wesley.

Table 1–D: Euler's method approximations for the solutions of $T' = K(M - T)$, $T(0) = 100$, with $K = 1$, $M = 70$, and $h = 0.1$.

n	t_n	T_n	n	t_n	T_n
1	0.1	97.000	11	1.1	79.414
2	0.2	94.300	12	1.2	78.473
3	0.3	91.870	13	1.3	77.626
4	0.4	89.683	14	1.4	76.863
5	0.5	87.715	15	1.5	76.177
6	0.6	85.943	16	1.6	75.559
7	0.7	84.349	17	1.7	75.003
8	0.8	82.914	18	1.8	74.503
9	0.9	81.623	19	1.9	74.053
10	1.0	80.460	20	2.0	73.647

FIGURES

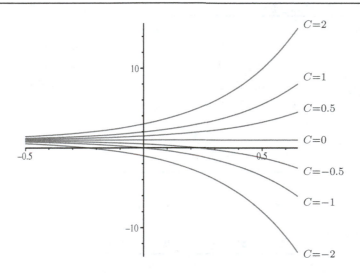

Figure 1–A: Graphs of the functions $y = Ce^{3x} + 1$ for $C = 0, \pm 0.5, \pm 1$, and ± 2.

 Copyright © 2012 Pearson Education, Inc. Publishing as Addison-Wesley.

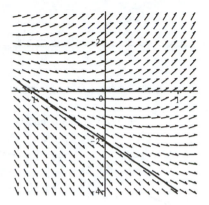

Figure 1–B: The solution curve in Problem 1(a).

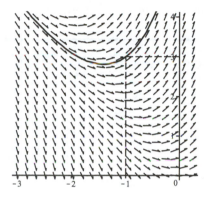

Figure 1–C: The solution curve in Problem 1(b).

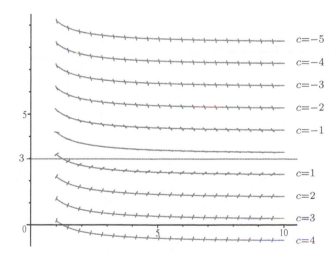

Figure 1–D: Isoclines and the direction field for Problem 17.

Copyright © 2012 Pearson Education, Inc. Publishing as Addison-Wesley.

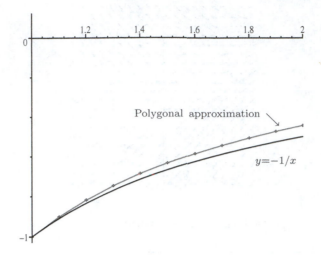

Figure 1–E: Polygonal line approximation and the actual solution for Problem 9.

 Copyright © 2012 Pearson Education, Inc. Publishing as Addison-Wesley.

CHAPTER 2: First-Order Differential Equations

EXERCISES 2.2:   Separable Equations

1. This equation is not separable because $\sin(x + y)$ cannot be expressed as a product $g(x)p(y)$.

3. This equation is separable because

$$\frac{ds}{dt} = t \ln\left(s^{2t}\right) + 8t^2 = t(2t) \ln|s| + 8t^2 = 2t^2(\ln|s| + 4).$$

5. Writing the equation in the form

$$\frac{dy}{dx} = \frac{2x}{xy^2 + 3y^2} = \frac{2x}{(x+3)y^2} = \frac{2x}{x+3} \cdot \frac{1}{y^2},$$

we see that the equation is separable.

7. To separate variables, we divide the equation by x and multiply by dt. Integrating yields

$$\frac{dx}{x} = 3t^2 dt \qquad \Rightarrow \qquad \ln|x| = t^3 + C_1 \qquad \Rightarrow \qquad |x| = e^{t^3 + C_1} = e^{C_1} e^{t^3}$$

$$\Rightarrow \qquad |x| = C_2 e^{t^3} \qquad \Rightarrow \qquad x = \pm C_2 e^{t^3} = C e^{t^3},$$

where C_1 is an arbitrary constant and, therefore, $C_2 := e^{C_1}$ is an arbitrary positive constant, $C = \pm C_2$ is any *nonzero* constant. Separating variables, we lost a solution $x \equiv 0$, which can be included in the above formula by taking $C = 0$. Thus, $x = C e^{t^3}$, C – arbitrary constant, is a general solution.

9. Separating variables yields

$$y^2 \, dy = \frac{x dx}{\sqrt{1+x}} \qquad \Rightarrow \qquad \int y^2 \, dy = \int \frac{x dx}{\sqrt{1+x}}.$$

The second integral can be evaluated using a substitution $u = 1 + x$.

$$\int \frac{x dx}{\sqrt{1+x}} = \int \frac{(u-1)du}{\sqrt{u}} = \int \left(u^{1/2} - u^{-1/2}\right) du = \frac{2}{3} u^{3/2} - 2u^{1/2} + C.$$

Thus,

$$\frac{y^3}{3} = \frac{2}{3}(1+x)^{3/2} - 2(1+x)^{1/2} + C \qquad \Rightarrow \qquad y = \sqrt[3]{2(1+x)^{3/2} - 6(1+x)^{1/2} + C}.$$

Copyright © 2012 Pearson Education, Inc. Publishing as Addison-Wesley.

Chapter 2

11. Separating variables, we obtain

$$\frac{dy}{\sec^2 y} = \frac{dx}{1 + x^2}.$$

Using the trigonometric identities $\sec y = 1/\cos y$ and $\cos^2 y = (1 + \cos 2y)/2$ and integrating, we get

$$\frac{dy}{\sec^2 y} = \frac{dx}{1 + x^2} \quad \Rightarrow \quad \frac{(1 + \cos 2y)dy}{2} = \frac{dx}{1 + x^2}$$

$$\Rightarrow \quad \int \frac{(1 + \cos 2y)dy}{2} = \int \frac{dx}{1 + x^2}$$

$$\Rightarrow \quad \frac{1}{2}\left(y + \frac{1}{2}\sin 2y\right) = \arctan x + C_1$$

$$\Rightarrow \quad 2y + \sin 2y = 4\arctan x + 4C_1 \quad \Rightarrow \quad 2y + \sin 2y = 4\arctan x + C.$$

The last equation defines implicit solutions to the given differential equation.

13. Writing the given equation in the form $dx/dt = x^3 + x$, we separate the variables to get

$$\frac{dx}{x^3 + x} = \frac{dx}{x(x^2 + 1)} = \int \left(\frac{1}{x^2} - \frac{1}{x^2 + 1}\right) x\, dx = dt.$$

Thus,

$$\frac{1}{2}\left[\ln\left(x^2\right) - \ln\left(x^2 + 1\right)\right] = t + C_1 \quad \Rightarrow \quad \frac{x^2}{x^2 + 1} = e^{2C_1}e^{2t} = Ce^{2t},$$

where $C > 0$ is an arbitrary constant. (Note that, separating variables, we lost a solution $x \equiv 0$, which can be included in the above formula by letting $C = 0$.) Solving for x, we get

$$x = \pm\sqrt{\frac{Ce^{2t}}{1 - Ce^{2t}}}, \quad C \geq 0.$$

15. To separate variables, we move the term, containing dx, to the right-hand side of the equation and divide both sides by y. This yields

$$y^{-1}dy = -ye^{\cos x}\sin x\, dx \quad \Rightarrow \quad y^{-2}dy = -e^{\cos x}\sin x\, dx.$$

Integrating the last equation, we obtain

$$\int y^{-2}dy = \int (-e^{\cos x}\sin x)\, dx \quad \Rightarrow \quad -y^{-1} + C = \int e^u du \quad (u = \cos x)$$

$$\Rightarrow \quad -\frac{1}{y} + C = e^u = e^{\cos x} \quad \Rightarrow \quad y = \frac{1}{C - e^{\cos x}},$$

where C is an arbitrary constant.

 Copyright © 2012 Pearson Education, Inc. Publishing as Addison-Wesley.

17. First we find a general solution to the equation. Separating variables and integrating, we get

$$\frac{dy}{dx} = x^3(1-y) \qquad \Rightarrow \qquad \frac{dy}{1-y} = x^3 dx$$

$$\Rightarrow \qquad \int \frac{dy}{1-y} = \int x^3 dx \qquad \Rightarrow \qquad -\ln|1-y| + C_1 = \frac{x^4}{4}$$

$$\Rightarrow \qquad |1-y| = \exp\left(C_1 - \frac{x^4}{4}\right) = Ce^{-x^4/4}.$$

We use the initial condition, $y(0) = 3$, to find C. Thus, substitution $y = 3$ and $x = 0$ into the last equation yields

$$|1-3| = Ce^{-0^4/4} \qquad \Rightarrow \qquad 2 = C.$$

Therefore, $|1-y| = 2e^{-x^4/4}$. Finally, since $1-y(0) = 1-3 < 0$, on an interval containing $x = 0$ one has $1 - y(x) < 0$ so that $|1 - y(x)| = y(x) - 1$. Thus, the answer

$$y - 1 = 2e^{-x^4/4} \qquad \text{or} \qquad y = 2e^{-x^4/4} + 1.$$

19. Separate variables to obtain
$$\frac{1}{2}\frac{dy}{\sqrt{1+y}} = \cos x \, dx.$$

Integrating, we have
$$(y+1)^{1/2} = \sin x + C.$$

Using the initial condition, that is $y(\pi) = 0$, we find
$$1 = \sin \pi + C \qquad \Rightarrow \qquad C = 1.$$

Thus,

$$(y+1)^{1/2} = \sin x + 1 \qquad \Rightarrow \qquad y = (\sin x + 1)^2 - 1 = \sin^2 x + 2\sin x.$$

21. For a general solution, separate variables and integrate:

$$\frac{1}{\theta}\frac{dy}{d\theta} = \frac{y\sin\theta}{y^2+1} \qquad \Rightarrow \qquad \frac{(y^2+1)\, dy}{y} = \theta\sin\theta\, d\theta$$

$$\Rightarrow \qquad \int \frac{(y^2+1)\, dy}{y} = \int \theta\sin\theta\, d\theta \qquad \Rightarrow \qquad \frac{1}{2}y^2 + \ln|y| = -\theta\cos\theta + \sin\theta + C,$$

Copyright © 2012 Pearson Education, Inc. Publishing as Addison-Wesley.

where C is an arbitrary constant. We substitute now the initial condition, $y(\pi) = 1$, and obtain

$$\frac{1}{2} = -\pi \cos \pi + \sin \pi + C = \pi + C \qquad \Rightarrow \qquad C = \frac{1}{2} - \pi.$$

Therefore, the solution $y(\theta)$ to the given initial value problem satisfies

$$\frac{y^2}{2} + \ln y = -\theta \cos \theta + \sin \theta + \frac{1}{2} - \pi.$$

(We have dropped the absolute value sigh in the logarithmic part thanks to positivity of the solution at the initial point.)

23. We have

$$t^{-1}\frac{dy}{dt} = 2\cos^2 y \qquad \Rightarrow \qquad \frac{dy}{\cos^2 y} = 2t\,dt \qquad \Rightarrow \qquad \sec^2 y\,dy = 2t\,dt$$

$$\Rightarrow \qquad \int \sec^2 y\,dy = \int 2t\,dt \qquad \Rightarrow \qquad \tan y = t^2 + C.$$

Since $y = \pi/4$ when $t = 0$, we get $\tan(\pi/4) = 0^2 + C$, and so $C = \tan(\pi/4) = 1$. Therefore, the solution is is

$$\tan y = t^2 + 1 \qquad \Leftrightarrow \qquad y = \arctan\left(t^2 + 1\right).$$

25. Separating variables, we obtain $(1 + y)^{-1}dy = x^2\,dx$. Integrating yields

$$\ln|1 + y| = \frac{x^3}{3} + C.$$

Substituting the initial condition, $y = 3$ when $x = 0$, we get $\ln 4 = 0 + C$ or $C = \ln 4$. Hence, the solution to the given initial value problem is

$$\ln|1 + y| = \frac{x^3}{3} + \ln 4$$
$$\Rightarrow \qquad |1 + y| = e^{(x^3/3)+\ln 4} = e^{x^3/3}e^{\ln 4} = 4e^{x^3/3}$$
$$\Rightarrow \qquad 1 + y = 4e^{x^3/3} \qquad \Rightarrow \qquad y = 4e^{x^3/3} - 1.$$

(We have dropped the absolute value sign because $1 + y(0) = 4 > 0$.)

27. (a) The differential equation $dy/dx = e^{x^2}$ separates if we multiply it by dx. Integrating the separated equation from $x = 0$ to $x = x_1$, we obtain

$$\int_0^{x_1} e^{x^2}\,dx = \int_{x=0}^{x=x_1} dy = y\,\Big|_{x=0}^{x=x_1} = y(x_1) - y(0).$$

 Copyright © 2012 Pearson Education, Inc. Publishing as Addison-Wesley.

If we let t be the variable of integration, replace x_1 by x, and use $y(0) = 0$, then we can express the solution to the initial value problem as

$$y(x) = \int_0^x e^{t^2}\, dt.$$

(b) The differential equation $dy/dx = e^{x^2}y^{-2}$ separates if we multiply both sides by $y^2\, dx$. We integrate the separated equation from $x = 0$ to $x = x_1$ to obtain

$$\int_0^{x_1} e^{x^2}\, dx = \int_0^{x_1} y^2\, dy = \frac{1}{3}\, y^3 \,\Big|_{x=0}^{x=x_1} = \frac{1}{3}\left[y(x_1)^3 - y(0)^3 \right].$$

If we let t be the variable of integration, replace x_1 by x, and put $y(0) = 1$, then we can express the solution to the given initial value problem as

$$\int_0^x e^{t^2}\, dt = \frac{1}{3}\left[y(x)^3 - 1 \right].$$

Solving for $y(x)$ we arrive at

$$y(x) = \left(1 + 3\int_0^x e^{t^2}\, dt \right)^{1/3}. \tag{2.1}$$

(c) The differential equation $dy/dx = \sqrt{1 + \sin x}(1 + y^2)$ separates if we divide by $(1 + y^2)$ and multiply by dx. We now integrate new equation from $x = 0$ to $x = x_1$ to get

$$\int_0^{x_1} \sqrt{1 + \sin x}\, dx = \int_{x=0}^{x=x_1} \frac{dy}{1 + y^2} = \tan^{-1} y(x_1) - \tan^{-1} y(0).$$

If we let t be the variable of integration, replace x_1 by x, and use the initial condition, $y(0) = 1$, then we can rewrite the solution to the initial value problem as

$$y(x) = \tan\left(\int_0^x \sqrt{1 + \sin t}\, dt + \frac{\pi}{4} \right).$$

(d) We will use Simpson's rule (Appendix C) to approximate the definite integral found in part (b). (Simpson's rule is implemented on the website for the text.) Simpson's rule requires an even number of intervals, but we don't know how many are required

Copyright © 2012 Pearson Education, Inc. Publishing as Addison-Wesley.

to obtain the desired three-place accuracy. Rather than make an error analysis, we compute the approximate value of $y(0.5)$ using 2, 4, 6, ... intervals for Simpson's rule until the approximate values for $y(0.5)$ change by less than five in the fourth place.

For $n = 2$, we divide $[0, 0.5]$ into 4 equal subintervals. Thus, each interval will be of length $(0.5 - 0)/4 = 1/8 = 0.125$. Therefore, the integral is approximated by

$$\int_0^{0.5} e^{x^2} dx = \frac{1}{24}\left[e^0 + 4e^{(0.125)^2} + 2e^{(0.25)^2} + 4e^{(0.325)^2} + e^{(0.5)^2} \right] \approx 0.544999.$$

Substituting this value into equation (2.1) from part (b) yields

$$y(0.5) \approx [1 + 3(0.544999)]^{1/3} \approx 1.3812.$$

Repeating these calculations for $n = 3$, 4, and 5 yields Table **2–A** on page 88. Since these values do not change in the fourth place, we can conclude that the first three places are accurate and that an approximate value of the solution is $y(0.5) \approx 1.381$.

29. **(a)** Separating variables and integrating yields

$$\frac{dy}{y^{1/3}} = dx \quad \Rightarrow \quad \int \frac{dy}{y^{1/3}} = \int dx$$

$$\Rightarrow \quad \frac{1}{2/3}\, y^{2/3} = x + C_1$$

$$\Rightarrow \quad y = \left(\frac{2}{3}x + \frac{2}{3}C_1 \right)^{3/2} = \left(\frac{2x}{3} + C \right)^{3/2}.$$

(b) Using the initial condition, $y(0) = 0$, we find that

$$0 = y(0) = \left[\frac{2(0)}{3} + C \right]^{3/2} = C^{3/2} \quad \Rightarrow \quad C = 0,$$

and so

$$y = (2x/3 + 0)^{3/2} = (2x/3)^{3/2}, \qquad x \ge 0,$$

is a solution to the initial value problem.

(c) The function $y(x) \equiv 0$, clearly, satisfies both, the differential equation $dy/dx = y^{1/3}$ and the initial condition $y(0) = 0$.

  Copyright © 2012 Pearson Education, Inc. Publishing as Addison-Wesley.

(d) In notation of Theorem 1, $f(x,y) = y^{1/3}$ and so

$$\frac{\partial f}{\partial y} = \frac{d}{dy}\left(y^{1/3}\right) = \frac{1}{3}\,y^{-2/3} = \frac{1}{3y^{2/3}}\,.$$

Since $\partial f/\partial y$ is not continuous when $y = 0$, there is no rectangle, containing the point $(0,0)$, in which both, f and $\partial f/\partial y$, are continuous. Therefore, Theorem 1 does not apply to this initial value problem.

31. (a) Separating variables and integrating yields

$$\frac{dy}{y^3} = x\,dx \qquad \Rightarrow \qquad \int \frac{dy}{y^3} = \int x\,dx$$

$$\Rightarrow \qquad \frac{y^{-2}}{-2} = \frac{x^2}{2} + C_1 \qquad \Rightarrow \qquad x^2 + y^{-2} = C, \qquad (2.2)$$

where $C := -2C_1$ is an arbitrary constant.

(b) To find the solution, satisfying the initial condition $y(0) = 1$, we substitute $x = 0$, $y = 1$ into (2.2) and obtain

$$0^2 + 1^{-2} = C \qquad \Rightarrow \qquad C = 1 \qquad \Rightarrow \qquad x^2 + y^{-2} = 1.$$

Solving for y yields

$$y = \pm\frac{1}{\sqrt{1 - x^2}}\,. \qquad (2.3)$$

Since, at the initial point, $x = 0$, $y(0) = 1 > 0$, we choose the positive sign in (2.3). Thus, the solution is

$$y = \frac{1}{\sqrt{1 - x^2}}\,.$$

Similarly we find solutions for the other two initial conditions:

$$y(0) = \frac{1}{2} \qquad \Rightarrow \qquad C = 4 \qquad \Rightarrow \qquad y = \frac{1}{\sqrt{4 - x^2}}\,;$$

$$y(0) = 2 \qquad \Rightarrow \qquad C = \frac{1}{4} \qquad \Rightarrow \qquad y = \frac{1}{\sqrt{(1/4) - x^2}}\,.$$

(c) For the solution to the first initial problem in (b), $y(0) = 1$, the domain is the set of all values of x satisfying two conditions

$$\begin{cases} 1 - x^2 \geq 0 & \text{(for existence of the square root)} \\ 1 - x^2 \neq 0 & \text{(for existence of the quotient)} \end{cases} \qquad \Rightarrow \qquad 1 - x^2 > 0.$$

Copyright © 2012 Pearson Education, Inc. Publishing as Addison-Wesley.

Solving for x, we get

$$x^2 < 1 \quad \Rightarrow \quad |x| < 1 \quad \text{or} \quad -1 < x < 1.$$

In the same manner, we find domains for solutions to the other two initial value problems.

$$y(0) = \frac{1}{2} \quad \Rightarrow \quad -2 < x < 2 \, ;$$
$$y(0) = 2 \quad \Rightarrow \quad -\frac{1}{2} < x < \frac{1}{2} .$$

(d) First, we find the solution to the initial value problem $y(0) = a$, $a > 0$, and its domain. Following the lines used in (b) and (c) for particular values of a, we conclude that

$$y(0) = a \quad \Rightarrow \quad 0^2 + a^{-2} = C \quad \Rightarrow \quad y = \frac{1}{\sqrt{a^{-2} - x^2}}$$

and so its domain is

$$a^{-2} - x^2 > 0 \quad \Rightarrow \quad x^2 < a^{-2} \quad \Rightarrow \quad |x| < \frac{1}{a} \quad \text{or} \quad -\frac{1}{a} < x < \frac{1}{a} .$$

As $a \to +0$, $1/a \to +\infty$, and the domain expands to the whole real line. As $a \to +\infty$, $1/a \to 0$, and the domain shrinks to $x = 0$.

(e) For the values $a = 1/2$, 1, and 2 the solutions are found in (b); for $a = -1$, we just have to choose the negative sign in (2.3); similarly, we reverse signs in the other two solutions in (b) to obtain the answers for $a = -1/2$ and -2. The graphs of these functions are shown in Figure **2–A** on page 90.

33. Let $A(t)$ be the number of kilograms of salt in the tank at t minutes after the process begins. Then we have

$$\frac{dA(t)}{dt} = \text{rate of salt in} \; - \; \text{rate of salt out.}$$

The rate of incoming salt is

$$\text{rate of salt in} \; = \; 10 \text{ L/min} \; \times 0.3 \text{ kg/L} \; = 3 \text{ kg/min.}$$

Since the tank is kept uniformly mixed, $A(t)/400$ is the mass of salt per liter that is flowing out of the tank at time t. Thus we have

$$\text{rate of salt out} = 10 \text{ L/min} \; \times \frac{A(t)}{400} \text{ kg/L} = \frac{A(t)}{40} \text{ kg/min.}$$

 Copyright © 2012 Pearson Education, Inc. Publishing as Addison-Wesley.

Therefore,

$$\frac{dA}{dt} = 3 - \frac{A}{40} = \frac{120 - A}{40}.$$

Separating this differential equation and integrating yields

$$\frac{40}{120 - A} dA = dt \qquad \Rightarrow \qquad -40 \ln |120 - A| = t + C_1$$

$$\Rightarrow \qquad \ln |120 - A| = -\frac{t}{40} + C_2 \qquad \left(C_2 = -\frac{C_1}{40} \right)$$

$$\Rightarrow \qquad 120 - A = Ce^{-t/40} \qquad \left(C = \pm e^{C_2} \right)$$

$$\Rightarrow \qquad A = 120 - Ce^{-t/40}.$$

There are 2 kg of salt in the tank initially, thus $A(0) = 2$. Using this initial condition, we find

$$2 = 120 - C \qquad \Rightarrow \qquad C = 118.$$

Substituting this value of C into the solution, we have

$$A(t) = 120 - 118e^{-t/40}.$$

Thus,

$$A(10) = 120 - 118e^{-10/40} \approx 28.1 \text{ kg}.$$

Note: For a detailed discussion of mixture problems see Section 3.2.

35. We saw in Problem 34 that the differential equation $dT/dt = k(M - T)$ can be solved by separation of variables, which yields

$$T = Ce^{kt} + M.$$

When the oven temperature is $120°$, we have $M = 120$. Also, $T(0) = 40$. Thus,

$$40 = C + 120 \qquad \Rightarrow \qquad C = -80.$$

Since $T(45) = 90$, we have

$$90 = -80e^{45k} + 120 \qquad \Rightarrow \qquad \frac{3}{8} = e^{45k} \qquad \Rightarrow \qquad 45k = \ln\left(\frac{3}{8}\right).$$

Thus, $k = \ln(3/8)/45 \approx -0.02180$. This k is independent of M. Therefore, we have a general solution

$$T(t) = Ce^{-0.02180t} + M.$$

Copyright © 2012 Pearson Education, Inc. Publishing as Addison-Wesley.

Chapter 2

(a) We are given that $M = 100$. To find C, we solve the equation $T(0) = 40 = C + 100$. This gives $C = -60$. Thus, the solution becomes

$$T(t) = -60e^{-0.02180t} + 100.$$

We now to solve for t the equation $T(t) = 90$. This gives us

$$90 = -60e^{-0.02180t} + 100 \qquad \Rightarrow \qquad \frac{1}{6} = e^{-0.02180t}$$

$$\Rightarrow \qquad -0.0218t = \ln\left(\frac{1}{6}\right) \qquad \Rightarrow \qquad 0.0218t = \ln 6.$$

Therefore, $t = \ln 6/0.0218 \approx 82.2$ min.

(b) Here, $M = 140$. So, we solve

$$T(0) = 40 = C + 140 \qquad \Rightarrow \qquad C = -100.$$

Thus,

$$T(t) = -100e^{-0.02180t} + 140.$$

Solving $T(t) = 90$ yields $t \approx 31.8$ min.

(c) With $M = 80$, we have $40 = C + 80$, yielding $C = -40$. Setting

$$T(t) = -40e^{-0.02180t} + 80 = 90,$$

we get $-1/4 = e^{-0.02180t}$. This equation has no solutions because an exponential function is never negative. Hence, the plasma never attains the desired temperature. The physical nature of this problem would lead us to expect this result. A further discussion of Newton's law of cooling is given in Section 3.3.

37. The differential equation

$$\frac{dP}{dt} = \frac{r}{100} P$$

separates if we divide it by P and multiply by dt.

$$\int \frac{1}{P}\, dP = \frac{r}{100} \int dt \qquad \Rightarrow \qquad \ln P = \frac{r}{100} t + C \qquad \Rightarrow \qquad P(t) = Ke^{rt/100},$$

where K is the initial amount of money in the savings account, $K = \$1000$, and $r = 5\%$ is the interest rate. These data result in

$$P(t) = 1000e^{5t/100}. \tag{2.4}$$

     Copyright © 2012 Pearson Education, Inc. Publishing as Addison-Wesley.

(a) To determine the amount of money in the account after 2 years, we substitute $t = 2$ into equation (2.4) and obtain

$$P(2) = 1000e^{10/100} = 1105.17 \text{ (dollars)}.$$

(b) To determine, when the account will reach \$4000, we solve equation (2.4) for t with $P = 4000$:

$$4000 = 1000e^{5t/100} \quad \Rightarrow \quad e^{5t/100} = 4 \quad \Rightarrow \quad t = 20\ln 4 \approx 27.73 \text{ (years)}.$$

(c) To determine the amount of money in the account after 3.5 years, we need to determine the value of each \$1000 deposit after 3.5 years has passed. This means that the initial amount of \$1000 is in the account for the entire 3.5 years and grows to the amount, which is given by $P_0 = 1000e^{5(3.5)/100}$. For the growth of the \$1000 deposited after 12 months, we take $t = 2.5$ in equation (2.4), because that is how long this \$1000 will be in the account. This gives $P_1 = 1000e^{5(2.5)/100}$. Using the above reasoning for the remaining deposits we arrive at $P_2 = 1000e^{5(1.5)/100}$ and $P_3 = 1000e^{5(0.5)/100}$. The total amount is determined by summing P_i's.

$$P = 1000\left[e^{5(3.5)/100} + e^{5(2.5)/100} + e^{5(1.5)/100} + e^{5(0.5)/100}\right] \approx 4,427.59 \text{ (dollars)}.$$

39. Let $s(t)$, $t > 0$, denote the distance traveled by driver A from the time $t = 0$, when he ran out of gas, to time t. Then driver A's velocity $v_A(t) = ds/dt$ is a solution to the initial value problem

$$\frac{dv_A}{dt} = -kv_A^2, \qquad v_A(0) = v_B,$$

where v_B is driver B's constant velocity, and $k > 0$ is a positive constant. Separating variables we get

$$\frac{dv_A}{v_A^2} = -k\,dt \quad \Rightarrow \quad \int \frac{dv_A}{v_A^2} = -\int k\,dt \quad \Rightarrow \quad \frac{1}{v_A(t)} = kt + C.$$

From the initial condition we find

$$\frac{1}{v_B} = \frac{1}{v_A(0)} = k \cdot 0 + C = C \quad \Rightarrow \quad C = \frac{1}{v_B}.$$

Thus,

$$v_A(t) = \frac{1}{kt + 1/v_B} = \frac{v_B}{v_B kt + 1}.$$

Copyright © 2012 Pearson Education, Inc. Publishing as Addison-Wesley.

The function $s(t)$, therefore, satisfies

$$\frac{ds}{dt} = \frac{v_B}{v_B kt + 1}, \qquad s(0) = 0.$$

Integrating, we obtain

$$s(t) = \int \frac{v_B}{v_B kt + 1}\, dt = \frac{1}{k} \ln\left(v_B kt + 1\right) + C_1.$$

To find C_1, we use the initial condition, $s(0) = 0$.

$$0 = s(0) = \frac{1}{k} \ln\left(v_B k \cdot 0 + 1\right) + C_1 = C_1 \qquad \Rightarrow \qquad C_1 = 0.$$

Hence,

$$s(t) = \frac{1}{k} \ln\left(v_B kt + 1\right).$$

At the moment $t = t_1$ when driver A's speed was halved, i.e., $v_A(t_1) = v_A(0)/2 = v_B/2$, we have

$$\frac{1}{2}v_B = v_A(t_1) = \frac{v_B}{v_B kt_1 + 1} \qquad \text{and} \qquad 1 = s(t_1) = \frac{1}{k} \ln\left(v_B kt_1 + 1\right)$$

$$\Rightarrow \qquad v_B kt_1 + 1 = 2 \qquad \text{and so} \qquad k = \ln\left(v_B kt_1 + 1\right) = \ln 2$$

$$\Rightarrow \qquad s(t) = \frac{1}{\ln 2} \ln\left(v_B t \ln 2 + 1\right).$$

Since driver B was 3 miles behind driver A at time $t = 0$, and his speed remained constant, he finished the race at time $t_B = (3 + 2)/v_B = 5/v_B$. At this moment, driver A had already passed

$$s(t_B) = \frac{1}{\ln 2} \ln\left(v_B t_B \ln 2 + 1\right) = \frac{1}{\ln 2} \ln\left(\frac{5}{v_B} v_B \ln 2 + 1\right)$$

$$= \frac{1}{\ln 2} \ln\left(5 \ln 2 + 1\right) \approx 2.1589 > 2 \text{ (miles)},$$

i.e., A wins the race.

EXERCISES 2.3:   Linear Equations

1. This equation is not linear since its right-hand side is not a linear function of x and/or x' but the left-hand side is. It is not separable either because, writing it in the form

$$\frac{dx}{dt} = e^x - xt,$$

we see that the right-hand side cannot be factored as $g(t)p(x)$.

    Copyright © 2012 Pearson Education, Inc. Publishing as Addison-Wesley.

3. Writing

$$\frac{dy}{dt} + e^{-t} \ln t \, y = 3te^{-t},$$

we see that this equation is linear (with $P(t) = e^{-t} \ln t$ and $Q(t) = 3te^{-t}$. It is not separable since the function $Q(t) - P(t)y = e^{-t}(3t - y \ln t)$ cannot be written as a product of two functions, $g(t)$ and $p(y)$, that depend only on t and y, resp.

5. This equation is equivalent to

$$\frac{dr}{d\theta} - 3r = \theta^3,$$

which fits the form of a linear equation with $P(\theta) = -3$ and $Q(\theta) = \theta^3$. It is not separable since the function $3r + \theta^3 (= dr/d\theta)$ cannot be represented as a product of functions of single variables, r and θ.

7. Writing the equation in standard form,

$$\frac{dy}{dx} - \frac{y}{x} = 2x + 1,$$

we see that

$$P(x) = -\frac{1}{x} \qquad \Rightarrow \qquad \mu(x) = \exp\left[\int \left(-\frac{1}{x}\right) dx\right] = \exp\left(-\ln|x|\right) = \frac{1}{|x|}.$$

Since the equation should be solved separately on each interval, where $P(x)$ is continuous (here, $(-\infty, 0)$ and $(0, \infty)$), and $\mu(x)$ is determined up to a constant multiple, we can take $\mu(x) = 1/x$.[1] Multiplying the given equation by $\mu(x)$, we get

$$\frac{d}{dx}\left(\frac{y}{x}\right) = 2 + \frac{1}{x} \qquad \Rightarrow \qquad y = x \int \left(2 + \frac{1}{x}\right) dx = x\left(2x + \ln|x| + C\right).$$

9. From the standard form of the given equation,

$$\frac{dy}{dx} + \frac{2}{x}y = x^{-4},$$

we find that

$$\mu(x) = \exp\left[\int (2/x)dx\right] = \exp\left(2\ln|x|\right) = x^2$$

$$\Rightarrow \quad \frac{d}{dx}\left(x^2 y\right) = x^{-2} \qquad \Rightarrow \qquad y = x^{-2}\int x^{-2}dx = x^{-2}\left(-x^{-1} + C\right) = \frac{Cx - 1}{x^3}.$$

---

[1]Similar arguments will be implicitly used in other problems when the absolute value sign appears in the integrating factor.

Copyright © 2012 Pearson Education, Inc. Publishing as Addison-Wesley.

11. Choosing t as the independent variable and y as the dependent variable, we put the equation in standard form:

$$t + y + 1 - \frac{dy}{dt} = 0 \qquad \Rightarrow \qquad \frac{dy}{dt} - y = t + 1. \tag{2.5}$$

Thus $P(t) \equiv -1$ and so $\mu(t) = \exp\left[\int(-1)dt\right] = e^{-t}$. We multiply both sides of the second equation in (2.5) by $\mu(t)$.

$$e^{-t}\frac{dy}{dt} - e^{-t}y = (t+1)e^{-t} \qquad \Rightarrow \qquad \frac{d}{dt}\left(e^{-t}y\right) = (t+1)e^{-t}.$$

Thus, integrating we get

$$e^{-t}y = \int (t+1)e^{-t}dt = -(t+1)e^{-t} + \int e^{-t}dt$$
$$= -(t+1)e^{-t} - e^{-t} + C = -(t+2)e^{-t} + C$$
$$\Rightarrow \qquad y = e^t\left[-(t+2)e^{-t} + C\right] = -t - 2 + Ce^t,$$

where we have used integration by parts to find $\int (t+1)e^{-t}dt$.

13. In this problem, the independent variable is y and the dependent variable is x. So, we divide the equation by y to rewrite it in standard form.

$$y\frac{dx}{dy} + 2x = 5y^3 \qquad \Rightarrow \qquad \frac{dx}{dy} + \frac{2}{y}x = 5y^2.$$

Therefore, $P(y) = 2/y$, and so the integrating factor $\mu(y)$ is

$$\mu(y) = \exp\left(\int \frac{2}{y}\,dy\right) = \exp\left(2\ln|y|\right) = y^2.$$

Multiplying the equation (in standard form) by y^2 and integrating yields

$$y^2\frac{dx}{dy} + 2y\,x = 5y^4 \qquad \Rightarrow \qquad \frac{d}{dy}\left(y^2x\right) = 5y^4$$
$$\Rightarrow \qquad y^2x = \int 5y^4\,dy = y^5 + C \qquad \Rightarrow \qquad x = y^{-2}\left(y^5 + C\right) = y^3 + Cy^{-2}.$$

15. First, we put this linear equation in standard form,

$$\frac{dy}{dx} + \frac{x}{x^2+1}\,y = \frac{x}{x^2+1}. \tag{2.6}$$

Here, $P(x) = x/(x^2+1)$, and so

$$\int P(x)\,dx = \int \frac{x}{x^2+1}\,dx = \frac{1}{2}\ln(x^2+1).$$

 Copyright © 2012 Pearson Education, Inc. Publishing as Addison-Wesley.

Thus, the integrating factor is

$$\mu(x) = e^{(1/2)\ln(x^2+1)} = e^{\ln\left[(x^2+1)^{1/2}\right]} = (x^2+1)^{1/2}.$$

Multiplying (2.6) by $\mu(x)$ yields

$$(x^2+1)^{1/2}\frac{dy}{dx} + \frac{x}{(x^2+1)^{1/2}}y = \frac{x}{(x^2+1)^{1/2}},$$

which becomes

$$\frac{d}{dx}\left[(x^2+1)^{1/2}y\right] = \frac{x}{(x^2+1)^{1/2}}.$$

We now integrate both sides and solve for y to find

$$(x^2+1)^{1/2}y = (x^2+1)^{1/2} + C \qquad \Rightarrow \qquad y = 1 + C(x^2+1)^{-1/2}.$$

This solution is valid for all x since $P(x)$ and $Q(x)$ are continuous for all x.

17. This is a linear equation with $P(x) = -1/x$ and $Q(x) = xe^x$, which are continuous on any interval not containing $x = 0$. Therefore, the integrating factor is given by

$$\mu(x) = \exp\left[\int\left(-\frac{1}{x}\right)dx\right] = e^{-\ln x} = \frac{1}{x}, \qquad x > 0.$$

Multiplying the equation by this integrating factor yields

$$\frac{1}{x}\frac{dy}{dx} - \frac{y}{x^2} = e^x \qquad \Rightarrow \qquad \frac{d}{dx}\left(\frac{y}{x}\right) = e^x.$$

Integrating gives us

$$\frac{y}{x} = e^x + C \qquad \Rightarrow \qquad y = xe^x + Cx.$$

Applying now the initial condition, $y(1) = e - 1$, we get

$$e - 1 = e + C \qquad \Rightarrow \qquad C = -1.$$

Thus, the solution is

$$y = xe^x - x, \quad 0 < x < \infty.$$

Note: This interval is the largest interval, containing the initial value $x = 1$, in which $P(x)$ and $Q(x)$ are continuous.

Copyright © 2012 Pearson Education, Inc. Publishing as Addison-Wesley.

19. First, we rewrite this equation in standard form. Namely,

$$\frac{dx}{dt} + \frac{3}{t}x = t^2 \ln t + \frac{1}{t^2}.$$

Thus we see that $P(t) = 3/t$ and $Q(t) = t^2 \ln t + t^{-2}$. The integrating factor $\mu(t)$ is then

$$\mu(t) = \exp\left[\int \left(\frac{3}{t}\right) dt\right] - \exp(3 \ln t) = t^3.$$

We next compute

$$\int \mu(t)Q(t)\, dt = \int t^3\left(t^2 \ln t + t^{-2}\right) dt = \int \left(t^5 \ln t + t\right) dt = \frac{t^6}{6}\left(\ln t - \frac{1}{6}\right) + \frac{t^2}{2} + C,$$

so that

$$x(t) = \frac{1}{t^3}\left[\frac{t^6}{6}\left(\ln t - \frac{1}{6}\right) + \frac{t^2}{2} + C\right] = \frac{t^3}{6}\left(\ln t - \frac{1}{6}\right) + \frac{1}{2t} + \frac{C}{t^3}.$$

We now use the initial condition and substitute $t = 1$ and $x = 0$ into this general solution to find C.

$$0 = \frac{1}{6}\cdot\left(-\frac{1}{6}\right) + \frac{1}{2} + C \qquad\Rightarrow\qquad C = -\frac{17}{36}$$

so that

$$x(t) = \frac{t^3}{6}\left(\ln t - \frac{1}{6}\right) + \frac{1}{2t} - \frac{17}{36t^3} = \frac{t^3}{6}\ln t - \frac{t^3}{36} + \frac{1}{2t} - \frac{17}{36t^3}.$$

21. Putting the equation in standard form yields

$$\frac{dy}{dx} + \frac{\sin x}{\cos x}y = 2x\cos x \qquad\Rightarrow\qquad \frac{dy}{dx} + (\tan x)y = 2x\cos x.$$

Therefore, $P(x) = \tan x$ and so

$$\mu(x) = \exp\left(\int \tan x\, dx\right) = \exp\left(-\ln|\cos x|\right) = |\sec x|.$$

We can take $\mu(x) = \sec x$. (See the footnote on page 35.) Multiplying the standard form of the given equation by $\mu(x)$ gives

$$\sec x\frac{dy}{dx} + (\sec x \tan x)\,y = 2x \qquad\Rightarrow\qquad \frac{d}{dx}(y\sec x) = 2x$$

$$\Rightarrow\qquad y\sec x = \int 2x\, dx = x^2 + C \qquad\Rightarrow\qquad y = \cos x\left(x^2 + C\right).$$

From the initial condition, we find that

$$\frac{-15\sqrt{2}\pi^2}{32} = y\left(\frac{\pi}{4}\right) = \cos\frac{\pi}{4}\left[\left(\frac{\pi}{4}\right)^2 + C\right] \qquad\Rightarrow\qquad C = -\pi^2.$$

Hence, the solution is given by $y = \cos x\left(x^2 - \pi^2\right) = x^2\cos x - \pi^2\cos x.$

 Copyright © 2012 Pearson Education, Inc. Publishing as Addison-Wesley.

23. We proceed similarly to Example 2 and obtain an analog of the initial value problem (13), that is,

$$\frac{dy}{dt} + 5y = 40e^{-20t}, \qquad y(0) = 10. \tag{2.7}$$

Thus, $P(t) \equiv 5$ and $\mu(t) = \exp\left(\int 5dt\right) = e^{5t}$. Multiplying the differential equation in (2.7) by $\mu(t)$ and integrating, we obtain

$$e^{5t}\frac{dy}{dt} + 5e^{5t}y = 40e^{-20t}e^{5t} = 40e^{-15t}$$

$$\Rightarrow \qquad \frac{d\left(e^{5t}y\right)}{dt} = 40e^{-15t} \qquad \Rightarrow \qquad e^{5t}y = \int 40e^{-15t}\,dt = \frac{40}{-15}e^{-15t} + C.$$

Therefore, a general solution to the differential equation in (2.7) is

$$y = e^{-5t}\left(\frac{40}{-15}e^{-15t} + C\right) = Ce^{-5t} - \frac{8}{3}e^{-20t}.$$

We find C using the initial condition, $y(0) = 10$.

$$10 = y(0) = Ce^{-5\cdot 0} - \frac{8}{3}e^{-20\cdot 0} = C - \frac{8}{3} \qquad \Rightarrow \qquad C = 10 + \frac{8}{3} = \frac{38}{3}.$$

Hence, the mass of RA_2 for $t \geq 0$ is given by

$$y(t) = \frac{38}{3}e^{-5t} - \frac{8}{3}e^{-20t}.$$

25. **(a)** This is a linear problem, and so we find an integrating factor $\mu(x)$.

$$\mu(x) = \exp\left[\int (2x)\,dx\right] = e^{x^2}.$$

Multiplying the equation by $\mu(x)$ yields

$$\frac{d}{dx}\left(ye^{x^2}\right) = e^{x^2} \qquad \Rightarrow \qquad \int_2^x \frac{d}{dt}\left(ye^{t^2}\right)dt = \int_2^x e^{t^2}\,dt,$$

where we have changed the dummy variable x to t and integrated with respect to t from $t = 2$ (since the initial value is $x = 2$) to $t = x$. Thus, since $y(2) = 1$,

$$ye^{x^2} - e^4 = \int_2^x e^{t^2}\,dt \qquad \Rightarrow \qquad y = e^{-x^2}\left(e^4 + \int_2^x e^{t^2}\,dt\right) = e^{4-x^2} + e^{-x^2}\int_2^x e^{t^2}\,dt.$$

Copyright © 2012 Pearson Education, Inc. Publishing as Addison-Wesley.

(b) We use Simpson's rule (Appendix C) to approximate the definite integral found in part (a) with upper bound $x = 3$. Simpson's rule requires an even number of intervals, but we do not know how many are required to obtain the desired three place accuracy. Thus, rather than make an error analysis, we will compute the approximate values of $y(3)$ using 4, 6, 8, 10, 12, ... intervals for Simpson's rule until the approximate values of $y(3)$ change by less than five in the fourth place.

For $n = 2$, we divide $[2, 3]$ into four equal subintervals. Thus, each subinterval will be of the length $(3 - 2)/4 = 1/4$. Therefore, the integral is approximated by

$$\int_2^3 e^{t^2}\, dt \approx \frac{1}{12}\left[e^{(2)^2} + e^{(2.25)^2} + e^{(2.5)^2} + e^{(2.75)^2} + e^{(3)^2} \right] \approx 1460.35435\,.$$

Dividing this result by $e^{(3)^2}$ and adding $e^{4-3^2} = e^{-5}$ gives

$$y(3) \approx 0.18696\,.$$

Calculations for $n = 3$, 4, 5, and 6 are shown in Table **2–B** on 89. Since the last three approximate values do not change by more than five in the fourth place, it appears that their first three places are accurate and the approximate value of the solution is $y(3) \approx 0.183$.

27. (a) The given differential equation is in standard form. Thus, $P(x) = \sqrt{1 + \sin^2 x}$. Since we cannot express $\int P(x)\, dx$ as an elementary function, we use the fundamental theorem of calculus to conclude that, with any fixed constant a,

$$\left(\int_a^x P(t) dt \right)' = P(x), \tag{2.8}$$

that is, the definite integral with variable upper bound is an antiderivative of $P(x)$. Since, in the formula for $\mu(x)$, one can choose any antiderivative of $P(x)$, we take $a = 0$ in (2.8). (Such a choice of a comes from the initial point $x = 0$ and makes it easy to satisfy the initial condition.) Therefore, the integrating factor $\mu(x)$ can be chosen as

$$\mu(x) = \exp\left(\int_0^x \sqrt{1 + \sin^2 t}\, dt \right).$$

 Copyright © 2012 Pearson Education, Inc. Publishing as Addison-Wesley.

Multiplying the differential equation by $\mu(x)$ and integrating from $x = 0$ to $x = s$, we obtain

$$\frac{d[\mu(x)y]}{dx} = \mu(x)x$$

$$\Rightarrow \quad \int_0^s d[\mu(x)y] = \int_0^s \mu(x)x\,dx$$

$$\Rightarrow \quad \mu(x)y(x)\Big|_{x=0}^{x=s} = \int_0^s \mu(x)x\,dx$$

$$\Rightarrow \quad \mu(s)y(s) - \mu(0)y(0) = \int_0^s \mu(x)x\,dx\,.$$

From the initial condition, $y(0) = 2$, and

$$\mu(0) = \exp\left(\int_0^0 \sqrt{1 + \sin^2 t}\,dt\right) = e^0 = 1$$

we obtain $\mu(0)y(0) = 2$, so that

$$\mu(s)y(s) = \int_0^s \mu(x)x\,dx + 2\,.$$

Dividing by $\mu(s)$ and interchanging x and s, we get the required solution.

(b) The values of $\mu(x)$ for $x = 0.1, 0.2, \ldots, 1.0$, approximated by using the Simpson's rule, are given in Table **2–C**, page 89.

We now use these values of $\mu(x)$ to approximate $\int_0^1 \mu(s)s\,ds$ by applying Simpson's rule again. With $n = 5$ and $h = (1 - 0)/(2n) = 0.1$ the Simpson's rule yields

$$\int_0^1 \mu(s)s\,ds \approx \frac{0.1}{3}[\mu(0)(0) + 4\mu(0.1)(0.1) + 2\mu(0.2)(0.2) + 4\mu(0.3)(0.3)$$

$$+ 2\mu(0.4)(0.4) + 4\mu(0.5)(0.5) + 2\mu(0.6)(0.6) + 4\mu(0.7)(0.7)$$

$$+ 2\mu(0.8)(0.8) + 4\mu(0.9)(0.9) + \mu(1.0)(1.0)] \approx 1.06454\,.$$

Therefore,

$$y(1) \approx \frac{1}{\mu(1)}\int_0^1 \mu(s)s\,ds + \frac{2}{\mu(1)} \approx \frac{1}{3.07672} \cdot 1.06454 + \frac{2}{3.07672} \approx 0.9960\,.$$

Copyright © 2012 Pearson Education, Inc. Publishing as Addison-Wesley.

(c) We rewrite the differential equation in the form used in Euler's method,

$$\frac{dy}{dx} = x - \sqrt{1 + \sin^2 x}\, y, \qquad y(0) = 2,$$

and conclude that $f(x, y) = x - \sqrt{1 + \sin^2 x}\, y$. Thus, the recursive formulas (2) and (3) become

$$x_{n+1} = x_n + h,$$
$$y_{n+1} = y_n + h\left(x_n - \sqrt{1 + \sin^2 x_n}\, y_n\right), \qquad n = 0, 1, \ldots,$$

$x_0 = 0$, $y_0 = 2$. With $h = 0.1$ we need $n = (1 - 0)/0.1 = 10$ steps to get an approximation at $x = 1$.

$$n = 0: \quad x_1 = 0.1, \quad y_1 = (2) + 0.1[(0) - \sqrt{1 + \sin^2(0)}\,(2)] = 1.8000;$$
$$n = 1: \quad x_2 = 0.2, \quad y_2 = (1.8) + 0.1[(0.1) - \sqrt{1 + \sin^2(0.1)}\,(1.8)] \approx 1.6291;$$
$$n = 2: \quad x_3 = 0.3, \quad y_3 = (1.6291) + 0.1[(0.2) - \sqrt{1 + \sin^2(0.2)}\,(1.6291)] \approx 1.4830;$$
$$\vdots$$

Results of these computations, rounded off to four decimal places, are given in Table **2–D**, page 89. Thus, Euler's method with step $h = 0.1$ gives $y(1) \approx 0.9486$.

Next, we take $h = 0.05$ and fill in Table **2–E** and get $y(1) \approx 0.9729$.

29. In the presented form, the equation

$$\frac{dy}{dx} = \frac{1}{e^{4y} + 2x}$$

is not linear. But, if we switch the roles of variables and consider y as the independent variable and x as the dependent variable (using the connection between derivatives of inverse functions, that is, the formula $y'(x) = 1/x'(y)$), then the equation transforms to

$$\frac{dx}{dy} = e^{4y} + 2x \qquad \Rightarrow \qquad \frac{dx}{dy} - 2x = e^{4y}.$$

This is a linear equation with $P(y) = -2$. Thus, the integrating factor is

$$\mu(y) = \exp\left[\int (-2)\,dy\right] = e^{-2y}$$

and so

$$\frac{d}{dy}\left(e^{-2y} x\right) = e^{-2y} e^{4y} = e^{2y} \qquad \Rightarrow \qquad e^{-2y} x = \int e^{2y}\,dy = \frac{e^{2y}}{2} + C.$$

Solving for x yields

$$x = e^{2y}\left(\frac{e^{2y}}{2} + C\right) = \frac{e^{4y}}{2} + Ce^{2y}.$$

 Copyright © 2012 Pearson Education, Inc. Publishing as Addison-Wesley.

31. (a) On the interval $0 \le x \le 2$, we have $P(x) = 1$. Thus, we are solving the equation

$$\frac{dy}{dx} + y = x.$$

An integrating factor is given by

$$\mu(x) = \exp\left(\int dx\right) = e^x.$$

Multiplying the equation by the integrating factor, we obtain

$$e^x \frac{dy}{dx} + e^x y = xe^x \quad \Rightarrow \quad \frac{d}{dx}(e^x y) = xe^x \quad \Rightarrow \quad e^x y = \int xe^x \, dx.$$

Integrating by parts and dividing by e^x yields a general solution

$$y = e^{-x}(xe^x - e^x + C) = x - 1 + Ce^{-x}.$$

(b) Using the initial condition, $y(0) = 1$, we see that

$$1 = y(0) = 0 - 1 + C = -1 + C \quad \Rightarrow \quad C = 2.$$

Thus, the solution becomes

$$y = x - 1 + 2e^{-x}.$$

(c) In the interval $x > 2$, we have $P(x) = 3$. Therefore, the integrating factor is given by

$$\mu(x) = \exp\left(\int 3 \, dx\right) = e^{3x}.$$

Multiplying the equation by $\mu(x)$ and solving yields

$$e^{3x} \frac{dy}{dx} + 3e^{3x} y = xe^{3x} \quad \Rightarrow \quad \frac{d}{dx}(e^{3x} y) = xe^{3x} \quad \Rightarrow \quad e^{3x} y = \int xe^{3x} \, dx.$$

Integrating by parts and dividing by e^{3x} gives us

$$y = e^{-3x}\left(\frac{1}{3} xe^{3x} - \frac{1}{9} e^{3x} + C\right) = \frac{x}{3} - \frac{1}{9} + Ce^{-3x}.$$

(d) We want the value of the initial point for the solution in part (c) to be the value of the solution found in part (b) at the point $x = 2$. This value is

$$y(2) = 2 - 1 + 2e^{-2} = 1 + 2e^{-2}.$$

Copyright © 2012 Pearson Education, Inc. Publishing as Addison-Wesley.

Thus, the initial point we seek for the solution in part (c) is

$$y(2) = 1 + 2e^{-2}.$$

Using this initial point, we find that

$$1 + 2e^{-2} = y(2) = \frac{2}{3} - \frac{1}{9} + Ce^{-6} \quad \Rightarrow \quad C = \frac{4e^6}{9} + 2e^4.$$

Thus, the solution of the equation on the interval $x > 2$ is given by

$$y = \frac{x}{3} - \frac{1}{9} + \left(\frac{4}{9}e^6 + 2e^4 \right) e^{-3x}.$$

Patching these two solutions together gives a continuous solution to the given equation on the interval $x \geq 0$.

$$y = \begin{cases} x - 1 + 2e^{-x}, & 0 \leq x \leq 2; \\ x/3 - 1/9 + (4e^6/9 + 2e^4) e^{-3x}, & 2 < x. \end{cases}$$

(e) The graph of the solution is given in Figure B.18 in the answers of the text.

33. (a) Writing the equation in standard form yields

$$\frac{dy}{dx} + \frac{2}{x} y = 3.$$

Therefore, $P(x) = 2/x$ and

$$\mu(x) = \exp \left(\int \frac{2}{x} \, dx \right) = \exp \left(2 \ln |x| \right) = |x|^2 = x^2.$$

Hence

$$\frac{d}{dx} \left(x^2 y \right) = 3x^2 \quad \Rightarrow \quad x^2 y = \int 3x^2 \, dx = x^3 + C \quad \Rightarrow \quad y = x + \frac{C}{x^2}$$

is a general solution to the given differential equation. Unless $C = 0$ (and so $y = x$), the function $y = x + C/x^2$ is not defined at $x = 0$. Therefore, among all solutions, the only function defined at $x = 0$ is $\phi(x) = x$, and the initial value problem with $y(0) = y_0$ has a (unique) solution if and only if

$$y_0 = \phi(x) \Big|_{x=0} = 0.$$

(b) Standard form of the equation $xy' - 2y = 3x$ is

$$\frac{dy}{dx} - \frac{2}{x}y = 3.$$

This gives $P(x) = -2/x$, $\mu(x) = \exp\left[\int(-2/x)dx\right] = x^{-2}$, and

$$\frac{d}{dx}\left(x^{-2}y\right) = 3x^{-2} \quad \Rightarrow \quad x^{-2}y = \int 3x^{-2}\,dx = -3x^{-1} + C \quad \Rightarrow \quad y = -3x + Cx^2.$$

Therefore, any solution is a polynomial and so is defined for all real numbers. Moreover, any solution satisfies the initial condition, $y(0) = 0$, because

$$-3x + Cx^2 \Big|_{x=0} = -3(0) + C(0)^2 = 0$$

for any C and, therefore, is a solution to the initial value problem. (This also implies that the initial value problem with $y(0) = y_0 \neq 0$ has no solution.)

35. (a) This part of the problem is similar to Problem 33 in Section 2.2. So, we proceed in the same way.

Let $A(t)$ denote the mass of salt in the tank at t minutes after the process begins. Then we have

$$\text{rate of input } = 5 \text{ L/min } \times 0.2 \text{ kg/L } = 1 \text{ kg/min},$$
$$\text{rate of exit} = 5 \text{ L/min } \times \frac{A(t)}{500} \text{ kg/L} = \frac{A(t)}{100} \text{ kg/min},$$
$$\frac{dA}{dt} = 1 - \frac{A}{100} = \frac{100 - A}{100}.$$

Separating variables in this differential equation yields

$$dA/(100 - A) = dt/100.$$

Integrating, we obtain

$$-\ln|100 - A| = \frac{t}{100} + C_1 \quad \Rightarrow \quad |100 - A| = e^{-t/100 - C_1} = e^{-C_1}e^{-t/100}$$
$$\Rightarrow \quad 100 - A = Ce^{-t/100} \quad \left(C = \pm e^{-C_1}\right) \quad \Rightarrow \quad A = 100 - Ce^{-t/100}.$$

The initial condition, $A(0) = 5$ (initially, there were 5 kg of salt in the tank) implies that

$$5 = A(0) = 100 - C \quad \Rightarrow \quad C = 95.$$

Copyright © 2012 Pearson Education, Inc. Publishing as Addison-Wesley.

Substituting this value of C into the general solution, we get

$$A(t) = 100 - 95e^{-t/100}.$$

Thus, the mass of salt in the tank after 10 min is

$$A(10) = 100 - 95e^{-10/100} \approx 14.04 \text{ kg},$$

which gives the concentration $14.04 \text{ kg}/500 \text{ L} \approx 0.0281 \text{ kg/L}$.

(b) After the leak develops, the system satisfies a new differential equation. While the rate of input remains the same, 1 kg/min, the rate of exit is now different. Since, every minute, 5 liters of the solution are coming in and $5 + 1 = 6$ liters are coming out, the volume of the solution in the tank decreases by $6 - 5 = 1$ liter per minute. Thus, for $t \geq 10$, the volume of the solution in the tank is $500 - (1)(t - 10) = 510 - t$ liters. This gives the concentration of salt in the tank

$$\frac{A(t)}{510 - t} \text{ kg/L} \tag{2.9}$$

and

$$\text{rate of exit} = 6 \text{ L/min} \times \frac{A(t)}{510 - t} \text{ kg/L} = \frac{6A(t)}{510 - t} \text{ kg/min}.$$

Hence, the differential equation, for $t > 10$, becomes

$$\frac{dA}{dt} = 1 - \frac{6A}{510 - t} \qquad \Rightarrow \qquad \frac{dA}{dt} + \frac{6A}{510 - t} = 1$$

with the initial condition $A(10) = 14.04$ (the value found in (a)). This equation is a linear equation. We have

$$\mu(t) = \exp\left(\int \frac{6}{510 - t} \, dt\right) = e^{-6 \ln |510 - t|} = (510 - t)^{-6}$$

$$\Rightarrow \qquad \frac{d}{dt}\left[(510 - t)^{-6} A\right] = (1)(510 - t)^{-6} = (510 - t)^{-6}.$$

Integrating yields

$$(510 - t)^{-6} A = \int (510 - t)^{-6} dt = \frac{1}{5}(510 - t)^{-5} + C$$

$$\Rightarrow \qquad A = \frac{1}{5}(510 - t) + C(510 - t)^6.$$

Using the initial condition, $A(10) = 14.04$, we compute C.

$$14.04 = A(10) = \frac{1}{5}(510 - 10) + C(510 - 10)^6 \qquad \Rightarrow \qquad C = -\frac{85.96}{(500)^6}.$$

Copyright © 2012 Pearson Education, Inc. Publishing as Addison-Wesley.

Therefore,

$$A(t) = \frac{1}{5}(510 - t) - \frac{85.96}{(500)^6}(510 - t)^6 = \frac{1}{5}(510 - t) - 85.96\left(\frac{510 - t}{500}\right)^6$$

and, according to (2.9), the concentration of salt is given by

$$\frac{A(t)}{510 - t} = \frac{1}{5} - \frac{85.96}{510 - t}\left(\frac{510 - t}{500}\right)^6.$$

20 minutes after the leak develops, that is, when $t = 30$, the concentration will be

$$\frac{1}{5} - \frac{85.96}{510 - 30}\left(\frac{510 - 30}{500}\right)^6 \approx 0.0598 \text{ (kg/L)}.$$

37. We are solving the initial value problem

$$\frac{dx}{dt} + 2x = 1 - \cos\left(\frac{\pi t}{12}\right), \qquad x(0) = 10.$$

This is a linear equation with dependent variable x and independent variable t. Thus, $P(t) \equiv 2$ and, to solve this equation, we find an integrating factor $\mu(t)$.

$$\mu(t) = \exp\left(\int 2\,dt\right) = e^{2t}.$$

Multiplying the equation by this factor yields

$$e^{2t}\frac{dx}{dt} + 2xe^{2t} = e^{2t}\left[1 - \cos\left(\frac{\pi t}{12}\right)\right] = e^{2t} - e^{2t}\cos\left(\frac{\pi t}{12}\right)$$

$$\Rightarrow \qquad xe^{2t} = \int e^{2t}\,dt - \int e^{2t}\cos\left(\frac{\pi t}{12}\right)dt = \frac{1}{2}e^{2t} - \int e^{2t}\cos\left(\frac{\pi t}{12}\right)dt.$$

The last integral can be found using integrating by parts twice, which leads back to an integral similar to the original. Combining these two similar integrals and simplifying, we obtain

$$\int e^{2t}\cos\left(\frac{\pi t}{12}\right)dt = \frac{e^{2t}\left[2\cos\left(\pi t/12\right) + (\pi/12)\sin\left(\pi t/12\right)\right]}{4 + (\pi/12)^2} + C.$$

Thus, we see that

$$x(t) = \frac{1}{2} - \frac{2\cos\left(\pi t/12\right) + (\pi/12)\sin\left(\pi t/12\right)}{4 + (\pi/12)^2} + Ce^{-2t}.$$

Using the initial condition, $x = 10$ when $t = 0$, we obtain

$$C = \frac{19}{2} + \frac{2}{4\,(\pi/12)^2}.$$

Therefore, the desired solution is

$$x(t) = \frac{1}{2} - \frac{2\cos\left(\pi t/12\right) + (\pi/12)\sin\left(\pi t/12\right)}{4 + (\pi/12)^2} + \left(\frac{19}{2} + \frac{2}{4\,(\pi/12)^2}\right)e^{-2t}.$$

Copyright © 2012 Pearson Education, Inc. Publishing as Addison-Wesley.

39. Let $T_j(t)$, $j = 0, 1, 2, \ldots$, denote the temperature in the classroom for $9 + j \leq t < 10 + j$, where $t = 13$ denotes $1:00$ P.M., $t = 14$ denotes $2:00$ P.M., etc. Then

$$T(9) = 0, \tag{2.10}$$

and the continuity of the temperature implies that

$$\lim_{t \to 10+j} = T_{j+1}(10 + j), \qquad j = 0, 1, 2, \ldots . \tag{2.11}$$

According to the work of the heating unit, the temperature satisfies the equation

$$\frac{dT_j}{dt} = \begin{cases} 1 - T_j, & \text{if } j = 2k \\ -T_j, & \text{if } j = 2k + 1 \end{cases}, \qquad 9 + j < t < 10 + j \quad k = 0, 1, \ldots .$$

For j even, we have

$$\frac{dT_j}{dt} = 1 - T_j \qquad \Rightarrow \qquad \frac{dT_j}{1 - T_j} = dt$$

$$\Rightarrow \qquad \ln|1 - T_j| = -t + c_j \qquad \Rightarrow \qquad T_j(t) = 1 - C_j e^{-t};$$

for j odd,

$$\frac{dT_j}{dt} = -T_j \qquad \Rightarrow \qquad \frac{dT_j}{-T_j} = dt$$

$$\Rightarrow \qquad \ln|T_j| = -t + c_j \qquad \Rightarrow \qquad T_j(t) = C_j e^{-t};$$

where $C_j \neq 0$ are constants. From (2.10) we have

$$0 = T_0(9) = \left(1 - C_0 e^{-t}\right)\Big|_{t=9} = 1 - C_0 e^{-9} \qquad \Rightarrow \qquad C_0 = e^9.$$

Also from (2.11), for even values of j (say, $j = 2k$) we get

$$\left(1 - C_{2k} e^{-t}\right)\Big|_{t=9+(2k+1)} = C_{2k+1} e^{-t}\Big|_{t=9+(2k+1)}$$

$$\Rightarrow \qquad 1 - C_{2k} e^{-(10+2k)} = C_{2k+1} e^{-(10+2k)}$$

$$\Rightarrow \qquad C_{2k+1} = e^{10+2k} - C_{2k}.$$

Similarly from (2.11) for odd values of j (say, $j = 2k + 1$) we get

$$C_{2k+1} e^{-t}\Big|_{t=9+(2k+2)} = \left(1 - C_{2k+2} e^{-t}\right)\Big|_{t=9+(2k+2)}$$

$$\Rightarrow \qquad C_{2k+1} e^{-(11+2k)} = 1 - C_{2k+2} e^{-(11+2k)}$$

 Copyright © 2012 Pearson Education, Inc. Publishing as Addison-Wesley.

$$\Rightarrow \qquad C_{2k+2} = e^{11+2k} - C_{2k+1}.$$

In general, we see that for any integer j (even or odd) the formula

$$C_j = e^{9+j} - C_{j-1}$$

holds. Using this recurrence formula we successively compute

$$C_1 = e^{10} - C_0 = e^{10} - e^9 = e^9(e-1)$$
$$C_2 = e^{11} - C_1 = e^{11} - e^{10} + e^9 = e^9(e^2 - e + 1)$$
$$\vdots$$
$$C_j = e^9 \sum_{k=0}^{j} (-1)^{j-k} e^k.$$

Therefore, the temperature at noon (when $t = 12$ and $j = 3$) is

$$T_3(12) = C_3 e^{-12} = e^{-12} e^9 \sum_{k=0}^{3} (-1)^{3-k} e^k = 1 - e^{-1} + e^{-2} - e^{-3} \approx 0.718 = 71.8\,(^\circ\text{F}).$$

At 5 P.M. (when $t = 17$ and $j = 8$), we find

$$T_8(17) = 1 - C_8 e^{-17} = 1 - e^{-17} e^9 \sum_{k=0}^{8} (-1)^{8-k} e^k = \sum_{k=1}^{8} (-1)^{k+1} e^{-k}$$
$$= e^{-1} \cdot \frac{1 - (-e^{-1})^8}{1 + e^{-1}} \approx 0.269 = 26.9\,(^\circ\text{F}).$$

EXERCISES 2.4: Exact Equations

1. This equation is not separable because the coefficient $x^{10/3} - 2y$ cannot be written as a product $p(x)q(y)$. Writing the equation in the form

$$\frac{dy}{dx} - 2x^{-1}y = -x^{7/3},$$

we see that the equation is linear. Since $M(x,y) = x^{10/3} - 2y$, $N(x,y) = x$,

$$\frac{\partial M}{\partial y} = -2 \neq \frac{\partial N}{\partial x} = 1,$$

and so the equation is not exact.

3. First we note that $M(x,y) = \sqrt{-2y - y^2}$ depends only on y and $N(x,y) = 3 + 2x - x^2$ depends only on x. So, the equation is separable. It is not linear with x as independent

Copyright © 2012 Pearson Education, Inc. Publishing as Addison-Wesley.

variable because $M(x, y)$ is not a linear function of y. Similarly, it is not linear with y as independent variable because $N(x, y)$ is not a linear function of x. Computing

$$\frac{\partial M}{\partial y} = \frac{1}{2}\left(-2y - y^2\right)^{-1/2}(-2 - 2y) = -\frac{1 + y}{\sqrt{-2y - y^2}},$$

$$\frac{\partial N}{\partial x} = 2 - 2x,$$

we see that the equation (5) in Theorem 2 is not satisfied. Therefore, the equation is not exact.

5. It is separable, linear with x as independent variable, and not exact because

$$\frac{\partial M}{\partial y} = x \neq \frac{\partial N}{\partial x} = 0.$$

7. Here, $M(x, y) = 2x + y\cos(xy)$, $N(x, y) = x\cos(xy) - 2y$. Since $M(x, y)/N(x, y)$ cannot be expressed as a product $f(x)g(y)$, the equation is not separable. We also conclude that it is not linear because $M(x, y)/N(x, y)$ is not a linear function of y and $N(x, y)/M(x, y)$ is not a linear function of x. Taking partial derivatives

$$\frac{\partial M}{\partial y} = \cos(xy) - xy\sin(xy) = \frac{\partial N}{\partial x},$$

we see that the equation is exact.

9. In this problem, $M(x, y) = 2x + y$, $N(x, y) = x - 2y$. Thus, $M_y = N_x = 1$, and the equation is exact. We find

$$F(x, y) = \int (2x + y)dx = x^2 + xy + g(y),$$

$$\frac{\partial F}{\partial y} = x + g'(y) = N(x, y) = x - 2y$$

$$\Rightarrow \quad g'(y) = -2y \quad \Rightarrow \quad g(y) = \int (-2y)dy = -y^2$$

$$\Rightarrow \quad F(x, y) = x^2 + xy - y^2,$$

and so $x^2 + xy - y^2 = C$ is a general solution.

11. Taking partial derivatives of $M(x, y) = \cos x\cos y + 2x$ and $N(x, y) = -(\sin x\sin y + 2y)$, we obtain

$$\frac{\partial M}{\partial y} = \frac{\partial}{\partial y}(\cos x\cos y + 2x) = -\cos x\sin y,$$

 Copyright © 2012 Pearson Education, Inc. Publishing as Addison-Wesley.

$$\frac{\partial N}{\partial x} = \frac{\partial}{\partial x} \left[-(\sin x \sin y + 2y) \right] = -\cos x \sin y,$$

and so the equation is exact.

Integrating $M(x, y)$ with respect to x yields

$$F(x, y) = \int M(x, y) dx = \int (\cos x \cos y + 2x) \, dx$$

$$= \cos y \int \cos x \, dx + \int 2x \, dx = \sin x \cos y + x^2 + g(y).$$

To find $g(y)$, we compute the partial derivative of $F(x, y)$ with respect to y and compare the result with $N(x, y)$.

$$\frac{\partial F}{\partial y} = \frac{\partial}{\partial y} \left[\sin x \cos y + x^2 + g(y) \right] = -\sin x \sin y + g'(y) = -(\sin x \sin y + 2y)$$

$$\Rightarrow \quad g'(y) = -2y \quad \Rightarrow \quad g(y) = \int (-2y) dy = -y^2.$$

(We have taken zero integration constant.) Therefore,

$$F(x, y) = \sin x \cos y + x^2 - y^2 = C$$

is a general solution to the given equation.

13. In this equation, the variables are y and t, $M(y, t) = t/y$, $N(y, t) = 1 + \ln y$. Since

$$\frac{\partial M}{\partial t} = \frac{\partial}{\partial t} \left(\frac{t}{y} \right) = \frac{1}{y} \quad \text{and} \quad \frac{\partial N}{\partial y} = \frac{\partial}{\partial y} (1 + \ln y) = \frac{1}{y},$$

the equation is exact.

Integrating $M(y, t)$ with respect to y, we get

$$F(y, t) = \int \frac{t}{y} dy = t \ln |y| + g(t) = t \ln y + g(t).$$

(From $N(y, t) = 1 + \ln y$ we conclude that $y > 0$.) Therefore,

$$\frac{\partial F}{\partial t} = \frac{\partial}{\partial t} \left[t \ln y + g(t) \right] = \ln y + g'(t) = 1 + \ln y$$

$$\Rightarrow \quad g'(t) = 1 \quad \Rightarrow \quad g(t) = t$$

$$\Rightarrow \quad F(y, t) = t \ln y + t,$$

and a general solution is given by $t \ln y + t = C$ (or, explicitly, $t = C/(\ln y + 1)$).

Copyright © 2012 Pearson Education, Inc. Publishing as Addison-Wesley.

15. This differential equation is expressed in variables r and θ. Since the variables x and y are dummy variables, this equation is solved in exactly the same way as an equation in x and y. We look for a solution with independent variable θ and dependent variable r.

$$M(r, \theta) = \cos\theta \quad \text{and} \quad N(r, \theta) = -r\sin\theta + e^\theta,$$

we find that

$$M_\theta(r, \theta) = -\sin\theta = N_r(r, \theta),$$

and so the equation is exact. Integrating $M(r, \theta)$ with respect to r yields

$$F(r, \theta) = \int \cos\theta\, dr = r\cos\theta + g(\theta)$$
$$\Rightarrow \quad F_\theta(r, \theta) = -r\sin\theta + g'(\theta) = N(r, \theta) = -r\sin\theta + e^\theta.$$

Thus, we have

$$g'(\theta) = e^\theta \quad \Rightarrow \quad g(\theta) = e^\theta,$$

where the constant of integration will be incorporated into the parameter of the solution. Substituting this expression for $g(\theta)$ into the formula we found for $F(r, \theta)$, we get

$$F(r, \theta) = r\cos\theta + e^\theta.$$

Therefore, the solution is given by a one parameter family $r\cos\theta + e^\theta = C$. Solving for r yields

$$r = \frac{C - e^\theta}{\cos\theta} = (C - e^\theta)\sec\theta.$$

17. Partial derivatives of $M(x, y) = 1/y$ and $N(x, y) = -(3y - x/y^2)$ are

$$\frac{\partial M}{\partial y} = \frac{\partial}{\partial y}\left(\frac{1}{y}\right) = -\frac{1}{y^2} \quad \text{and}$$
$$\frac{\partial N}{\partial x} = \frac{\partial}{\partial x}\left(-3y + \frac{x}{y^2}\right) = \frac{1}{y^2}.$$

Since $\partial M/\partial y \neq \partial N/\partial x$, the equation is not exact.

19. Taking partial derivatives of $M(x, y) = 2x + y/(1+x^2y^2)$ and $N(x, y) = -2y + x/(1+x^2y^2)$ with respect to y and x, respectively, we get

$$\frac{\partial M}{\partial y} = \frac{\partial}{\partial y}\left(2x + \frac{y}{1+x^2y^2}\right) = \frac{(1)(1+x^2y^2) - yx^2(2y)}{(1+x^2y^2)^2} = \frac{1 - x^2y^2}{(1+x^2y^2)^2},$$

Copyright © 2012 Pearson Education, Inc. Publishing as Addison-Wesley.

$$\frac{\partial N}{\partial x} = \frac{\partial}{\partial x}\left(-2y + \frac{x}{1 + x^2y^2}\right) = \frac{(1)(1 + x^2y^2) - xy^2(2x)}{(1 + x^2y^2)^2} = \frac{1 - x^2y^2}{(1 + x^2y^2)^2}.$$

Therefore, the equation is exact.

$$F(x, y) = \int\left(2x + \frac{y}{1 + x^2y^2}\right) dx = x^2 + \int \frac{d(xy)}{1 + (xy)^2} = x^2 + \arctan(xy) + g(y)$$

$$\frac{\partial F}{\partial y} = \frac{\partial}{\partial y}\left[x^2 + \arctan(xy) + g(y)\right] = \frac{x}{1 + (xy)^2} + g'(y) = -2y + \frac{x}{1 + x^2y^2}$$

$$\Rightarrow \qquad g'(y) = -2y \qquad \Rightarrow \qquad g(y) = -y^2$$

$$\Rightarrow \qquad F(x, y) = x^2 - y^2 + \arctan(xy)$$

and a general solution then is given implicitly by $x^2 - y^2 + \arctan(xy) = C$.

21. We check the equation for exactness. Since $M(x, y) = 1/x + 2y^2x$, $N(x, y) = 2yx^2 - \cos y$,

$$\frac{\partial M}{\partial y} = \frac{\partial}{\partial y}\left(\frac{1}{x} + 2y^2x\right) = 4yx,$$

$$\frac{\partial N}{\partial x} = \frac{\partial}{\partial x}\left(2yx^2 - \cos y\right) = 4yx.$$

Thus, $\partial M/\partial y = \partial N/\partial x$. Integrating $M(x, y)$ with respect to x yields

$$F(x, y) = \int\left(\frac{1}{x} + 2y^2x\right) dx = \ln|x| + x^2y^2 + g(y).$$

Therefore,

$$\frac{\partial F}{\partial y} = \frac{\partial}{\partial y}\left[\ln|x| + x^2y^2 + g(y)\right] = 2x^2y + g'(y) = N(x, y) = 2yx^2 - \cos y$$

$$\Rightarrow \qquad g'(y) = -\cos y \qquad \Rightarrow \qquad g(y) = \int(-\cos y)dy = -\sin y$$

$$\Rightarrow \qquad F(x, y) = \ln|x| + x^2y^2 - \sin y,$$

and a general solution to the given differential equation is

$$\ln|x| + x^2y^2 - \sin y = C.$$

Substituting the initial condition, $y = \pi$ when $x = 1$, we find C.

$$\ln|1| + 1^2\pi^2 - \sin \pi = C \qquad \Rightarrow \qquad C = \pi^2.$$

Therefore, the answer is given implicitly by $\ln x + x^2y^2 - \sin y = \pi^2$. (We used the fact that $x > 0$ at the initial point, $(1, \pi)$, to skip the absolute value sign in the logarithmic term.)

Copyright © 2012 Pearson Education, Inc. Publishing as Addison-Wesley.

23. Here, $M(t, y) = e^t y + te^t y$ and $N(t, y) = te^t + 2$. Thus, $M_y(t, y) = e^t + te^t = N_t(t, y)$ and so the equation is exact. To find $F(t, y)$, we first integrate $N(t, y)$ with respect to y to obtain

$$F(t, y) = \int (te^t + 2)\, dy = (te^t + 2)y + h(t).$$

(We have chosen to integrate the function $N(t, y)$ because this integration can be easily accomplished.) Thus,

$$F_t(t, y) = e^t y + te^t y + h'(t) = M(t, y) = e^t y + te^t y$$

$$\Rightarrow \quad h'(t) = 0 \quad \Rightarrow \quad h(t) = C.$$

We will incorporate this constant into the parameter of the solution. Combining these results yields $F(t, y) = te^t y + 2y$. Hence, a general solution is $te^t y + 2y = C$ or, solving for y, $y = C/(te^t + 2)$. We now use the initial condition, $y(0) = -1$, to find C.

$$y(0) = \frac{C}{0 + 2} = -1 \quad \Rightarrow \quad \frac{C}{2} = -1 \quad \Rightarrow \quad C = -2.$$

This gives us the answer

$$y = -\frac{2}{te^t + 2}.$$

25. One can check that the equation is not exact ($\partial M / \partial y \neq \partial N / \partial x$), but it is separable, because it can be written in the form

$$y^2 \sin x\, dx + \frac{1 - y}{x}\, dy = 0 \quad \Rightarrow \quad y^2 \sin x\, dx = \frac{y - 1}{x}\, dy$$

$$\Rightarrow \quad x \sin x\, dx = \frac{y - 1}{y^2}\, dy.$$

Integrating both sides yields

$$\int x \sin x\, dx = \int \frac{y - 1}{y^2}\, dy \quad \Rightarrow \quad x(-\cos x) - \int (-\cos x)\, dx = \int \left(\frac{1}{y} - \frac{1}{y^2} \right) dy$$

$$\Rightarrow \quad -x \cos x + \sin x = \ln |y| + \frac{1}{y} + C,$$

where we applied integration by parts to find $\int x \sin x\, dx$. Substitution of the initial condition, $y(\pi) = 1$, results

$$-\pi \cos \pi + \sin \pi = \ln |1| + 1 + C \quad \Rightarrow \quad C = \pi - 1.$$

So, the solution to the given initial value problem is

$$\sin x - x \cos x = \ln y + \frac{1}{y} + \pi - 1.$$

(Since $y(\pi) = 1 > 0$, we have removed the absolute value sign in the logarithmic term.)

 Copyright © 2012 Pearson Education, Inc. Publishing as Addison-Wesley.

27. (a) We want to find $M(x, y)$ such that, with $N(x, y) = \sec^2 y - x/y$, we have

$$M_y(x, y) = N_x(x, y) = -\frac{1}{y}.$$

Integrating this last expression with respect to y yields

$$M(x, y) = \int \left(-\frac{1}{y}\right) dy = -\ln|y| + f(x),$$

where $f(x)$, the "constant" of integration, is a function of x alone.

(b) We are looking for $M(x, y)$ such that for

$$N(x, y) = \sin x \cos y - xy - e^{-y}$$

there holds

$$M_y(x, y) = N_x(x, y) = \cos x \cos y - y.$$

We integrate this last expression with respect to y. That is,

$$\begin{aligned} M(x, y) &= \int (\cos x \cos y - y)\, dy = \cos x \int \cos y\, dy - \int y\, dy \\ &= \cos x \sin y - \frac{y^2}{2} + f(x), \end{aligned}$$

where $f(x)$, a function of x only, is the "constant" of integration.

29. (a) We have $M(x, y) = y^2 + 2xy$ and $N(x, y) = -x^2$. Therefore, $M_y(x, y) = 2y + 2x$ and $N_x(x, y) = -2x$. Thus, $M_y(x, y) \neq N_x(x, y)$, and so the differential equation is not exact.

(b) If we multiply $(y^2 + 2xy)dx - x^2 dy = 0$ by y^{-2}, we obtain

$$\left(1 + \frac{2x}{y}\right) dx - \frac{x^2}{y^2}\, dy = 0.$$

In this equation, we have $M(x, y) = 1 + 2xy^{-1}$ and $N(x, y) = -x^2 y^{-2}$. Therefore,

$$\frac{\partial M(x, y)}{\partial y} = -\frac{2x}{y^2} = \frac{\partial N(x, y)}{\partial x}.$$

So, the new differential equation is exact.

(c) Following the method for solving exact equations, we integrate $M(x, y)$ in part (b) with respect to x to obtain

$$F(x, y) = \int \left(1 + 2\frac{x}{y}\right) dx = x + \frac{x^2}{y} + g(y).$$

Copyright © 2012 Pearson Education, Inc. Publishing as Addison-Wesley.

To determine $g(y)$, we take the partial derivative of both sides of this equation with respect to y and get

$$\frac{\partial F}{\partial y} = -\frac{x^2}{y^2} + g'(y).$$

Substituting $N(x, y)$ (given in part (b)) for $\partial F/\partial y$, we can now solve for $g'(y)$ to obtain

$$N(x, y) = -\frac{x^2}{y^2} = -\frac{x^2}{y^2} + g'(y) \qquad \Rightarrow \qquad g'(y) = 0.$$

The integral of $g'(y)$ gives a constant, and the choice of the constant of integration is not important. So, we take $g(y) = 0$. Hence, we have $F(x, y) = x + x^2/y$, and an implicit solution to the given equation is

$$x + \frac{x^2}{y} = C.$$

Solving for y, we obtain

$$y = \frac{x^2}{C - x}.$$

(d) Dividing both sides by y^2, we lost the solution $y \equiv 0$.

31. Following the proof of Theorem 2, we come to the expression (10) for $g'(y)$, that is

$$g'(y) = N(x, y) - \frac{\partial}{\partial y} \int_{x_0}^{x} M(s, y)\, ds \qquad\qquad (2.12)$$

(where we have replaced the integration variable t by s). In other words, the function $g(y)$ is an antiderivative of the right-hand side in (2.12). Since an antiderivative is defined up to an additive constant and, in Theorem 2, such a constant can be chosen arbitrarily (that is, $g(y)$ can be *any* antiderivative), we choose $g(y)$ that vanishes at y_0. According to fundamental theorem of calculus, this function can be written in the form

$$g(y) = \int_{y_0}^{y} g'(t)\, dt \;=\; \int_{y_0}^{y} \left[N(x, t) - \frac{\partial}{\partial t} \int_{x_0}^{x} M(s, t)\, ds \right] dt$$

$$= \int_{y_0}^{y} N(x, t)\, dt - \int_{y_0}^{y} \frac{\partial}{\partial t} \left[\int_{x_0}^{x} M(s, t)\, ds \right] dt$$

$$= \int_{y_0}^{y} N(x, t)\, dt - \left[\int_{x_0}^{x} M(s, t)\, ds \right] \Bigg|_{t=y_0}^{t=y}$$

Copyright © 2012 Pearson Education, Inc. Publishing as Addison-Wesley.

$$= \int_{y_0}^{y} N(x,t)\,dt - \int_{x_0}^{x} M(s,y)\,ds + \int_{x_0}^{x} M(s,y_0)\,ds\,.$$

Substituting this function into the formula (9), we conclude that

$$F(x,y) = \int_{x_0}^{x} M(t,y)\,dt + \left[\int_{y_0}^{y} N(x,t)\,dt - \int_{x_0}^{x} M(s,y)\,ds + \int_{x_0}^{x} M(s,y_0)\,ds\right]$$

$$= \int_{y_0}^{y} N(x,t)\,dt + \int_{x_0}^{x} M(s,y_0)\,ds\,.$$

(a) In the differential form used in Example 1, $M(x,y) = 2xy^2 + 1$ and $N(x,y) = 2x^2y$. Thus, $N(x,t) = 2x^2t$ and $M(s,y_0) = 2s \cdot 0^2 + 1 = 1$, and (18) yields

$$F(x,y) = \int_{0}^{y} (2x^2t)\,dt + \int_{0}^{x} 1 \cdot ds = x^2 \int_{0}^{y} 2t\,dt + \int_{0}^{x} ds$$

$$= x^2 t^2 \Big|_{t=0}^{t=y} + s \Big|_{s=0}^{s=x} = x^2y^2 + x.$$

(b) Since $M(x,y) = 2xy - \sec^2 x$ and $N(x,y) = x^2 + 2y$, we have

$$N(x,t) = x^2 + 2t \quad \text{and} \quad M(s,y_0) = 2s \cdot 0 - \sec^2 s = -\sec^2 s,$$

$$F(x,y) = \int_{0}^{y} (x^2 + 2t)\,dt + \int_{0}^{x} (-\sec^2 s)\,ds$$

$$= (x^2 t + t^2) \Big|_{t=0}^{t=y} - \tan s \Big|_{s=0}^{s=x} = x^2 y + y^2 - \tan x.$$

(c) Here, $M(x,y) = 1 + e^x y + xe^x y$ and $N(x,y) = xe^x + 2$. Therefore,

$$N(x,t) = xe^x + 2 \quad \text{and} \quad M(s,y_0) = 1 + e^s \cdot 0 + se^s \cdot 0 = 1,$$

$$F(x,y) = \int_{0}^{y} (xe^x + 2)\,dt + \int_{0}^{x} (1)\,ds$$

$$= (xe^x + 2) t \Big|_{t=0}^{t=y} + s \Big|_{s=0}^{s=x} = (xe^x + 2) y + x,$$

which is identical to $F(x,y)$ obtained in Example 3.

33. We use notations and results of Problem 32. For a family of curves given by $F(x,y) = k$, the orthogonal trajectories satisfy the differential equation

$$\frac{\partial F(x,y)}{\partial y}\,dx - \frac{\partial F(x,y)}{\partial x}\,dy = 0\,. \tag{2.13}$$

Copyright © 2012 Pearson Education, Inc. Publishing as Addison-Wesley.

Chapter 2

(a) In this problem, $F(x, y) = 2x^2 + y^2$, and so the equation (2.13) becomes

$$\frac{\partial(2x^2 + y^2)}{\partial y} dx - \frac{\partial(2x^2 + y^2)}{\partial x} dy = 0 \quad \Rightarrow \quad 2y\, dx - 4x\, dy = 0. \quad (2.14)$$

Separating variables and integrating yields

$$2y\, dx = 4x\, dy \quad \Rightarrow \quad \frac{dx}{x} = \frac{2dy}{y} \quad \Rightarrow \quad \int \frac{dx}{x} = \int \frac{2dy}{y}$$

$$\Rightarrow \quad \ln|x| = 2\ln|y| + c_1 \quad \Rightarrow \quad e^{\ln|x|} = e^{2\ln|y|+c_1}$$

$$\Rightarrow \quad |x| = e^{c_1}|y|^2 = c_2 y^2 \quad \Rightarrow \quad x = \pm c_2 y^2 = cy^2,$$

where c as any nonzero constant.

Separating variables, we divided the equation (2.14) by xy. As a result, we lost two constant solutions $x \equiv 0$ and $y \equiv 0$ (see the discussion in Section 2.2 of the text). Thus, the orthogonal trajectories for the family $2x^2 + y^2 = k$ are $x = cy^2$, $c \neq 0$, and $x \equiv 0$, $y \equiv 0$. (Note that $x \equiv 0$ can be obtained from $x = cy^2$ by taking $c = 0$ while $y \equiv 0$ cannot.)

(b) First, we rewrite the equation, defining the family of curves, in the form $F(x, y) = k$. Dividing it by x^4 yields $yx^{-4} = k$. We use (2.13) to set up an equation for the orthogonal trajectories:

$$\frac{\partial F}{\partial x} = -4yx^{-5}, \quad \frac{\partial F}{\partial y} = x^{-4} \quad \Rightarrow \quad x^{-4} dx - \left(-4yx^{-5}\right) dy = 0.$$

Solving this separable equation yields

$$x^{-4} dx = -4yx^{-5} dy = 0 \quad \Rightarrow \quad x\, dx = -4y\, dy$$

$$\Rightarrow \quad \int x\, dx = \int (-4y) dy \quad \Rightarrow \quad \frac{x^2}{2} = -2y^2 + c_1 \quad \Rightarrow \quad x^2 + 4y^2 = c.$$

Thus, the family of orthogonal trajectories is $x^2 + 4y^2 = c$.

(c) Taking logarithm of both sides of the equation, we obtain

$$\ln y = kx \quad \Rightarrow \quad \frac{\ln y}{x} = k,$$

and so $F(x, y) = (\ln y)/x$, $\partial F/\partial x = -(\ln y)/x^2$, $\partial F/\partial y = 1/(xy)$. The equation (2.13) becomes

$$\frac{1}{xy} dx - \left(-\frac{\ln y}{x^2}\right) dy = 0 \quad \Rightarrow \quad \frac{1}{xy} dx = -\frac{\ln y}{x^2} dy.$$

Copyright © 2012 Pearson Education, Inc. Publishing as Addison-Wesley.

Separating variables and integrating, we obtain

$$x \, dx = -y \ln y \, dy \qquad \Rightarrow \qquad \int x \, dx = -\int y \ln y \, dy$$

$$\Rightarrow \qquad \frac{x^2}{2} = -\frac{y^2}{2} \ln y + \int \frac{y^2}{2} \cdot \frac{1}{y} \, dy = -\frac{y^2}{2} \ln y + \frac{y^2}{4} + c_1$$

$$\Rightarrow \qquad \frac{x^2}{2} + \frac{y^2}{2} \ln y - \frac{y^2}{4} = c_1 \qquad \Rightarrow \qquad 2y^2 \ln y - y^2 + 2x^2 = c,$$

where $c := 4c_1$, and we have used integration by parts to find $\int y \ln y \, dy$.

(d) Dividing the equation $y^2 = kx$ by x, we get $y^2/x = k$. Thus, $F(x, y) = y^2/x$ and

$$\frac{\partial F}{\partial x} = -\frac{y^2}{x^2}, \qquad \frac{\partial F}{\partial y} = \frac{2y}{x}$$

$$\Rightarrow \qquad \frac{2y}{x} \, dx - \left(-\frac{y^2}{x^2}\right) dy = 0 \qquad \Rightarrow \qquad \frac{2y}{x} \, dx = \left(-\frac{y^2}{x^2}\right) dy$$

$$\Rightarrow \qquad 2x \, dx = -y \, dy \qquad \Rightarrow \qquad x^2 = -\frac{y^2}{2} + c_1 \qquad \Rightarrow \qquad 2x^2 + y^2 = c.$$

35. Applying Leibniz's theorem, we switch the order of differentiation (with respect to y) and integration. This yields

$$g' = N(x, y) - \int_{x_0}^{x} \left(\frac{\partial}{\partial y} M(t, y)\right) dt.$$

Therefore, g' is differentiable (even continuously) with respect to x as a difference of two (continuously) differentiable functions ($N(x, y)$ and an integral with variable upper bound x of a continuous function $M_y'(t, y)$). Taking partial derivatives of both sides with respect to x and using the fundamental theorem of calculus, we obtain

$$\frac{\partial (g')}{\partial x} = \frac{\partial}{\partial x} \left[N(x, y) - \int_{x_0}^{x} \left(\frac{\partial}{\partial y} M(t, y)\right) dt \right]$$

$$= \frac{\partial}{\partial x} N(x, y) - \frac{\partial}{\partial x} \left[\int_{x_0}^{x} \left(\frac{\partial}{\partial y} M(t, y)\right) dt \right] = \frac{\partial}{\partial x} N(x, y) - \frac{\partial}{\partial y} M(x, y) = 0$$

due to (5). Thus, $\partial (g')/\partial x \equiv 0$, which implies that g' does not depend on x.

EXERCISES 2.5: Special Integrating Factors

1. This equation is neither separable, nor linear. Since

$$\frac{\partial M}{\partial y} = x^{-1} \neq \frac{\partial N}{\partial x} = y,$$

it is not exact either. But

$$\frac{M_y - N_x}{N} = \frac{x^{-1} - y}{xy - 1} = \frac{1 - xy}{x(xy - 1)} = -\frac{1}{x}$$

is a function of just x. So, there exists an integrating factor $\mu(x)$, which makes the equation exact.

3. This equation is also not separable and not linear. Computing

$$\frac{\partial M}{\partial y} = 1 = \frac{\partial N}{\partial x},$$

we see that it is exact.

5. It is not separable, but linear with x as independent variable. Since

$$\frac{\partial M}{\partial y} = 4 \neq \frac{\partial N}{\partial x} = 1,$$

this equation is not exact, but it has an integrating factor $\mu(x)$, because

$$\frac{M_y - N_x}{N} = \frac{3}{x}$$

depends on x only.

7. We find that

$$\frac{\partial M}{\partial y} = 2x, \qquad \frac{\partial N}{\partial x} = -6x \qquad \Rightarrow \qquad \frac{N_x - M_y}{M} = \frac{-8x}{2xy} = -\frac{4}{y}$$

depends just on y. So, an integrating factor is

$$\mu(y) = \exp\left[\int \left(-\frac{4}{y}\right) dy\right] = \exp\left(-4\ln y\right) = y^{-4}.$$

So, multiplying the given equation by y^{-4}, we get an exact equation

$$2xy^{-3}dx + \left(y^{-2} - 3x^2 y^{-4}\right) dy = 0.$$

Thus,

$$F(x, y) = \int 2xy^{-3}dx = x^2 y^{-3} + g(y),$$

$$\frac{\partial F}{\partial y} = -3x^2 y^{-4} + g'(y) = y^{-2} - 3x^2 y^{-4}$$

$$\Rightarrow \qquad g'(y) = y^{-2} \qquad \Rightarrow \qquad g(y) = -y^{-1}.$$

This yields a solution

$$F(x, y) = x^2 y^{-3} - y^{-1} = C,$$

which together with the lost solution $y \equiv 0$, gives a general solution to the given equation.

     Copyright © 2012 Pearson Education, Inc. Publishing as Addison-Wesley.

9. Since

$$\frac{\partial M}{\partial y} = 1, \quad \frac{\partial N}{\partial x} = -1, \quad \text{and} \quad \frac{M_y - N_x}{N} = \frac{2}{-x},$$

the equation has an integrating factor

$$\mu(x) = \exp\left[\int \left(-\frac{2}{x}\right) dx\right] = \exp\left(-2\ln x\right) = x^{-2}.$$

Therefore, the equation

$$x^{-2}\left[\left(x^4 - x + y\right) dx - x dy\right] = \left(x^2 - x^{-1} + x^{-2}y\right) dx - x^{-1} dy = 0$$

is exact. Therefore,

$$F(x, y) = \int \left(-x^{-1}\right) dy = -x^{-1}y + h(x),$$

$$\frac{\partial F}{\partial x} = x^{-2}y + h'(x) = x^2 - x^{-1} + x^{-2}y$$

$$\Rightarrow \quad h'(x) = x^2 - x^{-1} \quad \Rightarrow \quad h(x) = \frac{x^3}{3} - \ln|x|$$

$$\Rightarrow \quad -\frac{y}{x} + \frac{x^3}{3} - \ln|x| = C \quad \Rightarrow \quad y = \frac{x^4}{3} - x\ln|x| - Cx.$$

Together with the lost solution, $x \equiv 0$, this gives a general solution to the problem.

11. In this differential equation, $M(x, y) = y^2 + 2xy$, $N(x, y) = -x^2$. Therefore,

$$\frac{\partial M}{\partial y} = 2y + 2x, \quad \frac{\partial N}{\partial x} = -2x,$$

and so

$$\frac{\partial N/\partial x - \partial M/\partial y}{M} = \frac{-4x - 2y}{y^2 + 2xy} = -\frac{2}{y}$$

is a function of y. Thus,

$$\mu(y) = \exp\left[\int \left(-\frac{2}{y}\right) dy\right] = \exp\left(-2\ln|y|\right) = y^{-2}$$

is an integrating factor. Multiplying the given differential equation by $\mu(y)$ and solving the obtained exact equation, we get

$$y^{-2}\left(y^2 + 2xy\right) dx - y^{-2}x^2 dy = 0$$

$$\Rightarrow \quad F(x, y) = \int \left(-y^{-2}x^2\right) dy = x^2 y^{-1} + h(x)$$

$$\Rightarrow \quad \frac{\partial F}{\partial x} = \frac{\partial}{\partial x}\left[x^2 y^{-1} + h(x)\right] = 2xy^{-1} + h'(x) = y^{-2}\left(y^2 + 2xy\right) = 1 + 2xy^{-1}$$

Copyright © 2012 Pearson Education, Inc. Publishing as Addison-Wesley.

$$\Rightarrow \qquad h'(x) = 1 \qquad \Rightarrow \qquad h(x) = x \qquad \Rightarrow \qquad F(x,y) = x^2 y^{-1} + x.$$

Since we multiplied given equation by $\mu(y) = y^{-2}$ (in fact, divided by y^2) to get an exact equation, we could lose the solution $y \equiv 0$. This, indeed, happened: $y \equiv 0$ is, clearly, a solution to the original equation. Thus, a general solution is

$$x^2 y^{-1} + x = C. \qquad \text{and} \qquad y \equiv 0.$$

13. We multiply the equation by $x^n y^m$ and try to make it exact. Thus, we have

$$\left(2x^n y^{m+2} - 6x^{n+1} y^{m+1} \right) dx + \left(3x^{n+1} y^{m+1} - 4x^{n+2} y^m \right) dy = 0.$$

We want that $M_y(x,y) = N_x(x,y)$. Since

$$M_y(x,y) = 2(m+2)x^n y^{m+1} - 6(m+1)x^{n+1} y^m \,,$$
$$N_x(x,y) = 3(n+1)x^n y^{m+1} - 4(n+2)x^{n+1} y^m \,,$$

we need

$$2(m+2) = 3(n+1) \qquad \text{and} \qquad 6(m+1) = 4(n+2).$$

Solving these equations simultaneously, we obtain $n = 1$ and $m = 1$. So,

$$\mu(x,y) = xy.$$

With these choices for n and m, we obtain an exact equation

$$(2xy^3 - 6x^2 y^2)\, dx + (3x^2 y^2 - 4x^3 y)\, dy = 0.$$

Solving yields

$$F(x,y) = \int (2xy^3 - 6x^2 y^2)\, dx = x^2 y^3 - 2x^3 y^2 + g(y)$$
$$\Rightarrow \qquad F_y(x,y) = 3x^2 y^2 - 4x^3 y + g'(y) = N(x,y) = 3x^2 y^2 - 4x^3 y.$$

Therefore, $g'(y) = 0$. Since the constant of integration can be incorporated into the parameter C of the solution, we pick $g(y) \equiv 0$. Thus, we have

$$F(x,y) = x^2 y^3 - 2x^3 y^2 \,,$$

and the solution becomes

$$x^2 y^3 - 2x^3 y^2 = C.$$

Copyright © 2012 Pearson Education, Inc. Publishing as Addison-Wesley.

Since we have multiplied the original equation by xy we could have added the extraneous solutions $y \equiv 0$ or $x \equiv 0$. But, since $y \equiv 0$ implies that $dy/dx \equiv 0$ or $x \equiv 0$ implies that $dx/dy \equiv 0$, $y \equiv 0$ and $x \equiv 0$ are solutions to both the original equation and the transformed equation.

15. (a) Assume that, for a differential equation

$$M(x,y)dx + N(x,y)dy = 0, \tag{2.15}$$

the expression

$$\frac{\partial N/\partial x - \partial M/\partial y}{xM - yN} = H(xy) \tag{2.16}$$

is a function of $z = xy$ only. Denoting

$$\mu(z) = \exp\left(\int H(z)dz\right) \tag{2.17}$$

and multiplying (2.15) by $\mu(xy)$, we get a differential equation

$$\mu(xy)M(x,y)dx + \mu(xy)N(x,y)dy = 0. \tag{2.18}$$

Let us check it for exactness. First, we note that

$$\mu'(z) = \left[\exp\left(\int H(z)dz\right)\right]' = \exp\left(\int H(z)dz\right)\left(\int H(z)dz\right)' = \mu(z)H(z).$$

Next, using this fact, we compute partial derivatives of the coefficients in (2.18).

$$\frac{\partial}{\partial y}\{\mu(xy)M(x,y)\} = \mu'(xy)\frac{\partial(xy)}{\partial y}M(x,y) + \mu(xy)\frac{\partial M(x,y)}{\partial y}$$

$$= \mu(xy)H(xy)\,xM(x,y) + \mu(xy)\frac{\partial M(x,y)}{\partial y}$$

$$= \mu(xy)\left[H(xy)\,xM(x,y) + \frac{\partial M(x,y)}{\partial y}\right],$$

$$\frac{\partial}{\partial x}\{\mu(xy)N(x,y)\} = \mu'(xy)\frac{\partial(xy)}{\partial x}N(x,y) + \mu(xy)\frac{\partial N(x,y)}{\partial x}$$

$$= \mu(xy)H(xy)\,yN(x,y) + \mu(xy)\frac{\partial N(x,y)}{\partial x}$$

$$= \mu(xy)\left[H(xy)\,yN(x,y) + \frac{\partial N(x,y)}{\partial x}\right].$$

But (2.16) implies that

$$\frac{\partial N}{\partial x} - \frac{\partial M}{\partial y} = (xM - yN)H(xy) \quad \Leftrightarrow \quad yNH(xy) + \frac{\partial N}{\partial x} = xMH(xy) + \frac{\partial M}{\partial y},$$

Copyright © 2012 Pearson Education, Inc. Publishing as Addison-Wesley.

and, therefore,

$$\frac{\partial[\mu(xy)M(x,y)]}{\partial y} = \frac{\partial[\mu(xy)N(x,y)]}{\partial x}.$$

This means that the equation (2.18) is exact.

(b) Here, $M(x,y) = 3y + 2xy^2$, $N(x,y) = x + 2x^2y$. Thus,

$$\begin{aligned} M_y(x,y) &= 3 + 4xy \\ N_x(x,y) &= 1 + 4xy \end{aligned} \qquad \Rightarrow \qquad M_y - N_x = 2.$$

Clearly, $(M_y - N_x)/N = 2/(x + 2x^2y)$ is not a function depending only on x, and $(N_x - M_y)/M = -2/(3y + 2xy^2)$ is not a function of y. Therefore, this equation does not have an integrating factor $\mu(x)$ or $\mu(y)$. At the same time,

$$\frac{N_x - M_y}{xM - yN} = \frac{-2}{x(3y + 2xy^2) - y(x + 2x^2y)} = -\frac{1}{xy}$$

depends only on xy. So, we can find an integrating factor $\mu(xy)$ using (2.17) from part (a) with $H(z) = -1/z$.

$$\mu(z) = \exp\left[\int(-1/z)dz\right] = \frac{1}{|z|}.$$

Hence, $\mu(xy) = 1/|xy|$. We take, without loss of generality, $\mu(xy) = 1/xy$. Then, multiplying the given equation by $\mu(xy)$ yields an exact equation

$$\left(\frac{3}{x} + 2y\right)dx + \left(\frac{1}{y} + 2x\right)dy = 0.$$

Thus,

$$F(x,y) = \int\left(\frac{3}{x} + 2y\right)dx = 3\ln|x| + 2xy + g(y)$$

$$\Rightarrow \quad \frac{1}{y} + 2x = F_y(x,y) = 2x + g'(y) \quad \Rightarrow \quad g'(y) = \frac{1}{y} \quad \Rightarrow \quad g(y) = \ln|y|.$$

Therefore,

$$3\ln|x| + 2xy + \ln|y| = C = \ln\left|x^3y\right| + 2xy = C$$

is an implicit solution to the given equation. In addition, two constant solutions, $x \equiv 0$ and $y \equiv 0$ were lost in dividing the equation by xy. Substituting the initial condition, $y(1) = 1$, we find C.

$$\ln\left|(1^3)(1)\right| + 2(1)(1) = C \qquad \Rightarrow \qquad C = 2,$$

and so $\ln(x^3y) + 2xy = 2$ is the solution to the given initial value problem. (We can skip the absolute value sign because of positive initial values of variables.)

 Copyright © 2012 Pearson Education, Inc. Publishing as Addison-Wesley.

17. (a) A function $\mu(x^2 y)$ is an integrating factor if and only if the equation

$$\mu(x^2 y) M(x, y) dx + \mu(x^2 y) N(x, y) dy = 0$$

is exact. Thus,

$$\frac{\partial}{\partial y}\left[\mu(x^2 y) M(x, y)\right] = \frac{\partial}{\partial x}\left[\mu(x^2 y) N(x, y)\right]$$

$$\Rightarrow \quad x^2 \mu'(x^2 y) M(x, y) + \mu(x^2 y) M_y(x, y) = 2xy\mu'(x^2 y) N(x, y) + \mu(x^2 y) N_x(x, y)$$

$$\Rightarrow \quad \mu'(x^2 y)\left[x^2 M(x, y) - 2xy N(x, y)\right] = \mu(x^2 y)\left[N_x(x, y) - M_y(x, y)\right].$$

Therefore, the function

$$H(x, y) := \frac{N_x - M_y}{x^2 M - 2xy N} = \frac{\mu'(x^2 y)}{\mu(x^2 y)}$$

must depend on $z = x^2 y$ only. Integrating this equation yields

$$\mu(z) = \exp\left(\int H(z)dz\right). \tag{2.19}$$

(b) We have

$$\frac{N_x - M_y}{x^2 M - 2xy N} = \frac{(2 + 4x^3 + 6x^2 y) - (2 + 2x^3 + 8x^2 y)}{(2x^3 + 2x^2 y + 2x^5 y + 4x^4 y^2) - (4x^2 y + 2x^5 y + 4x^4 y^2)} \equiv 1.$$

Therefore, $H(z) \equiv 1$ and (2.19) imply that

$$\mu(z) = \exp\left(\int dz\right) = e^z.$$

Multiplying the given equation by $e^{x^2 y}$ yields an exact equation. Hence, we get

$$F(x, y) = \int e^{x^2 y} N(x, y) dy = \int e^{x^2 y}(2x + x^4 + 2x^3 y) dy$$

$$= \frac{2e^{x^2 y}}{x} + x^2 e^{x^2 y} + \frac{2}{x}\int z e^z dz \quad (z = x^2 y) = e^{x^2 y}(x^2 + 2xy) + h(x).$$

Differentiating $F(x, y)$ with respect to x and comparing the result with $e^{x^2 y} M(x, y)$ gives

$$\frac{\partial F}{\partial x} = e^{x^2 y}\left[2xy(x^2 + 2xy) + (2y + 2x)\right] + h'(x) = e^{x^2 y}\left(2x + 2y + 2x^3 y + 4x^2 y^2\right)$$

$$\Rightarrow \quad h'(x) = 0 \quad \Rightarrow \quad h(x) = \text{const.}$$

Thus, a general solution to the given equation is

$$e^{x^2 y}(x^2 + 2xy) = C.$$

Copyright © 2012 Pearson Education, Inc. Publishing as Addison-Wesley.

Chapter 2

19. (a) Expressing the family $y = x - 1 + ke^{-x}$ in the form $(y - x + 1)e^x = k$, we have (with notation of Problem 32, Exercises 2.4) $F(x, y) = (y - x + 1)e^x$. Computing

$$\frac{\partial F}{\partial x} = \frac{\partial}{\partial x}\left[(y - x + 1)e^x\right] = \frac{\partial(y - x + 1)}{\partial x}e^x + (y - x + 1)\frac{d(e^x)}{dx}$$
$$= -e^x + (y - x + 1)e^x = (y - x)e^x,$$

$$\frac{\partial F}{\partial y} = \frac{\partial}{\partial y}\left[(y - x + 1)e^x\right] = \frac{\partial(y - x + 1)}{\partial y}e^x = e^x,$$

we use the result of Problem 32 to derive an equation for the orthogonal trajectories (i.e., velocity potentials) of the given family of curves.

$$\frac{\partial F}{\partial y}\,dx - \frac{\partial F}{\partial x}\,dy = 0 \quad \Rightarrow \quad e^x\,dx - (y - x)e^x\,dy = 0 \quad \Rightarrow \quad dx + (x - y)dy = 0.$$

(b) In the differential equation $dx + (x - y)dy = 0$, $M = 1$ and $N = x - y$. Therefore,

$$\frac{\partial N/\partial x - \partial M/\partial y}{M} = \frac{\partial(x - y)/\partial x - \partial(1)/\partial y}{(1)} = 1,$$

and an integrating factor $\mu(y)$ is given by $\mu(y) = \exp\left[\int (1)dy\right] = e^y$. Multiplying the equation from part (a) by $\mu(y)$ yields an exact equation, and we look for its solutions of the form $G(x, y) = C$.

$$e^y\,dx + (x - y)e^y\,dy = 0$$
$$\Rightarrow \quad G(x, y) = \int e^y\,dx = xe^y + g(y)$$
$$\Rightarrow \quad \frac{\partial G}{\partial y} = xe^y + g'(y) = (x - y)e^y \quad \Rightarrow \quad g'(y) = -ye^y.$$

Integrating yields

$$g(y) = \int (-ye^y)dy = -\left(ye^y - \int e^y\,dy\right) = -ye^y + e^y.$$

Thus, the velocity potentials are given by

$$G(x, y) = xe^y - ye^y + e^y = C \quad \text{or} \quad x = y - 1 + Ce^{-y}.$$

EXERCISES 2.6: Substitutions and Transformations

1. We can write the equation in the form

$$\frac{dx}{dt} = \frac{x^2 - t^2}{2tx} = \frac{1}{2}\left(\frac{x}{t} - \frac{t}{x}\right),$$

Copyright © 2012 Pearson Education, Inc. Publishing as Addison-Wesley.

which shows that it is homogeneous. At the same time, it is a Bernoulli equation because it can be written as

$$\frac{dx}{dt} - \frac{1}{2t}x = -\frac{t}{2}x^{-1},$$

3. This is a Bernoulli equation $P(x) = 1/x$, $Q(x) = x^3$, and $n = 2$. Clearly. it does not fit any other form discussed in this section.

5. Dividing this equation by $\theta\,d\theta$, we obtain

$$\frac{dy}{d\theta} - \frac{1}{\theta}y = \frac{1}{\sqrt{\theta}}y^{1/2}.$$

Therefore, it is a Bernoulli equation. It can also be written in the form

$$\frac{dy}{d\theta} = \frac{y}{\theta} + \sqrt{\frac{y}{\theta}},$$

and so it is homogeneous as well.

7. We can rewrite the equation in the form

$$\frac{dy}{dx} = \frac{\sin(x+y)}{\cos(x+y)} = \tan(x+y).$$

Thus, it is of the form $dy/dx = G(ax + by)$ with $a = b = 1$ and $G(t) = \tan t$.

9. Writing the equation in the form

$$\frac{dy}{dx} = \frac{xy + y^2}{x^2} = \frac{y}{x} + \left(\frac{y}{x}\right)^2$$

and making the substitution $v = y/x$, we obtain

$$v + x\frac{dv}{dx} = v + v^2 \quad \Rightarrow \quad \frac{dv}{v^2} = \frac{dx}{x} \quad \Rightarrow \quad \int \frac{dv}{v^2} = \int \frac{dx}{x}$$

$$\Rightarrow \quad -\frac{1}{v} = \ln|x| + C \quad \Rightarrow \quad -\frac{x}{y} = \ln|x| + C \quad \Rightarrow \quad y = -\frac{x}{\ln|x| + C}.$$

In addition, separating variables, we lost a solution $v \equiv 0$, corresponding to $y \equiv 0$.

11. From

$$\frac{dx}{dy} = \frac{xy - y^2}{x^2} = \frac{y}{x} - \left(\frac{y}{x}\right)^2$$

we conclude that given equation is homogeneous. Let $u = y/x$. Then $y = xu$ and $y' = u + xu'$. Substitution yields

$$u + x\frac{du}{dx} = u - u^2 \quad \Rightarrow \quad x\frac{du}{dx} = -u^2 \quad \Rightarrow \quad -\frac{du}{u^2} = \frac{dx}{x}$$

Copyright © 2012 Pearson Education, Inc. Publishing as Addison-Wesley.

$$\Rightarrow \quad -\int \frac{du}{u^2} = \int \frac{dx}{x} \quad \Rightarrow \quad \frac{1}{u} = \ln|x| + C$$

$$\Rightarrow \quad \frac{x}{y} = \ln|x| + C \quad \Rightarrow \quad y = \frac{x}{\ln|x| + C}.$$

Note that, solving this equation, we have performed two divisions: by x^2 and u^2. In doing this, we lost two solutions, $x \equiv 0$ and $u \equiv 0$. (The latter gives $y \equiv 0$.) Therefore, a general solution to the given equation is

$$y = \frac{x}{\ln|x| + C}, \quad x \equiv 0, \quad \text{and} \quad y \equiv 0.$$

13. We can express $f(t, x)$ in the form $G(x/t)$ (dividing the numerator and denominator by t^2), that is,

$$\frac{x^2 + t\sqrt{t^2 + x^2}}{tx} = \frac{(x/t)^2 + \sqrt{(x/t)^2}}{(x/t)},$$

the equation is homogeneous. Substituting $v = x/t$ and $dx/dt = v + tdv/dt$ into the equation yields

$$v + t\frac{dv}{dt} = v + \frac{\sqrt{1 + v^2}}{v} \quad \Rightarrow \quad t\frac{dv}{dt} = \frac{\sqrt{1 + v^2}}{v}.$$

This transformed equation is separable. Thus, we have

$$\frac{v}{\sqrt{1 + v^2}}\,dv = \frac{1}{t}\,dt \quad \Rightarrow \quad \sqrt{1 + v^2} = \ln|t| + C,$$

where we have integrated with the integration on the left hand side being accomplished by the substitution $u = 1 + v^2$. Substituting back x/t for v in this formula gives the solution to the original equation, namely

$$\sqrt{1 + \frac{x^2}{t^2}} = \ln|t| + C.$$

15. This equation is homogeneous, because

$$\frac{dy}{dx} = \frac{x^2 - y^2}{3xy} = \frac{1 - (y/x)^2}{3(y/x)}.$$

Thus, we substitute $u = y/x$ (so that $y = xu$ and $y' = u + xu'$) to get

$$u + x\frac{du}{dx} = \frac{1 - u^2}{3u} \quad \Rightarrow \quad x\frac{du}{dx} = \frac{1 - 4u^2}{3u} \quad \Rightarrow \quad \frac{3u\,du}{1 - 4u^2} = \frac{dx}{x}$$

$$\Rightarrow \quad \int \frac{3u\,du}{1 - 4u^2} = \int \frac{dx}{x} \quad \Rightarrow \quad -\frac{3}{8}\ln\left|1 - 4u^2\right| = \ln|x| + C_1$$

 Copyright © 2012 Pearson Education, Inc. Publishing as Addison-Wesley.

$$\Rightarrow \quad -3\ln\left|1 - 4\left(\frac{y}{x}\right)^2\right| = 8\ln|x| + C_2$$

$$\Rightarrow \quad 3\ln(x^2) - 3\ln\left|x^2 - 4y^2\right| = 8\ln|x| + C_2,$$

which, after some algebra, yields $\left(x^2 - 4y^2\right)^3 x^2 = C$.

17. With the substitutions $z = x + y$ and $dz/dx = 1 + dy/dx$ or $dy/dx = dz/dx - 1$ this equation becomes a separable equation

$$\frac{dz}{dx} - 1 = \sqrt{z} - 1 \quad \Rightarrow \quad \frac{dz}{dx} = \sqrt{z}$$

$$\Rightarrow \quad z^{-1/2}\,dz = dx \quad \Rightarrow \quad 2z^{1/2} = x + C.$$

Substituting $z = x + y$ in this solution gives the solution of the original equation, i.e.,

$$2\sqrt{x + y} = x + C,$$

which, on solving for y, yields

$$y = \left(\frac{x}{2} + \frac{C}{2}\right)^2 - x.$$

Thus, we have

$$y = \frac{(x + C)^2}{4} - x.$$

Together with the lost solution $z = 0$ or $y = -x$, the last formula gives a general solution.

19. The right-hand side of this equation has the form $G(x - y)$ with

$$G(t) = (t + 5)^2.$$

Thus, we substitute

$$t = x - y \quad \Rightarrow \quad y = x - t \quad \Rightarrow \quad y' = 1 - t',$$

separate variables, and integrate.

$$1 - \frac{dt}{dx} = (t + 5)^2$$

$$\Rightarrow \quad \frac{dt}{dx} = 1 - (t + 5)^2 = (1 - t - 5)(1 + t + 5) = -(t + 4)(t + 6)$$

$$\Rightarrow \quad \frac{dt}{(t + 4)(t + 6)} = -dx \quad \Rightarrow \quad \int \frac{dt}{(t + 4)(t + 6)} = -\int dx$$

Copyright © 2012 Pearson Education, Inc. Publishing as Addison-Wesley.

$$\Rightarrow \quad \frac{1}{2} \int \left(\frac{1}{t+4} - \frac{1}{t+6} \right) dt = - \int dx \qquad \Rightarrow \qquad \ln \left| \frac{t+4}{t+6} \right| = -2x + C_1$$

$$\Rightarrow \quad \ln \left| \frac{x-y+4}{x-y+6} \right| = -2x + C_1 \qquad \Rightarrow \qquad \frac{x-y+6}{x-y+4} = C_2 e^{2x}$$

$$\Rightarrow \quad 1 + \frac{2}{x-y+4} = C_2 e^{2x} \qquad \Rightarrow \qquad y = x + 4 + \frac{2}{C e^{2x} + 1} = x + \frac{6 + 4C e^{2x}}{1 + C e^{2x}}.$$

Also, the solution

$$t + 4 \equiv 0 \qquad \Rightarrow \qquad y = x + 4$$

has been lost in separation of variables.

21. This is a Bernoulli equation with $n = 2$. So, we make a substitution $u = y^{1-n} = y^{-1}$, which gives $y = u^{-1}$, $y' = -u^{-2} u'$, and the equation becomes

$$-\frac{1}{u^2} \frac{du}{dx} + \frac{1}{ux} = \frac{x^2}{u^2} \qquad \Rightarrow \qquad \frac{du}{dx} - \frac{1}{x} u = -x^2.$$

The last equation is a linear equation with $P(x) = -1/x$. Following the procedure of solving linear equations, we find an integrating factor $\mu(x) = 1/x$ and multiply the equation by $\mu(x)$ to get

$$\frac{1}{x} \frac{du}{dx} - \frac{1}{x^2} u = -x \qquad \Rightarrow \qquad \frac{d}{dx} \left(\frac{1}{x} u \right) = -x$$

$$\Rightarrow \quad \frac{1}{x} u = \int (-x) dx = -\frac{1}{2} x^2 + C_1 \qquad \Rightarrow \qquad u = -\frac{1}{2} x^3 + C_1 x$$

$$\Rightarrow \quad y = \frac{1}{-(x^3/2) + C_1 x} = \frac{2}{Cx - x^3}.$$

Also, $y \equiv 0$ is a solution, which was lost when we multiplied the equation by u^2 (in terms of y, divided by y^2) to obtain a linear equation.

23. This is a Bernoulli equation with $n = 2$. Dividing it by y^2 and rewriting yields

$$y^{-2} \frac{dy}{dx} - 2x^{-1} y^{-1} = -x^2.$$

After the substitution $v = y^{-1}$ (and $dv/dx = -y^{-2} dy/dx$), the equation becomes

$$\frac{dv}{dx} + 2 \frac{v}{x} = x^2. \tag{2.20}$$

This is a linear equation in v and x. The integrating factor $\mu(x)$ is given by

$$\mu(x) = \exp \left(\int \frac{2}{x} dx \right) = e^{2 \ln |x|} = x^2.$$

 Copyright © 2012 Pearson Education, Inc. Publishing as Addison-Wesley.

Multiplying (2.20) by $\mu(x)$ and solving, we obtain

$$x^2 \frac{dv}{dx} + 2vx = x^4 \qquad \Rightarrow \qquad \frac{d}{dx}\left(x^2 v\right) = x^4$$

$$\Rightarrow \qquad x^2 v = \int x^4\, dx = \frac{x^5}{5} + C_1$$

$$\Rightarrow \qquad v = \frac{x^3}{5} + \frac{C_1}{x^2}.$$

Substituting y^{-1} for v in this solution gives the solution to the original equation. Thus, we find that

$$y^{-1} = \frac{x^3}{5} + \frac{C_1}{x^2} \qquad \Rightarrow \qquad y = \left(\frac{x^5 + 5C_1}{5x^2}\right)^{-1}.$$

Letting $C = 5C_1$ and simplifying yields

$$y = \frac{5x^2}{x^5 + C}.$$

Note: $y \equiv 0$ is also a solution to the original equation. It was lost in the first step when we divided by y^2.

25. In this Bernoulli equation, $n = 3$. Dividing the equation by x^3, we obtain

$$x^{-3} \frac{dx}{dt} + \frac{1}{t} x^{-2} = -t.$$

Now, we make a substitution $u = x^{-2}$ to obtain a linear equation. Since $u' = -2x^{-3}x'$, the equation becomes

$$-\frac{1}{2}\frac{du}{dt} + \frac{1}{t}u = -t \qquad \Rightarrow \qquad \frac{du}{dt} - \frac{2}{t}u = 2t.$$

This equation has an integrating factor

$$\mu(t) = \exp\left(-\int \frac{2}{t}\, dt\right) = t^{-2}.$$

Multiplying by $\mu(t)$ yields

$$\frac{d\left(t^{-2}u\right)}{dt} = \frac{2}{t} \qquad \Rightarrow \qquad t^{-2}u = \int \frac{2}{t}\, dt = 2\ln|t| + C$$

$$\Rightarrow \qquad u = 2t^2 \ln|t| + Ct^2 \qquad \Rightarrow \qquad x^{-2} = 2t^2 \ln|t| + Ct^2.$$

$x \equiv 0$ is also a solution, which we lost dividing the given equation by x^3.

Copyright © 2012 Pearson Education, Inc. Publishing as Addison-Wesley.

Chapter 2

27. This equation is a Bernoulli equation with $n = 2$, because it can be written in the form

$$\frac{dr}{d\theta} - \frac{2}{\theta} r = r^2 \theta^{-2}.$$

Dividing by r^2 and making the substitution $u = r^{-1}$, we obtain a linear equation.

$$r^{-2} \frac{dr}{d\theta} - \frac{2}{\theta} r^{-1} = \theta^{-2} \quad \Rightarrow \quad -\frac{du}{d\theta} - \frac{2}{\theta} u = \theta^{-2}$$

$$\Rightarrow \quad \frac{du}{d\theta} + \frac{2}{\theta} u = -\theta^{-2} \quad \Rightarrow \quad \mu(\theta) = \exp\left(\int \frac{2}{\theta} d\theta\right) = \theta^2$$

$$\Rightarrow \quad \frac{d\left(\theta^2 u\right)}{d\theta} = -1 \quad \Rightarrow \quad \theta^2 u = -\theta + C \quad \Rightarrow \quad u = \frac{-\theta + C}{\theta^2}.$$

Making back substitution (and adding the lost solution $r \equiv 0$), we obtain a general solution

$$r = \frac{\theta^2}{C - \theta} \quad \text{and} \quad r \equiv 0.$$

29. Solving the linear system

$$-3h + k - 1 = 0,$$
$$h + k + 3 = 0$$

for h and k gives $h = -1$, $k = -2$. Thus, we make the substitutions $x = u - 1$ and $y = v - 2$, so that $dx = du$ and $dy = dv$, to obtain

$$(-3u + v)\, du + (u + v)\, dv = 0.$$

This is the same transformed equation that we encountered in Example 4. There we found that its solution is

$$v^2 + 2uv - 3u^2 = C.$$

Substituting $x + 1$ for u and $y + 2$ for v gives the solution to the original equation

$$(y + 2)^2 + 2(x + 1)(y + 2) - 3(x + 1)^2 = C.$$

31. In this equation with linear coefficients, we make a substitution $x = u + h$, $y = v + k$, where h and k satisfy

$$\begin{aligned} 2h - k &= 0, \\ 4h + k &= 3 \end{aligned} \quad \Rightarrow \quad \begin{aligned} k &= 2h, \\ 4h + 2h &= 3 \end{aligned} \quad \Rightarrow \quad \begin{aligned} k &= 1, \\ h &= 1/2. \end{aligned}$$

Thus, $x = u + 1/2$, $y = v + 1$. As $dx = du$ and $dy = dv$, substitution yields

$$(2u - v)du + (4u + v)dv = 0 \quad \Rightarrow \quad \frac{du}{dv} = -\frac{4u + v}{2u - v} = -\frac{4(u/v) + 1}{2(u/v) - 1}$$

 Copyright © 2012 Pearson Education, Inc. Publishing as Addison-Wesley.

$$\Rightarrow \qquad z = \frac{u}{v} \qquad \Rightarrow \qquad u = vz \qquad \Rightarrow \qquad \frac{du}{dv} = z + v\frac{dz}{dv}$$

$$\Rightarrow \qquad z + v\frac{dz}{dv} = -\frac{4z+1}{2z-1} \qquad \Rightarrow \qquad v\frac{dz}{dv} = -\frac{4z+1}{2z-1} - z = -\frac{(2z+1)(z+1)}{2z-1}$$

$$\Rightarrow \qquad \frac{2z-1}{(2z+1)(z+1)}dz = -\frac{dv}{v} \qquad \Rightarrow \qquad \int \frac{2z-1}{(2z+1)(z+1)}dz = -\int \frac{dv}{v}\,.$$

To find the integral in the left-hand side of the above equation, we use the partial fractions decomposition

$$\frac{2z-1}{(2z+1)(z+1)} = -\frac{4}{2z+1} + \frac{3}{z+1}\,.$$

Therefore, the integration yields

$$-2\ln|2z+1| + 3\ln|z+1| = -\ln|v| + C_1 \qquad \Rightarrow \qquad |z+1|^3|v| = e^{C_1}|2z+1|^2$$

$$\Rightarrow \qquad \left(\frac{u}{v}+1\right)^3 v = C_2\left(2\frac{u}{v}+1\right)^2 \qquad \Rightarrow \qquad (u+v)^3 = C_2(2u+v)^2$$

$$\Rightarrow \qquad (x-1/2+y-1)^3 = C_2(2x-1+y-1)^2$$

$$\Rightarrow \qquad (2x+2y-3)^3 = C(2x+y-2)^2\,.$$

33. In Problem 1, we found that the given equation can be written as a Bernoulli equation,

$$\frac{dx}{dt} - \frac{1}{2t}x = -\frac{t}{2}x^{-1}\,.$$

Thus,

$$2x\frac{dx}{dt} - \frac{1}{t}x^2 = -t \qquad \Rightarrow \qquad (v = x^2) \qquad \frac{dv}{dt} - \frac{1}{t}v = -t.$$

The latter is a linear equation, which has an integrating factor

$$\mu(t) = \exp\left(-\int \frac{dt}{t}\right) = \frac{1}{t}\,.$$

Thus,

$$v = t\int(-1)dt = t(-t+C) = -t^2 + Ct$$

$$\Rightarrow \qquad x^2 + t^2 - Ct = 0,$$

where C is an arbitrary constant. We also note that a constant solution, $t \equiv 0$, was lost in writing the given equation as a Bernoulli equation.

35. Dividing the equation by y^2 yields

$$y^{-2}\frac{dy}{dx} + \frac{1}{x}y^{-1} = x^3 \qquad \Rightarrow \qquad v = y^{-1}, \; v' = -y^{-2}y'$$

Copyright © 2012 Pearson Education, Inc. Publishing as Addison-Wesley.

$$\Rightarrow \qquad -\frac{dv}{dx} + \frac{1}{x}v = x^3 \qquad \Rightarrow \qquad \frac{dv}{dx} - \frac{1}{x}v = -x^3$$

$$\Rightarrow \qquad \mu(x) = \exp\left(-\int \frac{dx}{x}\right) = \frac{1}{x}$$

$$\Rightarrow \qquad v = -x\int x^2 dx = -x\left(\frac{x^3}{3} + C_1\right) = -\frac{x^4 + Cx}{3},$$

where $C = 3C_1$ is an arbitrary constant. Thus,

$$y = v^{-1} = -\frac{3}{x^4 + Cx}.$$

Together with the constant (lost) solution $y \equiv 0$, this gives a general solution to the original equation.

37. Since this equation is a Bernoulli equation (see Problem 5), we make a substitution $v = y^{1/2}$ so that $2v' = y^{-1/2}y'$ and obtain a linear equation

$$2\frac{dv}{d\theta} - \frac{1}{\theta}v = \theta^{-1/2} \qquad \Rightarrow \qquad \frac{dv}{d\theta} - \frac{1}{2\theta}v = \frac{1}{2}\theta^{-1/2}.$$

An integrating factor for this equation is

$$\mu(\theta) = \exp\left(-\int \frac{d\theta}{2\theta}\right) = \theta^{-1/2}.$$

So,

$$v = \theta^{1/2}\int \left(\frac{1}{2}\theta^{-1/2}\theta^{-1/2}\right) d\theta = \frac{\theta^{1/2}}{2}(\ln|\theta| + C).$$

Therefore,

$$y = v^2 = \frac{\theta}{4}(\ln|\theta| + C)^2.$$

Dividing the given equation by $\theta\, d\theta$, we lost a constant solution $\theta \equiv 0$.

39. Using the conclusion made in Problem 7, we make a substitution $v = x + y$, $v' = 1 + y'$, and obtain a separable equation

$$\frac{dv}{dx} = \tan v + 1 \qquad \Rightarrow \qquad \frac{dv}{\tan v + 1} = dx.$$

The integral of the left-hand side can be found, for instance, as follows.

$$\int \frac{dv}{\tan v + 1} = \int \frac{\cos v\, dv}{\sin v + \cos v} = \frac{1}{2}\int \left(\frac{\cos v - \sin v}{\sin v + \cos v} + 1\right) dv$$

$$= \frac{1}{2}\left[\int \frac{d(\sin v + \cos v)}{\sin v + \cos v} + \int dv\right] = \frac{1}{2}(\ln|\sin v + \cos v| + v).$$

 Copyright © 2012 Pearson Education, Inc. Publishing as Addison-Wesley.

Therefore,

$$\frac{1}{2}\left(\ln|\sin v + \cos v| + v\right) = x + C_1$$

$$\Rightarrow \quad \ln|\sin(x+y) + \cos(x+y)| + x + y = 2x + C_2$$

$$\Rightarrow \quad \ln|\sin(x+y) + \cos(x+y)| = x - y + C_2$$

$$\Rightarrow \quad \sin(x+y) + \cos(x+y) = \pm e^{C_2} e^{x-y} = C e^{x-y},$$

where $C \neq 0$ is an arbitrary constant. Note that in procedure of separating variables we lost solutions corresponding to

$$\tan v + 1 = 0 \quad \Rightarrow \quad x + y = v = -\frac{\pi}{4} + k\pi, \quad k = 0, \pm 1, \pm 2, \ldots,$$

which can be included in the above formula by letting $C = 0$.

41. The right-hand side of (8) from Example 2 of the text can be written as

$$y - x - 1 + (x - y + 2)^{-1} = -(x - y + 2) + 1 + (x - y + 2)^{-1} = G(x - y + 2)$$

with $G(v) = -v + v^{-1} + 1$. With $v = x - y + 2$, we have $y' = 1 - v'$, and the equation becomes

$$1 - \frac{dv}{dx} = -v + v^{-1} + 1 \quad \Rightarrow \quad \frac{dv}{dx} = \frac{v^2 - 1}{v} \quad \Rightarrow \quad \frac{v}{v^2 - 1} dv = dx$$

$$\Rightarrow \quad \ln|v^2 - 1| = 2x + C_1 \quad \Rightarrow \quad v^2 - 1 = C e^{2x},$$

where $C \neq 0$ is an arbitrary constant. Dividing by $v^2 - 1$, we lost constant solutions $v = \pm 1$, which can be obtained by taking $C = 0$ in the above formula. Therefore, a general solution to the given equation is

$$(x - y + 2)^2 = C e^{2x} + 1,$$

where C is an arbitrary constant.

43. (a) If $f(tx, ty) = f(x, y)$ for any t, then, substituting $t = 1/x$, we obtain

$$f(tx, ty) = f\left(\frac{1}{x} \cdot x, \frac{1}{x} \cdot y\right) = f\left(1, \frac{y}{x}\right),$$

which shows that $f(x, y)$ depends, in fact, on y/x alone.

Copyright © 2012 Pearson Education, Inc. Publishing as Addison-Wesley.

(b) Since

$$\frac{dy}{dx} = -\frac{M(x,y)}{N(x,y)} =: f(x,y)$$

and the function $f(x,y)$ satisfies

$$f(tx, ty) = -\frac{M(tx, ty)}{N(tx, ty)} = -\frac{t^n M(x,y)}{t^n N(x,y)} = -\frac{M(x,y)}{N(x,y)} = f(x,y),$$

we apply part (a) to conclude that the equation $M(x,y)dx + N(x,y)dy = 0$ is homogeneous.

45. To obtain (17), we divide given equations:

$$\frac{dy}{dx} = -\frac{4x+y}{2x-y} = \frac{4+(y/x)}{(y/x)-2}.$$

Therefore, the equation is homogeneous, and the substitution $u = y/x$ yields

$$u + x\frac{du}{dx} = \frac{4+u}{u-2} \qquad \Rightarrow \qquad x\frac{du}{dx} = \frac{4+u}{u-2} - u = \frac{-u^2 + 3u + 4}{u-2}$$

$$\Rightarrow \qquad \frac{u-2}{u^2 - 3u - 4}\,du = -\frac{1}{x}\,dx \qquad \Rightarrow \qquad \int \frac{u-2}{u^2 - 3u - 4}\,du = -\int \frac{1}{x}\,dx\,.$$

Using partial fractions, we get

$$\frac{u-2}{u^2 - 3u - 4} = \frac{2}{5}\frac{1}{u-4} + \frac{3}{5}\frac{1}{u+1},$$

and so

$$\frac{2}{5}\ln|u-4| + \frac{3}{5}\ln|u+1| = -\ln|x| + C_1$$

$$\Rightarrow \qquad (u-4)^2(u+1)^3 x^5 = C$$

$$\Rightarrow \qquad \left(\frac{y}{x}-4\right)^2\left(\frac{y}{x}+1\right)^3 x^5 = C \qquad \Rightarrow \qquad (y-4x)^2(y+x)^3 = C.$$

47. (a) Substituting $y = u + 1/v$ into the Riccati equation yields

$$\frac{d}{dx}\left(u + \frac{1}{v}\right) = P(x)\left(u + \frac{1}{v}\right)^2 + Q(x)\left(u + \frac{1}{v}\right) + R(x)$$

$$\Rightarrow \qquad \frac{du}{dx} - \frac{1}{v^2}\frac{dv}{dx} = P(x)\left(u^2 + \frac{2u}{v} + \frac{1}{v^2}\right) + Q(x)\left(u + \frac{1}{v}\right) + R(x)$$

$$\Rightarrow \qquad \frac{du}{dx} - P(x)u^2 - Q(x)u - R(x) = \frac{1}{v^2}\frac{dv}{dx} + P(x)\frac{2uv+1}{v^2} + Q(x)\frac{1}{v}$$

$$\Rightarrow \qquad \frac{dv}{dx} + P(x)\left(2uv + 1\right) + Q(x)v = 0$$

Copyright © 2012 Pearson Education, Inc. Publishing as Addison-Wesley.

since $u(x)$ is a solution. Writing

$$\frac{dv}{dx} + [2P(x)u(x) + Q(x)]\,v = -P(x) \qquad \Rightarrow \qquad \frac{dv}{dx} + P_1(x)v = Q_1(x),$$

where $P_1(x) = 2P(x)u(x) + Q(x)$ and $Q_1(x) = -P(x)$, we see that this equation is linear in v.

(b) Writing

$$\frac{dy}{dx} = x^3\left(y^2 - 2xy + x^2\right) + \frac{y}{x} = x^3 y^2 + \left(\frac{1}{x} - 2x^4\right)y + x^5,$$

we see that in this Riccati equation $P(x) = x^3$, $Q(x) = (1/x) - 2x^4$, and $R(x) = x^5$. We compute $P_1(x) = 1/x$ and $Q_1(x) = -x^3$, so that the linear equation in v is

$$\frac{dv}{dx} + \frac{1}{x}v = -x^3.$$

The integrating factor

$$\mu(x) = \exp\left(\int \frac{dx}{x}\right) = x$$

implies that

$$v(x) = \frac{1}{x}\int(-x^3)x\,dx = \frac{1}{x}\left(-\frac{x^5}{5} + C_1\right) = \frac{C - x^5}{5x}.$$

Thus, a general solution to the given Riccati equation is

$$y = x + \frac{1}{v} = x + \frac{5x}{C - x^5}.$$

REVIEW PROBLEMS

1. Separation variables yields

$$\frac{y-1}{e^y}\,dy = e^x\,dx \qquad \Rightarrow \qquad (y-1)e^{-y}\,dy = e^x\,dx$$

$$\Rightarrow \qquad \int(y-1)e^{-y}\,dy = \int e^x\,dx \qquad \Rightarrow \qquad -(y-1)e^{-y} + \int e^{-y}\,dy = e^x + C$$

$$\Rightarrow \qquad -(y-1)e^{-y} - e^{-y} = e^x + C \qquad \Rightarrow \qquad e^x + ye^{-y} = -C,$$

and we can replace $-C$ by C.

3. The differential equation is an exact equation with $M = 2xy - 3x^2$ and $N = x^2 - 2y^{-3}$ because $M_y = 2x = N_x$. To solve this problem we will follow the procedure for solving exact equations given in Section 2.4. First we integrate $M(x, y)$ with respect to x to get

$$F(x, y) = \int\left(2xy - 3x^2\right)dx + g(y) \qquad \Rightarrow \qquad F(x, y) = x^2 y - x^3 + g(y). \qquad (2.21)$$

Copyright © 2012 Pearson Education, Inc. Publishing as Addison-Wesley.

To determine $g(y)$ take the partial derivative with respect to y of both sides and substitute $N(x, y)$ for $\partial F(x, y)/\partial y$ to obtain

$$N = x^2 - 2y^{-3} = x^2 + g'(y).$$

Solving for $g'(y)$ yields $g'(y) = -2y^{-3}$. Since the choice of the integration constant is arbitrary, we take $g(y) = y^{-2}$. Hence, from equation (2.21), $F(x, y) = x^2y - x^3 + y^{-2}$, and the solution to the differential equation is given implicitly by $x^2y - x^3 + y^{-2} = C$.

5. In this problem, $M(x, y) = \sin(xy) + xy\cos(xy)$ and $N(x, y) = 1 + x^2\cos(xy)$. Since

$$\frac{\partial M}{\partial y} = [x\cos(xy)] + [x\cos(xy) - xy\sin(xy)x] = 2x\cos(xy) - x^2y\sin(xy),$$

$$\frac{\partial N}{\partial x} = 0 + [2x\cos(xy) - x^2\sin(xy)y] = 2x\cos(xy) - x^2y\sin(xy),$$

we see that $\partial M/\partial y = \partial N/\partial x$ Therefore, the equation is exact. So, we use the method of solving exact equations and obtain

$$F(x, y) = \int N(x, y)dy = \int \left[1 + x^2\cos(xy)\right]dy = y + x\sin(xy) + h(x)$$

$$\Rightarrow \quad \frac{\partial F}{\partial x} = \sin(xy) + x\cos(xy)y + h'(x) = M(x, y) = \sin(xy) + xy\cos(xy)$$

$$\Rightarrow \quad h'(x) = 0 \quad \Rightarrow \quad h(x) \equiv 0,$$

and a general solution is given implicitly by $y + x\sin(xy) = C$.

7. This equation is separable. Separating variables and integrating, we get

$$t^3y^2\,dt = -t^4y^{-6}\,dy \quad \Rightarrow \quad \frac{dt}{t} = -\frac{dy}{y^8} \quad \Rightarrow \quad \ln|t| = \frac{1}{7y^7} + C_1.$$

Choosing y to be the independent variable yields

$$|t| = e^{1/(7y^7)+C_1} = e^{C_1}e^{1/(7y^7)} \quad \Rightarrow \quad t = \pm e^{C_1}e^{1/(7y^7)} = Ce^{1/(7y^7)},$$

where $C \neq 0$ is an arbitrary constant. Allowing $C = 0$, we include in this formula the solution $t \equiv 0$, which was lost in dividing the original equation by t^4.

9. The given differential equation can be written in the form

$$\frac{dy}{dx} + \frac{1}{3x}y = -\frac{x}{3}y^{-1}.$$

 Copyright © 2012 Pearson Education, Inc. Publishing as Addison-Wesley.

This is a Bernoulli equation with $n = -1$, $P(x) = 1/(3x)$, and $Q(x) = -x/3$. To transform this equation into a linear equation, we first multiply by y to obtain

$$y\frac{dy}{dx} + \frac{1}{3x}y^2 = -\frac{1}{3}x.$$

Next, we make the substitution $v = y^2$. Since $v' = 2yy'$, the transformed equation is

$$\frac{1}{2}v' + \frac{1}{3x}v = -\frac{1}{3}x, \qquad \Rightarrow \qquad v' + \frac{2}{3x}v = -\frac{2}{3}x. \tag{2.22}$$

The above equation is linear, so we can solve it for v using the method for solving linear equations discussed in Section 2.3. Following this procedure, the integrating factor $\mu(x)$ is found to be

$$\mu(x) = \exp\left(\int \frac{2}{3x}\,dx\right) = \exp\left(\frac{2}{3}\ln|x|\right) = x^{2/3}.$$

Multiplying equation (2.22) by $x^{2/3}$ yields

$$x^{2/3}v' + \frac{2}{3x^{1/3}}v = -\frac{2}{3}x^{5/3} \qquad \Rightarrow \qquad \left(x^{2/3}v\right)' = -\frac{2}{3}x^{5/3}.$$

We now integrate both sides and solve for v to find

$$x^{2/3}v = \int \frac{-2}{3}x^{5/3}\,dx = \frac{-1}{4}x^{8/3} + C_1 \qquad \Rightarrow \qquad v = \frac{-1}{4}x^2 + C_1 x^{-2/3}.$$

Substitution $v = y^2$ gives the solution

$$y^2 = -\frac{1}{4}x^2 + C_1 x^{-2/3} \qquad \Rightarrow \qquad (x^2 + 4y^2)x^{2/3} = 4C_1$$

or, cubing both sides, $(x^2 + 4y^2)^3 x^2 = C$, where $C := (4C_1)^3$ is an arbitrary constant.

11. The right-hand side of this equation is of the form $G(t - x)$ with $G(u) = 1 + \cos^2 u$. Thus, we make a substitution

$$t - x = u \qquad \Rightarrow \qquad x = t - u \qquad \Rightarrow \qquad x' = 1 - u',$$

which yields

$$1 - \frac{du}{dt} = 1 + \cos^2 u \qquad \Rightarrow \qquad \frac{du}{dt} = -\cos^2 u$$

$$\Rightarrow \qquad \sec^2 u\,du = -dt \qquad \Rightarrow \qquad \int \sec^2 u\,du = -\int dt$$

$$\Rightarrow \qquad \tan u = -t + C \qquad \Rightarrow \qquad \tan(t - x) + t = C.$$

Copyright © 2012 Pearson Education, Inc. Publishing as Addison-Wesley.

13. This is a linear equation with $P(x) = -1/x$. Following the method for solving linear equations, we find that an integrating factor is $\mu(x) = 1/x$, and so

$$\frac{d[(1/x)y]}{dx} = \frac{1}{x} x^2 \sin 2x = x \sin 2x$$

$$\Rightarrow \quad \frac{y}{x} = \int x \sin 2x \, dx = -\frac{1}{2} x \cos 2x + \frac{1}{2} \int \cos 2x \, dx = -\frac{1}{2} x \cos 2x + \frac{1}{4} \sin 2x + C$$

$$\Rightarrow \quad y = -\frac{x^2}{2} \cos 2x + \frac{x}{4} \sin 2x + Cx .$$

15. The right-hand side of the differential equation $y' = 2 - \sqrt{2x - y + 3}$ is a function of $2x - y$, and so it can be solved using the method for equations of the form $y' = G(ax + by)$. By letting $z = 2x - y$ we can transform the equation into a separable one. To make the substitution, we differentiate $z = 2x - y$ with respect to x to obtain

$$\frac{dz}{dx} = 2 - \frac{dy}{dx} \qquad \Rightarrow \qquad \frac{dy}{dx} = 2 - \frac{dz}{dx} .$$

Substituting $z = 2x - y$ and $y' = 2 - z'$ into the differential equation yields

$$2 - \frac{dz}{dx} = 2 - \sqrt{z + 3} \qquad \text{or} \qquad \frac{dz}{dx} = \sqrt{z + 3} .$$

To solve this equation, we divide it by $\sqrt{z + 3}$, multiply by dx, and integrate getting

$$\int (z + 3)^{-1/2} \, dz = \int dx \qquad \Rightarrow \qquad 2(z + 3)^{1/2} = x + C .$$

Thus,

$$z + 3 = \frac{(x + C)^2}{4} .$$

Finally, replacing z by $2x - y$ yields

$$2x - y + 3 = \frac{(x + C)^2}{4} .$$

Solving for y, we obtain

$$y = 2x + 3 - \frac{(x + C)^2}{4} .$$

17. This equation is a Bernoulli equation with $n = 2$. So, we divide it by y^2 and substitute $u = y^{-1}$ to get

$$-\frac{du}{d\theta} + 2u = 1 \quad \Rightarrow \quad \frac{du}{d\theta} - 2u = -1 \quad \Rightarrow \quad \mu(\theta) = \exp\left[\int (-2) d\theta\right] = e^{-2\theta}$$

$$\Rightarrow \quad \frac{d\left(e^{-2\theta} u\right)}{d\theta} = -e^{-2\theta} \quad \Rightarrow \quad e^{-2\theta} u = \int \left(-e^{-2\theta}\right) d\theta = \frac{e^{-2\theta}}{2} + C_1$$

   Copyright © 2012 Pearson Education, Inc. Publishing as Addison-Wesley.

$$\Rightarrow \qquad y^{-1} = \frac{1}{2} + C_1 e^{2\theta} = \frac{1 + Ce^{2\theta}}{2} \qquad \Rightarrow \qquad y = \frac{2}{1 + Ce^{2\theta}}.$$

This formula, together with $y \equiv 0$, gives a general solution to the given equation.

19. In the given differential equation, $M(x, y) = x^2 - 3y^2$ and $N(x, y) = 2xy$. The equation is not exact, because

$$\frac{\partial M}{\partial y} = -6y \neq 2x = \frac{\partial N}{\partial x}.$$

However, since $(\partial M/\partial y - \partial N/\partial x)/N = (-8y)/(2xy) = -4/x$ depends only on x, we can determine an integrating factor $\mu(x)$ from equation (8), Section 2.5. This gives

$$\mu(x) = \exp\left(\int \frac{-4}{x}\, dx\right) = x^{-4}.$$

Multiplying the given differential equation by $\mu(x) = x^{-4}$, we get an exact equation

$$(x^{-2} - 3x^{-4}y^2)\, dx + 2x^{-3}y\, dy = 0.$$

To find $F(x, y)$, we integrate $x^{-2} - 3x^{-4}y^2$ with respect to x and obtain

$$F(x, y) = \int (x^{-2} - 3x^{-4}y^2)\, dx = -x^{-1} + x^{-3}y^2 + g(y).$$

We take the partial derivative of F with respect to y and substitute $2x^{-3}y$ for $\partial F/\partial y$.

$$2x^{-3}y = 2x^{-3}y + g'(y).$$

Thus, $y'(y) = 0$ and, since the choice of the constant of integration is not important, we take $g(y) \equiv 0$. Hence, we have $F(x, y) = -x^{-1} + x^{-3}y^2$, and an implicit solution to the differential equation is

$$-x^{-1} + x^{-3}y^2 = C.$$

Solving for y^2 yields $y^2 = x^2 + Cx^3$.

Finally, we check if any solutions were lost in the process. We have multiplied the given equation by $\mu(x) = x^{-4}$, so we check if $x \equiv 0$ is a solution. Indeed, it is.

21. This equation has linear coefficients. Therefore, we are looking for substitutions of the form $x = u + h$ and $y = v + k$ with h and k satisfying

$$\begin{aligned} -2h + k - 1 &= 0, \\ h + k - 4 &= 0 \end{aligned} \qquad \Rightarrow \qquad \begin{aligned} h &= 1, \\ k &= 3. \end{aligned}$$

Copyright © 2012 Pearson Education, Inc. Publishing as Addison-Wesley.

So, $x = u + 1$ ($dx = du$) and $y = v + 3$ ($dy = dv$), and the equation becomes

$$(-2u + v)du + (u + v)dv = 0 \quad \Rightarrow \quad \frac{dv}{du} = \frac{2u - v}{u + v} = \frac{2 - (v/u)}{1 + (v/u)}.$$

With $z = v/u$, we have $v' = z + uz'$, and so

$$z + u\frac{dz}{du} = \frac{2 - z}{1 + z} \quad \Rightarrow \quad u\frac{dz}{du} = \frac{2 - z}{1 + z} - z = \frac{-z^2 - 2z + 2}{1 + z}$$

$$\Rightarrow \quad \frac{z + 1}{z^2 + 2z - 2}dz = -\frac{du}{u} \quad \Rightarrow \quad \int \frac{1 + z}{z^2 + 2z - 2}dz = -\int \frac{du}{u}$$

$$\Rightarrow \quad \frac{1}{2}\ln|z^2 + 2z - 2| = -\ln|u| + C_1 \quad \Rightarrow \quad (z^2 + 2z - 2)u^2 = C_2.$$

Back substitution, $z = v/u = (y - 3)/(x - 1)$, yields

$$v^2 + 2uv - 2u^2 = C_2 \quad \Rightarrow \quad (y - 3)^2 + 2(x - 1)(y - 3) - 2(x - 1)^2 = C_2$$

$$\Rightarrow \quad xy - x^2 - x + \frac{y^2}{2} - 4y = C.$$

23. Given equation is homogeneous because

$$\frac{dy}{dx} = \frac{x - y}{x + y} = \frac{1 - (y/x)}{1 + (y/x)}.$$

Therefore, substituting $u = y/x$, we obtain a separable equation.

$$u + x\frac{du}{dx} = \frac{1 - u}{1 + u} \quad \Rightarrow \quad x\frac{du}{dx} = \frac{-u^2 - 2u + 1}{1 + u}$$

$$\Rightarrow \quad \frac{u + 1}{u^2 + 2u - 1}du = -\frac{dx}{x} \quad \Rightarrow \quad \int \frac{1 + u}{u^2 + 2u - 1}du = -\int \frac{dx}{x}$$

$$\Rightarrow \quad \frac{1}{2}\ln|u^2 + 2u - 1| = -\ln|x| + C_1 \quad \Rightarrow \quad (u^2 + 2u - 1)x^2 = C,$$

and, substituting back $u = y/x$, after some algebra we get $y^2 + 2xy - x^2 = C$.

25. In this differential form, $M(x, y) = y(x - y - 2)$ and $N(x, y) = x(y - x + 4)$. Therefore,

$$\frac{\partial M}{\partial y} = x - 2y - 2, \qquad \frac{\partial N}{\partial x} = y - 2x + 4$$

$$\Rightarrow \quad \frac{\partial N/\partial x - \partial M/\partial y}{M} = \frac{(y - 2x + 4) - (x - 2y - 2)}{y(x - y - 2)} = \frac{-3(x - y - 2)}{y(x - y - 2)} = \frac{-3}{y},$$

which is a function of y alone. Therefore, the equation has a special integrating factor $\mu(y)$. We use formula (9), Section 2.5, to find that $\mu(y) = y^{-3}$. Multiplying the equation by $\mu(y)$ yields

$$y^{-2}(x - y - 2)\, dx + xy^{-3}(y - x + 4)\, dy = 0$$

Copyright © 2012 Pearson Education, Inc. Publishing as Addison-Wesley.

$$\Rightarrow \qquad F(x,y) = \int y^{-2}(x-y-2)\,dx = \frac{y^{-2}x^2}{2} - \left(y^{-1} + 2y^{-2}\right)x + g(y)$$

$$\Rightarrow \qquad \frac{\partial F}{\partial y} = -y^{-3}x^2 - \left(-y^{-2} - 4y^{-3}\right)x + g'(y) = N(x,y) = xy^{-3}\left(y - x + 4\right)$$

$$\Rightarrow \qquad g'(x) = 0 \qquad \Rightarrow \qquad g(y) \equiv 0\,,$$

and so

$$F(x,y) = \frac{y^{-2}x^2}{2} - x\left(y^{-1} + 2y^{-2}\right) = C_1 \qquad \Rightarrow \qquad x^2 y^{-2} - 2xy^{-1} - 4xy^{-2} = C$$

is a general solution. In addition, $y \equiv 0$ is a solution that we lost when multiplied the equation by $\mu(y) = y^{-3}$.

27. This equation has linear coefficients. Thus we make a substitution $x = u + h$, $y = v + k$ with h and k satisfying

$$\begin{aligned} 3h - k - 5 &= 0, \\ h - k + 1 &= 0 \end{aligned} \qquad \Rightarrow \qquad \begin{aligned} h &= 3, \\ k &= 4. \end{aligned}$$

With this substitution,

$$(3u - v)du + (u - v)dv = 0 \qquad \Rightarrow \qquad \frac{dv}{du} = -\frac{3u - v}{u - v} = -\frac{3 - (v/u)}{1 - (v/u)}$$

$$\Rightarrow \qquad z = \frac{v}{u}, \qquad v = uz, \qquad v' = z + uz'$$

$$\Rightarrow \qquad z + u\frac{dz}{du} = -\frac{3 - z}{1 - z} \qquad \Rightarrow \qquad u\frac{dz}{du} = -\frac{3 - z}{1 - z} - z = -\frac{z^2 - 3}{z - 1}$$

$$\Rightarrow \qquad \frac{z - 1}{z^2 - 3}dz = -\frac{du}{u} \qquad \Rightarrow \qquad \int \frac{z - 1}{z^2 - 3}dz = -\int \frac{du}{u}\,.$$

We use partial fractions decomposition to find the integral in the left-hand side. Namely, we write

$$\frac{z - 1}{z^2 - 3} = \frac{A}{z - \sqrt{3}} + \frac{B}{z + \sqrt{3}} \qquad \Rightarrow \qquad A = \frac{1}{2} - \frac{1}{2\sqrt{3}}, \qquad B = \frac{1}{2} + \frac{1}{2\sqrt{3}}\,.$$

Therefore, integration yields

$$A\ln\left|z - \sqrt{3}\right| + B\ln\left|z + \sqrt{3}\right| = -\ln|u| + C_1$$

$$\Rightarrow \qquad \left|z - \sqrt{3}\right|^{1 - 1/\sqrt{3}}\left|z + \sqrt{3}\right|^{1 + 1/\sqrt{3}}u^2 = C\,,$$

where $C := e^{2C_1} > 0$. Separating variables, we lost constant solutions $z = \pm\sqrt{3}$, which can be included in the above formula by letting $C = 0$. Back substitutions result

$$\left|v - u\sqrt{3}\right|^{1 - 1/\sqrt{3}}\left|v + u\sqrt{3}\right|^{1 + 1/\sqrt{3}} = C \quad \Rightarrow \quad \left|v^2 - 3u^2\right|\left|\frac{v + u\sqrt{3}}{v - u\sqrt{3}}\right|^{1/\sqrt{3}} = C$$

Copyright © 2012 Pearson Education, Inc. Publishing as Addison-Wesley.

$$\Rightarrow \qquad \left| (y-4)^2 - 3(x-3)^2 \right| \left| \frac{(y-4) + (x-3)\sqrt{3}}{(y-4) - (x-3)\sqrt{3}} \right|^{1/\sqrt{3}} = C, \quad C \geq 0.$$

We can eliminate the absolute value sign by allowing C to be negative. Thus

$$\left[(y-4)^2 - 3(x-3)^2 \right] \left| \frac{(y-4) + (x-3)\sqrt{3}}{(y-4) - (x-3)\sqrt{3}} \right|^{1/\sqrt{3}} = C,$$

where C is an arbitrary constant.

29. Here, $M(x,y) = 4xy^3 - 9y^2 + 4xy^2$ and $N(x,y) = 3x^2y^2 - 6xy + 2x^2y$. We compute

$$\frac{\partial M}{\partial y} = 12xy^2 - 18y + 8xy, \qquad \frac{\partial N}{\partial x} = 6xy^2 - 6y + 4xy,$$

$$\frac{\partial M/\partial y - \partial N/\partial x}{N} = \frac{(12xy^2 - 18y + 8xy) - (6xy^2 - 6y + 4xy)}{3x^2y^2 - 6xy + 2x^2y}$$

$$= \frac{2y(3xy - 6 + 2x)}{xy(3xy - 6 + 2x)} = \frac{2}{x},$$

which is a function of x alone. Thus, the equation has a special integrating factor

$$\mu(x) = \exp\left(\int \frac{2}{x}\, dx \right) = x^2.$$

Multiplying the equation by $\mu(x)$, we find that

$$F(x,y) = \int x^2 \left(4xy^3 - 9y^2 + 4xy^2 \right) dx = x^4 y^3 - 3x^3 y^2 + x^4 y^2 + g(y)$$

$$\Rightarrow \qquad \frac{\partial F}{\partial y} = 3x^4 y^2 - 6x^3 y + 2x^4 y + g'(y) = x^2 N(x,y) = x^2 \left(3x^2 y^2 - 6xy + 2x^2 y \right)$$

$$\Rightarrow \qquad g'(y) = 0 \qquad \Rightarrow \qquad g(y) \equiv 0$$

$$\Rightarrow \qquad F(x,y) = x^4 y^3 - 3x^3 y^2 + x^4 y^2 = C$$

is a general solution.

31. In this problem,

$$\frac{\partial M}{\partial y} = -1, \qquad \frac{\partial N}{\partial x} = 1, \qquad \text{and so} \qquad \frac{\partial M/\partial y - \partial N/\partial x}{N} = -\frac{2}{x}.$$

Therefore, the equation has a special integrating factor

$$\mu(x) = \exp\left[\int \left(\frac{-2}{x} \right) dx \right] = x^{-2}.$$

We multiply the given equation by $\mu(x)$ to get an exact equation.

$$\left(x - \frac{y}{x^2}\right) dx + \frac{1}{x} dy = 0$$

$$\Rightarrow \qquad F(x,y) = \int \left(\frac{1}{x}\right) dy = \frac{y}{x} + h(x)$$

$$\Rightarrow \qquad \frac{\partial F}{\partial x} = -\frac{y}{x^2} + h'(x) = x - \frac{y}{x^2} \qquad \Rightarrow \qquad h'(x) = x \qquad \Rightarrow \qquad h(x) = \frac{x^2}{2},$$

and a general solution is given by

$$F(x,y) = \frac{y}{x} + \frac{x^2}{2} = C \quad \text{and} \quad x \equiv 0.$$

(The latter has been lost in multiplication by $\mu(x)$.) Substitution the initial values yields

$$\frac{3}{1} + \frac{1^2}{2} = C \qquad \Rightarrow \qquad C = \frac{7}{2}.$$

Hence, the answer is

$$\frac{y}{x} + \frac{x^2}{2} = \frac{7}{2} \qquad \Rightarrow \qquad y = -\frac{x^3}{2} + \frac{7x}{2}.$$

33. Choosing x as the dependent variable, we transform the equation to

$$\frac{dx}{dt} + x = -(t + 3).$$

This equation is linear, $P(t) \equiv 1$. So, $\mu(t) = \exp\left(\int dt\right) = e^t$ and

$$\frac{d(e^t x)}{dt} = -(t + 3)e^t$$

$$\Rightarrow \qquad e^t x = -\int (t+3)e^t \, dt = -(t+3)e^t + \int e^t \, dt = -(t+2)e^t + C$$

$$\Rightarrow \qquad x = -(t+2) + Ce^{-t}.$$

Using the initial condition, $x(0) = 1$, we find that

$$1 = x(0) = -(0 + 2) + Ce^{-0} \qquad \Rightarrow \qquad C = 3,$$

and so $x = -t - 2 + 3e^{-t}$.

35. For $M(x,y) = 2y^2 + 4x^2$ and $N(x,y) = -xy$, we compute

$$\frac{\partial M}{\partial y} = 4y, \qquad \frac{\partial N}{\partial x} = -y \qquad \Rightarrow \qquad \frac{\partial M/\partial y - \partial N/\partial x}{N} = \frac{4y - (-y)}{-xy} = \frac{-5}{x},$$

Copyright © 2012 Pearson Education, Inc. Publishing as Addison-Wesley.

which is a function of x only. Using (8), Section 2.5, we find an integrating factor $\mu(x) = x^{-5}$ and multiply the equation by $\mu(x)$ to get an exact equation,

$$x^{-5}\left(2y^2 + 4x^2\right)dx - x^{-4}y\,dy = 0.$$

Hence,

$$F(x, y) = \int \left(-x^{-4}y\right)dy = -\frac{x^{-4}y^2}{2} + h(x)$$

$$\Rightarrow \quad \frac{\partial F}{\partial x} = \frac{4x^{-5}y^2}{2} + h'(x) = x^{-5}M(x, y) = 2x^{-5}y^2 + 4x^{-3}$$

$$\Rightarrow \quad h'(x) = 4x^{-3} \quad \Rightarrow \quad h(x) = -2x^{-2} \quad \Rightarrow \quad F(x, y) = -\frac{x^{-4}y^2}{2} - 2x^{-2} = C$$

is a general (implicit) solution to the given equation. We find C by substituting the initial condition, $y(1) = -2$.

$$-\frac{(1)^{-4}(-2)^2}{2} - 2(1)^{-2} = C \quad \Rightarrow \quad C = -4\,.$$

Thus, the solution is

$$-\frac{x^{-4}y^2}{2} - 2x^{-2} = -4 \quad \Rightarrow \quad y^2 + 4x^2 = 8x^4$$

$$\Rightarrow \quad y^2 = 8x^4 - 4x^2 = 4x^2\left(2x^2 - 1\right) \quad \Rightarrow \quad y = -2x\sqrt{2x^2 - 1}\,,$$

where, taking the square root, we have chosen the negative sign because of the initial negative value of y.

37. In this equation with linear coefficients we make a substitution $x = u + h$, $y = v + k$ with h and k such that

$$\begin{aligned}2h - k &= 0, \\ h + k &= 3\end{aligned} \quad \Rightarrow \quad \begin{aligned}k &= 2h \\ h + (2h) &= 3\end{aligned} \quad \Rightarrow \quad \begin{aligned}k &= 2\,, \\ h &= 1\,.\end{aligned}$$

Therefore,

$$(2u - v)du + (u + v)dv = 0$$

$$\Rightarrow \quad \frac{dv}{du} = \frac{v - 2u}{v + u} = \frac{(v/u) - 2}{(v/u) + 1} \quad \Rightarrow \quad z = v/u, \quad v = uz, \quad v' = z + uz'$$

$$\Rightarrow \quad z + u\frac{dz}{du} = \frac{z - 2}{z + 1} \quad \Rightarrow \quad u\frac{dz}{du} = -\frac{z^2 + 2}{z + 1} \quad \Rightarrow \quad \frac{z + 1}{z^2 + 2}dz = -\frac{du}{u}\,.$$

Integrating yields

$$\int \frac{z + 1}{z^2 + 2}dz = -\int \frac{du}{u} \quad \Rightarrow \quad \int \frac{z\,dz}{z^2 + 2} + \int \frac{dz}{z^2 + 2} = -\ln|u| + C_1\,.$$

 Copyright © 2012 Pearson Education, Inc. Publishing as Addison-Wesley.

The first integral on the left-hand side can be found by making a substitution $s = z^2 + 2$, and the second integral is a standard one. Thus, we obtain

$$\frac{1}{2} \ln\left(z^2 + 2\right) + \frac{1}{\sqrt{2}} \arctan\left(\frac{z}{\sqrt{2}}\right) = -\ln|u| + C_1$$

$$\Rightarrow \quad \ln\left[\left(z^2 + 2\right) u^2\right] + \sqrt{2} \arctan\left(\frac{z}{\sqrt{2}}\right) = C$$

$$\Rightarrow \quad \ln\left(v^2 + 2u^2\right) + \sqrt{2} \arctan\left(\frac{v}{u\sqrt{2}}\right) = C$$

$$\Rightarrow \quad \ln\left[(y-2)^2 + 2(x-1)^2\right] + \sqrt{2} \arctan\left[\frac{y-2}{(x-1)\sqrt{2}}\right] = C.$$

The initial condition, $y(0) = 2$, gives $C = \ln 2$, and so the answer is

$$\ln\left[(y-2)^2 + 2(x-1)^2\right] + \sqrt{2} \arctan\left[\frac{y-2}{\sqrt{2}(x-1)}\right] = \ln 2.$$

39. Multiplying the equation by y, we get

$$y \frac{dy}{dx} - \frac{2}{x} y^2 = \frac{1}{x}.$$

We substitute $u = y^2$ and obtain

$$\frac{1}{2} \frac{du}{dx} - \frac{2}{x} u = \frac{1}{x} \qquad \Rightarrow \qquad \frac{du}{dx} - \frac{4}{x} u = \frac{2}{x},$$

which is a linear equation having an integrating factor

$$\mu(x) = \exp\left[\int\left(-\frac{4}{x}\right) dx\right] = x^{-4}.$$

Hence,

$$\frac{d\left(x^{-4}u\right)}{dx} = 2x^{-5} \qquad \Rightarrow \qquad x^{-4}u = \int \left(2x^{-5}\right) dx = -\frac{x^{-4}}{2} + C$$

$$\Rightarrow \qquad x^{-4}y^2 = -\frac{x^{-4}}{2} + C \qquad \Rightarrow \qquad y^2 = -\frac{1}{2} + Cx^4.$$

Substitution $y(1) = 3$ yields

$$3^2 = -\frac{1}{2} + C(1)^4 \qquad \text{or} \qquad C = \frac{19}{2}.$$

Therefore, the solution to the given initial value problem is

$$y^2 = -\frac{1}{2} + \frac{19x^4}{2} \qquad \text{or} \qquad y = \sqrt{\frac{19x^4 - 1}{2}}.$$

Copyright © 2012 Pearson Education, Inc. Publishing as Addison-Wesley.

41. Writing the equation as

$$\frac{dy}{dt} + y = \frac{1}{1+t^2},$$

we see that $P(t) = 1$ and $Q(t) = 1/(1+t^2)$. Therefore,

$$\mu(t) = \exp\left(\int dt\right) = e^t$$

and

$$y = \frac{1}{e^t} \int \frac{e^t}{1+t^2}\, dt = e^{-t}\left(\int_2^t \frac{e^r}{1+r^2}\, dr + C\right),$$

where we have chosen the lower limit in the definite integral due to the initial value of $t = 2$. Since $y(2) = 3$, we obtain

$$3 = e^{-2}C \qquad \Rightarrow \qquad C = 3e^2.$$

Thus,

$$y = e^{-t}\left(\int_2^t \frac{e^r}{1+r^2}\, dr + 3e^2\right) = e^{-t}\int_2^t \frac{e^r}{1+r^2}\, dr + 3e^{-(t-2)}.$$

To approximate the value of $y(3)$, one can use the Simpson's method outlined in Appendix C. Results of computations are given in Table **2–F** on page 90. From this table we conclude that, within the required accuracy, $y(3) \approx 1.1883$.

TABLES

Table **2–A**: Successive approximations for $y(0.5)$ using Simpson's rule.

Number of Intervals	$y(0.5)$
6	1.38121
8	1.38121
10	1.38120

Copyright © 2012 Pearson Education, Inc. Publishing as Addison-Wesley.

Table 2–B: Successive approximations for $y(3)$ using Simpson's rule.

Number of Intervals	$y(3)$
6	0.18391
8	0.18329
10	0.18311
12	0.18304

Table 2–C: Approximations of $\nu(x) = \int_0^x \sqrt{1 + \sin^2 t}\, dt$ and $\mu(x) = e^{\nu(x)}$ using the Simpson's rule.

x	$\nu(x)$	$\mu(x)$	x	$\nu(x)$	$\mu(x)$
0.1	0.10017	1.10535	0.6	0.63202	1.88140
0.2	0.20132	1.22301	0.7	0.74890	2.11468
0.3	0.30436	1.35576	0.8	0.86992	2.38671
0.4	0.41010	1.50698	0.9	0.99498	2.70467
0.5	0.51917	1.68064	1.0	1.12387	3.07672

Table 2–D: Euler's method approximations for the solution of $y' + y\sqrt{1 + \sin^2 x} = x$, $y(0) = 2$, at $x = 1$ with $h = 0.1$.

k	x_k	y_k	k	x_k	y_k
0	0	2.0000	6	0.6	1.1637
1	0.1	1.8000	7	0.7	1.0900
2	0.2	1.6291	8	0.8	1.0304
3	0.3	1.4830	9	0.9	0.9836
4	0.4	1.3584	10	1.0	0.9486
5	0.5	1.2526			

Copyright © 2012 Pearson Education, Inc. Publishing as Addison-Wesley.

Table 2–E: Euler's method approximations for the solution of $y' + y\sqrt{1 + \sin^2 x} = x$, $y(0) = 2$, at $x = 1$ with $h = 0.05$.

n	x_n	y_n	n	x_n	y_n	n	x_n	y_n
0	0.00	2.0000	7	0.35	1.4368	14	0.70	1.1144
1	0.05	1.9000	8	0.40	1.3784	15	0.75	1.0831
2	0.10	1.8074	9	0.45	1.3244	16	0.80	1.0551
3	0.15	1.7216	10	0.50	1.2747	17	0.85	1.0301
4	0.20	1.6420	11	0.55	1.2290	18	0.90	1.0082
5	0.25	1.5683	12	0.60	1.1872	19	0.95	0.9892
6	0.30	1.5000	13	0.65	1.1490	20	1.00	0.9729

Table 2–F: Approximations to $y(3)$ in Problem 41 using Simpson's rule.

Number of Intervals	$y(3)$
4	1.188336292
6	1.188336279
8	1.188336277

FIGURES

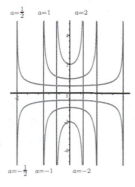

Figure 2–A: Solutions to the initial value problem $y' = xy^3$, $y(0) = a$, $a \pm 0.5$, ± 1, and ± 2.

 Copyright © 2012 Pearson Education, Inc. Publishing as Addison-Wesley.

CHAPTER 3: Mathematical Models and Numerical Methods Involving First-Order Equations

EXERCISES 3.2: Compartmental Analysis

1. Let $x(t)$ denote the mass of salt in the tank at time t with $t = 0$ denoting the moment when the process started. Thus, we have $x(0) = 0.5$ kg. We use the mathematical model described by equation (1) to find $x(t)$. Since the solution is entering the tank with rate 8 L/min and contains 0.05 kg/L of salt,

$$\text{input rate} = 8\,(\text{L/min}) \cdot 0.05\,(\text{kg/L}) = 0.4\,(\text{kg/min}).$$

We can determine the concentration of salt in the tank by dividing $x(t)$ by the volume of the solution, which remains constant, 100 L, because the flow rate in is the same as the flow rate out. Therefore, the concentration of salt at time t is $x(t)/100$ kg/L and

$$\text{output rate} = \frac{x(t)}{100}\,(\text{kg/L}) \cdot 8\,(\text{L/min}) = \frac{2x(t)}{25}\,(\text{kg/min}).$$

Then the equation (1) yields

$$\frac{dx}{dt} = 0.4 - \frac{2x}{25} \qquad \Rightarrow \qquad \frac{dx}{dt} + \frac{2x}{25} = 0.4\,, \qquad x(0) = 0.5\,.$$

This equation is linear, has integrating factor $\mu(t) = \exp\left[\int (2/25)dt\right] = e^{2t/25}$, and so

$$\frac{d\left(e^{2t/25}x\right)}{dt} = 0.4e^{2t/25}$$

$$\Rightarrow \qquad e^{2t/25}x = 0.4\left(\frac{25}{2}\right)e^{2t/25} + C = 5e^{2t/25} + C \qquad \Rightarrow \qquad x = 5 + Ce^{-2t/25}.$$

Using the initial condition, we find C.

$$0.5 = x(0) = 5 + C \qquad \Rightarrow \qquad C = -4.5\,.$$

So the mass of salt in the tank after t minutes is

$$x(t) = 5 - 4.5e^{-2t/25}\,(\text{kg}).$$

Copyright © 2012 Pearson Education, Inc. Publishing as Addison-Wesley.

If the concentration of salt in the tank is 0.02 kg/L, the mass of salt is $0.02 \times 100 = 2$ kg and, to find this instant in time, we solve

$$5 - 4.5e^{-2t/25} = 2 \qquad \Rightarrow \qquad e^{-2t/25} = \frac{2}{3} \qquad \Rightarrow \qquad t = \frac{25\ln(3/2)}{2} \approx 5.07 \,(\text{min}).$$

3. Let $x(t)$ be the volume of nitric acid in the tank at time t. The tank initially held 200 L of a 0.5% nitric acid solution; therefore, $x(0) = 200 \times 0.005 = 1$. Since 6 L of 20% nitric acid solution are flowing into the tank per minute, the rate at which nitric acid is entering is $6 \times 0.2 = 1.2$ L/min. Because the rate of flow out of the tank is 8 L/min and the rate of flow in the tank is only 6 L/min, there is a net loss in the tank of 2 L of solution every minute. Thus, at any time t, the tank will be holding $200 - 2t$ liters of solution. Combining this with the fact that the volume of nitric acid in the tank at time t is $x(t)$, we see that the concentration of nitric acid in the tank at time t is $x(t)/(200 - 2t)$. Here we are assuming that the solution is kept well stirred. The rate at which nitric acid flows out of the tank is, therefore, $8 \times [x(t)/(200 - 2t)]$ L/min. Thus, we see that

$$\text{input rate} = 1.2 \text{ L/min},$$
$$\text{output rate} = \frac{8x(t)}{200 - 2t} \text{ L/min}.$$

We know from (1) that

$$\frac{dx}{dt} = \text{input rate} - \text{output rate}.$$

Thus, we have to solve the initial value problem

$$\frac{dx}{dt} = 1.2 - \frac{4x(t)}{100 - t}, \qquad x(0) = 1.$$

The standard form of this linear equation is

$$\frac{dx}{dt} + \frac{4}{100 - t}x = 1.2,$$

and so it has an integrating factor

$$\mu(t) = \exp\left(\int \frac{4}{100 - t}\, dt\right) = e^{-4\ln(100-t)} = (100 - t)^{-4}.$$

Multiplying the previous equation by $\mu(t)$ yields

$$(100 - t)^{-4}\frac{dx}{dt} + 4x(100 - t)^{-5} = (1.2)(100 - t)^{-4}$$

 Copyright © 2012 Pearson Education, Inc. Publishing as Addison-Wesley.

$$\Rightarrow \qquad D_t\left[(100-t)^{-4}x\right] = (1.2)(100-t)^{-4}$$

$$\Rightarrow \qquad (100-t)^{-4}x = 1.2\int (100-t)^{-4}\,dt = \frac{1.2}{3}(100-t)^{-3} + C$$

$$\Rightarrow \qquad x(t) = (0.4)(100-t) + C(100-t)^4.$$

To find the value of C, we use the initial condition, $x(0) = 1$. So,

$$x(0) = (0.4)(100) + C(100)^4 = 1 \qquad \Rightarrow \qquad C = \frac{-39}{100^4} = -3.9 \times 10^{-7}.$$

This means that, at time t, there is

$$x(t) = (0.4)(100-t) - (3.9 \times 10^{-7})(100-t)^4 \qquad\qquad (3.1)$$

liters of nitric acid in the tank. When the percentage of nitric acid is 10%, its concentration equals 0.1. Thus, we to solve the equation

$$\frac{x(t)}{200 - 2t} = \frac{x(t)}{2(100-t)} = 0.1$$

for t. (3.1) yields

$$(0.2) - (1.95 \times 10^{-7})(100-t)^3 = 0.1$$

$$\Rightarrow \qquad t = -\left[0.1 \cdot \frac{10^7}{1.95}\right]^{1/3} + 100 \approx 19.96 \text{ (min)}.$$

5. Let $x(t)$ denote the volume of chlorine in the pool at time t. Then, in the formula

$$\text{rate of change} = \text{input rate} - \text{output rate},$$

we have

$$\text{input rate} = 5\,(\text{gal/min}) \cdot \frac{0.001\%}{100\%} = 5 \cdot 10^{-5}\,(\text{gal/min}),$$

$$\text{output rate} = 5\,(\text{gal/min}) \cdot \frac{x(t)\,(\text{gal})}{10,000\,(\text{gal})} = 5 \cdot 10^{-4}x(t)\,(\text{gal/min}),$$

and the equation for $x(t)$ becomes

$$\frac{dx}{dt} = 5 \cdot 10^{-5} - 5 \cdot 10^{-4}x \qquad \Rightarrow \qquad \frac{dx}{dt} + 5 \cdot 10^{-4}x = 5 \cdot 10^{-5}.$$

This is a linear equation. Solving yields

$$x(t) = 0.1 + Ce^{5 \cdot 10^{-4}t} = 0.1 + Ce^{-0.0005t}.$$

Copyright © 2012 Pearson Education, Inc. Publishing as Addison-Wesley.

Using the initial condition,

$$x(0) = 10,000\,(\text{gal}) \cdot \frac{0.01\%}{100\%} = 1\,(\text{gal}),$$

we find that

$$1 = 0.1 + Ce^{-0.0005 \cdot 0} \qquad \Rightarrow \qquad C = 0.9.$$

Therefore, $x(t) = 0.1 + 0.9e^{-0.0005t}$, and so the concentration of chlorine, say $c(t)$, in the pool at time t is

$$c(t) = \frac{x(t)}{10,000} \cdot 100\% = \frac{x(t)}{100}\,\% = 0.001 + 0.009e^{-0.0005t}\,\%.$$

After 1 hour (i.e., $t = 60$ min),

$$c(60) = 0.001 + 0.009e^{-0.0005 \cdot 60} = 0.001 + 0.009e^{-0.03} \approx 0.0097\,\%.$$

To answer the second question, we solve the equation

$$c(t) = 0.001 + 0.009e^{-0.0005t} = 0.002 \qquad \Rightarrow \qquad t = \frac{\ln(1/9)}{-0.0005} \approx 4394.45\,(\text{min}) \approx 73.24\,(\text{h}).$$

7. Let $x(t)$ denote the mass of salt in the first tank at time t. Assuming that the initial mass is $x(0) = x_0$, we use the mathematical model described by equation (1) to find $x(t)$. We can determine the concentration of salt in the first tank by dividing $x(t)$ by the volume of the tank: $x(t)/60$ (kg/gal). Note that the volume of the solution in this tank remains constant, because the flow rate in is the same as the flow rate out. Then

$$\text{output rate}_1 = (3\text{ gal/min}) \cdot \left(\frac{x(t)}{60}\text{ kg/gal}\right) = \frac{x(t)}{20}\text{ kg/min}.$$

Since the incoming liquid is pure water, we conclude that

$$\text{input rate}_1 = 0.$$

Therefore, $x(t)$ satisfies the initial value problem

$$\frac{dx}{dt} = \text{input rate}_1 - \text{output rate}_1 = -\frac{x}{20}, \qquad x(0) = x_0.$$

This equation is linear (and separable). Solving and using the initial condition, we find that

$$x(t) = x_0 e^{-t/20}.$$

 Copyright © 2012 Pearson Education, Inc. Publishing as Addison-Wesley.

Now, let $y(t)$ denote the mass of salt in the second tank at time t. Since initially this tank contained only pure water, we have $y(0) = 0$. For the function $y(t)$ we use the same mathematical model. Thus, we get

$$\text{input rate}_2 = \text{output rate}_1 = \frac{x(t)}{20} = \frac{x_0}{20} e^{-t/20} \text{ (kg/min)}.$$

Further, since the volume of the second tank also remains constant, we have

$$\text{output rate}_2 = (3 \text{ gal/min}) \cdot \left(\frac{y(t)}{60} \text{ kg/gal}\right) = \frac{y(t)}{20} \text{ kg/min}.$$

Therefore, $y(t)$ satisfies the initial value problem

$$\frac{dy}{dt} = \text{input rate}_2 - \text{output rate}_2 = \frac{x_0}{20} e^{-t/20} - \frac{y(t)}{20}, \qquad y(0) = 0$$

or

$$\frac{dy}{dt} + \frac{y(t)}{20} = \frac{x_0}{20} e^{-t/20}, \qquad y(0) = 0.$$

This is a linear equation in standard form. Using the method given in Section 2.3, we find a general solution

$$y(t) = \frac{x_0}{20} t e^{-t/20} + C e^{-t/20}.$$

The constant C can be found from the initial condition:

$$0 = y(0) = \frac{x_0}{20} \cdot 0 \cdot e^{-0/20} + C e^{-0/20} \qquad \Rightarrow \qquad C = 0.$$

Therefore, $y(t) = (x_0/20) t e^{-t/20}$. To investigate $y(t)$ for its maximum value, we calculate

$$\frac{dy}{dt} = \frac{x_0}{20} \left(e^{-t/20} - \frac{t e^{-t/20}}{20}\right) = \frac{x_0}{20} e^{-t/20} \left(1 - \frac{t}{20}\right).$$

Thus,

$$\frac{dy}{dt} = 0 \qquad \Leftrightarrow \qquad 1 - \frac{t}{20} = 0 \qquad \Leftrightarrow \qquad t = 20 \,(\text{min.}),$$

which is the point of global maximum (notice that $dy/dt > 0$ for $t < 20$ and $dy/dt < 0$ for $t > 20$). In other words, at this moment the water in the second tank will taste saltiest and, comparing the concentrations, it will be

$$\frac{y(20)/60}{x_0/60} = \frac{y(20)}{x_0} = \frac{1}{20} \cdot 20 \cdot e^{-20/20} = \frac{1}{e}$$

times as salty as the original brine.

Chapter 3

9. Let $p(t)$ be the population of splake in the lake at time t. We start counting the population in 1990. Thus, we let $t = 0$ correspond to the year 1990. By the Malthusian (exponential) law, we have

$$p(t) = p_0 e^{kt}.$$

Since $p_0 = p(0) = 1000$, we see that

$$p(t) = 1000 e^{kt}.$$

To find k, we use the fact that the population of splake was 3000 in 1997. Therefore,

$$p(7) = 3000 = 1000 e^{k \cdot 7} \quad \Rightarrow \quad 3 = e^{7k} \quad \Rightarrow \quad k = \frac{\ln 3}{7}.$$

Putting this value for k into the equation for $p(t)$ yields

$$p(t) = 1000 e^{(t \ln 3)/7} = 1000 \cdot 3^{t/7}.$$

To estimate the population in 2020 we plug in $t = 2020 - 1990 = 30$ into this formula and get

$$p(30) = 1000 \cdot 3^{30/7} \approx 110,868 \text{ (splakes)}.$$

11. In this problem, the dependent variable is p, the independent variable is t, and the function $f(t, p) = (a - bp)p$. Since $f(t, p) = f(p)$, i.e., it does not depend on t, the given equation is autonomous. To find equilibrium solutions, we solve

$$f(p) = 0 \quad \Rightarrow \quad (a - bp)p = 0 \quad \Rightarrow \quad p_1 = 0, \quad p_2 = \frac{a}{b}.$$

Thus, $p_1(t) \equiv 0$ and $p_2(t) \equiv a/b$ are equilibrium solutions. For $p_1 < p < p_2$, $f(p) > 0$, and $f(p) < 0$ when $p > p_2$. (Also, $f(p) < 0$ for $p < p_1$.) Thus, the phase line for the given equation is as it is shown in Figure **3–A** on page 149. From this picture, we conclude that the equilibrium $p = p_1$ is a source, while $p = p_2$ is a sink. Thus, regardless of an initial point $p_0 > 0$, the solution to the corresponding initial value problem will approach $p_2 = a/b$ as $t \to \infty$.

13. With the year 1990 corresponding to $t = 0$, the data given can be written as

$$t_0 = 0, \qquad\qquad p_0 = p(t_0) = 1000;$$
$$t_a = 1997 - 1990 = 7, \qquad p_a = p(t_a) = 3000;$$
$$t_b = 2004 - 1990 = 14, \quad p_b = p(t_b) = 5000.$$

    Copyright © 2012 Pearson Education, Inc. Publishing as Addison-Wesley.

Since $t_b = 2t_a$, we can use the formulas in Problem 12 to compute parameters p_1 and A in the logistic model (14).

$$p_1 = \frac{(3000)(5000) - 2(1000)(5000) + (1000)(3000)}{(3000)^2 - (1000)(5000)}(3000) = 6000\,,$$

$$A = \frac{1}{(6000)7}\ln\left[\frac{5000(3000 - 1000)}{1000(5000 - 3000)}\right] = \frac{\ln 5}{42000}\,.$$

Thus, the formula (15) becomes

$$p(t) = \frac{p_0 p_1}{p_0 + (p_1 - p_0)e^{-Ap_1 t}}$$

$$= \frac{(1000)(6000)}{(1000) + (6000 - 1000)e^{-(\ln 5/42000)6000t}} = \frac{6000}{1 + 5^{1-t/7}}\,. \qquad (3.2)$$

In the year 2020, $t = 2020 - 1990 = 30$, and the estimated population of splake is

$$p(30) = \frac{6000}{1 + 5^{1-30/7}} \approx 5970.$$

Taking the limit in (3.2), as $t \to \infty$, yields

$$\lim_{t\to\infty} p(t) = \lim_{t\to\infty}\frac{6000}{1 + 5^{1-t/7}} = \frac{6000}{1 + \lim_{t\to\infty} 5^{1-t/7}} = 6000.$$

Therefore, the predicted limiting population is 6000.

15. Counting time from the year 1980, we have the following data:

$$t_0 = 0, \qquad\qquad p_0 = p(t_0) = 1500;$$
$$t_a = 1993 - 1980 = 13, \quad p_a = p(t_a) = 4100;$$
$$t_b = 2006 - 1980 = 26, \quad p_b = p(t_b) = 6000.$$

Since $t_b = 2t_a$, we use the formulas in Problem 12 to find parameters in the logistic model.

$$p_1 = \left[\frac{(4100)(6000) - 2(1500)(6000) + (1500)(4100)}{(4100)^2 - (1500)(6000)}\right](4100) = \frac{5227500}{781};$$

$$Ap_1 = \frac{1}{13}\ln\left[\frac{(6000)(4100 - 1500)}{(1500)(6000 - 4100)}\right] = \frac{\ln(104/19)}{13}, \,.$$

Therefore,

$$p(t) = \frac{(1500)(5227500/781)}{1500 + [(5227500/781) - 1500]e^{-\ln(104/19)\,t/13}} = \frac{5227500}{781 + 2704(19/104)^{t/13}}\,.$$

Copyright © 2012 Pearson Education, Inc. Publishing as Addison-Wesley.

In the year 2020, $t = 2020 - 1980 = 40$, and so the estimated population of alligators is

$$p(40) = \frac{5227500}{781 + 2704(19/104)^{40/13}} \approx 6572.$$

Taking the limit of $p(t)$, as $t \to \infty$, we get the predicted limiting population

$$\lim_{t \to \infty} p(t) = p_1 = \frac{5227500}{781} \approx 6693.$$

17. **(a)** With the year of 1790 corresponding to $t = 0$, we compute

$$\frac{1}{p(0)} \frac{p(10) - p(0)}{10} = \frac{1}{3.93} \frac{5.31 - 3.93}{10} \approx 0.03511450382 \,;$$

$$\frac{1}{p(10)} \frac{p(20) - p(10)}{10} = \frac{1}{5.31} \frac{7.24 - 5.31}{10} \approx 0.03634651601 \,;$$

$$\vdots$$

$$\frac{1}{p(210)} \frac{p(220) - p(210)}{10} = \frac{1}{281.42} \frac{308.75 - 281.42}{10} \approx 0.00971146329 \,.$$

Results of these computations (rounded to four decimal places) are listed in the second column in Table **3–A** on page 146.

(b) We use the formulas for coefficients $\alpha(= Ap_1)$ and $\beta(= -A)$ given in Appendix E (with y_i's as new values of $(1/p)dp/dt$, x_i's as the values of p, and $N = 22$). Thus,

$$\sum_{i=1}^{22} x_i = 3.93 + 5.31 + \cdots + 281.42 = 2084.12 \,;$$

$$\sum_{i=1}^{22} x_i^2 = 3.93^2 + 5.31^2 + \cdots + 281.42^2 \approx 357204.0650 \,;$$

$$\sum_{i=1}^{22} y_i = 0.03511450382 + \cdots + 0.00971146329 = 0.4912692468 \,;$$

$$\sum_{i=1}^{22} x_i y_i = 3.93 \cdot 0.03511450382 + \cdots + 281.42 \cdot 0.00971146329 \approx 30.8420 \,.$$

Therefore,

$$\alpha = \frac{357204.0650 \cdot 0.4912692468 - 2084.12 \cdot 30.8420}{22 \cdot 357204.0650 - 2084.12^2} \approx 0.0318513093 \,,$$

$$\beta = \frac{22 \cdot 30.8420 - 2084.12 \cdot 0.4912692468}{22 \cdot 357204.0650 - 2084.12^2} \approx -0.0001005026 \,,$$

and so

$$A = -\beta \approx 0.0001005026 \approx 0.00010050 \,, \quad p_1 = \frac{\alpha}{A} \approx 316.9202518 \approx 316.920 \,.$$

 Copyright © 2012 Pearson Education, Inc. Publishing as Addison-Wesley.

(c) With $Ap_1 = \alpha = 0.0318513093$ and $p_1 = 316.9202508$, we compute p_0 as

$$p_0 = \frac{p_1}{23} \sum_{k=0}^{22} \frac{p(10k)e^{-\alpha(10k)}}{p_1 - p(10k)\left[1 - e^{-\alpha(10k)}\right]} = 3.287791117 \approx 3.28780.$$

(The Census result of 2010 is included.)

(d) We use p_1, Ap_1, and p_0 found in parts (b) and (c), and the formula

$$p(t) = \frac{p_0 p_1}{p_0 + (p_1 - p_0)\, e^{-Ap_1 t}}$$

for $t = 10k$, $k = 0, \ldots, 23$ to fill the third column in Table **3–A**. (Results are rounded to two decimal places.)

19. This problem can be regarded as a compartmental analysis problem for the population of fish. If we let $m(t)$ denote the mass in million tons of a certain species of fish, then the mathematical model for this process is given by

$$\frac{dm}{dt} = \text{increase rate} - \text{decrease rate},$$

where the increase rate of fish is $2m$ million tons/yr. The decrease rate of fish is 15 million tons/yr. Thus, $m(t)$ satisfies

$$\frac{dm}{dt} = 2m - 15, \qquad m(0) = 7.$$

This equation is linear (and also separable). Using the initial condition, $m(0) = 7$, to evaluate the constant of integration, we obtain

$$m(t) = -\frac{1}{2}\, e^{2t} + \frac{15}{2}\,.$$

We can now find the time when all the fish will be gone. Setting $m(t) = 0$ and solving for t yields

$$0 = -\frac{1}{2}\, e^{2t} + \frac{15}{2} \qquad \Rightarrow \qquad t = \frac{1}{2}\, \ln(15) \approx 1.354 \text{ (years)}.$$

To determine the fishing rate required to keep the fish mass constant, we solve

$$\frac{dm}{dt} = 2m - r, \qquad m(0) = 7,$$

with r as the fishing rate. Thus, we obtain

$$m(t) = Ke^{2t} + \frac{r}{2}\,.$$

Copyright © 2012 Pearson Education, Inc. Publishing as Addison-Wesley.

Substituting the initial condition into this equation, we find the constant K.

$$m(0) = 7 = K + \frac{r}{2} \quad \Rightarrow \quad K = 7 - \frac{r}{2} \quad \Rightarrow \quad m(t) = \left(7 - \frac{r}{2}\right)e^{2t} + \frac{r}{2}.$$

A fishing rate of $r = 14$ million tons/year will give a constant mass of fish by cancelling out the coefficient of the e^{2t} term.

21. Let $D = D(t)$, $S(t)$, and $V(t)$ denote the diameter, surface area, and volume of the snowball at time t, respectively. From geometry, we know that $V = \pi D^3/6$ and $S = \pi D^2$. Since we are given that $V'(t)$ is proportional to $S(t)$, the equation describing the melting process is

$$\frac{dV}{dt} = kS \quad \Rightarrow \quad \frac{d}{dt}\left(\frac{\pi}{6}D^3\right) = k\left(\pi D^2\right)$$

$$\Rightarrow \quad \frac{\pi}{2}D^2\frac{dD}{dt} = k\pi D^2 \quad \Rightarrow \quad \frac{dD}{dt} = 2k = \text{const.}$$

Solving, we get $D = 2kt + C$. Initially, $D(0) = 4$, and we also know that $D(30) = 3$. These data allow us to find k and C.

$$4 = D(0) = 2k \cdot 0 + C \quad \Rightarrow \quad C = 4;$$
$$3 = D(30) = 2k \cdot 30 + C = 2k \cdot 30 + 4 \quad \Rightarrow \quad 2k = -\frac{1}{30}.$$

Thus,

$$D(t) = -\frac{t}{30} + 4.$$

The diameter $D(t)$ of the snowball will be 2 inches after

$$-\frac{t}{30} + 4 = 2 \quad \Rightarrow \quad t = 60\,(\text{min}) = 1\,(\text{h}),$$

and the snowball will disappear after

$$-\frac{t}{30} + 4 = 0 \quad \Rightarrow \quad t = 120\,(\text{min}) = 2\,(\text{h}).$$

23. If $m(t)$ (with t measured in "days") denotes the mass of a radioactive substance, the law of decay says that

$$\frac{dm}{dt} = km(t),$$

with the decay constant k depending on the substance. Solving this equation yields

$$m(t) = Ce^{kt}.$$

     Copyright © 2012 Pearson Education, Inc. Publishing as Addison-Wesley.

If the initial mass of the substance is $m(0) = m_0$, then (similarly to the equation (11)) we find that

$$m(t) = m_0 e^{kt}. \tag{3.3}$$

In this problem, $m_0 = 50$ g, and we know that $m(3) = 10$ g. These data yields

$$10 = m(3) = 50 \cdot e^{k(3)} \qquad \Rightarrow \qquad k = -\frac{\ln 5}{3},$$

and so the decay is governed by the equation

$$m(t) = 50e^{-(\ln 5)t/3} = (50)5^{-t/3}.$$

After 4 days, the remaining amount will be $m(4) = (50)5^{-4/3}$ g, which is

$$\frac{(50)5^{-4/3}}{50} \cdot 100\% = 5^{-4/3} \cdot 100\% \approx 11.7\%$$

of the original amount.

25. Let $M(t)$ denote the mass of carbon-14 present in the burnt wood of the campfire. Then, since carbon-14 decays at a rate proportional to its mass, we have

$$\frac{dM}{dt} = -\alpha M,$$

where α is the proportionality constant. This equation is separable (and linear). Using the initial condition, $M(0) = M_0$, we obtain

$$M(t) = M_0 e^{-\alpha t}.$$

Given that the half-life of carbon-14 is 5600 years, we solve for α.

$$\frac{1}{2} M_0 = M_0 e^{-\alpha(5600)} \qquad \Rightarrow \qquad \frac{1}{2} = e^{-\alpha(5600)} \qquad \Rightarrow \qquad \alpha = \frac{\ln(0.5)}{-5600} \approx 0.000123776.$$

Thus,

$$M(t) = M_0 e^{-0.000123776t}.$$

We know that after t years 2% of the original amount of carbon-14 remains in the campfire. Solving for t yields

$$0.02 M_0 = M_0 e^{-0.000123776t} \qquad \Rightarrow \qquad 0.02 = e^{-0.000123776t}$$

$$\Rightarrow \qquad t = \frac{\ln 0.02}{-0.000123776} \approx 31,606 \text{ (years)}.$$

Copyright © 2012 Pearson Education, Inc. Publishing as Addison-Wesley.

27. The element Hh decays according to the general law of a radioactive decay, which is described by (3.3) (with t measured in "years"). Since the initial mass of Hh is $m_0 = 1$ kg and the decay constant $k = k_{\text{Hh}} = -2/\text{yr}$, we get

$$\text{Hh}(t) = e^{k_{\text{Hh}}t} = e^{-2t}. \tag{3.4}$$

For It, the process is more complicated: it has an incoming mass from the decay of Hh and, at the same, loses its mass decaying to Bu. (This process is similar to the "brine solution" models.) Thus, we use the general idea in getting a differential equation, describing this process, i.e.,

$$\text{rate of change} = \text{input rate} - \text{output rate}. \tag{3.5}$$

The "input rate" is the rate of mass coming from Hh's decay, which is opposite to the rate of decay of Hh (Hh loses the mass, but It gains it). So,

$$\text{input rate} = -\frac{d\text{Hh}}{dt} = 2e^{-2t}, \tag{3.6}$$

where we have used (3.4). The "output rate" is the rate with which It decays, which (again, according to the general law of a radioactive decay) is proportional to its current mass. Since the decay constant for It is $k = k_{\text{It}} = -1/\text{yr}$,

$$\text{output rate} = k_{\text{It}}\text{It}(t) = -\text{It}(t). \tag{3.7}$$

Therefore, combining (3.5)–(3.7), we get the equation for It, that is,

$$\frac{d\text{It}(t)}{dt} = 2e^{-2t} - \text{It}(t) \qquad \Rightarrow \qquad \frac{d\text{It}(t)}{dt} + \text{It}(t) = 2e^{-2t}.$$

This is a linear equation with $P(t) \equiv 1$. Multiplying the equation by an integrating factor $\mu(t) = \exp\left[\int (1)dt\right] = e^t$ yields

$$\frac{d\left[e^t\text{It}(t)\right]}{dt} = 2e^{-t} \qquad \Rightarrow \qquad e^t\text{It}(t) = -2e^{-t} + C \qquad \Rightarrow \qquad \text{It}(t) = -2e^{-2t} + Ce^{-t}.$$

Initially, there were no It, which means that $\text{It}(0) = 0$. With this initial condition, we find that

$$0 = \text{It}(0) = -2e^{-2(0)} + Ce^{-(0)} = -2 + C \qquad \Rightarrow \qquad C = 2,$$

and the mass of It remaining after t years is

$$\text{It}(t) = 2e^{-t} - 2e^{-2t}. \tag{3.8}$$

 Copyright © 2012 Pearson Education, Inc. Publishing as Addison-Wesley.

The element Bu only gains its mass from It, and the rate, with which it does this, is opposite to the rate with which It loses its mass. Hence (3.7) yields

$$\frac{d\text{Bu}(t)}{dt} = \text{It}(t) = 2\left(e^{-t} - e^{-2t}\right).$$

Integrating, we obtain

$$\text{Bu}(t) = 2\int \left(e^{-t} - e^{-2t}\right) dt = -2e^{-t} + e^{-2t} + C,$$

and the initial condition $\text{Bu}(0) = 0$ gives $C = 1$. Therefore,

$$\text{Bu}(t) = 1 - 2e^{-t} + e^{-2t}.$$

EXERCISES 3.3:   Heating and Cooling of Buildings

1. Let $T(t)$ denote the temperature of the coffee at time t (in minutes). According to the Newton's law of cooling (see (1)),

$$\frac{dT}{dt} = K[21 - T(t)],$$

where we have taken $H(t) \equiv U(t) \equiv 0$ and $M(t) \equiv 21°\,\text{C}$, with the initial condition $T(0) = 95°\,\text{C}$. Solving this initial value problem yields

$$\frac{dT}{21 - T} = K\,dt \quad \Rightarrow \quad -\ln|T - 21| = Kt + C_1 \quad \Rightarrow \quad T(t) = 21 + Ce^{-Kt};$$
$$95 = T(0) = 21 + Ce^{-K(0)} \quad \Rightarrow \quad C = 74 \quad \Rightarrow \quad T(t) = 21 + 74e^{-Kt}.$$

To find K, we use the fact that after $5\,\text{min}$ the temperature of the coffee was $80°\,\text{C}$.

$$80 = T(5) = 21 + 74e^{-K(5)} \quad \Rightarrow \quad K = \frac{\ln(74/59)}{5},$$

and so

$$T(t) = 21 + 74e^{-\ln(74/59)t/5} = 21 + 74\left(\frac{74}{59}\right)^{-t/5}.$$

Finally, we solve the equation $T(t) = 50$ to find the appropriate time for drinking coffee.

$$50 = 21 + 74\left(\frac{74}{59}\right)^{-t/5} \quad \Rightarrow \quad \left(\frac{74}{59}\right)^{-t/5} = \frac{29}{74} \quad \Rightarrow \quad t = \frac{5\ln(74/29)}{\ln(74/59)} \approx 20.7\,(\text{min}).$$

Copyright © 2012 Pearson Education, Inc. Publishing as Addison-Wesley.

Chapter 3

3. This problem is similar to one of cooling a building. In this problem, we have no additional heating or cooling, so we can say that the rate of change of the wine's temperature $T(t)$ is given by Newton's law of cooling

$$\frac{dT}{dt} = K[M(t) - T(t)],$$

where $M(t) = 32$ is the temperature of ice. This equation is linear and has standard form

$$\frac{dT}{dt} + KT(t) = 32K.$$

We find that the integrating factor is e^{Kt}. Multiplying both sides by e^{Kt} and integrating gives

$$e^{Kt}\frac{dT}{dt} + e^{Kt}KT(t) = 32Ke^{Kt} \qquad \Rightarrow \qquad e^{Kt}T(t) = \int 32Ke^{Kt}\,dt$$

$$\Rightarrow \qquad e^{Kt}T(t) = 32e^{Kt} + C \qquad \Rightarrow \qquad T(t) = 32 + Ce^{-Kt}.$$

By setting $t = 0$ and using the initial temperature, $70°\text{F}$, we find that

$$70 = 32 + C \qquad \Rightarrow \qquad C = 38.$$

Knowing that it takes 15 minutes for the wine to chill to $60°\text{F}$, we find K.

$$60 = 32 + 38e^{-K(15)} \qquad \Rightarrow \qquad K = \frac{-1}{15}\ln\left(\frac{60 - 32}{38}\right) \approx 0.02035.$$

Therefore,

$$T(t) = 32 + 38e^{-0.02035t}.$$

We can now determine how long it will take for the wine to reach $56°\text{F}$. Using our equation for $T(t)$, we solve

$$56 = 32 + 38e^{-0.02035t} \qquad \Rightarrow \qquad t = \frac{-1}{0.02035}\ln\left(\frac{56 - 32}{38}\right) \approx 22.6 \text{ (min)}.$$

5. This problem can be treated as a problem of cooling. If we assume the air surrounding the body has not changed since the death, we can say that the rate of change of the body's temperature, $T(t)$, satisfies Newton's law of cooling, that is,

$$\frac{dT}{dt} = K[M(t) - T(t)],$$

 Copyright © 2012 Pearson Education, Inc. Publishing as Addison-Wesley.

where $M(t)$ represents the surrounding temperature, which we have assumed to be constantly 16°C. This differential equation is linear, and so can be solved using an integrating factor $\mu(t) = e^{Kt}$. Rewriting the above equation in standard form, multiplying both sides by $\mu(t)$, and integrating yields

$$\frac{dT}{dt} + KT(t) = K(16) \quad \Rightarrow \quad e^{Kt}\frac{dT}{dt} + e^{Kt}KT(t) = 16Ke^{Kt}$$
$$\Rightarrow \quad e^{Kt}T(t) = 16e^{Kt} + C \quad \Rightarrow \quad T(t) = 16 + Ce^{-Kt}.$$

Let us set $t = 0$ for the time, when the person died. Then $T(0) = 37°C$ (normal body temperature), and so we have

$$37 = 16 + C \quad \Rightarrow \quad C = 21.$$

We know that after, say, X hours the body's temperature was 34.5°C, and that after $X + 1$ hours it was 33.7°C. Therefore, we have two equations,

$$34.5 = 16 + 21e^{-KX} \quad \text{and} \quad 33.7 = 16 + 21e^{-K(X+1)}.$$

Solving the first equation for KX, we arrive at

$$KX = -\ln\left(\frac{34.5 - 16}{21}\right) \approx 0.12675. \tag{3.9}$$

Substituting this value into the second equation, we find K.

$$33.7 = 16 + 21e^{-0.12675 - K}$$
$$\Rightarrow \quad K = -\left[0.12675 + \ln\left(\frac{33.7 - 16}{21}\right)\right] \approx 0.04421.$$

This results in an equation for the body temperature of

$$T(t) = 16 + 21e^{-0.04421t}.$$

From (3.9), we now find the number of hours X before 12 (noon), when the person died.

$$X = \frac{0.12675}{K} = \frac{0.12675}{0.04421} \approx 2.867 \text{ (hours)}.$$

Therefore, the time of death is approximately 2.867 hours (2 hours and 52 min) before noon or 9 : 08 A.M.

Copyright © 2012 Pearson Education, Inc. Publishing as Addison-Wesley.

Chapter 3

7. The temperature function $T(t)$ changes according to Newton's law of cooling. Similarly to Example 1, we conclude that, with $H(t) = U(t) \equiv 0$ and the outside temperature $M(t) \equiv 35°C$, a general solution formula (4) yields

$$T(t) = 35 + Ce^{-Kt}.$$

To find C, we use the initial condition,

$$T(0) = T(\text{at noon}) = 24°C,$$

and get

$$24 = T(0) = 35 + Ce^{-K(0)} \quad \Rightarrow \quad C = 24 - 35 = -11 \quad \Rightarrow \quad T(t) = 35 - 11e^{-Kt}.$$

The time constant for the building is $1/K = 4\,\mathrm{hr}$; so $K = 1/4$ and $T(t) = 35 - 11e^{-t/4}$.

At $2:00$ P.M., $t = 2$ and $t = 6$ at $6:00$ P.M. Substituting these values into the solution, we conclude that the temperature

$$\text{at } 2:00 \text{ P.M. will be} \quad T(2) = 35 - 11e^{-2/4} \approx 28.3°C;$$
$$\text{at } 6:00 \text{ P.M. will be} \quad T(6) = 35 - 11e^{-6/4} \approx 32.5°C.$$

Finally, we solve the equation

$$T(t) = 35 - 11e^{-t/4} = 27$$

to find the time when the temperature inside the building reaches $27°C$.

$$35 - 11e^{-t/4} = 27 \quad \Rightarrow \quad 11e^{-t/4} = 8 \quad \Rightarrow \quad t = 4\ln\left(\frac{11}{8}\right) \approx 1.27\,(\mathrm{hr}).$$

Thus, the temperature inside the building will be $27°C$ at 1.27 hours after noon, that is, approximately at $1:16$ P.M.

9. Since we are evaluating the temperature in a warehouse, we can assume that heat generated by people or equipment is negligible, so that $H(t) \equiv 0$. We also assume that there is no heating or air conditioning in the warehouse. Thus, we have $U(t) \equiv 0$. We are also given that the outside temperature has a sinusoidal fluctuation. Thus, as in Example 2, we see that

$$M(t) = M_0 - B\cos\omega t,$$

Copyright © 2012 Pearson Education, Inc. Publishing as Addison-Wesley.

where M_0 is the average outside temperature, B is a positive constant for the magnitude of the temperature shift from this average, and $\omega = \pi/2$ radians per hour. To find M_0 and B, we use the fact that $M(t)$ reaches a low of $16°C$ at $2:00$ A.M. and it reaches a high of $32°C$ at $2:00$ P.M. This gives

$$M_0 = \frac{16 + 32}{2} = 24°C.$$

By letting $t = 0$ at $2:00$ A.M. (so that low for the outside temperature corresponds to the low for the negative cosine function), we can compute the constant B. That is,

$$16 = 24 - B\cos 0 = 24 - B \qquad \Rightarrow \qquad B = 8.$$

Therefore, we see that

$$M(t) = 24 - 8\cos\omega t,$$

where $\omega = \pi/12$. As in Example 2, using that $B_0 = M_0 + H_0/K = M_0 + 0/K = M_0$, we see that

$$T(t) = 24 - 8F(t) + Ce^{-Kt},$$

where

$$F(t) = \frac{\cos\omega t + (\omega/K)\sin\omega t}{1 + (\omega/K)^2} = \left[1 + \left(\frac{\omega}{K}\right)^2\right]^{-1/2}\cos(\omega t - \alpha).$$

In the last equation, α is chosen such that $\tan\alpha = \omega/K$. Assuming that the exponential term dies off, we obtain

$$T(t) = 24 - 8\left[1 + \left(\frac{\omega}{K}\right)^2\right]^{-1/2}\cos(\omega t - \alpha).$$

This function will reach the minimum when $\cos(\omega t - \alpha) = 1$, and it will reach the maximum when $\cos(\omega t - \alpha) = -1$.

For the case, when the time constant of the building is 1, we see that $1/K = 1$, which implies that $K = 1$. Therefore, the temperature will reach the maximum of

$$T = 24 + 8\left[1 + \left(\frac{\pi}{12}\right)^2\right]^{-1/2} \approx 31.7°C$$

and the minimum of

$$T = 24 - 8\left[1 + \left(\frac{\pi}{12}\right)^2\right]^{-1/2} \approx 16.3°C.$$

Copyright © 2012 Pearson Education, Inc. Publishing as Addison-Wesley.

For the case, when the time constant of the building is 5, we have $1/K = 5$ or $K = 1/5$. Then, the temperature will reach the maximum of

$$T = 24 + 8\left[1 + \left(\frac{5\pi}{12}\right)^2\right]^{-1/2} \approx 28.9°\text{C}$$

and the minimum of

$$T = 24 - 8\left[1 + \left(\frac{5\pi}{12}\right)^2\right]^{-1/2} \approx 19.1°\text{C}.$$

11. As in Example 3 of the text, this problem involves a thermostat (to regulate the temperature in the van). Hence, we have

$$U(t) = K_U\left[T_D - T(t)\right],$$

where T_D is the desired temperature of $16°\text{C}$ and K_U is a proportionality constant. The time constant for the van is $1/K = 2$ hr, hence $K = 0.5$. Since the time constant for the van with its air conditioning system is $1/K_1 = 1/3$ hr, we have $K_1 = K + K_U = 3$. Therefore, $K_U = 3 - K = 2.5$.

We assume that the outside temperature $M(t)$ is constantly $35°\text{C}$. We also assume that outside objects (such as cars parked nearby, the asphalt, etc.) are heated to $55°\text{C}$ as well so that they generate an additional incoming heat at a rate

$$H(t) \equiv K(55 - 35) = 0.5(20) = 10.$$

Thus, the equation (1) in the text becomes

$$\frac{dT}{dt} = K\left[M(t) - T(t)\right] + H(t) + U(t) = 0.5(35 - T) + 10 + (2.5)(16 - T) = 67.5 - 3T.$$

Solving this separable equation yields $T(t) = 22.5 + Ce^{-3t}$. When $t = 0$, we have $T(0) = 55$. Solving for C yields $C = 32.5$. Hence, the van's temperature is given by

$$T(t) = 22.5 + 32.5e^{-3t}.$$

To find out, when the temperature in the van will reach $27°\text{C}$, we set $T(t) = 27$. Solving for t, we get

$$27 = 22.5 + 32.5e^{-3t} \qquad \Rightarrow \qquad e^{-3t} = \frac{4.5}{32.5} \approx 0.13846$$

$$\Rightarrow \qquad t \approx -\frac{\ln(0.13846)}{3} \approx 0.65906 \text{ (hr)} \qquad \text{or} \qquad 39.5 \text{ min.}$$

 Copyright © 2012 Pearson Education, Inc. Publishing as Addison-Wesley.

13. Since the time constant is 64, we have $K = 1/64$. The temperature in the tank increases at the rate of 2°F for every 1000 Btu. Furthermore, every hour of sunlight provides an input of 2000 Btu to the tank. Thus,

$$H(t) = 2 \times 2 = 4°F \,/\, \text{hour}.$$

We are given that $T(0) = 110$, and that the temperature $M(t)$ outside the tank is constantly 80°F. Hence, the temperature in the tank is governed by

$$\frac{dT}{dt} = \frac{1}{64}\,[80 - T(t)] + 4 = -\frac{1}{64}\,T(t) + 5.25\,, \qquad T(0) = 110.$$

Solving this separable equation gives

$$T(t) = 336 + Ce^{-t/64}\,.$$

To find C, we use the initial condition and find that

$$T(0) = 110 = 336 + C \qquad \Rightarrow \qquad C = -226.$$

This yields

$$T(t) = 336 - 226e^{-t/64}\,.$$

So, after 12 hours of sunlight, the temperature will be

$$T(12) = 336 - 226e^{-12/64} \approx 148.6°F.$$

15. The equation $dT/dt = k\,(M^4 - T^4)$ is separable. Separation variables yields

$$\frac{dT}{T^4 - M^4} = -k\,dt \qquad \Rightarrow \qquad \int \frac{dT}{T^4 - M^4} = -\int k\,dt = -kt + C_1. \qquad (3.10)$$

Since $T^4 - M^4 = (T^2 - M^2)\,(T^2 + M^2)$, we have

$$\frac{1}{T^4 - M^4} = \frac{1}{2M^2}\,\frac{(M^2 + T^2) + (M^2 - T^2)}{(T^2 - M^2)\,(T^2 + M^2)} = \frac{1}{2M^2}\left[\frac{1}{T^2 - M^2} - \frac{1}{T^2 + M^2}\right],$$

and the integral in the left-hand side of (3.10) becomes

$$\int \frac{dT}{T^4 - M^4} = \frac{1}{2M^2}\left[\int \frac{dT}{T^2 - M^2} - \int \frac{dT}{T^2 + M^2}\right]$$
$$= \frac{1}{4M^3}\left[\ln\frac{T - M}{T + M} - 2\arctan\left(\frac{T}{M}\right)\right].$$

Copyright © 2012 Pearson Education, Inc. Publishing as Addison-Wesley.

Thus, a general solution to Stefan's equation is given implicitly by

$$\frac{1}{4M^3}\left[\ln\frac{T-M}{T+M} - 2\arctan\left(\frac{T}{M}\right)\right] = -kt + C_1$$

or

$$T - M = C(T+M)\exp\left[2\arctan\left(\frac{T}{M}\right) - 4M^3kt\right].$$

When T is close to M,

$$M^4 - T^4 = (M-T)(M+T)\left(M^2+T^2\right) \approx (M-T)(2M)\left(2M^2\right) \approx 4M^3(M-T),$$

and so

$$\frac{dT}{dt} \approx k \cdot 4M^3(M-T) = k_1(M-T)$$

with $k_1 = 4M^3k$, which constitutes Newton's law of cooling.

EXERCISES 3.4:   Newtonian Mechanics

1. This problem is a particular case of Example 1. Therefore, we can use the general formula (6) with $m = 5$, $b = 50$, and $v_0 = v(0) = 0$. But let us follow the general idea of Section 3.4, to find an equation of the motion, and then solve it.

   From the given data, the force due to gravity is $F_1 = mg = 5g$ and the air resistance force is $F_2 = -50v$. Therefore, the velocity $v(t)$ satisfies

   $$m\frac{dv}{dt} = F_1 + F_2 = 5g - 50v \qquad \Rightarrow \qquad \frac{dv}{dt} = g - 10v \quad \text{with} \quad v(0) = 0.$$

   Separating variables yields

   $$\frac{dv}{10v - g} = -dt \qquad \Rightarrow \qquad \frac{1}{10}\ln|10v - g| = -t + C_1 \qquad \Rightarrow \qquad v(t) = \frac{g}{10} + Ce^{-10t}.$$

   Substituting the initial condition, $v(0) = 0$, we get $C = -g/10$, and so

   $$v(t) = \frac{g}{10}\left(1 - e^{-10t}\right).$$

   Integrating this equation, we obtain

   $$x(t) = \int v(t)\,dt = \int \frac{g}{10}\left(1 - e^{-10t}\right)dt = \frac{g}{10}\left(t + \frac{1}{10}e^{-10t}\right) + C,$$

   and we find C using the initial condition $x(0) = 0$.

   $$0 = \frac{g}{10}\left(0 + \frac{1}{10}e^{-10(0)}\right) + C \qquad \Rightarrow \qquad C = -\frac{g}{100}$$

$$\Rightarrow \qquad x(t) = \frac{g}{10}t + \frac{g}{100}\left(e^{-10t} - 1\right) = (0.981)t + (0.0981)e^{-10t} - 0.0981 \text{ (m)}.$$

When the object hits the ground, $x(t) = 1000$ (m). Thus, we solve

$$(0.981)t + (0.0981)e^{-10t} - 0.0981 = 1000,$$

which gives (since t is nonnegative) $t \approx 1019$ (sec).

3. In this problem, $m = 500$ kg, $v_0 = 0$, $g = 9.81$ m/sec^2, and $b = 50$ kg/sec. We also see that the object has 1000 m to fall before it hits the ground. Plugging in these values into (6) gives

$$x(t) = \frac{(500)(9.81)}{50}t + \frac{500}{50}\left(0 - \frac{(500)(9.81)}{50}\right)\left(1 - e^{-50t/500}\right)$$

$$\Rightarrow \qquad x(t) = 98.1t + 981e^{-t/10} - 981.$$

To find out, when the object hits the ground, we solve $x(t) = 1000$ for t. Thus,

$$1000 = 98.1t + 981e^{-t/10} - 981 \qquad \Rightarrow \qquad 98.1t + 981e^{-t/10} = 1981.$$

In this equation, if we ignore the term $981e^{-t/10}$, we find that $t \approx 20.2$. But this means that we have ignored the term like $981e^{-2} \approx 132.8$, which we find too large to ignore. Therefore, we use Newton's method (see Appendix B) to solve the equation

$$f(t) = 98.1t + 981e^{-t/10} - 1981 = 0.$$

(The positive root to this equation is the value of t we need.) Newton's method generates a sequence of approximations given by the formula

$$t_{n+1} = t_n - \frac{f(t_n)}{f'(t_n)}.$$

Since $f'(t) = 98.1 - 98.1e^{-t/10} = 98.1\left(1 - e^{-t/10}\right)$, the recursive formula becomes

$$t_{n+1} = t_n - \frac{t_n + 10e^{-t_n/10} - (1981/98.1)}{1 - e^{-t_n/10}}. \tag{3.11}$$

To start the process, let $t_0 = 1981/98.1 \approx 20.19$, which is about the approximation we obtained, when we neglected the exponential term. Using (3.11), we obtain

$$t_1 = 20.19368 - \frac{20.19368 + 10e^{-2.019368} - 20.19368}{1 - e^{-2.019368}} \approx 18.66.$$

To find t_2, we plug in this value for t_1 into (3.11). This gives $t_2 \approx 18.64$. Continuing this process, we find that $t_3 \approx 18.64$. Since t_2 and t_3 agree to two decimal places, an approximation for the time it takes the object to strike the ground is $t \approx 18.6$ sec.

Copyright © 2012 Pearson Education, Inc. Publishing as Addison-Wesley.

Chapter 3

5. We proceed similarly to the solution of Problem 1 to get

$$F_1 = 5g, \qquad F_2 = -10v$$

$$\Rightarrow \qquad 5\frac{dv}{dt} = F_1 + F_2 = 5g - 10v$$

$$\Rightarrow \qquad \frac{dv}{dt} = g - 2v, \qquad v(0) = 50.$$

Solving this initial value problem yields

$$v(t) = \frac{g}{2} + Ce^{-2t};$$

$$50 = v(0) = \frac{g}{2} + Ce^{-2(0)} \qquad \Rightarrow \qquad C = \frac{100 - g}{2}$$

$$\Rightarrow \qquad v(t) = \frac{g}{2} + \frac{100 - g}{2}e^{-2t}.$$

We now integrate $v(t)$ to obtain the equation of the motion of the object.

$$x(t) = \int v(t)\, dt = \int \left(\frac{g}{2} + \frac{100 - g}{2}e^{-2t} \right) dt = \frac{g}{2}t - \frac{100 - g}{4}e^{-2t} + C,$$

where C is such that $x(0) = 0$. Computing yields

$$0 = x(0) = \frac{g}{2}(0) - \frac{100 - g}{4}e^{-2(0)} + C \qquad \Rightarrow \qquad C = \frac{100 - g}{4}.$$

This gives the answer to the first question in this problem. Namely,

$$x(t) = \frac{g}{2}t - \frac{100 - g}{4}e^{-2t} + \frac{100 - g}{4} \approx 4.91t + 22.55 - 22.55\,e^{-2t}.$$

Answering the second question, we solve the equation $x(t) = 500$ to find time t, when the object passes $500\,\mathrm{m}$, and so strikes the ground.

$$4.91t + 22.55 - 22.55\,e^{-2t} = 500 \qquad \Rightarrow \qquad t \approx 97.3\,(\mathrm{sec}).$$

7. Since the air resistance force has different coefficients of proportionality for closed and for opened chute, we need two differential equations describing the motion. Let $x_1(t)$, $x_1(0) = 0$, denote the distance that the parachutist has fallen in t seconds, and we let $v_1(t) := dx_1/dt$ denote her velocity. With $m = 75$, $b = b_1 = 30$ N-sec/m, and $v_0 = 0$ the initial value problem (4) in the text becomes

$$75\frac{dv_1}{dt} = 75g - 30v_1 \qquad \Rightarrow \qquad \frac{dv_1}{dt} + \frac{2}{5}v_1 = g \quad \text{with} \quad v_1(0) = 0.$$

   Copyright © 2012 Pearson Education, Inc. Publishing as Addison-Wesley.

This is a linear equation. Solving yields

$$d\left(e^{2t/5}v_1\right) = e^{2t/5}g\,dt \qquad \Rightarrow \qquad v_1(t) = \frac{5g}{2} + C_1 e^{-2t/5};$$

$$0 = v_1(0) = \frac{5g}{2} + C_1 e^0 = \frac{5g}{2} + C_1 \qquad \Rightarrow \qquad C_1 = -\frac{5g}{2}$$

$$\Rightarrow \qquad v_1(t) = \frac{5g}{2}\left(1 - e^{-2t/5}\right)$$

$$\Rightarrow \qquad x_1(t) = \int_0^t v_1(s)\,ds = \frac{5g}{2}\left(s + \frac{5}{2}e^{-2s/5}\right)\Bigg|_{s=0}^{s=t} = \frac{5g}{2}\left(t + \frac{5}{2}e^{-2t/5} - \frac{5}{2}\right).$$

To find the time t_*, when the chute opens, we solve

$$20 = v_1(t_*) \qquad \Rightarrow \qquad 20 = \frac{5g}{2}\left(1 - e^{-2t_*/5}\right) \qquad \Rightarrow \qquad t_* = -\frac{5}{2}\ln\left(1 - \frac{8}{g}\right) \approx 4.23\,(\text{sec}).$$

By this time the parachutist has fallen

$$x_1(t_*) = \frac{5g}{2}\left(t_* + \frac{5}{2}e^{-2t_*/5} - \frac{5}{2}\right) \approx \frac{5g}{2}\left(4.225 + \frac{5}{2}e^{-2\cdot 4.225/5} - \frac{5}{2}\right) \approx 53.62\,(\text{m}),$$

and so she is $2000 - 53.62 = 1946.38$ m above the ground. Setting the second equation, we for convenience reset the time t. Denoting by $x_2(t)$ the distance passed by the parachutist from the moment when the chute opens, and by $v_2(t) := x_2'(t)$ – her velocity, we have

$$75\frac{dv_2}{dt} = 75g - 90v_2, \qquad v_2(0) = v_1(t_*) = 20, \qquad x_2(0) = 0.$$

Solving, we get

$$v_2(t) = \frac{5g}{6} + C_2 e^{-6t/5};$$

$$20 = v_2(0) = \frac{5g}{6} + C_2 \qquad \Rightarrow \qquad C_2 = 20 - \frac{5g}{6}$$

$$\Rightarrow \qquad v_2(t) = \frac{5g}{6} + \left(20 - \frac{5g}{6}\right)e^{-6t/5}$$

$$\Rightarrow \qquad x_2(t) = \int_0^t v_2(s)\,ds = \left[\frac{5g}{6}s - \frac{5}{6}\left(20 - \frac{5g}{6}\right)e^{-6s/5}\right]\Bigg|_{s=0}^{s=t}$$

$$= \frac{5g}{6}t + \frac{5}{6}\left(20 - \frac{5g}{6}\right)\left(1 - e^{-6t/5}\right).$$

With the chute open, the parachutist falls 1946.38 m. It takes t^* seconds, where t^* satisfies $x_2(t^*) = 1946.38$. Solving yields

$$\frac{5g}{6}t^* + \frac{5}{6}\left(20 - \frac{5g}{6}\right)\left(1 - e^{-6t^*/5}\right) = 1946.38 \qquad \Rightarrow \qquad t^* \approx 236.88\,(\text{sec}).$$

Therefore, the parachutist will hit the ground after $t^* + t_* \approx 241$ seconds.

Copyright © 2012 Pearson Education, Inc. Publishing as Addison-Wesley.

9. This problem is similar to that in Example 1 in the text with the addition of a buoyancy force of magnitude $mg/40$. If we let $x(t)$ be the downward distance of the object from the water level at time t and denote by $v(t)$ its velocity, then the total force acting on the object is

$$F = mg - bv - \frac{mg}{40}.$$

We are given that $m = 100$ kg, $g = 9.81$ m/sec^2, and $b = 10$ kg/sec. Applying Newton's second law yields

$$100\frac{dv}{dt} = (100)(9.81) - 10v - \frac{10}{4}(9.81)$$

$$\Rightarrow \quad \frac{dv}{dt} = 9.56 - (0.1)v.$$

Solving this equation using separation of variables, we obtain

$$v(t) = 95.65 + Ce^{-t/10}.$$

Since $v(0) = 0$, we find that $C = -95.65$, and so

$$v(t) = 95.65 - 95.65e^{-t/10}.$$

Integrating yields

$$x(t) = 95.65t + 956.5e^{-t/10} + C_1.$$

Using the fact that $x(0) = 0$, we find $C_1 = -956.5$. Therefore, the equation of motion of the object is given by

$$x(t) = 95.65t + 956.5e^{-t/10} - 956.5.$$

To determine when the object is traveling at the velocity of 70 m/sec, we solve $v(t) = 70$. That is,

$$70 = 95.65 - 95.65e^{-t/10} = 95.65\left(1 - e^{-t/10}\right)$$

$$\Rightarrow \quad t = -10\ln\left(1 - \frac{70}{95.65}\right) \approx 13.2 \text{ sec.}$$

11. Let $v(t) = V[x(t)]$. Then, using the chain rule, we get

$$\frac{dv}{dt} = \frac{dV}{dx}\frac{dx}{dt} = \frac{dV}{dx}V,$$

 Copyright © 2012 Pearson Education, Inc. Publishing as Addison-Wesley.

and so, for $V(x)$, the initial value problem (4) becomes

$$m \frac{dV}{dx} V = mg - bV, \qquad V(0) = V[x(0)] = v(0) = v_0 .$$

This differential equation is separable. Solving yields

$$\frac{V}{g - (b/m)V} \, dV = dx \qquad \Rightarrow \qquad \frac{m}{b} \left[\frac{g}{g - (b/m)V} - 1 \right] dV = dx$$

$$\Rightarrow \qquad \int \frac{m}{b} \left[\frac{g}{g - (b/m)V} - 1 \right] dV = \int dx$$

$$\Rightarrow \qquad \frac{m}{b} \left[-\frac{mg}{b} \ln |g - (b/m)V| - V \right] = x + C$$

$$\Rightarrow \qquad mg \ln |mg - bV| + bV = -\frac{b^2 x}{m} + C_1 .$$

Substituting the initial condition, $V(0) = v_0$, we find that $C_1 = mg \ln |mg - bv_0| + bv_0$, and hence (recall that $V(x) = v$) the answer is

$$mg \ln |bV - mg| + bV = -\frac{b^2 x}{m} + mg \ln |bv_0 - mg| + bv_0$$

$$\Rightarrow \qquad e^{bv} |bv - mg|^{mg} = e^{v_0 b} |bv_0 - mg|^{mg} e^{-b^2 x/m} .$$

13. There are two forces acting on the shell: a constant force due to the downward pull of gravity and a force due to air resistance that acts in opposition to the motion of the shell. All of the motion occurs along a vertical axis. On this axis, we choose the origin to be the point, where the shell was shot from, and let $x(t)$ denote the upward position of the shell at time t. The forces acting on the object can be expressed in terms of this choice. The force due to gravity is $F_1 = -mg$, where g is the acceleration due to gravity near Earth. Note that we have a negative force, because our coordinate system has positive upward direction, while gravity acts downward. The force due to air resistance is $F_2 = -(0.1)v^2$. The negative sign is present because air resistance acts in opposition to the motion of the object. Therefore, the net force acting on the shell is

$$F = F_1 + F_2 = -mg - (0.1)v^2.$$

We now apply Newton's second law to obtain

$$m \frac{dv}{dt} = - \left[mg + (0.1)v^2 \right] .$$

Because the initial velocity of the shell is 500 m/sec, a model for the velocity of the rising shell is expressed as an initial value problem

$$m \frac{dv}{dt} = - \left[mg + (0.1)v^2 \right] , \qquad v(0) = 500, \qquad (3.12)$$

Copyright © 2012 Pearson Education, Inc. Publishing as Addison-Wesley.

where $g \approx 9.81$. Separating variables, we get

$$\frac{dv}{10mg + v^2} = -\frac{dt}{10m},$$

and so

$$\int \frac{dv}{10mg + v^2} = -\int \frac{dt}{10m} \quad \Rightarrow \quad \frac{1}{\sqrt{10mg}} \tan^{-1}\left(\frac{v}{\sqrt{10mg}}\right) = -\frac{t}{10m} + C.$$

Setting $m = 3$, $g = 9.81$, and $v = 500$ when time $t = 0$, we find that

$$C = \frac{1}{\sqrt{10(3)(9.81)}} \tan^{-1}\left(\frac{500}{\sqrt{10(3)(9.81)}}\right) \approx 0.08956.$$

Thus, the equation for the velocity v (as a function of time t) is

$$\frac{1}{\sqrt{10mg}} \tan^{-1}\left(\frac{v}{\sqrt{10mg}}\right) = -\frac{t}{10m} + 0.08956.$$

From physics we know that, when the shell reaches the maximum height, its velocity is zero; therefore $t_{\max}$ can be found from

$$t_{\max} = -10(3)\left[\frac{1}{\sqrt{10(3)(9.81)}} \tan^{-1}\left(\frac{0}{\sqrt{10(3)(9.81)}}\right) - 0.08956\right]$$

$$\approx -(30)(-0.08956) \approx 2.69 \text{ (sec)}.$$

Using equation (3.12) and noting that $dv/dt = (dv/dx)(dx/dt) = (dv/dx)v$, we can determine the maximum height attained by the shell. With the above substitution, equation (3.12) becomes

$$mv\frac{dv}{dx} = -\left(mg + 0.1v^2\right), \qquad v(0) = 500.$$

Using separation of variables and integration, we get

$$\frac{v\,dv}{10mg + v^2} = -\frac{dx}{10m} \quad \Rightarrow \quad \frac{1}{2}\ln\left(10mg + v^2\right) = -\frac{x}{10m} + C$$

$$\Rightarrow \quad 10mg + v^2 = Ke^{-x/(5m)}.$$

Setting $v = 500$, when $x = 0$, we find that

$$K = e^0\left[(10)(3)(9.81) + (500)^2\right] = 250294.3.$$

Thus, the equation for the velocity, as a function of the distance, is

$$v^2 + 10mg = (250294.3)e^{-x/(5m)}.$$

Copyright © 2012 Pearson Education, Inc. Publishing as Addison-Wesley.

The maximum height will occur, when the shell's velocity is zero. Therefore, is

$$x_{\max} = -(5)(3) \ln \left[\frac{0 + (10)(3)(9.81)}{250294.3} \right] \approx 101.19 \text{ (m)}.$$

15. The total torque exerted on the flywheel is the sum of the torque exerted by the motor and the retarding torque due to friction. Thus, by Newton's second law for rotation, we have

$$I \frac{d\omega}{dt} = T - k\omega \qquad \text{with} \qquad \omega(0) = \omega_0,$$

where I is the moment of inertia of the flywheel, $\omega(t)$ is the angular velocity, $d\omega/dt$ is the angular acceleration, T is the constant torque exerted by the motor, and k is a positive constant of proportionality for the torque due to friction. Solving this separable equation gives

$$\omega(t) = \frac{T}{k} + Ce^{-kt/I}.$$

Using the initial condition, $\omega(0) = \omega_0$, we find that $C = (\omega_0 - T/k)$. Hence,

$$\omega(t) = \left(\omega_0 - \frac{T}{k} \right) e^{-kt/I} + \frac{T}{k}.$$

17. Since the motor is turned off, its torque is $T = 0$, and the only torque acting on the flywheel is the retarding one, $-5\sqrt{\omega}$. Then Newton's second law for rotational motion becomes

$$I \frac{d\omega}{dt} = -5\sqrt{\omega} \qquad \text{with} \qquad \omega(0) = \omega_0 = 225 \text{ (rad/sec)} \quad \text{and} \quad I = 50 \text{ (kg/m}^2)$$

A general solution to this separable equation is

$$\sqrt{\omega(t)} = -\frac{5}{2I} t + C = -0.05t + C.$$

Using the initial condition, we find that

$$\sqrt{\omega(0)} = -0.05 \cdot 0 + C \qquad \Rightarrow \qquad C = \sqrt{\omega(0)} = \sqrt{225} = 15.$$

Thus,

$$t = \frac{1}{0.05} \left[15 - \sqrt{\omega(t)} \right] = 20 \left[15 - \sqrt{\omega(t)} \right].$$

At the moment $t = t_{\text{stop}}$, when the flywheel stops rotating, we have $\omega(t_{\text{stop}}) = 0$ and so

$$t_{\text{stop}} = 20(15 - \sqrt{0}) = 300 \text{ (sec)}.$$

Copyright © 2012 Pearson Education, Inc. Publishing as Addison-Wesley.

19. There are three forces acting on the object: F_1 – the force due to gravity, F_2 – the air resistance force, and F_3 – the friction force. Using Figure 3.11 (with 30° replaced by 45°), we obtain

$$F_1 = mg \sin 45° = \frac{mg\sqrt{2}}{2},$$

$$F_2 = -3v,$$

$$F_3 = -\mu N = -\mu mg \cos 45° = -\frac{\mu mg\sqrt{2}}{2},$$

and so the equation describing the motion is

$$m\frac{dv}{dt} = \frac{mg\sqrt{2}}{2} - \frac{\mu mg\sqrt{2}}{2} - 3v \qquad \Rightarrow \qquad \frac{dv}{dt} = 0.475g\sqrt{2} - \frac{v}{20}$$

with the initial condition $v(0) = 0$. Solving yields

$$v(t) = 9.5g\sqrt{2} + Ce^{-t/20};$$

$$0 = v(0) = 9.5g\sqrt{2} + C \qquad \Rightarrow \qquad C = -9.5g\sqrt{2}$$

$$\Rightarrow \qquad v(t) = 9.5g\sqrt{2}\left(1 - e^{-t/20}\right).$$

Since $x(0) = 0$, integrating the above equation, we obtain

$$x(t) = \int_0^t v(s)ds = \int_0^t 9.5g\sqrt{2}\left(1 - e^{-s/20}\right)ds = 9.5g\sqrt{2}\left(s + 20e^{-s/20}\right)\Big|_{s=0}^{s=t}$$

$$= 9.5g\sqrt{2}\left(t + 20e^{-t/20} - 20\right) \approx 2636e^{-t/20} + 131.8t - 2636 \,(\text{m}).$$

The object reaches the end of the inclined plane when

$$x(t) = 131.8t + 2636e^{-t/20} - 2636 = 10 \qquad \Rightarrow \qquad t \approx 1.768 \,(\text{sec}).$$

21. In this problem there are two forces acting on the sailboat: a constant horizontal force due to the wind and a force due to the water resistance that acts in opposition to the motion of the sailboat. All of the motion occurs along the horizontal axis. On this axis, we choose the origin to be the point, where the hard blowing wind begins, let $x(t)$ denote the distance the sailboat travels during time t. The forces acting on the sailboat can be expressed in terms of this axis. The force due to the wind is

$$F_1 = 600 \text{ N}.$$

 Copyright © 2012 Pearson Education, Inc. Publishing as Addison-Wesley.

The force due to water resistance is

$$F_2 = -100v \text{ N.}$$

Applying Newton's second law, we obtain

$$m\frac{dv}{dt} = 600 - 100v.$$

Since the initial velocity of the sailboat is 1 m/sec, a model for the velocity of the moving sailboat is expressed as the following initial value problem

$$m\frac{dv}{dt} = 600 - 100v, \qquad v(0) = 1.$$

Using separation of variables, we get (with $m = 50$ kg)

$$\frac{dv}{6 - v} = 2dt \qquad \Rightarrow \qquad -\ln(6 - v) = 2t + C.$$

Therefore, the velocity is given by $v(t) = 6 - Ke^{-2t}$. Setting $v = 1$ when $t = 0$, we find that

$$1 = 6 - K \qquad \Rightarrow \qquad K = 5.$$

Thus, the equation for the velocity $v(t)$ is $v(t) = 6 - 5e^{-2t}$. The limiting velocity of the sailboat under the given conditions can be found by letting time t approach infinity, i.e.,

$$\lim_{t\to\infty} v(t) = \lim_{t\to\infty} \left(6 - 5e^{-2t}\right) = 6 \text{ (m/sec).}$$

To determine the equation of motion, we use the obtained velocity equation and substitute dx/dt for $v(t)$ to get

$$\frac{dx}{dt} = 6 - 5e^{-2t}, \qquad x(0) = 0.$$

Integrating this equation, we obtain

$$x(t) = \frac{5}{2}e^{-2t} + 6t + C_1.$$

Setting $x = 0$ when $t = 0$, we find that

$$0 = 0 + \frac{5}{2} + C_1 \qquad \Rightarrow \qquad C_1 = -\frac{5}{2}.$$

Thus, the equation of motion for the sailboat is given by

$$x(t) = \frac{5}{2}e^{-2t} + 6t - \frac{5}{2}.$$

Copyright © 2012 Pearson Education, Inc. Publishing as Addison-Wesley.

23. In this problem, there are two forces acting on the boat: the wind force F_1 and the water resistance force F_2. Since the proportionality constant in the water resistance force is different for the velocities below and above of a certain limit (5 m/sec for the boat A and 6 m/sec for the boat B), each boat has its own differential equation. (Compare with Problem 7.) Let $x_1^{(A)}(t)$ denote the distance passed by the boat A for the time t, $v_1^{(A)}(t) := dx_1^{(A)}(t)/dt$. Then the equation, describing the motion of the boat A before it reaches the velocity 5 m/sec, is

$$m\frac{dv_1^{(A)}}{dt} = F_1 + F_2 = 650 - b_1 v_1^{(A)} \qquad \Rightarrow \qquad \frac{dv_1^{(A)}}{dt} = \frac{65}{6} - \frac{4}{3}v_1^{(A)}. \qquad (3.13)$$

Solving this linear equation and using the initial condition, $v_1^{(A)}(0) = 2$, we get

$$v_1^{(A)}(t) = \frac{65}{8} - \frac{49}{8}e^{-4t/3},$$

and so

$$x_1^{(A)}(t) = \int_0^t \left(\frac{65}{8} - \frac{49}{8}e^{-4s/3}\right) ds = \frac{65}{8}t - \frac{147}{32}\left(e^{-4t/3} - 1\right).$$

The boat A will have the velocity 5 m/sec at $t = t_*$ satisfying

$$\frac{65}{8} - \frac{49}{8}e^{-4t_*/3} = 5 \qquad \Rightarrow \qquad t_* = -\frac{3\ln(25/49)}{4} \approx 0.5\,(\text{sec}),$$

and it will be

$$x_1^{(A)}(t_*) = \frac{65}{8}t_* - \frac{147}{32}\left(e^{-4t_*/3} - 1\right) \approx 1.85\,(\text{m})$$

away from the starting point or, equivalently, $500 - 1.85 = 498.15$ meters away from the finish. Similarly to (3.13), resetting the time, we obtain an equation of the motion of the boat A starting from the moment, when its velocity reaches 5 m/sec. Denoting by $x_2^{(A)}(t)$ the distance passed by the boat A and by $v_2^{(A)}(t)$ its velocity, we get $x_2^{(A)}(0) = 0$, $v_2^{(A)}(0) = 5$, and

$$m\frac{dv_2^{(A)}}{dt} = 650 - b_2 v_2^{(A)}$$

$$\Rightarrow \qquad \frac{dv_2^{(A)}}{dt} = \frac{65}{6} - v_2^{(A)} \qquad \Rightarrow \qquad v_2^{(A)}(t) = \frac{65}{6} - \frac{35}{6}e^{-t}$$

$$\Rightarrow \qquad x_2^{(A)}(t) = \int_0^t \left(\frac{65}{6} - \frac{35}{6}e^{-s}\right) ds = \frac{65}{6}t + \frac{35}{6}\left(e^{-t} - 1\right).$$

 Copyright © 2012 Pearson Education, Inc. Publishing as Addison-Wesley.

Solving the equation $x_2^{(A)}(t) = 498.15$, we find that the time (counting from the moment, when the boat A's velocity has reached 5 m/sec) is $t^* \approx 46.5$ sec, which is necessary to come to the end of the first leg. Therefore, the total time for the boat A is

$$t_* + t^* \approx 0.5 + 46.5 = 47 \text{ (sec)}.$$

Similarly, for the boat B, we find that

$$v_1^{(B)}(t) = \frac{65}{8} - \frac{49}{8} e^{-5t/3}, \quad x_1^{(B)}(t) = \frac{65}{8} t + \frac{147}{40} \left(e^{-5t/3} - 1 \right), \quad t_* = -\frac{3 \ln(17/49)}{5} \approx 0.6;$$

$$v_2^{(B)}(t) = \frac{65}{5} - \frac{35}{5} e^{-5t/6}, \quad x_2^{(B)}(t) = \frac{65}{5} t + \frac{42}{5} \left(e^{-5t/6} - 1 \right), \quad t^* \approx 38.9.$$

Thus, $t_* + t^* < 40$ sec, and so the boat B will be leading at the end of the first leg.

25. (a) From Newton's second law we have

$$m \frac{dv}{dt} = \frac{-GMm}{r^2}.$$

Dividing both sides by m, the mass of the rocket, and letting $g = GM/R^2$ we get

$$\frac{dv}{dt} = \frac{-gR^2}{r^2},$$

where g is the gravitational constant of Earth, R is the radius of Earth, and r is the distance between Earth and the projectile.

(b) Using the equation found in part (a), the chain rule

$$\frac{dv}{dt} = \frac{dv}{dr} \cdot \frac{dr}{dt},$$

and knowing that $dr/dt = v$, we get

$$v \frac{dv}{dr} = -\frac{gR^2}{r^2}.$$

(c) The differential equation found in part (b) is separable and can be written in the form

$$v \, dv = -\frac{gR^2}{r^2} \, dr.$$

If the projectile leaves Earth with a velocity of v_0, we have an initial value problem

$$v \, dv = -\frac{gR^2}{r^2} \, dr, \qquad v \Big|_{r=R} = v_0.$$

Copyright © 2012 Pearson Education, Inc. Publishing as Addison-Wesley.

Integrating, we get

$$\frac{v^2}{2} = \frac{gR^2}{r} + K,$$

where K is an arbitrary constant. We can find K by using the initial condition.

$$K = \frac{v_0^2}{2} - \frac{gR^2}{R} = \frac{v_0^2}{2} - gR.$$

Substituting this formula for K and solving for the velocity, we obtain

$$v^2 = \frac{2gR^2}{r} + v_0^2 - 2gR.$$

(d) In order for the velocity of the projectile to always remain positive, the expression $(2gR^2/r) + v_0^2$ must be greater than $2gR$, as r approaches infinity. This means that

$$\lim_{r \to \infty} \left(\frac{2gR^2}{r} + v_0^2 \right) > 2gR \qquad \Rightarrow \qquad v_0^2 > 2gR.$$

Therefore,

$$v_0^2 - 2gR > 0.$$

(e) Using the equation

$$v_e = \sqrt{2gR}$$

for the escape velocity and converting meters to kilometers, we get

$$v_e = \sqrt{2gR} = \sqrt{2 \cdot 9.81 \cdot (1/1000)(6370)} \approx 11.18 \text{ (km/sec)}.$$

(f) Similarly to (e), we find that

$$v_e = \sqrt{2(g/6)R} = \sqrt{2(9.81/6)(1/1000)(1738)} = 2.38 \text{ (km/sec)}.$$

EXERCISES 3.5: Electrical Circuits

1. In this problem, $R = 5\,\Omega$, $L = 0.05\,\text{H}$, and the voltage function is $E(t) = 5\cos 120t\,\text{V}$. Substituting these data into a general solution (3) to the Kirchhoff's equation (2) yields

$$I(t) = e^{-Rt/L} \left(\int e^{Rt/L} \frac{E(t)}{L} \, dt + K \right)$$

$$= e^{-5t/0.05} \left(\int e^{5t/0.05} \frac{5\cos 120t}{0.05} \, dt + K \right) = e^{-100t} \left(100 \int e^{100t} \cos 120t \, dt + K \right).$$

Using the integral tables, we evaluate the integral in the right-hand side and obtain

$$I(t) = e^{-100t} \left[100 \frac{e^{100t}(100\cos 120t + 120\sin 120t)}{(100)^2 + (120)^2} + K \right]$$

$$= \frac{\cos 120t + 1.2\sin 120t}{2.44} + Ke^{-100t}.$$

The initial condition, $I(0) = 1$, implies that

$$1 = I(0) = \frac{\cos(120(0)) + 1.2\sin(120(0))}{2.44} + Ke^{-100(0)} \frac{1}{2.44} + K$$

$$\Rightarrow \quad K = 1 - \frac{1}{2.44} = \frac{1.44}{2.44}$$

and so

$$I(t) = \frac{1.44e^{-100t} + \cos 120t + 1.2\sin 120t}{2.44}.$$

The subsequent inductor voltage is then determined by

$$E_L(t) = L\frac{dI}{dt} = 0.05\frac{d}{dt}\left(\frac{1.44e^{-100t} + \cos 120t + 1.2\sin 120t}{2.44}\right)$$

$$= \frac{-7.2e^{-100t} - 6\sin 120t + 7.2\cos 120t}{2.44}.$$

3. In this RC circuit, $R = 100\,\Omega$, $C = 10^{-12}\,\text{F}$, the initial charge of the capacitor is $Q = q(0) = 0$ coulombs, and the applied constant voltage is $V = 5$ volts. Thus, we can use a general equation for the charge $q(t)$ of the capacitor derived in Example 2. Substitution of given data yields

$$q(t) = CV + [Q - CV]e^{-t/RC} = 10^{-12}(5)\left(1 - e^{-t/(100\cdot 10^{-12})}\right) = 5\cdot 10^{-12}\left(1 - e^{-10^{10}t}\right),$$

and so

$$E_C(t) = \frac{q(t)}{C} = 5\left(1 - e^{-10^{10}t}\right).$$

Solving the equation $E_C(t) = 3$, we get

$$5\left(1 - e^{-10^{10}t}\right) = 3 \quad \Rightarrow \quad e^{-10^{10}t} = 0.4 \quad \Rightarrow \quad t = -\frac{\ln 0.4}{10^{10}} \approx 9.2 \times 10^{-11} \text{ (sec)}.$$

Thus, it will take about 9.2×10^{-11} sec. for the voltage to reach 3 volts at the receiving gate.

5. Let $V(t)$ denote the voltage across an element, and let $I(t)$ be the current through this element. Then for the power, say $P = P(t)$, generated or dissipated by the element we have

$$P = I(t)V(t). \tag{3.14}$$

Copyright © 2012 Pearson Education, Inc. Publishing as Addison-Wesley.

We use formulas given in (a), (b), and (c) in the text to find P for the resistor, inductor, and capacitor.

(a) *Resistor.* In this case,

$$V(t) = E_R(t) = RI(t),$$

and substitution into (3.14) yields

$$P_R = I(t)\,[RI(t)] = I(t)^2 R.$$

(b) *Inductor.* We have

$$V(t) = E_L(t) = L\,\frac{dI(t)}{dt}$$

$$\Rightarrow \quad P_L = I(t)\left[L\,\frac{dI(t)}{dt}\right] = \frac{L}{2}\left[2I(t)\frac{dI(t)}{dt}\right] = \frac{L}{2}\frac{d\,[I(t)^2]}{dt} = \frac{d}{dt}\frac{LI(t)^2}{2}.$$

(c) *Capacitor.* Here, with $q(t)$ denoting the electrical charge on the capacitor,

$$V(t) = E_C(t) = \frac{1}{C}\,q(t)$$

$$\Rightarrow \quad q(t) = CE_C(t) \qquad \Rightarrow \qquad I(t) = \frac{dq(t)}{dt} = \frac{d\,[CE_C(t)]}{dt}$$

and so

$$P_C = \frac{d\,[CE_C(t)]}{dt}\,E_C(t) = \frac{C}{2}\left[2E_C(t)\frac{dE_C(t)}{dt}\right] = \frac{C}{2}\frac{d\,[E_C(t)^2]}{dt} = \frac{d}{dt}\frac{CE_C(t)^2}{2}.$$

7. First, we find a formula for the current $I(t)$. Given that $R = 3\,\Omega$, $L = 10\,\mathrm{H}$, and the voltage function $E(t)$ is a constant, say V, the formula (3) (which describes currents in RL circuits) becomes

$$I(t) = e^{-3t/10}\left(\int e^{3t/10}\,\frac{V}{10}\,dt + K\right) = e^{-3t/10}\left(\frac{V}{3}\,e^{3t/10} + K\right) = \frac{V}{3} + Ke^{-3t/10}.$$

The initial condition, $I(0) = 0$ (there was no current in the electromagnet before the voltage source was applied), yields

$$0 = \frac{V}{3} + Ke^{-3(0)/10} \qquad \Rightarrow \qquad K = -\frac{V}{3} \qquad \Rightarrow \qquad I(t) = \frac{V}{3}\left(1 - e^{-3t/10}\right).$$

Next, we find the limiting value I_∞ of $I(t)$, that is,

$$I_\infty = \lim_{t\to\infty}\left[\frac{V}{3}\left(1 - e^{-3t/10}\right)\right] = \frac{V}{3}(1 - 0) = \frac{V}{3}.$$

 Copyright © 2012 Pearson Education, Inc. Publishing as Addison-Wesley.

Therefore, we are looking for the moment t when $I(t) = (0.9)I_\infty = (0.9)V/3$. Solving yields

$$\frac{0.9V}{3} = \frac{V}{3}\left(1 - e^{-3t/10}\right) \qquad \Rightarrow \qquad c^{-3t/10} = 0.1 \qquad \Rightarrow \qquad l = -\frac{10 \ln 0.1}{3} \approx 7.68\,.$$

Thus, it takes approximately 7.68 seconds for the electromagnet to reach 90% of its final value.

EXERCISES 3.6: Improved Euler's Method

1. Given the step size h and considering equally spaced points we have

$$x_{n+1} = x_n + nh, \qquad n = 0, 1, 2, \ldots\,.$$

Euler's method, given by equation (4), says that

$$y_{n+1} = y_n + hf(x_n, y_n), \qquad n = 0, 1, 2, \ldots\,,$$

where $f(x, y) = 5y$. Starting with the given value $y_0 = 1$, we compute

$$y_1 = y_0 + h(5y_0) = 1 + 5h.$$

We now use this value to find that

$$y_2 = y_1 + h(5y_1) = (1 + 5h)y_1 = (1 + 5h)^2.$$

Proceeding in this manner, we find a general formula for y_n, that is,

$$y_n = (1 + 5h)^n.$$

Referring back to our equation for x_n and using the given values of $x_0 = 0$ and $x_1 = 1$, we find that

$$1 = nh \qquad \Rightarrow \qquad n = \frac{1}{h}\,.$$

Substituting this value back into the formula for y_n, we find that an approximation to our initial value problem, $y' = 5y$, $y(0) = 1$, at $x = 1$ is $(1 + 5h)^{1/h}$.

3. In this initial value problem, $f(x, y) = y$, $x_0 = 0$, and $y_0 = 1$. Thus, formula (8) becomes

$$y_{n+1} = y_n + \frac{h}{2}\left(y_n + y_{n+1}\right).$$

Copyright © 2012 Pearson Education, Inc. Publishing as Addison-Wesley.

Chapter 3

Solving this equation for y_{n+1} yields

$$\left(1 - \frac{h}{2}\right) y_{n+1} = \left(1 + \frac{h}{2}\right) y_n$$

$$\Rightarrow \qquad y_{n+1} = \left(\frac{1 + h/2}{1 - h/2}\right) y_n, \qquad n = 0, 1, \dots. \qquad (3.15)$$

For $n \geq 1$, we can use (3.15) to express y_n in terms of y_{n-1} and substitute this expression into the right-hand side of (3.15). Continuing this process, we get

$$y_{n+1} = \left(\frac{1 + h/2}{1 - h/2}\right) \left[\left(\frac{1 + h/2}{1 - h/2}\right) y_{n-1}\right] = \left(\frac{1 + h/2}{1 - h/2}\right)^2 y_{n-1} = \cdots = \left(\frac{1 + h/2}{1 - h/2}\right)^{n+1} y_0.$$

In order to approximate the solution $\phi(x) = e^x$ at the point $x = 1$ with N steps, we take $h = (x - x_0)/N = 1/N$, and so $N = 1/h$. Then the above formula becomes

$$y_N = \left(\frac{1 + h/2}{1 - h/2}\right)^N y_0 = \left(\frac{1 + h/2}{1 - h/2}\right)^N = \left(\frac{1 + h/2}{1 - h/2}\right)^{1/h}$$

and hence

$$e = \phi(1) \approx y_N = \left(\frac{1 + h/2}{1 - h/2}\right)^{1/h}.$$

Substituting $h = 10^{-k}$, $k = 0, 1, 2, 3$, and 4, we fill in Table **3–D** on page 147. These approximations are better then those in Tables 3.4 and 3.5 of the text.

5. In this problem, we have $f(x, y) = 4y$. Thus,

$$f(x_n, y_n) = 4y_n \quad \text{and} \quad f(x_n + h, y_n + hf(x_n, y_n)) = 4\left[y_n + h(4y_n)\right] = 4y_n + 16hy_n.$$

By equation (9), we have

$$y_{n+1} = y_n + \frac{h}{2}(4y_n + 4y_n + 16hy_n) = \left(1 + 4h + 8h^2\right) y_n. \qquad (3.16)$$

Since the initial condition, $y(0) = 1/3$, implies that $x_0 = 0$ and $y_0 = 1/3$, (3.16) yields

$$y_1 = \left(1 + 4h + 8h^2\right) y_0 = \frac{1}{3}\left(1 + 4h + 8h^2\right),$$

$$y_2 = \left(1 + 4h + 8h^2\right) y_1 = \left(1 + 4h + 8h^2\right)\left(\frac{1}{3}\right)\left(1 + 4h + 8h^2\right) = \frac{1}{3}\left(1 + 4h + 8h^2\right)^2,$$

$$y_3 = \left(1 + 4h + 8h^2\right) y_2 = \left(1 + 4h + 8h^2\right)\left(\frac{1}{3}\right)\left(1 + 4h + 8h^2\right)^2 = \frac{1}{3}\left(1 + 4h + 8h^2\right)^3.$$

Continuing this way, we see that

$$y_n = \frac{1}{3}\left(1 + 4h + 8h^2\right)^n. \qquad (3.17)$$

Copyright © 2012 Pearson Education, Inc. Publishing as Addison-Wesley.

(This can be proved by induction using (3.16).) We are looking for an approximation to our solution at the point $x = 1/2$. Therefore, we have

$$h = \frac{1/2 - x_0}{n} = \frac{1/2 - 0}{n} = \frac{1}{2n} \qquad \Rightarrow \qquad n = \frac{1}{2h}.$$

Substituting this value for n into equation (3.17) yields

$$y_n = \frac{1}{3}\left(1 + 4h + 8h^2\right)^{1/(2h)}.$$

7. In this problem, $f(x, y) = x - y^2$. We need to approximate the solution on the interval $[1, 1.5]$ using a step size of $h = 0.1$. Thus, the number of steps needed is $N = 5$. The inputs to the improved Euler's method subroutine are $x_0 = 1$, $y_0 = 0$, $c = 1.5$, and $N = 5$. For Step 3 of the subroutine we have

$$F = f(x, y) = x - y^2,$$
$$G = f(x + h, y + hF) = (x + h) - (y + hF)^2 = (x + h) - \left[y + h(x - y^2)\right]^2.$$

Starting with $x = x_0 = 1$ and $y = y_0 = 0$, we get $h = 0.1$ (as specified) and

$$F = 1 - 0^2 = 1,$$
$$G = (1 + 0.1) - \left[0 + 0.1(1 - 0^2)\right]^2 = 1.1 - (0.1)^2 = 1.09.$$

Hence, in Step 4 we compute

$$x = 1 + 0.1 = 1.1,$$
$$y = 0 + 0.05(1 + 1.09) = 0.1045.$$

Thus, the approximate value of the solution at 1.1 is 0.1045. Next, we repeat Step 3 with $x = 1.1$ and $y = 0.1045$ to obtain

$$F = 1.1 + (0.1045)^2 \approx 1.0891,$$
$$G = (1.1 + 0.1) - \left[0.1045 + 0.1\left(1.1 - (0.1045)^2\right)\right]^2 \approx 1.1545.$$

Hence, in Step 4 we compute

$$x = 1.1 + 0.1 = 1.2,$$
$$y = 0.1045 + 0.05(1.0891 + 1.1545) \approx 0.21668.$$

Thus, the approximate value of the solution at $x = 1.2$ is $y = 0.21668$. Continuing this way, we fill in Table 3–E on page 147. (The reader can also use the software provided with the text.)

Copyright © 2012 Pearson Education, Inc. Publishing as Addison-Wesley.

9. In this initial value problem,

$$f(x, y) = x + 3\cos(xy), \quad x_0 = 0, \text{ and } y_0 = 0.$$

To approximate the solution on $[0, 2]$ with a step size $h = 0.2$, we need $N = 10$ steps. The functions F and G in the improved Euler's method subroutine are

$$\begin{aligned} F &= f(x, y) = x + 3\cos(xy); \\ G &= f(x + h, y + hF) = x + h + 3\cos[(x + h)(y + hF)] \\ &= x + 0.2 + 3\cos[(x + 0.2)(y + 0.2\{x + 3\cos(xy)\})]. \end{aligned}$$

Starting with $x = x_0 = 0$ and $y = y_0 = 0$, we compute

$$\begin{aligned} F &= 0 + 3\cos(0 \cdot 0) = 3\,; \\ G &= 0 + 0.2 + 3\cos[(0 + 0.2)(0 + 0.2\{0 + 3\cos(0 \cdot 0)\})] \approx 3.178426\,. \end{aligned}$$

Using these values, we find on Step 4 that

$$\begin{aligned} x &= 0 + 0.2 = 0.2\,, \\ y &= 0 + 0.1(3 + 3.178426) \approx 0.617843\,. \end{aligned}$$

With these new values of x and y, we repeat Step 3 and obtain

$$F = 0.2 + 3\cos(0.2 \cdot 0.617843) \approx 3.177125\,;$$
$$G = 0.2 + 0.2 + 3\cos[(0.2 + 0.2)(0.617843 + 0.2\{0.2 + 3\cos(0.2 \cdot 0.617843)\})]$$
$$\approx 3.030865\,.$$

Step 4 then yields an approximation of the solution at $x = 0.4$, namely,

$$\begin{aligned} x &= 0.2 + 0.2 = 0.4\,, \\ y &= 0.617843 + 0.1(3.177125 + 3.030865) \approx 1.238642\,. \end{aligned}$$

Continuing, we get Table **3–F** on page 148.

The polygonal line, approximating the graph of the solution to the given initial value problem, which has vertices at points (x, y) from Table **3–F**, is sketched in Figure **3–B** on page 150.

 Copyright © 2012 Pearson Education, Inc. Publishing as Addison-Wesley.

11. In this problem, we have $f(t,x) = 1 + t\sin(tx)$, $t_0 = 0$, $x_0 = 0$, $c = 1$. Starting with the number of steps $N = 2^0 = 1$ (so that $h = (c - t_0)/N = 1$), we set $t = t_0 = 0$, $x = x_0 = 0$ and compute

$$F = f(t,x) = 1 + t\sin(tx) = 1 + (0)\sin(0) = 1,$$

$$G = f(t + h, x + hF) = f(1,1) = 1 + \sin(1),$$

so that for $t = t_0 + h = 1$ we have

$$x(1;1) = x_0 + \frac{h}{2}[F + G] = 0 + \frac{1}{2}[1 + (1 + \sin(1))] \approx 1.42074.$$

Next we take $N = 2^1 = 2$ steps with $h = (c - t_0)/N = 2^{-1}$ and use the improved Euler's method to find

$$x\left(0.5; 2^{-1}\right) \approx 0.53093, \quad x\left(1; 2^{-1}\right) \approx 1.28613 \quad \Rightarrow \quad \phi(1) \approx 1.28613.$$

Continuing this way, we double the number of steps taking $N = 2^2$, 2^3, etc. (so halving the step size h), but stop when the difference between two successive approximations is within $\varepsilon = 0.01$. (We assume that the reader has a programmable calculator or a computer available, and so can transform the step-by-step subroutine into an executable program. Alternatively, the reader can use the software provided with the text.) Since

$$x\left(1; 2^{-2}\right) \approx 1.26026, \quad x\left(1; 2^{-3}\right) \approx 1.25494,$$

and $|1.26026 - 1.25494| = 0.00532 < 0.01$, we take $\phi(1) \approx 1.25494$.

13. We want to approximate the solution $\phi(x)$ to $y' = 1 - y + y^3$, $y(0) = 0$, at the point $x = 1$. (In other words, we want to find an approximate value for $\phi(1)$.) To do this, we use the improved Euler's method with tolerance. (We suggest to use a programmable calculator or a computer, and so transform the step-by-step subroutine into an executable program or use the software provided with the text.)

The inputs to the program are $x_0 = 0$, $y_0 = 0$, $c = 1$, $\varepsilon = 0.003$, and, say, $M = 100$. Notice that by Step 6 of the improved Euler's method with tolerance, the computations should terminate when two successive approximations differ by less that 0.003. The initial value for h in Step 1 is

$$h = (1 - 0)2^{-0} = 1.$$

Copyright © 2012 Pearson Education, Inc. Publishing as Addison-Wesley.

For the given equation, we have $f(x, y) = 1 - y + y^3$, and so the functions F and G in Step 3 of the improved Euler's method subroutine are

$$F = f(x, y) = 1 - y + y^3\,,$$
$$G = f(x + h, y + hF) = 1 - (y + hF) + (y + hF)^3\,.$$

From Step 4 of the improved Euler's method subroutine with $x = 0$, $y = 0$, and $h = 1$, we get

$$x = x + h = 0 + 1 = 1\,,$$
$$y = y + \frac{h}{2}(F + G) = 0 + \frac{1}{2}\left[1 + (1 - 1 + 1^3)\right] = 1\,.$$

Thus,

$$\phi(1) \approx y(1; 1) = 1\,.$$

The algorithm (Step 1 of the improved Euler's method subroutine) next sets $h = 2^{-1}$. The inputs to the subroutine are $x = 0$, $y = 0$, $c = 1$, and $N = 2$. For Step 3 of the subroutine we have

$$F = 1 - 0 + 0 = 1\,,$$
$$G = 1 - [0 + 0.5(1)] + [0 + 0.5(1)]^3 = 0.625\,.$$

Hence, in Step 4 we compute

$$x = 0 + 0.5 = 0.5\,,$$
$$y = 0 + 0.25(1 + 0.625) \approx 0.4063\,.$$

Thus, the approximate value of the solution at 0.5 is 0.4063. Next, we repeat Step 3 with $x = 0.5$ and $y = 0.4063$ to obtain

$$F = 1 - 0.4063 + (0.4063)^3 = 0.6608\,,$$
$$G = 1 - [0.4063 + 0.5(0.6608)] + [0.4063 + 0.5(0.6608)]^3 \approx 0.6631\,.$$

In Step 4, we compute

$$x = 0.5 + 0.5 = 1\,,$$
$$y = 0.4063 + 0.25(0.6608 + 0.6631) \approx 0.7372\,.$$

Copyright © 2012 Pearson Education, Inc. Publishing as Addison-Wesley.

Thus, the approximate value of the solution at $x = 1$ is $y = 0.7372$. Further outputs of the algorithm are given in Table **3–G**. Since

$$\left| y(1; 2^{-3}) - y(1; 2^{-2}) \right| = |0.7170 - 0.7194| < 0.003 \,,$$

the algorithm stops (see Step 6 of the improved Euler's method with tolerance) and prints out that $\phi(1)$ is approximately 0.7170.

15. For this problem, $f(x, y) = (x + y + 2)^2$. We want to approximate the solution on the interval $[0, 1.4]$ that satisfies $y(0) = -2$, and find the point (with two decimal places of accuracy), where its graph crosses the x-axis, that is, where $y = 0$. Our approach is to use a step size of 0.005 and look for a change in the sign of y. This procedure requires 280 steps. The inputs in the improved Euler's method subroutine are $x_0 = 0$, $y_0 = -2$, $c = 1.4$, and $N = 280$. We will stop computations, when we see a sign change in the value of y. (The subroutine is implemented on the software package provided with the text.)

For Step 3 of the subroutine we have

$$F = f(x, y) = (x + y + 2)^2 \,,$$
$$G = f(x + h, y + hF) = (x + h + y + hF + 2)^2 = [x + y + 2 + h(1 + F)]^2 \,.$$

Starting with the inputs $x = x_0 = 0$, $y = y_0 = -2$, and $h = 0.005$, we obtain

$$F = (0 - 2 + 2)^2 = 0,$$
$$G = [0 - 2 + 2 + 0.005(1 + 0)]^2 = 0.000025 \,.$$

Thus, in Step 4 we compute

$$x = 0 + 0.005 = 0.005 \,, \quad y = -2 + 0.005(0 + 0.000025)(1/2) \approx -2.$$

Therefore, the approximate value of the solution at $x = 0.005$ is $y = -2$. We continue with Steps 3 and 4 of the improved Euler's method subroutine until we arrive at $x = 1.270$ and $y \approx -0.04658$. The next iteration, with $x = 1.275$, yields $y \approx 0.00630$. This tells us that $y = 0$ occurs somewhere between $x = 1.270$ and $x = 1.275$. Therefore, rounding off to two decimal places yields $x \approx 1.27$.

Copyright © 2012 Pearson Education, Inc. Publishing as Addison-Wesley.

17. In this initial value problem, $f(x, y) = -20y$, $x_0 = 0$, and $y_0 = 1$. Applying the formula (4), we find a general expression for y_n in terms of h. Indeed,

$$y_n = y_{n-1} + h(-20y_{n-1}) = (1 - 20h)y_{n-1} = \cdots = (1 - 20h)^n y_0 = (1 - 20h)^n = [c(h)]^n,$$

where $c(h) = 1 - 20h$. For suggested values of h, we have

$$h = 0.1 \quad \Rightarrow \quad c(0.1) = -1 \quad \Rightarrow \quad x_n = 0.1n, \quad y_n = (-1)^n, \quad n = 1, \ldots, 10;$$
$$h = 0.025 \quad \Rightarrow \quad c(0.025) = 0.5 \quad \Rightarrow \quad x_n = 0.025n, \quad y_n = (0.5)^n, \quad n = 1, \ldots, 40;$$
$$h = 0.2 \quad \Rightarrow \quad c(0.2) = -3 \quad \Rightarrow \quad x_n = 0.2n, \quad y_n = (-3)^n, \quad n = 1, \ldots, 5.$$

These values are shown in Table **3–H** on page 148. Thus, for $h = 0.1$ we have alternating $y_n = \pm 1$; for $h = 0.2$, y_n's have an increasing magnitude and alternating sign; $h = 0.025$ is a good step size. From this example, we conclude that in Euler's method one should be very careful in choosing a step size. Wrong choice can even lead to a diverging process.

19. We will use the improved Euler's method with $h = 2/3$ to approximate the solution to the problem

$$\left\{ \left[75 - 20 \cos \left(\frac{\pi t}{12} \right) \right] - T(t) \right\} + 0.1 + 1.5[70 - T(t)], \qquad T(0) = 65,$$

with $K = 0.2$. Since $h = 2/3$, it will take 36 steps to go from $t = 0$ to $t = 24$. By simplifying the above expression, we obtain

$$\frac{dT}{dt} = (75K + 105.1) - 20K \cos \left(\frac{\pi t}{12} \right) - (K + 1.5)T(t), \qquad T(0) = 65.$$

(Note that here t takes the place of x and T takes the place of y.) Therefore, with $K = 0.2$, the inputs to the subroutine are $t_0 = 0$, $T_0 = 65$, $c = 24$, and $N = 36$. For Step 3 of the subroutine we have

$$F = f(t, T) = (75K + 105.1) - 20K \cos \left(\frac{\pi t}{12} \right) - (K + 1.5)T, \qquad (3.18)$$
$$G = f(t + h, T + hF)$$
$$= (75K + 105.1) - 20K \cos \left(\frac{\pi(t + h)}{12} \right) - (K + 1.5)\{T + hF\}. \qquad (3.19)$$

For Step 4 in the subroutine, we have

$$t = t + h,$$
$$T = T + \frac{h}{2}(F + G).$$

 Copyright © 2012 Pearson Education, Inc. Publishing as Addison-Wesley.

Now, starting with $t = t_0 = 0$ and $T = T_0 = 65$, and $h = 2/3$ (as specified), we complete Step 3 of the subroutine and get

$$F = [75(0.2) + 105.1] - 20(0.2)\cos 0 - [(0.2) + 1.5](65) = 5.6\,,$$

$$G = [75(0.2) + 105.1] - 20(0.2)\cos\left[\frac{\pi(0.6667)}{12}\right] - [(0.2) + 1.5][65 + (0.6667)(5.6)]$$
$$\approx -0.6862\,.$$

Hence, in Step 4 we compute

$$t = 0 + 0.6667 = 0.6667\,,$$

$$T = 65 + 0.3333(5.6 - 0.6862) \approx 66.638\,.$$

Recalling that t_0 is midnight, we see that these results imply that at 0.6667 hours after midnight (or 12 : 40 A.M.) the temperature is approximately 66.638. Continuing with this process for $n = 1, 2, \ldots, 35$ gives us the approximate temperatures in a building with $K = 0.2$ over a 24 hr period. These results are given in Table **3–I** on page 149. (This is just a partial table.)

The next step is to redo the computations with $K = 0.4$. That is, we substitute $K = 0.4$ and $h = 2/3 \approx 0.6667$ into equations (3.18) and (3.19) above. This yields

$$F = 135.1 - 8\cos\left(\frac{\pi t}{12}\right) - 1.9T,$$

$$G = 135.1 - 8\cos\left[\frac{\pi(t + 0.6667)}{12}\right] - 1.9(T + 0.6667F),$$

and

$$T = T + (0.3333)(F + G).$$

Then, using these equations, we go through the process of finding F first, then find G and, finally, using both results determine T. (This process must be done for $n = 0, 1, \ldots, 35$.) We redo these calculations with $K = 0.6$ and $h = 2/3$. In so doing, we obtain the results given in the table in the answers of the text.

EXERCISES 3.7: Higher-Order Numerical Methods: Taylor and Runge-Kutta

1. In this problem, $f(x, y) = \cos(x + y)$. Applying the formula (4), we compute

$$\frac{\partial f(x, y)}{\partial x} = \frac{\partial}{\partial x}\left[\cos(x + y)\right] = -\sin(x + y)\frac{\partial}{\partial x}(x + y) = -\sin(x + y);$$

Copyright © 2012 Pearson Education, Inc. Publishing as Addison-Wesley.

$$\frac{\partial f(x,y)}{\partial y} = \frac{\partial}{\partial y}\left[\cos(x+y)\right] = -\sin(x+y)\frac{\partial}{\partial y}(x+y) = -\sin(x+y);$$

$$f_2(x,y) = \frac{\partial f(x,y)}{\partial x} + \left[\frac{\partial f(x,y)}{\partial y}\right]f(x,y) = -\sin(x+y) + \left[-\sin(x+y)\right]\cos(x+y)$$
$$= -\sin(x+y)\left[1 + \cos(x+y)\right],$$

and so, with $p = 2$, (5) and (6) give

$$x_{n+1} = x_n + h,$$
$$y_{n+1} = y_n + h\cos(x_n + y_n) - \frac{h^2}{2}\sin(x_n + y_n)\left[1 + \cos(x_n + y_n)\right].$$

3. Here we have $f(x,y) = x - y$ and so

$$f_2(x,y) = \frac{\partial(x-y)}{\partial x} + \frac{\partial(x-y)}{\partial y}(x-y) = 1 + (-1)(x-y) = 1 - x + y.$$

To obtain $f_3(x,y)$ and then $f_4(x,y)$, we differentiate the equation $y'' = f_2(x,y)$ twice. This yields

$$y'''(x) = [f_2(x,y)]' = (1 - x + y)' = -1 + y' = -1 + x - y =: f_3(x,y);$$

$$y^{(4)}(x) = [f_3(x,y)]' = (-1 + x - y)' = 1 - y' = 1 - x + y =: f_4(x,y).$$

Therefore, the recursive formulas of order 4 for the Taylor method are

$$x_{n+1} = x_n + h,$$
$$y_{n+1} = y_n + h(x_n - y_n) + \frac{h^2}{2}(1 - x_n + y_n) + \frac{h^3}{3!}(-1 + x_n - y_n) + \frac{h^4}{4!}(1 - x_n + y_n)$$
$$= y_n + h(x_n - y_n) + \frac{h^2}{2}(1 - x_n + y_n) - \frac{h^3}{6}(1 - x_n + y_n) + \frac{h^4}{24}(1 - x_n + y_n)$$
$$= y_n + h + (1 - x_n + y_n)\left(-h + \frac{h^2}{2} - \frac{h^3}{6} + \frac{h^4}{24}\right)$$
$$= y_n + h(x_n - y_n) + (1 - x_n + y_n)\left(\frac{h^2}{2} - \frac{h^3}{6} + \frac{h^4}{24}\right).$$

5. For the Taylor method of *order* 2, we need to find (see the equation (4))

$$f_2(x,y) = \frac{\partial f(x,y)}{\partial x} + \left[\frac{\partial f(x,y)}{\partial y}\right]f(x,y)$$

Copyright © 2012 Pearson Education, Inc. Publishing as Addison-Wesley.

for $f(x,y) = x + 1 - y$. Thus, we have

$$f_2(x,y) = 1 + (-1)(x + 1 - y) = y - x.$$

Therefore, from equations (5) and (6) we see that the recursive formulas (with $h = 0.25$) become

$$x_{n+1} = x_n + 0.25 \,,$$

$$y_{n+1} = y_n + 0.25 \, (x_n + 1 - y_n) + \frac{(0.25)^2}{2} \, (y_n - x_n) \,.$$

By starting with $x_0 = 0$ and $y_0 = 1$ (the initial values for the problem), we find

$$y_1 = 1 + \frac{0.0625}{2} \approx 1.03125 \,.$$

Plugging this value into the recursive formulas yields

$$y_2 = 1.03125 + 0.25(0.25 + 1 - 1.03125) + \left(\frac{0.0625}{2} \right)(1.03125 - 0.25) \approx 1.11035 \,.$$

Continuing, we fill in the first three columns in Table **3–B**, page 146.

For the Taylor method of *order* 4, we need to find f_3 and f_4. Thus, we obtain

$$f_3(x,y) = \frac{\partial f_2(x,y)}{\partial x} + \left[\frac{\partial f_2(x,y)}{\partial y} \right] f(x,y) = -1 + 1 \cdot (x + 1 - y) = x - y,$$

$$f_4(x,y) = \frac{\partial f_3(x,y)}{\partial x} + \left[\frac{\partial f_3(x,y)}{\partial y} \right] f(x,y) = 1 + (-1) \cdot (x + 1 - y) = y - x.$$

Hence, equation (6) gives a recursive formula for y_{n+1} in Taylor's method of order 4 with $h = 0.25$. We have

$$y_{n+1} = y_n + 0.25 \, (x_n + 1 - y_n) + \frac{(0.25)^2}{2} \, (y_n - x_n) + \frac{(0.25)^3}{6} \, (x_n - y_n) + \frac{(0.25)^4}{24} \, (y_n - x_n) \,.$$

By starting with $x_0 = 0$ and $y_0 = 1$, we can fill in the fourth column of Table **3–B**.

Thus, the approximation (rounded to five decimal places) of the solution by the Taylor method at the point $x = 1$ is given by $\phi_2(1) = 1.37253$ if we use order 2 and by $\phi_4(1) = 1.36789$ if we use order 4. The actual solution, $y = x + e^{-x}$, has the value $y(1) = 1 + e^{-1} \approx 1.36788$ at $x = 1$. Comparing these results, we see that

$$|y(1) - \phi_2(1)| = 0.00465 \qquad \text{and} \qquad |y(1) - \phi_4(1)| = 0.00001 \,.$$

Copyright © 2012 Pearson Education, Inc. Publishing as Addison-Wesley.

Chapter 3

7. We will use the 4th order Runge-Kutta subroutine. Since $x_0 = 0$ and $h = 0.25$, we need $N = 4$ steps to approximate the solution at $x = 1$. With $f(x,y) = 2y - 6$, we set

$$x = x_0 = 0 , \quad y = y_0 = 1$$

and go to Step 3 to compute k_j's.

$$k_1 = hf(x,y) = 0.25[2(1) - 6] = -1 \, ;$$
$$k_2 = hf(x + h/2, y + k_1/2) = 0.25[2(1 + (-1)/2) - 6] = -1.25 \, ;$$
$$k_3 = hf(x + h/2, y + k_2/2) = 0.25[2(1 + (-1.25)/2) - 6] = -1.3125 \, ;$$
$$k_4 = hf(x + h, y + k_3) = 0.25[2(1 + (-1.3125)) - 6] = -1.65625 \, .$$

Step 4 then yields

$$x = 0 + 0.25 = 0.25 \, ,$$
$$y = 1 + \frac{1}{6}(k_1 + 2k_2 + 2k_3 + k_4) = 1 + \frac{1}{6}(-1 - 2 \cdot 1.25 - 2 \cdot 1.3125 - 1.65625)$$
$$\approx -0.29688 \, .$$

Now we go back to Step 3 and recalculate k_j's for new values of x and y.

$$k_1 = 0.25[2(-0.29688) - 6] = -1.64844 \, ;$$
$$k_2 = 0.25[2(-0.29688 + (-1.64844)/2) - 6] = -2.06055 \, ;$$
$$k_3 = 0.25[2(-0.29688 + (-2.06055)/2) - 6] = -2.16358 \, ;$$
$$k_4 = 0.25[2(-0.29688 + (-2.16358)) - 6] = -2.73022 \, .$$

Thus, Step 4 gives

$$x = 0.25 + 0.25 = 0.5 \, ,$$
$$y = -0.29688 + \frac{1}{6}(-1.64844 - 2 \cdot 2.06055 - 2 \cdot 2.16358 - 2.73022) \approx -2.43470 \, .$$

We now repeat the cycle twice.

$$k_1 = 0.25[2(-2.43470) - 6] = -2.71735 \, ;$$
$$k_2 = 0.25[2(-2.43470 + (-2.71735)/2) - 6] = -3.39670 \, ;$$
$$k_3 = 0.25[2(-2.43470 + (-3.39670)/2) - 6] = -3.56652 \, ;$$
$$k_4 = 0.25[2(-2.43470 + (-3.56652)) - 6] = -4.50060 \, ;$$

 Copyright © 2012 Pearson Education, Inc. Publishing as Addison-Wesley.

$$x = 0.5 + 0.25 = 0.75, \qquad y \approx -5.95876$$

and

$$k_1 = 0.25[2(-5.95876) - 6] = -4.47938;$$

$$k_2 = 0.25[2(-5.95876 + (-4.47938)/2) - 6] = -5.59922;$$

$$k_3 = 0.25[2(-5.95876 + (-5.59922)/2) - 6] = -5.87918;$$

$$k_4 = 0.25[2(-5.95876 + (-5.87918)) - 6] = -7.41895;$$

$$x = 0.75 + 0.25 = 1.00, \qquad y \approx -11.7679.$$

Thus $\phi(1) \approx -11.7679$. The actual solution, $\phi(x) = 3 - 2e^{2x}$, evaluated at $x = 1$, gives

$$\phi(1) = 3 - 2e^{2(1)} = 3 - 2e^2 \approx -11.7781.$$

9. In this problem, we use the 4th order Runge-Kutta subroutine with

$$f(x, y) = x + 1 - y.$$

Using the step size of $h = 0.25$, we find that the number of steps needed to approximate the solution at $x = 1$ is $N = 4$. For Step 3, we have

$$k_1 = hf(x, y) = 0.25(x + 1 - y);$$

$$k_2 = hf\left(x + \frac{h}{2}, y + \frac{k_1}{2}\right) = 0.25(0.875x + 1 - 0.875y);$$

$$k_3 = hf\left(x + \frac{h}{2}, y + \frac{k_2}{2}\right) = 0.25(0.890625x + 1 - 0.890625y);$$

$$k_4 = hf(x + h, y + k_3) \approx 0.25(0.777344x + 1 - 0.777344y).$$

Hence, in Step 4 we have

$$x = x + 0.25, \qquad y = y + \frac{1}{6}(k_1 + 2k_2 + 2k_3 + k_4).$$

Using the initial conditions $x_0 = 0$ and $y_0 = 1$, $c = 1$, and $N = 4$ for Step 3 we obtain

$$k_1 = 0.25(0 + 1 - 1) = 0;$$

$$k_2 = 0.25(0.875(0) + 1 - 0.875(1)) = 0.03125;$$

$$k_3 = 0.25(0.890625(0) + 1 - 0.890625(1)) \approx 0.027344;$$

$$k_4 = 0.25(0.777344(0) + 1 - 0.777344(1)) \approx 0.055664.$$

Copyright © 2012 Pearson Education, Inc. Publishing as Addison-Wesley.

Thus, Step 4 gives

$$x = 0 + 0.25 = 0.25,$$

$$y \approx 1 + \frac{1}{6}\left[0 + 2(0.03125) + 2(0.027344) + 0.055664\right] \approx 1.02881.$$

Thus the approximate value of the solution at 0.25 is 1.02881. By repeating Steps 3 and 4 of the algorithm we fill in the following Table **3–C** on page 147.

Thus, our approximation at $x = 1$ is approximately 1.36789. Comparing this with Problem 5, we see that we have obtained accuracy to four decimal places, as we did with the Taylor method of order four, but without computing partial derivatives.

11. In this problem, $f(x, y) = 2x^{-4} - y^2$. To find the root of the solution within two decimal places of accuracy, we choose a step size $h = 0.005$ in 4th order Runge-Kutta subroutine. It will require $(2 - 1)/0.005 = 200$ steps to approximate the solution on $[1, 2]$. With the initial input $x = x_0 = 1$, $y = y_0 = -0.414$, we get

$$k_1 = hf(x, y) = 0.005[2(1)^{-4} - (-0.414)^2] \approx 0.009143;$$

$$k_2 = hf(x + h/2, y + k_1/2) = 0.005[2(1 + 0.005/2)^{-4} - (-0.414 + 0.009143/2)^2]$$
$$\approx 0.009062;$$

$$k_3 = hf(x + h/2, y + k_2/2) = 0.005[2(1 + 0.005/2)^{-4} - (-0.414 + 0.009062/2)^2]$$
$$\approx 0.009062;$$

$$k_4 = hf(x + h, y + k_3) = 0.005[2(1 + 0.005)^{-4} - (-0.414 + 0.009062)^2] = 0.008983.$$

Therefore,

$$x = 1 + 0.005 = 1.005,$$

$$y = -0.414 + (0.009143 + 2 \cdot 0.009062 + 2 \cdot 0.009062 + 0.008983)/6 \approx -0.404937.$$

Continuing, on 82nd step we get

$$x = 1.405 + 0.005 = 1.410,$$

$$y = -0.004425 + \frac{1}{6}(0.002566 + 2 \cdot 0.002548 + 2 \cdot 0.002548 + 0.002530) \approx -0.001876,$$

and the next step gives

$$k_1 = 0.005[2(1.410)^{-4} - (-0.001876)^2] = 0.002530;$$

 Copyright © 2012 Pearson Education, Inc. Publishing as Addison-Wesley.

$$k_2 = 0.005[2(1.410 + 0.005/2)^{-4} - (-0.001876 + 0.002530/2)^2] = 0.002512 \, ;$$

$$k_3 = 0.005[2(1.410 + 0.005/2)^{-4} - (-0.001876 + 0.002512/2)^2] = 0.002512 \, ;$$

$$k_4 = 0.005[2(1.410 + 0.005)^{-4} - (-0.001876 + 0.002512)^2] = 0.002494 \, ;$$

$$x = 1.410 + 0.005 = 1.415 \, , \qquad y = \approx 0.000636 \, .$$

Since $y(1.41) < 0$ and $y(1.415) > 0$ we conclude that the root of the solution is on $(1.41, 1.415)$.

As a check, we apply the 4th order Runge-Kutta subroutine to approximate the solution to the given initial value problem on $[1, 1.5]$ with a step size $h = 0.001$, which requires $N = (1.5 - 1)/0.001 = 500$ steps. This yields

$$y(1.413) \approx -0.000367 \, , \quad y(1.414) \approx 0.000134 \, ,$$

and so, within two decimal places of accuracy, $x \approx 1.41$.

13. For this problem,

$$f(x, y) = y^2 - 2e^x y + e^{2x} + e^x.$$

We want to find the vertical asymptote located in the interval $[0, 2]$ within two decimal places of accuracy using the 4th order Runge-Kutta subroutine. One approach is to use a step size of 0.005 and look for y to approach infinity. This requires 400 steps. We will stop computations, when the value of y "blows up" (becomes very large). For Step 3, we have

$$k_1 = hf(x, y) = 0.005 \left(y^2 - 2e^x y + e^{2x} + e^x \right),$$

$$k_2 = hf \left(x + \frac{h}{2}, y + \frac{k_1}{2} \right)$$

$$= 0.005 \left[\left(y + \frac{k_1}{2} \right)^2 - 2e^{(x+h/2)} \left(y + \frac{k_1}{2} \right) + e^{2(x+h/2)} + e^{(x+h/2)} \right],$$

$$k_3 = hf \left(x + \frac{h}{2}, y + \frac{k_2}{2} \right)$$

$$= 0.005 \left[\left(y + \frac{k_2}{2} \right)^2 - 2e^{(x+h/2)} \left(y + \frac{k_2}{2} \right) + e^{2(x+h/2)} + e^{(x+h/2)} \right],$$

$$k_4 = hf(x + h, y + k_3) = 0.005 \left[(y + k_3)^2 - 2e^{(x+h)}(y + k_3) + e^{2(x+h)} + e^{(x+h)} \right].$$

Hence, Step 4 yields

$$x = x + 0.005 \, ,$$

Copyright © 2012 Pearson Education, Inc. Publishing as Addison-Wesley.

$$y = y + \frac{1}{6} \left(k_1 + 2k_2 + 2k_3 + k_4 \right).$$

Using the initial conditions $x_0 = 0$, $y_0 = 3$, $c = 2$, and $N = 400$ on Step 3, we obtain

$$k_1 = 0.005 \left(3^2 - 2e^0(3) + e^{2(0)} + e^0 \right) = 0.025,$$

$$k_2 = 0.005 \left[(3 + 0.0125)^2 - 2e^{(0+0.0025)}(3 + 0.0125) + e^{2(0+0.0025)} + e^{(0+0.0025)} \right]$$
$$\approx 0.02522,$$

$$k_3 = 0.005 \left[(3 + 0.01261)^2 - 2e^{(0+0.0025)}(3 + 0.01261) + e^{2(0+0.0025)} + e^{(0+0.0025)} \right]$$
$$\approx 0.02522,$$

$$k_4 = 0.005 \left[(3 + 0.02522)^2 - 2e^{(0+0.0025)}(3 + 0.02522) + e^{2(0+0.0025)} + e^{(0+0.0025)} \right]$$
$$\approx 0.02543.$$

Thus, Step 4 gives us

$$x = 0 + 0.005 = 0.005,$$

$$y \approx 3 + \tfrac{1}{6} (0.025 + 2(0.02522) + 2(0.02522) + 0.02543) \approx 3.02522.$$

Thus, the approximate value at $x = 0.005$ is 3.02522. By repeating Steps 3 and 4 of the subroutine we find that, at $x = 0.505$, $y = 2.0201 \cdot 10^{13}$. The next iteration gives a floating point overflow. This would lead one to think the asymptote occurs at $x = 0.51$.

As a check, let apply the 4th order Runge-Kutta subroutine with the initial conditions $x_0 = 0$, $y_0 = 3$, $c = 1$, and $N = 400$. This gives a finer step size of $h = 0.0025$. With these inputs, we find $y(0.5025) \approx 4.0402 \cdot 10^{13}$.

Repeating the subroutine one more time with a step size of 0.00125, we obtain the value $y(0.50125) \approx 8.0804 \cdot 10^{13}$. Therefore, we conclude that the vertical asymptote occurs at $x = 0.50$, not at $x = 0.51$ as was earlier thought.

15. Here $f(x, y) = \cos(5y) - x$, $x_0 = 0$, and $y_0 = 0$. With a step size $h = 0.1$, we take $N = 30$ in order to approximate the solution on $[0, 3]$. Setting $x = x_0 = 0$, $y = y_0 = 0$, we compute

$$k_1 = hf(x, y) = 0.1[\cos(5 \cdot 0) - 0] = 0.1;$$
$$k_2 = hf(x + h/2, y + k_1/2) = 0.1[\cos(5(0 + 0.1/2)) - (0 + 0.1/2)] = 0.091891;$$
$$k_3 = hf(x + h/2, y + k_2/2) = 0.1[\cos(5(0 + 0.091891/2)) - (0 + 0.1/2)] = 0.092373;$$
$$k_4 = hf(x + h, y + k_3) = 0.1[\cos(5(0 + 0.092373)) - (0 + 0.1)] = 0.079522;$$
$$x = 0 + 0.1 = 0.1,$$
$$y = 0 + (0.1 + 2 \cdot 0.091891 + 2 \cdot 0.092373 + 0.079522)/6 \approx 0.091342.$$

    Copyright © 2012 Pearson Education, Inc. Publishing as Addison-Wesley.

The results of further computations are shown in Table **3–J**, page 149.

Using these values, we sketch the polygonal line, approximating the graph of the solution on $[0, 3]$, in Figure **3–C** on page 150.

17. Taylor method of order 2 has recursive formulas given by equations (5) and (6).

$$x_{j+1} = x_j + h \qquad \text{and} \qquad y_{j+1} = y_j + hf(x_j, y_j) + \frac{h^2}{2!} f_2(x_j, y_j).$$

In this problem, $f(x, y) = y$ and so (4) yields

$$f_2(x, y) = \frac{\partial f(x, y)}{\partial x} + \left[\frac{\partial f(x, y)}{\partial y} \right] f(x, y) = 0 + 1 \cdot (y) = y.$$

Therefore, since $h = 1/n$, the recursive formula for y_{j+1} is given by the equation

$$y_{j+1} = y_j + \frac{1}{n} y_j + \frac{1}{2n^2} y_j = \left(1 + \frac{1}{n} + \frac{1}{2n^2} \right) y_j.$$

We start the process with $x_0 = 0$, $y_0 = 1$ and take steps of size $1/n$ until we reach $x = 1$. This means that we need n steps, and that y_n will be an approximation of the solution to the differential equation at $x = 1$. Since the actual solution is $y = e^x$, $y_n \approx e$. Using the recursive formula yields

$$y_1 = \left(1 + \frac{1}{n} + \frac{1}{2n^2} \right) y_0 = \left(1 + \frac{1}{n} + \frac{1}{2n^2} \right),$$

$$y_2 = \left(1 + \frac{1}{n} + \frac{1}{2n^2} \right) y_1 = \left(1 + \frac{1}{n} + \frac{1}{2n^2} \right)^2,$$

$$y_3 = \left(1 + \frac{1}{n} + \frac{1}{2n^2} \right) y_2 = \left(1 + \frac{1}{n} + \frac{1}{2n^2} \right)^3,$$

$$y_4 = \left(1 + \frac{1}{n} + \frac{1}{2n^2} \right) y_3 = \left(1 + \frac{1}{n} + \frac{1}{2n^2} \right)^4,$$

$$\vdots$$

$$y_n = \left(1 + \frac{1}{n} + \frac{1}{2n^2} \right) y_{n-1} = \left(1 + \frac{1}{n} + \frac{1}{2n^2} \right)^n.$$

(This can be proved rigorously by mathematical induction.) Thus,

$$e \approx \left(1 + \frac{1}{n} + \frac{1}{2n^2} \right)^n.$$

19. In this initial value problem, the independent variable is u and the dependent variable is v, $u_0 = 2$ and $v_0 = 0.1$. Also,

$$f(u, v) = u \left(\frac{u}{2} + 1 \right) v^3 + \left(u + \frac{5}{2} \right) v^2.$$

Copyright © 2012 Pearson Education, Inc. Publishing as Addison-Wesley.

Chapter 3

We use the classical 4th order Runge-Kutta algorithm with tolerance but, since the stopping criteria should be based on the relative error, we replace the condition $|z-v| < \varepsilon$ in Step 6 by

$$\frac{z-v}{v} < \varepsilon$$

(see Step 6').

We start with $m = 0$, $N = 2^m = 1$, and a step size $h = (3-2)/N = 1$. Setting

$$u = u_0 = 2, \quad v = v_0 = 0.1,$$

on Step 4 we compute

$$k_1 = hf(u, v) = (1)\left[2\left(\frac{2}{2}+1\right)(0.1)^3 + \left(2+\frac{5}{2}\right)(0.1)^2\right] = 0.049;$$

$$k_2 = hf(u+h/2, v+k_1/2) = (1)\left[(2+1/2)\left(\frac{2+1/2}{2}+1\right)(0.1+0.049/2)^3\right.$$
$$\left. + \left((2+1/2)+\frac{5}{2}\right)(0.1+0.049/2)^2\right] = 0.088356;$$

$$k_3 = hf(u+h/2, v+k_2/2) = (1)\left[(2+1/2)\left(\frac{2+1/2}{2}+1\right)(0.1+0.088356/2)^3\right.$$
$$\left. + \left((2+1/2)+\frac{5}{2}\right)(0.1+0.088356/2)^2\right] = 0.120795;$$

$$k_4 = hf(u+h, v+k_3) = (1)\left[(2+1)\left(\frac{2+1}{2}+1\right)(0.1+0.120795)^3\right.$$
$$\left. + \left((2+1)+\frac{5}{2}\right)(0.1+0.120795)^2\right] = 0.348857.$$

Therefore,

$$u = u+h = 2+1 = 3,$$
$$v = v + \frac{1}{6}(0.049 + 2\cdot0.088356 + 2\cdot0.120795 + 0.348857) \approx 0.236027.$$

Because the relative error between two successive approximations, $v(3; 2^0) = 0.236027$ and $v = 0.1$ is

$$\varepsilon = \left|\frac{0.236027 - 0.1}{0.236027}\right| \approx 0.576320 > 0.0001,$$

we go back to Step 2, set $m = 1$, take $N = 2^m = 2$ on Step 3, compute $h = 1/N = 0.5$, and use the 4th order Runge-Kutta subroutine again to find $v(3; 2^{-1})$. This takes two steps and yields

$$k_1 = (0.5)\left[2\left(\frac{2}{2}+1\right)(0.1)^3 + \left(2+\frac{5}{2}\right)(0.1)^2\right] = 0.0245;$$

     Copyright © 2012 Pearson Education, Inc. Publishing as Addison-Wesley.

$$k_2 = (0.5) \left[(2 + 0.5/2) \left(\frac{2 + 0.5/2}{2} + 1 \right) (0.1 + 0.0245/2)^3 \right.$$
$$\left. + \left((2 + 0.5/2) + \frac{5}{2} \right) (0.1 + 0.0245/2)^2 \right] = 0.033306\,;$$

$$k_3 = (0.5) \left[(2 + 0.5/2) \left(\frac{2 + 0.5/2}{2} + 1 \right) (0.1 + 0.033306/2)^3 \right.$$
$$\left. + \left((2 + 0.5/2) + \frac{5}{2} \right) (0.1 + 0.033306/2)^2 \right] = 0.036114\,;$$

$$k_4 = (0.5) \left[(2 + 0.5) \left(\frac{2 + 0.5}{2} + 1 \right) (0.1 + 0.036114)^3 \right.$$
$$\left. + \left((2 + 0.5) + \frac{5}{2} \right) (0.1 + 0.036114)^2 \right] = 0.053410\,;$$

$$u = 2 + 0.5 = 2.5\,,$$

$$v = 0.1 + \frac{1}{6} (0.0245 + 2 \cdot 0.033306 + 2 \cdot 0.036114 + 0.053410) \approx 0.136125\,.$$

We compute k_j's one more time and find an approximate value of $v(3)$.

$$k_1 = (0.5) \left[2.5 \left(\frac{2.5}{2} + 1 \right) (0.136125)^3 + \left(2.5 + \frac{5}{2} \right) (0.136125)^2 \right] = 0.053419;$$

$$k_2 = (0.5) \left[(2.5 + 0.5/2) \left(\frac{2.5 + 0.5/2}{2} + 1 \right) (0.136125 + 0.053419/2)^3 \right.$$
$$\left. + \left((2.5 + 0.5/2) + \frac{5}{2} \right) (0.136125 + 0.053419/2)^2 \right] = 0.083702\,;$$

$$k_3 = (0.5) \left[(2.5 + 0.5/2) \left(\frac{2.5 + 0.5/2}{2} + 1 \right) (0.136125 + 0.083702/2)^3 \right.$$
$$\left. + \left((2.5 + 0.5/2) + \frac{5}{2} \right) (0.136125 + 0.083702/2)^2 \right] = 0.101558\,;$$

$$k_4 = (0.5) \left[(2.5 + 0.5) \left(\frac{2.5 + 0.5}{2} + 1 \right) (0.136125 + 0.101558)^3 \right.$$
$$\left. + \left((2.5 + 0.5) + \frac{5}{2} \right) (0.136125 + 0.101558)^2 \right] = 0.205709\,;$$

$$u = 2.5 + 0.5 = 3.0\,,$$

$$v = 0.136125 + \frac{1}{6} (0.053419 + 2 \cdot 0.083702 + 2 \cdot 0.101558 + 0.205709) \approx 0.241066\,.$$

This time the relative error is

$$\varepsilon = \left| \frac{v(3; 2^{-1}) - v(3; 2^0)}{v(3; 2^{-1})} \right| = \frac{0.241066 - 0.236027}{0.241066} \approx 0.020903 > 0.0001\,.$$

Thus, we set $m = 2$, $N = 2^m = 4$, $h = 1/N = 0.25$, repeat computations with this new step, and find

$$v(3; 2^{-2}) \approx 0.241854\,,$$

Copyright © 2012 Pearson Education, Inc. Publishing as Addison-Wesley.

so that

$$\varepsilon = \left| \frac{v(3; 2^{-2}) - v(3; 2^{-1})}{v(3; 2^{-2})} \right| = \frac{0.241854 - 0.241066}{0.241854} \approx 0.003258 > 0.0001 \, .$$

We continue increasing m and get

$$m = 3, \quad h = 0.125, \quad v(3; 2^{-3}) = 0.241924 \, ,$$
$$\varepsilon = \left| \frac{v(3; 2^{-3}) - v(3; 2^{-2})}{v(3; 2^{-3})} \right| = \left| \frac{0.241924 - 0.241854}{0.241924} \right| \approx 0.00029 \, ;$$

$$m = 4, \quad h = 0.0625, \quad v(3; 2^{-4}) = 0.241929 \, ,$$
$$\varepsilon = \left| \frac{v(3; 2^{-4}) - v(3; 2^{-3})}{v(3; 2^{-4})} \right| = \left| \frac{0.241929 - 0.241924}{0.241929} \right| \approx 0.00002 \, .$$

Therefore, within relative accuracy of 0.0001,

$$v(3) \approx 0.24193 \, .$$

21. In this initial value problem, the independent variable is x and the dependent variable is z, $x_0 = 0$ and $z_0 = 1$. Writing the given equation in standard form

$$\frac{dz}{dx} = f(x) - g(x)z^2 = 5x + 2 - x^2 z^2 \, ,$$

we see that

$$f(x, z) = 5x + 2 - x^2 z^2 \, .$$

We use the classical 4th order Runge-Kutta algorithm with tolerance. The required accuracy here is 10^{-4}.

We start with $m = 0$, $N = 2^m = 1$, and a step size $h = (1 - 0)/N = 1$. Setting

$$x = x_0 = 0, \quad z = z_0 = 1 \, ,$$

on Step 4 we compute

$$k_1 = 2; \qquad k_2 = 3.5;$$
$$k_3 = 2.609375; \quad k_4 = -6.027588;$$
$$x = 1.0, \qquad z = 2.365194 \, .$$

Because the error between two successive approximations, $z(1; 2^0) = 2.365194$ and $z = 1$ is

$$\varepsilon = 2.365194 - 1 > 0.0001,$$

 Copyright © 2012 Pearson Education, Inc. Publishing as Addison-Wesley.

we go back to Step 2, set $m = 1$, take $N = 2^m = 2$ on Step 3, compute $h = 1/N = 0.5$, and use the 4th order Runge-Kutta subroutine to find $z\left(1; 2^{-1}\right)$. This takes two more steps.

$$k_1 = 1; \qquad k_2 = 1.554688;$$
$$k_3 = 1.526283; \quad k_4 = 1.452237;$$
$$x = 0.5, \qquad z = 2.435696$$

and

$$k_1 = 1.508423; \quad k_2 = 0.013137;$$
$$k_3 = 1.197440; \quad k_4 = -3.099840;$$
$$x = 1.0, \qquad z = 2.573986.$$

This time, the error

$$\left| z\left(1; 2^{-1}\right) - z\left(1; 2^0\right) \right| = 2.573986 - 2.365194 > 0.0001.$$

We repeat our computations, taking $m = 2, 3, \ldots$ until two successive approximations differ by less than 10^{-4}.

$$m = 2, \quad z\left(1; 2^{-2}\right) = 2.854553;$$
$$m = 3, \quad z\left(1; 2^{-3}\right) = 2.870202;$$
$$m = 4, \quad z\left(1; 2^{-4}\right) = 2.870805;$$
$$m = 5, \quad z\left(1; 2^{-5}\right) = 2.870833.$$

Since

$$\left| z\left(1; 2^{-5}\right) - z\left(1; 2^{-4}\right) \right| = 2.870833 - 2.870805 < 0.0001,$$

we stop and take

$$z(1) \approx z\left(1; 0.03125\right) = 2.870833.$$

TABLES

Table 3–A: Computation results in Problem 17, Sec. 3.2.

Year	$(1/p)dp/dt$	Logistic (L. S.)
1790	0.0351	3.29
1800	0.0363	4.51
1810	0.0331	6.16
1820	0.0335	8.41
1830	0.0326	11.45
1840	0.0359	15.53
1850	0.0356	20.97
1860	0.0267	28.14
1870	0.0260	37.45
1880	0.0255	49.31
1890	0.0210	64.07
1900	0.0210	81.89
1910	0.0150	102.65
1920	0.0162	125.87
1930	0.0073	150.64
1940	0.0145	175.80
1950	0.0185	200.10
1960	0.0134	222.47
1970	0.0114	242.16
1980	0.0098	258.81
1990	0.0132	272.44
2000	0.0097	283.28
2010		291.73
2020		298.18

Table 3–B: Taylor approximations of order 2 and 4 for the equation $y' = x + 1 - y$.

n	x_n	y_n (order 2)	y_n (order 4)
1	0.25	1.03125	1.02881
2	0.50	1.11035	1.10654
3	0.75	1.22684	1.22238
4	1.00	1.37253	1.36789

 Copyright © 2012 Pearson Education, Inc. Publishing as Addison-Wesley.

Table 3–C: 4th order Runge-Kutta subroutine approximations for $y' = x + 1 - y$ at $x = 1$ with $h = 0.25$.

x	0	0.25	0.50	0.75	1.0
y	1	1.02881	1.10654	1.22238	1.36789

Table 3–D: Approximations $\left(\dfrac{1 + h/2}{1 + h/2}\right)^{1/h}$ to $e \approx 2.71828\ldots$.

h	Approximation	Error
1	3	0.28172
10^{-1}	2.72055	0.00227
10^{-2}	2.71830	0.00002
10^{-3}	2.71828	0.00000
10^{-4}	2.71828	0.00000

Table 3–E: Improved Euler's method to approximate the solution of $y' = x - y^2$, $y(1) = 0$, with $h = 0.1$.

n	x_n	y_n
1	1.1	0.10450
2	1.2	0.21668
3	1.3	0.33382
4	1.4	0.45300
5	1.5	0.57135

Copyright © 2012 Pearson Education, Inc. Publishing as Addison-Wesley.

Table 3–F: Improved Euler's method approximations to $y' = x + 3\cos(xy)$, $y(0) = 0$, on $[0, 2]$ with $h = 0.2$.

n	x_n	y_n	n	x_n	y_n
1	0.2	0.61784	6	1.2	1.88461
2	0.4	1.23864	7	1.4	1.72447
3	0.6	1.73653	8	1.6	1.56184
4	0.8	1.98111	9	1.8	1.41732
5	1.0	1.99705	10	2.0	1.29779

Table 3–G: Improved Euler's method approximations to $\phi(1)$, where $\phi(x)$ is the solution to $y' = 1 - y + y^3$, $y(0) = 0$.

h	$y(1; h) \approx \phi(1)$
1	1.0
2^{-1}	0.7372
2^{-2}	0.7194
2^{-3}	0.7170

Table 3–H: Euler's method approximations to the solution of $y' = -20y$, $y(0) = 1$, on $[0, 1]$ with $h = 0.1$, 0.2, and 0.025.

x_n	y_n ($h = 0.2$)	y_n ($h = 0.1$)	y_n ($h = 0.025$)
0.1		-1	0.06250
0.2	-3	1	0.00391
0.3		-1	0.00024
0.4	9	1	0.00002
0.5		-1	0.00000
0.6	-27	1	0.00000
0.7		-1	0.00000
0.8	81	1	0.00000
0.9		-1	0.00000
1.0	-243	1	0.00000

 Copyright © 2012 Pearson Education, Inc. Publishing as Addison-Wesley.

Table 3–I: Improved Euler's method to approximate the temperature in a building over a 24-hour period (with $K = 0.2$).

Time	t_n	T_n
Midnight	0	65
12:40 A.M.	0.6667	66.6380
1:20 A.M.	1.3333	67.5291
2:00 A.M.	2.0000	68.0727
2:40 A.M.	2.6667	68.4696
3:20 A.M.	3.3333	68.8181
4:00 A.M.	4.0000	69.1639
8:00 A.M.	8.0000	71.4836
Noon	12.000	72.9089
4:00 P.M.	16.000	72.0714
8:00 P.M.	20.000	69.8095
Midnight	24.000	68.3852

Table 3–J: 4th order Runge-Kutta approximations to the solution of $y' = \cos(5y) - x$, $y(0) = 0$, on $[0, 3]$ with $h = 0.1$.

x_n	y_n	x_n	y_n	x_n	y_n
0.1	0.09134	1.1	0.11439	2.1	−1.02887
0.2	0.15663	1.2	0.08686	2.2	−1.17307
0.3	0.19458	1.3	0.05544	2.3	−1.30020
0.4	0.21165	1.4	0.01855	2.4	−1.45351
0.5	0.21462	1.5	−0.02668	2.5	−1.69491
0.6	0.20844	1.6	−0.85748	2.6	−2.03696
0.7	0.19629	1.7	−0.17029	2.7	−2.30917
0.8	0.18006	1.8	−0.30618	2.8	−2.50088
0.9	0.16079	1.9	−0.53517	2.9	−2.69767
1.0	0.13890	2.0	−0.81879	3.0	−2.99510

FIGURES

Figure 3–A: The phase line for $p' = (a - bp)p$.

Copyright © 2012 Pearson Education, Inc. Publishing as Addison-Wesley.

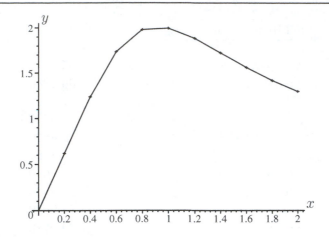

Figure 3–B: Polygonal line approximation to the solution of $y' = x + 3\cos(xy)$, $y(0) = 0$.

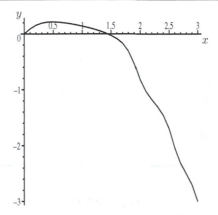

Figure 3–C: Polygonal line approximation to the solution of $y' = \cos(5y) - x$, $y(0) = 0$, on $[0, 3]$.

 Copyright © 2012 Pearson Education, Inc. Publishing as Addison-Wesley.

CHAPTER 4: Linear Second-Order Equations

EXERCISES 4.1:   Introduction: The Mass-Spring Oscillator

1. With $b = 0$ and $F_{ext} = 0$, equation (3) becomes

$$my'' + ky = 0.$$

Substitution $y = \cos \omega t$, where $\omega = \sqrt{k/m}$, yields

$$
\begin{aligned}
m(\cos \omega t)'' + k(\cos \omega t) &= -m\omega^2 \cos \omega t + k \cos \omega t \\
&= \left(-m\omega^2 + k\right) \cos \omega t = (-m(k/m) + k) \cos \omega t = 0.
\end{aligned}
$$

Thus, $y = \cos \omega t$ is indeed a solution.

3. With $F_{ext} = 0$, $m = 1$, $k = 9$, and $b = 6$ equation (3) becomes

$$y'' + 6y' + 9y = 0 \,.$$

Substitution $y_1 = e^{-3t}$ and $y_2 = te^{-3t}$ yields

$$\left(e^{-3t}\right)'' + 6\left(e^{-3t}\right)' + 9\left(e^{-3t}\right) = 9e^{-3t} - 18e^{-3t} + 9e^{-3t} = 0,$$

$$\left(te^{-3t}\right)'' + 6\left(te^{-3t}\right)' + 9\left(te^{-5t}\right) = (9t - 6)e^{-3t} + 6(1 - 3t)e^{-3t} + 9te^{-3t} = 0.$$

Thus, $y_1 = e^{-3t}$ and $y_2 = te^{-3t}$ are solutions to the given equation.

Both solutions approach zero as $t \to \infty$.

5. We differentiate $y(t)$ twice and obtain

$$
\begin{aligned}
y(t) &= e^{-3t} \sin(\sqrt{3}t) \\
y'(t) &= e^{-3t}[(-3) \sin(\sqrt{3}t) + \sqrt{3} \cos(\sqrt{3}t)] \\
y''(t) &= e^{-3t} \left[(-3)^2 \sin(\sqrt{3}t) + (-3)\sqrt{3} \cos(\sqrt{3}t) + (-3)\sqrt{3} \cos(\sqrt{3}t) - (\sqrt{3})^2 \sin(\sqrt{3}t) \right] \\
&= e^{-3t} \left[6 \sin(\sqrt{3}t) - 6\sqrt{3} \cos(\sqrt{3}t) \right].
\end{aligned}
$$

Copyright © 2012 Pearson Education, Inc. Publishing as Addison-Wesley.

Chapter 4

Substituting these functions into the differential equation yields

$$my'' + by' + ky = y'' + 6y' + 12y = e^{-3t}\left[6\sin(\sqrt{3}t) - 6\sqrt{3}\cos(\sqrt{3}t)\right]$$
$$+6e^{-3t}[(-3)\sin(\sqrt{3}t) + \sqrt{3}\cos(\sqrt{3}t)] + 12e^{-3t}\sin(\sqrt{3}t)$$
$$= e^{-3t}\left[(6 - 18 + 12)\sin(\sqrt{3}t) + (-6\sqrt{3} + 6\sqrt{3})\cos(\sqrt{3}t)\right] = 0.$$

Therefore, $y = e^{-3t}\sin(\sqrt{3}t)$ is a solution. As $t \to \infty$, $e^{-2t} \to 0$ and $\sin(\sqrt{3}t)$ remains bounded. Therefore, $\lim_{t\to\infty} y(t) = 0$.

7. For $y = A\cos 3t + B\sin 3t$,

$$y' = -3A\sin 3t + 3B\cos 3t, \qquad y'' = -9A\cos 3t - 9B\sin 3t.$$

Inserting y, y', and y'' into the given equation and matching coefficients yield

$$y'' + 2y' + 4y = 5\sin 3t$$
$$\Rightarrow \quad (-9A\cos 3t - 9B\sin 3t) + 2(-3A\sin 3t + 3B\cos 3t) + 4(A\cos 3t + B\sin 3t)$$
$$= (-5A + 6B)\cos 3t + (-6A - 5B)\sin 3t = 5\sin 3t$$
$$\Rightarrow \quad \begin{matrix} -5A + 6B = 0 \\ -6A - 5B = 5 \end{matrix} \quad \Rightarrow \quad \begin{matrix} A = -30/61 \\ B = -25/61. \end{matrix}$$

Thus,

$$y = -(30/61)\cos 3t - (25/61)\sin 3t$$

is a synchronous solution to $y'' + 2y' + 4y = 5\sin 3t$.

9. We differentiate $y = A\cos 2t + B\sin 2t$ twice to get

$$y' = -2A\sin 2t + 2B\cos 2t \quad \text{and} \quad y'' = -4A\cos 2t - 4B\sin 2t.$$

Substituting y, y', and y'' into the given equation and comparing coefficients yields

$$y'' + 2y' + 4y = (-4A\cos 2t - 4B\sin 2t) + 2(-2A\sin 2t + 2B\cos 2t)$$
$$+4(A\cos 2t + B\sin 2t)$$
$$= 4B\cos 2t - 4A\sin 2t = 6\cos 2t + 8\sin 2t$$
$$\Rightarrow \quad \begin{matrix} 4B = 6, \\ -4A = 8 \end{matrix} \quad \Rightarrow \quad \begin{matrix} A = -2, \\ B = 3/2 \end{matrix} \quad \Rightarrow \quad y = -2\cos 2t + (3/2)\sin 2t.$$

 Copyright © 2012 Pearson Education, Inc. Publishing as Addison-Wesley.

EXERCISES 4.2: Homogeneous Linear Equations: The General Solution

1. The auxiliary equation, $r^2 + 6r + 9 = (r+3)^2 = 0$, has a double root $r = -3$. Therefore, e^{-3t} and te^{-3t} are two linearly independent solutions for this differential equation, and a general solution is given by

$$y(t) = c_1 e^{-3t} + c_2 t e^{-3t},$$

where c_1 and c_2 are arbitrary constants.

3. The auxiliary equation for this problem is $r^2 - r - 2 = (r-2)(r+1) = 0$, which has the roots $r = 2$ and $r = -1$. Thus $\{e^{2t}, e^{-t}\}$ is a set of two linearly independent solutions to this differential equation. Therefore, a general solution is given by

$$y(t) = c_1 e^{2t} + c_2 e^{-t},$$

where c_1 and c_2 are arbitrary constants.

5. The auxiliary equation for this problem is $r^2 - 5r + 6 = 0$ with roots $r = 2, 3$. Therefore, a general solution is

$$y(t) = c_1 e^{2t} + c_2 e^{3t}.$$

7. Solving the auxiliary equation, $6r^2 + r - 2 = 0$, yields $r = -2/3, 1/2$. Thus a general solution is given by

$$y(t) = c_1 e^{t/2} + c_2 e^{-2t/3},$$

where c_1 and c_2 are arbitrary constants.

9. Solving the auxiliary equation, $4r^2 - 4r + 1 = (2r-1)^2 = 0$, we conclude that $r = 1/2$ is its double root. Therefore, a general solution to the given differential equation is

$$y(t) = c_1 e^{t/2} + c_2 t e^{t/2}.$$

11. Solving the auxiliary equation, $4r^2 + 20r + 25 = (2r+5)^2 = 0$, we conclude that $r = -5/2$ is its double root. Therefore, a general solution to the given differential equation is

$$w(t) = c_1 e^{-5t/2} + c_2 t e^{-5t/2}.$$

Copyright © 2012 Pearson Education, Inc. Publishing as Addison-Wesley.

13. The auxiliary equation for this problem is $r^2 + 2r - 8 = 0$, which has roots $r = -4, 2$. Thus, a general solution is given by

$$y(t) = c_1 e^{-4t} + c_2 e^{2t},$$

where c_1, c_2 are arbitrary constants. To satisfy the initial conditions, $y(0) = 3$ and $y'(0) = -12$, we find the derivative $y'(t) = -4c_1 e^{-4t} + 2c_2 e^{2t}$ and solve the system

$$
\begin{aligned}
y(0) &= c_1 e^{-4\cdot 0} + c_2 e^{2\cdot 0} = c_1 + c_2 = 3, \\
y'(0) &= -4c_1 e^{-4\cdot 0} + 2c_2 e^{2\cdot 0} = -4c_1 + 2c_2 = -12
\end{aligned}
\quad\Rightarrow\quad
\begin{aligned}
c_1 &= 3, \\
c_2 &= 0.
\end{aligned}
$$

Therefore, the solution to the given initial value problem is

$$y(t) = (3)e^{-4t} + (0)e^{2t} = 3e^{-4t}.$$

15. Here, the auxiliary equation is $r^2 - 4r - 5 = (r-5)(r+1) = 0$, which has roots $r = 5$, -1. Consequently, a general solution to the differential equation is $y(t) = c_1 e^{5t} + c_2 e^{-t}$, where c_1 and c_2 are arbitrary constants. To find the solution that satisfies the initial conditions, $y(-1) = 3$ and $y'(-1) = 9$, we first differentiate the solution found above and then plug into y and y' these initial conditions. Thus,

$$y(-1) = 3 = c_1 e^{-5} + c_2 e$$

$$y'(-1) = 9 = 5c_1 e^{-5} - c_2 e.$$

Solving this system yields $c_1 = 2e^5$, $c_2 = e^{-1}$. Therefore, $y(t) = 2e^{5(t+1)} + e^{-(t+1)}$ is the desired solution.

17. The auxiliary equation for this problem, $r^2 - 2r - 2 = 0$, has roots $r = 1 \pm \sqrt{3}$. Thus, a general solution is given by $z(t) = c_1 e^{(1+\sqrt{3})t} + c_2 e^{(1-\sqrt{3})t}$. Differentiating, we find that $z'(t) = c_1(1+\sqrt{3})e^{(1+\sqrt{3})t} + c_2(1-\sqrt{3})e^{(1-\sqrt{3})t}$. Substitution of $z(t)$ and $z'(t)$ into the initial conditions yields the system

$$
\begin{aligned}
z(0) &= c_1 + c_2 = 0, \\
z'(0) &= c_1(1+\sqrt{3}) + c_2(1-\sqrt{3}) = \sqrt{3}(c_1 - c_2) = 3
\end{aligned}
\quad\Rightarrow\quad
\begin{aligned}
c_1 &= \sqrt{3}/2, \\
c_2 &= -\sqrt{3}/2.
\end{aligned}
$$

Thus, the solution satisfying the given initial conditions is

$$z(t) = \frac{\sqrt{3}}{2} e^{(1+\sqrt{3})t} - \frac{\sqrt{3}}{2} e^{(1-\sqrt{3})t} = \frac{\sqrt{3}}{2} \left[e^{(1+\sqrt{3})t} - e^{(1-\sqrt{3})t} \right].$$

Copyright © 2012 Pearson Education, Inc. Publishing as Addison-Wesley.

19. The auxiliary equation for this differential equation is $r^2 + 2r + 1 = (r+1)^2 = 0$. We see that $r = -1$ is a repeated root. Thus, two linearly independent solutions are $y_1(t) = e^{-t}$ and $y_2(t) = te^{-t}$. This means that a general solution is given by $y(t) = c_1 e^{-t} + c_2 te^{-t}$.

To find the constants c_1 and c_2, we substitute the initial conditions into the general solution and its derivative, $y'(t) = -c_1 e^{-t} + c_2 \left(e^{-t} - te^{-t} \right)$, and obtain

$$1 = y(0) = c_1 e^0 + c_2 \cdot 0 = c_1,$$
$$-3 = y'(0) = -c_1 e^0 + c_2 \left(e^0 - 0 \right) = -c_1 + c_2.$$

Hence, $c_1 = 1$, $c_2 = -2$, and the solution that satisfies the initial conditions is given by

$$y(t) = e^{-t} - 2te^{-t}.$$

21. (a) With $y(t) = e^{rt}$, $y'(t) = re^{rt}$, the equation becomes

$$are^{rt} + be^{rt} = (ar + b)e^{rt} = 0.$$

Since the function e^{rt} is never zero on $(-\infty, \infty)$, to satisfy the above equation we must have

$$ar + b = 0.$$

(b) Solving the characteristic equation, $ar + b = 0$, obtained in part (a), we find that $r = -b/a$. So, $y(t) = e^{rt} = e^{-bt/a}$, and a general solution is given by $y = ce^{-bt/a}$, where c is an arbitrary constant.

23. We form the characteristic equation, $5r + 4 = 0$, and find its root, $r = -4/5$. Therefore, $y(t) = ce^{-4t/5}$ is a general solution to the given equation.

25. The characteristic equation, $6r - 13 = 0$, has the root $r = 13/6$. Therefore, a general solution is given by $w(t) = ce^{13t/6}$.

27. Since $\sin 2t = 2 \sin t \cos t$ (the double angle identity), we have $y_2(t) = 2y_1(t)$ on $(-\infty, \infty)$ (and so on $(0,1)$). Thus, $y_1(t)$ and $y_2(t)$ are linearly dependent.

29. These functions are linearly independent, because the equality $y_1(t) \equiv cy_2(t)$ would imply that, for some constant c,

$$te^{2t} \equiv ce^{2t} \qquad \Rightarrow \qquad t \equiv c$$

on $(0,1)$, which is wrong.

Copyright © 2012 Pearson Education, Inc. Publishing as Addison-Wesley.

31. Using the trigonometric identity $1 + \tan^2 t = \sec^2 t$, we conclude that

$$y_1(t) = \tan^2 t - \sec^2 t = -1 \qquad \Rightarrow \qquad y_2(t) = 3 = (-3)y_1(t),$$

and so $y_1(t)$ and $y_2(t)$ are linearly dependent on $(0, 1)$ (even on $(-\infty, \infty)$).

33. If two functions, $y_1(t)$ and $y_2(t)$, are linearly dependent on I in the sense of Definition 1, then there is a constant c such that, say,

$$y_1(t) = cy_2(t) \qquad \Rightarrow \qquad y_1(t) - cy_2(t) = 0 \quad \text{for all } t \text{ in } I.$$

Thus, taking $c_1 = 1$ and $c_2 = -c$, we get

$$c_1 y_1(t) + c_2 y_2(t) = 0 \quad \text{for all } t \text{ in } I. \tag{4.1}$$

On the other hand, if (4.1) holds and one of the coefficients, say, $c_1 \neq 0$, then dividing (4.1) by c_1 and separating the terms yields

$$y_1(t) = (-c_2/c_1)\, y_2 = cy_2(t) \quad \text{for all } t \text{ in } I,$$

i.e., $y_1(t)$ and $y_2(t)$ are linearly dependent on I in the sense of Definition 1.

35. (a) A linear combination $c_1 + c_2 t + c_3 t^2$ of $y_1(t) = 1$, $y_2(t) = t$, and $y_3(t) = t^2$ is a polynomial of degree at most two and so can have at most two real roots, unless it is a zero polynomial, i.e., has all zero coefficients. Therefore, the above linear combination vanishes on $(-\infty, \infty)$ if and only if $c_1 = c_2 = c_3 = 0$, and $y_1(t)$, $y_2(t)$, and $y_3(t)$ are linearly independent on $(-\infty, \infty)$.

(b) Since

$$5y_1(t) + 3y_2(t) + 15y_3(t) = -15 + 15\sin^2 t + 15\cos^2 t = 15(-1 + \sin^2 t + \cos^2 t) \equiv 0$$

on $(-\infty, \infty)$ (the Pythagorean identity), given functions are linearly dependent.

(c) These functions are linearly independent. Indeed, since the function e^t does not vanish on $(-\infty, \infty)$,

$$c_1 y_1 + c_2 y_2 + c_3 y_3 = c_1 e^t + c_2 t e^t + c_3 t^2 e^t = \left(c_1 + c_2 t + c_3 t^2\right) e^t \equiv 0$$

if and only if $c_1 + c_2 t + c_3 t^2 \equiv 0$. But functions 1, t, and t^2 are linearly independent on $(-\infty, \infty)$ (see (a)), and so their linear combination is identically zero if and only if $c_1 = c_2 = c_3 = 0$.

 Copyright © 2012 Pearson Education, Inc. Publishing as Addison-Wesley.

(d) By the definition of the hyperbolic cosine function,

$$y_3(t) = \cosh t = \frac{e^t + e^{-t}}{2} = \frac{1}{2}e^t + \frac{1}{2}e^{-t} = \frac{1}{2}y_1(t) + \frac{1}{2}y_2(t),$$

and so given functions are linearly dependent on $(-\infty, \infty)$.

37. The auxiliary equation for this problem is $r^3 + r^2 - 6r + 4 = 0$. Factoring yields

$$
\begin{aligned}
r^3 + r^2 - 6r + 4 &= \left(r^3 - r^2\right) + \left(2r^2 - 2r\right) + (-4r + 4) \\
&= r^2(r-1) + 2r(r-1) - 4(r-1) = (r-1)(r^2 + 2r - 4).
\end{aligned}
$$

Thus, the roots of the auxiliary equation are

$$r = 1 \quad \text{and} \quad r = \frac{-2 \pm \sqrt{(-2)^2 - 4(1)(-4)}}{2} = -1 \pm \sqrt{5}.$$

Therefore, the functions e^t, $e^{(-1-\sqrt{5})t}$, and $e^{(-1+\sqrt{5})t}$ are solutions to the given equation, and they are linearly independent on $(-\infty, \infty)$ (see Problem 36). Hence, a general solution to $y''' + y'' - 6y' + 4y = 0$ is given by

$$y(t) = c_1 e^t + c_2 e^{(-1-\sqrt{5})t} + c_3 e^{(-1+\sqrt{5})t}.$$

39. Factoring the auxiliary polynomial yields

$$
\begin{aligned}
r^3 + 2r^2 - 4r - 8 &= \left(r^3 + 2r^2\right) - (4r + 8) \\
&= r^2(r+2) - 4(r+2) = (r+2)\left(r^2 - 4\right) = (r+2)(r+2)(r-2).
\end{aligned}
$$

Therefore, the auxiliary equation has a double root $r = -2$ and a simple root $r = 2$. The functions e^{-2t}, te^{-2t}, and e^{2t} form a linearly independent solution set. Therefore, the answer to this problem is

$$z(t) = c_1 e^{-2t} + c_2 t e^{-2t} + c_3 e^{2t}.$$

41. By inspection, we see that $r = 2$ is a root of the auxiliary equation, $r^3 + 3r^2 - 4r - 12 = 0$. Factoring yields

$$r^3 + 3r^2 - 4r - 12 = (r-2)\left(r^2 + 5r + 6\right) = (r-2)(r+2)(r+3).$$

Hence, two other roots of the auxiliary equation are $r = -2$ and $r = -3$. The functions e^{-3t}, e^{-2t}, and e^{2t} are three linearly independent solutions to the given equation, and a general solution is given by

$$y(t) = c_1 e^{-3t} + c_2 e^{-2t} + c_3 e^{2t}.$$

Copyright © 2012 Pearson Education, Inc. Publishing as Addison-Wesley.

43. First, we find a general solution to the equation $y''' - y' = 0$. Its characteristic equation, $r^3 - r = 0$, has roots $r = 0$, -1, and 1, and so a general solution is given by

$$y(t) = c_1 e^{(0)t} + c_2 e^{(-1)t} + c_3 e^{(1)t} = c_1 + c_2 e^{-t} + c_3 e^t.$$

Differentiating $y(t)$ twice yields

$$y'(t) = -c_2 e^{-t} + c_3 e^t, \qquad y''(t) = c_2 e^{-t} + c_3 e^t.$$

Now we substitute y, y', and y'' into the initial conditions and find c_1, c_2, and c_3.

$$
\begin{aligned}
y(0) &= c_1 + c_2 + c_3 = 2, & c_1 &= 3, \\
y'(0) &= -c_2 + c_3 = 3, & \Rightarrow \qquad c_2 &= -2, \\
y''(0) &= c_2 + c_3 = -1 & c_3 &= 1.
\end{aligned}
$$

Therefore, the solution to the given initial value problem is

$$y(t) = 3 + e^t - 2e^{-t}.$$

45. (a) To find the roots of the auxiliary equation, $p(r) := 3r^3 + 18r^2 + 13r - 19 = 0$, one can use Newton's method or intermediate value theorem. We note that

$$
\begin{aligned}
p(-5) &= -9 < 0, & p(-4) &= 25 > 0, \\
p(-2) &= 3 > 0, & p(-1) &= -17 < 0, \\
p(0) &= -19 < 0, & p(1) &= 15 > 0.
\end{aligned}
$$

Therefore, the roots of $p(r)$ belong to the intervals $[-5, -4]$, $[-2, -1]$, and $[0, 1]$, and we can take $r = -5$, $r = -2$, and $r = 0$ as initial guesses. Approximation yields $r_1 \approx -4.832$, $r_2 \approx -1.869$, and $r_3 \approx 0.701$. So, a general solution is given by

$$y(t) = c_1 e^{r_1 t} + c_2 e^{r_2 t} + c_3 e^{r_3 t} = c_1 e^{-4.832t} + c_2 e^{-1.869t} + c_3 e^{0.701t}.$$

(b) The auxiliary equation, $r^4 - 5r^2 + 5 = 0$, is of quadratic type. The substitution $s = r^2$ yields

$$s^2 - 5s + 5 = 0 \quad \Rightarrow \quad s = \frac{5 \pm \sqrt{5}}{2} \quad \Rightarrow \quad r = \pm\sqrt{s} = \pm\sqrt{\frac{5 \pm \sqrt{5}}{2}}.$$

Therefore,

$$r_1 = \sqrt{\frac{5 - \sqrt{5}}{2}} \approx 1.176, \quad r_2 = \sqrt{\frac{5 + \sqrt{5}}{2}} \approx 1.902, \quad r_3 = -r_1, \quad \text{and} \quad r_4 = -r_2$$

are the roots of the auxiliary equation, and a general solution to $y^{(\mathrm{iv})} - 5y'' + 5y = 0$ is given by $y(t) = c_1 e^{r_1 t} + c_2 e^{-r_1 t} + c_3 e^{r_2 t} + c_4 e^{-r_2 t}$.

 Copyright © 2012 Pearson Education, Inc. Publishing as Addison-Wesley.

(c) We can use numerical tools to find the roots of the auxiliary fifth degree polynomial equation $r^5 - 3r^4 - 5r^3 + 15r^2 + 4r - 12 = 0$. Alternatively, one can involve the rational root theorem and examine the divisors of the free coefficient, -12. These divisors are $\pm 1, \pm 2, \pm 3, \pm 4, \pm 6$, and ± 12. By inspection, $r = \pm 1, \pm 2$, and 3 satisfy the equation. Thus, a general solution is $y(t) = c_1 e^{-t} + c_2 e^t + c_3 e^{-2t} + c_4 e^{2t} + c_5 e^{3t}$.

EXERCISES 4.3:   Auxiliary Equations with Complex Roots

1. The auxiliary equation in this problem is $r^2 + 1 = 0$, which has roots $r = \pm i$. We see that $\alpha = 0$ and $\beta = 1$. Thus, a general solution to the differential equation is given by

$$y(t) = c_1 e^{(0)t} \cos t + c_2 e^{(0)t} \sin t = c_1 \cos t + c_2 \sin t.$$

3. The auxiliary equation, $r^2 - 10r + 26 = 0$, has roots $r = 5 \pm i$. So, $\alpha = 5$, $\beta = 1$, and

$$y(t) = c_1 e^{5t} \cos t + c_2 e^{5t} \sin t$$

is a general solution.

5. This differential equation has the auxiliary equation $r^2 - 4r + 7 = 0$. The roots of this auxiliary equation are $r = \left(4 \pm \sqrt{16 - 28}\right)/2 = 2 \pm \sqrt{3}\, i$. We see that $\alpha = 2$ and $\beta = \sqrt{3}$. Thus, a general solution to the differential equation is given by

$$w(t) = c_1 e^{2t} \cos\left(\sqrt{3t}\right) + c_2 e^{2t} \sin\left(\sqrt{3t}\right).$$

7. The auxiliary equation for this problem is given by

$$4r^2 + 4r + 6 = 0 \quad \Rightarrow \quad 2r^2 + 2r + 3 = 0 \quad \Rightarrow \quad r = \frac{-2 \pm \sqrt{4 - 24}}{4} = -\frac{1}{2} \pm \frac{\sqrt{5}}{2} i.$$

Therefore, $\alpha = -1/2$ and $\beta = \sqrt{5}/2$, and a general solution is given by

$$y(t) = c_1 e^{-t/2} \cos\left(\frac{\sqrt{5}t}{2}\right) + c_2 e^{-t/2} \sin\left(\frac{\sqrt{5}t}{2}\right).$$

9. The associated auxiliary equation, $r^2 + 4r + 8 = 0$, has two complex roots, $r = -2 \pm 2i$. Thus the answer is

$$y(t) = c_1 e^{-2t} \cos 2t + c_2 e^{-2t} \sin 2t.$$

Copyright © 2012 Pearson Education, Inc. Publishing as Addison-Wesley.

Chapter 4

11. The auxiliary equation for this problem is $r^2 + 10r + 25 = (r+5)^2 = 0$. We see that $r = -5$ is a repeated root. Thus, two linearly independent solutions are $z_1(t) = e^{-5t}$ and $z_2(t) = te^{-5t}$. This means that a general solution is given by

$$z(t) = c_1 e^{-5t} + c_2 t e^{-5t},$$

where c_1 and c_2 are arbitrary constants.

13. Solving the auxiliary equation yields complex roots. Indeed,

$$r^2 + 2r + 5 = 0 \quad \Rightarrow \quad r = \frac{-2 \pm \sqrt{2^2 - 4(1)(5)}}{2} = -1 \pm 2i.$$

So, $\alpha = -1$, $\beta = 2$, and a general solution is given by

$$y(t) = c_1 e^{-t} \cos 2t + c_2 e^{-t} \sin 2t.$$

15. First, we find the roots of the auxiliary equation.

$$r^2 + 10r + 41 = 0 \quad \Rightarrow \quad r = \frac{-10 \pm \sqrt{10^2 - 4(1)(41)}}{2} = -5 \pm 4i.$$

These are complex numbers with $\alpha = -5$ and $\beta = 4$. Hence, a general solution to the given differential equation is

$$y(t) = c_1 e^{-5t} \cos 4t + c_2 e^{-5t} \sin 4t.$$

17. The auxiliary equation in this problem, $r^2 - r + 7 = 0$, has the roots

$$r = \frac{1 \pm \sqrt{1^2 - 4(1)(7)}}{2} = \frac{1 \pm \sqrt{-27}}{2} = \frac{1}{2} \pm \frac{3\sqrt{3}}{2} i.$$

Therefore, a general solution is

$$y(t) = c_1 e^{t/2} \cos\left(\frac{3\sqrt{3}}{2} t\right) + c_2 e^{t/2} \sin\left(\frac{3\sqrt{3}}{2} t\right).$$

19. The auxiliary equation, $r^3 + r^2 + 3r - 5 = 0$, is a cubic one. Since any cubic equation has a real root, first we examine the divisors of the free coefficient to find integer real roots (if any). By inspection, $r = 1$ satisfies the equation. Dividing $r^3 + r^2 + 3r - 5$ by $r - 1$ yields

$$r^3 + r^2 + 3r - 5 = (r - 1)(r^2 + 2r + 5).$$

 Copyright © 2012 Pearson Education, Inc. Publishing as Addison-Wesley.

Therefore, the other two roots of the auxiliary equation are the roots of the quadratic equation $r^2 + 2r + 5 = 0$, which are $r = -1 \pm 2i$. A general solution to the given equation is then given by

$$y(t) = c_1 e^t + c_2 e^{-t} \cos 2t + c_3 e^{-t} \sin 2t.$$

21. The auxiliary equation for this problem is $r^2 + 2r + 2 = 0$, which has roots

$$r = \frac{-2 \pm \sqrt{4-8}}{2} = -1 \pm i.$$

So, a general solution is given by

$$y(t) = c_1 e^{-t} \cos t + c_2 e^{-t} \sin t,$$

where c_1 and c_2 are arbitrary constants. To find the solution that satisfies the initial conditions, $y(0) = 2$ and $y'(0) = 1$, we first differentiate the solution found above and then plug in these initial conditions into y and y'. We have

$$y'(t) = c_1 e^{-t}(-\cos t - \sin t) + c_2 e^{-t}(\cos t - \sin t)$$

and

$$y(0) = c_1 = 2,$$
$$y'(0) = -c_1 + c_2 = 1.$$

Thus, $c_1 = 2$, $c_2 = 3$, and the answer is given by $y(t) = 2e^{-t} \cos t + 3e^{-t} \sin t$.

23. The auxiliary equation for this problem is $r^2 - 4r + 2 = 0$. The roots of this equation are

$$r = \frac{4 \pm \sqrt{16-8}}{2} = 2 \pm \sqrt{2},$$

which are real numbers. A general solution is given by $w(t) = c_1 e^{(2+\sqrt{2})t} + c_2 e^{(2-\sqrt{2})t}$, where c_1 and c_2 are arbitrary constants. To find the solution that satisfies the initial conditions, $w(0) = 0$ and $w'(0) = 1$, we first differentiate the solution found above and then solve the system

$$w(0) = c_1 + c_2 = 0,$$
$$w'(0) = \left(2 + \sqrt{2}\right) c_1 + \left(2 - \sqrt{2}\right) c_2 = 1.$$

Copyright © 2012 Pearson Education, Inc. Publishing as Addison-Wesley.

Chapter 4

This yields $c_1 = \sqrt{2}/4$ and $c_2 = -\sqrt{2}/4$. Thus,

$$w(t) = \frac{\sqrt{2}}{4}e^{(2+\sqrt{2})t} - \frac{\sqrt{2}}{4}e^{(2-\sqrt{2})t} = \frac{\sqrt{2}}{4}\left[e^{(2+\sqrt{2})t} - e^{(2-\sqrt{2})t}\right]$$

is the desired solution.

25. The auxiliary equation, $r^2 - 2r + 2 = 0$, has roots $r = 1 \pm i$. Thus, a general solution is

$$y(t) = c_1 e^t \cos t + c_2 e^t \sin t,$$

where c_1 and c_2 are arbitrary constants. To find the solution that satisfies the initial conditions, $y(\pi) = e^\pi$ and $y'(\pi) = 0$, we find $y'(t) = c_1 e^t(\cos t - \sin t) + c_2 e^t(\sin t + \cos t)$ and solve the system

$$e^\pi = y(\pi) = -c_1 e^\pi,$$
$$0 = y'(\pi) = -c_1 e^\pi - c_2 e^\pi.$$

This yields $c_1 = -1$, $c_2 = 1$. So, the answer is $y(t) = e^t \sin t - e^t \cos t$.

27. To solve the auxiliary equation, $r^3 - 4r^2 + 7r - 6 = 0$, which is of the third order, we find its real root first. Examining the divisors of -6, that is, ± 1, ± 2, ± 3, and ± 6, we find that $r = 2$ satisfies the equation. Next, we divide $r^3 - 4r^2 + 7r - 6$ by $r - 2$ and obtain

$$r^3 - 4r^2 + 7r - 6 = (r - 2)\left(r^2 - 2r + 3\right).$$

Therefore, the other two roots of the auxiliary equation are

$$r = \frac{2 \pm \sqrt{4 - 12}}{2} = 1 \pm \sqrt{2}i,$$

and a general solution to the given differential equation is given by

$$y(t) = c_1 e^{2t} + c_2 e^t \cos \sqrt{2}t + c_3 e^t \sin \sqrt{2}t.$$

Next, we find the derivatives

$$y'(t) = 2c_1 e^{2t} + c_2 e^t \left(\cos \sqrt{2}t - \sqrt{2}\sin \sqrt{2}t\right) + c_3 e^t \left(\sin \sqrt{2}t + \sqrt{2}\cos \sqrt{2}t\right),$$
$$y''(t) = 4c_1 e^{2t} + c_2 e^t \left(-\cos \sqrt{2}t - 2\sqrt{2}\sin \sqrt{2}t\right) + c_3 e^t \left(-\sin \sqrt{2}t + 2\sqrt{2}\cos \sqrt{2}t\right),$$

 Copyright © 2012 Pearson Education, Inc. Publishing as Addison-Wesley.

and substitute y, y', and y'' into the initial conditions. This yields

$$c_1 + c_2 = 1,$$
$$2c_1 + c_2 + \sqrt{2}c_3 = 0, \qquad \Rightarrow \qquad \begin{aligned} c_1 &= 1, \\ c_2 &= 0, \end{aligned}$$
$$4c_1 - c_2 + 2\sqrt{2}c_3 = 0 \qquad \qquad c_3 = -\sqrt{2}.$$

With these values of the constants c_1, c_2, and c_3, the solution becomes

$$y(t) = e^{2t} - \sqrt{2}e^t \sin\left(\sqrt{2}t\right).$$

29. **(a)** As it was stated in Section 4.2, linear homogeneous differential equations with constant coefficients of any order can be handled in the same way as second order equations. Thus, we look for the roots of the auxiliary equation $r^3 - r^2 + r + 3 = 0$. By the rational root theorem, the only possible integer roots are $r = \pm 1$ and ± 3. Checking these values, we find that one of the roots of the auxiliary equation is $r = -1$. Factoring yields

$$r^3 - r^2 + r + 3 = (r+1)(r^2 - 2r + 3).$$

Using the quadratic formula, we find that the other two roots are

$$r = \frac{2 \pm \sqrt{4-12}}{2} = 1 \pm \sqrt{2}\,i.$$

A general solution is, therefore,

$$y(t) = c_1 e^{-t} + c_2 e^t \cos\left(\sqrt{2}t\right) + c_3 e^t \sin\left(\sqrt{2}t\right).$$

(b) By inspection, $r = 2$ is a root of the auxiliary equation, $r^3 + 2r^2 + 5r - 26 = 0$. Since

$$r^3 + 2r^2 + 5r - 26 = (r-2)\left(r^2 + 4r + 13\right),$$

the other two roots are the roots of $r^2 + 4r + 13 = 0$, which are $r = -2 \pm 3i$. Therefore, a general solution to the given equation is

$$y(t) = c_1 e^{2t} + c_2 e^{-2t} \cos 3t + c_3 e^{-2t} \sin 3t.$$

(c) The fourth order auxiliary equation $r^4 + 13r^2 + 36 = 0$ can be reduced to a quadratic equation by making a substitution $s = r^2$. This yields

$$s^2 + 13r + 36 = 0 \qquad \Rightarrow \qquad s = \frac{-13 \pm \sqrt{169 - 144}}{2} = \frac{-13 \pm 5}{2}.$$

Copyright © 2012 Pearson Education, Inc. Publishing as Addison-Wesley.

Thus, $s = (-13 + 5)/2 = -4$ or $s = (-13 - 5)/2 = -9$, and the solutions to the auxiliary equation are $r = \pm\sqrt{-4} = \pm 2i$ and $r = \pm\sqrt{-9} = \pm 3i$. A general solution, therefore, has the form

$$y(t) = c_1 \cos 2t + c_2 \sin 2t + c_3 \cos 3t + c_4 \sin 3t.$$

31. (a) Comparing the equation $y'' + 16y = 0$ with the mass-spring model (16) in Example 4, we conclude that the damping coefficient $b = 0$ and the positive stiffness $k = 16$. Thus, solutions should have an oscillatory behavior.

Indeed, the auxiliary equation, $r^2 + 16 = 0$, has roots $r = \pm 4i$, and a general solution is given by

$$y(t) = c_1 \cos 4t + c_2 \sin 4t.$$

Evaluating $y'(t)$ and substituting the initial conditions, we get

$$\begin{aligned} y(0) &= c_1 = 2, \\ y'(0) &= 4c_2 = 0 \end{aligned} \quad \Rightarrow \quad \begin{aligned} c_1 &= 2, \\ c_2 &= 0 \end{aligned} \quad \Rightarrow \quad y(t) = 2\cos 4t.$$

(b) Positive damping $b = 100$ and stiffness $k = 1$ imply that the displacement $y(t)$ tends to zero as $t \to \infty$.

To confirm this prediction, we solve the given initial value problem explicitly. The roots of the associated equation are

$$r = \frac{-100 \pm \sqrt{100^2 - 4}}{2} = -50 \pm \sqrt{2499}.$$

Thus, the roots $r_1 = -50 - \sqrt{2499}$ and $r_2 = -50 + \sqrt{2499}$ are both negative. A general solution is given by

$$y(t) = c_1 e^{r_1 t} + c_2 e^{r_2 t} \quad \Rightarrow \quad y'(t) = c_1 r_1 e^{r_1 t} + c_2 r_2 e^{r_2 t}.$$

Solving the initial value problem yields

$$\begin{aligned} y(0) &= 1 = c_1 + c_2, \\ y'(0) &= 0 = c_1 r_1 + c_2 r_2 \end{aligned} \quad \Rightarrow \quad \begin{aligned} c_1 &= r_2/(r_2 - r_1), \\ c_2 &= r_1/(r_1 - r_2), \end{aligned}$$

and so the desired solution is

$$y(t) = \frac{r_2}{r_2 - r_1} e^{r_1 t} + \frac{r_1}{r_1 - r_2} e^{r_2 t}.$$

Since $r_1 t$ and $r_2 t$ tend to $-\infty$ as $t \to \infty$, we have $y(t) \to 0$.

 Copyright © 2012 Pearson Education, Inc. Publishing as Addison-Wesley.

(c) The corresponding mass-spring model has negative damping $b = -6$ and positive stiffness $k = 8$. Thus, the magnitude $|y(t)|$ of the displacement $y(t)$ will increase without bound, as $t \to \infty$. Moreover, because of the positive initial displacement and initial zero velocity, the mass will move in the negative direction. Therefore, our guess is that $y(t) \to -\infty$ as $t \to \infty$.

Now we find the actual solution. Since the roots of the auxiliary equation are $r = 2$ and $r = 4$, a general solution to the given equation is $y(t) = c_1 e^{2t} + c_2 e^{4t}$. Next, we find c_1 and c_2 satisfying the initial conditions.

$$
\begin{aligned}
y(0) &= 1 = c_1 + c_2, \\
y'(0) &= 0 = 2c_1 + 4c_2
\end{aligned}
\qquad \Rightarrow \qquad
\begin{aligned}
c_1 &= 2, \\
c_2 &= -1.
\end{aligned}
$$

Thus, the desired solution is $y(t) = 2e^{2t} - e^{4t}$, and it approaches $-\infty$ as $t \to \infty$.

(d) In this problem, the stiffness $k = -3$ is negative. In the mass-spring model, this means that the spring force acts in the same direction (positive or negative) as the sign of the displacement is. Initially, the displacement $y(0) = -2$ is negative, and the mass has no initial velocity. Thus, the mass, when released, will move in the negative direction, and the spring will enforce this motion. So, we expect that $y(t) \to -\infty$ as $t \to \infty$.

To find the actual solution, we solve the auxiliary equation $r^2 + 2r - 3 = 0$ and obtain $r = -3, 1$. Therefore, a general solution is given by $y(t) = c_1 e^{-3t} + c_2 e^t$. We find c_1 and c_2 from the initial conditions

$$
\begin{aligned}
y(0) &= -2 = c_1 + c_2, \\
y'(0) &= 0 = -3c_1 + c_2
\end{aligned}
\qquad \Rightarrow \qquad
\begin{aligned}
c_1 &= -1/2, \\
c_2 &= -3/2.
\end{aligned}
$$

Thus, the solution to the initial value problem is

$$
y(t) = -\frac{e^{-3t}}{2} - \frac{3e^t}{2},
$$

and, as $t \to \infty$, it approaches $-\infty$.

(e) As in the previous problem, we have negative stiffness $k = -6$, but, this time, the initial displacement, $y(0) = 1$ is positive, as well as the initial velocity, $y'(0) = 1$. So, the mass will start moving in the positive direction, and will continue doing this (due to the negative stiffness) with increasing velocity. Thus, our prediction is that $y(t) \to \infty$ as $t \to \infty$.

Copyright © 2012 Pearson Education, Inc. Publishing as Addison-Wesley.

Indeed, the roots of the characteristic equation in this problem are $r = -2$ and 3, and so a general solution has the form $y(t) = c_1 e^{-2t} + c_2 e^{3t}$. To satisfy the initial conditions, we solve the system

$$y(0) = 1 = c_1 + c_2, \qquad\qquad c_1 = 2/5,$$
$$y'(0) = 1 = -2c_1 + 3c_2 \qquad \Rightarrow \qquad c_2 = 3/5.$$

Thus, the solution to the initial value problem is

$$y(t) = \frac{2e^{-2t}}{5} + \frac{3e^{3t}}{5},$$

and it approaches ∞ as $t \to \infty$.

33. From Example 3 we know that, in the study of a vibrating spring with damping, we have the initial value problem

$$my''(t) + by'(t) + ky(t) = 0; \qquad y(0) = y_0, \qquad y'(0) = v_0,$$

where m is the mass of the spring system, b is the damping constant, k denotes the spring constant, $y(t)$ is the displacement of the mass from the equilibrium at time t, $y(0)$ is the initial displacement, and $y'(0)$ is the initial velocity.

(a) We want to determine the equation of motion for a spring system with $m = 10$ kg, $b = 60$ kg/sec, $k = 250$ kg/sec^2, $y(0) = 0.3$ m, and $y'(0) = -0.1$ m/sec. That is, we seek the solution to the initial value problem

$$10y''(t) + 60y'(t) + 250y(t) = 0; \qquad y(0) = 0.3, \qquad y'(0) = -0.1.$$

The auxiliary equation for this differential equation is

$$10r^2 + 60r + 250 = 0 \qquad \Rightarrow \qquad r^2 + 6r + 25 = 0,$$

which has roots

$$r = \frac{-6 \pm \sqrt{36 - 100}}{2} = \frac{-6 \pm 8i}{2} = -3 \pm 4i.$$

Hence $\alpha = -3$ and $\beta = 4$, and the displacement function $y(t)$ has the form

$$y(t) = c_1 e^{-3t} \cos 4t + c_2 e^{-3t} \sin 4t.$$

 Copyright © 2012 Pearson Education, Inc. Publishing as Addison-Wesley.

We find c_1 and c_2 by using the initial conditions. Differentiating $y(t)$, we get

$$y'(t) = (-3c_1 + 4c_2)e^{-3t}\cos 4t + (-4c_1 - 3c_2)e^{-3t}\sin 4t.$$

Substituting y and y' into the initial conditions, we obtain the system

$$y(0) = 0.3 = c_1,$$
$$y'(0) = -0.1 = -3c_1 + 4c_2.$$

Solving yields $c_1 = 0.3$ and $c_2 = 0.2$. Therefore, the equation of motion is given by

$$y(t) = 0.3e^{-3t}\cos 4t + 0.2e^{-3t}\sin 4t \text{ (m)}.$$

(b) From Problem 32, we know that the frequency of oscillation is given by $\beta/(2\pi)$. In part (a) we found that $\beta = 4$. So, the frequency of oscillation is $4/(2\pi) = 2/\pi$.

(c) We see a decrease in the frequency of oscillation. We also have the factor e^{-3t}, which causes the solution to decay to zero. This is a result of energy loss due to damping.

35. The equation of the motion of a swinging door is similar to that for mass-spring model (with the mass m replaced by the moment of inertia I and the displacement $y(t)$ replaced by the angle θ that the door is open). So, from the discussion following Example 3, we conclude that the door will not continually swing back and forth (that is, the solution $\theta(t)$ will not oscillate) if $b \geq \sqrt{4Ik} = 2\sqrt{Ik}$.

37. (a) The auxiliary equation for this problem is

$$r^4 + 2r^2 + 1 = (r^2 + 1)^2 = 0.$$

This equation has the roots $r_1 = r_2 = -i$, $r_3 = r_4 = i$. Thus, $y_1(t) = \cos t$ and $y_2(t) = \sin t$ are solutions and, since the roots are repeated, we get two more solutions by multiplying $y_1(t)$ and $y_2(t)$ by t, that is, $y_3(t) = t\cos t$ and $y_4(t) = t\sin t$ are also solutions. This gives a general solution of the form

$$y(t) = c_1\cos t + c_2\sin t + c_3 t\cos t + c_4 t\sin t.$$

Copyright © 2012 Pearson Education, Inc. Publishing as Addison-Wesley.

(b) The auxiliary equation in this problem is

$$r^4 + 4r^3 + 12r^2 + 16r + 16 = (r^2 + 2r + 4)^2 = 0.$$

The roots of the quadratic equation $r^2 + 2r + 4 = 0$ are

$$r = \frac{-2 \pm \sqrt{4 - 16}}{2} = -1 \pm \sqrt{3}i.$$

Hence, the roots of the auxiliary equation are

$$r_{1,2} = -1 - \sqrt{3}i \quad \text{and} \quad r_{3,4} = -1 + \sqrt{3}i\,.$$

So, two linearly independent solutions are

$$y_1(t) = e^{-t}\cos(\sqrt{3}t) \quad \text{and} \quad y_2(t) = e^{-t}\sin(\sqrt{3}t),$$

and we get two more linearly independent solutions by multiplying $y_1(t)$ and $y_2(t)$ by t. This gives a general solution

$$y(t) = (c_1 + c_2 t)e^{-t}\cos(\sqrt{3}t) + (c_3 + c_4 t)e^{-t}\sin(\sqrt{3}t).$$

EXERCISES 4.4: Nonhomogeneous Equations: The Method of Undetermined Coefficients

1. We cannot use the method of undetermined coefficients to find a particular solution because of the t^{-1} factor, which is not a polynomial.

3. Writing

$$\frac{\sin x}{e^{4x}} = e^{-4x}\sin x\,,$$

we see that the right-hand side is of the form, for which (15) applies.

5. The nonhomogeneous term simplifies to

$$f(x) = 4x\sin^2 x + 4x\cos^2 x = 4x\left(\sin^2 x + \cos^2 x\right) = 4x\,.$$

Therefore, the method of undetermined coefficients can be used, and a particular solution has the form (14) with $m = 1$ and $r = 0$.

7. The given equation is not an equation with constant coefficients. Thus, the method of undetermined coefficients cannot be applied.

 Copyright © 2012 Pearson Education, Inc. Publishing as Addison-Wesley.

9. Since $r = 0$ is not a root of the auxiliary equation, $r^2 + 2r - 1 = 0$, we choose $s = 0$ in (14) and seek a particular solution of the form $y_p(t) \equiv A_0$. Substitution into the original equation yields

$$(A_0)'' + 2(A_0)' - A_0 = 10 \qquad \Rightarrow \qquad A_0 = -10.$$

Thus, $y_p(t) \equiv -10$ is a particular solution to the given nonhomogeneous equation.

11. The auxiliary equation in this problem, $r^2 + 1 = 0$, has complex roots. Therefore, $2^x = e^{(\ln 2)x}$ is not a solution to the corresponding homogeneous equation, and a particular solution to the original nonhomogeneous equation has the form $y_p(x) = A2^x$. Substituting this expression into the equation, we find the constant A.

$$(A2^x)'' + A2^x = A2^x(\ln 2)^2 + A2^x = A2^x\left[(\ln 2)^2 + 1\right] = 2^x \qquad \Rightarrow \qquad A = \frac{1}{(\ln 2)^2 + 1}.$$

Hence, $y_p(x) = \left[(\ln 2)^2 + 1\right]^{-1} 2^x$.

13. The right-hand side of the original nonhomogeneous equation suggests the form

$$y_p(t) = t^s(A\cos 3t + B\sin 3t)$$

for a particular solution. Since the roots of the auxiliary equation, $r^2 - r + 9 = 0$, are different from $3i$, neither $\cos 3t$ nor $\sin 3t$ is a solution to the corresponding homogeneous equation. Therefore, we can choose $s = 0$, and so

$$y_p(t) = A\cos 3t + B\sin 3t,$$
$$y_p'(t) = -3A\sin 3t + 3B\cos 3t,$$
$$y_p''(t) = -9A\cos 3t - 9B\sin 3t.$$

Substituting these expressions into the original equation and equating the corresponding coefficients, we conclude that

$$(-9A\cos 3t - 9B\sin 3t) - (-3A\sin 3t + 3B\cos 3t) + 9(A\cos 3t + B\sin 3t) = 3\sin 3t$$

$$\Rightarrow \qquad -3B\cos 3t + 3A\sin 3t = 3\sin 3t \qquad \Rightarrow \qquad A = 1, \; B = 0.$$

Hence, the answer is $y_p(t) = \cos 3t$.

Copyright © 2012 Pearson Education, Inc. Publishing as Addison-Wesley.

15. For this problem, the corresponding homogeneous equation is $y'' - 5y' + 6y = 0$, which has the associated auxiliary equation $r^2 - 5r + 6 = 0$. The roots of this equation are $r = 3$ and $r = 2$. Therefore, neither $y = e^x$ nor $y = xe^x$ satisfies the homogeneous equation, and in the expression $y_p(x) = x^s(Ax + B)e^x$ for a particular solution we can take $s = 0$. So,

$$y_p(x) = (Ax + B)e^x$$
$$\Rightarrow \quad y_p'(x) = (Ax + B + A)e^x$$
$$\Rightarrow \quad y_p''(x) = (Ax + B + 2A)e^x$$
$$\Rightarrow \quad (Ax + B + 2A)e^x - 5(Ax + B + A)e^x + 6(Ax + B)e^x = xe^x$$
$$\Rightarrow \quad (2Ax - 3A + 2B)e^x = xe^x \quad \Rightarrow \quad \begin{matrix} 2A = 1, \\ -3A + 2B = 0 \end{matrix} \quad \Rightarrow \quad \begin{matrix} A = 1/2, \\ B = 3/4, \end{matrix}$$

and $y_p(x) = (x/2 + 3/4)e^x = xe^x/2 + 3e^x/4$.

17. The right-hand side of the original equation suggests that a particular solution should be of the form $y_p(t) = At^s e^t$. Since $r = 1$ is a double root of the corresponding auxiliary equation, $r^2 - 2r + 1 = (r - 1)^2 = 0$, we take $s = 2$. Hence

$$y_p(t) = At^2 e^t \quad \Rightarrow \quad y_p'(t) = A\left(t^2 + 2t\right)e^t \quad \Rightarrow \quad y_p''(t) = A\left(t^2 + 4t + 2\right)e^t.$$

Substituting these expressions into the original equation, we find the constant A.

$$A\left(t^2 + 4t + 2\right)e^t - 2A\left(t^2 + 2t\right)e^t + At^2 e^t = 8e^t \quad \Rightarrow \quad 2Ae^t = 8e^t \quad \Rightarrow \quad A = 4.$$

Thus, $y_p(t) = 4t^2 e^t$.

19. According to the right-hand side of the given equation, a particular solution has the form $y_p(t) = t^s(A_1 t + A_0)e^{-3t}$. To choose s, we solve the auxiliary equation, $4r^2 + 11r - 3 = 0$, and find that $r = -3$ is its simple root. Therefore, we take $s = 1$, and so

$$y_p(t) = t\left(A_1 t + A_0\right)e^{-3t} = \left(A_1 t^2 + A_0 t\right)e^{-3t}.$$

Differentiating yields

$$y_p'(t) = \left[-3A_1 t^2 + (2A_1 - 3A_0)t + A_0\right]e^{-3t},$$
$$y_p''(t) = \left[9A_1 t^2 + (9A_0 - 12A_1)t + 2A_1 - 6A_0\right]e^{-3t}.$$

 Copyright © 2012 Pearson Education, Inc. Publishing as Addison-Wesley.

Substituting y, y', and y'' into the original equation, after some algebra we get

$$[-26A_1t + (8A_1 - 13A_0)]e^{-3t} = -2te^{-3t} \quad \Rightarrow \quad \begin{aligned} -26A_1 &= -2, \\ 8A_1 - 13A_0 &= 0. \end{aligned}$$

Solving this system, we obtain $A_0 = 8/169$, $A_1 = 1/13$. Therefore,

$$y_p(t) = \left(\frac{t}{13} + \frac{8}{169} \right) te^{-3t}.$$

21. The nonhomogeneous term of the original equation is te^{2t}. Therefore, a particular solution has the form $x_p(t) = t^s (A_1t + A_0) e^{2t}$. The corresponding homogeneous differential equation has the auxiliary equation $r^2 - 4r + 4 = (r-2)^2 = 0$. Since $r = 2$ is its double root, we choose $s = 2$. Thus, a particular solution to the nonhomogeneous equation has the form

$$x_p(t) = t^2 (A_1t + A_0) e^{2t} = \left(A_1t^3 + A_0t^2\right) e^{2t}.$$

We compute

$$x_p' = \left(3A_1t^2 + 2A_0t\right) e^{2t} + 2 \left(A_1t^3 + A_0t^2\right) e^{2t},$$
$$x_p'' = (6A_1t + 2A_0) e^{2t} + 4 \left(3A_1t^2 + 2A_0t\right) e^{2t} + 4 \left(A_1t^3 + A_0t^2\right) e^{2t}.$$

Substituting these expressions into the original differential equation yields

$$\begin{aligned} x_p'' - 4x_p' + 4x_p &= (6A_1t + 2A_0) e^{2t} + 4 \left(3A_1t^2 + 2A_0t\right) c^{2t} + 4 \left(A_1t^3 + A_0t^2\right) e^{2t} \\ &\quad - 4 \left(3A_1t^2 + 2A_0t\right) e^{2t} - 8 \left(A_1t^3 + A_0t^2\right) e^{2t} + 4 \left(A_1t^3 + A_0t^2\right) e^{2t} \\ &= (6A_1t + 2A_0) e^{2t} = te^{2t}. \end{aligned}$$

Equating coefficients gives $A_0 = 0$ and $A_1 = 1/6$. Therefore, $x_p(t) = t^3 e^{2t}/6$ is a particular solution to the given nonhomogeneous equation.

23. The right-hand side of this equation suggests that $y_p(\theta) = \theta^s(A_2\theta^2 + A_1\theta + A_0)$. We choose $s = 1$ because $r = 0$ is a simple root of the auxiliary equation, $r^2 - 7r = 0$. Therefore,

$$y_p(\theta) = \theta(A_2\theta^2 + A_1\theta + A_0) = A_2\theta^3 + A_1\theta^2 + A_0\theta$$
$$\Rightarrow \quad y_p'(\theta) = 3A_2\theta^2 + 2A_1\theta + A_0$$
$$\Rightarrow \quad y_p''(\theta) = 6A_2\theta + 2A_1.$$

Copyright © 2012 Pearson Education, Inc. Publishing as Addison-Wesley.

So,

$$y_p'' - 7y_p' = (6A_2\theta + 2A_1) - 7\left(3A_2\theta^2 + 2A_1\theta + A_0\right)$$
$$= -21A_2\theta^2 + (6A_2 - 14A_1)\theta + 2A_1 - 7A_0 = \theta^2.$$

Comparing the corresponding coefficients, we find A_2, A_1, and A_0.

$$-21A_2 = 1, \qquad\qquad A_2 = -1/21,$$
$$6A_2 - 14A_1 = 0, \qquad \Rightarrow \qquad A_1 = 3A_2/7 = -1/49,$$
$$2A_1 - 7A_0 = 0 \qquad\qquad A_0 = 2A_1/7 = -2/343.$$

Hence,

$$y_p(\theta) = -\frac{1}{21}\theta^3 - \frac{1}{49}\theta^2 - \frac{2}{343}\theta.$$

25. We look for a particular solution of the form $y_p(t) = t^s(A\cos 3t + B\sin 3t)e^{2t}$. Since $r = 2 + 3i$ is not a root of the auxiliary equation (which is $r^2 + 2r + 4 = 0$), we take $s = 0$. Thus,

$$y_p(t) = (A\cos 3t + B\sin 3t)e^{2t}$$
$$\Rightarrow \quad y_p'(t) = [(2A + 3B)\cos 3t + (-3A + 2B)\sin 3t]e^{2t}$$
$$\Rightarrow \quad y_p''(t) = [(-5A + 12B)\cos 3t + (-12A - 5B)\sin 3t]e^{2t}.$$

Next, we substitute y_p, y_p', and y_p'' into the original equation and compare the corresponding coefficients.

$$y_p'' + 2y_p' + 4y_p = [(3A + 18B)\cos 3t + (-18A + 3B)\sin 3t]e^{2t} = 111e^{2t}\cos 3t$$
$$\Rightarrow \quad \begin{aligned} 3A + 18B &= 111, \\ -18A + 3B &= 0. \end{aligned}$$

This system has the solution $A = 1$, $B = 6$. So,

$$y_p(t) = (\cos 3t + 6\sin 3t)e^{2t}.$$

27. The right-hand side of this equation suggests that

$$y_p(t) = t^s\left(A_3t^3 + A_2t^2 + A_1t + A_0\right)\cos 3t + t^s\left(B_3t^3 + B_2t^2 + B_1t + B_0\right)\sin 3t.$$

To choose s, we find the roots of the characteristic equation, which is $r^2 + 9 = 0$. Since $r = \pm 3i$ are its simple roots, we take $s = 1$. Thus

$$\begin{aligned} y_p(t) &= t\left(A_3t^3 + A_2t^2 + A_1t + A_0\right)\cos 3t + t\left(B_3t^3 + B_2t^2 + B_1t + B_0\right)\sin 3t \\ &= \left(A_3t^4 + A_2t^3 + A_1t^2 + A_0t\right)\cos 3t + \left(B_3t^4 + B_2t^3 + B_1t^2 + B_0t\right)\sin 3t. \end{aligned}$$

 Copyright © 2012 Pearson Education, Inc. Publishing as Addison-Wesley.

29. The characteristic equation $r^2 - 6r + 9 = (r-3)^2 = 0$ has a double root $r = 3$. Therefore, a particular solution is of the form

$$
\begin{aligned}
y_p(t) &= t^2 \left(A_6 t^6 + A_5 t^5 + A_4 t^4 + A_3 t^3 + A_2 t^2 + A_1 t + A_0 \right) e^{3t} \\
&= \left(A_6 t^8 + A_5 t^7 + A_4 t^6 + A_3 t^5 + A_2 t^4 + A_1 t^3 + A_0 t^2 \right) e^{3t}.
\end{aligned}
$$

31. From the right-hand side, we conclude that a particular solution should be of the form

$$
y_p(t) = t^s \left[\left(A_3 t^3 + A_2 t^2 + A_1 t + A_0 \right) \cos t + \left(B_3 t^3 + B_2 t^2 + B_1 t + B_0 \right) \sin t \right] e^{-t}.
$$

Since $r = -1 \pm i$ are simple roots of the characteristic equation, $r^2 + 2r + 2 = 0$, we should take $s = 1$. Therefore,

$$
y_p(t) = \left[\left(A_3 t^4 + A_2 t^3 + A_1 t^2 + A_0 t \right) \cos t + \left(B_3 t^4 + B_2 t^3 + B_1 t^2 + B_0 t \right) \sin t \right] e^{-t}.
$$

33. The right-hand side of the equation suggests that $y_p(t) = t^s (A \cos t + B \sin t)$. By inspection, we see that $r = i$ is not a root of the auxiliary equation, $r^3 - r^2 + 1 = 0$. Thus, with $s = 0$,

$$
\begin{aligned}
y_p(t) &= A \cos t + B \sin t, \\
y_p'(t) &= -A \sin t + B \cos t, \\
y_p''(t) &= -A \cos t - B \sin t, \\
y_p'''(t) &= A \sin t - B \cos t,
\end{aligned}
$$

and substituting into the original equation yields

$$
(A \sin t - B \cos t) - (-A \cos t - B \sin t) + (A \cos t + B \sin t) = \sin t
$$
$$
\Rightarrow \quad (2A - B) \cos t + (A + 2B) \sin t = \sin t
$$
$$
\Rightarrow \quad
\begin{aligned}
2A - B &= 0, \\
A + 2B &= 1
\end{aligned}
\quad \Rightarrow \quad
\begin{aligned}
A &= 1/5, \\
B &= 2/5
\end{aligned}
\quad \Rightarrow \quad
y_p(t) = \frac{1}{5} \cos t + \frac{2}{5} \sin t.
$$

35. We look for a particular solution of the form $y_p(t) = t^s (A_1 t + A_0) e^t$, and choose $s = 1$ because the auxiliary equation, $r^3 + r^2 - 2 = (r-1)(r^2 + 2r + 2) = 0$, has a simple root $r = 1$.

$$
y_p(t) = t(A_1 t + A_0) e^t = (A_1 t^2 + A_0 t) e^t
$$

Copyright © 2012 Pearson Education, Inc. Publishing as Addison-Wesley.

$$\Rightarrow \quad y_p'(t) = \left[A_1 t^2 + (2A_1 + A_0)t + A_0 \right] e^t$$

$$\Rightarrow \quad y_p''(t) = \left[A_1 t^2 + (4A_1 + A_0)t + (2A_1 + 2A_0) \right] e^t$$

$$\Rightarrow \quad y_p'''(t) = \left[A_1 t^2 + (6A_1 + A_0)t + (6A_1 + 3A_0) \right] e^t$$

$$\Rightarrow \quad y''' + y'' - 2y = \left[10A_1 t + (8A_1 + 5A_0) \right] e^t = te^t .$$

Equating the corresponding coefficients, we find that

$$\begin{aligned} 10A_1 &= 1, \\ 8A_1 + 5A_0 &= 0 \end{aligned} \quad \Rightarrow \quad \begin{aligned} A_1 &= 1/10, \\ A_0 &= -8A_1/5 = -4/25 \end{aligned} \quad \Rightarrow \quad y_p(t) = \left(\frac{1}{10} t^2 - \frac{4}{25} t \right) e^t.$$

EXERCISES 4.5: The Superposition Principle and Undetermined Coefficients Revisited

1. Let $g_1(t) := \cos 2t$ and $g_2(t) := t$. Then $y_1(t) = (1/4)\sin 2t$ is a solution to

$$y'' + 2y' + 4y = g_1(t)$$

and $y_2(t) = t/4 - 1/8$ is a solution to

$$y'' + 2y' + 4y = g_2(t).$$

(a) The right-hand side of the given equation equals $g_2(t) + g_1(t)$. Therefore, the function $y(t) = y_2(t) + y_1(t) = t/4 - 1/8 + (1/4)\sin 2t$ is a solution to the equation $y'' + 2y' + 4y = t + \cos 2t$.

(b) We can express $2t - 3\cos 2t = 2g_2(t) - 3g_1(t)$. So, by the superposition principle, the desired solution is $y(t) = 2y_2(t) - 3y_1(t) = t/2 - 1/4 - (3/4)\sin 2t$.

(c) Since $11t - 12\cos 2t = 11g_2(t) - 12g_1(t)$, the function

$$y(t) = 11y_2(t) - 12y_1(t) = 11t/4 - 11/8 - 3\sin 2t$$

is a solution to the given equation.

3. The corresponding homogeneous equation, $y'' + y' = 0$, has the associated auxiliary equation $r^2 + r = r(r+1) = 0$. This gives $r = 0, -1$, and a general solution to the homogeneous equation is $y_h(t) = c_1 + c_2 e^{-t}$. Combining this solution with the particular solution, $y_p(t) = t$, we find that a general solution is given by

$$y(t) = y_p(t) + y_h(t) = t + c_1 + c_2 e^{-t} .$$

   Copyright © 2012 Pearson Education, Inc. Publishing as Addison-Wesley.

5. The corresponding auxiliary equation, $r^2 + 5r + 6 = 0$, has the roots $r = -3, -2$. Therefore, a general solution to the corresponding homogeneous equation has the form $y_h(x) = c_1 e^{-2x} + c_2 e^{-3x}$. By the superposition principle, a general solution to the original nonhomogeneous equation is

$$y(x) = y_p(x) + y_h(x) = e^x + x^2 + c_1 e^{-2x} + c_2 e^{-3x}.$$

7. First, we rewrite the equation in standard form, that is,

$$y'' - 2y = 2 \tan^3 x.$$

The corresponding homogeneous equation, $y'' - 2y = 0$, has the associated auxiliary equation $r^2 - 2 = 0$. Thus $r = \pm\sqrt{2}$, and a general solution to the homogeneous equation is

$$y_h(x) = c_1 e^{\sqrt{2}x} + c_2 e^{-\sqrt{2}x}.$$

Combining this with the particular solution, $y_p(x) = \tan x$, we find that a general solution is given by

$$y(x) = y_p(x) + y_h(x) = \tan x + c_1 e^{\sqrt{2}x} + c_2 e^{-\sqrt{2}x}.$$

9. We can write the nonhomogeneous term as a difference

$$\left(e^t + t\right)^2 = e^{2t} + 2te^t + t^2 = g_1(t) + g_2(t) + g_3(t).$$

The functions $g_1(t)$, $g_2(t)$, and $g_3(t)$ have a form suitable for the method of undetermined coefficients. Therefore, we can apply this method to find particular solutions $y_{p,1}(t)$, $y_{p,2}(t)$, and $y_{p,3}(t)$ to

$$y'' - y' + y = g_k(t), \quad k = 1, 2, 3,$$

respectively. Then, by the superposition principle, $y_p(t) = y_{p,1}(t) + y_{p,2}(t) + y_{p,3}(t)$ is a particular solution to the given equation.

11. The answer is "no" because the term $1/t$ in the right-hand side.

13. In the original form, the function $\sin^2 t$ does not fit any of the cases, given in the method of undetermined coefficients, but it can be written as $\sin^2 t = (1 - \cos 2t)/2$, and so

$$2t + \sin^2 t + 3 = 2t + \frac{1 - \cos 2t}{2} + 3 = \left(2t + \frac{7}{2}\right) - \left(\frac{1}{2}\cos 2t\right).$$

Copyright © 2012 Pearson Education, Inc. Publishing as Addison-Wesley.

Now, the method of undetermined coefficients can be applied to each term in the above difference to find a particular solution to the corresponding nonhomogeneous equation, and the difference of these particular solutions, by the superposition principle, is a particular solution to the original equation. Thus, the answer is "yes".

15. "No", because the given equation is not an equation with constant coefficients.

17. The auxiliary equation in this problem is $r^2 - 1 = 0$ with roots $r = \pm 1$. Hence,

$$y_h(t) = c_1 e^t + c_2 e^{-t}$$

is a general solution to the corresponding homogeneous equation. Our next step is to find a particular solution $y_p(t)$ to the original nonhomogeneous equation. Applying the method of undetermined coefficients yields

$$y_p(t) = At + B \quad \Rightarrow \quad y_p'(t) \equiv A \quad \Rightarrow \quad y_p''(t) \equiv 0;$$
$$y_p'' - y_p = 0 - (At + B) = -At - B = -11t + 1$$
$$\Rightarrow \quad A = 11, \ B = -1$$
$$\Rightarrow \quad y_p(t) = 11t - 1.$$

By the superposition principle, a general solution is given by

$$y(t) = y_p(t) + y_h(t) = 11t - 1 + c_1 e^t + c_2 e^{-t}.$$

19. Solving the auxiliary equation, $r^2 - 3r + 2 = 0$, we find that $r = 1, 2$. Therefore, a general solution to the corresponding homogeneous equation, $y'' - 3y' + 2y = 0$, is

$$y_h(x) = c_1 e^x + c_2 e^{2x}.$$

By the method of undetermined coefficients, a particular solution $y_p(x)$ to the original equation has the form $y_p(x) = x^s(A\cos x + B\sin x)e^x$. We choose $s = 0$ because $r = 1+i$ is not a root of the auxiliary equation. So,

$$y_p(x) = (A\cos x + B\sin x)e^x$$
$$\Rightarrow \quad y_p'(x) = [(A + B)\cos x + (B - A)\sin x]e^x$$
$$\Rightarrow \quad y_p''(x) = (2B\cos x - 2A\sin x)e^x.$$

 Copyright © 2012 Pearson Education, Inc. Publishing as Addison-Wesley.

Substituting these expressions into the equation, we compare the corresponding coefficients and find A and B.

$$\{(2B\cos x - 2A\sin x) - 3[(A+B)\cos x + (B-A)\sin x]$$

$$+2(A\cos x + B\sin x)\}\, e^x = e^x \sin x$$

$$\Rightarrow \quad -(A+B)\cos x + (A-B)\sin x = \sin x \quad \Rightarrow \quad \begin{matrix} A+B=0, \\ A-B=1 \end{matrix} \quad \Rightarrow \quad \begin{matrix} A=1/2, \\ B=-1/2. \end{matrix}$$

Therefore,

$$y_p(x) = \frac{(\cos x - \sin x)e^x}{2},$$

and so

$$y(x) = \frac{(\cos x - \sin x)e^x}{2} + c_1 e^x + c_2 e^{2x}$$

is a general solution to the given nonhomogeneous equation.

21. Since the roots of the auxiliary equation, which is $r^2 + 2r + 2 = 0$, are $r = -1 \pm i$, we have a general solution to the corresponding homogeneous equation

$$y_h(\theta) = c_1 e^{-\theta}\cos\theta + c_2 e^{-\theta}\sin\theta = (c_1\cos\theta + c_2\sin\theta)\,e^{-\theta},$$

and we look for a particular solution of the form

$$y_p(\theta) = \theta^s(A\cos\theta + B\sin\theta)e^{-\theta} \qquad \text{with} \qquad s=1.$$

Differentiating $y_p(\theta)$, we get

$$y_p'(\theta) = (A\cos\theta + B\sin\theta)e^{-\theta} + \theta\left[(A\cos\theta + B\sin\theta)e^{-\theta}\right]',$$

$$y_p''(\theta) = 2\left[(A\cos\theta + B\sin\theta)e^{-\theta}\right]' + \theta\left[(A\cos\theta + B\sin\theta)e^{-\theta}\right]''$$

$$= 2\left[(B-A)\cos\theta - (B+A)\sin\theta\right]e^{-\theta} + \theta\left[(A\cos\theta + B\sin\theta)e^{-\theta}\right]''.$$

(Note that we did not evaluate the terms containing the factor θ because they give zero result when substituted into the original equation.) Therefore,

$$y_p'' + 2y_p' + 2y_p = 2\left[(B-A)\cos\theta - (B+A)\sin\theta\right]e^{-\theta} + 2(A\cos\theta + B\sin\theta)e^{-\theta}$$

$$= 2\left(B\cos\theta - A\sin\theta\right)e^{-\theta} = e^{-\theta}\cos\theta.$$

Hence $A=0$, $B=1/2$, $y_p(\theta) = (1/2)\theta e^{-\theta}\sin\theta$, and a general solution is given by

$$y(\theta) = \frac{1}{2}\theta e^{-\theta}\sin\theta + (c_1\cos\theta + c_2\sin\theta)\,e^{-\theta}.$$

Copyright © 2012 Pearson Education, Inc. Publishing as Addison-Wesley.

23. The corresponding homogeneous equation, $y' - y = 0$, is separable. Solving yields

$$\frac{dy}{dt} = y \quad \Rightarrow \quad \frac{dy}{y} = dt \quad \Rightarrow \quad \ln|y| = t + c \quad \Rightarrow \quad y = \pm e^c e^t = C e^t,$$

where $C \neq 0$ is an arbitrary constant. By inspection, $y \equiv 0$ is also a solution. Therefore, $y_h(t) = C e^t$, where C is an arbitrary constant, is a general solution to the homogeneous equation. (Alternatively, one can apply the method of solving first-order linear equations in Section 2.3 or the method discussed in Problem 21, Section 4.2.) A particular solution has the form $y_p(t) = A$. Substituting y_p into the original equation yields

$$(A)' - A = 1 \quad \Rightarrow \quad A = -1.$$

Thus, $y(t) = C e^t - 1$ is a general solution. To satisfy the initial condition, $y(0) = 0$, we find

$$0 = y(0) = C e^0 - 1 = C - 1 \quad \Rightarrow \quad C = 1.$$

So, the answer is $y(t) = e^t - 1$.

25. The auxiliary equation, $r^2 + 1 = 0$, has roots $r = \pm i$. Therefore, a general solution to the corresponding homogeneous equation is $z_h(x) = c_1 \cos x + c_2 \sin x$, and a particular solution to the original equation has the form $z_p(x) = A e^{-x}$. Substituting this function into the given equation, we find the constant A.

$$z'' + z = \left(A e^{-x}\right)'' + A e^{-x} = 2A e^{-x} = 2e^{-x} \quad \Rightarrow \quad A = 1,$$

and a general solution to the given nonhomogeneous equation is

$$z(x) = e^{-x} + c_1 \cos x + c_2 \sin x.$$

Next, since $z'(x) = -e^{-x} - c_1 \sin x + c_2 \cos x$, from the initial conditions we get a system for determining constants c_1 and c_2.

$$\begin{aligned} 0 &= z(0) = 1 + c_1, \\ 0 &= z'(0) = -1 + c_2 \end{aligned} \quad \Rightarrow \quad \begin{aligned} c_1 &= -1, \\ c_2 &= 1. \end{aligned}$$

Hence, $z(x) = e^{-x} - \cos x + \sin x$ is the solution to the given initial value problem.

27. The roots of the auxiliary equation, $r^2 - r - 2 = 0$, are $r = -1$ and $r = 2$. This gives a general solution to the corresponding homogeneous equation of the form

$$y_h(x) = c_1 e^{-x} + c_2 e^{2x}.$$

 Copyright © 2012 Pearson Education, Inc. Publishing as Addison-Wesley.

We use the superposition principle to find particular solutions to the nonhomogeneous equations.

(i) For the equation

$$y'' - y' - 2y = \cos x,$$

a particular solution has the form $y_{p,1}(x) = A\cos x + B\sin x$. Substituting $y_{p,1}$ into the differential equation yields

$$(-A\cos x - B\sin x) - (-A\sin x + B\cos x) - 2(A\cos x + B\sin x)$$

$$= (-3A - B)\cos x + (A - 3B)\sin x = \cos x$$

$$\Rightarrow \quad \begin{array}{l} -3A - B = 1, \\ A - 3B = 0 \end{array} \quad \Rightarrow \quad \begin{array}{l} A = -3/10\,, \\ B = -1/10\,. \end{array}$$

So, $y_{p,1}(x) = -(3/10)\cos x - (1/10)\sin x$.

(ii) For the equation

$$y'' - y' - 2y = \sin 2x,$$

a particular solution has the form $y_{p,2}(x) = A\cos 2x + B\sin 2x$. Substituting yields

$$(-4A\cos 2x - 4B\sin 2x) - (-2A\sin 2x + 2B\cos 2x) - 2(A\cos 2x + B\sin 2x)$$

$$= (-6A - 2B)\cos 2x + (2A - 6B)\sin 2x = \sin 2x$$

$$\Rightarrow \quad \begin{array}{l} -6A - 2B = 0, \\ 2A - 6B = 1 \end{array} \quad \Rightarrow \quad \begin{array}{l} A = 1/20, \\ B = -3/20. \end{array}$$

Thus, we conclude that

$$y_{p,2}(x) = \frac{1}{20}\cos 2x - \frac{3}{20}\sin 2x.$$

Therefore, a general solution to the original equation is

$$\begin{aligned} y(x) &= y_{p,1}(x) - y_{p,2}(x) + y_h(x) \\ &= -\frac{3}{10}\cos x - \frac{1}{10}\sin x - \frac{1}{20}\cos 2x + \frac{3}{20}\sin 2x + c_1 e^{-x} + c_2 e^{2x}. \end{aligned}$$

Next, we find c_1 and c_2 such that the initial conditions are satisfied.

$$\begin{array}{l} -7/20 = y(0) = -3/10 - 1/20 + c_1 + c_2\,, \\ 1/5 = y'(0) = -1/10 + 2(3/20) - c_1 + 2c_2 \end{array} \quad \Rightarrow \quad \begin{array}{l} c_1 + c_2 = 0\,, \\ -c_1 + 2c_2 = 0\,. \end{array}$$

Solving yields $c_1 = c_2 = 0$. With these constants, the solution becomes

$$y(x) = -\frac{3}{10}\cos x - \frac{1}{10}\sin x - \frac{1}{20}\cos 2x + \frac{3}{20}\sin 2x\,.$$

Copyright © 2012 Pearson Education, Inc. Publishing as Addison-Wesley.

Chapter 4

29. The roots of the auxiliary equation, $r^2 - 1 = 0$, are $r = \pm 1$. Therefore, a general solution to the corresponding homogeneous equation is

$$y_h(\theta) = c_1 e^{\theta} + c_2 e^{-\theta}.$$

(i) For the equation

$$y'' - y = \sin\theta,$$

a particular solution has the form $y_{p,1}(x) = A\cos\theta + B\sin\theta$. Substituting into the given differential equation yields

$$(-A\cos\theta - B\sin\theta) - (A\cos\theta + B\sin\theta) = -2A\cos\theta - 2B\sin\theta = \sin\theta$$

$$\Rightarrow \quad \begin{matrix} -2A = 0, \\ -2B = 1 \end{matrix} \quad \Rightarrow \quad \begin{matrix} A = 0, \\ B = -1/2. \end{matrix}$$

So, $y_{p,1}(\theta) = -(1/2)\sin\theta$.

(ii) For the equation

$$y'' - y = e^{2\theta},$$

a particular solution has the form $y_{p,2}(\theta) = Ae^{2\theta}$. Substituting yields

$$\left(Ae^{2\theta}\right)'' - \left(Ae^{2\theta}\right) = 3Ae^{2\theta} = e^{2\theta} \quad \Rightarrow \quad A = 1/3,$$

and $y_{p,2}(\theta) = (1/3)e^{2\theta}$.

By the superposition principle, a particular solution to the original nonhomogeneous equation is given by

$$y_p(\theta) = y_{p,1}(\theta) - y_{p,2}(\theta) = -(1/2)\sin\theta - (1/3)e^{2\theta},$$

and a general solution is

$$y(\theta) = y_p(\theta) + y_h(\theta) = -(1/2)\sin\theta - (1/3)e^{2\theta} + c_1 e^{\theta} + c_2 e^{-\theta}.$$

Next, we satisfy the initial conditions.

$$\begin{matrix} 1 = y(0) = -1/3 + c_1 + c_2, \\ -1 = y'(0) = -1/2 - 2/3 + c_1 - c_2 \end{matrix} \quad \Rightarrow \quad \begin{matrix} c_1 + c_2 = 4/3, \\ c_1 - c_2 = 1/6 \end{matrix} \quad \Rightarrow \quad \begin{matrix} c_1 = 3/4, \\ c_2 = 7/12. \end{matrix}$$

Therefore, the solution to the given initial value problem is

$$y(\theta) = -\frac{1}{2}\sin\theta - \frac{1}{3}e^{2\theta} + \frac{3}{4}e^{\theta} + \frac{7}{12}e^{-\theta}.$$

Copyright © 2012 Pearson Education, Inc. Publishing as Addison-Wesley.

31. For the nonhomogeneous term $\sin t + t \cos t$, a particular solution has the form

$$y_{p,1}(t) = (A_1 t + A_0)t^s \cos t + (B_1 t + B_0)t^s \sin t.$$

For $10^t = e^{t \ln 10}$, a particular solution should be of the form

$$y_{p,2}(t) = Ct^p e^{t \ln 10} = Ct^p 10^t.$$

Since the roots of the auxiliary equation, $r^2 + 1 = 0$, are $r = \pm i$, we choose $s = 1$ and $p = 0$. Thus, by the superposition principle,

$$y_p(t) = y_{p,1}(t) + y_{p,2}(t) = (A_1 t + A_0)t \cos t + (B_1 t + B_0)t \sin t + C \cdot 10^t.$$

33. To apply the method of undetermined coefficients, we first express

$$\cos^3 t = \frac{\cos 2t + 1}{2} \cos t = \frac{\cos 2t \cos t}{2} + \frac{\cos t}{2} = \frac{1}{4} \cos 3t + \frac{3}{4} \cos t.$$

Thus, the given equation can be written as

$$x'' - x' - 2x = e^t \cos t - t^2 + \frac{1}{4} \cos 3t + \frac{3}{4} \cos t$$

so that, by the superposition principle, a particular solution has has the form

$$x_p = x_{p,1} + x_{p,2} + x_{p,3} + x_{p,4},$$

where $x_{p,1}$ corresponds to $e^t \cos t$, $x_{p,2}$ corresponds to $-t^2$, etc.

The roots of the auxiliary equation, which is $r^2 - r - 2 = 0$, are $r = -1, 2$, and they are different from $1 + i$, 0, $3i$, and i. Therefore,

$$x_p(t) = (A \cos t + B \sin t)e^t + C_2 t^2 + C_1 t + C_0 + D_1 \cos 3t + D_2 \sin 3t + E_1 \cos t + E_2 \sin t.$$

35. Since the roots of the auxiliary equation are

$$r = \frac{4 \pm \sqrt{16 - 20}}{2} = 2 \pm i,$$

which are different from 5 and $3i$, a particular solution has the form

$$y_p(t) = (A_1 t + A_0) \cos 3t + (B_1 t + B_0) \sin 3t + Ce^{5t}.$$

(The last term corresponds to e^{5t} in the right-hand side of the original equation, and the first two come from $t \sin 3t - \cos 3t$.)

Copyright © 2012 Pearson Education, Inc. Publishing as Addison-Wesley.

37. Clearly, $r = 0$ is not a root of the auxiliary equation, $r^3 - 2r^2 - r + 2 = 0$. (One can find the roots, say, using the factorization $r^3 - 2r^2 - r + 2 = (r - 2)(r - 1)(r + 1)$, but they are not needed for the form of a particular solution: the only important thing is that they are different from zero.) Therefore, a particular solution has the form

$$y_p(t) = A_2 t^2 + A_1 t + A_0.$$

Substitution into the original equation yields

$$
\begin{aligned}
y_p''' - 2y_p'' - y_p' + 2y_p &= (0) - 2(2A_2) - (2A_2 t + A_1) + 2(A_2 t^2 + A_1 t + A_0) \\
&= 2A_2 t^2 + (A_1 - 2A_2)t + (A_0 - A_1 - 4A_2) = 2t^2 + 4t - 9.
\end{aligned}
$$

Equating the coefficients, we obtain

$$
\begin{array}{llll}
2A_2 &= 2, & & A_2 = 1, \\
2A_1 - 2A_2 &= 4, & \Rightarrow & A_1 = 3, \\
2A_0 - A_1 - 4A_2 &= -9 & & A_0 = -1.
\end{array}
$$

Therefore, $y_p(t) = t^2 + 3t - 1$.

39. The auxiliary equation in this problem is $r^3 + r^2 - 2 = 0$. By inspection, we see that $r = 0$ is not a root. Next, we find that $r = 1$ is a simple root because

$$\left. \left(r^3 + r^2 - 2 \right) \right|_{r=1} = 0$$

and

$$\left. \left(r^3 + r^2 - 2 \right)' \right|_{r=1} = \left. \left(3r^2 + 2r \right) \right|_{r=1} \neq 0.$$

Therefore, by the superposition principle, a particular solution has the form

$$y_p(t) = t(A_1 t + A_0)e^t + B = (A_1 t^2 + A_0 t)e^t + B.$$

Differentiating, we get

$$
\begin{aligned}
y_p'(t) &= \left[A_1 t^2 + (A_0 + 2A_1)t + A_0 \right] e^t, \\
y_p''(t) &= \left[A_1 t^2 + (A_0 + 4A_1)t + 2A_0 + 2A_1 \right] e^t, \\
y_p'''(t) &= \left[A_1 t^2 + (A_0 + 6A_1)t + 3A_0 + 6A_1 \right] e^t.
\end{aligned}
$$

Substituting y_p and its derivatives into the original equation, and equating the coefficients at like-terms, after some algebra yields

 Copyright © 2012 Pearson Education, Inc. Publishing as Addison-Wesley.

$$\left\{ \left[A_1 t^2 + (A_0 + 6A_1)t + 3A_0 + 6A_1 \right] + \left[A_1 t^2 + (A_0 + 4A_1)t + 2A_0 + 2A_1 \right] \right.$$
$$\left. -2 \left[A_1 t^2 + A_0 t \right] \right\} e^t - 2B = te^t + 1$$
$$\Rightarrow \quad \left[10A_1 t + 8A_1 + 5A_0 \right] e^t - 2B = te^t + 1$$

$$10A_1 = 1, \qquad\qquad A_1 = 1/10,$$
$$\Rightarrow \quad 8A_1 + 5A_0 = 0, \qquad \Rightarrow \qquad A_0 = -4/25,$$
$$-2B = 1 \qquad\qquad B = -1/2.$$

Hence, a particular solution is

$$y_p(t) = \left(\frac{t}{10} - \frac{4}{25} \right) te^t - \frac{1}{2}.$$

41. The characteristic equation in this problem is $r^2 + 2r + 5 = 0$, which has roots $r = -1 \pm 2i$. Therefore, a general solution to the corresponding homogeneous equation is given by

$$y_h(t) = (c_1 \cos 2t + c_2 \sin 2t) e^{-t}. \tag{4.2}$$

(a) For $0 \le t \le 3\pi/2$, $g(t) \equiv 10$, and so the equation becomes

$$y'' + 2y' + 5y = 10.$$

Hence, a particular solution has the form $y_p(t) \equiv A$. Substitution into the equation yields

$$(A)'' + 2(A)' + 5(A) = 10 \qquad \Rightarrow \qquad 5A = 10 \qquad \Rightarrow \qquad A = 2,$$

and so, on $[0, 3\pi/2]$, a general solution to the original equation is

$$y_1(t) = (c_1 \cos 2t + c_2 \sin 2t) e^{-t} + 2.$$

We find c_1 and c_2 by substituting this function into the initial conditions.

$$0 = y_1(0) = c_1 + 2, \qquad\qquad c_1 = -2,$$
$$0 = y_1'(0) = -c_1 + 2c_2 \qquad \Rightarrow \qquad c_2 = -1$$
$$\Rightarrow \quad y_1(t) = -(2 \cos 2t + \sin 2t) e^{-t} + 2.$$

(b) For $t > 3\pi/2$, $g(t) \equiv 0$, and so the given equation becomes homogeneous. Thus, a general solution, $y_2(t)$, is given by (4.2), i.e.,

$$y_2(t) = y_h(t) = (c_1 \cos 2t + c_2 \sin 2t) e^{-t}.$$

Copyright © 2012 Pearson Education, Inc. Publishing as Addison-Wesley.

(c) We want to satisfy the conditions

$$y_1(3\pi/2) = y_2(3\pi/2),$$
$$y_1'(3\pi/2) = y_2'(3\pi/2).$$

Evaluating y_1, y_2, and their derivatives at $t = 3\pi/2$, we solve the system

$$2e^{-3\pi/2} + 2 = -c_1 e^{-3\pi/2},$$
$$0 = (c_1 - 2c_2)e^{3\pi/2}$$

$$\Rightarrow \quad c_1 = -2\left(e^{3\pi/2} + 1\right),$$
$$c_2 = -\left(e^{3\pi/2} + 1\right).$$

43. Recall that the motion of a mass-spring system is governed by the equation

$$my'' + by' + ky = g(t),$$

where m is the mass, b is the damping coefficient, k is the spring constant, and $g(t)$ is the external force. Thus, we have an initial value problem

$$y'' + 4y' + 3y = 5\sin t, \qquad y(0) = \frac{1}{2}, \quad y'(0) = 0.$$

The roots of the auxiliary equation, $r^2 + 4r + 3 = 0$, are $r = -3, -1$, and a general solution to the corresponding homogeneous equation is

$$y_h(t) = c_1 e^{-3t} + c_2 e^{-t}.$$

A particular solution to the original equation has the form $y_p(t) = A\cos t + B\sin t$. Substituting this function into the equation, we get

$$y_p'' + 4y_p' + 3y_p = (-A\cos t - B\sin t) + 4(-A\sin t + B\cos t) + 3(A\cos t + B\sin t)$$
$$= (2A + 4B)\cos t + (2B - 4A)\sin t = 5\sin t$$

$$\Rightarrow \quad \begin{matrix} 2A + 4B = 0, \\ 2B - 4A = 5 \end{matrix} \quad \Rightarrow \quad \begin{matrix} A = -1, \\ B = 1/2. \end{matrix}$$

Thus, a general solution to the equation, describing the motion, is

$$y(t) = -\cos t + \frac{1}{2}\sin t + c_1 e^{-3t} + c_2 e^{-t}.$$

Differentiating, we find that $y'(t) = \sin t + (1/2)\cos t - 3c_1 e^{-3t} - c_2 e^{-t}$. Applying the initial conditions yields

$$y(0) = -1 + c_1 + c_2 = 1/2, \qquad \Rightarrow \quad c_1 = -1/2,$$
$$y'(0) = 1/2 - 3c_1 - c_2 = 0 \qquad \qquad c_2 = 2.$$

Hence, the equation of motion is

$$y(t) = -\cos t + \frac{1}{2}\sin t - \frac{1}{2}e^{-3t} + 2e^{-t}.$$

Copyright © 2012 Pearson Education, Inc. Publishing as Addison-Wesley.

45. **(a)** With $m = k = 1$, $L = \pi$, and $F_0 = 1$ given initial value problem becomes

$$y(t) = 0, \qquad t \leq -\frac{\pi}{2V},$$

$$y'' + y = \begin{cases} \cos(Vt), & -\pi/(2V) < t < \pi/(2V), \\ 0, & t \geq \pi/(2V). \end{cases}$$

The corresponding homogeneous equation, $y'' + y = 0$, is the simple harmonic equation, whose general solution is

$$y_h(t) = C_1 \cos t + C_2 \sin t. \tag{4.3}$$

First, we find the solution to the given problem for $-\pi/(2V) < t < \pi/(2V)$. We have to consider two cases.

(1) Assuming that $V \neq 1$, the nonhomogeneous term, $\cos(Vt)$, suggests a particular solution of the form

$$y_p(t) = A \cos(Vt) + B \sin(Vt).$$

Substituting $y_p(t)$ into the equation yields

$$[A \cos(Vt) + B \sin(Vt)]'' + [A \cos(Vt) + B \sin(Vt)] = \cos(Vt)$$

$$\Rightarrow \quad \left[-V^2 A \cos(Vt) - V^2 B \sin(Vt)\right] + [A \cos(Vt) + B \sin(Vt)] = \cos(Vt)$$

$$\Rightarrow \quad \left(1 - V^2\right) A \cos(Vt) + \left(1 - V^2\right) B \sin(Vt) = \cos(Vt).$$

Equating coefficients, we get

$$A = \frac{1}{1 - V^2}, \qquad B = 0,$$

Thus, a general solution on $(-\pi/(2V), \pi/(2V))$ is

$$y_1(t) = y_h(t) + y_p(t) = C_1 \cos t + C_2 \sin t + \frac{1}{1 - V^2} \cos(Vt). \tag{4.4}$$

Since $y(t) \equiv 0$ for $t \leq -\pi/(2V)$, the initial conditions for the above solution are

$$y_1\left(-\frac{\pi}{2V}\right) = y_1'\left(-\frac{\pi}{2V}\right) = 0.$$

From (4.4) we obtain

$$y_1\left(-\frac{\pi}{2V}\right) = C_1 \cos\left(-\frac{\pi}{2V}\right) + C_2 \sin\left(-\frac{\pi}{2V}\right) = 0$$

$$y_1'\left(-\frac{\pi}{2V}\right) = -C_1 \sin\left(-\frac{\pi}{2V}\right) + C_2 \cos\left(-\frac{\pi}{2V}\right) + \frac{V}{1-V^2} = 0.$$

Solving the system yields

$$C_1 = \frac{V}{V^2-1} \sin\left(\frac{\pi}{2V}\right), \qquad C_2 = \frac{V}{V^2-1} \cos\left(\frac{\pi}{2V}\right),$$

and so

$$\begin{aligned} y_1(t) &= \frac{V}{V^2-1} \sin\left(\frac{\pi}{2V}\right) \cos t + \frac{V}{V^2-1} \cos\left(\frac{\pi}{2V}\right) \sin t + \frac{1}{1-V^2} \cos(Vt) \\ &= \frac{V}{V^2-1} \sin\left(t + \frac{\pi}{2V}\right) - \frac{1}{V^2-1} \cos(Vt), \qquad -\frac{\pi}{2V} < t < \frac{\pi}{2V}. \end{aligned}$$

(2) In the case $V = 1$, similar arguments lead to

$$y_p = \frac{t}{2} \sin t$$

$$\Rightarrow \quad y_1(t) = \frac{1}{2} \cos t + \frac{\pi}{4} \sin t + \frac{t}{2} \sin t,$$

$-\pi/(2V) < t < \pi/(2V)$.

The given equation is homogeneous for $t > \pi/(2V)$, and its general solution, $y_2(t)$, is given by (4.3). That is,

$$y_2(t) = C_3 \cos t + C_4 \sin t.$$

From the initial conditions,

$$y_2\left(\frac{\pi}{2V}\right) = y_1\left(\frac{\pi}{2V}\right) \quad \text{and} \quad y_2'\left(\frac{\pi}{2V}\right) = y_1'\left(\frac{\pi}{2V}\right),$$

we conclude that, for $V \neq 1$,

$$C_3 \cos\left(\frac{\pi}{2V}\right) + C_4 \sin\left(\frac{\pi}{2V}\right) = \frac{V}{V^2-1} \sin\left(\frac{\pi}{V}\right),$$

$$\begin{aligned} -C_3 \sin\left(\frac{\pi}{2V}\right) + C_4 \cos\left(\frac{\pi}{2V}\right) &= \frac{V}{V^2-1} \cos\left(\frac{\pi}{V}\right) + \frac{V}{V^2-1} \\ &= \frac{2V}{V^2-1} \cos^2\left(\frac{\pi}{2V}\right). \end{aligned}$$

Solving these equations simultaneously yields

$$C_3 = 0, \qquad C_4 = \frac{2V}{V^2-1} \cos\left(\frac{\pi}{2V}\right).$$

Hence,

$$y_2(t) = \frac{2V}{V^2-1} \cos\left(\frac{\pi}{2V}\right) \sin t. \tag{4.5}$$

 Copyright © 2012 Pearson Education, Inc. Publishing as Addison-Wesley.

For $V = 1$, similar computations give $C_3 = 0$ and $C_4 = \pi/2$, so that

$$y_2(t) = \frac{\pi}{2} \sin t. \tag{4.6}$$

We remark that the solution (4.6) for the case $V = 1$ can be, formally, obtained from (4.5) by letting $V \to 1$. Applying, say, the L'Hospital's rule, one gets

$$\lim_{V \to 1} \frac{2V}{V^2 - 1} \cos\left(\frac{\pi}{2V}\right) = \lim_{V \to 1} \frac{\cos\left[\pi/(2V)\right]}{V - 1} = \lim_{V \to 1} \frac{-\sin\left[\pi/(2V)\right]\left[-\pi/(2V^2)\right]}{1} = \frac{\pi}{2}.$$

Of course, this formal approach must be rigorously justified.

(b) The graph of the function

$$|A(V)| = \left|\frac{2V}{V^2 - 1} \cos\left(\frac{\pi}{2V}\right)\right|$$

is given in Figure **4–A** on page 241. From this graph, we find that the most violent shaking of the vehicle (the maximum of $|A(V)|$) happens when the speed $V \approx 0.73$.

47. The auxiliary equation in this problem is $r^2 + 9 = 0$ with roots $r = \pm 3i$. So, a general solution to the corresponding homogeneous equation is

$$y_h = c_1 \cos 3t + c_2 \sin 3t.$$

The form of a particular solution, corresponding to the right-hand side, is

$$y_p(t) = A \cos 6t + B \sin 6t.$$

Substituting into the original equation yields

$$-27(A \cos 6t + B \sin 6t) = 27 \cos 6t \quad \Rightarrow \quad A = -1, B = 0 \quad \Rightarrow \quad y_p(t) = -\cos 6t.$$

Therefore, a general solution has the form

$$y(t) = c_1 \cos 3t + c_2 \sin 3t - \cos 6t.$$

In (a)–(c), we have the same boundary condition at $t = 0$, that is, $y(0) = -1$. This yields

$$-1 = y(0) = c_1 - 1 \quad \Rightarrow \quad c_1 = 0.$$

Hence, all the solutions satisfying this condition are given by

$$y(t) = c_2 \sin 3t - \cos 6t. \tag{4.7}$$

Copyright © 2012 Pearson Education, Inc. Publishing as Addison-Wesley.

(a) The second boundary condition gives

$$3 = y(\pi/6) = c_2 + 1 \qquad \Rightarrow \qquad c_2 = 2,$$

and the answer is $y = 2\sin 3t - \cos 6t$.

(b) This time we have

$$5 = y(\pi/3) = c_2 \cdot 0 - 1 \qquad \Rightarrow \qquad 5 = -1,$$

and so there is no solution of the form (4.7) satisfying second boundary condition.

(c) Now we have

$$-1 = y(\pi/3) = c_2 \cdot 0 - 1 \qquad \Rightarrow \qquad -1 = -1,$$

which is an identity. This means that any function in (4.7), where c_2 is an arbitrary constant, satisfies both boundary conditions.

EXERCISES 4.6:   Variation of Parameters

1. From Example 1 in the text, we know that functions $y_1(t) = \cos t$ and $y_2(t) = \sin t$ are two linearly independent solutions to the corresponding homogeneous equation, and so its general solution is given by $y_h(t) = c_1 \cos t + c_2 \sin t$. Now we apply the method of variation of parameters to find a particular solution to the original equation. By the formula (3) in the text, $y_p(t)$ has the form

$$y_p(t) = v_1(t)y_1(t) + v_2(t)y_2(t).$$

Since

$$y_1'(t) = (\cos t)' = -\sin t, \qquad y_2'(t) = (\sin t)' = \cos t,$$

the system (9) becomes

$$v_1'(t)\cos t + v_2'(t)\sin t = 0$$
$$-v_1'(t)\sin t + v_2'(t)\cos t = \sec t.$$

Multiplying the first equation by $\sin t$ and the second equation by $\cos t$ yields

$$v_1'(t)\sin t \cos t + v_2'(t)\sin^2 t = 0$$
$$-v_1'(t)\sin t \cos t + v_2'(t)\cos^2 t = 1.$$

     Copyright © 2012 Pearson Education, Inc. Publishing as Addison-Wesley.

Adding these equations together, we obtain

$$v_2'(t) \left(\cos^2 t + \sin^2 t \right) = 1 \qquad \text{or} \qquad v_2'(t) = 1.$$

From the first equation in the system, we can now find $v_1'(t)$.

$$v_1'(t) = -v_2'(t) \frac{\sin t}{\cos t} = -\tan t.$$

So,

$$\begin{array}{ll} v_1'(t) = -\tan t & \qquad v_1(t) = -\int \tan t \, dt = \ln|\cos t| + c_3 \\ v_2'(t) = 1 & \Rightarrow \qquad v_2(t) = \int dt = t + c_4. \end{array}$$

Since we are looking for a particular solution, we can take $c_3 = c_4 = 0$ and get

$$y_p(t) = (\cos t) \ln|\cos t| + t \sin t.$$

Thus, a general solution to the given equation is

$$y(t) = y_p(t) + y_h(t) = (\cos t) \ln|\cos t| + t \sin t + c_1 \cos t + c_2 \sin t.$$

3. This equation has associated homogeneous equation $y'' + 2y' + y = 0$. Its auxiliary equation, $r^2 + 2r + 1 = 0$, has a double root $r = -1$. Thus, a general solution to the homogeneous equation is $y_h(t) = c_1 e^{-t} + c_2 t e^{-t}$. For the variation of parameters method, we let

$$y_p(t) = v_1(t) y_1(t) + v_2(t) y_2(t), \qquad \text{where} \qquad y_1(t) = e^{-t} \quad \text{and} \quad y_2(t) = t e^{-t}.$$

Thus, $y_1'(t) = -e^{-t}$ and $y_2'(t) = (1-t)e^{-t}$. This means that we have to solve the system (see system (9) in text)

$$e^{-t} v_1' + t e^{-t} v_2' = 0$$
$$-e^{-t} v_1' + (1-t) e^{-t} v_2' = e^{-t}.$$

Adding these two equations yields

$$e^{-t} v_2' = c^{-t} \qquad \Rightarrow \qquad v_2' = 1 \qquad \Rightarrow \qquad v_2 = \int (1) dt = t.$$

Also, from the first equation of the system we have

$$v_1' = -t v_2' = -t \qquad \Rightarrow \qquad v_1 = -\int t \, dt = -\frac{t^2}{2}.$$

Copyright © 2012 Pearson Education, Inc. Publishing as Addison-Wesley.

Therefore,

$$y_p(t) = -\frac{t^2}{2} e^{-t} + t \cdot t e^{-t} = \frac{t^2}{2} e^{-t}$$

$$\Rightarrow \quad y(t) = y_p(t) + y_h(t) = \frac{t^2}{2} e^{-t} + c_1 e^{-t} + c_2 t e^{-t}.$$

5. In this problem, the corresponding homogeneous equation is $r^2 + 9 = 0$ with roots $r = \pm 3i$. Hence, $y_1(t) = \cos 3t$ and $y_2(t) = \sin 3t$ are two linearly independent solutions, and a general solution to the corresponding homogeneous equation is given by

$$y_h(t) = c_1 \cos 3t + c_2 \sin 3t,$$

and, in the method of variation of parameters, a particular solution has the form

$$y_p(t) = v_1(t) \cos 3t + v_2(t) \sin 3t,$$

where $v_1'(t)$, $v_2'(t)$ satisfy the system

$$v_1'(t) \cos 3t + v_2'(t) \sin 3t = 0$$

$$-3v_1'(t) \sin 3t + 3v_2'(t) \cos 3t = \sec^2 3t.$$

Multiplying the first equation by $3 \sin 3t$ and the second equation by $\cos 3t$, and adding the resulting equations, we get

$$3v_2' \left(\sin^2 3t + \cos^2 3t \right) = \sec 3t \quad \Rightarrow \quad v_2' = \frac{1}{3} \sec 3t$$

$$\Rightarrow \quad v_2 = \frac{1}{3} \int \sec 3t \, dt = \frac{1}{9} \ln | \sec 3t + \tan 3t |.$$

From the first equation in the system we also find that

$$v_1'(t) = -v_2'(t) \tan 3t = -\frac{1}{3} \sec 3t \tan 3t$$

$$\Rightarrow \quad v_1(t) = -\frac{1}{3} \int \sec 3t \tan 3t \, dt = -\frac{1}{9} \sec 3t \, .$$

Therefore,

$$\begin{aligned} y_p(t) &= -\frac{1}{9} \sec 3t \cos 3t + \frac{1}{9} \sin 3t \ln | \sec 3t + \tan 3t | \\ &= -\frac{1}{9} + \frac{1}{9} \sin 3t \ln | \sec 3t + \tan 3t | \end{aligned}$$

and

$$y(t) = -\frac{1}{9} + \frac{1}{9} \sin 3t \ln | \sec 3t + \tan 3t | + c_1 \cos 3t + c_2 \sin 3t$$

is a general solution to the given equation.

 Copyright © 2012 Pearson Education, Inc. Publishing as Addison-Wesley.

7. This equation has associated homogeneous equation $y'' + 4y' + 4y = 0$. Its auxiliary equation, $r^2 + 4r + 4 = 0$, has a double root $r = -2$. Thus, a general solution to the homogeneous equation is

$$y_h(t) = c_1 e^{-2t} + c_2 t e^{-2t}.$$

We look for a particular solution to the given equation in the form

$$y_p(t) = v_1(t)y_1(t) + v_2(t)y_2(t), \qquad \text{where} \qquad y_1(t) = e^{-2t} \quad \text{and} \quad y_2(t) = te^{-2t}.$$

Since $y_1' = -2e^{-2t}$ and $y_2' = (1 - 2t)e^{-2t}$, v_1' and v_2' satisfy the system

$$e^{-2t}v_1' + te^{-2t}v_2' = 0$$
$$-2e^{-2t}v_1' + (1 - 2t)e^{-2t}v_2' = e^{-2t}\ln t.$$

Multiplying the first equation by 2 and then adding them together yields

$$e^{-2t}v_2' = e^{-2t}\ln t \qquad \Rightarrow \qquad v_2' = \ln t \qquad \Rightarrow \qquad v_2 = \int \ln t\, dt = t(\ln t - 1).$$

Since $v_1' = -tv_2' = -t\ln t$, we find that

$$v_1 = -\int t\ln t\, dt = -\left(\frac{1}{2}t^2 \ln t - \frac{1}{4}t^2\right).$$

So,

$$y_p(t) = -\left(\frac{1}{2}t^2 \ln t - \frac{1}{4}t^2\right)e^{-2t} + t(\ln t - 1) \cdot te^{-2t} = \frac{2\ln t - 3}{4}t^2 o^{-2t},$$

and a general solution is given by

$$y(t) = \frac{2\ln t - 3}{4}t^2 e^{-2t} + c_1 e^{-2t} + c_2 te^{-2t}.$$

9. This equation has associated homogeneous equation $y'' - y = 0$. The roots of the associated auxiliary equation, $r^2 - 1 = 0$, are $r = \pm 1$. Therefore, a general solution to the homogeneous equation is

$$y_h(t) = c_1 e^t + c_2 e^{-t}.$$

(1) We first apply the method of undetermined coefficients to find a particular solution. Since $r = 0$ is not a root of the auxiliary equation, we have $y_p = At + B$. Substituting this function into the given equation yields

$$-(At + B) = 2t + 4 \qquad \Rightarrow \qquad A = -2,\ B = -4 \qquad \Rightarrow \qquad y_p(t) = -(2t + 4).$$

Copyright © 2012 Pearson Education, Inc. Publishing as Addison-Wesley.

(2) For the variation of parameters method, we let

$$y_p(t) = v_1(t)y_1(t) + v_2(t)y_2(t), \qquad \text{where} \qquad y_1(t) = e^t \quad \text{and} \quad y_2(t) = e^{-t}.$$

Thus, $y_1'(t) = e^t$ and $y_2'(t) = -e^{-t}$. This means that we have to solve the system

$$e^t v_1' + e^{-t} v_2' = 0$$
$$e^t v_1' - e^{-t} v_2' = 2t + 4.$$

Adding these two equations yields

$$2e^t v_1' = 2t + 4 \qquad \Rightarrow \qquad v_1' = (t+2)e^{-t}.$$

Integration yields

$$v_1(t) = \int (t+2)e^{-t} dt = -(t+3)e^{-t}.$$

Substituting v_1' into the first equation, we get

$$v_2' = -v_1' e^{2t} = -(t+2)e^t \qquad \Rightarrow \qquad v_2(t) = -\int (t+2)e^t dt = -(t+1)e^t.$$

Therefore,

$$y_p(t) = -(t+3)e^{-t}e^t - (t+1)e^t e^{-t} = -(2t+4).$$

Comparing these two methods, we see that (in this problem) the method of undetermined coefficients leads to $y_p(t)$ much faster than the method of variation of parameters.

11. This equation is similar to that discussed in Example 1. Only the nonhomogeneous term is different. Thus, we will follow steps of Example 1. Two linearly independent solutions to the corresponding homogeneous equation are $y_1(t) = \cos t$ and $y_2(t) = \sin t$. A particular solution to the original equation is of the form

$$y_p(t) = v_1(t) \cos t + v_2(t) \sin t,$$

where $v_1(t)$ and $v_2(t)$ satisfy

$$v_1'(t) \cos t + v_2'(t) \sin t = 0,$$
$$-v_1'(t) \sin t + v_2'(t) \cos t = \tan^2 t.$$

Multiplying the first equation by $\sin t$ and the second equation by $\cos t$, and adding them together yields

$$v_2'(t) = \tan^2 t \cos t = (\sec^2 t - 1) \cos t = \sec t - \cos t.$$

 Copyright © 2012 Pearson Education, Inc. Publishing as Addison-Wesley.

We find $v_1'(t)$ from the first equation in the system.

$$v_1'(t) = -v_2'(t) \tan t = -(\sec t - \cos t) \tan t = \sin t - \frac{\sin t}{\cos^2 t}.$$

Integrating, we get

$$v_1(t) = \int \left(\sin t - \frac{\sin t}{\cos^2 t} \right) dt = -\cos t - \sec t,$$

$$v_2(t) = \int (\sec t - \cos t) \, dt = \ln|\sec t + \tan t| - \sin t,$$

where we have taken zero integration constants. Therefore,

$$y_p(t) = -(\cos t + \sec t) \cos t + (\ln|\sec t + \tan t| - \sin t) \sin t = \sin t \ln|\sec t + \tan t| - 2,$$

and a general solution is given by

$$y(t) = c_1 \cos t + c_2 \sin t + (\sin t) \ln|\sec t + \tan t| - 2.$$

13. The corresponding homogeneous equation in this problem is the same as that in Problem 1 (with y replaced by v). Similarly to the solution in Problem 1, we conclude that $v_1(t) = \cos 2t$ and $v_2(t) = \sin 2t$ are two linearly independent solutions of the corresponding homogeneous equation, and a particular solution to the original equation can be found as

$$v_p(t) = u_1(t) \cos 2t + u_2(t) \sin 2t,$$

where $u_1(t)$ and $u_2(t)$ satisfy

$$u_1'(t) \cos 2t + u_2'(t) \sin 2t = 0,$$

$$-2u_1'(t) \sin 2t + 2u_2'(t) \cos 2t = \sec^4 2t.$$

Multiplying the first equation by $\sin 2t$ and the second equation by $(1/2) \cos 2t$, and adding the results together, we get

$$u_2'(t) = \frac{1}{2} \sec^3 2t.$$

From the first equation in the above system, we also obtain

$$u_1'(t) = -u_2'(t) \tan 2t = -\frac{1}{2} \sec^4 2t \sin 2t.$$

Integrating yields

$$u_1(t) = -\frac{1}{2} \int \sec^4 2t \sin 2t \, dt = -\frac{1}{2} \int \cos^{-4} 2t \sin 2t \, dt = -\frac{1}{12} \sec^3 2t,$$

$$u_2(t) = \frac{1}{2} \int \sec^3 2t \, dt = \frac{1}{8} \left(\sec 2t \tan 2t + \ln | \sec 2t + \tan 2t | \right).$$

Thus,

$$
\begin{aligned}
v_p(t) &= -\frac{1}{12} \sec^3 2t \cos 2t + \frac{1}{8} \left(\sec 2t \tan 2t + \ln | \sec 2t + \tan 2t | \right) \sin 2t \\
&= -\frac{1}{12} \sec^2 2t + \frac{1}{8} \tan^2 2t + \frac{1}{8} \sin 2t \ln | \sec 2t + \tan 2t | \\
&= \frac{1}{24} \sec^2 2t - \frac{1}{8} + \frac{1}{8} \sin 2t \ln | \sec 2t + \tan 2t |,
\end{aligned}
$$

and a general solution to the given equation is

$$v(t) = c_1 \cos 2t + c_2 \sin 2t + \frac{1}{24} \sec^2 2t - \frac{1}{8} + \frac{1}{8} (\sin 2t) \ln | \sec 2t + \tan 2t |.$$

15. The corresponding homogeneous equation is $y'' + y = 0$. Its auxiliary equation has the roots $r = \pm i$. Hence, a general solution to the homogeneous problem is given by

$$y_h(t) = c_1 \cos t + c_2 \sin t.$$

We find a particular solution to the original equation by first finding a particular solution to each of two problems – the one with the nonhomogeneous term $g_1(t) = 3 \sec t$ and the other one with the nonhomogeneous term $g_2(t) = -t^2 + 1$. Then we will use the superposition principle to obtain a particular solution for the original equation. The term $3 \sec t$ is not in a form that allows us to use the method of undetermined coefficients. Therefore, we will use the method of variation of parameters. To this end, let $y_1(t) = \cos t$ and $y_2(t) = \sin t$ (linearly independent solutions to the corresponding homogeneous problem). Then a particular solution $y_{p,1}$ to $y'' + y = 3 \sec t$ has the form

$$y_{p,1}(t) = v_1(t) y_1(t) + v_2(t) y_2(t) = v_1(t) \cos t + v_2(t) \sin t,$$

where $v_1(t)$ and $v_2(t)$ are determined by the system

$$v_1' \cos t + v_2' \sin t = 0,$$

$$-v_1' \sin t + v_2' \cos t = 3 \sec t.$$

    Copyright © 2012 Pearson Education, Inc. Publishing as Addison-Wesley.

Multiplying the first equation by $\cos t$ and the second equation by $\sin t$ and subtracting the results, we get

$$v_1' = -3 \sec t \sin t = -3 \tan t.$$

Hence,

$$v_1(t) = -3 \int \tan t \, dt = 3 \ln |\cos t|.$$

To find $v_2'(t)$, we multiply the first equation of the above system by $\sin t$, the second by $\cos t$, and add the results to obtain

$$v_2' = 3 \sec t \cos t = 3 \qquad \Rightarrow \qquad v_2(t) = 3t.$$

Therefore, for this first equation (with $g_1(t) = 3 \sec t$), we have a particular solution given by

$$y_{p,1}(t) = 3(\cos t) \ln |\cos t| + 3t \sin t.$$

The nonhomogeneous term $g_2(t) = -t^2 + 1$ is of a form that allows us to use the method of undetermined coefficients. Thus, a particular solution to this nonhomogeneous equation has the form

$$y_{p,2}(t) = A_2 t^2 + A_1 t + A_0 \qquad \Rightarrow \qquad y_{p,2}'(t) = 2A_2 t + A_1 \qquad \Rightarrow \qquad y_{p,2}''(t) = 2A_2.$$

Plugging in these expressions into the equation $y'' + y = -t^2 + 1$ yields

$$y_{p,2}'' + y_{p,2} = 2A_2 + A_2 t^2 + A_1 t + A_0 = A_2 t^2 + A_1 t + (2A_2 + A_0) = -t^2 + 1.$$

By equating coefficients, we obtain

$$A_2 = -1, \qquad A_1 = 0, \qquad 2A_2 + A_0 = 1 \quad \Rightarrow \quad A_0 = 3.$$

Therefore,

$$y_{p,2}(t) = -t^2 + 3.$$

By the superposition principle, we see that a particular solution to the original problem is

$$y_p(t) = y_{p,1}(t) + y_{p,2}(t) = 3(\cos t) \ln |\cos t| + 3t \sin t - t^2 + 3.$$

Combining this solution with the general solution to the homogeneous equation yields a general solution to the original differential equation,

$$y(t) = c_1 \cos t + c_2 \sin t - t^2 + 3 + 3t \sin t + 3(\cos t) \ln |\cos t|.$$

Copyright © 2012 Pearson Education, Inc. Publishing as Addison-Wesley.

17. Multiplying the given equation by 2, we get

$$y'' + 4y = 2\tan 2t - e^t.$$

The nonhomogeneous term can be written as a linear combination $2g_1(t) - g_2(t)$, where $g_1(t) = \tan 2t$ and $g_2(t) = e^t$. A particular solution to the equation

$$y'' + 4y = \tan 2t$$

is found in Problem 1, that is,

$$y_{p,1}(t) = -\frac{1}{4}\cos 2t \ln|\sec 2t + \tan 2t|.$$

A particular solution to

$$y'' + 4y = e^t$$

can be found using the method of undetermined coefficients. We look for $y_{p,2}$ of the form $y_{p,2}(t) = Ae^t$. Substitution yields

$$\left(Ae^t\right)'' + 4\left(Ae^t\right) = e^t \qquad \Rightarrow \qquad 5Ae^t = e^t \qquad \Rightarrow \qquad A = \frac{1}{5},$$

and so $y_{p,2} = (1/5)e^t$. By the superposition principle, a particular solution to the original equation is

$$y_p(t) = 2y_{p,1} - y_{p,2} = -\frac{1}{2}(\cos 2t)\ln|\sec 2t + \tan 2t| - \frac{1}{5}e^t.$$

Adding a general solution to the homogeneous equation to $y_p(t)$, we get

$$y(t) = c_1\cos 2t + c_2\sin 2t - \frac{1}{5}e^t - \frac{1}{2}(\cos 2t)\ln|\sec 2t + \tan 2t|.$$

19. A general solution to the corresponding homogeneous equation is given by

$$y_h(t) = c_1 e^{-t} + c_2 e^t.$$

We will try to find a particular solution to the original nonhomogeneous equation of the form $y_p(t) = v_1(t)y_1(t) + v_2(t)y_2(t)$, where $y_1(t) = e^{-t}$ and $y_2(t) = e^t$. We apply formulas (10), but replace the indefinite integrals by definite integrals. Note that the Wronskian

$$y_1(t)y_2'(t) - y_1'(t)y_2(t) = e^{-x}e^x - \left(-e^{-x}\right)e^x = 2.$$

 Copyright © 2012 Pearson Education, Inc. Publishing as Addison-Wesley.

With $g(t) = 1/t$ and integration limits from $u = 1$ to $u = t$, formulas (10) become

$$v_1(t) = \int_1^t \frac{-g(u)y_2(u)}{2}\, du = -\frac{1}{2}\int_1^t \frac{e^u}{u}\, du\,,$$

$$v_2(t) = \int_1^t \frac{g(u)y_1(u)}{2}\, du = \frac{1}{2}\int_1^t \frac{e^{-u}}{u}\, du\,.$$

(Notice that we have chosen the lower limit of integration $u = 1$ because the initial conditions are given at $t = 1$. We could have chosen any other value for the lower limit, but the choice of $u = 1$ will make finding the constants c_1 and c_2 easier.) Thus,

$$y_p(t) = -\frac{e^{-t}}{2}\int_1^t \frac{e^u}{u}\, du + \frac{e^t}{2}\int_1^t \frac{e^{-u}}{u}\, du\,,$$

and so a general solution to the original differential equation is

$$y(t) = c_1 e^{-t} + c_2 e^t - \frac{e^{-t}}{2}\int_1^t \frac{e^u}{u}\, du + \frac{e^t}{2}\int_1^t \frac{e^{-u}}{u}\, du\,.$$

By plugging in the first initial condition (and using the fact that the integral of a function from a to a is zero (which was the reason for choosing the lower limit of integration $u = 1$), we find that

$$y(1) = c_1 e^{-1} + c_2 e = 0.$$

Differentiating $y(t)$ yields

$$y'(t) = -c_1 e^{-t} + c_2 e^t + \frac{e^{-t}}{2}\int_1^t \frac{e^u}{u}\, du - \left(\frac{e^{-t}}{2}\right)\left(\frac{e^t}{t}\right) + \frac{e^t}{2}\int_1^t \frac{e^{-u}}{u}\, du + \left(\frac{e^t}{2}\right)\left(\frac{e^{-t}}{t}\right),$$

where we have used the product rule and the fundamental theorem of calculus to differentiate the last two terms. We plug in the second initial condition into $y'(t)$ to get

$$-2 = y'(1) = -c_1 e^{-1} + c_2 e^1 - \left(\frac{e^{-1}}{2}\right)\left(\frac{e^1}{1}\right) + \left(\frac{e^1}{2}\right)\left(\frac{e^{-1}}{1}\right) = -c_1 e^{-1} + c_2 e\,.$$

Solving the system

$$c_1 e^{-1} + c_2 e = 0,$$

$$-c_1 e^{-1} + c_2 e = -2$$

yields $c_2 = -e^{-1}$ and $c_1 = e$. Therefore, the solution to our problem is given by

$$y(t) = e^{1-t} - e^{t-1} + \frac{e^t}{2} \int\limits_1^t \frac{e^{-u}}{u}\, du - \frac{e^{-t}}{2} \int\limits_1^t \frac{e^u}{u}\, du. \qquad (4.8)$$

Simpson's rule is implemented on the software package provided with the text (see also the discussion of the solution to Problem 25 in Exercises 2.3). Simpson's rule requires an even number of intervals, but we don't know how many are required to obtain the *two*-place accuracy. We will compute the approximate value of $y(t)$ at $t = 2$ using 2, 4, 6, ... intervals until the approximate value changes by less than five in the third place. For $n = 2$, we divide $[1, 2]$ into four subintervals of equal width of $(2 - 1)/4 = 1/4$. Therefore, the integrals in (4.8) have approximate values

$$\int\limits_1^2 \frac{e^u}{u}\, du \approx \frac{1}{12}\left[\frac{e^1}{1} + 4\frac{e^{1.25}}{1.25} + 2\frac{e^{1.5}}{1.5} + 4\frac{e^{1.75}}{1.75} + \frac{e^2}{2}\right] \approx 3.0592,$$

$$\int\limits_1^2 \frac{e^{-u}}{u}\, du \approx \frac{1}{12}\left[\frac{e^{-1}}{1} + 4\frac{e^{-1.25}}{1.25} + 2\frac{e^{-1.5}}{1.5} + 4\frac{e^{-1.75}}{1.75} + \frac{e^{-2}}{2}\right] \approx 0.1706.$$

Substituting these values into (4.8), we obtain

$$y(2) \approx e^{1-2} - e^{2-1} - \frac{e^{-2}}{2}(3.0592) + \frac{e^2}{2}(0.1706) = -1.9271.$$

Repeating these calculations for $n = 3$, 4, and 5 yields the approximations in Table **4–A**, page 241. Since these values do not change in the fourth place, we conclude that the first three places are accurate, and so an approximate solution is $y(2) \approx -1.93$.

21. The corresponding auxiliary equation, $r^2 + 10r + 25 = (r + 5)^2 = 0$, has a double root $r = -5$. Thus, the functions $y_1(t) = e^{-5t}$ and $y_2(t) = te^{-5t}$ form a fundamental solution set, and so

$$y_h(t) = C_1 y_1(t) + C_2 y_2(t) = C_1 e^{-5t} + C_2 te^{-5t}$$

is a general solution to the corresponding homogeneous equation.

Next, we compute the Wronskian of these two functions.

$$W\left[y_1, y_2\right](t) = y_1(t)y_2'(t) - y_1'(t)y_2(t) = e^{-5t}\left[(1 - 5t)e^{-5t}\right] - \left[-5e^{-5t}\right]te^{-5t} = e^{-10t}.$$

Copyright © 2012 Pearson Education, Inc. Publishing as Addison-Wesley.

We now can use formulas (10) to find $v_1(t)$ and $v_2(t)$. For convenience (looking at the initial conditions), we choose the following antiderivatives.

$$v_1(t) = \int_0^t \frac{-e^{u^3}ue^{-5u}du}{e^{-10u}} = -\int_0^t ue^{u^3+5u}\,du\,, \quad v_2(t) = \int_0^t \frac{e^{u^3}e^{-5u}du}{e^{-10u}} = -\int_0^t e^{u^3+5u}\,du\,.$$

Then a particular solution $y_p(t) = y_1(t)v_1(t) + y_2(t)v_2(t)$ to the given equation satisfies

$$
\begin{aligned}
y_p(0) &= y_1(0)v_1(0) + y_2(0)v_2(0) = (1)(0) + (0)(0) = 0\,, \\
y_p'(0) &= y_1'(0)v_1(0) + y_1(0)v_1'(0) + y_2'(0)v_2(0) + y_2(0)v_2'(0) \\
&= (-5)(0) + (1)(0) + (1)(0) + (0)(1) = 0\,.
\end{aligned}
$$

By the superposition principle, a general solution to the given equation is

$$y(t) = y_h(t) + y_p(t) = C_1 e^{-5t} + C_2 te^{-5t} + y_p(t).$$

Next, we find constants C_1 and C_2 such that $y(t)$ satisfies the initial conditions.

$$
\begin{aligned}
1 &= y(0) = y_h(0) + y_p(0) = y_h(0) = C_1\,, \\
-5 &= y'(0) = y_h'(0) + y_p'(0) = y_h'(0) = C_2 - 5C_1\,.
\end{aligned}
$$

Thus, $C_1 = 1$, $C_2 = 0$, and

$$y(t) = e^{-5t} - e^{-5t} \int_0^t ue^{u^3+5u}\,du + te^{-5t} \int_0^t e^{u^3+5u}\,du\,.$$

solves the given initial value problem. Hence,

$$y(0.2) = e^{-1} - e^{-1} \int_0^{0.2} ue^{u^3+5u}\,du + 0.2e^{-1} \int_0^{0.2} e^{u^3+5u}\,du \approx 0.3785\,.$$

EXERCISES 4.7:   Variable-Coefficient Equations

1. Writing the equation in standard form,

$$y'' + \frac{2}{t-3}y' - \frac{1}{t(t-3)}y = \frac{t}{t-3}\,,$$

we see that the coefficients $p(t) = 2/(t-3)$ and $q(t) = 1/[t(t-3)]$, and $g(t) = t/(t-3)$ are simultaneously continuous on $(-\infty, 0)$, $(0, 3)$, and $(3, \infty)$. Since the initial value of t belongs to $(0, 3)$, Theorem 5 applies, and so there exists a unique solution to the given initial value problem on $(0, 3)$ (with any choice of Y_0 and Y_1).

Copyright © 2012 Pearson Education, Inc. Publishing as Addison-Wesley.

3. The standard form for this equation is

$$y'' + \frac{1}{t^2}y = \frac{\cos t}{t^2}.$$

The function $p(t) \equiv 0$ is continuous everywhere, $q(t) = t^{-2}$, and $g(t) = t^{-2}\cos t$ are simultaneously continuous on $(-\infty, 0)$ and $(0, \infty)$. Thus, the given initial value problem has a unique solution on $(0, \infty)$.

5. Theorem 5 does not apply to this initial value problem since the initial point, $t = 0$, is a point of discontinuity of (say) $p(t) = t^{-1}$ (actually, $q(t)$ and $g(t)$ are also discontinuous at this point).

7. Theorem 5 does not apply because the given problem is not an initial value problem.

9. In this a homogeneous Cauchy-Euler equation with

$$a = 1, \quad b = 2, \quad c = -6.$$

Thus, substituting $y = t^r$, we get its characteristic equation (see (7))

$$ar^2 + (b - a)r + c = r^2 + r - 6 = 0 \qquad \Rightarrow \qquad r = -3, 2.$$

Therefore, $y_1(t) = t^{-3}$ and $y_2(t) = t^2$ are two linearly independent solutions to the given differential equation, and a general solution has the form

$$y(t) = c_1 t^{-3} + c_2 t^2.$$

11. Multiplying by t^2, we get a homogeneous Cauchy-Euler equation

$$t^2 w'' + 6tw' + 4w = 0.$$

Substituting $w = t^r$, we find that r must satisfy the equation

$$r(r - 1) + 6r + 4 = r^2 + 5r + 4 = 0 \qquad \Rightarrow \qquad r = -1, -4.$$

Thus $w_1(t) = t^{-1}$ and $w_2(t) = t^{-4}$ form a fundamental solution set. So,

$$w(t) = c_1 t^{-1} + c_2 t^{-4}$$

gives a general solution to the given equation.

 Copyright © 2012 Pearson Education, Inc. Publishing as Addison-Wesley.

13. Comparing this equation with (6), we see that $a = 9$, $b = 15$, and $c = 1$. Therefore, the corresponding auxiliary equation,

$$ar^2 + (b-a)r + c = 9r^2 + 6r + 1 = (3r+1)^2 = 0$$

has a double root $r = -1/3$. Therefore, $y_1(t) = t^{-1/3}$ and $y_2(t) = t^{-1/3}\ln t$ represent two linearly independent solutions, and so

$$y(t) = c_1 t^{-1/3} + c_2 t^{-1/3}\ln t$$

is a general solution to the given equation.

15. A substitution $t = -\tau$, $y(-\tau) = z(\tau)$, $\tau > 0$, so that

$$z'(\tau) = -y'(-\tau), \quad z''(\tau) = y''(-\tau),$$

leads to

$$z'' - \frac{1}{(-\tau)}(-z') + \frac{5}{(-\tau)^2}z = 0 \quad \Rightarrow \quad z'' - \frac{1}{\tau}z' + \frac{5}{\tau^2}z = 0.$$

We multiply the equation through by τ^2 to get a Cauchy-Euler equation

$$\tau^2 z'' - \tau z' + 5z = 0.$$

Substituting $z = \tau^r$ yields the characteristic equation

$$r(r-1) - r + 5 = r^2 - 2r + 5 = (r-1)^2 + 4 = 0 \quad \Rightarrow \quad r = 1 \pm 2i.$$

Therefore, a fundamental solution set for this differential equation is $z_1(\tau) = \tau\cos(2\ln\tau)$ and $z_2(\tau) = \tau\sin(2\ln\tau)$, which results in a general solution

$$z(\tau) = C_1\tau\cos(2\ln\tau) + C_2\tau\sin(2\ln\tau), \quad \tau > 0.$$

Making now the back substitutions $\tau = -t$ and $z(-t) = y(t)$ yields

$$y(t) = c_1 t\cos[2\ln(-t)] + c_2 t\sin[2\ln(-t)], \quad t < 0,$$

where $c_1 = -C_1$, $c_2 = -C_2$ are arbitrary constants.

Copyright © 2012 Pearson Education, Inc. Publishing as Addison-Wesley.

17. Similarly to Problem 15, we let $t = -\tau$, $y(-\tau) = z(\tau)$, $\tau > 0$, and obtain a Cauchy-Euler equation

$$\tau^2 z'' + 9\tau z' + 17z = 0, \quad \tau > 0.$$

The substitution $z = \tau^r$ leads the characteristic equation

$$r(r-1) + 9r + 17 = 0 \quad \Rightarrow \quad r^2 + 8r + 17 = 0.$$

Solving yields

$$r = \frac{-8 \pm \sqrt{8^2 - 4(1)(17)}}{2} = \frac{-8 \pm 2i}{2} = -4 \pm i.$$

Thus, the roots are complex numbers $\alpha \pm \beta i$ with $\alpha = -4$, $\beta = 1$. According to the text, the functions

$$z_1(\tau) = \tau^{-4} \cos(\ln \tau), \quad z_2(\tau) = \tau^{-4} \sin(\ln \tau)$$

are two linearly independent solutions to the given homogeneous equation. Thus, a general solution is given by

$$z(\tau) = c_1 z_1(\tau) + c_2 z_2(\tau) = \tau^{-4} \left[c_1 \cos(\ln \tau) + c_2 \sin(\ln \tau) \right], \quad \tau > 0.$$

The back substitutions $\tau = -t$ and $z(-t) = y(t)$ result

$$y(t) = t^{-4} \left\{ c_1 \cos[\ln(-t)] + c_2 \sin[\ln(-t)] \right\}, \quad t < 0.$$

19. First, we find a general solution to the given Cauchy-Euler equation. Substitution $y = t^r$ leads to the characteristic equation

$$r(r-1) - 4r + 4 = r^2 - 5r + 4 = 0 \quad \Rightarrow \quad r = 1, 4.$$

Thus, $y = c_1 t + c_2 t^4$ is a general solution. We now find constants c_1 and c_2 such that the initial conditions are satisfied.

$$\begin{array}{ll} -2 = y(1) = c_1 + c_2, \\ -11 = y'(1) = c_1 + 4c_2 \end{array} \quad \Rightarrow \quad \begin{array}{l} c_1 = 1, \\ c_2 = -3, \end{array}$$

and, therefore, $y = t - 3t^4$ is the solution to the given initial value problem.

21. We will look for solutions to the given equation of the form

$$y(t) = (t-2)^r \quad \Rightarrow \quad y'(t) = r(t-2)^{r-1} \quad \Rightarrow \quad y''(t) = r(r-1)(t-2)^{r-2}.$$

 Copyright © 2012 Pearson Education, Inc. Publishing as Addison-Wesley.

Substituting these formulas into the differential equation yields

$$[r(r-1) - 7r + 7](t-2)^r = 0 \quad \Rightarrow \quad r^2 - 8r + 7 = 0 \quad \Rightarrow \quad r = 1, 7.$$

Therefore, $y_1 = t - 2$ and $y_2 = (t-2)^7$ are two linearly independent solutions on $(2, \infty)$. Taking their linear combination, we obtain a general solution of the form

$$y = c_1(t-2) + c_2(t-2)^7.$$

23. (a) Let $t = e^x$. Then $y(t) = y(e^x) =: Y(x)$ can be treated as a function of new variable x. The chain rule yields

$$\frac{dY}{dx} = \frac{dy}{dt}\frac{dt}{dx} = \frac{dy}{dt}e^x = t\frac{dy}{dt}. \tag{4.9}$$

Differentiating one more time, we get

$$\begin{aligned}
\frac{d^2Y}{dx^2} &= \frac{d}{dx}\left(\frac{dY}{dx}\right) = \frac{d}{dt}\left(\frac{dY}{dx}\right)\frac{dt}{dx} = \frac{d}{dt}\left(t\frac{dy}{dt}\right)e^x \\
&= t\left(\frac{dy}{dt} + t\frac{d^2y}{dt^2}\right) = t\frac{dy}{dt} + t^2\frac{d^2y}{dt^2} = \frac{dY}{dx} + t^2\frac{d^2y}{dt^2}.
\end{aligned}$$

Thus,

$$t^2\frac{d^2y}{dt^2} = \frac{d^2Y}{dx^2} - \frac{dY}{dx}. \tag{4.10}$$

(b) Substituting $t = e^x$ into the equation

$$at^2\frac{d^2y}{dt^2} + bt\frac{dy}{dt} + cy = f(t) \tag{4.11}$$

and using relations (4.9) and (4.10), we obtain

$$a\left(\frac{d^2Y}{dx^2} - \frac{dY}{dx}\right) + b\frac{dY}{dx} + cY = f(e^x)$$

$$\Rightarrow \quad a\frac{d^2Y}{dx^2} + (b-a)\frac{dY}{dx} + cY = f(e^x), \tag{4.12}$$

which is a linear equation with constant coefficients.

(c) The characteristic equation for (4.12), $ar^2 + (b-a)r + c = 0$, is the same as that for the original Cauchy-Euler equation given in (7) in the text.

If this equation has complex roots, $r = \alpha \pm i\beta$, then

$$Y_1(x) = e^{\alpha x}\cos(\beta x) \quad \text{and} \quad Y_2(x) = e^{\alpha x}\sin(\beta x)$$

Copyright © 2012 Pearson Education, Inc. Publishing as Addison-Wesley.

form a fundamental solution set for the homogeneous equation corresponding to (4.12). Since $x = \ln t$, back substitution yields

$$y_1(t) = Y_1(\ln t) = e^{\alpha \ln t} \cos(\beta \ln t) = t^\alpha \cos(\beta \ln t),$$
$$y_2(t) = Y_2(\ln t) = e^{\alpha \ln t} \sin(\beta \ln t) = t^\alpha \sin(\beta \ln t).$$

Similarly, if the characteristic equation has a double root r, then two linearly independent solutions to the homogeneous equation, corresponding to (4.12), are $Y_1(x) = e^{rx}$ and $Y_2(x) = xe^{rx}$, and the back substitution yields

$$y_1(t) = e^{r \ln t} = t^r, \qquad y_2(t) = (\ln t)e^{r \ln t} = t^r \ln t.$$

25. (a) True. Since $y_1(t)$ and $y_2(t)$ are linearly dependent on $[a, b]$, there exists a constant c such that $y_1(t) = cy_2(t)$ (or $y_2(t) = cy_1(t)$) for all t in $[a, b]$. In particular, this equality is satisfied on any smaller interval $[c, d]$, and so $y_1(t)$ and $y_2(t)$ are linearly dependent on $[c, d]$.

(b) False. As an example, consider $y_1(t) = t$ and $y_2(t) = |t|$ on $[-1, 1]$. For t in $[0, 1]$, $y_2(t) = t = y_1(t)$, and so $y_2(t) \equiv c_1 y_1(t)$ with $c_1 = 1$. For t in $[-1, 0]$, we have that $y_2(t) = -t = -y_1(t)$, and so $y_2(t) \equiv c_2 y_1(t)$ with $c_2 = -1$. Therefore, these two functions are linearly dependent on $[0, 1]$ and on $[-1, 0]$. Since $c_1 \neq c_2$, there is no such a constant c that $y_1(t) \equiv cy_2(t)$ on $[-1, 1]$. So, $y_1(t)$ and $y_2(t)$ are linearly independent on $[-1, 1]$.

27. (a) First we verify that $y_1(t) = t$ and $y_2(t) = t^3$ are solutions.

$$t^2 y_1'' - 3t y_1' + 3y_1 = t^2(0) - 3t(1) + 3t = 0 \, ;$$
$$t^2 y_2'' - 3t y_2' + 3y_2 = t^2(6t) - 3t(3t^2) + 3t^3 = 0 \, .$$

Computing the Wronskian at $t_0 = 1$ yields

$$W[y_1, y_2](t_0) = y_1(t_0)y_2'(t_0) - y_1'(t_0)y_2(t_0) = (1)3(1)^2 - (1)(1)^3 = 2 \, .$$

(b) The linear combination, $y = c_1 y_1 + c_2 y_2 = c_1 t + c_2 t^3$ is a polynomial. By the fundamental theorem of algebra, it has a finite number of real zeros (not exceeding its degree), and so cannot vanish on an interval unless it is the *zero* polynomial, i.e., $c_1 = c_2 = 0$.

 Copyright © 2012 Pearson Education, Inc. Publishing as Addison-Wesley.

(c) Since

$$y_3(t) = \begin{cases} -t^3 = -y_2(t), & t < 0, \\ t^3 = y_2(t), & t > 0, \end{cases}$$

$y_3(t)$ is a solution to the given linear homogeneous equation on $(-\infty, 0)$ and on $(0, \infty)$ since $y_2(t)$ is. The point $t = 0$ needs a special consideration due to the fact that $y_3(t)$ is a piecewise defined function. We find

$$y_3'(0) = \lim_{h \to 0} \frac{y_3(h) - y_3(0)}{h} = \lim_{h \to 0} \frac{|h|^3}{h} = \lim_{h \to 0} h|h| = 0.$$

Note that, for $t \neq 0$,

$$y_3'(t) = \begin{cases} 3t^2, & t > 0 \\ -3t^2, & t < 0 \end{cases} = 3t|t|.$$

Thus,

$$y_3''(0) = \lim_{h \to 0} \frac{y_3'(h) - y_3'(0)}{h} = \lim_{h \to 0} \frac{3h|h|}{h} = 3 \lim_{h \to 0} |h| = 0.$$

Hence, $y_3(0) = y_3'(0) = y_3''(0) = 0$, and the given equation is trivially satisfied at $t = 0$.

(d) Assuming, to the contrary, that

$$y_3 = |t|^3 = c_1 y_1 + c_2 y_2 = c_1 t + c_2 t^3 \quad \text{for all } t \text{ in } (-\infty, \infty),$$

we get

(i) for $t < 0$, $-t^3 = c_1 t + c_2 t^3$ and so $c_1 t + (c_2 + 1)t^3 = 0$;

(ii) for $t > 0$, $t^3 = c_1 t + c_2 t^3$ and so $c_1 t + (c_2 - 1)t^3 = 0$.

Since functions t and t^3 are linearly independent on any interval, (i) gives $c_2 = -1$ while (ii) is true only for $c_2 = 1$. Thus, there is no unique constant c_2 such that for all t in $(-\infty, \infty)$ there holds $y_3(t) = c_1 y_1(t) + c_2 y_2(t)$.

(e) No. Writing the given equation in standard form,

$$y'' - \frac{3}{t} y' + \frac{3}{t^2} y = 0,$$

we see that the coefficients are not continuos on $(-\infty, \infty)$.

29. Assuming that $y_1(t)$ and $y_2(t)$ are two linearly independent solutions to $y'' - py' + qy = 0$ on (a, b) and that $y_1(t_0) = y_2(t_0) = 0$ at some point t_0 on (a, b), we would conclude that

$$W[y_1, y_2](t_0) = y_1(t_0) y_2'(t_0) - y_1'(t_0) y_2(t_0) = 0$$

and so, by Lemma 2, $y_1(t)$ and $y_2(t)$ must be linearly dependent.

Copyright © 2012 Pearson Education, Inc. Publishing as Addison-Wesley.

31. (a) Yes. Actually, it is the Wronskian of solutions $y_1(t) = e^{5t}$ and $y_2(t) = e^{-t}$ to the differential equation $y'' - 4y' - 5y = 0$.

(b) No. Since $w(0) = 0$ and the point $t = 0$ belongs to $(-1, 1)$, this function cannot be the Wronskian of two independent solutions to the given equation thanks to Lemma 2. But it also cannot be the Wronskian of two linearly dependent solutions since it is not identically zero on $(-1, 1)$.

(c) Yes, since this function does not vanish at any point on the *open* interval $(-1, 1)$. In fact, it is the Wronskian of two linearly independent solutions, $y_1(t) \equiv 1$ and $y_2(t) = \ln(1 + t)$ to the equation $(1 + t)y'' + y' = 0$.

(d) Yes. As an example, consider any two linearly dependent solutions to the given equation, say, $y(t)$ and $2y(t)$.

33. Writing the equation in standard form,

$$y'' + \frac{t-1}{t}\, y' + \frac{3}{t}\, y = 0,$$

we see that $p(t) = (t-1)/t$. So, applying Abel's formula yields

$$
\begin{aligned}
W\left[y_1, y_2\right](t) &= C_1 \exp\left(-\int_{t_0}^{t} \frac{\tau - 1}{\tau}\, d\tau\right) = C_1 \exp\left[(\ln|\tau| - \tau)\,\big|_{t_0}^{t}\right] \\
&= C_1 \exp\left[(\ln t - t) - (\ln t_0 - t_0)\right] = C e^{\ln t - t} = C t e^{-t},
\end{aligned}
$$

where $C = C_1 e^{t_0}/t_0$.

35. Using the superposition principle, discussed in Problem 30 (for the constant-coefficient case, see Section 4.5, Theorem 3), we conclude that the functions

$$y_1(t) = (1 + 2t) - (1 + t) = t \quad \text{and} \quad y_2(t) = \frac{1}{3}\left[(1 + 3t^2) - (1 + t) + y_1(t)\right] = t^2$$

are solutions to the corresponding homogeneous equation. Also,

$$y_p(t) = (1 + t) - y_1(t) = 1$$

is a particular solution to the given nonhomogeneous equation. Since y_1 and y_2 are linearly independent (on any interval), we get a general solution

$$y(t) = 1 + C_1 t + C_2 t^2.$$

Copyright © 2012 Pearson Education, Inc. Publishing as Addison-Wesley.

Substituting the initial conditions, we find that

$$2 = y(1) = 1 + C_1 + C_2, \qquad C_1 + C_2 = 1,$$
$$0 = y'(1) = C_1 + 2C_2 \qquad \Rightarrow \qquad C_1 + 2C_2 = 0 \qquad \Rightarrow \qquad C_1 = 2, \ C_2 = -1.$$

Therefore, the answer is $y(t) = 1 + 2t - t^2$.

37. In standard form, the equation becomes

$$y'' - \frac{t+1}{t} y' + \frac{1}{t} y = t.$$

Thus, $g(t) = t$. We are also given two linearly independent solutions to the corresponding homogeneous equation, $y_1(t) = e^t$ and $y_2(t) = t + 1$. Computing their Wronskian

$$W[y_1, y_2](t) = e^t(t+1)' - \left(e^t\right)'(t+1) = -te^t,$$

we can use Theorem 7 to find $v_1(t)$ and $v_2(t)$.

$$v_1(t) = \int \frac{-t(t+1)}{-te^t} \, dt = \int (t+1)e^{-t} dt = -(t+2)e^{-t},$$
$$v_2(t) = \int \frac{te^t}{-te^t} \, dt = -\int dt = -t.$$

Therefore,

$$y_p = y_1 v_1 + y_2 v_2 = -e^t(t+2)e^{-t} - (t+1)t = -t^2 - 2t - 2 = -t^2 - 2y_2$$

is a particular solution to the given equation. By Superposition Principle, $-t^2$ is also a particular solution. Thus, a general solution to the given equation is

$$y(t) = c_1 e^t + c_2(t+1) - t^2.$$

39. Writing the equation in standard form,

$$y'' + \frac{5t-1}{t} y' - \frac{5}{t} y = te^{-5t},$$

we see that $g(t) = te^{-5t}$. We also have two linearly independent solutions to the corresponding homogeneous equation, $y_1(t) = 5t - 1$ and $y_2(t) = e^{-5t}$. Computing their Wronskian

$$W[y_1, y_2](t) = (5t-1)\left(e^{-5t}\right)' - (5t-1)'e^{-5t} = -25te^{-5t},$$

Copyright © 2012 Pearson Education, Inc. Publishing as Addison-Wesley.

we use Theorem 7 to find $v_1(t)$ and $v_2(t)$.

$$v_1(t) = \int \frac{-te^{-5t}e^{-5t}}{-25te^{-5t}} \, dt = \frac{1}{25} \int e^{-5t} dt = -\frac{1}{125} e^{-5t},$$

$$v_2(t) = \int \frac{te^{-5t}(5t-1)}{-25te^{-5t}} \, dt = -\frac{1}{25} \int (5t-1) dt = -\frac{t^2}{10} + \frac{t}{25}.$$

Therefore,

$$y_p = v_1 y_1 + v_2 y_2 = -\frac{1}{125} e^{-5t}(5t-1) + \left(-\frac{t^2}{10} + \frac{t}{25}\right) e^{-5t} = -\frac{t^2 e^{-5t}}{10} + \frac{y_2}{125}$$

is a particular solution to the given equation. By the superposition principle, $-t^2 e^{-5t}/10$ is also a particular solution. Thus, a general solution to the given equation is

$$y(t) = c_1(5t-1) + c_2 e^{-5t} - \frac{t^2 e^{-5t}}{10}.$$

41. First, we find a fundamental solution set for the corresponding homogeneous Cauchy-Euler equation. Using (7) yields

$$r^2 + 9 = 0 \quad \Rightarrow \quad r = \pm 3i.$$

Thus, $y_1(t) = \cos(3\ln t)$ and $y_2(t) = \sin(3\ln t)$ are two linearly independent solutions to the homogeneous equation.

Next, we compute their Wronskian.

$$W[y_1, y_2](t) = \begin{vmatrix} \cos(3\ln t) & \sin(3\ln t) \\ -3\sin(3\ln t)/t & 3\cos(3\ln t)/t \end{vmatrix} = \frac{3}{t}.$$

We now use Theorem 7 to find a particular solution to the given equation. Putting it in standard form (11) yields

$$g(t) = -\frac{\tan(3\ln t)}{t^2}.$$

Thus, applying formulas (12), we get

$$
\begin{aligned}
v_1(t) &= \int \frac{[\tan(3\ln t)/t^2]\sin(3\ln t)}{(3/t)} \, dt = \frac{1}{3} \int \frac{\tan(3\ln t)\sin(3\ln t)}{t} \, dt \\
&= \frac{1}{9} \left[\ln|\sec(3\ln t) + \tan(3\ln t)| - \sin(3\ln t)\right], \\
v_2(t) &= \int \frac{[-\tan(3\ln t)/t^2]\cos(3\ln t)}{(3/t)} \, dt = -\frac{1}{3} \int \frac{\sin(3\ln t)}{t} \, dt \\
&= \frac{1}{9} \cos(3\ln t).
\end{aligned}
$$

 Copyright © 2012 Pearson Education, Inc. Publishing as Addison-Wesley.

(Evaluating integrals, we used a substitution $u = 3\ln t$.) Thus, a particular solution to the given equation is

$$
\begin{aligned}
y_p(t) &= y_1(t)v_1(t) + v_2(t)y_2(t) \\
&= \frac{1}{9}\left[\ln|\sec(3\ln t) + \tan(3\ln t)| - \sin(3\ln t)\right]\cos(3\ln t) + \frac{1}{9}\cos(3\ln t)\sin(3\ln t) \\
&= \frac{1}{9}\cos(3\ln t)\ln|\sec(3\ln t) + \tan(3\ln t)|,
\end{aligned}
$$

and so a general solution is given by

$$
y = y_h + y_p = c_1\cos(3\ln t) + c_2\sin(3\ln t) + \frac{1}{9}\cos(3\ln t)\ln|\sec(3\ln t) + \tan(3\ln t)|.
$$

43. In this problem, the auxiliary equation (7) becomes

$$
r^2 - 2r + 1 = 0 \qquad \Rightarrow \qquad r = 1
$$

is a double root. Hence $z_1(t) = t$ and $z_2(t) = t\ln t$ are two linearly independent solutions to the corresponding homogeneous equation. Computing their Wronskian yields

$$
\begin{vmatrix} t & t\ln t \\ 1 & \ln t + 1 \end{vmatrix} = t.
$$

The standard form of the given equation,

$$
z'' - t^{-1}z' + t^{-2}z = t^{-1}\left(1 + \frac{3}{\ln t}\right),
$$

says that $g(t) = t^{-1}(1 + 3/\ln t)$. We now apply formulas (12) to find a particular solution.

$$
\begin{aligned}
v_1(t) &= \int \frac{[-t^{-1}(1 + 3/\ln t)]\,t\ln t}{t}\,dt = -\int \frac{\ln t + 3}{t}\,dt = -\left(\frac{1}{2}\ln^2 t + 3\ln t\right), \\
v_2(t) &= \int \frac{[t^{-1}(1 + 3/\ln t)]\,t}{t}\,dt = \int \left(1 + \frac{3}{\ln t}\right)\frac{dt}{t} = \ln t + 3\ln|\ln t|.
\end{aligned}
$$

Thus,

$$
\begin{aligned}
z_p = v_1 z_1 + v_2 z_2 &= -\left[\frac{1}{2}\ln^2 t + 3\ln t\right]t + [\ln t + 3\ln|\ln t|]\,t\ln t \\
&= \frac{t\ln^2 t}{2} - 3t\ln t + 3t(\ln t)\ln|\ln t|.
\end{aligned}
$$

Clearly, the middle term in z_p can be incorporated into the general solution to the corresponding homogeneous equation. Hence, a general solution to the given equation is

$$
z = z_h + z_p = c_1 t + c_2 t\ln t + \frac{1}{2}t\ln^2 t + 3t(\ln t)\ln|\ln t|.
$$

Copyright © 2012 Pearson Education, Inc. Publishing as Addison-Wesley.

45. In standard form, the equation becomes

$$y'' - \frac{2}{t}\,y' - \frac{4}{t^2}\,y = 0\,.$$

Thus, $p(t) = -2/t$. We also have a nontrivial solution $y_1(t) = t^{-1}$. We compute

$$\exp\left[-\int p(t)dt\right] = \exp\left(\int \frac{2dt}{t}\right) = \exp\left(2\ln t\right) = t^2\,.$$

Hence, Theorem 8 yields a second linearly independent solution

$$y_2(t) = t^{-1}\int \frac{t^2 dt}{(t^{-1})^2} = t^{-1}\int t^4 dt = \frac{t^4}{5}\,.$$

Note that one can take $y_2(t) = t^4$, because the given equation is linear and homogeneous.

47. Here, we have

$$x'' - \frac{t+1}{t}\,x' + \frac{1}{t}\,x = 0, \qquad x_1(t) = e^t\,.$$

Hence, $p(t) = -(t+1)/t$ and so

$$\exp\left[-\int p(t)dt\right] = \exp\left[\int \frac{(t+1)dt}{t}\right] = \exp\left(t + \ln t\right) = te^t\,.$$

Therefore, by Theorem 8, a second linearly independent solution is

$$x_2(t) = e^t\int \frac{te^t dt}{(e^t)^2} = e^t\int te^{-t}dt = e^t(-t-1)e^{-t} = -(t+1)\,.$$

Clearly, $-x_2(t) = t+1$ is also a second linearly independent solution.

49. The Hermite's equation is already in standard form with $p(t) = -2t$. Thus,

$$\exp\left[-\int p(t)dt\right] = \exp\left[\int (2t)dt\right] = e^{t^2}\,,$$

and the formula in Theorem 8 for the second linearly independent solution yields

(a)

$$y_2(t) = \left(1 - 2t^2\right)\int \frac{e^{t^2}}{(1-2t^2)^2}\,dt = \left(1 - 2t^2\right)\int \left(1 - 2t^2\right)^{-2} e^{t^2}\,dt\,;$$

(b)

$$y_2(t) = \left(3t - 2t^3\right)\int \frac{e^{t^2}}{(3t-2t^3)^2}\,dt = \left(3t - 2t^3\right)\int \left(3t - 2t^3\right)^{-2} e^{t^2}\,dt\,.$$

Copyright © 2012 Pearson Education, Inc. Publishing as Addison-Wesley.

51. For $y(t) = v(t)f(t) = v(t)e^t$, we find

$$y' = v'e^t + v\left(e^t\right)' = \left(v' + v\right)e^t\,,$$

$$y'' = \left(v' + v\right)'e^t + \left(v' + v\right)\left(e^t\right)' = \left(v'' + 2v' + v\right)e^t\,,$$

$$y''' = \left(v'' + 2v' + v\right)'e^t + \left(v'' + 2v' + v\right)\left(e^t\right)' = \left(v''' + 3v'' + 3v' + v\right)e^t\,.$$

Substituting y and its derivatives into the given equation and collecting like-terms, we get

$$t\left(v''' + 3v'' + 3v' + v\right)e^t - t\left(v'' + 2v' + v\right)e^t + \left(v' + v\right)e^t - ve^t$$

$$= \left[tv''' + 2tv'' + (t+1)v'\right]e^t = 0$$

$$\Rightarrow \qquad tv''' + 2tv'' + (t+1)v' = 0\,.$$

Hence, denoting $v' = w$ (so that $v'' = w'$ and $v''' = w''$) yields

$$tw'' + 2tw' + (t+1)w = 0\,,$$

which is a second order linear homogeneous equation in w.

53. Let $\phi(t)$ be a nontrivial solution to $y'' + py' + qy = 0$ on (a, b), where the coefficients $p(t)$ and $q(t)$ are continuous functions on (a, b).

(a) Assume, to the contrary, that zeros of $\phi(t)$ on (a, b) are not isolated, i.e., there exists t_0 in (a, b) such that $\phi(t_0) = 0$ and, for any $\delta > 0$, the exists t in (a, b), satisfying $\phi(t) = 0$ and $0 < |t - t_0| < \delta$.

We take a sequence $\delta_n = 1/n$, $n = 1, 2, \ldots$, and denote by t_n a zero of ϕ satisfying $0 < |t_n - t_0| < 1/n$. Thus, $\phi(t_n) = 0$ for all n and the sequence of t_n's converges to t_0, as $n \to \infty$.

The function $\phi(t)$ is differentiable (even twice) on (a, b). Let us compute $\phi'(t_0)$ using the definition of the derivative..

$$\phi'(t_0) = \lim_{t \to t_0} \frac{\phi(t) - \phi(t_0)}{t - t_0}$$

$$= \lim_{n \to \infty} \frac{\phi(t_n) - \phi(t_0)}{t_n - t_0} = \lim_{n \to \infty} \frac{0 - 0}{t_n - t_0} = 0\,.$$

(b) From part (a), we have $\phi(t_0) = \phi'(t_0) = 0$. But we assumed that $p(t)$ and $q(t)$ are continuous on (a, b), and so the existence and uniqueness theorem (Theorem 5) applies.

Copyright © 2012 Pearson Education, Inc. Publishing as Addison-Wesley.

Since the trivial solution, $y(t) \equiv 0$ also satisfies these initial conditions, we conclude that $\phi(t) \equiv 0$ on (a, b), which is a contradiction.

EXERCISES 4.8: Qualitative Considerations for Variable-Coefficient and Nonlinear Equations

1. Let $Y(t) := y(-t)$. Then, using the chain rule, we get

$$\frac{dY}{dt} = y'(-t)\frac{d(-t)}{dt} = -y'(-t),$$

$$\frac{d^2Y}{dt^2} = \frac{d[-y'(-t)]}{dt} = -y''(-t)\frac{d(-t)}{dt} = y''(-t).$$

Therefore, denoting $-t = s$, we obtain

$$Y''(t) + tY(t) = y''(-t) + ty(-t) = y''(s) - sy(s) = 0.$$

3. As in Problem 2, this equation describes the motion of the mass-spring system with unit mass, no damping, and stiffness "k" $= -6y$. The initial displacement $y(0) = -1$ is negative as well as the initial velocity $y'(0) = -1$. So, starting from $t = 0$, $y(t)$ will decrease for a while. This will result increasing positive stiffness, $-6y$, i.e., "the spring will become stiffer and stiffer". Eventually, the spring will become so strong that the mass will stop and then go in the positive direction. While $y(t)$ is negative, the positive stiffness will force the mass to approach zero displacement point, $y = 0$. Thereafter, with $y(t) > 0$, the stiffness becomes negative, which means that the spring itself will push the mass further away from $y = 0$ in the positive direction with force, which increases with y. Thus, the curve $y(t)$ will go up unboundedly. Figure 4.23 confirms our prediction.

5. (a) Comparing the equation $y'' = 2y^3$ with the equation (7) of Lemma 3, we conclude that $f(y) = 2y^3$, and so

$$F(y) = \int 2y^3 \, dy = \frac{1}{2}y^4 + C,$$

where C is a constant. We can choose any particular value for C, say, $C = 0$. Thus, $F(y) = (1/2)y^4$. Next, with constant $K = 0$ and sign "$-$" in front of the integral, the equation (11) becomes

$$t = -\int \frac{dy}{\sqrt{2(1/2)y^4}} = -\int y^{-2} dy = \frac{1}{y} + c,$$

Copyright © 2012 Pearson Education, Inc. Publishing as Addison-Wesley.

or, equivalently,

$$y = \frac{1}{t - c},$$

where c is an arbitrary constant.

(b) A linear combination of $y_1(t) := 1/(t - c_1)$ and $y_2(t) := 1/(t - c_2)$,

$$C_1 y_1(t) + C_2 y_2(t) = \frac{C_1}{t - c_1} + \frac{C_2}{t - c_2} = \frac{(C_1 + C_2)t - (C_1 c_2 + C_2 c_1)}{(t - c_1)(t - c_2)},$$

is identically zero in a neighborhood of $t = 0$ if and only if

$$(C_1 + C_2)t - (C_1 c_2 + C_2 c_1) \equiv 0.$$

Thus, the numerator must be the zero polynomial, i.e., C_1 and C_2 must satisfy

$$\begin{aligned} C_1 + C_2 &= 0, \\ C_1 c_2 + C_2 c_1 &= 0 \end{aligned} \qquad \Rightarrow \qquad \begin{aligned} C_2 &= -C_1, \\ C_1(c_2 - c_1) &= 0. \end{aligned}$$

Since $c_1 \neq c_2$, the second equation implies that $C_1 = 0$, and then $C_2 = 0$ from the first equation. Thus, only the trivial linear combination of $y_1(t)$ and $y_2(t)$ vanishes identically near the origin, and so these functions are linearly independent.

(c) For any function of the form $y_c(t) := 1/(t - c)$, the equality

$$y_c'(t) = -\frac{1}{(t - c)^2} = -[y_c(t)]^2$$

holds for all $t \neq c$. In particular, at $t = 0$,

$$y_c'(0) = -[y_c(0)]^2.$$

(We assume that $c \neq 0$; otherwise, $t = 0$ is not in the domain.) Obviously, this equality fails for any positive initial velocity $y'(0)$; in particular, it fails for the given data, which are $y(0) = 1$ and $y'(0) = 2$.

7. (a) Since, for a point moving along a circle of radius ℓ, the magnitude v of its linear velocity $\vec{v}$ and the angular velocity $\omega = d\theta/dt$ are connected by $v = \omega\ell = (d\theta/dt)\ell$, and the velocity vector $\vec{v}$ is tangent to the circle (so that it is perpendicular to the radius), we have

$$\text{angular momentum} = \ell \cdot mv = \ell \cdot m \cdot \frac{d\theta}{dt}\ell = m\ell^2 \frac{d\theta}{dt}.$$

Copyright © 2012 Pearson Education, Inc. Publishing as Addison-Wesley.

(b) From Figure 4.18, we see that the component of the gravitational force mg, which is perpendicular to the level arm, has the magnitude $|mg \sin \theta|$ and is directed toward decreasing of θ. Thus,

$$\text{torque} = \ell \cdot (-mg \sin \theta) = -\ell mg \sin \theta.$$

(c) According to Newton's law of rotational motion,

$$\text{torque} = \frac{d}{dt}(\text{angular momentum}) \qquad \Rightarrow \qquad -\ell mg \sin \theta = \frac{d}{dt}\left(m\ell^2 \frac{d\theta}{dt}\right)$$

$$\Rightarrow \qquad -\ell mg \sin \theta = m\ell^2 \frac{d^2\theta}{dt^2} \qquad \Rightarrow \qquad \frac{d^2\theta}{dt^2} + \frac{g}{\ell}\sin \theta = 0.$$

9. According to Problem 8, with $\ell = g$, the function $\theta(t)$ satisfies the identity

$$\frac{(\theta')^2}{2} - \cos \theta = C = \text{const.} \tag{4.13}$$

Our first purpose is to determine the constant C. Let t_a denote the moment when pendulum is in the apex point, i.e., $\theta(t_a) = \pi$. Since it doesn't cross the apex over, we also have $\theta'(t_a) = 0$. Substituting these two values into (4.13), we obtain

$$\frac{0^2}{2} - \cos \pi = C \qquad \Rightarrow \qquad C = 1.$$

Thus, (4.13) becomes

$$\frac{(\theta')^2}{2} - \cos \theta = 1.$$

In particular, at the initial moment, $t = 0$,

$$\frac{[\theta'(0)]^2}{2} - \cos[\theta(0)] = 1.$$

Since $\theta(0) = 0$, we get

$$\frac{[\theta'(0)]^2}{2} - \cos 0 = 1 \qquad \Rightarrow \qquad [\theta'(0)]^2 = 4$$

$$\Rightarrow \qquad \theta'(0) = 2 \qquad \text{or} \qquad \theta'(0) = -2.$$

11. The "damping coefficient" in the Rayleigh equation is $b = (y')^2 - 1$. Thus, for low velocities y', we have $b < 0$, and $b > 0$ for high velocities. Therefore, the low velocities are boosted, while high velocities are slowed, and so one should expect a limit cycle.

 Copyright © 2012 Pearson Education, Inc. Publishing as Addison-Wesley.

13. Qualitative features of solutions to Airy, Duffing, and van der Pol equations are discussed after Example 3 and in Examples 6 and 7, respectively. Comparing curves in Figure 4.26 with graphs depicted in Figures 4.13, 4.16, and 4.17, we conclude that the answers are

    (a) Airy;

    (b) Duffing;

    (c) van der Pol.

15. **(a)** Yes, because the "stiffness" t^2 is positive and no damping.

    (b) No, because of the negative "stiffness" $-t^2$.

    (c) Writing $y'' + y^5 = y'' + (y^4)y$, we conclude that the mass-spring model, corresponding to this equation, has positive "stiffness" y^4 and no damping. Thus, the answer is "yes".

    (d) Here, the "stiffness" is y^5, which is negative for $y < 0$. So, "no".

    (e) Yes, because the "stiffness" $4 + 2\cos t \geq 2 > 0$ and no damping.

    (f) Since both the "damping" t and the stiffness 1 are positive, all solutions are bounded.

    (g) No, because the "stiffness", -1, is negative.

17. For the radius, $r(t)$, we have the initial value problem

$$r''(t) = -GMr^{-2}, \qquad r(0) = a, \qquad r'(0) = 0.$$

Thus, in the energy integral lemma, $f(r) = -GMr^{-2}$. Since

$$\int f(r)\,dr = \int \left(-GMr^{-2}\right) dr = GMR^{-1} + C,$$

we can take $F(r) = GMR^{-1}$, and the energy integral lemma yields

$$\frac{1}{2}\left[r'(t)\right]^2 - \frac{GM}{r(t)} = C_1 = \text{const}.$$

To find the constant C_1, we use the initial conditions.

$$C_1 = \frac{1}{2}\left[r'(0)\right]^2 - \frac{GM}{r(0)} = \frac{1}{2} \cdot 0^2 - \frac{GM}{a} = -\frac{GM}{a}.$$

Copyright © 2012 Pearson Education, Inc. Publishing as Addison-Wesley.

Therefore, $r(t)$ satisfies

$$\frac{1}{2}[r'(t)]^2 - \frac{GM}{r(t)} = -\frac{GM}{a} \quad \Rightarrow \quad \frac{1}{2}(r')^2 = \frac{GM}{r} - \frac{GM}{a} \quad \Rightarrow \quad r' = -\sqrt{\frac{2GM}{a}}\sqrt{\frac{a-r}{r}}.$$

(Remember, $r(t)$ is decreasing, and so $r'(t) < 0$.) Separating variables and integrating, we get

$$\int \sqrt{\frac{r}{a-r}}\, dr = \int \left(-\sqrt{\frac{2GM}{a}}\right) dt$$

$$\Rightarrow \quad a\left(\arctan\sqrt{\frac{r}{a-r}} - \frac{\sqrt{r(a-r)}}{a}\right) = -\sqrt{\frac{2GM}{a}}\, t + C_2.$$

We apply the initial condition, $r(0) = a$, once again to find the constant C_2. But this time we have to be careful because the argument of "arctan" function becomes infinite at $r = a$. So, we take the limit of both sides rather than making simple substitution.

$$\lim_{t \to +0} a\left(\arctan\sqrt{\frac{r(t)}{a-r(t)}} - \frac{\sqrt{r(t)[a-r(t)]}}{a}\right)$$

$$= a\left(\lim_{t\to+0}\arctan\sqrt{\frac{r(t)}{a-r(t)}} - \lim_{t\to+0}\frac{\sqrt{r(t)[a-r(t)]}}{a}\right) = a\left(\frac{\pi}{2} - 0\right) = \frac{a\pi}{2},$$

and, in the right-hand side,

$$\lim_{t\to+0}\left(-\sqrt{\frac{2GM}{a}}\, t + C_2\right) = -\sqrt{\frac{2GM}{a}}(0) + C_2 = C_2.$$

Thus $C_2 = a\pi/2$ and $r(t)$ satisfies

$$a\left(\arctan\sqrt{\frac{r(t)}{a-r(t)}} - \frac{\sqrt{r(t)[a-r(t)]}}{a}\right) = -\sqrt{\frac{2GM}{a}}\, t + \frac{a\pi}{2}.$$

At the moment $t = T_0$, when Earth splashes into the sun, we have $r(T_0) = 0$. Substituting this condition into the last equation yields

$$a\left(\arctan\sqrt{\frac{0}{a-0}} - \frac{\sqrt{0(a-0)}}{a}\right) = -\sqrt{\frac{2GM}{a}}\, T_0 + \frac{a\pi}{2}$$

$$\Rightarrow \quad 0 = -\sqrt{\frac{2GM}{a}}\, T_0 + \frac{a\pi}{2}$$

$$\Rightarrow \quad T_0 = \frac{a\pi}{2}\sqrt{\frac{a}{2GM}} = \frac{\pi}{2\sqrt{2}}\sqrt{\frac{a^3}{GM}}.$$

Then the required ratio is

$$\frac{T_0}{T} = \left(\frac{\pi}{2\sqrt{2}}\sqrt{\frac{a^3}{GM}}\right) \Big/ \left(2\pi\sqrt{\frac{a^3}{GM}}\right) = \frac{1}{4\sqrt{2}}.$$

 Copyright © 2012 Pearson Education, Inc. Publishing as Addison-Wesley.

EXERCISES 4.9:   A Closer Look at Free Mechanical Vibrations

1. In this problem, we have undamped free vibration case governed by equation (2) in the text. With $m = 2$ and $k = 50$, the equation becomes

$$2y'' + 50y = 0$$

with the initial conditions $y(0) = -1/4$, $y'(0) = -1$.

The angular velocity of the motion is

$$\omega = \sqrt{\frac{k}{m}} = \sqrt{\frac{50}{2}} = 5 \,.$$

It follows that

$$\text{period} = \frac{2\pi}{\omega} = \frac{2\pi}{5}$$

$$\text{natural frequency} = \frac{\omega}{2\pi} = \frac{5}{2\pi} \,.$$

A general solution, given in (4) in the text, becomes

$$y(t) = C_1 \cos \omega t + C_2 \sin \omega t = C_1 \cos 5t + C_2 \sin 5t.$$

We find C_1 and C_2 from the initial conditions.

$$y(0) = (C_1 \cos 5t + C_2 \sin 5t)\,\big|_{t=0} = C_1 = -1/4$$
$$y'(0) = (-5C_1 \sin 5t + 5C_2 \cos 5t)\,\big|_{t=0} = 5C_2 = -1 \qquad \Rightarrow \qquad \begin{aligned} C_1 &= -1/4 \\ C_2 &= 1/5. \end{aligned}$$

Thus, the solution to the initial value problem is

$$y(t) = -\frac{1}{4}\cos 5t - \frac{1}{5}\sin 5t.$$

The amplitude of the motion therefore is

$$A = \sqrt{C_1^2 + C_2^2} = \sqrt{\frac{1}{16} + \frac{1}{25}} = \frac{\sqrt{41}}{20} \,.$$

Setting $y = 0$ in the above solution, we find values of t when the mass passes through the point of equilibrium.

$$-\frac{1}{4}\cos 5t - \frac{1}{5}\sin 5t = 0 \qquad \Rightarrow \qquad \tan 5t = -\frac{5}{4}$$

$$\Rightarrow \qquad t = \frac{\pi k - \arctan(5/4)}{5} \,, \qquad k = 1, 2, \dots .$$

Copyright © 2012 Pearson Education, Inc. Publishing as Addison-Wesley.

(Time t is nonnegative.) The first moment when this happens, i.e., the smallest value of t, corresponds to $k = 1$. So,

$$t = \frac{\pi - \arctan(5/4)}{5} \approx 0.45 \text{ (sec)}.$$

3. The characteristic equation in this problem, $r^2 + br + 16 = 0$, has roots

$$r = \frac{-b \pm \sqrt{b^2 - 64}}{2}. \tag{4.14}$$

Substituting given particular values of b into (4.14), we find roots of the characteristic equation and solutions to the initial value problems in each case.

$b = 0$.

$$r = \frac{\pm\sqrt{-64}}{2} = \pm 4i.$$

A general solution has the form $y = C_1 \cos 4t + C_2 \sin 4t$. Constants C_1 and C_2 can be found from the initial conditions.

$$\begin{aligned} y(0) &= (C_1 \cos 4t + C_2 \sin 4t)\,\big|_{t=0} = C_1 = 1, \\ y'(0) &= (-4C_1 \sin 4t + 4C_2 \cos 4t)\,\big|_{t=0} = 4C_2 = 0 \end{aligned} \quad \Rightarrow \quad \begin{aligned} C_1 &= 1, \\ C_2 &= 0, \end{aligned}$$

and so $y(t) = \cos 4t$.

$b = 6$.

$$r = \frac{-6 \pm \sqrt{36 - 64}}{2} = -3 \pm \sqrt{7}i.$$

A general solution has the form $y = (C_1 \cos \sqrt{7}t + C_2 \sin \sqrt{7}t)e^{-3t}$. For constants C_1 and C_2, we have the system

$$\begin{aligned} y(0) &= \left(C_1 \cos \sqrt{7}t + C_2 \sin \sqrt{7}t\right) e^{-3t}\,\big|_{t=0} = C_1 = 1, \\ y'(0) &= \left[(\sqrt{7}C_2 - 3C_1) \cos \sqrt{7}t - (\sqrt{7}C_1 + 3C_2) \sin \sqrt{7}t\right] e^{-3t}\,\big|_{t=0} \\ &= \sqrt{7}C_2 - 3C_1 = 0. \end{aligned}$$

Thus, we conclude that $C_1 = 1$, $C_2 = 3/\sqrt{7}$, and so

$$y(t) = \left(\cos \sqrt{7}t + \frac{3}{\sqrt{7}} \sin \sqrt{7}t\right) e^{-3t} = \frac{4}{\sqrt{7}} e^{-3t} \sin\left(\sqrt{7}t + \phi\right),$$

where $\phi = \arctan(\sqrt{7}/3) \approx 0.723$.

 Copyright © 2012 Pearson Education, Inc. Publishing as Addison-Wesley.

$b = 8$.

$$r = \frac{-8 \pm \sqrt{64 - 64}}{2} = -4.$$

Thus, $r = -4$ is a double root of the characteristic equation. So, a general solution has the form $y = (C_0 + C_1 t)e^{-4t}$. For constants C_0 and C_1, we obtain the system

$$
\begin{aligned}
y(0) &= (C_0 + C_1 t)\, e^{-4t}\,\big|_{t=0} = C_0 = 1, \\
y'(0) &= (C_1 - 4C_0 - 4C_1 t)\, e^{-4t}\,\big|_{t=0} = C_1 - 4C_0 = 0
\end{aligned}
\qquad \Rightarrow \qquad
\begin{aligned}
C_0 &= 1, \\
C_1 &= 4
\end{aligned}
$$

so that $y(t) = (4t + 1)e^{-4t}$.

$b = 10$.

$$r = \frac{-10 \pm \sqrt{100 - 64}}{2} = -5 \pm 3.$$

Thus, $r = -2, -8$, and a general solution is given by $y = C_1 e^{-2t} + C_2 e^{-8t}$. Initial conditions imply that

$$
\begin{aligned}
y(0) &= \left(C_1 e^{-2t} + C_2 e^{-8t} \right)\big|_{t=0} = C_1 + C_2 = 1, \\
y'(0) &= \left(-2C_1 e^{-2t} - 8C_2 e^{-8t} \right)\big|_{t=0} = -2C_1 - 8C_2 = 0
\end{aligned}
\qquad \Rightarrow \qquad
\begin{aligned}
C_1 &= 4/3, \\
C_2 &= -1/3
\end{aligned}
$$

and, therefore, $y(t) = (4/3)e^{-2t} - (1/3)e^{-8t}$ is the solution to the initial value problem.

The graphs of the solutions are depicted in Figures B.19–B.22 in the answers of the text.

5. The auxiliary equation associated with the given differential equation is $r^2 + 10r + k = 0$, and its roots are $r = -5 \pm \sqrt{25 - k}$.

   $k = 20$. In this case, $r = -5 \pm \sqrt{25 - 20} = -5 \pm \sqrt{5}$. Thus, a general solution is given by $y = C_1 e^{(-5+\sqrt{5})t} + C_2 e^{(-5-\sqrt{5})t}$. The initial conditions imply that

$$
\begin{aligned}
y(0) &= \left[C_1 e^{(-5+\sqrt{5})t} + C_2 e^{(-5-\sqrt{5})t} \right]\Big|_{t=0} = C_1 + C_2 = 1, \\
y'(0) &= \left[(-5+\sqrt{5})C_1 e^{(-5+\sqrt{5})t} + (-5-\sqrt{5})C_2 e^{(-5-\sqrt{5})t} \right]\Big|_{t=0} \\
&= (-5+\sqrt{5})C_1 + (-5-\sqrt{5})C_2 = 0.
\end{aligned}
$$

Solving yields $C_1 = \left(1 + \sqrt{5}\right)/2$, $C_2 = \left(1 - \sqrt{5}\right)/2$ and, therefore,

$$y(t) = \frac{\left(1 + \sqrt{5}\right) e^{(-5+\sqrt{5})t} + \left(1 - \sqrt{5}\right) e^{(-5-\sqrt{5})t}}{2}$$

is the solution to the problem.

Copyright © 2012 Pearson Education, Inc. Publishing as Addison-Wesley.

$k = 25$. Then $r = -5 \pm \sqrt{25 - 25} = -5$. Thus, $r = -5$ is a double root of the characteristic equation. So, a general solution has the form $y = (C_0 + C_1 t)e^{-5t}$. For constants C_0 and C_1, using the initial conditions, we obtain the system

$$y(0) = (C_1 t + C_0) e^{-5t} \big|_{t=0} = C_0 = 1,$$
$$y'(0) = (-5C_1 t - 5C_0 + C_1) e^{-5t} \big|_{t=0} = C_1 - 5C_0 = 0 \qquad \Rightarrow \qquad \begin{array}{l} C_0 = 1, \\ C_1 = 5, \end{array}$$

and so $y(t) = (1 + 5t)e^{-5t}$.

$k = 30$. In this case, $r = -5 \pm \sqrt{25 - 30} = -5 \pm \sqrt{5}i$. A general solution has the form $y = (C_1 \cos \sqrt{5}t + C_2 \sin \sqrt{5}t)e^{-5t}$. For constants C_1 and C_2, we have the system

$$y(0) = \left(C_1 \cos \sqrt{5}t + C_2 \sin \sqrt{5}t\right) e^{-5t} \big|_{t=0} = C_1 = 1,$$
$$y'(0) = \left[(\sqrt{5}C_2 - 5C_1) \cos \sqrt{5}t - (\sqrt{5}C_1 + 5C_2) \sin \sqrt{5}t\right] e^{-5t} \big|_{t=0}$$
$$= \sqrt{5}C_2 - 5C_1 = 0$$

$$\Rightarrow \qquad \begin{array}{l} C_1 = 1, \\ C_2 = \sqrt{5} \end{array}$$

and so

$$y(t) = \left(\cos \sqrt{5}t + \sqrt{5} \sin \sqrt{5}t\right) e^{-5t} = \sqrt{6}e^{-5t} \sin\left(\sqrt{5}t + \phi\right),$$

where $\phi = \arctan(1/\sqrt{5}) \approx 0.421$.

Graphs of the solutions for $k = 20$, 25, and 30 are shown in Figures B.23–B.25 in the answers of the text.

7. The motion of this mass-spring system is governed by equation (12). With $m = 1/8$, $b = 2$, and $k = 16$ this equation becomes

$$\frac{1}{8} y'' + 2y' + 16y = 0, \tag{4.15}$$

and the initial conditions are $y(0) = -3/4$, $y'(0) = -2$. Since

$$b^2 - 4mk = 4 - 4(1/8)16 = -4 < 0,$$

we have a case of underdamped motion. A general solution to (4.15) is given in (16), that is, with $\alpha = -b/(2m) = -8$ and $\beta = (1/2m)\sqrt{4mk - b^2} = 8$, we have

$$y = (C_1 \cos 8t + C_2 \sin 8t) e^{-8t}.$$

 Copyright © 2012 Pearson Education, Inc. Publishing as Addison-Wesley.

Using the initial conditions, we find the constants C_1 and C_2.

$$y(0) = (C_1 \cos 8t + C_2 \sin 8t)\, e^{-8t} \big|_{t=0} = C_1 = -3/4\,,$$
$$y'(0) = 8\left[(C_2 - C_1)\cos 8t - (C_2 + C_1)\sin 8t\right] e^{-8t} \big|_{t=0} = 8\,(C_2 - C_1) = -2$$

$$\Rightarrow \qquad \begin{aligned} C_1 &= -3/4\,, \\ C_2 &= -1\,. \end{aligned}$$

Therefore,

$$y(t) = \left(-\frac{3}{4}\cos 8t - \sin 8t\right) e^{-8t} = \frac{5}{4}\, e^{-8t} \sin(8t + \phi),$$

where $\tan\phi = (-3/4)/(-1) = 3/4$ and $\cos\phi = -1 < 0$. Thus,

$$\phi = \pi + \arctan(3/4) \approx 3.785\,.$$

The damping factor is $(5/4)e^{-8t}$, the quasiperiod is $P = 2\pi/8 = \pi/4$, and the quasifrequency is $1/P = 4/\pi$.

9. Substituting the values $m = 2$, $k = 40$, and $b = 8\sqrt{5}$ into the equation (12) and using the initial conditions, we obtain the following initial value problem.

$$2\frac{d^2y}{dt^2} + 8\sqrt{5}\,\frac{dy}{dt} + 40y = 0, \qquad y(0) = 0.1\,, \quad y'(0) = 2\,.$$

The initial conditions are positive, reflecting the fact that we have chosen *positive downward* direction in our coordinate system. The auxiliary equation for this system is

$$2r^2 + 8\sqrt{5}r + 40 = 0 \qquad \text{or} \qquad r^2 + 4\sqrt{5}r + 20 = 0.$$

This equation has a double root $r = -2\sqrt{5}$. Therefore, this system is critically damped and the equation of motion has the form

$$y(t) = (C_1 + C_2 t)\, e^{-2\sqrt{5}t}\,.$$

To find the constants C_1 and C_2, we use the initial conditions $y(0) = 0.1$ and $y'(0) = 2$. Thus, we have

$$0.1 = y(0) = C_1 \qquad \Rightarrow \qquad C_1 = 0.1\,,$$
$$2 = y'(0) = C_2 - 2\sqrt{5}C_1 \qquad \Rightarrow \qquad C_2 = 2 + 0.2\sqrt{5}\,.$$

Hence, we obtain

$$y(t) = \left[0.1 + \left(2 + 0.2\sqrt{5}\right)t\right] e^{-2\sqrt{5}t}\,.$$

Copyright © 2012 Pearson Education, Inc. Publishing as Addison-Wesley.

The maximum displacement of the mass is found by determining the first time, when the velocity of the mass becomes zero. Therefore, we have

$$y'(t) = 0 = \left(2 + 0.2\sqrt{5}\right) e^{-2\sqrt{5}t} - 2\sqrt{5} \left[0.1 + \left(2 + 0.2\sqrt{5}\right) t\right] e^{-2\sqrt{5}t},$$

which gives

$$t = \frac{2}{2\sqrt{5}(2 + 0.2\sqrt{5})} = \frac{1}{2\sqrt{5} + 1}.$$

Thus, the maximum displacement is

$$y\left(\frac{1}{2\sqrt{5} + 1}\right) = \left[0.1 + \left(2 + 0.2\sqrt{5}\right) \frac{1}{2\sqrt{5} + 1}\right] e^{-2\sqrt{5}/(2\sqrt{5}+1)} \approx 0.242 \,(\text{m}).$$

11. The equation of the motion of this mass-spring system is

$$y'' + 0.2y' + 100y = 0 \quad \text{with} \quad y(0) = 0, \quad y'(0) = 1.$$

Clearly, this is an underdamped motion, because

$$b^2 - 4mk = (0.2)^2 - 4(1)(100) = -399.96 < 0.$$

So, we use use the equation (16) for a general solution. With

$$\alpha = -\frac{b}{2m} = -\frac{0.2}{2} = -0.1 \quad \text{and} \quad \beta = \frac{1}{2m}\sqrt{4mk - b^2} = \frac{1}{2}\sqrt{399.96} = \sqrt{99.99},$$

the solution (16) becomes

$$y(t) = \left(C_1 \cos \sqrt{99.99}\,t + C_2 \sin \sqrt{99.99}\,t\right) e^{-0.1t}.$$

From the initial condiions,

$$\begin{aligned} y(0) &= \left(C_1 \cos \sqrt{99.99}\,t + C_2 \sin \sqrt{99.99}\,t\right) e^{-0.1t}\Big|_{t=0} = C_1 = 0, \\ y'(0) &= \left[\left(\sqrt{99.99}\,C_2 - 0.1C_1\right) \cos \sqrt{99.99}\,t - \left(0.1C_2 + \sqrt{99.99}\,C_1\right) \sin \sqrt{99.99}\,t\right] e^{-0.1t}\Big|_{t=0} \\ &= \sqrt{99.99}\,C_2 - 0.1C_1 = 1 \end{aligned}$$

$$\Rightarrow \quad \begin{aligned} C_1 &= 0, \\ C_2 &= 1/\sqrt{99.99}. \end{aligned}$$

Therefore, the equation of motion is given by

$$y(t) = \frac{1}{\sqrt{99.99}} e^{-0.1t} \sin \sqrt{99.99}\,t.$$

Copyright © 2012 Pearson Education, Inc. Publishing as Addison-Wesley.

The maximum displacement to the right occurs at the first point of local maximum of $y(t)$. The critical points of $y(t)$ are solutions to

$$y'(t) = \frac{e^{-0.1t}}{\sqrt{99.99}} \left(\sqrt{99.99} \cos \sqrt{99.99}t - 0.1 \sin \sqrt{99.99}t \right) = 0$$

$$\Rightarrow \quad \sqrt{99.99} \cos \sqrt{99.99}t - 0.1 \sin \sqrt{99.99}t = 0$$

$$\Rightarrow \quad \tan \sqrt{99.99}t = 10\sqrt{99.99} = \sqrt{9999}.$$

Solving for t, we conclude that the first point of local maximum is at

$$t = (1/\sqrt{99.99}) \arctan \sqrt{9999} \approx 0.156 \, \text{sec}.$$

13. In Example 3, the solution was found to be

$$y(t) = \sqrt{\frac{7}{12}} \, e^{-2t} \sin\left(2\sqrt{3}t + \phi\right), \tag{4.16}$$

where $\phi = \pi + \arctan(\sqrt{3}/2)$. Therefore, we have

$$y'(t) = -\sqrt{\frac{7}{3}} \, e^{-2t} \sin\left(2\sqrt{3}t + \phi\right) + \sqrt{7} \, e^{-2t} \cos\left(2\sqrt{3}t + \phi\right).$$

Thus, to find the relative extrema for $y(t)$, we set

$$y'(t) = -\sqrt{\frac{7}{3}} \, e^{-2t} \sin\left(2\sqrt{3}t + \phi\right) + \sqrt{7} \, e^{-2t} \cos\left(2\sqrt{3}t + \phi\right) = 0$$

$$\Rightarrow \quad \frac{\sin\left(2\sqrt{3}t + \phi\right)}{\cos\left(2\sqrt{3}t + \phi\right)} = \frac{\sqrt{7}}{\sqrt{7/3}} = \sqrt{3}$$

$$\Rightarrow \quad \tan\left(2\sqrt{3}t + \phi\right) = \sqrt{3}.$$

Since $\tan\theta = \sqrt{3}$, when $\theta = (\pi/3) + n\pi$, where n is an integer, we see that the relative extrema will occur at the points t_n, where

$$2\sqrt{3}t_n + \phi = \frac{\pi}{3} + n\pi \quad \Rightarrow \quad t_n = \frac{(\pi/3) + n\pi - \phi}{2\sqrt{3}}.$$

Substituting $\pi + \arctan\left(\sqrt{3}/2\right)$ for ϕ into the last equation and requiring $t > 0$, we obtain

$$t_n = \frac{(\pi/3) + n\pi - \arctan\left(\sqrt{3}/2\right)}{2\sqrt{3}}, \qquad n = 0, 1, 2, \dots .$$

Thus, we see that the solution curve, given by the equation (4.16) will touch the exponential curves $y(t) = \pm\left(\sqrt{7/12}\right) e^{-2t}$ when we have

$$\sqrt{\frac{7}{12}} \, e^{-2t} \sin\left(2\sqrt{3}t + \phi\right) = \pm\sqrt{\frac{7}{12}} \, e^{-2t},$$

Copyright © 2012 Pearson Education, Inc. Publishing as Addison-Wesley.

where $\phi = \pi + \arctan\left(\sqrt{3}/2\right)$. This occurs, when $\sin\left(2\sqrt{3}t + \phi\right) = \pm 1$. Since $\sin\theta = \pm 1$, when $\theta = (\pi/2) + m\pi$ for any integer m, we see that times T_m, when the solution curve touches the exponential curves, satisfy

$$2\sqrt{3}T_m + \phi = \frac{\pi}{2} + m\pi \qquad \Rightarrow \qquad T_m = \frac{(\pi/2) + m\pi - \phi}{2\sqrt{3}},$$

where $\phi = \pi + \arctan\left(\sqrt{3}/2\right)$ and m is an integer. Again, requiring that t is positive, we see that $y(t)$ touches the exponential curve when

$$T_m = \frac{(\pi/2) + m\pi - \arctan\left(\sqrt{3}/2\right)}{2\sqrt{3}}, \qquad m = 0, 1, 2, \ldots .$$

From these facts it follows that, for $y(t)$ to have an extremum and, at the same time, touch the curves $y(t) = \pm\sqrt{7/12}\,e^{-2t}$, there must be integers m and n such that

$$\frac{(\pi/3) + n\pi - \arctan\left(\sqrt{3}/2\right)}{2\sqrt{3}} = \frac{(\pi/2) + m\pi - \arctan\left(\sqrt{3}/2\right)}{2\sqrt{3}}$$

$$\Rightarrow \qquad \frac{\pi}{3} + n\pi = \frac{\pi}{2} + m\pi$$

$$\Rightarrow \qquad n - m = \frac{1}{2} - \frac{1}{3} = \frac{1}{6}.$$

But, since m and n are integers, their difference is an integer and never $1/6$. Thus, the extrema of $y(t)$ do not occur on the exponential curves.

15. Since the exponential function is never zero, from the equation of motion (16) we conclude that the mass passes the equilibrium position, that is, $y(t) = 0$, if and only if

$$\sin(\omega t + \phi) = 0.$$

Therefore, the time between two successive crossings of the equilibrium position is π/ω, which is a half of the quasiperiod P. So, we can find the quasiperiod P by multiplying the time between two successive crossings of the equilibrium position by two. Whenever P is computed, we can measure the displacement $y(t)$ at any moment t (with $y(t) \neq 0$) and then at the moment $t + P$. Taking the quotient

$$\frac{y(t + P)}{y(t)} = \frac{Ae^{-(b/2m)(t+P)}\sin[\omega(t + P) + \phi]}{Ae^{-(b/2m)t}\sin(\omega t + \phi)} = e^{-(b/2m)P},$$

we find that the damping coefficient b is

$$b = -\frac{2m\ln[y(t + P)/y(t)]}{P}.$$

 Copyright © 2012 Pearson Education, Inc. Publishing as Addison-Wesley.

17. According to (20), the equation of a critically damped motion is given by

$$y(t) = (C_1 + C_2 t) \, e^{-(b/2m)t} .$$

Differentiating, we find the velocity of the mass.

$$y'(t) = \left(C_2 - \frac{b}{2m} C_1 - \frac{b}{2m} C_2 t \right) e^{-(b/2m)t} .$$

Given initial conditions, we compute the values of C_1 and C_2.

$$
\begin{aligned}
y_0 = y(0) = C_1 , & & C_1 = y_0 , \\
v_0 = y'(0) = C_2 - \frac{b}{2m} C_1 & \quad \Rightarrow \quad & C_2 = v_0 + \frac{b}{2m} C_1 = \frac{2mv_0 + by_0}{2m} .
\end{aligned}
$$

Therefore,

$$y(t) = \left(y_0 + \frac{2mv_0 + by_0}{2m} t \right) e^{-(b/2m)t} . \tag{4.17}$$

(a) At the equilibrium position, $y(t) = 0$. With $y(t)$, defined in (4.17), solving for t yields a *unique* solution

$$
\left(y_0 + \frac{2mv_0 + by_0}{2m} t \right) e^{-(b/2m)t} = 0 \quad \Leftrightarrow \quad y_0 + \frac{2mv_0 + by_0}{2m} t = 0
$$

$$
\Leftrightarrow \quad t = \frac{-2my_0}{2mv_0 + by_0}
$$

(since the exponential function is never zero). This solution is positive when

$$-\frac{2my_0}{2mv_0 + by_0} > 0 \quad \Leftrightarrow \quad \frac{2my_0}{2mv_0 + by_0} < 0 . \tag{4.18}$$

(b) Taking $m = 1$, $k = 1$, and $b = 2$, we find that $b^2 - 4mk = (2)^2 - 4(1)(1) = 0$. Thus, these values of parameters result a critically damped motion. If we choose $y_0 = 2$ and $v_0 = -3$, then (4.18) will be satisfied. Substituting these data into (4.17) yields

$$y(t) = (2 - t)e^{-t} .$$

The graph of this function is shown in Figure 4–B, page 242.

EXERCISES 4.10: A Closer Look at Forced Mechanical Vibrations

1. The frequency response curve (13), with $m = 4$, $k = 1$, and $b = 2$, becomes

$$M(\gamma) = \frac{1}{\sqrt{(k - m\gamma^2)^2 + b^2\gamma^2}} = \frac{1}{\sqrt{(1 - 4\gamma^2)^2 + 4\gamma^2}} .$$

The graph of this function is shown in Figure B.26 in the answers of the text.

Copyright © 2012 Pearson Education, Inc. Publishing as Addison-Wesley.

Chapter 4

3. The auxiliary equation in this problem is $r^2 + 9 = 0$, which has roots $r = \pm 3i$. Thus, a general solution to the corresponding homogeneous equation has the form

$$y_h(t) = C_1 \cos 3t + C_2 \sin 3t.$$

We look for a particular solution to the original nonhomogeneous equation of the form

$$y_p(t) = t^s(A \cos 3t + B \sin 3t),$$

where we take $s = 1$ because $r = 3i$ is a simple root of the auxiliary equation. Computing the derivatives

$$y'(t) = A \cos 3t + B \sin 3t + t(-3A \sin 3t + 3B \cos 3t),$$
$$y''(t) = 6B \cos 3t - 6A \sin 3t + t(-9A \cos 3t - 9B \sin 3t),$$

and substituting $y(t)$ and $y''(t)$ into the original equation, we get

$$6B \cos 3t - 6A \sin 3t + t(-9A \cos 3t - 9B \sin 3t) + 9t(A \cos 3t + B \sin 3t) = 2 \cos 3t$$

$$\Rightarrow \quad 6B \cos 3t - 6A \sin 3t = 2 \cos 3t \quad \Rightarrow \quad \begin{aligned} A &= 0, \\ B &= 1/3. \end{aligned}$$

So, $y_p(t) = (1/3)t \sin 3t$, and $y(t) = C_1 \cos 3t + C_2 \sin 3t + (1/3)t \sin 3t$ is a general solution. To satisfy the initial conditions, we solve

$$\begin{aligned} y(0) &= = C_1 = 1, \\ y'(0) &= 3C_2 = 0 \end{aligned} \quad \Rightarrow \quad \begin{aligned} C_1 &= 1, \\ C_2 &= 0. \end{aligned}$$

Therefore, the solution to the given initial value problem is

$$y(t) = \cos 3t + \frac{1}{3} t \sin 3t.$$

The graph of $y(t)$ is depicted in Figure B.27 in the answers of the text.

5. (a) The corresponding homogeneous equation, $my'' + ky = 0$, is the equation of a simple harmonic motion, and so its general solution is given by

$$y_h(t) = C_1 \cos \omega t + C_2 \sin \omega t, \qquad \omega = \sqrt{k/m}.$$

Since $\gamma \neq \omega$, we look for a particular solution of the form

$$y_p(t) = A \cos \gamma t + B \sin \gamma t$$

 Copyright © 2012 Pearson Education, Inc. Publishing as Addison-Wesley.

$$\Rightarrow \qquad y_p'(t) = -A\gamma \sin \gamma t + B\gamma \cos \gamma t$$

$$\Rightarrow \qquad y_p''(t) = -A\gamma^2 \cos \gamma t - B\gamma^2 \sin \gamma t.$$

Substitution into the original equation yields

$$m\left(-A\gamma^2 \cos \gamma t - B\gamma^2 \sin \gamma t\right) + k\left(A \cos \gamma t + B \sin \gamma t\right) = F_0 \cos \gamma t$$

$$\Rightarrow \qquad A\left(-m\gamma^2 + k\right) \cos \gamma t + B\left(-m\gamma^2 + k\right) \sin \gamma t = F_0 \cos \gamma t$$

$$\Rightarrow \qquad \begin{aligned} A &= F_0/\left(k - m\gamma^2\right), \\ B &= 0 \end{aligned} \qquad \Rightarrow \qquad y_p(t) = \frac{F_0}{k - m\gamma^2} \cos \gamma t.$$

Therefore, a general solution to the original equation is

$$y(t) = C_1 \cos \omega t + C_2 \sin \omega t + \frac{F_0}{k - m\gamma^2} \cos \gamma t.$$

With the initial conditions, $y(0) = y'(0) = 0$, we get

$$\begin{aligned} y(0) &= C_1 + F_0/\left(k - m\gamma^2\right) = 0, \\ y'(0) &= \omega C_2 = 0 \end{aligned} \qquad \Rightarrow \qquad \begin{aligned} C_1 &= -F_0/\left(k - m\gamma^2\right), \\ C_2 &= 0. \end{aligned}$$

Therefore,

$$y(t) = -\frac{F_0}{k - m\gamma^2} \cos \omega t + \frac{F_0}{k - m\gamma^2} \cos \gamma t,$$

which can also be written in the form

$$y(t) = \frac{F_0}{k - m\gamma^2}\left(\cos \gamma t - \cos \omega t\right) = \frac{F_0}{m(\omega^2 - \gamma^2)}\left(\cos \gamma t - \cos \omega t\right).$$

(b) Here, one can apply the "difference-to-product" identity

$$\cos A - \cos B = 2 \sin\left(\frac{B + A}{2}\right) \sin\left(\frac{B - A}{2}\right)$$

with $A = \gamma t$ and $B = \omega t$ to get

$$y(t) = \frac{2F_0}{m(\omega^2 - \gamma^2)} \sin\left(\frac{\omega + \gamma}{2} t\right) \sin\left(\frac{\omega - \gamma}{2} t\right).$$

(c) For $F_0 = 32$, $m = 2$, $\omega = 9$, and $\gamma = 7$, the solution in part (b) becomes

$$y(t) = \frac{2(32)}{2(9^2 - 7^2)} \sin\left(\frac{9 + 7}{2} t\right) \sin\left(\frac{9 - 7}{2} t\right) = \sin 8t \sin t.$$

The graph of this function is shown in Figure B.28.

Copyright © 2012 Pearson Education, Inc. Publishing as Addison-Wesley.

Chapter 4

7. The auxiliary equation to (1), $mr^2 + br + k = 0$, has roots

$$r = \frac{-b \pm \sqrt{b^2 - 4mk}}{2m},$$

which are both real ($b^2 > 4mk$) and negative, because $\sqrt{b^2 - 4mk} < b$. Let

$$r_1 := \frac{-b - \sqrt{b^2 - 4mk}}{2m},$$

$$r_2 := \frac{-b + \sqrt{b^2 - 4mk}}{2m}.$$

Then a general solution to the homogeneous equation, corresponding to (1), is

$$y_h(t) = c_1 e^{r_1 t} + c_2 e^{r_2 t}.$$

A particular solution to the equation (1) is still given by (7). Thus,

$$y(t) = c_1 e^{r_1 t} + c_2 e^{r_2 t} + \frac{F_0 \sin(\gamma t + \theta)}{\sqrt{(k - m\gamma^2)^2 + b^2 \gamma^2}},$$

where $\tan\theta = (k - m\gamma^2)/(b\gamma)$, is a general solution to the forced overdamped equation.

9. If a mass of $m = 8\,\text{kg}$ stretches the spring by $\ell = 1.96\,\text{m}$, then the spring stiffness must be

$$k = \frac{mg}{\ell} = \frac{8 \cdot 9.8}{1.96} = 40\,(\text{N/m}).$$

Substitution $m = 8$, $b = 3$, $k = 40$, and the external force $F(t) = \cos 2t$ into the equation (23) yields

$$8y'' + 3y' + 40y = \cos 2t.$$

The steady-state (a particular) solution to this equation is given in (6) and (7), that is,

$$
\begin{aligned}
y_p(t) &= \frac{F_0}{(k - m\gamma^2)^2 + b^2\gamma^2} \left\{ (k - m\gamma^2)\cos\gamma t + b\gamma \sin\gamma t \right\} \\
&= \frac{1}{[40 - (8)(2)^2]^2 + (3)^2(2)^2} \left\{ (40 - 8(2)^2)\cos 2t + (3)(2)\sin 2t \right\} \\
&= 0.08\cos 2t + 0.06\sin 2t = 0.1\sin(2t + \theta),
\end{aligned}
$$

where $\theta = \arctan(0.8/0.6) = \arctan(4/3) \approx 0.927$.

11. First, we find the mass

$$m = \frac{8\,\text{lb}}{32\,\text{ft/sec}^2} = \frac{1}{4}\,\text{slug}.$$

    Copyright © 2012 Pearson Education, Inc. Publishing as Addison-Wesley.

Thus the equation (23), describing the motion, with $m = 1/4$, $b = 1$, $k = 10$, and the external force $F(t) = 2\cos 2t$ becomes

$$\frac{1}{4}y'' + y' + 10y = 2\cos 2t, \tag{4.19}$$

with the initial conditions are $y(0) = y'(0) = 0$. A general solution to the corresponding homogeneous equation is given in Section 4.9, formula (16). That is,

$$y_h(t) = e^{\alpha t}\left(C_1 \cos\beta t + C_2 \sin\beta t\right).$$

We compute

$$\alpha = -\frac{b}{2m} = -\frac{1}{2(1/4)} = -2 \quad \text{and} \quad \beta = \frac{1}{2(1/4)}\sqrt{4(1/4)(10) - 1^2} = 6.$$

So, $y_h(t) = e^{-2t}\left(C_1 \cos 6t + C_2 \sin 6t\right)$. For a particular solution, we use formula (7).

$$
\begin{aligned}
y_p(t) &= \frac{F_0}{\sqrt{(k - m\gamma^2)^2 + b^2\gamma^2}}\sin(\gamma t + \theta) \\
&= \frac{2}{\sqrt{[10 - (1/4)(2)^2]^2 + (1)^2(2)^2}}\sin(2t + \theta) = \frac{2}{\sqrt{85}}\sin(2t + \theta),
\end{aligned}
$$

where $\theta = \arctan[(k - m\gamma^2)/(b\gamma)] = \arctan(9/2) \approx 1.352$. A general solution to (4.19) is then given by

$$y(t) = e^{-2t}\left(C_1 \cos 6t + C_2 \sin 6t\right) + \frac{2}{\sqrt{85}}\sin(2t + \theta).$$

From the initial conditions, we find

$$
\begin{aligned}
y(0) &= C_1 + (2/\sqrt{85})\sin\theta = 0, \\
y'(0) &= -2C_1 + 6C_2 + (4/\sqrt{85})\cos\theta = 0
\end{aligned}
$$

$$\Rightarrow \quad
\begin{aligned}
C_1 &= -(2/\sqrt{85})\sin\theta = -18/85, \\
C_2 &= \left[C_1 - (2/\sqrt{85})\cos\theta\right]/3 = -22/255.
\end{aligned}
$$

$$\Rightarrow \quad y(t) = e^{-2t}\left(-\frac{18}{85}\cos 6t - \frac{22}{255}\sin 6t\right) + \frac{2}{\sqrt{85}}\sin(2t + \theta).$$

The resonance frequency for the system is

$$\frac{\gamma_r}{2\pi} = \frac{\sqrt{(k/m) - (b^2)/(2m^2)}}{2\pi} = \frac{\sqrt{40 - 8}}{2\pi} = \frac{2\sqrt{2}}{\pi} \quad (\text{cycles/sec}),$$

where we have used the formula (15) for γ_r.

Copyright © 2012 Pearson Education, Inc. Publishing as Addison-Wesley.

13. The mass attached to the spring is

$$m = \frac{32 \text{ lb}}{32 \text{ ft/sec}^2} = 1 \text{ slug}.$$

Thus, the equation governing the motion, $my'' + by' + ky = F_{\text{ext}}$, with $m = 1$, $b = 2$, $k = 5$, and $F_{\text{ext}}(t) = 3\cos 4t$ becomes

$$y'' + 2y' + 5y = 3\cos 4t.$$

This is an underdamped motion, because $b^2 - 4mk = (2)^2 - 4(1)(5) = -16 < 0$. For the steady-state solution to this equation, we use the formula (6). Since $F_{\text{ext}}(t) = 3\cos 4t$, we have $F_0 = 3$, and $\gamma = 4$. Substituting m, b, k, F_0, and γ into (6), we obtain

$$
\begin{aligned}
y_p(t) &= \frac{3}{[5 - (1)(4)^2]^2 + (2)^2(4)^2} \left\{ [5 - (1)(4)^2]\cos 4t + (2)(4)\sin 4t \right\} \\
&= \frac{3}{185}(8\sin 4t - 11\cos 4t).
\end{aligned}
$$

15. In the equation, governing this motion, $my'' + by' + ky = F_{\text{ext}}$, we have $m = 8$, $b = 3$, $k = 40$, and $F_{\text{ext}}(t) = 2\sin 2t \cos 2t = \sin 4t$. Thus, the equation becomes

$$8y'' + 3y' + 40y = \sin 4t.$$

Clearly, this is an underdamped motion, and the steady-state solution has the form

$$
\begin{aligned}
y_p(t) &= A\sin 4t + B\cos 4t \\
\Rightarrow \quad y_p'(t) &= 4A\cos 4t - 4B\sin 4t \\
\Rightarrow \quad y_p''(t) &= -16A\sin 4t - 16B\cos 4t.
\end{aligned}
$$

Substituting these formulas into the equation and collecting similar terms yields

$$(-88A - 12B)\sin 4t + (12A - 88B)\cos 4t = \sin 4t \quad \Rightarrow \quad \begin{aligned} -88A - 12B &= 1, \\ 12A - 88B &= 0. \end{aligned}$$

Solving this system, we get $A = -11/986$, $B = -3/1972$. Thus, the steady-state solution of the given motion is

$$y(t) = -\frac{11}{986}\sin 4t - \frac{3}{1972}\cos 4t.$$

This function has the amplitude

$$\sqrt{A^2 + B^2} = \sqrt{\left(-\frac{11}{986}\right)^2 + \left(-\frac{3}{1972}\right)^2} \approx 0.01 \text{ (m)}$$

and the frequency $4/(2\pi) = 2/\pi$ (sec^{-1}).

 Copyright © 2012 Pearson Education, Inc. Publishing as Addison-Wesley.

REVIEW PROBLEMS

1. Solving the auxiliary equation, $r^2 + 8r - 9 = 0$, we find $r_1 = -9$, $r_2 = 1$. Thus, a general solution is given by

$$y(t) = c_1 e^{r_1 t} + c_2 e^{r_2 t} = c_1 e^{-9t} + c_2 e^t.$$

3. The auxiliary equation, $4r^2 - 4r + 10 = 0$, has roots $r_{1,2} = (1 \pm 3i)/2$. Therefore, a general solution is

$$y(t) = \left[c_1 \cos\left(\frac{3t}{2}\right) + c_2 \sin\left(\frac{3t}{2}\right) \right] e^{t/2}.$$

5. The roots of the auxiliary equation, $6r^2 - 11r + 3 = 0$, are $r_1 = 3/2$ and $r_2 = 1/3$. Thus,

$$y(t) = c_1 e^{r_1 t} + c_2 e^{r_2 t} = c_1 e^{3t/2} + c_2 e^{t/3}$$

is a general solution.

7. Solving the auxiliary equation, $36r^2 + 24r + 5 = 0$, we find that

$$r = \frac{-24 \pm \sqrt{24^2 - 4(36)(5)}}{2(36)} = -\frac{1}{3} \pm \frac{1}{6} i.$$

Thus, a general solution is given by

$$y(t) = \left[c_1 \cos\left(\frac{t}{6}\right) + c_2 \sin\left(\frac{t}{6}\right) \right] e^{-t/3}.$$

9. The auxiliary equation, $16r^2 - 56r + 49 = (4r - 7)^2 = 0$, has a double root $r = 7/4$. Therefore, $e^{7t/4}$ and $te^{7t/4}$ are two linearly independent solutions, and a general solution is given by

$$y(t) = c_1 e^{7t/4} + c_2 t e^{7t/4} = (c_1 + c_2 t) e^{7t/4}.$$

11. This equation is a Cauchy-Euler equation. Using the characteristic equation (7), Section 4.7, we obtain

$$r^2 - r + 5 = 0 \qquad \Rightarrow \qquad r = \frac{1 \pm \sqrt{1^2 - 4(1)(5)}}{2} = \frac{1 \pm \sqrt{19}i}{2}.$$

Thus,

$$y(t) = t^{1/2} \left[c_1 \cos\left(\frac{\sqrt{19}}{2} \ln t\right) + c_2 \sin\left(\frac{\sqrt{19}}{2} \ln t\right) \right].$$

Copyright © 2012 Pearson Education, Inc. Publishing as Addison-Wesley.

13. The roots of the auxiliary equation, $r^2 + 16 = 0$, are $r = \pm 4i$. Thus a general solution to the corresponding homogeneous equation is given by

$$y_h(t) = c_1 \cos 4t + c_2 \sin 4t.$$

The method of undetermined coefficients suggests the form $y_p(t) = (A_1 t + A_0)e^t$ for a particular solution to the original equation. We compute

$$y_p'(t) = (A_1 t + A_0 + A_1)e^t, \qquad y_p''(t) = (A_1 t + A_0 + 2A_1)e^t,$$

and substitute $y_p''(t)$ and $y_p(t)$ into the given equation. This yields

$$y_p'' + 16y_p = \left[(A_1 t + A_0 + 2A_1)e^t\right] + 16\left[(A_1 t + A_0)e^t\right] = te^t$$
$$\Rightarrow \quad (17A_1 t + 17A_0 + 2A_1)\,e^t = te^t \quad \Rightarrow \quad A_1 = \frac{1}{17}, \ A_0 = -\frac{2}{289}.$$

Therefore,

$$y_p(t) = \left(\frac{t}{17} - \frac{2}{289}\right)e^t$$
$$\Rightarrow \quad y(t) = y_h(t) + y_p(t) = c_1 \cos 4t + c_2 \sin 4t + \left(\frac{t}{17} - \frac{2}{289}\right)e^t.$$

15. This is a third order homogeneous linear differential equation with constant coefficients. Its auxiliary equation is $3r^3 + 10r^2 + 9r + 2 = 0$. Factoring yields

$$3r^3 + 10r^2 + 9r + 2 = (3r^3 + 3r^2) + (7r^2 + 7r) + (2r + 2) = (3r^2 + 7r + 2)(r + 1).$$

Thus, the roots of the auxiliary equation are

$$r = -1 \quad \text{and} \quad r = \frac{-7 \pm \sqrt{7^2 - 4(3)(2)}}{6} = -2, \ -\frac{1}{3},$$

and a general solution is given by

$$y(t) = c_1 e^{-2t} + c_2 e^{-t} + c_3 e^{-t/3}.$$

17. To solve the auxiliary equation, $r^3 + 10r - 11 = 0$, we note that $r_1 = 1$ is a root. Dividing the polynomial $r^3 + 10r - 11$ by $r - 1$, we get

$$r^3 + 10r - 11 = (r - 1)(r^2 + r + 11),$$

 Copyright © 2012 Pearson Education, Inc. Publishing as Addison-Wesley.

and so the other two roots are

$$r_{2,3} = \frac{-1 \pm \sqrt{1 - 4(1)(11)}}{2} = \frac{-1}{2} \pm \frac{\sqrt{43}}{2} i.$$

A general solution is then given by

$$y(t) = c_1 e^t + e^{-t/2} \left[c_2 \cos\left(\frac{\sqrt{43}t}{2}\right) + c_3 \sin\left(\frac{\sqrt{43}t}{2}\right) \right].$$

19. By inspection, we find that the auxiliary equation, $4r^3 + 8r^2 - 11r + 3 = 0$, has a root $r = -3$. Factoring yields

$$4r^3 + 8r^2 - 11r + 3 = (r + 3)(4r^2 - 4r + 1) = (r + 3)(2r - 1)^2.$$

Thus, in addition, $r = 1/2$ is a double root of the auxiliary equation. A general solution then has the form

$$y(t) = c_1 e^{-3t} + c_2 e^{t/2} + c_3 t e^{t/2}.$$

21. First, we solve the corresponding homogeneous equation,

$$y'' - 3y' + 7y = 0.$$

Since the roots of the auxiliary equation, $r^2 - 3r + 7 = 0$, are

$$r = \frac{3 \pm \sqrt{9 - 28}}{2} = \frac{3 \pm \sqrt{19}i}{2},$$

a general solution to the homogeneous equation is

$$y_h(t) = \left[c_1 \cos\left(\frac{\sqrt{19}t}{2}\right) + c_2 \sin\left(\frac{\sqrt{19}t}{2}\right) \right] e^{3t/2}.$$

We now use the superposition principle to find a particular solution to the nonhomogeneous equation.

A particular solution $y_{p,1}(t)$ to $y'' - 3y' + 7y = 7t^2$ has the form

$$y_{p,1}(t) = A_2 t^2 + A_1 t + A_0.$$

Substituting yields

$$y_{p,1}'' - 3y_{p,1}' + 7y_{p,1} = 2A_2 - 3(2A_2 t + A_1) + 7(A_2 t^2 + A_1 t + A_0) = 7t^2$$

Copyright © 2012 Pearson Education, Inc. Publishing as Addison-Wesley.

$$\Rightarrow \qquad (7A_2)t^2 + (7A_1 - 6A_2)t + (7A_0 - 3A_1 + 2A_2) = 7t^2$$

$$7A_2 = 7, \qquad\qquad A_2 = 1,$$
$$\Rightarrow \qquad 7A_1 - 6A_2 = 0, \qquad \Rightarrow \qquad A_1 = 6/7,$$
$$7A_0 - 3A_1 + 2A_2 = 0 \qquad\qquad A_0 = 4/49,$$

and so

$$y_{p,1}(t) = t^2 + \frac{6}{7}t + \frac{4}{49}.$$

The other nonhomogeneous term in the right-hand side of the original equation is e^t. So, a particular solution to $y'' - 3y' + 7y = e^t$ has the form $y_{p,2}(t) = Ae^t$. Substituting, we obtain

$$y_{p,2}'' - 3y_{p,2}' + 7y_{p,2} = 5Ae^t = e^t \qquad \Rightarrow \qquad A = \frac{1}{5} \qquad \Rightarrow \qquad y_{p,2}(t) = \frac{1}{5}e^t.$$

By the superposition principle, a general solution to the original equation is

$$\begin{aligned} y(t) &= y_h(t) - y_{p,2}(t) + y_{p,1}(t) \\ &= \left[c_1 \cos\left(\frac{\sqrt{19}t}{2}\right) + c_2 \sin\left(\frac{\sqrt{19}t}{2}\right) \right] e^{3t/2} - \frac{e^t}{5} + t^2 + \frac{6t}{7} + \frac{4}{49}. \end{aligned}$$

23. The corresponding homogeneous equation in this problem is similar to that in Problem 13. Thus, $y_1(t) = \cos 4\theta$ and $y_2(t) = \sin 4\theta$ are its two linearly independent solutions, and a general solution is given by

$$y_h(\theta) = c_1 \cos 4\theta + c_2 \sin 4\theta.$$

For a particular solution to the original nonhomogeneous equation, we use the variation of parameters method. Letting

$$y_p(\theta) = v_1(\theta)\cos 4\theta + v_2(\theta)\sin 4\theta,$$

we get the following system for v_1' and v_2' (see the system (9) in Section 4.6).

$$v_1'(\theta)\cos 4\theta + v_2'(\theta)\sin 4\theta = 0$$
$$-4v_1'(\theta)\sin 4\theta + 4v_2'(\theta)\cos 4\theta = \tan 4\theta.$$

Multiplying the first equation by $\sin 4\theta$ and the second equation by $(1/4)\cos 4\theta$, and adding the results together, we get

$$v_2'(\theta) = \frac{1}{4}\sin 4\theta \qquad \Rightarrow \qquad v_2 = -\frac{1}{16}\cos 4\theta.$$

 Copyright © 2012 Pearson Education, Inc. Publishing as Addison-Wesley.

From the first equation in the above system we also obtain

$$v_1'(\theta) = -v_2'(\theta)\tan 4\theta = -\frac{1}{4}\frac{\sin^2 4\theta}{\cos 4\theta} = -\frac{1}{4}(\sec 4\theta - \cos 4\theta)$$

$$\Rightarrow \quad v_1(\theta) = -\frac{1}{4}\int(\sec 4\theta - \cos 4\theta)\,d\theta = -\frac{1}{16}\ln|\sec 4\theta + \tan 4\theta| + \frac{1}{16}\sin 4\theta.$$

Thus,

$$\begin{aligned}
y_p(\theta) &= \left(-\frac{1}{16}\ln|\sec 4\theta + \tan 4\theta| + \frac{1}{16}\sin 4\theta\right)\cos 4\theta + \left(-\frac{1}{16}\cos\theta\right)\sin 4\theta \\
&= -\frac{1}{16}(\cos 4\theta)\ln|\sec 4\theta + \tan 4\theta|,
\end{aligned}$$

and a general solution to the original equation is

$$y(\theta) = c_1\cos 4\theta + c_2\sin 4\theta - \frac{1}{16}(\cos 4\theta)\ln|\sec 4\theta + \tan 4\theta|.$$

25. Since the auxiliary equation, $4r^2 - 12r + 9 = (2r-3)^2 = 0$, has a double root $r = 3/2$, a general solution to the corresponding homogeneous equation is

$$y_h(t) = c_1 e^{3t/2} + c_2 t e^{3t/2}.$$

By the superposition principle, a particular solution to the original equation has the form

$$y_p(t) = Ae^{3t} + Be^{5t}.$$

Substituting this expression into the given nonhomogeneous equation, we get

$$\begin{aligned}
4y_p'' - 12y_p' + 9y_p &= 4\left(9Ae^{3t} + 25Be^{5t}\right) - 12\left(3Ae^{3t} + 5Be^{5t}\right) + 9\left(Ae^{3t} + Be^{5t}\right) \\
&= 9Ae^{3t} + 49Be^{5t} = e^{5t} + e^{3t} \quad \Rightarrow \quad A = 1/9,\ B = 1/49.
\end{aligned}$$

Therefore, $y_p(t) = (1/9)e^{3t} + (1/49)e^{5t}$, and a general solution to the original equation is

$$y(t) = c_1 e^{3t/2} + c_2 t e^{3t/2} + \frac{1}{9}e^{3t} + \frac{1}{49}e^{5t}.$$

27. This is a Cauchy-Euler equation. Thus we make the substitution $x = e^t$ and get

$$x^2\frac{d^2y}{dx^2} + 2x\frac{dy}{dx} - 2y = 6x^{-2} + 3x$$

$$\Rightarrow \quad \left(\frac{d^2y}{dt^2} - \frac{dy}{dt}\right) + 2\frac{dy}{dt} - 2y = 6(e^t)^{-2} + 3(e^t)$$

Copyright © 2012 Pearson Education, Inc. Publishing as Addison-Wesley.

$$\Rightarrow \qquad \frac{d^2y}{dt^2} + \frac{dy}{dt} - 2y = 6e^{-2t} + 3e^t. \qquad (4.20)$$

The auxiliary equation, $r^2 + r - 2 = 0$, has the roots $r = -2, 1$. Therefore, a general solution to the corresponding homogeneous equation is

$$y_h(t) = c_1 e^t + c_2 e^{-2t}.$$

A particular solution to (4.20) has the form

$$y_p(t) = Ate^{-2t} + Bte^t.$$

(The factor t appeared in both terms because e^t and e^{-2t} are both solutions to the corresponding homogeneous equation.) Differentiating, we find

$$y_p(t) = Ate^{-2t} + Bte^t$$
$$\Rightarrow \qquad y_p'(t) = A(1 - 2t)e^{-2t} + B(t + 1)e^t$$
$$\Rightarrow \qquad y_p'(t) = A(4t - 4)e^{-2t} + B(t + 2)e^t.$$

Substituting into (4.20) yields

$$-3Ae^{-2t} + 3Be^t = 6e^{-2t} + 3e^t \qquad \Rightarrow \qquad A = -2, \; B = 1.$$

Thus, a general solution to (4.20) is given by

$$y(t) = y_h(t) + y_p(t) = c_1 e^t + c_2 e^{-2t} - 2te^{-2t} + te^t.$$

The back substitution, $t = \ln x$, results

$$y(x) = c_1 x + c_2 x^{-2} - 2x^{-2} \ln x + x \ln x.$$

29. The roots of the auxiliary equation in this problem are

$$r = \frac{-4 \pm \sqrt{4^2 - 4(1)(7)}}{2} = -2 \pm \sqrt{3}i.$$

Therefore, a general solution is given by

$$y(t) = \left(c_1 \cos \sqrt{3}t + c_2 \sin \sqrt{3}t\right) e^{-2t}.$$

Substituting the initial conditions, we obtain

$$y(0) = \left(c_1 \cos \sqrt{3}t + c_2 \sin \sqrt{3}t\right) e^{-2t} \Big|_{t=0} = c_1 = 1,$$
$$y'(0) = \left[(-2c_1 + \sqrt{3}c_2) \cos \sqrt{3}t - (\sqrt{3}c_1 + 2c_2) \sin \sqrt{3}t\right] e^{-2t} \Big|_{t=0} = -2c_1 + \sqrt{3}c_2 = -2.$$

Solving yields $c_1 = 1$, $c_2 = 0$. Hence, the solution to the given initial value problem is

$$y(t) = e^{-2t} \cos\left(\sqrt{3}t\right).$$

 Copyright © 2012 Pearson Education, Inc. Publishing as Addison-Wesley.

31. First, we solve the corresponding homogeneous equation. Its auxiliary equation, which is $r^2 - 2r + 10 = 0$, has the roots $r = 1 \pm 3i$. Thus,

$$y_h(t) = (c_1 \cos 3t + c_2 \sin 3t)\, e^t$$

is a general solution.

Now, we apply the method of undetermined coefficients and look for a particular solution to the original nonhomogeneous equation of the form $y_p(t) = A\cos 3t + B\sin 3t$. Differentiating $y_p(t)$ twice, we obtain $y_p'(t) = -3A\sin 3t + 3B\cos 3t$, $y_p'' = -9A\cos 3t - 9B\sin 3t$ and substitute these expressions into the original equation. We get

$$(-9A\cos 3t - 9B\sin 3t) - 2(-3A\sin 3t + 3B\cos 3t) + 10(A\cos 3t + B\sin 3t)$$
$$= 6\cos 3t - \sin 3t$$

$$\Rightarrow \quad (A - 6B)\cos 3t + (6A + B)\sin 3t = 6\cos 3t - \sin 3t$$

$$\Rightarrow \quad \begin{aligned} A - 6B &= 6, \\ 6A + B &= -1 \end{aligned} \quad \Rightarrow \quad \begin{aligned} A &= 0, \\ B &= -1. \end{aligned}$$

So, $y_p(t) = -\sin 3t$, and $y(t) = (c_1\cos 3t + c_2\sin 3t)\, e^t - \sin 3t$ is a general solution to the given equation.

Next, we satisfy the initial conditions.

$$\begin{aligned} y(0) &= c_1 = 2, \\ y'(0) &= c_1 + 3c_2 - 3 = -8 \end{aligned} \quad \Rightarrow \quad \begin{aligned} c_1 &= 2, \\ c_2 &= -7/3. \end{aligned}$$

Hence, the answer is

$$y(t) = \left(2\cos 3t - \frac{7}{3}\sin 3t\right) e^t - \sin 3t.$$

33. The associated characteristic equation in this problem is $r^3 - 12r^2 + 27r + 40 = 0$, which is a third order equation. Using the rational root theorem, we look for its integer roots among the divisors of 40, which are ± 1, ± 2, ± 4, ± 8, ± 10, ± 20, and ± 40. By inspection, $r = -1$ is a root. Dividing $r^3 - 12r^2 + 27r + 40$ by $r + 1$, we get

$$r^3 - 12r^2 + 27r + 40 = (r^2 - 13r + 40)(r + 1),$$

and so the other two roots of the auxiliary equation are the roots of $r^2 - 13r + 40 = 0$, which are $r = 5, 8$. Therefore, a general solution to the given equation is

$$y(t) = c_1 e^{-t} + c_2 e^{5t} + c_3 e^{8t}.$$

Copyright © 2012 Pearson Education, Inc. Publishing as Addison-Wesley.

We find the values of c_1, c_2, and c_3 from the initial conditions.

$$y(0) = (c_1 e^{-t} + c_2 e^{5t} + c_3 e^{8t})\big|_{t=0} = c_1 + c_2 + c_3 = -3\,,$$
$$y'(0) = (-c_1 e^{-t} + 5c_2 e^{5t} + 8c_3 e^{8t})\big|_{t=0} = -c_1 + 5c_2 + 8c_3 = -6\,,$$
$$y''(0) = (c_1 e^{-t} + 25c_2 e^{5t} + 64c_3 e^{8t})\big|_{t=0} = c_1 + 25c_2 + 64c_3 = -12\,.$$

Solving this system yields $c_1 = -1$, $c_2 = -3$, and $c_3 = 1$. Therefore, the function $y(t) = -e^{-t} - 3e^{5t} + e^{8t}$ is the solution to the given initial value problem.

35. Since the roots of the corresponding auxiliary equation, $r^2 + 1 = 0$, are $r = \pm i$, the functions $y_1(\theta) = \cos\theta$ and $y_2(\theta) = \sin\theta$ are two linearly independent solutions to the associated homogeneous equation, and so its general solution is given by

$$y_h(\theta) = c_1 \cos\theta + c_2 \sin\theta\,.$$

We apply the method of variation of parameters to find a particular solution to the original equation. We look for a particular solution of the form

$$y_p(\theta) = v_1(\theta)\cos\theta + v_2(\theta)\sin\theta,$$

where $v_1(\theta)$ and $v_2(\theta)$ satisfy the system (9), Section 4.6. That is,

$$v_1' \cos\theta + v_2' \sin\theta = 0\,,$$
$$-v_1' \sin\theta + v_2' \cos\theta = \sec\theta\,.$$

Multiplying the first equation by $\sin\theta$, the second equation by $\cos\theta$, and adding them together yields

$$v_2' \sin^2\theta + v_2' \cos^2\theta = \sec\theta\cos\theta \quad\Rightarrow\quad v_2' = 1 \quad\Rightarrow\quad v_2(\theta) = \theta.$$

From the first equation in the above system, we also get

$$v_1' = -v_2' \tan\theta = -\tan\theta \quad\Rightarrow\quad v_1(\theta) = -\int \tan\theta\, d\theta = \ln|\cos\theta|,$$

Therefore,

$$y_p(\theta) = (\cos\theta)\ln|\cos\theta| + \theta\sin\theta\,,$$

and so

$$y(\theta) = c_1 \cos\theta + c_2 \sin\theta + (\cos\theta)\ln|\cos\theta| + \theta\sin\theta$$

Copyright © 2012 Pearson Education, Inc. Publishing as Addison-Wesley.

is a general solution to the original equation. Differentiating, we find that

$$y'(\theta) = -c_1 \sin\theta + c_2 \cos\theta - \sin\theta \ln|\cos\theta| + \theta\cos\theta.$$

Substituting of $y(\theta)$ and $y'(\theta)$ into the initial conditions yields

$$
\begin{aligned}
y(0) &= c_1 = 1, \\
y'(0) &= c_2 = 2
\end{aligned}
\qquad \Rightarrow \qquad
\begin{aligned}
c_1 &= 1, \\
c_2 &= 2.
\end{aligned}
$$

Hence, the answer is

$$y(\theta) = \cos\theta + 2\sin\theta + (\cos\theta)\ln|\cos\theta| + \theta\sin\theta.$$

37. Comparing the given homogeneous equations with mass-spring oscillator equation (13) in Section 4.8,

$$[\text{inertia}]\, y'' + [\text{damping}]\, y' + [\text{stiffness}]\, y = 0,$$

we see that in equations (a) through (d) the damping coefficient is 0. So, the behavior of solutions, as $t \to +\infty$, depends on the sign of the stiffness coefficient "k".

(a) "k" $= t^4 > 0$. This implies that all the solutions remain bounded.

(b) "k" $= -t^4 < 0$. The stiffness of the system is negative and increases unboundedly as $t \to +\infty$. It reinforces the displacement, $y(t)$, with magnitude increasing with time. Hence some solutions grow rapidly with time.

(c) "k" $= y^6 > 0$. Similarly to (a), we conclude that all the solutions are bounded.

(d) "k" $= y^7$. The function $f(y) = y^7$ is positive for positive y and negative if y is negative. Hence, we can expect that some of the solutions (say, ones satisfying negative initial conditions) are unbounded.

(e) "k" $= 3 + \sin t$. Since $|\sin t| \le 1$ for any t, we conclude that

$$\text{"k"} \ge 3 + (-1) = 2 > 0,$$

and all the solutions are bounded.

(f) Here there is positive damping "b" $= t^2$ increasing with time, which results an increasing drain of energy from the system, and positive constant stiffness $k = 1$. Thus, all the solutions are bounded.

(g) Negative damping "b" $= -t^2$ increases (in absolute value) with time, which imparts energy to the system instead of draining it. Note that the stiffness $k = -1$ is also negative. Thus, we should expect that some of the solutions increase unboundedly as $t \to +\infty$.

39. If a weight of $w = 32\,\mathrm{lb}$ stretches the spring by $\ell = 6\,\mathrm{in} = 0.5\,\mathrm{ft}$, the spring stiffness must be

$$k = \frac{w}{\ell} = \frac{32}{0.5} = 64 \ (\mathrm{lb/ft}).$$

Also, the mass m of the weight is

$$m = w/g = 32/32 = 1 \ (\mathrm{slug}),$$

and the damping constant $b = 2\,\mathrm{lb\text{-}sec/ft}$. The external force is given to be

$$F(t) = F_0 \cos \gamma t$$

with $F_0 = 4$ and $\gamma = 8$.

Clearly, we have an underdamped motion because

$$b^2 - 4mk = 4 - 256 < 0.$$

So, we can use formula (6) in Section 4.10 to find the steady-state solution. This formula yields

$$
\begin{aligned}
y_p(t) &= \frac{F_0}{(k - m\gamma^2)^2 + b^2\gamma^2} \left\{ (k - m\gamma^2) \cos \gamma t + b\gamma \sin \gamma t \right\} \\
&= \frac{4}{(64 - 8^2)^2 + 2^2 8^2} \left\{ (64 - 8^2) \cos 8t + (2)(8) \sin 8t \right\} = \frac{1}{4} \sin 8t.
\end{aligned}
$$

The resonant frequency for the system is $\gamma_r/(2\pi)$, where γ_r is given in (15), Section 4.10. Applying this formula, we get

$$\text{resonant frequency} = \frac{1}{2\pi} \sqrt{\frac{k}{m} - \frac{b^2}{2m^2}} = \frac{1}{2\pi} \sqrt{\frac{64}{1} - \frac{2^2}{2(1)^2}} = \frac{\sqrt{62}}{2\pi}.$$

Copyright © 2012 Pearson Education, Inc. Publishing as Addison-Wesley.

TABLES

Table 4–A: Successive approximations for $y(2)$ in Problem 19 using Simpson's rule.

Intervals	$y(2) \approx$
6	-1.9275
8	-1.9275
10	-1.9275

FIGURES

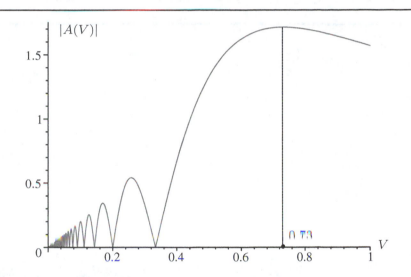

Figure 4–A: The graph of the function $|A(V)|$.

Copyright © 2012 Pearson Education, Inc. Publishing as Addison-Wesley.

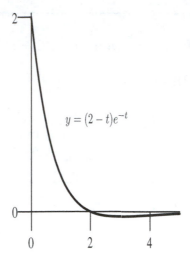

Figure 4–B: The graph of a critically damped motion in Problem 17(b).

 Copyright © 2012 Pearson Education, Inc. Publishing as Addison-Wesley.

CHAPTER 5: Introduction to Systems and Phase Plane Analysis

EXERCISES 5.2:   Elimination Method for Systems with Constant Coefficients

1. Since $D = d/dt$ and $D^2 = d^2/dt^2$, we have

(a) $A[y] = (D-1)[y] = \dfrac{dy}{dt} - y = \left(3t^2\right) - \left(t^3 - 8\right) = -t^3 + 3t^2 + 8;$

(b) $B[A[y]] = (D+2)[-t^3 + 3t^2 + 8] = \left(-3t^2 + 6t\right) + 2\left(-t^3 + 3t^2 + 8\right)$

$\qquad = -2t^3 + 3t^2 + 6t + 16;$

(c) $B[y] = (D+2)[y] = \dfrac{dy}{dt} + 2y = \left(3t^2\right) + 2\left(t^3 - 8\right) = 2t^3 + 3t^2 - 16;$

(d) $A[B[y]] = (D-1)[2t^3 + 3t^2 - 16] = \left(6t^2 + 6t\right) - \left(2t^3 + 3t^2 - 16\right)$

$\qquad = -2t^3 + 3t^2 + 6t + 16;$

(e) $C[y] = (D^2 + D - 2)[y] = \dfrac{d^2y}{dt^2} + \dfrac{dy}{dt} - 2y = (6t) + \left(3t^2\right) - 2\left(t^3 - 8\right)$

$\qquad = -2t^3 + 3t^2 + 6t + 16.$

3. We eliminate x by subtracting the second equation from the first one. This yields

$$y' + 2y = 0 \quad \Rightarrow \quad \frac{dy}{y} = -2dt \quad \Rightarrow \quad \ln|y| = -2t + c \quad \Rightarrow \quad y(t) = c_2 e^{-2t}.$$

From the second equation, we get

$$x' - y' = 0 \quad \Rightarrow \quad (x - y)' = 0 \quad \Rightarrow \quad x(t) - y(t) = c_1 \quad \Rightarrow \quad x(t) = c_1 + c_2 e^{-2t},$$

and a general solution is given by

$$x(t) = c_1 + c_2 e^{-2t}, \qquad y(t) = c_2 e^{-2t}.$$

5. Writing this system in operator notation yields

$$(D-1)[x] + D[y] = 5,$$
$$D[x] + (D+1)[y] = 1. \tag{5.1}$$

Copyright © 2012 Pearson Education, Inc. Publishing as Addison-Wesley.

We first eliminate the function $x(t)$, although we also could proceed by eliminating the function $y(t)$. Thus, we apply the operator D to the first equation and the operator $-(D-1)$ to the second equation to obtain

$$D(D-1)[x] + D^2[y] = D[5] = 0,$$
$$-(D-1)D[x] - (D-1)(D+1)[y] = -(D-1)[1] = 1.$$

Adding these two equations yields

$$\{D(D-1) - (D-1)D\}[x] + \{D^2 - (D^2 - 1)\}[y] = 1$$
$$\Rightarrow \quad 0 \cdot x + 1 \cdot y = 1 \quad \Rightarrow \quad y(t) = 1.$$

To find the function $x(t)$, we eliminate y from the system (5.1). Thus, we multiply the first equation in (5.1) by $(D+1)$ and the second one – by $(-D)$ to get

$$(D+1)(D-1)[x] + (D+1)D[y] = (D+1)[5] = 5,$$
$$-D^2[x] - D(D+1)[y] = D[1] = 0.$$

Adding these two equations, we obtain

$$\{(D^2 - 1) - D^2\}[x] = 5 \quad \Rightarrow \quad -x = 5 \quad \Rightarrow \quad x(t) = -5.$$

Therefore, $x(t) \equiv -5$, $y(t) \equiv 1$ is the solution to the given system of equations.

7. To eliminate u, we multiply the first equation by $(D-1)$, the second one – by $(D+1)$, and subtract the results.

$$(D-1)\{(D+1)[u] - (D+1)[v]\} = (D-1)[e^t] = (e^t)' - e^t = 0,$$
$$(D+1)\{(D-1)[u] + (2D+1)[v]\} = (D+1)[5] = (5)' + 5 = 5$$
$$\Rightarrow \quad \begin{aligned} (D^2 - 1)[u] - (D^2 - 1)[v] &= 0, \\ (D^2 - 1)[u] + \{(D+1)(2D+1)\}[v] &= 5 \end{aligned}$$
$$\Rightarrow \quad \{(D+1)(2D+1) + (D^2 - 1)\}[v] = 5 \quad \Rightarrow \quad \{D(D+1)\}[v] = \frac{5}{3}. \quad (5.2)$$

The corresponding homogeneous equation, $\{D(D+1)\}[v] = 0$, has the characteristic equation

$$r(r+1) = 0 \quad \Rightarrow \quad r = 0, -1,$$

and so its general solution is

$$v_h(t) = c_1 + c_2 e^{-t}.$$

 Copyright © 2012 Pearson Education, Inc. Publishing as Addison-Wesley.

Applying the method of undetermined coefficients, we look for a particular solution to (5.2) of the form $v_p(t) = ct^s$, where we choose $s = 1$ (since the homogeneous equation has constant solutions and does not have solutions of the form $v(t) = ct$). Substituting $v = ct$ into (5.2) yields

$$\{D(D+1)\}[ct] = (D+1)[c] = c = \frac{5}{3} \quad \Rightarrow \quad v_p(t) = \frac{5}{3}t.$$

Therefore, a general solution to (5.2) is

$$v(t) = v_h(t) + v_p(t) = c_1 + c_2 e^{-t} + \frac{5}{3}t.$$

We now take the original system and subtract the second equation from the first one.

$$2u - (3D+2)[v] = e^t - 5$$
$$\Rightarrow \quad u = \left(\frac{3}{2}D + 1\right)[v] + \frac{1}{2}e^t - \frac{5}{2}$$
$$\Rightarrow \quad u = \frac{3}{2}\left(c_1 + c_2 e^{-t} + \frac{5}{3}t\right)' + \left(c_1 + c_2 e^{-t} + \frac{5}{3}t\right) + \frac{1}{2}e^t - \frac{5}{2}$$
$$\Rightarrow \quad u(t) = c_1 - \frac{1}{2}c_2 e^{-t} + \frac{1}{2}e^t + \frac{5}{3}t.$$

Thus, a general solution to the given system is

$$u(t) = c_1 - \frac{1}{2}c_2 e^{-t} + \frac{1}{2}e^t + \frac{5}{3}t,$$
$$v(t) = c_1 + c_2 e^{-t} + \frac{5}{3}t.$$

9. Expressed in operator notation, this system becomes

$$(D+2)[x] + D[y] = 0,$$
$$(D-1)[x] + (D-1)[y] = \sin t.$$

In order to eliminate the function $y(t)$, we apply the operator $(D-1)$ to the first equation and the operator $(-D)$ to the second one. Thus, we have

$$(D-1)(D+2)[x] + (D-1)D[y] = (D-1)[0] = 0,$$
$$-D(D-1)[x] - D(D-1)[y] = -D[\sin t] = -\cos t.$$

Adding these two equations yields a differential equation involving $x(t)$ alone.

$$\{(D^2 + D - 2) - (D^2 - D)\}[x] = -\cos t$$

Copyright © 2012 Pearson Education, Inc. Publishing as Addison-Wesley.

$$\Rightarrow \qquad 2(D-1)[x] = -\cos t. \qquad\qquad (5.3)$$

This is a linear first-order differential equation with constant coefficients, and so it can be solved by methods of Chapter 2, Section 2.3. However, we will use methods of Chapter 4. We see that the auxiliary equation, associated with the corresponding homogeneous equation, is given by $2(r-1) = 0$, which has the root $r = 1$. Thus, a general solution to the homogeneous equation is

$$x_h(t) = c_1 e^t.$$

We now use the method of undetermined coefficients to find a particular solution to the given nonhomogeneous equation. To this end, we note that a particular solution has the form

$$x_p(t) = A\cos t + B\sin t \qquad \Rightarrow \qquad x_p'(t) = -A\sin t + B\cos t.$$

Substituting these expressions into the nonhomogeneous equation, given in (5.3), yields

$$\begin{aligned}
2x_p' - 2x_p &= 2(-A\sin t + B\cos t) - 2(A\cos t + B\sin t)\\
&= (2B - 2A)\cos t + (-2A - 2B)\sin t = -\cos t.
\end{aligned}$$

By equating coefficients, we obtain

$$2B - 2A = -1 \qquad \text{and} \qquad -2A - 2B = 0.$$

Solving these equations simultaneously, we see that

$$A = \frac{1}{4} \qquad \text{and} \qquad B = -\frac{1}{4}.$$

Thus, a particular solution to the nonhomogeneous equation given in (5.3) will be

$$x_p(t) = \frac{1}{4}\cos t - \frac{1}{4}\sin t$$

and a general solution to the nonhomogeneous equation (5.3) is

$$x(t) = x_h(t) + x_p(t) = c_1 e^t + \frac{1}{4}\cos t - \frac{1}{4}\sin t.$$

We now find $y(t)$. Subtracting the second equation in the system from the first one, to obtain

$$3x + y = -\sin t \qquad \Rightarrow \qquad y = -3x - \sin t.$$

Copyright © 2012 Pearson Education, Inc. Publishing as Addison-Wesley.

Therefore, we see that

$$y(t) = -3\left(c_1 e^t + \frac{1}{4}\cos t - \frac{1}{4}\sin t\right) - \sin t$$

$$\Rightarrow \qquad y(t) = -3c_1 e^t - \frac{3}{4}\cos t - \frac{1}{4}\sin t.$$

Hence, this system of differential equations has a general solution

$$x(t) = c_1 e^t + \frac{1}{4}\cos t - \frac{1}{4}\sin t \qquad \text{and} \qquad y(t) = -3c_1 e^t - \frac{3}{4}\cos t - \frac{1}{4}\sin t.$$

11. From the second equation, we obtain $u = -(D^2 + 2)[v]/2$. Substitution into the first equation eliminates u and gives

$$(D^2 - 1)\left\{-\frac{1}{2}(D^2 + 2)[v]\right\} + 5v = e^t$$

$$\Rightarrow \qquad \left[(D^2 - 1)(D^2 + 2) - 10\right][v] = -2e^t$$

$$\Rightarrow \qquad (D^4 + D^2 - 12)[v] = -2e^t. \tag{5.4}$$

Solving the characteristic equation,

$$r^4 + r^2 - 12 = 0 \qquad \Rightarrow \qquad (r^2 + 4)(r^2 - 3) = 0 \qquad \Rightarrow \qquad r = \pm 2i,\, \pm\sqrt{3},$$

we conclude that a general solution to the corresponding homogeneous equation is

$$v_h(t) = c_1 \cos 2t + c_2 \sin 2t + c_3 e^{\sqrt{3}t} + c_4 e^{-\sqrt{3}t}.$$

A particular solution to (5.4) has the form $v_p(t) = ce^t$. Substituting yields

$$(D^4 + D^2 - 12)\left[ce^t\right] = ce^t + ce^t - 12ce^t = -10ce^t = -2e^t \qquad \Rightarrow \qquad c = \frac{1}{5}.$$

Therefore,

$$v = v_h + v_p = c_1 \cos 2t + c_2 \sin 2t + c_3 e^{\sqrt{3}t} + c_4 e^{-\sqrt{3}t} + \frac{e^t}{5}$$

and

$$u = -\frac{1}{2}(D^2 + 2)[v] = -\frac{1}{2}\left(c_1 \cos 2t + c_2 \sin 2t + c_3 e^{\sqrt{3}t} + c_4 e^{-\sqrt{3}t} + \frac{1}{5}e^t\right)''$$

$$-\left(c_1 \cos 2t + c_2 \sin 2t + c_3 e^{\sqrt{3}t} + c_4 e^{-\sqrt{3}t} + \frac{1}{5}e^t\right)$$

$$= c_1 \cos 2t + c_2 \sin 2t - \frac{5}{2}c_3 e^{\sqrt{3}t} - \frac{5}{2}c_4 e^{-\sqrt{3}t} - \frac{3}{10}e^t.$$

Replacing $(-5/2)c_3$ by c_3 and $(-5/2)c_4$ by c_4, we obtain the answer given in the text.

Copyright © 2012 Pearson Education, Inc. Publishing as Addison-Wesley.

13. Expressing x from the second equation and substituting the result into the first equation, we obtain

$$x = y' - y \qquad \Rightarrow \qquad \frac{d(y' - y)}{dt} = (y' - y) - 4y \qquad \Rightarrow \qquad y'' - 2y' + 5y = 0.$$

This homogeneous linear equation with constant coefficients has the characteristic equation $r^2 - 2r + 5 = 0$, whose roots are $r = 1 \pm 2i$. Thus, a general solution is

$$y = c_1 e^t \cos 2t + c_2 e^t \sin 2t \, .$$

Therefore,

$$
\begin{aligned}
x &= \left(c_1 e^t \cos 2t + c_2 e^t \sin 2t\right)' - \left(c_1 e^t \cos 2t + c_2 e^t \sin 2t\right) \\
&= \left(c_1 e^t \cos 2t - 2c_1 e^t \sin 2t + c_2 e^t \sin 2t + 2c_2 e^t \cos 2t\right) - \left(c_1 e^t \cos 2t + c_2 e^t \sin 2t\right) \\
&= 2c_2 e^t \cos 2t - 2c_1 e^t \sin 2t.
\end{aligned}
$$

15. In operator form, the system becomes

$$
\begin{aligned}
-2z + (D - 5)[w] &= 5t, \\
(D - 4)[z] - 3w &= 17t.
\end{aligned}
$$

We multiply the first equation by 3, the second equation – by $(D - 5)$, and add the results.

$$\{-6 + (D - 5)(D - 4)\}\,[z] = 3(5t) + (D - 5)[17t] = -70t + 17$$

$$\Rightarrow \quad \left(D^2 - 9D + 14\right)[z] = -70t + 17.$$

Solving the characteristic equation, $r^2 - 9r + 14 = 0$, we obtain $r = 2, 7$. Hence, a general solution to the corresponding homogeneous equation is $z_h(t) = c_1 e^{2t} + c_2 e^{7t}$. A particular solution has the form $z_p(t) = At + B$. Substituting yields

$$\left(D^2 - 9D + 14\right)[At + B] = (At + B)'' - 9(At + B)' + 14(At + B)$$

$$= 14At - 9A + 14B = -70t + 17$$

$$\Rightarrow \quad A = \frac{-70}{14} = -5, \quad B = \frac{17 + 9A}{14} = -2 \, .$$

Therefore,

$$z(t) = z_h(t) + z_p(t) = c_1 e^{2t} + c_2 e^{7t} - 5t - 2 \, .$$

We now use the second equation in the original system to find w.

$$w = \frac{1}{3} \left(z' - 4z - 17t\right) = -\frac{2}{3} c_1 e^{2t} + c_2 e^{7t} + t + 1.$$

 Copyright © 2012 Pearson Education, Inc. Publishing as Addison-Wesley.

17. Expressed in operator notation, this system becomes

$$(D^2 + 5)\,[x] - 4[y] = 0,$$
$$-[x] + (D^2 + 2)\,[y] = 0.$$

In order to eliminate the function $x(t)$, we apply the operator $(D^2 + 5)$ to the second equation. Thus, we have

$$(D^2 + 5)\,[x] - 4[y] = 0,$$
$$-(D^2 + 5)\,[x] + (D^2 + 5)\,(D^2 + 2)\,[y] = 0.$$

Adding these two equations together yields a differential equation involving $y(t)$ only.

$$\{(D^2 + 5)(D^2 + 2) - 4\}\,[y] = 0 \quad\Rightarrow\quad (D^4 + 7D^2 + 6)\,[y] = 0.$$

The auxiliary equation for this homogeneous equation, $r^4 + 7r^2 + 6 = (r^2 + 1)(r^2 + 6) = 0$, has roots $r = \pm i,\ \pm i\sqrt{6}$. Thus, a general solution is given by

$$y(t) = c_1 \cos t + c_2 \sin t + c_3 \cos \sqrt{6}\,t + c_4 \sin \sqrt{6}\,t.$$

Now, we find a function $x(t)$ that satisfies the system of differential equations given in the problem. To do this, we solve the second equation in the system for $x(t)$ to obtain

$$x(t) = (D^2 + 2)\,[y].$$

Substituting the expression found for $y(t)$, we see that

$$x(t) = -c_1 \cos t - c_2 \sin t - 6c_3 \cos \sqrt{6}\,t - 6c_4 \sin \sqrt{6}\,t$$
$$+ 2\left(c_1 \cos t + c_2 \sin t + c_3 \cos \sqrt{6}\,t + c_4 \sin \sqrt{6}\,t\right)$$
$$\Rightarrow \quad x(t) = c_1 \cos t + c_2 \sin t - 4c_3 \cos \sqrt{6}\,t - 4c_4 \sin \sqrt{6}\,t.$$

Hence, this system of differential equations has a general solution

$$x(t) = c_1 \cos t + c_2 \sin t - 4c_3 \cos \sqrt{6}\,t - 4c_4 \sin \sqrt{6}\,t$$
$$y(t) = c_1 \cos t + c_2 \sin t + c_3 \cos \sqrt{6}\,t + c_4 \sin \sqrt{6}\,t.$$

19. From the first equation, we conclude that $y = x' - 4x$. Substituting into the second equation yields

$$(x' - 4x)' = -2x + (x' - 4x) \quad\Rightarrow\quad x'' - 5x' + 6x = 0.$$

Copyright © 2012 Pearson Education, Inc. Publishing as Addison-Wesley.

Chapter 5

The characteristic equation, $r^2 - 5r + 6 = 0$, has roots $r = 2, 3$, and so a general solution is

$$x(t) = c_1 e^{2t} + c_2 e^{3t}$$

$$\Rightarrow \quad y(t) = \left(c_1 e^{2t} + c_2 e^{3t}\right)' - 4\left(c_1 e^{2t} + c_2 e^{3t}\right) = -2c_1 e^{2t} - c_2 e^{3t}.$$

We find constants c_1 and c_2 from the initial conditions.

$$1 = x(0) = c_1 e^{2(0)} + c_2 e^{3(0)} = c_1 + c_2,$$
$$0 = y(0) = -2c_1 e^{2(0)} - c_2 e^{3(0)} = -2c_1 - c_2 \qquad \Rightarrow \qquad c_1 = -1,$$
$$c_2 = 2.$$

Therefore, the answer to this problem is

$$x(t) = 2e^{3t} - e^{2t}, \qquad y(t) = -2e^{3t} + 2e^{2t}.$$

21. To apply the elimination method, we write the system using operator notation.

$$D^2[x] - y = 0,$$
$$-x + D^2[y] = 0. \tag{5.5}$$

Eliminating y by applying D^2 to the first equation and adding to the second equation gives

$$\left(D^2 D^2 - 1\right)[x] = 0,$$

which reduces to

$$\left(D^4 - 1\right)[x] = 0. \tag{5.6}$$

The corresponding auxiliary equation, $r^4 - 1 = 0$, has roots ± 1, $\pm i$. Thus, the general solution to (5.6) is given by

$$x(t) = C_1 e^t + C_2 e^{-t} + C_3 \cos t + C_4 \sin t. \tag{5.7}$$

Substituting $x(t)$ into the first equation in (5.5) yields

$$y(t) = x''(t) = C_1 e^t + C_2 e^{-t} - C_3 \cos t - C_4 \sin t. \tag{5.8}$$

We use initial conditions to determine constants C_1, C_2, C_3, and C_4. Differentiating (5.7) and (5.8), we get

$$3 = x(0) = C_1 e^0 + C_2 e^{-0} + C_3 \cos 0 + C_4 \sin 0 = C_1 + C_2 + C_3,$$
$$1 = x'(0) = C_1 e^0 - C_2 e^{-0} - C_3 \sin 0 + C_4 \cos 0 = C_1 - C_2 + C_4,$$
$$1 = y(0) = C_1 e^0 + C_2 e^{-0} - C_3 \cos 0 - C_4 \sin 0 = C_1 + C_2 - C_3,$$
$$-1 = y'(0) = C_1 e^0 - C_2 e^{-0} + C_3 \sin 0 - C_4 \cos 0 = C_1 - C_2 - C_4.$$

Copyright © 2012 Pearson Education, Inc. Publishing as Addison-Wesley.

This yields

$$C_1 + C_2 + C_3 = 3\,,$$
$$C_1 - C_2 + C_4 = 1\,,$$
$$C_1 + C_2 - C_3 = 1\,,$$
$$C_1 - C_2 - C_4 = -1\,.$$

Solving, we obtain $C_1 = C_2 = C_3 = C_4 = 1$. So, the answer is

$$x(t) = e^t + e^{-t} + \cos t + \sin t,$$

$$y(t) = e^t + e^{-t} - \cos t - \sin t.$$

23. We will attempt to solve this system by eliminating the function $y(t)$. Thus, we multiply the first equation by $(D+2)$ and the second one – by $(1-D)$. Thus, we obtain

$$(D+2)(D-1)[x] + (D+2)(D-1)D[y] = (D+2)\left[-3e^{-2t}\right] = 6e^{-2t} - 6e^{-2t} = 0,$$
$$-(D-1)(D+2)[x] - (D-1)(D+2)[y] = -(D-1)\left[3e^t\right] = -3e^t + 3e^t = 0.$$

Adding these two equations yields

$$0 \cdot x + 0 \cdot y = 0,$$

which is true for any two functions $x(t)$ and $y(t)$. (But *not* every pair of functions satisfies the given system of differential equations.) Thus, this is a degenerate system, and has infinitely many linearly independent solutions. To see if we can find them, let us examine the system more closely. Notice that we can write

$$(D-1)[x+y] = -3e^{-2t},$$
$$(D+2)[x+y] = 3e^t.$$

Therefore, let us try the substitution $z(t) = x(t) + y(t)$. We want to find a function $z(t)$ that satisfies two equations,

$$z'(t) - z(t) = -3e^{-2t} \qquad \text{and} \qquad z'(t) + 2z(t) = 3e^t \tag{5.9}$$

simultaneously. We start by solving the first equation given in (5.9). This is a linear differential equation with constant coefficients, which has the auxiliary equation $r - 1 = 0$. Hence, a general solution to the corresponding homogeneous equation is

$$z_h(t) = Ce^t.$$

Copyright © 2012 Pearson Education, Inc. Publishing as Addison-Wesley.

By the method of undetermined coefficients, we see that a particular solution has the form

$$z_p(t) = Ae^{-2t} \qquad \Rightarrow \qquad z_p' = -2Ae^{-2t}.$$

Substituting these expressions into the first differential equation in (5.9) yields

$$z_p'(t) - z_p(t) = -2Ae^{-2t} - Ae^{-2t} = -3Ae^{-2t} = -3e^{-2t} \qquad \Rightarrow \qquad A = 1.$$

Thus, the first equation, given in (5.9), has a general solution

$$z(t) = Ce^t + e^{-2t}.$$

Now, substituting $z(t)$ into the second equation in (5.9) gives

$$Ce^t - 2e^{-2t} + 2\left(Ce^t + e^{-2t}\right) = 3e^t \qquad \Rightarrow \qquad 3Ce^t = 3e^t.$$

Hence, $C = 1$. Therefore, $z(t) = e^t + e^{-2t}$ is the only solution that satisfies both differential equations given in (5.9). Thus, we conclude that any two differentiable functions, satisfying the equation $x(t) + y(t) = e^t + e^{-2t}$, satisfy the original system.

25. Writing the system in operator form yields

$$
\begin{aligned}
(D-1)[x] - 2y + z &= 0, \\
-x + D[y] - z &= 0, \\
-4x + 4y + (D-5)[z] &= 0.
\end{aligned}
$$

We use the second equation to express z in terms of x and y.

$$z = -x + D[y]. \tag{5.10}$$

Substituting this expression into the other two equations, we obtain

$$
\begin{aligned}
(D-1)[x] - 2y + [-x + D[y]] &= 0, \\
-4x + 4y + (D-5)\left[-x + D[y]\right] &= 0.
\end{aligned}
$$

This system can be written as

$$
\begin{aligned}
(D-2)[x] + (D-2)[y]) &= 0, \\
-(D-1)[x] + (D^2 - 5D + 4)[y] &= 0.
\end{aligned}
\tag{5.11}
$$

Now, we eliminate x by multiplying the first equation by $(D-1)$, the second equation – by $(D-2)$, and adding the results together. This yields

$$\{(D-1)(D-2) + (D-2)(D^2 - 5D + 4)\}[y] = 0$$

Copyright © 2012 Pearson Education, Inc. Publishing as Addison-Wesley.

$$\Rightarrow \quad \{(D-2)(D^2-4D+3)\}[y]=0 \quad \Rightarrow \quad \{(D-2)(D-1)(D-3)\}[y]=0.$$

The roots of the characteristic equation, $(r-2)(r-1)(r-3)=0$, are $r=1, 2$, and 3. Thus, a general solution for $y(t)$ is

$$y(t) = c_1 e^t + c_2 e^{2t} + c_3 e^{3t}.$$

With $h(t) := x(t) + y(t)$, the first equation in (5.11) can be written in the form

$$(D-2)[h]=0 \quad \text{or} \quad h' - 2h = 0,$$

which has a general solution $h(t) = Ke^{2t}$. Therefore,

$$x(t) = h(t) - y(t) = -c_1 e^t + (K-c_2)e^{2t} - c_3 e^{3t}.$$

To find K, we substitute the above solutions, $x(t)$ and $y(t)$ (with $c_1 = c_3 = 0$), into the second equation in (5.11). Thus, we get

$$-(D-1)\left[(K-c_2)e^{2t}\right] + \left(D^2-5D+4\right)\left[c_2 e^{2t}\right] = 0$$
$$\Rightarrow \quad -(K-c_2)e^{2t} + (4(c_2) - 5(2c_2) + 4(c_2))e^{2t} = 0$$
$$\Rightarrow \quad -K - c_2 = 0 \quad \Rightarrow \quad K = -c_2.$$

Hence,

$$x(t) = -c_1 e^t - 2c_2 e^{2t} - c_3 e^{3t}.$$

Finally, we find $z(t)$ using (5.10).

$$z(t) = -\left(-c_1 e^t - 2c_2 e^{2t} - c_3 e^{3t}\right) + \left(c_1 e^t + c_2 e^{2t} + c_3 e^{3t}\right)' = 2c_1 e^t + 4c_2 e^{2t} + 4c_3 e^{3t}.$$

27. We eliminate z by expressing

$$z = \frac{1}{4}\left(-x' + 4x\right) = -\frac{1}{4}(D-4)[x] \tag{5.12}$$

from the first equation and substituting (5.12) into the second and third equations. We obtain

$$2\left\{-\frac{1}{4}(D-4)[x]\right\} + (D-4)[y] = 0,$$
$$2x + 4y + D\left\{-\frac{1}{4}(D-4)[x]\right\} - 4\left\{-\frac{1}{4}(D-4)[x]\right\} = 0.$$

Copyright © 2012 Pearson Education, Inc. Publishing as Addison-Wesley.

After some algebra, the above system simplifies to

$$-(D-4)[x] + 2(D-4)[y] = 0,$$
$$\left(D^2 - 8D + 8\right)[x] - 16y = 0.$$

We use the second equation to find that

$$y = \frac{1}{16}\left(D^2 - 8D + 8\right)[x]. \tag{5.13}$$

Then the first equation becomes

$$-(D-4)[x] + 2(D-4)\left\{\frac{1}{16}\left(D^2 - 8D + 8\right)[x]\right\} = 0$$

$$\Rightarrow \quad (D-4)\left\{-1 + \frac{1}{8}\left(D^2 - 8D + 8\right)\right\}[x] = 0 \quad \Rightarrow \quad (D-4)D(D-8)[x] = 0.$$

Solving the characteristic equation, we get $r = 0, 4$, and 8. So,

$$x = c_1 e^{8t} + c_2 e^{4t} + c_3.$$

Substitution of this solution into (5.12) and (5.13) yields

$$z = \frac{1}{4}\left(-x' + 4x\right) = -c_1 e^{8t} + c_3,$$

$$y = \frac{1}{16}\left(x'' - 8x' + 8x\right) = \frac{1}{2}\left(c_1 e^{8t} - c_2 e^{4t} + c_3\right).$$

29. We begin by expressing the system in operator notation.

$$(D - \lambda)[x] + y = 0,$$
$$-3x + (D-1)[y] = 0.$$

We eliminate y by applying $(D-1)$ to the first equation and subtracting the second equation from the result. This yields

$$\{(D-1)(D-\lambda) - (-3)\}[x] = 0$$

$$\Rightarrow \quad \left\{D^2 - (\lambda + 1)D + (\lambda + 3)\right\}[x] = 0. \tag{5.14}$$

Note that, since the given system is homogeneous, $y(t)$ also satisfies this equation (compare (7) and (8)). So, we can investigate $x(t)$ only. The auxiliary equation, $r^2 - (\lambda + 1)r + (\lambda + 3) = 0$, has roots

$$r_1 = \frac{(\lambda + 1) - \sqrt{\Delta}}{2}, \qquad r_2 = \frac{(\lambda + 1) + \sqrt{\Delta}}{2},$$

where the discriminant $\Delta := (\lambda + 1)^2 - 4(\lambda + 3)$. We consider two cases:

 Copyright © 2012 Pearson Education, Inc. Publishing as Addison-Wesley.

i) If $\lambda + 3 < 0$, i.e., $\lambda < -3$, then $\Delta > (\lambda + 1)^2$, and so the root

$$r_2 > \frac{(\lambda + 1) + |\lambda + 1|}{2} = 0$$

because $\lambda + 1 < 0$. Therefore, the solution $x(t) = e^{r_2 t}$ is unbounded, as $t \to +\infty$.

ii) If $\lambda + 3 \geq 0$, i.e., $\lambda \geq -3$, then $\Delta \leq (\lambda + 1)^2$. If $\Delta < 0$, then a fundamental solution set to (5.14) is

$$\left\{ e^{(\lambda+1)t/2} \cos\left(\frac{\sqrt{-\Delta}\,t}{2}\right), \; e^{(\lambda+1)t/2} \sin\left(\frac{\sqrt{-\Delta}\,t}{2}\right) \right\}. \tag{5.15}$$

If $\Delta \geq 0$, then $\sqrt{\Delta} < |\lambda + 1|$ and a fundamental solution set is

$$\begin{array}{ll} \{e^{r_1 t}, \, e^{r_2 t}\}, & \text{if } \Delta > 0, \\ \{e^{r_1 t}, \, t e^{r_1 t}\}, & \text{if } \Delta = 0, \end{array} \tag{5.16}$$

where both roots r_1, r_2 are nonpositive if and only if $\lambda \leq -1$. For $\lambda = -1$, we have $\Delta = (-1+1)^2 - 4(-1+3) < 0$, and get a particular case of the fundamental solution set (5.15) (without exponential term), consisting of bounded functions. Finally, if $\lambda < -1$, then $r_1 < 0$, $r_2 \leq 0$, and all the functions listed in (5.15), (5.16) are bounded.

Any solution $x(t)$ is a linear combination of fundamental solutions and, therefore, all solutions $x(t)$ are bounded if and only if $-3 \leq \lambda \leq -1$.

31. Solving this problem, we follow the arguments, described in Section 5.1, i.e., we let $x(t)$ be the mass of salt in the tank A and $y(t)$ be the mass of salt in the tank B. These functions satisfy the system

$$\begin{aligned} x' &= \text{input}_A - \text{output}_A, \\ y' &= \text{input}_B - \text{output}_B, \end{aligned} \tag{5.17}$$

with initial conditions $x(0) = 0$, $y(0) = 20$. It is important to notice that the volume of each tank stays at 100 L, because the net flow rate into each tank is the same as the net outflow. Next we observe that "input$_A$" consists of the salt coming from outside, which is

$$0.2 \, (\text{kg/L}) \cdot 6 \, (\text{L/min}) = 1.2 \, (\text{kg/min}),$$

and the salt coming from the tank B, which is given by

$$\frac{y}{100} \, (\text{kg/L}) \cdot 1 \, (\text{L/min}) = \frac{y}{100} \, (\text{kg/min}).$$

Copyright © 2012 Pearson Education, Inc. Publishing as Addison-Wesley.

Thus,

$$\text{input}_A = \left(1.2 + \frac{y}{100}\right) \text{ (kg/min)}.$$

The "output$_A$" consists of two flows: one is going out of the system and the other one is going to the tank B. So,

$$\text{output}_A = \frac{x}{100} \text{ (kg/L)} \cdot (4+3) \text{ (L/min)} = \frac{7x}{100} \text{ (kg/min)},$$

and the first equation in (5.17) becomes

$$\frac{dx}{dt} = 1.2 + \frac{y}{100} - \frac{7x}{100}.$$

Similarly, the second equation in (5.17) can be written as

$$\frac{dy}{dt} = \frac{3x}{100} - \frac{3y}{100}.$$

Rewriting this system in operator form, we obtain

$$(D + 0.07)[x] - 0.01y = 1.2,$$
$$-0.03x + (D + 0.03)[y] = 0. \tag{5.18}$$

Eliminating y yields

$$\{(D + 0.07)(D + 0.03) - (-0.01)(-0.03)\}[x] = (D + 0.03)[1.2] = 0.036,$$

which simplifies to

$$\left(D^2 + 0.1D + 0.0018\right)[x] = 0.036. \tag{5.19}$$

The auxiliary equation, $r^2 + 0.1r + 0.0018 = 0$, has roots

$$r_1 = -\frac{1}{20} - \sqrt{\frac{1}{400} - \frac{18}{10000}} = -\frac{1}{20} - \frac{\sqrt{7}}{100} = \frac{-5 - \sqrt{7}}{100} \approx -0.0765,$$
$$r_2 = \frac{-5 + \sqrt{7}}{100} \approx -0.0235.$$

Therefore, the general solution the corresponding homogeneous equation is

$$x_h(t) = C_1 e^{r_1 t} + C_2 e^{r_2 t}.$$

Since the nonhomogeneous term in (5.19) is a constant, we are looking for a particular solution of the form $x_p(t) = A$. Substituting into (5.19) yields

$$0.0018A = 0.036 \qquad \Rightarrow \qquad A = 20,$$

 Copyright © 2012 Pearson Education, Inc. Publishing as Addison-Wesley.

and a general solution is

$$x(t) = x_h(t) + x_p(t) = C_1 e^{r_1 t} + C_2 e^{r_2 t} + 20.$$

From the first equation in (5.18) we find that

$$\begin{aligned}
y(t) &= 100 \cdot \{(D + 0.07)[x] - 1.2\} = 100 \frac{dx}{dt} + 7x(t) - 120 \\
&= 100 \left\{ r_1 C_1 e^{r_1 t} + r_2 C_2 e^{r_2 t} \right\} + 7 \left\{ C_1 e^{r_1 t} + C_2 e^{r_2 t} + 20 \right\} - 120 \\
&= \left(2 - \sqrt{7}\right) C_1 e^{r_1 t} + \left(2 + \sqrt{7}\right) C_2 e^{r_2 t} + 20.
\end{aligned}$$

The initial conditions imply

$$\begin{aligned}
0 &= x(0) = C_1 + C_2 + 20, \\
20 &= y(0) = \left(2 - \sqrt{7}\right) C_1 + \left(2 + \sqrt{7}\right) C_2 + 20
\end{aligned}$$

$$\Rightarrow \quad \begin{aligned}
C_1 + C_2 &= -20, \\
\left(2 - \sqrt{7}\right) C_1 + \left(2 + \sqrt{7}\right) C_2 &= 0
\end{aligned}$$

$$\Rightarrow \quad C_1 = -\left(10 + \frac{20}{\sqrt{7}}\right), \qquad C_2 = -\left(10 - \frac{20}{\sqrt{7}}\right).$$

Thus, the solution to the problem is

$$\begin{aligned}
x(t) &= -\left(10 + \frac{20}{\sqrt{7}}\right) e^{r_1 t} - \left(10 - \frac{20}{\sqrt{7}}\right) e^{r_2 t} + 20 \ (\text{kg}), \\
y(t) &= \frac{30}{\sqrt{7}} e^{r_1 t} - \frac{30}{\sqrt{7}} e^{r_2 t} + 20 \ (\text{kg}).
\end{aligned}$$

33. Since no solution flows in or out of the system from the tank B, we conclude that the solution flows from the tank B to the tank A with the same rate as it does from A to B, that is, 1 L/min. Furthermore, the solution flows in and out of the tank A with the same rate (4 L/min), and so the volume of the solution in the tank A (as well as in the tank B) remains constant, namely, 100 L. Thus, with $x(t)$ and $y(t)$ denoting the amount of salt in the tanks A and B, respectively, the law "rate of change = input rate − output rate" becomes

Tank A:

$$\begin{aligned}
x' = &\left[4 \, (\text{L/min}) \cdot 0.2 \, (\text{kg/L}) + 1 \, (\text{L/min}) \cdot \frac{y}{100} \, (\text{kg/L}) \right] \\
&- \frac{x}{100} \, (\text{kg/L}) \cdot [1 \, (\text{L/min}) + 4 \, (\text{L/min})] \, ;
\end{aligned}$$

Copyright © 2012 Pearson Education, Inc. Publishing as Addison-Wesley.

Tank B:

$$y' = 1 \, (\text{L/min}) \cdot \frac{x}{100} \, (\text{kg/L}) - 1 \, (\text{L/min}) \cdot \frac{y}{100} \, (\text{kg/L}).$$

Hence, we obtain the system

$$x' = 0.8 - \frac{x}{20} + \frac{y}{100},$$
$$y' = \frac{x}{100} - \frac{y}{100}.$$

From the second equation, we find that $x = 100y' + y$. Substituting into the first equation yields

$$(100y' + y)' = 0.8 - \frac{100y' + y}{20} + \frac{y}{100}$$

$$\Rightarrow \quad 100y'' + 6y' + \frac{1}{25}y = 0.8 \quad \Rightarrow \quad y'' + 0.06y' + 0.0004y = 0.008 . (5.20)$$

The characteristic equation, $r^2 + 0.06r + 0.0004 = 0$, of the corresponding homogeneous equation has roots

$$r = \frac{-0.06 \pm \sqrt{(0.06)^2 - 4(1)(0.0004)}}{2} = \frac{-3 \pm \sqrt{5}}{100},$$

and so

$$y_h(t) = c_1 e^{(-3-\sqrt{5})t/100} + c_2 e^{(-3+\sqrt{5})t/100}$$

is a general solution to the homogeneous equation. We now look for a particular solution of the form $y_p(t) = c$. Substituting $y_p(t)$ into (5.20) yields

$$0.0004c = 0.008 \quad \Rightarrow \quad c = \frac{0.008}{0.0004} = 20.$$

Thus,

$$y(t) = y_p(t) + y_h(t) = 20 + c_1 e^{(-3-\sqrt{5})t/100} + c_2 e^{(-3+\sqrt{5})t/100} \tag{5.21}$$

is a general solution to (5.20). Then

$$x(t) = y + 100y' = 20 + (1 - 3 - \sqrt{5})c_1 e^{(-3-\sqrt{5})t/100} + (1 - 3 + \sqrt{5})c_2 e^{(-3+\sqrt{5})t/100}$$

$$= 20 - (2 + \sqrt{5})c_1 e^{(-3-\sqrt{5})t/100} + (-2 + \sqrt{5})c_2 e^{(-3+\sqrt{5})t/100} . \tag{5.22}$$

Next, we use the initial conditions, $x(0) = 0$ and $y(0) = 20$, to find values of c_1 and c_2.

$$20 - (2 + \sqrt{5})c_1 + (-2 + \sqrt{5})c_2 = 0, \qquad\qquad c_1 = 10/\sqrt{5},$$
$$\Rightarrow$$
$$20 + c_1 + c_2 = 20 \qquad\qquad c_2 = -10/\sqrt{5}.$$

 Copyright © 2012 Pearson Education, Inc. Publishing as Addison-Wesley.

With these values, the solution, given in (5.21) and (5.22), becomes

$$x(t) = 20 - \left(\frac{20 + 10\sqrt{5}}{\sqrt{5}}\right) e^{(-3-\sqrt{5})t/100} + \left(\frac{20 - 10\sqrt{5}}{\sqrt{5}}\right) e^{(-3+\sqrt{5})t/100},$$

$$y(t) = 20 + \left(\frac{10}{\sqrt{5}}\right) e^{(-3-\sqrt{5})t/100} - \left(\frac{10}{\sqrt{5}}\right) e^{(-3+\sqrt{5})t/100}.$$

35. Let $x(t)$ and $y(t)$ denote the temperatures at time t in zones A and B, respectively. Then the rate of change of temperature in zone A is $x'(t)$, and in zone B it is $y'(t)$. We can apply Newton's law of cooling to express these rates of change in an alternate manner. Thus, we observe that the rate of change of the temperature in zone A due to the outside temperature is $k_1[100 - x(t)]$ and that, due to the temperature in zone B, it is $k_2[y(t) - x(t)]$. Since the time constant for heat transfer between zone A and the outside is 2 hrs $(= 1/k_1)$, we see that $k_1 = 1/2$. Similarly, we see that $1/k_2 = 4$, which implies that $k_2 = 1/4$. Therefore, since there is no heating or cooling source in zone A, we can write the equation for the rate of change of the temperature in the attic as

$$x'(t) = \frac{1}{2}[100 - x(t)] + \frac{1}{4}[y(t) - x(t)].$$

In the same way, we see that the rate of change of the temperature in zone B due to the temperature of the attic is $k_3[x(t) - y(t)]$, where $1/k_3 = 4$, and that the rate of change of the temperature in this zone due to the outside temperature is $k_4[100 - y(t)]$, where $1/k_4 = 4$. In this zone, however, we must consider the cooling due to the air conditioner. Since the heat capacity of zone B is $(1/2)°$F per thousand Btu and the air conditioner has the cooling capacity of 24 thousand Btu per hr, we see that the air conditioner removes heat from this zone at the rate of $(1/2) \times 24° = 12°$ F/hr. (Since heat is *removed* from the house, this rate will be negative.) Combining these observations, we see that the rate of change of the temperature in zone B is given by

$$y'(t) = -12 + \frac{1}{4}[x(t) - y(t)] + \frac{1}{4}[100 - y(t)].$$

Simplifying these equations, we find that the given cooling problem satisfies

$$4x'(t) + 3x(t) - y(t) = 200,$$

$$-x'(t) + 4y'(t) + 2y(t) = 52.$$

In operator notation, this system becomes

$$(4D + 3)[x] - [y] = 200,$$

$$-[x] + (4D + 2)[y] = 52.$$

Copyright © 2012 Pearson Education, Inc. Publishing as Addison-Wesley.

Since we are interested in the temperature $x(t)$ in the attic, we eliminate the function $y(t)$ in the system above by applying $(4D+2)$ to the first equation and adding the result to the second one.

$$\{(4D+2)(4D+3)-1\}[x] = (4D+2)[200]+52 = 452$$
$$\Rightarrow \quad \left(16D^2 + 20D + 5\right)[x] = 452\,. \tag{5.23}$$

The corresponding homogeneous equation to this linear equation with constant coefficients has the auxiliary equation $16r^2 + 20r + 5 = 0$. Solving yields

$$r_1 = \frac{-5+\sqrt{5}}{8} \approx -0.345 \quad \text{and} \quad r_2 = \frac{-5-\sqrt{5}}{8} \approx -0.905\,.$$

Therefore,

$$x_h(t) = c_1 e^{r_1 t} + c_2 e^{r_2 t}$$

is a general solution to the homogeneous equation. The method of undetermined coefficients suggests a particular solution to (5.23) of the form

$$x_p(t) = A \quad \Rightarrow \quad x_p'(t) = 0 \quad \Rightarrow \quad x_p''(t) = 0.$$

Substituting these expressions into (5.23) yields

$$16x_p'' + 20x_p' + 5x_p = 5A = 452 \quad \Rightarrow \quad A = 90.4\,.$$

Thus, a particular solution to the differential equation given in (5.23) is $x_p(t) = 90.4$, and a general solution to this equation is given by

$$x(t) = c_1 e^{r_1 t} + c_2 e^{r_2 t} + 90.4\,,$$

where $r_1 = (-5+\sqrt{5})/8$ and $r_2 = (-5-\sqrt{5})/8$. To determine the maximum temperature in the attic, we assume that zones A and B have sufficiently cool initial temperatures (so that, for example, c_1 and c_2 are negative). Since r_1 and r_2 are negative, $c_1 e^{r_1 t}$ and $c_2 e^{r_2 t}$ go to zero, as t approaches to infinity. Therefore, the maximum temperature that can be attained in the attic is

$$\lim_{t\to\infty} x(t) = 90.4^\circ\,\text{F}.$$

Copyright © 2012 Pearson Education, Inc. Publishing as Addison-Wesley.

37. In this problem, we combine the idea exploded in interconnected tanks problems,

$$\text{rate of change} = \text{rate in} - \text{rate out}, \tag{5.24}$$

with the Newton's law of cooling

$$\frac{dT}{dt} = K(T - M). \tag{5.25}$$

Let $x(t)$ and $y(t)$ denote temperatures in rooms A and B, respectively.

Room A. It gets temperature only from the heater with a rate

$$\text{rate in} = 80,000\,\text{Btu/h} \cdot \frac{1/4°}{1000\,\text{Btu}} = 20°/\text{h}.$$

Temperature goes out of the room A into the room B and outside with coefficients of proportionality in (5.25) equal to $K_1 = 1/2$ and $K_2 = 1/4$, respectively. Therefore,

$$
\begin{aligned}
\text{rate out} &= \text{rate into B} + \text{rate outside} \\
&= \frac{1}{2}(x - y) + \frac{1}{4}(x - 0) = \frac{3}{4}x - \frac{1}{2}y.
\end{aligned}
$$

Thus, (5.24) implies that

$$x' = 20 - \left(\frac{3}{4}x - \frac{1}{2}y\right) = 20 - \frac{3}{4}x + \frac{1}{2}y.$$

Room B. Similarly, we obtain

$$y' = \left[1000 \cdot \frac{2}{1000} + \frac{1}{2}(x - y)\right] - \frac{1}{5}(y - 0) = 2 + \frac{1}{2}x - \frac{7}{10}y.$$

Hence, the system governing the temperature exchange is

$$
\begin{aligned}
x' &= 20 - (3/4)x + (1/2)y, \\
y' &= 2 + (1/2)x - (7/10)y.
\end{aligned}
$$

We find the critical points of this system by solving

$$
\begin{aligned}
20 - (3/4)x + (1/2)y &= 0, \\
2 + (1/2)x - (7/10)y &= 0
\end{aligned}
\quad \Rightarrow \quad
\begin{aligned}
3x - 2y &= 80, \\
-5x + 7y &= 20
\end{aligned}
\quad \Rightarrow \quad
\begin{aligned}
x &= 600/11, \\
y &= 460/11.
\end{aligned}
$$

Therefore, $(600/11, 460/11)$ is the only critical point of the system. Analyzing the direction field, we conclude that $(600/11, 460/11)$ is an asymptotically stable node. Hence,

$$\lim_{t \to \infty} y(t) = \frac{460}{11} \approx 41.8°\text{F}.$$

Copyright © 2012 Pearson Education, Inc. Publishing as Addison-Wesley.

One can also find an explicit solution

$$y(t) = 460/11 + c_1 e^{r_1 t} + c_2 e^{r_2 t},$$

where $r_1 < 0$, $r_2 < 0$, to conclude that $y(t) \to 460/11$, as $t \to \infty$.)

39. Let y be an arbitrary function, which is differentiable as many times as necessary. Note that, for a differential operator A, $A[y]$ is a function, and so we can use commutative, associative, and distributive laws dealing with such functions.

(a) It is straightforward that

$$(A + B)[y] := A[y] + B[y] = B[y] + A[y] =: (B + A)[y].$$

To prove commutativity of the multiplication, we will use the linearity of the differential operator D, that is,

$$D[\alpha x + \beta y] = \alpha D[x] + \beta D[y]$$

and the fact that $D^i D^j = D^{i+j} = D^j D^i$. Indeed,

$$\left(D^i D^j\right)[y] := D^i\left[D^j[y]\right] = \left(y^{(j)}\right)^{(i)} = y^{(i+j)} = \left(y^{(i)}\right)^{(j)} = D^j\left[D^i[y]\right] =: \left(D^j D^i\right)[y].$$

Thus, we have

$$(AB)[y] := A\left[B[y]\right] = \left(\sum_{j=0}^{2} a_j D^j\right)\left[\left(\sum_{i=0}^{2} b_i D^i\right)[y]\right]$$

$$:= \left(\sum_{j=0}^{2} a_j D^j\right)\left[\sum_{i=0}^{2} b_i D^i[y]\right] := \sum_{j=0}^{2} a_j D^j\left[\sum_{i=0}^{2} b_i D^i[y]\right]$$

$$= \sum_{j=0}^{2}\sum_{i=0}^{2}\left(a_j D^j b_i D^i\right)[y] = \sum_{i=0}^{2}\sum_{j=0}^{2}\left(b_i D^i a_j D^j\right)[y]$$

$$= \sum_{i=0}^{2} b_i D^i\left[\sum_{j=0}^{2} a_j D^j[y]\right] =: \left(\sum_{i=0}^{2} b_i D^i\right)\left[\sum_{j=0}^{2} a_j D^j[y]\right]$$

$$=: \left(\sum_{i=0}^{2} b_i D^i\right)\left[\left(\sum_{j=0}^{2} a_j D^j\right)[y]\right] = B\left[A[y]\right] =: (BA)[y].$$

(b) We have

$$\{(A + B) + C\}[y] := (A + B)[y] + C[y] := (A[y] + B[y]) + C[y]$$

 Copyright © 2012 Pearson Education, Inc. Publishing as Addison-Wesley.

$$= A[y] + (B[y] + C[y]) =: A[y] + (B + C)[y]$$
$$:= \{A + (B + C)\}[y]$$

and

$$\{(AB)C\}[y] := (AB)[C[y]] := A\left[B\left[C[y]\right]\right] =: A\left[(BC)[y]\right] =: \{A(BC)\}[y].$$

(c) Using the linearity of differential operators, we obtain

$$\{A(B + C)\}[y] := A\left[(B + C)[y]\right] := A\left[B[y] + C[y]\right]$$
$$= A\left[B[y]\right] + A\left[C[y]\right] =: (AB)[y] + (AC)[y] =: \{(AB) + (AC)\}[y].$$

41. As it was noticed in Example 1, performing arithmetic operations, we can treat a "polynomial" in D, that is, an expression of the form $p(D) = \sum_{i=0}^{n} a_i D^i$, as a regular algebraic polynomial $p(r) = \sum_{i=0}^{n} a_i r^i$. Hence, the factorization problem for $p(D)$ is equivalent to the factorization problem for $p(r)$, which reduces to finding its roots.

(a) $r = \dfrac{-3 \pm \sqrt{3^2 - 4(-4)}}{2} = \dfrac{-3 \pm 5}{2} = -4, 1 \quad \Rightarrow \quad D^2 + 3D - 4 = (D + 4)(D - 1).$

(b) $r = \dfrac{-1 \pm \sqrt{1^2 - 4(-6)}}{2} = \dfrac{-1 \pm 5}{2} = -3, 2 \quad \Rightarrow \quad D^2 + D - 6 = (D + 3)(D - 2).$

(c) $r = \dfrac{-9 \pm \sqrt{9^2 - 4(-5)2}}{4} = \dfrac{-9 \pm 11}{4} = -5, 1/2 \quad \Rightarrow \quad 2D^2 + 9D - 5 = (D + 5)(2D - 1)$

(d) $r = \pm\sqrt{2} \quad \Rightarrow \quad D^2 - 2 = (D + \sqrt{2})(D - \sqrt{2}).$

EXERCISES 5.3: Solving Systems and Higher–Order Equations Numerically

1. First, we isolate $y''(t)$ and get an equivalent equation, $y''(t) = 3y(t) - ty'(t) + t^2$. Denoting $x_1 := y$, $x_2 := y'$, we conclude that

$$x_1' = y' = x_2,$$
$$x_2' = (y')' = y'' = 3y - ty' + t^2 = 3x_1 - tx_2 + t^2,$$

with initial conditions

$$x_1(0) = y(0) = 3, \quad x_2(0) = y'(0) = -6.$$

Copyright © 2012 Pearson Education, Inc. Publishing as Addison-Wesley.

Therefore, the given initial value problem is equivalent to

$$x_1' = x_2 \,,$$
$$x_2' = 3x_1 - tx_2 + t^2 \,,$$
$$x_1(0) = 3, \quad x_2(0) = -6.$$

3. Isolating $y^{(4)}(t)$, we get

$$y^{(4)}(t) = y^{(3)}(t) - 7y(t) + \cos t \,.$$

In this problem, we need four new variables – for $y(t)$, $y'(t)$, $y''(t)$, and $y'''(t)$. Thus, we denote

$$x_1 = y, \qquad x_2 = y', \qquad x_3 = y'', \quad \text{and} \quad x_4 = y''' \,.$$

The initial conditions then become

$$x_1(0) = y(0) = 1 \,, \qquad x_2(0) = y'(0) = 1 \,,$$
$$x_3(0) = y''(0) = 0, \quad x_4(0) = y'''(0) = 2 \,.$$

We have

$$x_1' = y' = x_2 \,,$$
$$x_2' = (y')' = y'' = x_3 \,,$$
$$x_3' = (y'')' = y''' = x_4 \,,$$
$$x_4' = (y''')' = y^{(4)} = y''' - 7y + \cos t = x_4 - 7x_1 + \cos t \,.$$

Hence, the required initial value problem for a system in normal form is

$$x_1' = x_2 \,,$$
$$x_2' = x_3 \,,$$
$$x_3' = x_4 \,,$$
$$x_4' = x_4 - 7x_1 + \cos t,$$
$$x_1(0) = x_2(0) = 1, \quad x_3(0) = 0, \quad x_4(0) = 2 \,.$$

5. First, we express the given system as

$$x'' = x' - y + 2t \,,$$
$$y'' = x - y - 1 \,.$$

 Copyright © 2012 Pearson Education, Inc. Publishing as Addison-Wesley.

Setting $x_1 = x$, $x_2 = x'$, $x_3 = y$, $x_4 = y'$ we obtain

$$
\begin{aligned}
x_1' &= x' = x_2, \\
x_2' &= x'' = x_2 - x_3 + 2t, \\
x_3' &= y' = x_4, \\
x_4' &= y'' = x_1 - x_3 - 1
\end{aligned}
\qquad \Rightarrow \qquad
\begin{aligned}
x_1' &= x_2, \\
x_2' &= x_2 - x_3 + 2t, \\
x_3' &= x_4, \\
x_4' &= x_1 - x_3 - 1
\end{aligned}
$$

with initial conditions $x_1(3) = 5$, $x_2(3) = 2$, $x_3(3) = 1$, and $x_4(3) = -1$.

7. In equivalent form, we have a system

$$
\begin{aligned}
x''' &= y + t, \\
y'' &= \frac{2y - 2x'' + 1}{5}.
\end{aligned}
$$

Setting

$$
x_1 = x, \quad x_2 = x', \quad x_3 = x'', \quad x_4 = y, \quad x_5 = y',
$$

we obtain a system in normal form

$$
\begin{aligned}
x_1' &= x_2, \\
x_2' &= x_3, \\
x_3' &= x_4 + t, \\
x_4' &= x_5, \\
x_5' &= \frac{1}{5}(2x_4 - 2x_3 + 1)
\end{aligned}
$$

with initial conditions

$$
x_1(0) = x_2(0) = x_3(0) = 4, \quad x_4(0) = x_5(0) = 1.
$$

9. To see how the improved Euler's method can be extended, let us recall (see Section 3.6) the improved Euler's method. For the initial value problem

$$
x' = f(t, x), \qquad x(t_0) = x_0,
$$

the recursive formulas for the improved Euler's method are

$$
\begin{aligned}
t_{n+1} &= t_n + h, \\
x_{n+1} &= x_n + \frac{h}{2}\left[f(t_n, x_n) + f(t_n + h, x_n + hf(t_n, x_n)) \right],
\end{aligned}
$$

Copyright © 2012 Pearson Education, Inc. Publishing as Addison-Wesley.

where h is the step size. Now, suppose we want to approximate a solution $x_1(t)$, $x_2(t)$ to the system

$$x_1' = f_1(t, x_1, x_2),$$
$$x_2' = f_2(t, x_1, x_2)$$

satisfying the initial conditions

$$x_1(t_0) = a_1, \quad x_2(t_0) = a_2.$$

Let $x_{1;n}$ and $x_{2;n}$ denote approximations to $x_1(t_n)$ and $x_2(t_n)$, resp., where $t_n = t_0 + nh$ for $n = 0, 1, 2, \ldots$. The recursive formulas for the improved Euler's method can be obtained by forming the vector analogue of the scalar formula. We have

$$t_{n+1} = t_n + h,$$
$$x_{1;n+1} = x_{1;n} + \frac{h}{2} \left[f_1(t_n, x_{1;n}, x_{2;n}) \right.$$
$$\left. + f_1(t_n + h, x_{1;n} + h f_1(t_n, x_{1;n}, x_{2;n}), x_{2;n} + h f_2(t_n, x_{1;n}, x_{2;n})) \right],$$
$$x_{2;n+1} = x_{2;n} + \frac{h}{2} \left[f_2(t_n, x_{1;n}, x_{2;n}) \right.$$
$$\left. + f_2(t_n + h, x_{1;n} + h f_1(t_n, x_{1;n}, x_{2;n}), x_{2;n} + h f_2(t_n, x_{1;n}, x_{2;n})) \right].$$

This approach can be used, in general, for systems of m equations in normal form. Indeed, suppose we want to approximate the solution $x_1(t)$, $x_2(t)$, ..., $x_m(t)$ to the system

$$x_1' = f_1(t, x_1, x_2, \ldots, x_m),$$
$$x_2' = f_2(t, x_1, x_2, \ldots, x_m),$$
$$\vdots$$
$$x_m' = f_m(t, x_1, x_2, \ldots, x_m),$$

with the initial conditions $x_1(t_0) = a_1$, $x_2(t_0) = a_2, \ldots$, $x_m(t_0) = a_m$.

We adapt these recursive formulas and obtain for $n = 0, 1, 2, \ldots$

$$t_{n+1} = t_n + h \, ;$$
$$x_{1;n+1} = x_{1;n} + \frac{h}{2} \left[f_1(t_n, x_{1;n}, \ldots, x_{m;n}) + f_1(t_n + h, x_{1;n} + h f_1(t_n, x_{1;n}, \ldots, x_{m;n}), \right.$$
$$\left. x_{2;n} + h f_2(t_n, x_{1;n}, \ldots, x_{m;n}), \ldots, x_{m;n} + h f_m(t_n, x_{1;n}, \ldots, x_{m;n})) \right],$$
$$x_{2;n+1} = x_{2;n} + \frac{h}{2} \left[f_2(t_n, x_{1;n}, \ldots, x_{m;n}) + f_2(t_n + h, x_{1;n} + h f_1(t_n, x_{1;n}, \ldots, x_{m;n}), \right.$$

Copyright © 2012 Pearson Education, Inc. Publishing as Addison-Wesley.

$$x_{2;n} + hf_2(t_n, x_{1;n}, \ldots, x_{m;n}), \ldots, x_{m;n} + hf_m(t_n, x_{1;n}, \ldots, x_{m;n}))],$$

$$\vdots$$

$$x_{m;n+1} = x_{m;n} + \frac{h}{2} \left[f_m(t_n, x_{1,n}, \ldots, x_{m;n}) + f_m(t_n + h, x_{1;n} + hf_1(t_n, x_{1;n}, \ldots, x_{m;n}), \right.$$
$$\left. x_{2;n} + hf_2(t_n, x_{1;n}, \ldots, x_{m;n}), \ldots, x_{m;n} + hf_m(t_n, x_{1;n}, \ldots, x_{m;n})) \right].$$

11. See the answer in the text.

13. See the answer in the text.

15. See the answer in the text.

17. Let $x_1 := u$, $x_2 := v$, and denote the independent variable by t (in order to be consistent with formulas in Section 5.3). In new notation, we have an initial value problem

$$x_1' = 3x_1 - 4x_2,$$
$$x_2' = 2x_1 - 3x_2,$$
$$x_1(0) = x_2(0) = 1$$

for a system in normal form. Here, $f_1(t, x_1, x_2) = 3x_1 - 4x_2$, $f_2(t, x_1, x_2) = 2x_1 - 3x_2$. Thus, formulas for $k_{i,j}$'s in vectorized Runge-Kutta algorithm become

$$k_{1,1} = h(3x_{1;n} - 4x_{2;n}),$$

$$k_{2,1} = h(2x_{1;n} - 3x_{2;n}),$$

$$k_{1,2} = h \left[3 \left(x_{1;n} + \frac{k_{1,1}}{2} \right) - 4 \left(x_{2;n} + \frac{k_{2,1}}{2} \right) \right],$$

$$k_{2,2} = h \left[2 \left(x_{1;n} + \frac{k_{1,1}}{2} \right) - 3 \left(x_{2;n} + \frac{k_{2,1}}{2} \right) \right],$$

$$k_{1,3} = h \left[3 \left(x_{1;n} + \frac{k_{1,2}}{2} \right) - 4 \left(x_{2;n} + \frac{k_{2,2}}{2} \right) \right],$$

$$k_{2,3} = h \left[2 \left(x_{1;n} + \frac{k_{1,2}}{2} \right) - 3 \left(x_{2;n} + \frac{k_{2,2}}{2} \right) \right],$$

$$k_{1,4} = h \left[3 \left(x_{1;n} + k_{1,3} \right) - 4 \left(x_{2;n} + k_{2,3} \right) \right],$$

$$k_{2,4} = h \left[2 \left(x_{1;n} + k_{1,3} \right) - 3 \left(x_{2;n} + k_{2,3} \right) \right].$$

With inputs $t_0 = 0$, $x_{1;0} = x_{2;0} = 1$, and the step size $h = 1$, we compute

$$k_{1,1} = h(3x_{1;0} - 4x_{2;0}) = 3(1) - 4(1) = -1,$$
$$k_{2,1} = h(2x_{1;0} - 3x_{2;0}) = 2(1) - 3(1) = -1,$$

Copyright © 2012 Pearson Education, Inc. Publishing as Addison-Wesley.

$$k_{1,2} = h\left[3\left(x_{1;0} + \frac{k_{1,1}}{2}\right) - 4\left(x_{2;0} + \frac{k_{2,1}}{2}\right)\right] = 3\left(1 + \frac{-1}{2}\right) - 4\left(1 + \frac{-1}{2}\right) = -\frac{1}{2},$$

$$k_{2,2} = h\left[2\left(x_{1;0} + \frac{k_{1,1}}{2}\right) - 3\left(x_{2;0} + \frac{k_{2,1}}{2}\right)\right] = 2\left(1 + \frac{-1}{2}\right) - 3\left(1 + \frac{-1}{2}\right) = -\frac{1}{2},$$

$$k_{1,3} = h\left[3\left(x_{1;0} + \frac{k_{1,2}}{2}\right) - 4\left(x_{2;0} + \frac{k_{2,2}}{2}\right)\right]$$

$$= 3\left(1 + \frac{-1/2}{2}\right) - 4\left(1 + \frac{-1/2}{2}\right) = -\frac{3}{4},$$

$$k_{2,3} = h\left[2\left(x_{1;0} + \frac{k_{1,2}}{2}\right) - 3\left(x_{2;0} + \frac{k_{2,2}}{2}\right)\right]$$

$$= 2\left(1 + \frac{-1/2}{2}\right) - 3\left(1 + \frac{-1/2}{2}\right) = -\frac{3}{4},$$

$$k_{1,4} = h\left[3\left(x_{1;0} + k_{1,3}\right) - 4\left(x_{2;0} + k_{2,3}\right)\right] = 3\left(1 + \frac{-3}{4}\right) - 4\left(1 + \frac{-3}{4}\right) = -\frac{1}{4},$$

$$k_{2,4} = h\left[2\left(x_{1;0} + k_{1,3}\right) - 3\left(x_{2;0} + k_{2,3}\right)\right] = 2\left(1 + \frac{-3}{4}\right) - 3\left(1 + \frac{-3}{4}\right) = -\frac{1}{4}.$$

Using the recursive formulas, we find $t_1 = t_0 + h = 0 + 1 = 1$ and

$$x_{1;1} = x_{1;0} + \frac{1}{6}\left(k_{1,1} + 2k_{1,2} + 2k_{1,3} + k_{1,4}\right)$$

$$= 1 + \frac{(-1) + 2(-1/2) + 2(-3/4) + (-1/4)}{6} = \frac{3}{8},$$

$$x_{2;1} = x_{2;0} + \frac{1}{6}\left(k_{2,1} + 2k_{2,2} + 2k_{2,3} + k_{2,4}\right)$$

$$= 1 + \frac{(-1) + 2(-1/2) + 2(-3/4) + (-1/4)}{6} = \frac{3}{8}$$

as approximations to $x_1(1)$ and $x_2(1)$ with the step size $h = 1$.

We repeat the algorithm with $h = 2^{-m}$, $m = 1, 2, \ldots$. The results of these computations are shown in Table **5–A** on page 317. We stopped at $m = 2$, since

$$\left|x_1(1; 2^{-1}) - x_1(1; 2^{-2})\right| = \left|x_2(1; 2^{-1}) - x_2(1; 2^{-2})\right| = 0.36817 - 0.36789 < 0.001.$$

Hence $u(1) = v(1) \approx 0.36789$.

19. See the answer in the text.

21. We convert the given initial value problem to an initial value problem for a normal system. Let $x_1(t) = H(t)$, $x_2(t) = H'(t)$. Then, $H''(t) = x_2'(t)$, $x_1(0) = H(0) = 0$,

 Copyright © 2012 Pearson Education, Inc. Publishing as Addison-Wesley.

$x_2(0) = H'(0) = 0$, and so we get

$$x_1' = x_2\,,$$
$$60 - x_1 = (77.7)x_2' + (19.42)\,(x_2)^2\,, \qquad \Rightarrow \qquad x_2' = \left[60 - x_1 - (19.42)\,(x_2)^2\right]/77.7\,,$$
$$x_1(0) = x_2(0) = 0 \qquad\qquad\qquad\qquad x_1(0) = x_2(0) = 0.$$

Thus, $f_1(t, x_1, x_2) = x_2$, $f_2(t, x_1, x_2) = \left[60 - x_1 - (19.42)\,(x_2)^2\right]/77.7$, $t_0 = 0$, $x_{1;0} = 0$, and $x_{2;0} = 0$. With $h = 0.5$, we need $(5 - 0)/0.5 = 10$ steps to approximate the solution on the interval $[0, 5]$. Taking $n = 0$ in the vectorized Runge-Kutta algorithm, we approximate the solution at $t = 0.5$.

$$k_{1,1} = hx_{2;0} = 0.5(0) = 0,$$

$$k_{2,1} = h\left[60 - x_{1;0} - (19.42)\,(x_{2;0})^2\right]/77.7 = 0.38610\,,$$

$$k_{1,2} = h\left(x_{2;0} + \frac{k_{2,1}}{2}\right) = 0.5\left[(0) + \frac{0.38610}{2}\right] = 0.09653\,,$$

$$k_{2,2} = h\left[60 - \left(x_{1;0} + \frac{k_{1,1}}{2}\right) - (19.42)\left(x_{2;0} + \frac{k_{2,1}}{2}\right)^2\right]/77.7 = 0.38144\,,$$

$$k_{1,3} = h\left(x_{2;0} + \frac{k_{2,2}}{2}\right) = 0.5\left[(0) + \frac{0.38144}{2}\right] = 0.09536\,,$$

$$k_{2,3} = h\left[60 - \left(x_{1;0} + \frac{k_{1,2}}{2}\right) - (19.42)\left(x_{2;0} + \frac{k_{2,2}}{2}\right)^2\right]/77.7 = 0.38124\,,$$

$$k_{1,4} = h\left(x_{2;0} + k_{2,3}\right) = 0.5\left((0) + 0.38124\right) = 0.19062\,,$$

$$k_{2,4} = h\left[60 - (x_{1;0} + k_{1,3}) - (19.42)\,(x_{2;0} + k_{2,3})^2\right]/77.7 = 0.36732$$

Using the recursive formulas, we find that

$$t_1 = t_0 + h = 0 + 0.5 = 0.5$$
$$x_1(0.5) \approx x_{1;1} = x_{1;0} + \frac{1}{6}\left(k_{1,1} + 2k_{1,2} + 2k_{1,3} + k_{1,4}\right) = 0.09573\,,$$
$$x_2(0.5) \approx x_{2;1} = x_{2;0} + \frac{1}{6}\left(k_{2,1} + 2k_{2,2} + 2k_{2,3} + k_{2,4}\right) = 0.37980\,.$$

Repeating this procedure with $n = 1, 2, \ldots, 9$, we get the results listed in Table 5–B on page 317.

23. We let $x_1 = y$ and $x_2 = y'$ to get an initial value problem

$$x_1' = f_1(t, x_1, x_2) = x_2\,,$$
$$x_2' = f_2(t, x_1, x_2) = -x_1\left(1 + rx_1^2\right)$$

Copyright © 2012 Pearson Education, Inc. Publishing as Addison-Wesley.

with

$$x_1(0) = a, \quad x_2(0) = 0.$$

Using the definitions of t_n, $x_{i;n}$, $k_{i,1}$, $k_{i,2}$, $k_{i,3}$, and $k_{i,4}$, we obtain

$$k_{1,1} = h f_1 \left(t_n, x_{1;n}, x_{2;n} \right) = h x_{2;n},$$

$$k_{2,1} = h f_2 \left(t_n, x_{1;n}, x_{2;n} \right) = -h x_{1;n} \left(1 + r x_{1;n}^2 \right),$$

$$k_{1,2} = h f_1 \left(t_n + \frac{h}{2}, x_{1;n} + \frac{k_{1,1}}{2}, x_{2;n} + \frac{k_{2,1}}{2} \right) = h \left(x_{2;n} + \frac{k_{2,1}}{2} \right),$$

$$k_{2,2} = h f_2 \left(t_n + \frac{h}{2}, x_{1;n} + \frac{k_{1,1}}{2}, x_{2;n} + \frac{k_{2,1}}{2} \right)$$

$$= -h \left(x_{1;n} + \frac{k_{1,1}}{2} \right) \left[1 + r \left(x_{1;n} + \frac{k_{1,1}}{2} \right)^2 \right],$$

$$k_{1,3} = h f_1 \left(t_n + \frac{h}{2}, x_{1;n} + \frac{k_{1,2}}{2}, x_{2;n} + \frac{k_{2,2}}{2} \right) = h \left(x_{2;n} + \frac{k_{2,2}}{2} \right),$$

$$k_{2,3} = h f_2 \left(t_n + \frac{h}{2}, x_{1;n} + \frac{k_{1,2}}{2}, x_{2;n} + \frac{k_{2,2}}{2} \right)$$

$$= -h \left(x_{1;n} + \frac{k_{1,2}}{2} \right) \left[1 + r \left(x_{1;n} + \frac{k_{1,2}}{2} \right)^2 \right],$$

$$k_{1,4} = h f_1 \left(t_n + h, x_{1;n} + k_{1,3}, x_{2;n} + k_{2,3} \right) = h \left(x_{2;n} + k_{2,3} \right),$$

$$k_{2,4} = h f_2 \left(t_n + h, x_{1;n} + k_{1,3}, x_{2;n} + k_{2,3} \right) = -h \left(x_{1;n} + k_{1,3} \right) \left[1 + r \left(x_{1;n} + k_{1,3} \right)^2 \right].$$

Using these formulas, we find that

$$t_{n+1} = t_n + h = t_n + 0.1,$$

$$x_{1;n+1} = x_{1;n} + \frac{1}{6} \left(k_{1,1} + 2k_{1,2} + 2k_{1,3} + k_{1,4} \right),$$

$$x_{2;n+1} = x_{2;n} + \frac{1}{6} \left(k_{2,1} + 2k_{2,2} + 2k_{2,3} + k_{2,4} \right).$$

In Table **5–C**, page 317, we give the approximate period for $r = 1$ and $r = 2$ with $a = 1$, 2 and 3. From this table, we see that the period varies as r is varied or as a is varied.

25. With $x_1 = y$, $x_2 = y'$, and $x_3 = y''$, the the initial value problem can be expressed as

$$\begin{aligned} x_1' &= x_2, & x_1(0) &= 1, \\ x_2' &= x_3, & x_2(0) &= 1, \\ x_3' &= t - x_3 - x_1^2, & x_3(0) &= 1. \end{aligned}$$

Here,

$$f_1(t, x_1, x_2, x_3) = x_2,$$

     Copyright © 2012 Pearson Education, Inc. Publishing as Addison-Wesley.

$$f_2(t, x_1, x_2, x_3) = x_3\,,$$

$$f_3(t, x_1, x_2, x_3) = t - x_3 - x_1^2\,.$$

Since we are computing the approximations for $c = 1$, the initial value for h in Step 1 of the algorithm in Appendix F of the text is

$$h = (1 - 0)2^{-0} = 1\,.$$

Equations in Step 3 then become

$$k_{1,1} = hf_1(t, x_1, x_2, x_3) = hx_2\,,$$

$$k_{2,1} = hf_2(t, x_1, x_2, x_3) = hx_3\,,$$

$$k_{3,1} = hf_3(t, x_1, x_2, x_3) = h\left(t - x_3 - x_1^2\right),$$

$$k_{1,2} = hf_1\left(t + \frac{h}{2}, x_1 + \frac{k_{1,1}}{2}, x_2 + \frac{k_{2,1}}{2}, x_3 + \frac{k_{3,1}}{2}\right) = h\left(x_2 + \frac{k_{2,1}}{2}\right),$$

$$k_{2,2} = hf_2\left(t + \frac{h}{2}, x_1 + \frac{k_{1,1}}{2}, x_2 + \frac{k_{2,1}}{2}, x_3 + \frac{k_{3,1}}{2}\right) = h\left(x_3 + \frac{k_{3,1}}{2}\right),$$

$$k_{3,2} = hf_3\left(t + \frac{h}{2}, x_1 + \frac{k_{1,1}}{2}, x_2 + \frac{k_{2,1}}{2}, x_3 + \frac{k_{3,1}}{2}\right)$$

$$= h\left[t + \frac{h}{2} - x_3 - \frac{k_{3,1}}{2} - \left(x_1 + \frac{k_{1,1}}{2}\right)^2\right],$$

$$k_{1,3} = hf_1\left(t + \frac{h}{2}, x_1 + \frac{k_{1,2}}{2}, x_2 + \frac{k_{2,2}}{2}, x_3 + \frac{k_{3,2}}{2}\right) = h\left(x_2 + \frac{k_{2,2}}{2}\right),$$

$$k_{2,3} = hf_2\left(t + \frac{h}{2}, x_1 + \frac{k_{1,2}}{2}, x_2 + \frac{k_{2,2}}{2}, x_3 + \frac{k_{3,2}}{2}\right) = h\left(x_3 + \frac{k_{3,2}}{2}\right),$$

$$k_{3,3} = hf_3\left(t + \frac{h}{2}, x_1 + \frac{k_{1,2}}{2}, x_2 + \frac{k_{2,2}}{2}, x_3 + \frac{k_{3,2}}{2}\right)$$

$$= h\left[t + \frac{h}{2} - x_3 - \frac{k_{3,2}}{2} - \left(x_1 + \frac{k_{1,2}}{2}\right)^2\right],$$

$$k_{1,4} = hf_1(t + h, x_1 + k_{1,3}, x_2 + k_{2,3}, x_3 + k_{3,3}) = h(x_2 + k_{2,3})\,,$$

$$k_{2,4} = hf_2(t + h, x_1 + k_{1,3}, x_2 + k_{2,3}, x_3 + k_{3,3}) = h(x_3 + k_{3,3})\,,$$

$$k_{3,4} = hf_3(t + h, x_1 + k_{1,3}, x_2 + k_{2,3}, x_3 + k_{3,3}) = h\left[t + h - x_3 - k_{3,3} - (x_1 + k_{1,3})^2\right].$$

Using the values $t_0 = 0$, $a_1 = 1$, $a_2 = 0$, and $a_3 = 1$, we get the first approximations.

$$x_1(1; 1) = 1.29167\,, \qquad x_2(1; 1) = 0.28125\,, \qquad x_3(1; 1) = 0.03125\,.$$

Copyright © 2012 Pearson Education, Inc. Publishing as Addison-Wesley.

Repeating the algorithm with $h = 2^{-1}$, 2^{-2}, and 2^{-3}, we obtain the approximations shown in Table **5–D**, page 319. We stopped at $n = 3$ since

$$\left| \frac{x_1(1; 2^{-3}) - x_1(1; 2^{-2})}{x_1(1; 2^{-3})} \right| = \left| \frac{1.25958 - 1.25960}{1.25958} \right| = 0.00002 < 0.01 \,,$$

$$\left| \frac{x_2(1; 2^{-3}) - x_2(1; 2^{-2})}{x_2(1; 2^{-3})} \right| = \left| \frac{0.34704 - 0.34696}{0.34704} \right| = 0.00023 < 0.01 \,,$$

$$\left| \frac{x_3(1; 2^{-3}) - x_3(1; 2^{-2})}{x_3(1; 2^{-3})} \right| = \left| \frac{-0.06971 + 0.06957}{-0.06971} \right| = 0.00201 < 0.01 \,.$$

Hence, with tolerance 0.01, $y(1) \approx x_1\left(1; 2^{-3}\right) = 1.25958$.

27. See the answer in the text.

29. See the answer in the text.

EXERCISES 5.4:   Introduction to the Phase Plane

1. Substitution of $x(t) = e^{3t}$, $y(t) = e^t$ into the system yields

$$\frac{dx}{dt} = \frac{d}{dt}\left(e^{3t}\right) = 3e^{3t} = 3\left(e^t\right)^3 = 3y^3 \,,$$

$$\frac{dy}{dt} = \frac{d}{dt}\left(e^t\right) = e^t = y.$$

Thus, the given pair of functions is a solution. To sketch the trajectory of this solution, we express x as a function of y.

$$x = e^{3t} = \left(e^t\right)^3 = y^3 \qquad \text{for} \qquad y = e^t > 0.$$

Since $y = e^t$ is an increasing function, the flow arrows are directed away from the origin. See Figure B.29 in the answers of the text.

3. In this problem, $f(x, y) = x - y$, $g(x, y) = x^2 + y^2 - 1$. To find the critical point set, we solve the system

$$\begin{aligned} x - y &= 0, \\ x^2 + y^2 - 1 &= 0 \end{aligned} \qquad \Rightarrow \qquad \begin{aligned} x &= y, \\ x^2 + y^2 &= 1. \end{aligned}$$

Eliminating y yields

$$2x^2 = 1 \qquad \Rightarrow \qquad x = \pm\frac{1}{\sqrt{2}} \,.$$

Substituting x into the first equation, we find the corresponding value for y. Thus the critical points of the given system are $(1/\sqrt{2}, 1/\sqrt{2})$ and $(-1/\sqrt{2}, -1/\sqrt{2})$.

   Copyright © 2012 Pearson Education, Inc. Publishing as Addison-Wesley.

5. In this problem, $f(x, y) = x^2 - 2xy$, $g(x, y) = 3xy - y^2$, and so we find critical points by solving the system

$$\begin{aligned} x^2 - 2xy &= 0, \\ 3xy - y^2 &= 0 \end{aligned} \quad \Rightarrow \quad \begin{aligned} x(x - 2y) &= 0, \\ y(3x - y) &= 0. \end{aligned}$$

From the first equation we conclude that either $x = 0$ or $x = 2y$. Substituting these values into the second equation, we get

$$x = 0 \quad \Rightarrow \quad y[3(0) - y] = 0 \quad \Rightarrow \quad -y^2 = 0 \quad \Rightarrow \quad y = 0\,;$$

$$x = 2y \quad \Rightarrow \quad y[3(2y) - y] = 0 \quad \Rightarrow \quad 5y^2 = 0 \quad \Rightarrow \quad x = 2(0) = 0\,.$$

Therefore, $(0, 0)$ is the only critical point.

7. Here, $f(x, y) = y - 1$, $g(x, y) = e^{x+y}$. Thus, the phase plane equation becomes

$$\frac{dy}{dx} = \frac{e^{x+y}}{y - 1} = \frac{e^x e^y}{y - 1}.$$

Separating variables yields

$$(y - 1)e^{-y}\, dy = e^x\, dx \quad \Rightarrow \quad \int (y - 1)e^{-y}\, dy = \int e^x\, dx$$

$$\Rightarrow \quad -ye^{-y} + c = e^x \quad \Rightarrow \quad e^x + ye^{-y} = c.$$

9. The phase plane equation for this system is

$$\frac{dy}{dx} = \frac{g(x, y)}{f(x, y)} = \frac{e^x + y}{2y - x}.$$

We rewrite this equation in symmetric form,

$$-(e^x + y)\, dx + (2y - x)\, dy = 0,$$

and check it for exactness.

$$\frac{\partial M}{\partial y} = \frac{\partial}{\partial y}\left[-(e^x + y)\right] = -1,$$

$$\frac{\partial N}{\partial x} = \frac{\partial}{\partial x}(2y - x) = -1.$$

Therefore, the equation is exact, and we have

$$F(x, y) = \int N(x, y)\, dy = \int (2y - x)\, dy = y^2 - xy + g(x)\,;$$

Copyright © 2012 Pearson Education, Inc. Publishing as Addison-Wesley.

Chapter 5

$$M(x,y) = \frac{\partial}{\partial x} F(x,y) = \frac{\partial}{\partial x}\left(y^2 - xy + g(x)\right) = -y + g'(x) = -\left(e^x + y\right)$$

$$\Rightarrow \quad g'(x) = -e^x \quad \Rightarrow \quad g(x) = \int \left(-e^x\right) dx = -e^x.$$

Hence, a general solution to the phase plane equation is given implicitly by

$$F(x,y) = y^2 - xy - e^x = c_1 \qquad \text{or} \qquad e^x + xy - y^2 = c,$$

where $c = -c_1$ is an arbitrary constant.

11. In this problem, $f(x,y) = 2y$ and $g(x,y) = 2x$. Therefore, the phase plane equation for given system is

$$\frac{dy}{dx} = \frac{2x}{2y} = \frac{x}{y}.$$

Separating variables and integrating yields

$$y\,dy = x\,dx \qquad \Rightarrow \qquad \int y\,dy = \int x\,dx$$

$$\Rightarrow \qquad \frac{1}{2}y^2 = \frac{1}{2}x^2 + c_1 \qquad \Rightarrow \qquad y^2 - x^2 = c.$$

Thus, the trajectories are hyperbolas if $c \neq 0$ and, for $c = 0$, the lines $y = \pm x$.

In the upper half-plane, $y > 0$, we have $x' = 2y > 0$ and, therefore, $x(t)$ increases. In the lower half-plane, $x' < 0$ and so $x(t)$ decreases. This implies that solutions flow from the left to the right in the upper half-plane and from the right to the left in the lower half-plane. See Figure B.30 in the answers of the text.

13. First, we find the critical points of this system solving

$$(y - x)(y - 1) = 0,$$
$$(x - y)(x - 1) = 0.$$

Notice that both of these equations are satisfied if $y = x$. Thus, $x = C$ and $y = C$, for any fixed constant C, is a solution to the given system of differential equations and one family of critical points is the line $y = x$. We also see that there is a critical point at $(1,1)$. (This critical point belongs to the line $y = x$.)

Next we find the integral curves. Solving the first order separable differential equation

$$\frac{dy}{dx} = \frac{dy/dt}{dx/dt} = \frac{(x-y)(x-1)}{(y-x)(y-1)} \qquad \Rightarrow \qquad \frac{dy}{dx} = \frac{1-x}{y-1}$$

Copyright © 2012 Pearson Education, Inc. Publishing as Addison-Wesley.

yields

$$\int (y-1)\,dy = \int (1-x)\,dx$$

$$\Rightarrow \quad \frac{y^2}{2} - y = x - \frac{x^2}{2} + c_1 \quad \Rightarrow \quad x^2 - 2x + y^2 - 2y = 2c_1.$$

Completing squares, we obtain

$$(x-1)^2 + (y-1)^2 = c,$$

where $c = 2c_1 + 2$. Therefore, the integral curves are concentric circles with centers at $(1,1)$. The trajectories, associated with constants $c = 1, 4$, and 9, are sketched in Figure B.31 in the answers of the text.

Finally, we determine the flow along the trajectories. Notice that the variable t imparts a flow to the trajectories of a solution to a system of differential equations in the same manner as the parameter t imparts a direction to a curve written in parametric form. We find this flow by determining the regions in the xy-plane, where $x(t)$ is increasing (moving from the left to the right on each trajectory) and the regions, where $x(t)$ is decreasing (moving from the right to the left on each trajectory). Therefore, we consider four cases in studying $dx/dt = (y-x)(y-1)$, the first equation in our system.

Case 1 : $y > x$ and $y < 1$. (This region is above the line $y = x$, but below the line $y = 1$.) In this case, $y - x > 0$ and $y - 1 < 0$. Thus, $dx/dt = (y-x)(y-1) < 0$. Hence, $x(t)$ is decreasing here, and the flow along the trajectories goes from the right to the left. Therefore, the motion is clockwise.

Case 2 : $y > x$ and $y > 1$. (This region is above the lines $y = x$ and $y = 1$.) In this case, we see that $y - x > 0$ and $y - 1 > 0$. Hence, $dx/dt = (y-x)(y-1) > 0$. Thus, $x(t)$ is increasing and the flow along the trajectories in this region is still going clockwise.

Case 3 : $y < x$ and $y < 1$. (This region is below the lines $y = x$ and $y = 1$.) In this case, $y - x < 0$ and $y - 1 < 0$. Thus, $dx/dt > 0$, and so $x(t)$ is increasing. Thus, the movement is from the left to the right, and the flow along the trajectories is counterclockwise.

Case 4 : $y < x$ and $y > 1$. (This region is below the line $y = x$ but above the line $y = 1$.) In this case, $y - x < 0$ and $y - 1 > 0$. Thus, $dx/dt < 0$, so that $x(t)$ is a

Copyright © 2012 Pearson Education, Inc. Publishing as Addison-Wesley.

decreasing function. Therefore, the flow goes from the right to the left, and so it is counterclockwise.

Therefore, the flow is clockwise above the line $y = x$, and it is counterclockwise below this line. See Figure B.31 in the answers of the text.

15. According to Definition 1, we have to solve the system of equations

$$2x + y + 3 = 0,$$
$$-3x - 2y - 4 = 0.$$

Eliminating y in the first equation, we obtain

$$x + 2 = 0$$

and, eliminating x in the first equation, we get

$$-y + 1 = 0 \,.$$

We see that $x = -2$, $y = 1$ satisfy both equations. Therefore, $(-2, 1)$ is a critical point.

From Figure B.32 in the answers of the text, we conclude that all solutions going near the point $(-2, 1)$ do not stay close to it, which implies that the critical point $(-2, 1)$ is unstable.

17. For critical points, we solve the system

$$\begin{aligned} f(x,y) &= 0, \\ g(x,y) &= 0 \end{aligned} \Rightarrow \begin{aligned} 2x + 13y &= 0, \\ -x - 2y &= 0 \end{aligned} \Rightarrow \begin{aligned} 2(-2y) + 13y &= 0, \\ x &= 2y \end{aligned} \Rightarrow \begin{aligned} y &= 0, \\ x &= 0. \end{aligned}$$

Therefore, the system has just one critical point, namely, $(0, 0)$. The direction field is shown in Figure B.33 in the text. From this picture, we conclude that $(0, 0)$ is a center (stable).

19. We set $v = y'$. Then $y'' = (y')' = v'$, and so given equation is equivalent to the system

$$\begin{aligned} y' &= v, \\ v' - y &= 0 \end{aligned} \Rightarrow \begin{aligned} y' &= v, \\ v' &= y \,. \end{aligned}$$

In this system, $f(y, v) = v$ and $g(y, v) = y$. For critical points we solve

$$\begin{aligned} f(y, v) &= v = 0, \\ g(y, v) &= y = 0 \end{aligned} \Rightarrow \begin{aligned} y &= 0, \\ v &= 0 \end{aligned}$$

 Copyright © 2012 Pearson Education, Inc. Publishing as Addison-Wesley.

and conclude that, in yv-plane, the system has only one critical point, $(0,0)$. In the upper half-plane, $y' = v > 0$ and, therefore, y increases and solutions flow to the right; similarly, solutions flow to the left in the lower half-plane. See Figure B.34 in the answers of the text.

The phase plane equation for the system is

$$\frac{dv}{dy} = \frac{dv/dx}{dy/dx} = \frac{y}{v} \quad \Rightarrow \quad v\,dv = y\,dy \quad \Rightarrow \quad v^2 - y^2 = c.$$

Thus, the integral curves are hyperbolas for $c \neq 0$ and the lines $v = \pm y$ for $c = 0$. On the line $v = -y$, the solutions flow into the critical point $(0,0)$, whereas solutions flow away from $(0,0)$ on $v = y$. So, $(0,0)$ is a saddle point (unstable).

21. We convert the given equation into a system of first order equations involving the functions $y(t)$ and $v(t)$ by using the substitution

$$v(t) = y'(t) \quad \Rightarrow \quad v'(t) = y''(t).$$

This yields

$$y' = v,$$
$$v' = -y - y^5 = -y\left(1 + y^4\right).$$

To find the critical points, we solve the equations $v = 0$ and $-y\left(1 + y^4\right) = 0$ simultaneously. This system is satisfied, when $v = 0$ and $y = 0$ only. Thus, the critical point is $(0,0)$. To find the integral curves, we solve the first order equation

$$\frac{dv}{dy} = \frac{dv/dt}{dy/dt} = \frac{-y - y^5}{v}.$$

This is a separable equation, which can be solved as follows.

$$v\,dv = \left(-y - y^5\right) dy \quad \Rightarrow \quad \frac{v^2}{2} = -\frac{y^2}{2} - \frac{y^6}{6} + c_1$$
$$\Rightarrow \quad 3v^2 + 3y^2 + y^6 = c \qquad (c = 6c_1).$$

Therefore, the integral curves for this system are given implicitly by $3v^2 + 3y^2 + y^6 = c$, where c is a positive constant.

To determine the flow direction along the trajectories, we examine the equation $y' = v$. Thus, we see that

$$\frac{dy}{dt} > 0 \quad \text{for} \quad v > 0 \quad \text{and} \quad \frac{dy}{dt} < 0 \quad \text{for} \quad v < 0.$$

Copyright © 2012 Pearson Education, Inc. Publishing as Addison-Wesley.

Chapter 5

Therefore, $y(t)$ is increasing, when $v > 0$, and it is decreasing, when $v < 0$. This means that above the y-axis the flow is going from the left to the right, and that below the y-axis the flow is going from the right to the left. Thus, the flow on these trajectories is directed clockwise (see Figure B.35 in the answers of the text). Therefore, $(0,0)$ is a center (stable).

23. With $v = y'$, $v' = y''$, the equation transforms to the system

$$
\begin{aligned}
y' &= v, \\
v' + y - y^4 &= 0
\end{aligned}
\qquad \Rightarrow \qquad
\begin{aligned}
y' &= v, \\
v' &= y^4 - y.
\end{aligned}
\tag{5.26}
$$

Therefore, $f(y,v) = v$ and $g(y,v) = y^4 - y = y(y^3 - 1)$. We find critical points by solving

$$
\begin{aligned}
v &= 0, \\
y(y^3 - 1) &= 0
\end{aligned}
\qquad \Rightarrow \qquad
\begin{aligned}
v &= 0, \\
y &= 0 \quad \text{or} \quad y = 1.
\end{aligned}
$$

Hence, system (5.26) has two critical points, $(0,0)$ and $(1,0)$.

In the upper half plane, $y' = v > 0$ and so solutions flow to the right; similarly, solutions flow to the left in the lower half-plane. See Figure B.36 in the text for the direction field. This figure indicates that $(0,0)$ is a stable critical point (center), whereas $(1,0)$ is a saddle point (unstable).

25. This system has two critical points, $(0,0)$ and $(1,0)$, which are solutions to

$$
\begin{aligned}
y &= 0, \\
-x + x^3 &= 0.
\end{aligned}
$$

The direction field for this system is shown in Figure B.37. From this figure we conclude:

(a) the solution $(x(t), y(t))$, passing through the point $(0.25, 0.25)$, flows around $(0,0)$, and so it is periodic;

(b) for the solution $(x(t), y(t))$, passing through the point $(2,2)$, $y(t) \to \infty$ as $t \to \infty$, and so this solution is not periodic;

(c) the solution $(x(t), y(t))$, passing through the critical point $(1,0)$, is a constant solution (equilibrium), and so it is periodic.

27. The direction field for given system is shown in Figure B.38 in the answers of the text. From the starting point, which is $(1,1)$, following the direction arrows, the solution flows

 Copyright © 2012 Pearson Education, Inc. Publishing as Addison-Wesley.

down and to the left, crosses the x-axis, has a turning point in the fourth quadrant, and then does to the left and up toward the critical point $(0, 0)$. Thus, we predict that, as $t \to \infty$, the solution $(x(t), y(t))$ approaches $(0, 0)$.

29. **(a)** The phase plane equation for this system is

$$\frac{dy}{dx} = \frac{3y}{x} .$$

Separating variables and integrating, we get

$$\frac{dy}{y} = \frac{3dx}{x} \qquad \Rightarrow \qquad \ln|y| = 3\ln|x| + c_1 \qquad \Rightarrow \qquad y = cx^3 .$$

So, integral curves are cubic curves. Since, in the right half-plane, $x' = x > 0$ and, in the left half-plane, $x' < 0$, the solutions flow to the right in the right half-plane and to the left in the left half-plane. Solutions starting on the y-axis stay on it ($x' = 0$) – they flow up if the initial point is in the upper half-plane (because $y' = y > 0$) and flow down if the initial point in the lower half-plane. This matches the description of an unstable node. (See Figure 5.7.)

(b) Solving the phase plane equation for this system, we get

$$\frac{dy}{dx} = \frac{-4x}{y} \qquad \Rightarrow \qquad y\,dy = -4x\,dx \qquad \Rightarrow \qquad y^2 + 4x^2 = c .$$

Thus, the integral curves are ellipses. (Also, notice that the solutions flow along these ellipses in clockwise direction, because x increases in the upper half-plane and decreases in the lower half-plane.) Therefore, here we have a center (stable).

(c) Solving $-5x + 2y > 0$ and $x - 4y > 0$ simultaneously, we find that x increases in the half-plane $y > 5x$ and decreases in the half-plane $y < 5x$, and that y increases in the half-plane $y < x/4$ and decreases in the half-plane $y > x/4$. This leads us to the scheme ⟍⟋⟋ for the solutions' flow. Thus, all solutions approach the critical point $(0, 0)$ as $t \to \infty$, which corresponds to a stable node.

(d) An analysis, similar to that in (c), shows that all the solutions flow away from $(0, 0)$. Among pictures shown in Figure 5.7, only the unstable node and the unstable spiral have this feature. Since the unstable node is the answer to (a), we have the unstable spiral in this case.

Copyright © 2012 Pearson Education, Inc. Publishing as Addison-Wesley.

(e) The phase plane equation, $dy/dx = (4x - 3y)/(5x - 3y)$, has two linear solutions, $y = 2x$ and $y = 2x/3$. (One can find them by substituting $y = ax$ into the above phase plane equation and solving for a.) Solutions, that start at a point on the line $y = 2x$ in the first quadrant, have $x' = 5x - 3(2x) = -x < 0$, and so flow toward $(0, 0)$; similarly, solutions, that start at a point on this line in the third quadrant, have $x' = -x > 0$ and, again, flow to $(0, 0)$. On the other line, $y = 2x/3$, the picture is opposite: in the first quadrant, $x' = 5x - 3(2x/3) = 3x > 0$, and $x' < 0$ in the third quadrant. Therefore, there are two lines, passing through the critical point $(0, 0)$, such that solutions to the system flow into $(0, 0)$ on one of them and flow away from $(0, 0)$ on the other. This is the case of a saddle (unstable) point.

(f) The only remaining picture is the asymptotically stable spiral. (One can also get a diagram $\nwarrow\!\!\!\!\!\diagup\!\!\!\!\!\swarrow$ for solutions' flow with just one matching picture in Figure 5.7.)

31. (a) Setting $y' = v$ (so that $y'' = v'$), we transform given equation to a first order system

$$\frac{dy}{dx} = v,$$
$$\frac{dv}{dx} = f(y).$$

(b) By the chain rule,

$$\frac{dv}{dy} = \frac{dv}{dx} \cdot \frac{dx}{dy} = \frac{dv/dx}{dy/dx} = \frac{f(y)}{v} \qquad \Rightarrow \qquad \frac{dv}{dy} = \frac{f(y)}{v}.$$

This equation is separable. Separating variables and integrating yields

$$v \, dv = f(y) \, dy \qquad \Rightarrow \qquad \int v \, dv = \int f(y) \, dy$$

$$\Rightarrow \qquad \frac{1}{2} v^2 = F(y) + K,$$

where $F(y)$ is an antiderivative of $f(y)$. Substituting back $v = y'$ gives the required.

33. (a) We denote $v(t) = x'(t)$ to transform the equation

$$\frac{d^2 x}{dt^2} = -x + \frac{1}{\lambda - x}$$

to an equivalent system of two first order differential equations, that is,

$$\frac{dx}{dt} = v,$$
$$\frac{dv}{dt} = -x + \frac{1}{\lambda - x}.$$

Copyright © 2012 Pearson Education, Inc. Publishing as Addison-Wesley.

(b) The phase plane equation in xv-plane for the system in (a) is

$$\frac{dv}{dx} = \frac{-x + 1/(\lambda - x)}{v}.$$

This equation is separable. Separating variables and integrating, we obtain

$$v\,dv = \left(-x + \frac{1}{\lambda - x}\right) dx \qquad \Rightarrow \qquad \int v\,dv = \int \left(-x + \frac{1}{\lambda - x}\right) dx$$

$$\Rightarrow \qquad \frac{1}{2}v^2 = -\frac{1}{2}x^2 - \ln|\lambda - x| + C_1 \qquad \Rightarrow \qquad v^2 = C - x^2 - 2\ln|\lambda - x|$$

$$\Rightarrow \qquad v = \pm\sqrt{C - x^2 - 2\ln(\lambda - x)}.$$

(The absolute value sign is not necessary since $x < \lambda$.)

(c) To find critical points for the system in (a), we solve

$$\begin{array}{ccccc} v = 0 & & v = 0 & & v = 0, \\ -x + \dfrac{1}{\lambda - x} = 0 & \Rightarrow & x^2 - \lambda x + 1 = 0 & \Rightarrow & x = \dfrac{\lambda \pm \sqrt{\lambda^2 - 4}}{2}. \end{array}$$

For $0 < \lambda < 2$, we have $\lambda^2 - 4 < 0$, and so both roots are complex. However, for $\lambda > 2$, there are two distinct real solutions,

$$x_1 = \frac{\lambda - \sqrt{\lambda^2 - 4}}{2} \qquad \text{and} \qquad x_2 = \frac{\lambda + \sqrt{\lambda^2 - 4}}{2},$$

and the critical points are

$$\left(\frac{\lambda - \sqrt{\lambda^2 - 4}}{2}, 0\right) \qquad \text{and} \qquad \left(\frac{\lambda + \sqrt{\lambda^2 - 4}}{2}, 0\right).$$

(d) The phase plane diagrams for $\lambda = 1$ and $\lambda = 3$ are shown in Figures B.39 and B.40 in the answers of the text.

(e) From Figure B.39 we conclude that, for $\lambda = 1$, all solution curves approach the vertical line $x = 1$. This means that the bar is attracted to the magnet. The case $\lambda = 3$ is more complicated. The behavior of the bar depends on the initial displacement $x(0)$ and the initial velocity $v(0) = x'(0)$. From Figure B.40 we see that (with $v(0) = 0$) if $x(0)$ is small enough, then the bar will oscillate about the position $x = x_1$; if $x(0)$ is close enough to λ, then the bar will be attracted to the magnet. It is also possible, with an appropriate combination of $x(0)$ and $v(0)$, that the bar will come to rest at the saddle point $(x_2, 0)$.

Copyright © 2012 Pearson Education, Inc. Publishing as Addison-Wesley.

Chapter 5

35. (a) Denoting $y' = v$, we have $y'' = v'$, and (with $m = \mu = k = 1$) (16) can be written as

$$y' = v,$$

$$v' = -y + \begin{cases} y, & |y| < 1, v = 0, \\ \text{sign}(y), & |y| \geq 1, v = 0, \\ -\text{sign}(v), & v \neq 0 \end{cases} = \begin{cases} 0, & |y| < 1, v = 0, \\ -y + \text{sign}(y), & |y| \geq 1, v = 0, \\ -y - \text{sign}(v), & v \neq 0. \end{cases} \quad (5.27)$$

(b) The condition $v \neq 0$ corresponds to the third case in (5.27), i.e., the system has the form

$$y' = v,$$

$$v' = -y - \text{sign}(v).$$

The phase plane equation for this system is

$$\frac{dv}{dy} = \frac{dv/dt}{dy/dt} = \frac{-y - \text{sign}(v)}{v}.$$

We consider two cases.

1) $v > 0$. In this case $\text{sign}(v) = 1$ and we have

$$\frac{dv}{dy} = \frac{-y - 1}{v} \quad \Rightarrow \quad v\,dv = -(y+1)dy$$

$$\Rightarrow \quad \int v\,dv = -\int (y+1)dy$$

$$\Rightarrow \quad \frac{1}{2}v^2 = -\frac{1}{2}(y+1)^2 + c_1 \quad \Rightarrow \quad v^2 + (y+1)^2 = c,$$

where $c = 2c_1$.

2) $v < 0$. In this case, $\text{sign}(v) = -1$, and so we have

$$\frac{dv}{dy} = \frac{-y + 1}{v} \quad \Rightarrow \quad v\,dv = -(y-1)dy$$

$$\Rightarrow \quad \int v\,dv = -\int (y-1)dy$$

$$\Rightarrow \quad \frac{1}{2}v^2 = -\frac{1}{2}(y-1)^2 + c_1 \quad \Rightarrow \quad v^2 + (y-1)^2 = c.$$

(c) The equation $v^2 + (y+1)^2 = c$ defines a circle in the yv-plane centered at $(-1, 0)$ and of radius $\sqrt{c}$ if $c > 0$, and it is the empty set if $c < 0$. The condition $v > 0$ means that we have to take only the half of these circles lying in the upper half plane.

 Copyright © 2012 Pearson Education, Inc. Publishing as Addison-Wesley.

Moreover, the first equation implies that trajectories flow from the left to the right. Similarly, in the lower half plane, we have concentric semicircles $v^2 + (y-1)^2 = c$, $c \geq 0$, centered at $(1,0)$ and flowing from the right to the left.

(d) For the system found in (a),

$$f(y,v) = v,$$

$$g(y,v) = \begin{cases} 0, & |y| < 1, \; v = 0, \\ -y + \text{sign}(y), & |y| \geq 1, \; v = 0, \\ -y - \text{sign}(v), & v \neq 0. \end{cases}$$

Since $f(y,v) = 0 \iff v = 0$ and

$$g(y,0) = \begin{cases} 0, & |y| < 1, \\ -y + \text{sign}(y), & |y| \geq 1, \end{cases}$$

we consider two cases. If $y < 1$, then $g(y,0) \equiv 0$. This means that any point of the interval $-1 < y < 1$ is a critical point. If $|y| \geq 1$, then $g(y,0) = -y + \text{sign}(y)$, which is zero if $y = \pm 1$. Thus, the critical point set is the segment $v = 0$, $-1 \leq y \leq 1$.

(e) According to (c), the mass released at $(7.5, 0)$ goes in the lower half plane from the right to the left along a semicircle centered at $(1,0)$. The radius of this semicircle is $7.5 - 1 = 6.5$, and its other end is $(1 - 6.5, 0) = (-5.5, 0)$. From this point, the mass goes from the left to the right in the upper half plane along the semicircle centered at $(-1, 0)$ and of radius $-1 - (-5.5) = 4.5$, and comes to the point $(-1 + 4.5, 0) = (3.5, 0)$. Then the mass again goes from the right to the left in the lower half plane along the semicircle centered at $(1, 0)$ and of the radius $3.5 - 1 = 2.5$, and comes to the point $(1 - 2.5, 0) = (-1.5, 0)$. From this point, the mass goes in the upper half plane from the left to the right along the semicircle centered at $(-1, 0)$ and of radius $-1 - (-1.5) = 0.5$, and comes to the point $(-1 + 0.5, 0) = (-0.5, 0)$. Here it comes to rest because $|-0.5| < 1$, and there is no a lower semicircle, starting at this point. See the colored curve in Figure B.41.

Copyright © 2012 Pearson Education, Inc. Publishing as Addison-Wesley.

Chapter 5

EXERCISES 5.5: **Applications to Biomathematics: Epidemic and Tumor Growth Models**

1. In this problem, the variables are t and p. With suggested values of parameters, the initial value problem (23) becomes

$$\frac{dp}{dt} = 3p - p^r, \qquad p(0) = 1.$$

Thus, the inputs into the 4th order Runge-Kutta subroutine are $t_0 = 0$, $p_0 = 1$, and $N = 20$ (so that $h = 0.25$). Since $f(t,p) = 3p - p^r$, the formulas in Step 3 become

$$k_1 = 0.25 \left(3p - p^r\right),$$
$$k_2 = 0.25 \left[3 \left(p + \frac{k_1}{2} \right) - \left(p + \frac{k_1}{2} \right)^r \right],$$
$$k_3 = 0.25 \left[3 \left(p + \frac{k_2}{2} \right) - \left(p + \frac{k_2}{2} \right)^r \right],$$
$$k_4 = 0.25 \left[3 \left(p + k_3 \right) - \left(p + k_3 \right)^r \right].$$

The results of computations are shown in Table **5–E** on page 319. These results indicate that the limiting populations for $r = 1.5$, $r = 2$, and $r = 3$ are $p_\infty = 9$, $p_\infty = 3$, and $p_\infty = \sqrt{3}$, respectively.

Since the right-hand side of the given logistic equation, $f(t,p) = 3p - p^r$, does not depend on t, we conclude that this equation is autonomous. Therefore, its equilibrium solutions (if any) can be found by solving

$$f(p) = 3p - p^r = 0 \quad \Leftrightarrow \quad p\left(3 - p^{r-1}\right) = 0 \quad \Leftrightarrow \quad p = 0 \quad \text{or} \quad p = 3^{1/(r-1)}.$$

The condition $r > 1$ implies that $f(p) > 0$ on $\left(0, 3^{1/(r-1)}\right)$ and $f(p) < 0$ on $\left(3^{1/(r-1)}, \infty\right)$. Therefore, $p = 3^{1/(r-1)}$ is a sink and, regardless of the initial value, $p(0) = p_0 > 0$,

$$\lim_{t \to \infty} p(t) = 3^{1/(r-1)}.$$

3. Substituting the given values of parameters into the equation governing the hormone level in the blood yields a nonhomogeneous linear equation, which can be written in standard form

$$\frac{dx}{dt} + 2x = 1 - \cos\left(\frac{\pi t}{12}\right).$$

 Copyright © 2012 Pearson Education, Inc. Publishing as Addison-Wesley.

We now use the technique discussed in Section 2.3 to conclude that

$$\mu(t) = \exp\left(\int(2)dt\right) = e^{2t} \qquad \Rightarrow \qquad x(t) = \frac{1}{e^{2t}}\int\left[1 - \cos\left(\frac{\pi t}{12}\right)e^{2t}\right]dt.$$

Computing the integral yields

$$x(t) = \frac{1}{e^{2t}}\left[\frac{1}{2}e^{2t} - \frac{2\cos(\pi t/12) + (\pi/12)\sin(\pi t/12)}{2^2 + (\pi/12)^2}e^{2t} + C\right]$$

$$= \frac{1}{2} - \frac{2\cos(\pi t/12) + (\pi/12)\sin(\pi t/12)}{4 + (\pi/12)^2} + Ce^{-2t}.$$

To find the constant C, we substitute the initial condition, $x(0) = 10$, into this general solution and obtain

$$10 = \frac{1}{2} - \frac{2}{2^2 + (\pi/12)^2} + C \qquad \Rightarrow \qquad C = \frac{19}{2} + \frac{2}{4 + (\pi/12)^2}.$$

Hence, the answer is

$$x(t) = \frac{1}{2} - \frac{2\cos(\pi t/12) + (\pi/12)\sin(\pi t/12)}{4 + (\pi/12)^2} + \left[\frac{19}{2} + \frac{2}{4 + (\pi/12)^2}\right]e^{-2t}.$$

5. Since $S(t)$ and $I(t)$ represent population and we cannot have a negative population, we are only interested in the first quadrant of the SI-plane.

(a) In order to find the trajectory, corresponding to the initial conditions $I(0) = 1$ and $S(0) = 700$, we must solve the first order equation

$$\frac{dI}{dS} = \frac{dI/dt}{dS/dt} = \frac{aSI - bI}{-aSI} = -\frac{aS - b}{aS}$$

$$\Rightarrow \qquad \frac{dI}{dS} = -1 + \frac{b}{aS}. \tag{5.28}$$

By integrating both sides of equation (5.28) with respect to S, we obtain the integral curves given by

$$I(S) = -S + \frac{b}{a}\ln S + C.$$

A sketch of this curve for $a = 0.003$ and $b = 0.5$ is shown in Figure B.42 in the answers of the text.

(b) From the sketch in Figure B.42, we see that the peak number of infected people is approximately 295.

Copyright © 2012 Pearson Education, Inc. Publishing as Addison-Wesley.

From Calculus, we know that a function can attain its maximum only at a critical point, i.e., when $dI/dS = 0$ or does not exist. Thus, from (5.28) we have

$$-1 + \frac{b}{aS} = 0.$$

Solving for S, we obtain

$$S = \frac{b}{a} = \frac{0.5}{0.003} \approx 167 \text{ (people)},$$

which is consistent with the graph in Figure B.42.

7. For $P(t)$ (see (15)), we have an initial value problem

$$\frac{dP}{dt} = cP - r(N)P = [c - r(N)]\, P, \quad P(0) = 1.$$

According to the Gompertz law from Exapmle 3 in the text,

$$N = N(t) = \exp\left[\frac{c\left(1 - e^{-bt}\right)}{b}\right] \quad \Rightarrow \quad r(N) = b\left(1 + \ln N\right) = b + c\left(1 - e^{-bt}\right),$$

and so $P(t)$ satisfies the equation

$$\frac{dP}{dt} = \left(ce^{-bt} - b\right) P.$$

Integrating this separable equation yields

$$\frac{dP}{P} = \left(ce^{-bt} - b\right) dt \quad \Rightarrow \quad \ln P = -(c/b)e^{-bt} - bt + C$$

$$\Rightarrow \quad P = \exp\left(-\frac{ce^{-bt}}{b} - bt + C\right).$$

Since $P(0) = 1$, we find that

$$1 = \exp\left(-\frac{ce^{-b(0)}}{b} - b(0) + C\right) \quad \Rightarrow \quad C = \frac{c}{b}.$$

Thus,

$$P(t) = \exp\left(-\frac{ce^{-bt}}{b} - bt + \frac{c}{b}\right) = \exp\left[\frac{c\left(1 - e^{-bt}\right)}{b} - bt\right].$$

We now use the equation (16) to get an initial value problem for $Q(t)$.

$$\frac{dQ}{dt} = r(N)P = \left[b + c\left(1 - e^{-bt}\right)\right] \exp\left[\frac{c\left(1 - e^{-bt}\right)}{b} - bt\right], \quad Q(0) = 0.$$

 Copyright © 2012 Pearson Education, Inc. Publishing as Addison-Wesley.

Integrating yields

$$Q(t) = \exp\left[\frac{c\left(1 - e^{-bt}\right)}{b}\right] - \exp\left[\frac{c\left(1 - e^{-bt}\right)}{b} - bt\right] + C$$

$$= \exp\left[\frac{c\left(1 - e^{-bt}\right)}{b} - bt\right]\left(e^{bt} - 1\right) + C.$$

Finally, the initial condition, $Q(0) = 0$, implies that $C = 0$.

9. From (20) we conclude that, with $r(N) = s(2N - 1)$,

$$\frac{dN}{dt} = cN - \int_1^N r(x)dx = cN - \int_1^N s(2x - 1)dx = cN - s\left(x^2 - x\right)\Big|_1^N = cN - 2s\frac{N(N - 1)}{2}.$$

Thus,

$$\frac{dN}{dt} = cN - 2s\frac{N(N - 1)}{2},$$

which has the form of the logistic model equation (14), Section 3.2, with

$$k_1 = c, \quad k_3 = 2s \quad \Rightarrow \quad A = s, \quad p_1 = c/s + 1.$$

Thus, the general formula (15) yields

$$N(t) = \frac{N_0\left(c/s + 1\right)}{N_0 + \left[(c/s + 1) - N_0\right]e^{-s(c/s+1)t}} = \frac{N_0\left(c + s\right)}{N_0 s + (c + s - N_0 s)\,e^{-(c+s)t}},$$

where $N_0 = N(0)$.

EXERCISES 5.6: Coupled Mass-Spring Systems

1. For the mass m_1 there is only one force acting on it, that is, the force due to the spring with the spring constant k_1; so, it equals to $-k_1(x - y)$. Hence, we get

$$m_1 x'' = -k_1(x - y).$$

For the mass m_2, there are two forces: the force due to the spring with the spring constant k_2, which is $-k_2 y$; and the force due to the spring with the spring constant k_1, which is $k_1(y - x)$. Thus, we get

$$m_2 y'' = k_1(x - y) - k_2 y.$$

Copyright © 2012 Pearson Education, Inc. Publishing as Addison-Wesley.

Therefore, the system, describing the motion, is

$$m_1 x'' = k_1(y - x),$$
$$m_2 y'' = -k_1(y - x) - k_2 y,$$

or, in operator form,

$$\left(m_1 D^2 + k_1\right)[x] - k_1 y = 0,$$
$$-k_1 x + \left\{m_2 D^2 + (k_1 + k_2)\right\}[y] = 0.$$

With $m_1 = 1$, $m_2 = 2$, $k_1 = 4$, and $k_2 = 10/3$, we get

$$\left(D^2 + 4\right)[x] - 4y = 0,$$
$$-4x + (2D^2 + 22/3)[y] = 0,$$

(5.29)

with initial conditions

$$x(0) = -1, \quad x'(0) = 0, \quad y(0) = 0, \quad y'(0) = 0.$$

Multiplying the second equation of the system (5.29) by 4, applying $(2D^2 + 22/3)$ to the first equation of this system, and adding the results together, we get

$$\left(D^2 + 4\right)\left(2D^2 + \frac{22}{3}\right)[x] - 16x = 0$$
$$\Rightarrow \quad \left(2D^4 + \frac{46}{3}D^2 + \frac{40}{3}\right)[x] = 0$$
$$\Rightarrow \quad \left(3D^4 + 23D^2 + 20\right)[x] = 0.$$

The characteristic equation for this fourth order differential equation is

$$3r^4 + 23r^2 + 20 = 0,$$

which is a quadratic in r^2. So,

$$r^2 = \frac{-23 \pm \sqrt{529 - 240}}{6} = \frac{-23 \pm 17}{6}.$$

Since both, $-20/3$ and -1, are negative, the roots of the characteristic equation are $\pm i\beta_1$ and $\pm i\beta_2$, where

$$\beta_1 = \sqrt{\frac{20}{3}}, \quad \beta_2 = 1.$$

    Copyright © 2012 Pearson Education, Inc. Publishing as Addison-Wesley.

Hence,

$$x(t) = c_1 \cos \beta_1 t + c_2 \sin \beta_1 t + c_3 \cos \beta_2 t + c_4 \sin \beta_2 t.$$

Solving the first equation in the system (5.29) for y, we get

$$y(t) = \frac{1}{4}\left(D^2 + 4\right)[x] = \frac{1}{4}\Big[\left(-\beta_1^2 + 4\right)c_1 \cos \beta_1 t + \left(-\beta_1^2 + 4\right)c_2 \sin \beta_1 t$$
$$+ \left(-\beta_2^2 + 4\right)c_3 \cos \beta_2 t + \left(-\beta_2^2 + 4\right)c_4 \sin \beta_2 t\Big].$$

Next, we substitute the initial conditions. Setting $x(0) = -1$, $x'(0) = 0$ yields

$$-1 = c_1 + c_3\,,$$
$$0 = c_2\beta_1 + c_4\beta_2\,.$$

From $y(0) = 0$, $y'(0) = 0$, we also get

$$0 = \frac{1}{4}\Big[\left(-\beta_1^2 + 4\right)c_1 + \left(-\beta_2^2 + 4\right)c_3\Big],$$
$$0 = \frac{1}{4}\Big[\beta_1\left(-\beta_1^2 + 4\right)c_2 + \beta_2\left(-\beta_2^2 + 4\right)c_4\Big].$$

The solution to the above system is

$$c_2 = c_4 = 0, \qquad c_1 = -\frac{9}{17}, \qquad c_3 = -\frac{8}{17},$$

which gives us

$$x(t) = -\frac{8}{17}\cos t + \frac{9}{17}\cos\sqrt{\frac{20}{3}}\,t\,,$$
$$y(t) = -\frac{6}{17}\cos t + \frac{6}{17}\cos\sqrt{\frac{20}{3}}\,t\,.$$

3. We define the displacements $x(t)$, $y(t)$, and $z(t)$ of masses from the equilibrium at time t as we did in Example 2. For each mass, there are two forces, acting on it due to the Hook's law.

For the mass on the left,

$$F_{11} = -kx \qquad \text{and} \qquad F_{12} = k(y - x);$$

for the mass in the middle,

$$F_{21} = -k(y - x) \qquad \text{and} \qquad F_{22} = k(z - y);$$

Copyright © 2012 Pearson Education, Inc. Publishing as Addison-Wesley.

for the mass on the right,

$$F_{31} = -k(z - y) \qquad \text{and} \qquad F_{32} = -kz.$$

Applying Newton's second law for each mass, we obtain the following system

$$
\begin{aligned}
mx'' &= -kx + k(y - x), \\
my'' &= -k(y - x) + k(z - y), \\
mz'' &= -k(z - y) - kz
\end{aligned}
$$

or, in operator form,

$$
\begin{aligned}
\left(mD^2 + 2k\right)[x] - ky &= 0, \\
-kx + \left(mD^2 + 2k\right)[y] - kz &= 0, \\
-ky + \left(mD^2 + 2k\right)[z] &= 0.
\end{aligned}
$$

From the first equation, we express

$$y = \frac{1}{k}\left(mD^2 + 2k\right)[x] \tag{5.30}$$

and substitute this expression into the other two equations to get

$$
\begin{aligned}
-kx + \left(mD^2 + 2k\right)\left[\frac{1}{k}\left(mD^2 + 2k\right)[x]\right] - kz &= 0, \\
-\left(mD^2 + 2k\right)[x] + \left(mD^2 + 2k\right)[z] &= 0.
\end{aligned}
$$

The first equation yields

$$z = -x + \left\{\frac{1}{k}\left(mD^2 + 2k\right)\right\}^2 [x] = \left\{\frac{1}{k^2}\left(mD^2 + 2k\right)^2 - 1\right\}[x], \tag{5.31}$$

and so

$$
-\left(mD^2 + 2k\right)[x] + \left(mD^2 + 2k\right)\left[\left\{\frac{1}{k^2}\left(mD^2 + 2k\right)^2 - 1\right\}[x]\right]
$$
$$
= \left(mD^2 + 2k\right)\left\{\frac{1}{k^2}\left(mD^2 + 2k\right)^2 - 2\right\}[x] = 0.
$$

The characteristic equation for this linear equation with constant coefficients is

$$\left(mr^2 + 2k\right)\left[\frac{1}{k^2}\left(mr^2 + 2k\right)^2 - 2\right] = 0,$$

 Copyright © 2012 Pearson Education, Inc. Publishing as Addison-Wesley.

which splits onto two equations,

$$mr^2 + 2k = 0 \quad \Rightarrow \quad r = \pm i\sqrt{\frac{2k}{m}} \tag{5.32}$$

and

$$\frac{1}{k^2}\left(mr^2 + 2k\right)^2 - 2 = 0$$

$$\Rightarrow \quad \left(mr^2 + 2k\right)^2 - 2k^2 = 0$$

$$\Rightarrow \quad \left(mr^2 + 2k - \sqrt{2}k\right)\left(mr^2 + 2k + \sqrt{2}k\right) = 0$$

$$\Rightarrow \quad r = \pm i\sqrt{\frac{(2 - \sqrt{2})k}{m}}, \quad r = \pm i\sqrt{\frac{(2 + \sqrt{2})k}{m}}. \tag{5.33}$$

Solutions (5.32) and (5.33) give normal frequences

$$\omega_1 = \frac{1}{2\pi}\sqrt{\frac{2k}{m}}, \quad \omega_2 = \frac{1}{2\pi}\sqrt{\frac{(2 - \sqrt{2})k}{m}}, \quad \omega_3 = \frac{1}{2\pi}\sqrt{\frac{(2 + \sqrt{2})k}{m}}.$$

Thus, a general solution $x(t)$ has the form $x(t) = x_1(t) + x_2(t) + x_3(t)$, where functions

$$x_j(t) = c_{1j}\cos(2\pi\omega_j t) + c_{2j}\sin(2\pi\omega_j t).$$

Note that x_j's satisfy the following differential equations:

$$\left(mD^2 + 2k\right)[x_1] = 0,$$

$$\left(mD^2 + 2k - \sqrt{2}k\right)[x_2] = 0, \tag{5.34}$$

$$\left(mD^2 + 2k + \sqrt{2}k\right)[x_3] = 0.$$

For normal modes, we find solutions $y_j(t)$ and $z_j(t)$, corresponding to $x_j(t)$, $j = 1, 2,$ and 3, by using (5.30), (5.31), and identities (5.34).

ω_1:

$$y_1 = \frac{1}{k}\left(mD^2 + 2k\right)[x_1] \equiv 0,$$

$$z_1 = \left\{\frac{1}{k}\left(mD^2 + 2k\right)^2 - 1\right\}[x_1] = -x_1;$$

ω_2:

$$y_2 = \frac{1}{k}\left(mD^2 + 2k\right)[x_2] = \left\{\frac{1}{k}\left(mD^2 + 2k - \sqrt{2}k\right) + \sqrt{2}\right\}[x_2] = \sqrt{2}x_2,$$

$$z_2 = \left\{\frac{1}{k}\left(mD^2 + 2k\right)^2 - 1\right\}[x_2] = \left\{\left[\frac{1}{k}\left(mD^2 + 2k\right)^2 - 2\right] + 1\right\}[x_2] = x_2;$$

Copyright © 2012 Pearson Education, Inc. Publishing as Addison-Wesley.

ω_3:

$$y_3 = \frac{1}{k}\left(mD^2 + 2k\right)[x_3] = \left\{\frac{1}{k}\left(mD^2 + 2k + \sqrt{2}k\right) - \sqrt{2}\right\}[x_3] = -\sqrt{2}x_3,$$

$$z_3 = \left\{\frac{1}{k}\left(mD^2 + 2k\right)^2 - 1\right\}[x_3] = \left\{\left[\frac{1}{k}\left(mD^2 + 2k\right)^2 - 2\right] + 1\right\}[x_3] = x_3;$$

5. This spring system is similar to the system in Example 2, except the fact that the middle spring has been replaced by a dashpot. We proceed as in Example 1. Let x and y represent the displacements of masses m_1 and m_2 to the right of their respective equilibrium positions. The mass m_1 has a force F_1 acting on its left side due to the left spring and a force F_2 acting on its right side due to the dashpot. Applying Hooke's law, we see that

$$F_1 = -k_1 x.$$

Assuming, as we did in Section 4.1, that the damping force due to the dashpot is proportional to the magnitude of the velocity, but opposite in direction, we have

$$F_2 = b\left(y' - x'\right),$$

where b is the damping constant. Notice that velocity of the arm of the dashpot is the difference between the velocities of mass m_2 and mass m_1. The mass m_2 has a force F_3 acting on its left side due to the dashpot and a force F_4 acting on its right side due to the right spring. Using similar arguments, we find

$$F_3 = -b\left(y' - x'\right) \qquad \text{and} \qquad F_4 = -k_2 y.$$

Applying Newton's second law to each mass gives

$$m_1 x''(t) = F_1 + F_2 = -k_1 x(t) + b\left[y'(t) - x'(t)\right],$$
$$m_2 y''(t) = F_3 + F_4 = -b\left[y'(t) - x'(t)\right] - k_2 y.$$

Plugging in the constants $m_1 = m_2 = 1$, $k_1 = k_2 = 1$, and $b = 1$, and simplifying yields

$$\begin{aligned} x''(t) + x'(t) + x(t) - y'(t) &= 0, \\ -x'(t) + y''(t) + y'(t) + y(t) &= 0. \end{aligned} \tag{5.35}$$

The initial conditions for the system are $y(0) = 0$ (m_2 was held in its equilibrium position), $x(0) = -2$ (m_1 was pushed 2 ft to the left), and $x'(0) = y'(0) = 0$ (the masses

Copyright © 2012 Pearson Education, Inc. Publishing as Addison-Wesley.

were simply released at time $t = 0$ with no initial velocity). In operator notation, this system becomes

$$(D^2 + D + 1)[x] - D[y] = 0,$$
$$-D[x] + y''(t) + (D^2 + D + 1)[y] = 0.$$

By multiplying the first equation by D, the second one – by $(D^2 + D + 1)$, and adding the resulting equations, we eliminate $y(t)$. Thus, we have

$$\left\{ (D^2 + D + 1)^2 - D^2 \right\}[x] = 0$$
$$\Rightarrow \quad \left\{ \left[(D^2 + D + 1) - D \right] \cdot \left[(D^2 + D + 1) + D \right] \right\}[x] = 0$$
$$\Rightarrow \quad \left\{ (D^2 + 1)(D + 1)^2 \right\}[x] = 0.$$

This equation is a fourth order linear differential equation with constant coefficients, whose associated auxiliary equation has roots $r = -1, -1, i,$ and $-i$. Therefore, a general solution is given by

$$x(t) = c_1 e^{-t} + c_2 t e^{-t} + c_3 \cos t + c_4 \sin t$$
$$\Rightarrow \quad x'(t) = (-c_1 + c_2)e^{-t} - c_2 t e^{-t} - c_3 \sin t + c_4 \cos t$$
$$\Rightarrow \quad x''(t) = (c_1 - 2c_2)e^{-t} + c_2 t e^{-t} - c_3 \cos t - c_4 \sin t.$$

To find $y(t)$, we note that the first equation in the system (5.35) implies

$$y'(t) = x''(t) + x'(t) + x(t),$$

Substituting $x(t)$, $x'(t)$, and $x''(t)$ into this equation yields

$$y'(t) = (c_1 - 2c_2)e^{-t} + c_2 t e^{-t} - c_3 \cos t - c_4 \sin t$$
$$+ (-c_1 + c_2)e^{-t} - c_2 t e^{-t} - c_3 \sin t + c_4 \cos t + c_1 e^{-t} + c_2 t e^{-t} + c_3 \cos t + c_4 \sin t$$
$$\Rightarrow \quad y'(t) = (c_1 - c_2)e^{-t} + c_2 t e^{-t} - c_3 \sin t + c_4 \cos t.$$

By integrating both sides with respect to t, we obtain

$$y(t) = -(c_1 - c_2)e^{-t} - c_2 t e^{-t} - c_2 e^{-t} + c_3 \cos t + c_4 \sin t + c_5.$$

Simplifying yields

$$y(t) = -c_1 e^{-t} - c_2 t e^{-t} + c_3 \cos t + c_4 \sin t + c_5.$$

Copyright © 2012 Pearson Education, Inc. Publishing as Addison-Wesley.

To determine the five constants, we use the four initial conditions and the second equation in system (5.35). (We used the first equation to determine $y(t)$). Substituting into the second equation in (5.35) gives

$$- \left[(-c_1 + c_2)e^{-t} - c_2 t e^{-t} - c_3 \sin t + c_4 \cos t \right]$$
$$+ \left[(-c_1 + 2c_2)e^{-t} - c_2 t e^{-t} - c_3 \cos t - c_4 \sin t \right]$$
$$+ \left[(c_1 - c_2)e^{-t} + c_2 t e^{-t} - c_3 \sin t + c_4 \cos t \right]$$
$$+ \left[-c_1 e^{-t} - c_2 t e^{-t} + c_3 \cos t + c_4 \sin t + c_5 \right] = 0,$$

which reduces to $c_5 = 0$. Thus, the initial conditions imply

$$x(0) = c_1 + c_3 = -2, \qquad x'(0) = (-c_1 + c_2) + c_4 = 0,$$
$$y(0) = -c_1 + c_3 = 0, \qquad y'(0) = (c_1 - c_2) + c_4 = 0.$$

Solving these equations simultaneously, we find that

$$c_1 = -1, \quad c_2 = -1, \quad c_3 = -1, \quad \text{and} \quad c_4 = 0.$$

Therefore, the solution to this mass-spring-dashpot system is

$$x(t) = -e^{-t} - te^{-t} - \cos t, \qquad y(t) = e^{-t} + te^{-t} - \cos t.$$

7. In operator notation, we have

$$\left(D^2 + 5 \right) [x] - 2y = 0,$$
$$-2x + \left(D^2 + 2 \right) [y] = 3 \sin 2t.$$

Multiplying the first equation by $(D^2 + 2)$, the second equation – by 2, and adding the results, we obtain

$$\left\{ \left(D^2 + 2 \right) \left(D^2 + 5 \right) - 4 \right\} [x] = 6 \sin 2t$$
$$\Rightarrow \qquad \left(D^4 + 7D^2 + 6 \right) [x] = 6 \sin 2t$$
$$\Rightarrow \qquad \left(D^2 + 1 \right) \left(D^2 + 6 \right) [x] = 6 \sin 2t. \qquad (5.36)$$

Since the characteristic equation, $(r^2 + 1)(r^2 + 6) = 0$, has roots $r = \pm i$ and $r = \pm i\sqrt{6}$, a general solution to the corresponding homogeneous equation for $x(t)$ is given by

$$x_h(t) = c_1 \cos t + c_2 \sin t + c_3 \cos \sqrt{6}t + c_4 \sin \sqrt{6}t.$$

 Copyright © 2012 Pearson Education, Inc. Publishing as Addison-Wesley.

Due to the right-hand side in (5.36), a particular solution has the form

$$x_p(t) = A \cos 2t + B \sin 2t.$$

In order to simplify computations, we note that both functions, $\cos 2t$ and $\sin 2t$ (and so $x_p(t)$) satisfy the differential equation $(D^2 + 4)[x] = 0$. Thus,

$$\left(D^2 + 1\right)\left(D^2 + 6\right)[x_h] = \left\{(D^2 + 4) - 3\right\}\left\{(D^2 + 4) + 2\right\}[x_h] = 2\left\{(D^2 + 4) - 3\right\}[x_h]$$

$$= -6x_h = -6A \cos 2t - 6B \sin 2t = 6 \sin 2t$$

$$\Rightarrow \quad A = 0, \quad B = -1 \quad \Rightarrow \quad x_h(t) = -\sin 2t$$

and

$$x(t) = x_h(t) + x_p(t) = c_1 \cos t + c_2 \sin t + c_3 \cos \sqrt{6}t + c_4 \sin \sqrt{6}t - \sin 2t.$$

From the first equation in the original system, we have

$$y(t) = \frac{x'' + 5x}{2} = 2c_1 \cos t + 2c_2 \sin t - \frac{1}{2}c_3 \cos \sqrt{6}t - \frac{1}{2}c_4 \sin \sqrt{6}t - \frac{1}{2}\sin 2t.$$

We determine constants c_1 and c_3 using the initial conditions $x(0) = 0$ and $y(0) = 1$.

$$\begin{aligned} 0 &= x(0) = c_1 + c_3, \\ 1 &= y(0) = 2c_1 - c_3/2 \end{aligned} \quad \Rightarrow \quad \begin{aligned} c_3 &= -c_1, \\ 2c_1 - (-c_1)/2 = 1 \end{aligned} \quad \Rightarrow \quad \begin{aligned} c_3 &= -2/5, \\ c_1 &= 2/5. \end{aligned}$$

To find c_2 and c_4, we compute $x'(t)$ and $y'(t)$, evaluate these functions at $t = 0$, and use the other two initial conditions, $x'(0) = y'(0) = 0$. This yields

$$\begin{aligned} 0 &= x'(0) = c_2 + \sqrt{6}c_4 - 2, \\ 0 &= y'(0) = 2c_2 - \sqrt{6}c_4/2 - 1 \end{aligned} \quad \Rightarrow \quad \begin{aligned} c_4 &= \sqrt{6}/5, \\ c_2 &= 4/5. \end{aligned}$$

Therefore, the required solution is

$$x(t) = \frac{2}{5}\cos t + \frac{4}{5}\sin t - \frac{2}{5}\cos \sqrt{6}t + \frac{\sqrt{6}}{5}\sin \sqrt{6}t - \sin 2t,$$

$$y(t) = \frac{4}{5}\cos t + \frac{8}{5}\sin t + \frac{1}{5}\cos \sqrt{6}t - \frac{\sqrt{6}}{10}\sin \sqrt{6}t - \frac{1}{2}\sin 2t.$$

9. Writing the equations of this system in operator form, we obtain

$$\left\{mD^2 + \left(\frac{mg}{l} + k\right)\right\}[x_1] - kx_2 = 0,$$

$$-kx_1 + \left\{mD^2 + \left(\frac{mg}{l} + k\right)\right\}[x_2] = 0.$$

Copyright © 2012 Pearson Education, Inc. Publishing as Addison-Wesley.

Applying $\{mD^2 + (mg/l + k)\}$ to the first equation, multiplying the second equation by k, and then adding the results together yields

$$\left\{ \left[mD^2 + \left(\frac{mg}{l} + k \right) \right]^2 - k^2 \right\} [x_1] = 0.$$

This differential equation has the auxiliary equation

$$\left(mr^2 + \frac{mg}{l} + k \right)^2 - k^2 = \left(mr^2 + \frac{mg}{l} \right) \left(mr^2 + \frac{mg}{l} + 2k \right) = 0$$

with roots $\pm i\sqrt{g/l}$ and $\pm i\sqrt{(g/l) + (2k/m)}$. As it was discussed in the text, $\sqrt{g/l}$ and $\sqrt{(g/l) + (2k/m)}$ are the normal angular frequencies. To find the normal frequencies, we divide each angular frequency by 2π and obtain

$$\left(\frac{1}{2\pi} \right) \sqrt{\frac{g}{l}} \qquad \text{and} \qquad \left(\frac{1}{2\pi} \right) \sqrt{\frac{g}{l} + \frac{2k}{m}} .$$

EXERCISES 5.7: Electrical Systems

1. In this problem, $R = 100\,\Omega$, $L = 4\,\text{H}$, $C = 0.01\,\text{F}$, and $E(t) = 20\,\text{V}$. Therefore, the equation (4) in the text becomes

$$4\frac{d^2 I}{dt^2} + 100\frac{dI}{dt} + 100I = \frac{d(20)}{dt} = 0 \qquad \Rightarrow \qquad \frac{d^2 I}{dt^2} + 25\frac{dI}{dt} + 25I = 0.$$

The roots of the characteristic equation, $r^2 + 25r + 25 = 0$, are

$$r = \frac{-25 \pm \sqrt{(25)^2 - 4(25)(1)}}{2} = \frac{-25 \pm 5\sqrt{21}}{2},$$

and so a general solution is

$$I(t) = c_1 e^{(-25-5\sqrt{21})t/2} + c_2 e^{(-25+5\sqrt{21})t/2} .$$

To determine constants c_1 and c_2, we first find the initial value $I'(0)$ using $I(0) = 0$ and $q(0) = 4$. Substituting $t = 0$ into the equation (3) (with dq/dt replaced by $I(t)$), we obtain

$$L\frac{d[I(t)]}{dt} + RI(t) + \frac{1}{C}q(t) = E(t)$$

$$\Rightarrow \qquad 4I'(0) + 100(0) + \frac{1}{0.01}(4) = 20$$

$$\Rightarrow \qquad I'(0) = -95.$$

 Copyright © 2012 Pearson Education, Inc. Publishing as Addison-Wesley.

Thus, $I(t)$ satisfies $I(0) = 0$, $I'(0) = -95$. Next, we compute

$$I'(t) = \frac{c_1(-25 - 5\sqrt{21})}{2} e^{(-25 - 5\sqrt{21})t/2} + \frac{c_2(-25 + 5\sqrt{21})}{2} e^{(-25 + 5\sqrt{21})t/2},$$

substitute $t = 0$ into formulas for $I(t)$ and $I'(t)$, and obtain the system

$$\begin{aligned} 0 &= I(0) = c_1 + c_2, \\ -95 &= I'(0) = c_1(-25 - 5\sqrt{21})/2 + c_2(-25 + 5\sqrt{21})/2 \end{aligned} \quad \Rightarrow \quad \begin{aligned} c_1 &= 19/\sqrt{21}, \\ c_2 &= -19/\sqrt{21}. \end{aligned}$$

So, the solution is

$$I(t) = \frac{19}{\sqrt{21}} \left[e^{(-25 - 5\sqrt{21})t/2} - e^{(-25 + 5\sqrt{21})t/2} \right].$$

3. In this problem, $L = 4$, $R = 120$, $C = (2200)^{-1}$, and $E(t) = 10 \cos 20t$. Therefore, we see that $1/C = 2200$ and $E'(t) = -200 \sin 20t$. Substituting these expressions into the equation (4), we obtain

$$4 \frac{d^2 I}{dt^2} + 120 \frac{dI}{dt} + 2200 I = -200 \sin 20t.$$

Simplifying yields

$$\frac{d^2 I}{dt^2} + 30 \frac{dI}{dt} + 550 I = -50 \sin 20t. \tag{5.37}$$

The auxiliary equation, associated with the homogeneous equation corresponding to (5.37), is $r^2 + 30r + 550 = 0$. Solving, we get

$$r = -15 \pm 5\sqrt{13}i.$$

Therefore, the transient current, that is, $I_h(t)$ is given by

$$I_h(t) = e^{-15t} \left[C_1 \cos \left(5\sqrt{13}t \right) + C_2 \sin \left(5\sqrt{13}t \right) \right].$$

By the method of undetermined coefficients, a particular solution $I_p(t)$ to (5.37) has the form $I_p(t) = t^s(A \cos 20t + B \sin 20t)$. Since neither $y(t) = \cos 20t$ nor $y(t) = \sin 20t$ is a solution to the homogeneous equation (that is, the system is not at resonance), we can let $s = 0$ in $I_p(t)$. Thus, we see that $I_p(t)$, the steady-state current, has the form

$$I_p(t) = A \cos 20t + B \sin 20t.$$

To find the steady-state current, we must compute A and B. Observe that

$$I_p'(t) = -20A \sin 20t + 20B \cos 20t,$$

Copyright © 2012 Pearson Education, Inc. Publishing as Addison-Wesley.

$$I_p''(t) = -400A \cos 20t - 400B \sin 20t.$$

Plugging in these expressions into (5.37) yields

$$I_p''(t) + 30I_p'(t) + 550I(t) = -400A \cos 20t - 400B \sin 20t$$

$$-600A \sin 20t + 600B \cos 20t + 550A \cos 20t + 550B \sin 20t = -50 \sin 20t$$

$$\Rightarrow \quad (150A + 600B) \cos 20t + (150B - 600A) \sin 20t = -50 \sin 20t.$$

By equating coefficients, we obtain the system

$$15A + 60B = 0,$$

$$-60A + 15B = -5.$$

Solving these equations simultaneously for A and B yields $A = 4/51$ and $B = -1/51$. Thus, the steady-state current is given by

$$I_p(t) = \frac{4}{51} \cos 20t - \frac{1}{51} \sin 20t.$$

As was observed (see the discussion after Example 2), there is a correlation between RLC series circuits and mechanical vibration. So, we associate the variable L with m, R with b, and $1/C$ with k. Thus, we see that the resonance frequency for an RLC series circuit is given by $\gamma_r/(2\pi)$, where

$$\gamma_r = \sqrt{\frac{1}{CL} - \frac{R^2}{2L^2}},$$

provided $R^2 < 2L/C$. For this problem

$$R^2 = 14,400 < 2L/C = 17,600.$$

Therefore, we find the resonance frequency of this circuit to be

$$\gamma_r = \sqrt{\frac{1}{CL} - \frac{R^2}{2L^2}} = \sqrt{\frac{2200}{4} - \frac{14400}{32}} = 10.$$

Hence, the resonance frequency of this circuit is $10/(2\pi) = 5/\pi$.

5. In this problem, $C = 0.01$ F, $L = 4$ H, and $R = 10\,\Omega$. Hence, the equation, governing the RLC circuit, is

$$4\frac{d^2I}{dt^2} + 10\frac{dI}{dt} + \frac{1}{0.01}I = \frac{d}{dt}(E_0 \cos \gamma t) = -\frac{E_0\gamma}{4} \sin \gamma t.$$

     Copyright © 2012 Pearson Education, Inc. Publishing as Addison-Wesley.

The frequency response curve $M(\gamma)$ for an RLC curcuit is determined by

$$M(\gamma) = \frac{1}{\sqrt{[(1/C) - L\gamma^2]^2 + R^2\gamma^2}},$$

which comes from the comparison Table 5.3 and the equation (13) of Section 4.10. Thus,

$$M(\gamma) = \frac{1}{\sqrt{[(1/0.01) - 4\gamma^2]^2 + (10)^2\gamma^2}} = \frac{1}{\sqrt{(100 - 4\gamma^2)^2 + 100\gamma^2}}.$$

The graph of this function is shown in Figure B.43 in the answers of the text. $M(\gamma)$ has its maximal value at the point $\gamma_0 = \sqrt{x_0}$, where x_0 is the point where the quadratic function $(100 - 4x)^2 + 100x$ attains its minimum (the first coordinate of the vertex). Thus, we find

$$\gamma_0 = \sqrt{\frac{175}{8}} \qquad \text{and} \qquad M(\gamma_0) = \frac{2}{25\sqrt{15}} \approx 0.02.$$

7. This mass-spring system satisfies the differential equation

$$7\frac{d^2x}{dt^2} + 2\frac{dx}{dt} + 3x = 10\cos 10t. \tag{5.38}$$

Since we want to find an RLC series circuit with $R = 10\,\Omega$, which is an analog for the mass-spring system, we need L, $1/C$, and $E(t)$ such that the differential equation

$$L\frac{d^2q}{dt^2} + 10\frac{dq}{dt} + \frac{1}{C}q = E(t)$$

is equivalent to (5.38). Multiplying (5.38) by 5 yields

$$35\frac{d^2x}{dt^2} + 10\frac{dx}{dt} + 15x = 50\cos 10t.$$

Therefore, $E(t) = 50\cos 10t\,\text{V}$, $L = 35\,\text{H}$, and $C = 1/15\,\text{F}$.

11. For this electric network, there are three loops. Loop 1 is through a 10V battery, a $10\,\Omega$ resistor, and a 20 H inductor. Loop 2 is through a 10V battery, a $10\,\Omega$ resistor, a $5\,\Omega$ resistor, and a $(1/30)\,\text{F}$ capacitor. Loop 3 is through a $5\,\Omega$ resistor, a $(1/30)\,\text{F}$ capacitor, and a 20 H inductor. Therefore, applying Kirchhoff's second law to this network yields three equations,

$$\text{Loop 1:} \qquad 10I_1 + 20\frac{dI_2}{dt} = 10,$$

$$\text{Loop 2:} \qquad 10I_1 + 5I_3 + 30q_3 = 10,$$

$$\text{Loop 3:} \qquad 5I_3 + 30q_3 - 20\frac{dI_2}{dt} = 0.$$

Copyright © 2012 Pearson Education, Inc. Publishing as Addison-Wesley.

Since the equation for Loop 2 minus the equation for Loop 1 yields the equation for Loop 3, we use the first two equations in our calculations. By examining a junction point, we see that we also have the equation $I_1 = I_2 + I_3$. Thus, we have $I_1' = I_2' + I_3'$. We begin by dividing the equation for Loop 1 by 10 and the equation for Loop 2 by 5. Differentiating the equation for Loop 2 yields the system

$$I_1 + 2\frac{dI_2}{dt} = 1\,,$$

$$2\frac{dI_1}{dt} + \frac{dI_3}{dt} + 6I_3 = 0\,,$$

where $I_3 = q_3'$. Since $I_1 = I_2 + I_3$ and $I_1' = I_2' + I_3'$, we can rewrite the system, using operator notation, in the form

$$(2D + 1)[I_2] + I_3 = 1\,,$$

$$(2D)[I_2] + (3D + 6)[I_3] = 0\,.$$

If we multiply the first equation above by $(3D+6)$ and then subtract the second equation, we obtain

$$\{(3D + 6)(2D + 1) - 2D\}\,[I_2] = 6 \qquad \Rightarrow \qquad \left(6D^2 + 13D + 6\right)[I_2] = 6. \qquad (5.39)$$

This equation is a linear nonhomogeneous equation with constant coefficients, whose auxiliary equation, $6r^2 + 13r + 6 = 0$, has roots $-3/2$, $-2/3$. Therefore, the solution to the homogeneous equation, corresponding to the given equation is given by

$$I_{2h}(t) = c_1 e^{-3t/2} + c_2 e^{-2t/3}\,.$$

By the method of undetermined coefficients, a particular solution to (5.39) has the form $I_{2p}(t) = A$. Substituting this function into the differential equation, we see that a particular solution is given by $I_{2p}(t) = 1$. Thus, the current, $I_2(t)$ satisfies

$$I_2(t) = c_1 e^{-3t/2} + c_2 e^{-2t/3} + 1.$$

As we noticed above, $I_3(t)$ can now be found from the first equation.

$$I_3(t) = -(2D + 1)[I_2] + 1$$

$$= -2\left(-\frac{3}{2}c_1 e^{-3t/2} - \frac{2}{3}c_2 e^{-2t/3}\right) - \left(c_1 e^{-3t/2} + c_2 e^{-2t/3} + 1\right) + 1$$

$$\Rightarrow \qquad I_3(t) = 2c_1 e^{-3t/2} + \frac{1}{3}c_2 e^{-2t/3}\,.$$

 Copyright © 2012 Pearson Education, Inc. Publishing as Addison-Wesley.

To find $I_1(t)$, we use the relation $I_1 = I_2 + I_3$. Thus, we have

$$I_1(t) = c_1 e^{-3t/2} + c_2 e^{-2t/3} + 1 + 2c_1 e^{-3t/2} + \frac{1}{3} c_2 e^{-2t/3}$$

$$\Rightarrow \quad I_1(t) = 3c_1 e^{-3t/2} + \frac{4}{3} c_2 e^{-2t/3} + 1.$$

We will use the initial conditions, $I_2(0) = I_3(0) = 0$, to find the constants c_1 and c_2.

$$I_2(0) = c_1 + c_2 + 1 = 0 \quad \text{and} \quad I_3(0) = 2c_1 + \frac{1}{3} c_2 = 0.$$

Solving these two equations simultaneously yields $c_1 = 1/5$ and $c_2 = -6/5$. Therefore, the equations for the currents for this electric network are given by

$$I_1(t) = \frac{3}{5} e^{-3t/2} - \frac{8}{5} e^{-2t/3} + 1,$$

$$I_2(t) = \frac{1}{5} e^{-3t/2} - \frac{6}{5} e^{-2t/3} + 1,$$

$$I_3(t) = \frac{2}{5} e^{-3t/2} - \frac{2}{5} e^{-2t/3}.$$

13. In this problem, there are three loops. Loop 1 is through a $0.5\,\mathrm{H}$ inductor and a $1\,\Omega$ resistor; Loop 2 is through a $0.5\,\mathrm{H}$ inductor, a $0.5\,\mathrm{F}$ capacitor, and a voltage source supplying the voltage $\cos 3t\,\mathrm{V}$ at time t; Loop 3 is through a $1\,\Omega$ resistor, a $0.5\,\mathrm{F}$ capacitor, and the voltage source. We apply Kirchhoff's voltage law, $E_L + E_R + E_C = E(t)$, to Loop 1 and Loop 2 to get two equations connecting currents in the network. (Similarly to Example 2 and Problem 11, there is no need to apply Kirchhoff's voltage law to Loop 3, because the resulting equation is just a linear combination of those for other two loops.)

Loop 1:

$$E_L + E_R = 0 \quad \Rightarrow \quad 0.5 \frac{dI_1}{dt} + 1 \cdot I_2 = 0 \quad \Rightarrow \quad \frac{dI_1}{dt} + 2I_2 = 0. \quad (5.40)$$

Loop 2:

$$E_L + E_C = \cos 3t \quad \Rightarrow \quad 0.5 \frac{dI_1}{dt} + \frac{q_3}{0.5} = \cos 3t \quad \Rightarrow \quad \frac{dI_1}{dt} + 4q_3 = 2\cos 3t. \quad (5.41)$$

Additionally, at joint points, by Kirchhoff's current law,

$$-I_1 + I_2 + I_3 = 0 \quad \Rightarrow \quad -I_1 + I_2 + \frac{dq_3}{dt} = 0. \quad (5.42)$$

Copyright © 2012 Pearson Education, Inc. Publishing as Addison-Wesley.

Putting (5.40)–(5.42) together yields the following system.

$$\frac{dI_1}{dt} + 2I_2 = 0,$$

$$\frac{dI_1}{dt} + 4q_3 = 2\cos 3t,$$

$$-I_1 + I_2 + \frac{dq_3}{dt} = 0$$

or, in operator form,

$$D[I_1] + 2I_2 = 0,$$

$$D[I_1] + 4q_3 = 2\cos 3t,$$

$$-I_1 + I_2 + D[q_3] = 0$$

with the initial condition $I_1(0) = I_2(0) = I_3(0) = 0$ (since $I_3 = dq_3/dt$).

From the first equation, $I_2 = -(1/2)D[I_1]$, which (when substituted into the third equation) leads to the system

$$D[I_1] + 4q_3 = 2\cos 3t,$$

$$-(D+2)[I_1] + 2D[q_3] = 0.$$

Multiplying the first equation by D, the second equation – by 2, and subtracting the results, we eliminate q_3.

$$\{D^2 + 2(D+2)\}[I_1] = -6\sin 3t \qquad \Rightarrow \qquad (D^2 + 2D + 4)[I_1] = -6\sin 3t. \quad (5.43)$$

The roots of the characteristic equation

$$r^2 + 2r + 4 = 0,$$

are $r = -1 \pm \sqrt{3}i$, and so a general solution to the corresponding homogeneous equation is

$$I_{1h} = C_1 e^{-t} \cos \sqrt{3}t + C_2 e^{-t} \sin \sqrt{3}t.$$

A particular solution to (5.43) has the form $I_{1p} = A\cos 3t + B\sin 3t$. Substituting I_{1p} into the equation (5.43) yields

$$(-5A + 6B)\cos 3t + (-6A - 5B)\sin 3t = -6\sin 3t$$

$$\Rightarrow \quad \begin{array}{l} -5A + 6B = 0, \\ -6A - 5B = -6 \end{array} \quad \Rightarrow \quad \begin{array}{l} A = 36/61, \\ B = 30/61. \end{array}$$

 Copyright © 2012 Pearson Education, Inc. Publishing as Addison-Wesley.

Therefore,

$$I_1 = I_{1h} + I_{1p} = C_1 e^{-t} \cos\sqrt{3}t + C_2 e^{-t} \sin\sqrt{3}t + \frac{36}{61}\cos 3t + \frac{30}{61}\sin 3t\,.$$

Substituting this solution into (5.40), we find that

$$I_2 = -\frac{1}{2}\frac{dI_1}{dt} = \frac{C_1 - C_2\sqrt{3}}{2}e^{-t}\cos\sqrt{3}t + \frac{C_1\sqrt{3}+C_2}{2}e^{-t}\sin\sqrt{3}t - \frac{45}{61}\cos 3t + \frac{54}{61}\sin 3t\,.$$

The initial condition, $I_1(0) = I_2(0) = 0$, yields

$$\begin{aligned} C_1 + 36/61 &= 0, \\ (C_1 - C_2\sqrt{3})/2 - 45/61 &= 0 \end{aligned} \quad\Rightarrow\quad \begin{aligned} C_1 &= -36/61, \\ C_2 &= -42\sqrt{3}/61. \end{aligned}$$

Thus,

$$I_1 = -\frac{36}{61}e^{-t}\cos\sqrt{3}t - \frac{42\sqrt{3}}{61}e^{-t}\sin\sqrt{3}t + \frac{36}{61}\cos 3t + \frac{30}{61}\sin 3t,$$

$$I_2 = \frac{45}{61}e^{-t}\cos\sqrt{3}t - \frac{39\sqrt{3}}{61}e^{-t}\sin\sqrt{3}t - \frac{45}{61}\cos 3t + \frac{54}{61}\sin 3t,$$

$$I_3 = I_1 - I_2 = -\frac{81}{61}e^{-t}\cos\sqrt{3}t - \frac{3\sqrt{3}}{61}e^{-t}\sin\sqrt{3}t + \frac{81}{61}\cos 3t - \frac{24}{61}\sin 3t.$$

EXERCISES 5.8: Dynamical Systems, Poincarè Maps, and Chaos

1. Let $\omega = 3/2$. Using system (3) with $A = F = 1$, $\phi = 0$, and $\omega = 3/2$, we define the Poincaró map

$$x_n = \sin(3\pi n) + \frac{1}{(9/4)-(4/4)} = \sin(3\pi n) + \frac{4}{5} = \frac{4}{5},$$

$$v_n = \frac{3}{2}\cos(3\pi n) = (-1)^n\left(\frac{3}{2}\right), \qquad n = 0,1,2,\dots.$$

Calculating the first few values of (x_n, v_n), we find that they alternate between $(4/5, 3/2)$ and $(4/5, -3/2)$. Consequently, we can deduce that there is a subharmonic solution of period 4π. Let $\omega = 3/5$. Using system (3) with $A = F = 1$, $\phi = 0$, and $\omega = 3/5$, we find that the Poincaré map is

$$x_n = \sin\left(\frac{6\pi n}{5}\right) + \frac{1}{(9/25)-1} = \sin\left(\frac{6\pi n}{5}\right) - 1.5625,$$

$$v_n = \frac{3}{5}\cos\left(\frac{6\pi n}{5}\right) = (0.6)\cos\left(\frac{6\pi n}{5}\right),$$

Copyright © 2012 Pearson Education, Inc. Publishing as Addison-Wesley.

for $n = 0, 1, 2, \ldots$. Calculating the first few values of (x_n, v_n), we find that the Poincaré map cycles through the points

$$
\begin{array}{ll}
(-1.5625, 0.6), & n = 0, 5, 10, \ldots, \\
(-2.1503, -0.4854), & n = 1, 6, 11, \ldots, \\
(-0.6114, 0.1854), & n = 2, 7, 12, \ldots, \\
(-2.5136, 0.1854), & n = 3, 8, 13, \ldots, \\
(-0.9747, -0.4854), & n = 4, 9, 14, \ldots.
\end{array}
$$

Thus, we conclude that there is a subharmonic solution of period 10π.

3. With $A = F = 1$, $\phi = 0$, $\omega = 1$, $b = -0.1$, and $\theta = 0$ (because $\tan\theta = (\omega^2 - 1)/b = 0$), the solution (5) to the equation (4) becomes

$$
x(t) = e^{0.05t} \sin\left(\frac{\sqrt{3.99}}{2}\, t\right) + 10\sin t.
$$

Thus,

$$
v(t) = x'(t) = e^{0.05t}\left[0.05\sin\left(\frac{\sqrt{3.99}}{2}\,t\right) + \frac{\sqrt{3.99}}{2}\cos\left(\frac{\sqrt{3.99}}{2}\,t\right)\right] + 10\cos t
$$

and, therefore,

$$
x_n = x(2\pi n) \approx e^{0.1\pi n}\sin(1.997498\pi n),
$$
$$
v_n = v(2\pi n) \approx e^{0.1\pi n}\left(0.05\sin(1.997498\pi n) + 0.998749\cos(1.997498\pi n)\right) + 10.
$$

The values of x_n and v_n for $n = 0, 1, \ldots, 20$ are listed in Table **5–F**, page 320, and points (x_n, v_n) are shown in Figure **5–A** on page 318. As $n \to \infty$, the points (x_n, v_n) become unbounded thanks to the factor $e^{0.1\pi n}$.

5. We want to construct the Poincaré map using $t = 2\pi n$ for $x(t)$ given in equation (5) with $A = F = 1$, $\phi = 0$, $\omega = 1/3$, and $b = 0.22$. Since

$$
\tan\theta = \frac{\omega^2 - 1}{b} = -4.040404,
$$

we take $\theta = \tan^{-1}(-4.040404) = -1.328172$ and get

$$
x_n = x(2\pi n) = e^{-0.22\pi n}\sin(0.629321\pi n) - (1.092050)\sin(1.328172),
$$
$$
v_n = x'(2\pi n) = -0.11e^{-0.22\pi n}\sin(0.629321\pi n) + (1.258642)e^{-0.22\pi n}\cos(0.629321\pi n)
$$

 Copyright © 2012 Pearson Education, Inc. Publishing as Addison-Wesley.

$$+(1.092050)\cos(1.328172).$$

In Table **5–G** on page 320, we have listed the first twenty one values of the Poincaré map.

As n is getting large, we see that

$$x_n \approx -(1.092050)\sin(1.328172) \approx -1.0601\,,$$

$$v_n \approx (1.092050)\cos(1.328172) \approx 0.2624\,.$$

Hence, as $n \to \infty$, the Poincaré map approaches the point $(-1.0601, 0.2624)$.

7. Let A, ϕ and A^*, ϕ^* denote the values of constants A, ϕ in solution formula (2), corresponding to initial values (x_0, v_0) and (x_0^*, v_0^*), respectively.

(i) From recursive formulas (3) we conclude that

$$x_n - F/(\omega^2 - 1) = A\sin(2\pi\omega n + \phi),$$

$$v_n/\omega = A\cos(2\pi\omega n + \phi),$$

and so $(A, 2\pi\omega n + \phi)$ are polar coordinates of the point $(v_n/\omega, x_n - F/(\omega^2 - 1))$ in xv-plane. Similarly, $(A^*, 2\pi\omega n + \phi^*)$ represent polar coordinates of the point $(v_n^*/\omega, x_n^* - F/(\omega^2 - 1))$. Therefore,

$$(v_n^*/\omega, x_n^* - F/(\omega^2 - 1)) \to (v_n/\omega, x_n - F/(\omega^2 - 1))$$

as $A^* \to A$ and $\phi^* \to \phi$ if $A \neq 0$ or as $A^* \to 0$ (regardless of ϕ^*) if $A = 0$. Note that the convergence is uniform with respect to n. (One can easily see this from the distance formula in polar coordinates.) This is equivalent to

$$x_n^* - F/(\omega^2 - 1) \to x_n - F/(\omega^2 - 1), \qquad\qquad x_n^* \to x_n\,,$$
$$v_n^*/\omega \to v_n/\omega \qquad\qquad\qquad\qquad \Leftrightarrow \qquad v_n^* \to v_n$$

uniformly with respect to n.

(ii) On the other hand, A^* and ϕ^* satisfy

$$A^*\sin\phi^* + F/(\omega^2 - 1) = x_0^*, \qquad\quad A^* = \sqrt{(x_0^* - F/(\omega^2 - 1))^2 + (v_0^*/\omega)^2}\,,$$
$$\omega A^*\cos\phi^* = v_0^* \qquad\Rightarrow\qquad \cos\phi^* = v_0^*/(\omega A^*)\,.$$

Therefore, A^* is a continuous function of (x_0^*, v_0^*), implying that $A^* \to A$ as $(x_0^*, v_0^*) \to (x_0, v_0)$. If (x_0, v_0) is such that $A \neq 0$, then ϕ^*, as a function of (x_0^*, v_0^*), is also continuous at (x_0, v_0) and, therefore, $\phi^* \to \phi$ as $(x_0^*, v_0^*) \to (x_0, v_0)$.

Copyright © 2012 Pearson Education, Inc. Publishing as Addison-Wesley.

Combining (i) and (ii), we conclude that

$$(x_n^*, v_n^*) \to (x_n, v_n) \quad \text{as} \quad (x_0^*, v_0^*) \to (x_0, v_0)$$

(uniformly with respect to n). Thus, if (x_0^*, v_0^*) is close to (x_0, v_0), (x_n^*, v_n^*) is close to (x_n, v_n) for all n.

9. **(a),(b)** When $x_0 = 1/7$, the doubling modulo 1 map gives

$$x_1 = \frac{2}{7} \, (\mathrm{mod}\, 1) = \frac{2}{7}, \qquad\qquad x_2 = \frac{4}{7} \, (\mathrm{mod}\, 1) = \frac{4}{7},$$

$$x_3 = \frac{8}{7} \, (\mathrm{mod}\, 1) = \frac{1}{7}, \qquad\qquad x_4 = \frac{2}{7} \, (\mathrm{mod}\, 1) = \frac{2}{7},$$

$$x_5 = \frac{4}{7} \, (\mathrm{mod}\, 1) = \frac{4}{7}, \qquad\qquad x_6 = \frac{8}{7} \, (\mathrm{mod}\, 1) = \frac{1}{7},$$

$$x_7 = \frac{2}{7} \, (\mathrm{mod}\, 1) = \frac{2}{7}, \qquad\qquad \text{etc.}$$

This is the sequence $\left\{ \dfrac{1}{7}, \dfrac{2}{7}, \dfrac{4}{7}, \dfrac{1}{7}, \dots \right\}$. For $x_0 = \dfrac{k}{7}$, $k = 2, \dots, 6$, we obtain

$$\left\{ \frac{2}{7}, \frac{4}{7}, \frac{1}{7}, \frac{2}{7}, \dots \right\}, \quad \left\{ \frac{3}{7}, \frac{6}{7}, \frac{5}{7}, \frac{3}{7}, \dots \right\},$$

$$\left\{ \frac{4}{7}, \frac{1}{7}, \frac{2}{7}, \frac{4}{7}, \dots \right\}, \quad \left\{ \frac{5}{7}, \frac{3}{7}, \frac{6}{7}, \frac{5}{7}, \dots \right\},$$

$$\left\{ \frac{6}{7}, \frac{5}{7}, \frac{3}{7}, \frac{6}{7}, \dots \right\}.$$

These sequences fall into two classes. The first class has a repeating sequence $\dfrac{1}{7}, \dfrac{2}{7}, \dfrac{4}{7}$, and the second one has a repeating sequence $\dfrac{3}{7}, \dfrac{6}{7}, \dfrac{5}{7}$.

(c) To see what happens, when $x_0 = \dfrac{k}{2^j}$, we consider a special case, when $x_0 = \dfrac{3}{2^2} = \dfrac{3}{4}$.

$$x_1 = 2 \left(\frac{3}{4} \right) (\mathrm{mod}\, 1) = \frac{3}{2} \, (\mathrm{mod}\, 1) = \frac{1}{2},$$

$$x_2 = 2 \left(\frac{1}{2} \right) (\mathrm{mod}\, 1) = 1 \, (\mathrm{mod}\, 1) = 0,$$

$$x_3 = 0,$$

$$x_4 = 0,$$

$$\vdots$$

 Copyright © 2012 Pearson Education, Inc. Publishing as Addison-Wesley.

Observe that

$$x_2 = 2x_1 \,(\text{mod } 1) = 2^2\left(\frac{3}{2^2}\right)(\text{mod } 1) = 3\,(\text{mod } 1) = 0 \quad \Rightarrow \quad x_n = 0 \text{ for } n \geq 2.$$

In general, if $x_0 = k/2^j$, then

$$x_j = 2^j\left(\frac{k}{2^j}\right)(\text{mod } 1) = k\,(\text{mod } 1) = 0.$$

Consequently, $x_n = 0$ for $n \geq j$.

11. (a) A general solution to equation (6) is given by $x(t) = x_h(t) + x_p(t)$, where

$$x_h(t) = Ae^{-0.11t}\sin\left(\sqrt{9879}t + \phi\right)$$

is the transient term (a general solution to the corresponding homogeneous equation),

$$x_p(t) = \frac{1}{0.22}\sin t + \frac{1}{\sqrt{1 + 2(0.22)^2}}\sin\left(\sqrt{2}t + \psi\right), \quad \tan\psi = -\frac{1}{0.22\sqrt{2}},$$

is the steady-state term (a particular solution to (6)), which can be found by applying the formula (7) and using the superposition principle discussed in Section 4.5. Differentiating $x(t)$, we get

$$v(t) = x_h'(t) + x_p'(t) = x_h'(t) + \frac{1}{0.22}\cos t + \frac{\sqrt{2}}{\sqrt{1 + 2(0.22)^2}}\cos\left(\sqrt{2}t + \psi\right).$$

The steady-state solution does not depend on initial values x_0 and v_0; these values affect only constants A and ϕ in the transient part. But, as $t \to \infty$, $x_h(t)$ and $x_h'(t)$ tend to zero, and so the values of $x(t)$ and $v(t)$ approach the values of $x_p(t)$ and $x_p'(t)$, respectively. Thus, the limit set of points $(x(t), v(t))$ is the same as that of $(x_p(t), x_p'(t))$, which is independent of initial values.

(b) Substitution $t = 2\pi n$ into $x_p(t)$ and $x_p'(t)$ yields

$$x_n = x(2\pi n) = x_h(2\pi n) + \frac{1}{\sqrt{1 + 2(0.22)^2}}\sin\left(\sqrt{2}(2\pi n) + \psi\right),$$

$$v_n = v(2\pi n) = x_h'(2\pi n) + \frac{1}{0.22} + \frac{\sqrt{2}}{\sqrt{1 + 2(0.22)^2}}\cos\left(\sqrt{2}(2\pi n) + \psi\right).$$

As $n \to \infty$, $x_h(2\pi n) \to 0$ and $x_h'(2\pi n) \to 0$. Therefore, for n large,

$$x_n \approx \frac{1}{\sqrt{1 + 2(0.22)^2}}\sin\left(\sqrt{2}(2\pi n) + \psi\right) = a\sin\left(2\sqrt{2}\pi n + \psi\right),$$

$$v_n \approx \frac{1}{0.22} + \frac{\sqrt{2}}{\sqrt{1 + 2(0.22)^2}}\cos\left(\sqrt{2}(2\pi n) + \psi\right) = c + \sqrt{2}a\cos\left(2\sqrt{2}\pi n + \psi\right).$$

Copyright © 2012 Pearson Education, Inc. Publishing as Addison-Wesley.

(c) From part (b), we conclude that, for n large,

$$x_n^2 \approx a^2 \sin^2\left(2\sqrt{2}\pi n + \psi\right) \quad \text{and} \quad (v_n - c)^2 \approx 2a^2 \cos^2\left(2\sqrt{2}\pi n + \psi\right)$$

$$\Rightarrow \quad x_n^2 + \frac{(v_n - c)^2}{2} \approx a^2\left[\sin^2\left(2\sqrt{2}\pi n + \psi\right) + \cos^2\left(2\sqrt{2}\pi n + \psi\right)\right] = a^2,$$

and the error (coming from the transient part) tends to zero as $n \to \infty$. Thus, any limiting point of the sequence (x_n, v_n) satisfies the equation

$$x^2 + \frac{(v - c)^2}{2} = a^2,$$

which defines an ellipse centered at $(0, c)$ with semiaxes a and $a\sqrt{2}$.

13. A Poincaré section for the Duffing equation

$$x'' + 0.3x' - x + x^3 = 0.31\sin(1.2t)$$

is shown in Figure B.44 in the answers of the text. From the picture, we conclude that the solution exhibits a chaotic behavior.

REVIEW PROBLEMS

1. Expressing the system in the operator notation gives

$$D[x] + \left(D^2 + 1\right)[y] = 0,$$
$$D^2[x] + D[y] = 0.$$

We eliminate x by applying D to the first equation and then subtracting the second equation.

$$\{D\left(D^2 + 1\right) - D\}[y] = 0 \quad \Rightarrow \quad D^3[y] = 0.$$

Thus, integrating three times, we get

$$y(t) = c_3 + c_2 t + c_1 t^2.$$

We now substitute this solution into the first equation of the given system to get

$$x' = -(y'' + y) = -\left[(2c_1) + (c_3 + c_2 t + c_1 t^2)\right] = -\left[(c_3 + 2c_1) + c_2 t + c_1 t^2\right].$$

Integrating yields

$$x(t) = -\int \left[(c_3 + 2c_1) + c_2 t + c_1 t^2\right] dt = c_4 - (c_3 + 2c_1)t - \frac{1}{2}c_2 t^2 - \frac{1}{3}c_1 t^3.$$

Copyright © 2012 Pearson Education, Inc. Publishing as Addison-Wesley.

Thus, a general solution to the given system is

$$x(t) = c_4 - (c_3 + 2c_1)\,t - \frac{1}{2}\,c_2 t^2 - \frac{1}{3}\,c_1 t^3\,,$$
$$y(t) = c_3 + c_2 t + c_1 t^2\,.$$

3. Writing the system in operator form yields

$$(2D - 3)[x] - (D + 1)[y] = e^t\,,$$
$$(-4D + 15)[x] + (3D - 1)[y] = e^{-t}\,.$$

(5.44)

We eliminate y by multiplying the first equation by $(3D-1)$, the second one – by $(D+1)$, and summing the results.

$$\{(2D - 3)(3D - 1) + (-4D + 15)(D + 1)\}\,[x] = (3D - 1)[e^t] + (D + 1)[e^{-t}]$$
$$\Rightarrow \quad (D^2 + 9)[x] = e^t\,.$$

Since the characterictic equation,

$$r^2 + 9 = 0\,,$$

has roots $r = \pm 3i$, a general solution to the corresponding homogeneous equation is

$$x_h(t) = c_1 \cos 3t + c_2 \sin 3t.$$

We look for a particular solution of the form $x_p(t) = Ae^t$. Substituting this function into the equation, we obtain

$$Ae^t + 9Ae^t = e^t \quad \Rightarrow \quad A = \frac{1}{10} \quad \Rightarrow \quad x_p(t) = \frac{e^t}{10}\,,$$

and so

$$x(t) = x_h(t) + x_p(t) = c_1 \cos 3t + c_2 \sin 3t + \frac{e^t}{10}\,.$$

To find $y(t)$, we multiply the first equation in (5.44) by 3 and add to the second equation.

$$2(D + 3)[x] - 4y = 3e^t + e^{-t}$$
$$\Rightarrow \quad y(t) = \frac{1}{2}(D + 3)[x] - \frac{3}{4}\,e^t - \frac{1}{4}\,e^{-t}$$
$$= \frac{3(c_1 + c_2)}{2}\,\cos 3t - \frac{3(c_1 - c_2)}{2}\,\sin 3t - \frac{11}{20}\,e^t - \frac{1}{4}\,e^{-t}\,.$$

Copyright © 2012 Pearson Education, Inc. Publishing as Addison-Wesley.

5. Differentiating the second equation, we obtain $y'' = z'$. We eliminate z from the first and third equations by substituting y' for z and y'' for z'. This yields

$$\begin{aligned} x' &= y' - y, \\ y'' &= y' - x \end{aligned} \qquad \Rightarrow \qquad \begin{aligned} x' - y' + y &= 0, \\ y'' - y' + x &= 0 \end{aligned} \qquad (5.45)$$

or, in operator notation,

$$D[x] - (D-1)[y] = 0\,,$$
$$x + (D^2 - D)[y] = 0\,.$$

We eliminate y by applying D to the first equation and adding the result to the second one.

$$\left\{ D^2[x] - D(D-1)[y] \right\} + \left\{ x + (D^2 - D)[y] \right\} = 0 \qquad \Rightarrow \qquad \left(D^2 + 1 \right)[x] = 0.$$

This equation is the simple harmonic equation, and its general solution is given by

$$x(t) = C_1 \cos t + C_2 \sin t.$$

Substituting $x(t)$ into the first equation in (5.45) yields

$$y' - y = -C_1 \sin t + C_2 \cos t. \qquad (5.46)$$

A general solution to the corresponding homogeneous equation, $y' - y = 0$, is

$$y_h(t) = C_3 e^t\,.$$

We look for a particular solution to (5.46) of the form

$$y_p(t) = C_4 \cos t + C_5 \sin t\,.$$

Differentiating, we obtain $y_p'(t) = -C_4 \sin t + C_5 \cos t$. Thus, the equation (5.46) becomes

$$\begin{aligned} -C_1 \sin t + C_2 \cos t = y_p' - y &= (-C_4 \sin t + C_5 \cos t) - (C_4 \cos t + C_5 \sin t) \\ &= (C_5 - C_4) \cos t - (C_5 + C_4) \sin t. \end{aligned}$$

Equating coefficients yields

$$\begin{aligned} C_5 - C_4 &= C_2, \\ C_5 + C_4 &= C_1 \end{aligned} \qquad (5.47)$$

 Copyright © 2012 Pearson Education, Inc. Publishing as Addison-Wesley.

$$\Rightarrow \qquad 2C_5 = C_1 + C_2$$

$$\Rightarrow \qquad C_5 = \frac{C_1 + C_2}{2}.$$

From the second equation in (5.47), we find that

$$C_4 = C_1 - C_5 = \frac{C_1 - C_2}{2}.$$

Therefore, a general solution to (5.46) is

$$y(t) = y_h(t) + y_p(t) = C_3 e^t + \frac{C_1 - C_2}{2} \cos t + \frac{C_1 + C_2}{2} \sin t.$$

Finally, we find $z(t)$ by differentiating $y(t)$.

$$z(t) = y'(t) = \left(C_3 e^t + \frac{C_1 - C_2}{2} \cos t + \frac{C_1 + C_2}{2} \sin t \right)'$$

$$= C_3 e^t - \frac{C_1 - C_2}{2} \sin t + \frac{C_1 + C_2}{2} \cos t.$$

Hence, a general solution to the given system is

$$x(t) = C_1 \cos t + C_2 \sin t,$$

$$y(t) = C_3 e^t + \frac{C_1 - C_2}{2} \cos t + \frac{C_1 + C_2}{2} \sin t,$$

$$z(t) = C_3 e^t - \frac{C_1 - C_2}{2} \sin t + \frac{C_1 + C_2}{2} \cos t.$$

To find constants C_1, C_2, and C_3, we use the initial conditions. So, we get

$$0 = x(0) = C_1 \cos 0 + C_2 \sin 0 = C_1,$$

$$0 = y(0) = C_3 e^0 + \frac{C_1 - C_2}{2} \cos 0 + \frac{C_1 + C_2}{2} \sin 0 = C_3 + \frac{C_1 - C_2}{2},$$

$$2 = z(0) = C_3 e^0 - \frac{C_1 - C_2}{2} \sin 0 + \frac{C_1 + C_2}{2} \cos 0 = C_3 + \frac{C_1 + C_2}{2},$$

which simplifies to

$$C_1 = 0,$$

$$C_1 - C_2 + 2C_3 = 0,$$

$$C_1 + C_2 + 2C_3 = 4.$$

Solving yields $C_1 = 0$, $C_2 = 2$, $C_3 = 1$ and so

$$x(t) = 2 \sin t, \qquad y(t) = e^t - \cos t + \sin t, \qquad z(t) = e^t + \cos t + \sin t.$$

Copyright © 2012 Pearson Education, Inc. Publishing as Addison-Wesley.

7. Let $x(t)$ and $y(t)$ denote the mass of salt in tanks A and B, respectively. The only difference between this problem and the problem in Section 5.1 is that a brine solution flows in tank A instead of pure water. This change affects the input rate for tank A only, adding

$$6\,\text{L/min} \times 0.2\,\text{kg/L} = 1.2\,\text{kg/min}$$

to the original $(y/12)\,\text{kg/min}$. Thus, the system (1), Section 5.1, becomes

$$x' = -\frac{1}{3}x + \frac{1}{12}y + 1.2\,,$$
$$y' = \frac{1}{3}x - \frac{1}{3}y.$$

Following the given solution, we express $x = 3y' + y$ from the second equation and substitute it into the first one.

$$(3y' + y)' = -\frac{1}{3}\,(3y' + y) + \frac{1}{12}y + 1.2 \qquad \Rightarrow \qquad 3y'' + 2y' + \frac{1}{4}y = 1.2\,.$$

A general solution to the corresponding homogeneous equation is given in formula (3), Section 5.1.

$$y_h(t) = c_1 e^{-t/2} + c_2 e^{-t/6}\,.$$

A particular solution has the form $y_p(t) \equiv C$, which results

$$3(C)'' + 2(C)' + \frac{1}{4}C = 1.2 \qquad \Rightarrow \qquad C = 4.8\,.$$

Therefore, $y_p(t) \equiv 4.8$, and a general solution to the system is

$$x(t) = 3y'(t) + y(t) = -\frac{c_1}{2}e^{-t/2} + \frac{c_2}{2}e^{-t/6} + 4.8\,,$$
$$y(t) = y_h(t) + y_p(t) = c_1 e^{-t/2} + c_2 e^{-t/6} + 4.8\,.$$

We find constants c_1, c_2 from the initial conditions $x(0) = 0.1$, $y(0) = 0.3$. Substituting yields

$$-\frac{c_1}{2} + \frac{c_2}{2} + 4.8 = 0.1\,,$$
$$c_1 + c_2 + 4.8 = 0.3\,.$$

Solving this system, we obtain $c_1 = 49/20$, $c_2 = -139/20$, and so

$$x(t) = -\frac{13.9}{4}e^{-t/6} - \frac{4.9}{4}e^{-t/2} + 4.8\,,$$
$$y(t) = -\frac{13.9}{2}e^{-t/6} + \frac{4.9}{2}e^{-t/2} + 4.8\,.$$

 Copyright © 2012 Pearson Education, Inc. Publishing as Addison-Wesley.

9. First, we rewrite the given differential equation in an equivalent form as

$$y''' = \frac{1}{3}\left(5 + e^t y - 2y'\right).$$

Denoting $x_1(t) = y(t)$, $x_2(t) = y'(t)$, and $x_3(t) = y''(t)$, we conclude that

$$x_1' = y' = x_2\,,$$

$$x_2' = (y')' = y'' = x_3\,,$$

$$x_3' = (y'')' = y''' = \frac{1}{3}\left(5 + e^t x_1 - 2x_2\right).$$

11. This system is equivalent to

$$x''' = t - y' - y''\,,$$

$$y''' = x' - x''\,.$$

Next, we introduce, as additional unknowns, derivatives of $x(t)$ and $y(t)$ by setting

$$x_1(t) := x(t), \quad x_2(t) := x'(t), \quad x_3(t) := x''(t),$$

$$x_4(t) := y(t), \quad x_5(t) := y'(t), \quad x_6(t) := y''(t).$$

With these new variables, the system becomes

$$x''' = (x'')' =: x_3' = t - y' - y'' =: t - x_5 - x_6\,,$$
$$y''' = (y'')' =: x_6' = x' - x'' =: x_2 - x_3\,.$$

Also, we have four new equations, connecting x_j's:

$$x_1' = x' =: x_2\,,$$

$$x_2' = (x')' = x'' =: x_3\,,$$

$$x_4' = y' =: x_5\,,$$

$$x_5' = (y')' = y'' =: x_6\,.$$

Therefore, the answer is

$$x_1' = x_2\,,$$

$$x_2' = x_3\,,$$

$$x_3' = t - x_5 - x_6\,,$$

$$x_4' = x_5\,,$$

$$x_5' = x_6\,,$$

$$x_6' = x_2 - x_3\,.$$

13. With notation used in (1), Section 5.4,

$$f(x, y) = 4 - 4y,$$

$$g(x, y) = -4x,$$

and the phase plane equation (see the equation (2), Section 5.4) can be written as

$$\frac{dy}{dx} = \frac{g(x, y)}{f(x, y)} = \frac{-4x}{4 - 4y} = \frac{x}{y - 1}.$$

This equation is separable. Solving yields

$$(y - 1)\,dy = x\,dx \quad \Rightarrow \quad \int (y - 1)\,dy = \int x\,dx \quad \Rightarrow \quad (y - 1)^2 + C = x^2$$

or $x^2 - (y - 1)^2 = C$, where C is an arbitrary constant. We find the critical points by solving

$$\begin{aligned} f(x, y) = 4 - 4y = 0, \\ g(x, y) = -4x = 0 \end{aligned} \quad \Rightarrow \quad \begin{aligned} y = 1, \\ x = 0. \end{aligned}$$

Thus, $(0, 1)$ is the unique critical point. For $y > 1$,

$$\frac{dx}{dt} = 4(1 - y) < 0,$$

which implies that trajectories flow to the left. Similarly, for $y < 1$, trajectories flow to the right. Comparing the phase plane diagram with those in Figure 5.12, we conclude that the critical point $(0, 1)$ is a saddle point (unstable).

15. Some integral curves and the direction field for the given system are shown in Figure 5–B on page 318. Comparing this picture with Figure 5.12, we conclude that the origin is an asymptotically stable spiral point.

17. A trajectory is a path traced by an actual solution pair $(x(t), y(t))$, as t increases; thus, it is a directed (oriented) curve. An integral curve is the graph of a solution to the phase plane equation; it has no direction. All trajectories lie along (parts of) integral curves. A given integral curve can be the underlying point set for several different trajectories.

19. We apply Kirchhoff's voltage law to Loops 1 and 2.

Loop 1 contains a capacitor C and a resistor R_2; note that the direction of the loop is opposite to that of I_2. Thus, we have

$$\frac{q}{C} - R_2 I_2 = 0 \quad \Rightarrow \quad \frac{q}{C} = R_2 I_2,$$

where q denotes the charge of the capacitor.

 Copyright © 2012 Pearson Education, Inc. Publishing as Addison-Wesley.

Loop 2 consists of an inductor L and two resistors R_1 and R_2; note that the loop direction is opposite to the direction of I_3. Therefore,

$$R_2 I_2 - R_1 I_3 - L I_3' = 0 \quad \Rightarrow \quad R_2 I_2 = R_1 I_3 + L I_3'.$$

For the top juncture, all the currents flow out, and the Kirchhoff's current law gives

$$-I_1 - I_2 - I_3 = 0 \quad \Rightarrow \quad I_1 + I_2 + I_3 = 0.$$

Therefore, the system, describing the current in RLC, is

$$\frac{q}{C} = R_2 I_2,$$
$$R_2 I_2 = R_1 I_3 + L I_3',$$
$$I_1 + I_2 + I_3 = 0.$$

With $R_1 = R_2 = 1\,\Omega$, $L = 1\,\mathrm{H}$, and $C = 1\,\mathrm{F}$, and the relation $I_1 = dq/dt$, this system becomes

$$q = I_2,$$
$$I_2 = I_3 + I_3',$$
$$q' + I_2 + I_3 = 0.$$

Replacing in the last two equations I_2 by q, we get

$$I_3' + I_3 - q = 0,$$
$$q' + q + I_3 = 0.$$

We eliminate q by substituting $q = I_3' + I_3$ into the second equation.

$$I_3'' + 2I_3' + 2I_3 = 0.$$

The characteristic equation,

$$r^2 + 2r + 2 = 0,$$

has roots $r = -1 \pm i$, and so a general solution to this linear homogeneous equation is

$$I_3 = e^{-t}(A\cos t + B\sin t).$$

Copyright © 2012 Pearson Education, Inc. Publishing as Addison-Wesley.

Therefore,

$$I_2 = q = I_3' + I_3$$
$$= -e^{-t}(A\cos t + B\sin t) + e^{-t}(-A\sin t + B\cos t) + e^{-t}(A\cos t + B\sin t)$$
$$= e^{-t}(B\cos t - A\sin t)$$

and

$$I_1 = \frac{dq}{dt} = -e^{-t}(B\cos t - A\sin t) + e^{-t}(-B\sin t - A\cos t)$$
$$= e^{-t}[(A - B)\sin t - (A + B)\cos t].$$

 Copyright © 2012 Pearson Education, Inc. Publishing as Addison-Wesley.

TABLES

Table 5 A: Approximations to the solution in Problem 17.

m	$h = 2^{-m}$	$x_1(1; h)$	$x_2(1; h)$
0	1.0	0.375	0.375
1	0.5	0.36817	0.36817
2	0.25	0.36789	0.36789

Table 5–B: Approximations of the solution to Problem 21.

n	t_n	$x_{1;n} \approx H(t_n)$	n	t_n	$x_{1;n} \approx H(t_n)$
1	0.5	0.09573	6	3.0	2.75497
2	1.0	0.37389	7	3.5	3.52322
3	1.5	0.81045	8	4.0	4.31970
4	2.0	1.37361	9	4.5	5.13307
5	2.5	2.03111	10	5.0	5.95554

Table 5–C: Approximate period of the solution to Problem 23.

r	$a = 1$	$a = 2$	$a = 3$
1	4.8	3.3	2.3
2	4.0	2.4	1.7

FIGURES

Copyright © 2012 Pearson Education, Inc. Publishing as Addison-Wesley.

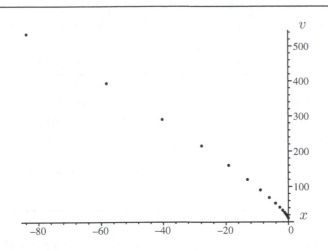

Figure 5–A: Poincaré section for Problem 3.

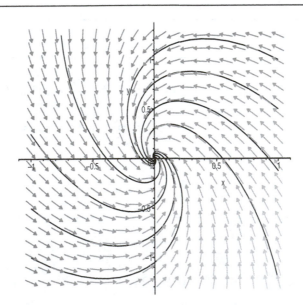

Figure 5–B: Integral curves and the direction field for Problem 15.

 Copyright © 2012 Pearson Education, Inc. Publishing as Addison-Wesley.

Table 5–D: Approximations of the solution in Problem 25.

n	h	$y(1) \approx x_1(1; 2^{-n})$	$x_2(1; 2^{-n})$	$x_3(1; 2^{-n})$
0	1.0	1.29167	0.28125	0.03125
1	0.5	1.26039	0.34509	−0.06642
2	0.25	1.25960	0.34696	−0.06957
3	0.125	1.25958	0.34704	−0.06971

Table 5–E: Runge-Kutta method approximations to the solution of $p' = 3p - p^r$, $p(0) = 1$, on $[0, 5]$ with $h = 0.25$ for $r = 1.5$, 2, and 3.

x_n	y_n ($r = 1.5$)	y_n ($r = 2$)	y_n ($r = 3$)
0.25	1.59600	1.54249	1.43911
0.5	2.37945	2.07410	1.64557
0.75	3.30824	2.47727	1.70801
1.0	4.30243	2.72769	1.72545
1.25	5.27054	2.86458	1.73024
1.5	6.13869	2.93427	1.73156
1.75	6.86600	2.96848	1.73192
2.0	7.44350	2.98498	1.73201
2.25	7.88372	2.99286	1.73204
2.5	8.20933	2.99661	1.73205
2.75	8.44497	2.99839	1.73205
3.0	8.61286	2.99924	1.73205
3.25	8.73117	2.99964	1.73205
3.5	8.81392	2.99983	1.73205
3.75	8.87147	2.99992	1.73205
4.0	8.91136	2.99996	1.73205
4.25	8.93893	2.99998	1.73205
4.5	8.95796	2.99999	1.73205
4.75	8.97107	3.00000	1.73205
5.0	8.98010	3.00000	1.73205

Copyright © 2012 Pearson Education, Inc. Publishing as Addison-Wesley.

Table 5–F: Poincaré map for Problem 3.

n	x_n	v_n	n	x_n	v_n
0	0	10.998749	11	−2.735915	41.387469
1	−0.010761	11.366815	12	−4.085318	52.925111
2	−0.029466	11.870407	13	−6.057783	68.700143
3	−0.060511	12.559384	14	−8.929255	90.267442
4	−0.110453	13.501933	15	−13.09442	119.75193
5	−0.189009	14.791299	16	−19.11674	160.05736
6	−0.310494	16.554984	17	−27.79923	215.15152
7	−0.495883	18.967326	18	−40.28442	290.45581
8	−0.775786	22.266682	19	−58.19561	393.37721
9	−1.194692	26.778923	20	−83.83579	534.03491
10	−1.817047	32.949532			

Table 5–G: Poincaré map for Problem 5.

n	x_n	v_n	n	x_n	v_n
0	−1.060065	1.521008	11	−1.059944	0.261743
1	−0.599847	0.037456	12	−1.060312	0.262444
2	−1.242301	0.065170	13	−1.059997	0.262491
3	−1.103418	0.415707	14	−1.060030	0.262297
4	−0.997156	0.251142	15	−1.060096	0.262362
5	−1.074094	0.228322	16	−1.060061	0.262385
6	−1.070300	0.278664	17	−1.060058	0.262360
7	−1.052491	0.264458	18	−1.060068	0.262364
8	−1.060495	0.257447	19	−1.060065	0.262369
9	−1.061795	0.263789	20	−1.060064	0.262366
10	−1.059271	0.263037			

 Copyright © 2012 Pearson Education, Inc. Publishing as Addison-Wesley.

CHAPTER 6: Theory of Higher-Order Linear Differential Equations

EXERCISES 6.1: Basic Theory of Linear Differential Equations

1. Putting the equation in standard form,

$$y''' - \frac{3}{x}\,y' + \frac{e^x}{x}\,y = \frac{x^2-1}{x}\,,$$

we find that

$$p_1(x) \equiv 0, \quad p_2(x) = -\frac{3}{x}\,, \quad p_3(x) = \frac{e^x}{x}\,, \quad \text{and} \quad q(x) = \frac{x^2-1}{x}\,.$$

Functions $p_2(x)$, $p_3(x)$, and $q(x)$ have only one point of discontinuity, $x = 0$, while $p_1(x)$ is continuous everywhere. Therefore, these functions are continuous simultaneously on $(-\infty, 0)$ and $(0, \infty)$. Since the initial point, $x_0 = -2$, belongs to $(-\infty, 0)$, Theorem 1 guarantees the existence of a unique solution to the given initial value problem on $(-\infty, 0)$.

3. In this problem, $p_1(x) \equiv -1$, $p_2(x) = \sqrt{x-1}$, and $q(x) = \tan x$. We note that $p_1(x)$ is continuous everywhere, $p_2(x)$ is continuous for $x \geq 1$, and $g(x)$ is continuous everywhere except odd multiples of $\pi/2$. Therefore, these three functions are continuous simultaneously on the intervals

$$\left[1, \frac{\pi}{2}\right), \quad \left(\frac{\pi}{2}, \frac{3\pi}{2}\right), \quad \left(\frac{3\pi}{2}, \frac{5\pi}{2}\right), \ldots .$$

Since $x = 5$, the initial point, is on the interval $(3\pi/2, 5\pi/2)$, Theorem 1 guarantees a unique solution to the given initial value problem on this interval.

5. Dividing the equation by $x\sqrt{x+1}$, we obtain

$$y''' - \frac{1}{x\sqrt{x+1}}\,y' + \frac{1}{\sqrt{x+1}}\,y = 0.$$

Thus, $p_1(x) \equiv 0$, $p_2(x) = 1/(x\sqrt{x+1})$, $p_3(x) = 1/\sqrt{x+1}$, and $g(x) \equiv 0$. Functions $p_1(x)$ and $q(x)$ are continuous on the whole real line; $p_3(x)$ is defined and continuous for $x > -1$; $p_2(x)$ is defined and continuous for $x > -1$ and $x \neq 0$. Therefore, these functions are continuous simultaneously on $(-1, 0)$ and $(0, \infty)$. The initial point lies on $(0, \infty)$ and so, by Theorem 1, the initial value problem has a unique solution on the interval $(0, \infty)$.

7. Let us assume that c_1, c_2, and c_3 are constants, for which

$$c_1 e^{3x} + c_2 e^{5x} + c_3 e^{-x} \equiv 0 \quad \text{on} \quad (-\infty, \infty). \tag{6.1}$$

If we show that this is possible only if $c_1 = c_2 = c_3 = 0$, then linear independence will follow. Evaluating the linear combination in (6.1) at $x = 0$, $x = \ln 2$, and $x = -\ln 2$, we find that constants c_1, c_2, and c_3 must satisfy

$$c_1 + c_2 + c_3 = 0,$$
$$8c_1 + 32c_2 + \frac{1}{2} c_3 = 0,$$
$$\frac{1}{8} c_1 + \frac{1}{32} c_2 + 2c_3 = 0.$$

This system is a homogeneous system of linear equations, whose determinant

$$\begin{vmatrix} 1 & 1 & 1 \\ 8 & 32 & 1/2 \\ 1/8 & 1/32 & 2 \end{vmatrix} = \begin{vmatrix} 32 & 1/2 \\ 1/32 & 2 \end{vmatrix} - \begin{vmatrix} 8 & 1/2 \\ 1/8 & 2 \end{vmatrix} + \begin{vmatrix} 8 & 32 \\ 1/8 & 1/32 \end{vmatrix} = \frac{2827}{64} \neq 0.$$

Hence, it has only the trivial solution, that is, $c_1 = c_2 = c_3 = 0$.

9. Let

$$y_1 = \sin^2 x, \qquad y_2 = \cos^2 x, \qquad y_3 = 1.$$

We want to find c_1, c_2, and c_3, not all zero, such that

$$c_1 y_1 + c_2 y_2 + c_3 y_3 = c_1 \sin^2 x + c_2 \cos^2 x + c_3 \cdot 1 = 0,$$

for all x in the interval $(-\infty, \infty)$. Since $\sin^2 x + \cos^2 x = 1$ for all real x, we can choose $c_1 = 1$, $c_2 = 1$, and $c_3 = -1$. Thus, these functions are linearly dependent.

11. Let $y_1 = x^{-1}$, $y_2 = x^{1/2}$, and $y_3 = x$. We want to find constants c_1, c_2, and c_3 such that

$$c_1 y_1 + c_2 y_2 + c_3 y_3 = c_1 x^{-1} + c_2 x^{1/2} + c_3 x = 0,$$

Copyright © 2012 Pearson Education, Inc. Publishing as Addison-Wesley.

for all x on the interval $(0, \infty)$. This equation then holds for $x = 1$, 4, and 9. Plugging in these values for x into the above equation, we see that c_1, c_2, and c_3 must satisfy the system

$$c_1 + c_2 + c_3 = 0,$$
$$\frac{c_1}{4} + 2c_2 + 4c_3 = 0,$$
$$\frac{c_1}{9} + 3c_2 + 9c_3 = 0.$$

Solving these equations simultaneously yields $c_1 = c_2 = c_3 = 0$. Thus, the only way to have $c_1 x^{-1} + c_2 x^{1/2} + c_3 x = 0$ for all x on $(0, \infty)$ is $c_1 = c_2 = c_3 = 0$. Therefore, x^{-1}, $x^{1/2}$, and x are linearly independent on $(0, \infty)$.

13. A linear combination, $c_1 x + c_2 x^2 + c_3 x^3 + c_4 x^4$, is a polynomial of degree at most four and so, by the fundamental theorem of algebra, it cannot have more than four zeros unless it is the zero polynomial (that is, it has all zero coefficients). Thus, if this linear combination vanishes on *an interval*, then $c_1 = c_2 = c_3 = c_4 = 0$. Therefore, the functions x, x^2, x^3, and x^4 are linearly independent on any interval, in particular, on $(-\infty, \infty)$.

15. Since, by inspection, $r = 3$, $r = -1$, and $r = -4$ are the roots of the characteristic equation, $r^3 + 2r^2 - 11r - 12 = 0$, the functions e^{3x}, e^{-x}, and e^{-4x} form a solution set. Next, we check that these functions are linearly independent by showing that their Wronskian is never zero.

$$W\left[e^{3x}, e^{-x}, e^{-4x}\right](x) = \begin{vmatrix} e^{3x} & e^{-x} & e^{-4x} \\ 3e^{3x} & -e^{-x} & -4e^{-4x} \\ 9e^{3x} & e^{-x} & 16e^{-4x} \end{vmatrix} = e^{3x}e^{-x}e^{-4x} \begin{vmatrix} 1 & 1 & 1 \\ 3 & -1 & -4 \\ 9 & 1 & 16 \end{vmatrix} = -84e^{-2x},$$

which does not vanish. Therefore, $\{e^{3x}, e^{-x}, e^{-4x}\}$ is a *fundamental* solution set and, by Theorem 4, a general solution to the given differential equation is

$$y = c_1 e^{3x} + c_2 e^{-x} + c_3 e^{-4x}.$$

17. Writing the given differential equation,

$$x^3 y''' - 3x^2 y'' + 6xy' - 6y = 0,$$

in standard form (17), we see that its coefficients, $-3/x$, $6/x^2$, and $-6/x^3$ are continuous on the specified interval, which is $x > 0$.

Next, substituting x, x^2, and x^3 into the differential equation, we verify that these functions are indeed solutions.

$$x^3(x)''' - 3x^2(x)'' + 6x(x)' - 6(x) = 0 - 0 + 6x - 6x = 0,$$

$$x^3(x^2)''' - 3x^2(x^2)'' + 6x(x^2)' - 6(x^2) = 0 - 6x^2 + 12x^2 - 6x^2 = 0,$$

$$x^3(x^3)''' - 3x^2(x^3)'' + 6x(x^3)' - 6(x^3) = 6x^3 - 18x^3 + 18x^3 - 6x^3 = 0.$$

Evaluating the Wronskian by expanding along the first column yields

$$W\left[x, x^2, x^3\right](x) = \begin{vmatrix} x & x^2 & x^3 \\ 1 & 2x & 3x^2 \\ 0 & 2 & 6x \end{vmatrix}$$

$$= x \begin{vmatrix} 2x & 3x^2 \\ 2 & 6x \end{vmatrix} - \begin{vmatrix} x^2 & x^3 \\ 2 & 6x \end{vmatrix} = x\left(6x^2\right) - \left(4x^3\right) = 2x^3.$$

Thus $W\left[x, x^2, x^3\right](x) \neq 0$ on $(0, \infty)$ and so $\{x, x^2, x^3\}$ is a fundamental solution set for the given differential equation. By Theorem 2, $y = c_1 x + c_2 x^2 + c_3 x^3$ is a general solution.

19. (a) Since $\{e^x, e^{-x}\cos 2x, e^{-x}\sin 2x, \}$ is a fundamental solution set for the corresponding homogeneous differential equation and $y_p = x^2$ is a solution to the nonhomogeneous equation, by the superposition principle, we have a general solution given by

$$y(x) = c_1 e^x + c_2 e^{-x}\cos 2x + c_3 e^{-x}\sin 2x + x^2.$$

(b) To find the solution that satisfies the initial conditions, we differentiate the general solution $y(x)$ from part (a) twice. Thus, we have

$$y'(x) = c_1 e^x - c_2 e^{-x}\cos 2x - 2c_2 e^{-x}\sin 2x - c_3 e^{-x}\sin 2x + 2c_3 e^{-x}\cos 2x + 2x$$

$$= c_1 e^x + (-c_2 + 2c_3)e^{-x}\cos 2x + (-2c_2 - c_3)e^{-x}\sin 2x + 2x,$$

$$y''(x) = c_1 e^x + (c_2 - 2c_3)e^{-x}\cos 2x - 2(-c_2 + 2c_3)e^{-x}\sin 2x$$

$$\qquad - (-2c_2 - c_3)e^{-x}\sin 2x + 2(-2c_2 - c_3)e^{-x}\cos 2x + 2$$

$$= c_1 e^x + (-3c_2 - 4c_3)e^{-x}\cos 2x + (4c_2 - 3c_3)e^{-x}\sin 2x + 2.$$

Plugging the initial conditions into these formulas yields the equations

 Copyright © 2012 Pearson Education, Inc. Publishing as Addison-Wesley.

$$y(0) = c_1 + c_2 = -1,$$

$$y'(0) = c_1 - c_2 + 2c_3 = 1,$$

$$y''(0) = c_1 - 3c_2 - 4c_3 + 2 = -3.$$

By solving these equations simultaneously, we obtain $c_1 = -1$, $c_2 = 0$, and $c_3 = 1$. Therefore, the solution to the given initial value problem is

$$y(x) = -e^x + e^{-x} \sin 2x + x^2.$$

21. In standard form, the given equation becomes

$$y''' + \frac{1}{x^2} y' - \frac{1}{x^3} y = \frac{3 - \ln x}{x^3}.$$

Since its coefficients are continuous on $(0, \infty)$, we can apply Theorems 2 and 4 to conclude that a general solution to the corresponding homogeneous equation is

$$y_h(x) = c_1 x + c_2 x \ln x + c_3 x (\ln x)^2,$$

and so a general solution to the given nonhomogeneous equation is

$$y(x) = y_h(x) + y_p(x) = c_1 x + c_2 x \ln x + c_3 x (\ln x)^2 + \ln x.$$

To satisfy the initial conditions, we find

$$y'(x) = \frac{1}{x} + c_1 + c_2(\ln x + 1) + c_3 \left[(\ln x)^2 + 2 \ln x\right],$$

$$y''(x) = -\frac{1}{x^2} + \frac{c_2}{x} + c_3 \left(\frac{2 \ln x}{x} + \frac{2}{x}\right).$$

Substituting the initial conditions, $y(1) = 3$, $y'(1) = 3$, and $y''(1) = 0$, we get the system

$$\begin{aligned}
3 &= y(1) = c_1, \\
3 &= y'(1) = 1 + c_1 + c_2, \\
0 &= y''(1) = -1 + c_2 + 2c_3
\end{aligned} \qquad \Rightarrow \qquad \begin{aligned}
c_1 &= 3, \\
c_1 + c_2 &= 2, \\
c_2 + 2c_3 &= 1
\end{aligned} \qquad \Rightarrow \qquad \begin{aligned}
c_1 &= 3, \\
c_2 &= -1, \\
c_3 &= 1.
\end{aligned}$$

Thus,

$$y(x) = 3x - x \ln x + x (\ln x)^2 + \ln x$$

is the desired solution.

Copyright © 2012 Pearson Education, Inc. Publishing as Addison-Wesley.

23. Substituting $y_1(x) = \sin x$ and $y_2(x) = x$ into the given differential operator yields

$$L[\sin x] = (\sin x)''' + (\sin x)' + x(\sin x) = -\cos x + \cos x + x \sin x = x \sin x,$$
$$L[x] = (x)''' + (x)' + x(x) = 0 + 1 + x^2 = x^2 + 1.$$

Note that $L[y]$ is a linear operator of the form (7). So, we can use the superposition principle.

(a) Since $2x \sin x - x^2 - 1 = 2(x \sin x) - (x^2 + 1)$, by the superposition principle,

$$y(x) = 2y_1(x) - y_2(x) = 2 \sin x - x$$

is a solution to $L[y] = 2x \sin x - x^2 - 1$.

(b) We can express $4x^2 + 4 - 6x \sin x = 4(x^2 + 1) - 6(x \sin x)$. Hence,

$$y(x) = 4y_2(x) - 6y_1(x) = 4x - 6 \sin x$$

is a solution to $L[y] = 4x^2 + 4 - 6x \sin x$.

25. Clearly, it is sufficient to prove (9) just for two functions, y_1 and y_2. Using the linear property of differentiation, we have

$$
\begin{aligned}
L[y_1 + y_2] &= (y_1 + y_2)^{(n)} + p_1 (y_1 + y_2)^{(n-1)} + \cdots + p_n (y_1 + y_2) \\
&= \left(y_1^{(n)} + y_2^{(n)} \right) + p_1 \left(y_1^{(n-1)} + y_2^{(n-1)} \right) + \cdots + p_n (y_1 + y_2) \\
&= \left(y_1^{(n)} + p_1 y_1^{(n-1)} + \cdots + p_n y_1 \right) + \left(y_2^{(n)} + p_1 y_2^{(n-1)} + \cdots + p_n y_2 \right) \\
&= L[y_1] + L[y_1].
\end{aligned}
$$

Next, we verify (10).

$$
\begin{aligned}
L[cy] &= (cy)^{(n)} + p_1 (cy)^{(n-1)} + \cdots + p_n (cy) = cy^{(n)} + p_1 cy^{(n-1)} + \cdots + p_n cy \\
&= c \left(y^{(n)} + p_1 y^{(n-1)} + \cdots + p_n y \right) = cL[y].
\end{aligned}
$$

27. A linear combination $c_0 + c_1 x + c_2 x^2 + \cdots + c_n x^n$ of the functions from the given set is a polynomial of degree at most n and so, by the fundamental theorem of algebra, it cannot have more than n zeros unless it is the zero polynomial, i.e., it has all zero coefficients. Thus, if this linear combination vanishes on a whole interval (a, b), then it follows that the coefficients $c_0 = c_1 = \ldots = c_n = 0$. Therefore, the functions $1, x, x^2, \ldots, x^n$ are linearly independent on any interval (a, b).

 Copyright © 2012 Pearson Education, Inc. Publishing as Addison-Wesley.

29. (a) Assuming that functions $f_1, f_2, \ldots, f_m$ are linearly dependent on $(-\infty, \infty)$, we can find their nontrivial linear combination, vanishing identically on $(-\infty, \infty)$, i.e.,

$$c_1 f_1 + c_2 f_2 + \cdots + c_m f_m \equiv 0 \qquad \text{on} \qquad (-\infty, \infty),$$

where not all c_j's are zero. In particular, this linear combination vanishes on $(-1, 1)$, which contradicts the assumption that $f_1, f_2, \ldots, f_m$ are linearly independent on $(-1, 1)$.

(b) Let

$$f_1(x) := |x - 1|, \qquad f_2(x) := x - 1.$$

On $(-1, 1)$ (even on $(-\infty, 1)$) we have $f_1(x) = -f_2(x)$ or, equivalently, $f_1(x) + f_2(x) \equiv 0$ and so these functions are linearly dependent on $(-1, 1)$. At the same time, their linear combination

$$c_1 f_1(x) + c_2 f_2(x) = \begin{cases} (c_2 - c_1)(x - 1), & x \le 1; \\ (c_1 + c_2)(x - 1), & x > 1 \end{cases}$$

cannot vanish identically on $(-\infty, \infty)$ unless $c_1 - c_2 = 0$ and $c_1 + c_2 = 0$, which implies $c_1 = c_2 = 0$.

31. (a) The linearity of differentiation and the product rule imply that

$$y'(x) = [v(x)e^x]' = v'(x)e^x + v(x)(e^x)' = [v'(x) + v(x)]e^x,$$

$$y''(x) = [v'(x) + v(x)]'e^x + [v'(x) + v(x)](e^x)' = [v''(x) + 2v'(x) + v(x)]e^x,$$

$$y'''(x) = [v''(x) + 2v'(x) + v(x)]'e^x + [v''(x) + 2v'(x) + v(x)](e^x)'$$
$$= [v'''(x) + 3v''(x) + 3v'(x) + v(x)]e^x.$$

(b) Substituting y, y', y'', and y''' into the differential equation (35), we obtain

$$(v''' + 3v'' + 3v' + v)e^x - 2(v'' + 2v' + v)e^x - 5(v' + v)e^x + 6ve^x = 0$$

$$\Rightarrow \qquad [(v''' + 3v'' + 3v' + v) - 2(v'' + 2v' + v) - 5(v' + v) + 6v]e^x = 0$$

$$\Rightarrow \qquad v''' + v'' - 6v' = 0,$$

where we have used the fact that the function e^x is never zero. Let $w := v'$. Then $w' = v''$, $w'' = v'''$, and so the above equation becomes

$$w'' + w' - 6w = 0. \tag{6.2}$$

Copyright © 2012 Pearson Education, Inc. Publishing as Addison-Wesley.

(c) The auxiliary equation for (6.2), $r^2 + r - 6 = 0$, has roots $r = -3, 2$. Thus, a general solution to this differential equation is $w(x) = c_1 e^{-3x} + c_2 e^{2x}$, where c_1 and c_2 are arbitrary constants. Choosing, say, $c_1 = -3$, $c_2 = 0$ and $c_1 = 0$, $c_2 = 2$, we find two linearly independent solutions,

$$w_1(x) = -3e^{-3x} \qquad \text{and} \qquad w_2(x) = 2e^{2x} \,.$$

Integrating yields

$$v_1(x) = \int w_1(x)\,dx = \int \left(-3e^{-3x}\right) dx = e^{-3x} \,,$$

$$v_2(x) = \int w_2(x)\,dx = \int \left(2e^{2x}\right) dx = e^{2x} \,,$$

where we have chosen zero integration constants.

(d) With functions $v_1(x)$ and $v_2(x)$ obtained in (c), we have

$$y_1(x) = v_1(x)e^x = e^{-3x}e^x = e^{-2x} \,, \quad y_2(x) = v_2(x)e^x = e^{2x}e^x = e^{3x} \,.$$

To show that the functions e^x, e^{-2x}, and e^{3x} are linearly independent on $(-\infty, \infty)$, we can use the approach similar to that in Problem 7. Alternatively, since these functions are solutions to the differential equation (35), one can apply Theorem 3, as we did in Problem 15. To this end,

$$W\left[e^x, e^{-2x}, e^{3x}\right](x) = \begin{vmatrix} e^x & e^{-2x} & e^{3x} \\ e^x & -2e^{-2x} & 3e^{3x} \\ e^x & 4e^{-2x} & 9e^{3x} \end{vmatrix} = e^x e^{-2x} e^{3x} \begin{vmatrix} 1 & 1 & 1 \\ 1 & -2 & 3 \\ 1 & 4 & 9 \end{vmatrix} = -30e^{2x} \neq 0$$

on $(-\infty, \infty)$, and so the functions e^x, e^{-2x}, and e^{3x} are linearly independent.

33. Let $y(x) = v(x)e^{2x}$. Differentiating $y(x)$, we obtain

$$y'(x) = \left[v'(x) + 2v(x)\right]e^{2x} \,,$$

$$y''(x) = \left[v''(x) + 4v'(x) + 4v(x)\right]e^{2x} \,,$$

$$y'''(x) = \left[v'''(x) + 6v''(x) + 12v'(x) + 8v(x)\right]e^{2x} \,.$$

Substituting these expressions into the given differential equation yields

$$\left[(v''' + 6v'' + 12v' + 8v) - 2\left(v'' + 4v' + 4v\right) + (v' + 2v) - (2v)\right]e^{2x} = 0$$

 Copyright © 2012 Pearson Education, Inc. Publishing as Addison-Wesley.

$$\Rightarrow \qquad \left[v''' + 4v'' + 5v'\right]e^{2x} = 0 \qquad \Rightarrow \qquad v''' + 4v'' + 5v' = 0.$$

With $w(x) := v'(x)$, the above equation becomes $w''(x) + 4w'(x) + 5w(x) = 0$. The roots of the auxiliary equation, $r^2 + 4r + 5 = 0$, for this second order equation are $r = -2 \pm i$. Therefore,

$$\{w_1(x), w_2(x)\} = \left\{e^{-2x}\cos x, e^{-2x}\sin x\right\}$$

is a fundamental solution set. Integrating, we get

$$v_1(x) = \int w_1(x) = \int e^{-2x}\cos x\, dx = \frac{e^{-2x}(\sin x - 2\cos x)}{5},$$

$$v_2(x) = \int w_2(x) = \int e^{-2x}\sin x\, dx = -\frac{e^{-2x}(2\sin x + \cos x)}{5},$$

where we have chosen zero integration constants. Thus, functions

$$f(x) = e^{2x},$$

$$y_1(x) = v_1(x)f(x) = \frac{e^{-2x}(\sin x - 2\cos x)}{5}e^{2x} = \frac{\sin x - 2\cos x}{5},$$

$$y_2(x) = v_2(x)f(x) = -\frac{e^{-2x}(2\sin x + \cos x)}{5}e^{2x} = -\frac{2\sin x + \cos x}{5}$$

are three linearly independent solutions to the given differential equation.

35. In notation of Problem 34,

$$f_1(x) = x, \quad f_2(x) = \sin x, \quad \text{and} \quad f_3(x) = \cos x.$$

These functions are analytic. Let us evaluate their Wronskian to make sure that the result of Problem 34 can be applied. Expanding along the first column we get

$$W\left[x, \sin x, \cos x\right](x) = \begin{vmatrix} x & \sin x & \cos x \\ 1 & \cos x & -\sin x \\ 0 & -\sin x & -\cos x \end{vmatrix}$$

$$= x\begin{vmatrix} \cos x & -\sin x \\ -\sin x & -\cos x \end{vmatrix} - \begin{vmatrix} \sin x & \cos x \\ -\sin x & -\cos x \end{vmatrix}$$

$$= x\left(-\cos^2 x - \sin^2 x\right) - \left(-\sin x\cos x + \sin x\cos x\right) = -x.$$

Thus,

$$W\left[x, \sin x, \cos x\right](x) \neq 0 \quad \text{on } (-\infty, 0) \text{ and } (0, \infty).$$

Copyright © 2012 Pearson Education, Inc. Publishing as Addison-Wesley.

Hence, on either of these two intervals, $\{x, \sin x, \cos x\}$ is a fundamental solution set for the third order linear differential equation given in Problem 34.

Expanding the determinant in Problem 34 over its last column and then expanding third-order determinants over their first column yields

$$
\begin{vmatrix}
x & \sin x & \cos x & y \\
1 & \cos x & -\sin x & y' \\
0 & -\sin x & -\cos x & y'' \\
0 & -\cos x & \sin x & y'''
\end{vmatrix}
= y''' W\left[x, \sin x, \cos x\right](x) - y''
\begin{vmatrix}
x & \sin x & \cos x \\
1 & \cos x & -\sin x \\
0 & -\cos x & \sin x
\end{vmatrix}
$$

$$
+ y'
\begin{vmatrix}
x & \sin x & \cos x \\
0 & -\sin x & -\cos x \\
0 & -\cos x & \sin x
\end{vmatrix}
- y
\begin{vmatrix}
1 & \cos x & -\sin x \\
0 & -\sin x & -\cos x \\
0 & -\cos x & \sin x
\end{vmatrix}
$$

$$
= -xy''' - y''\left(x
\begin{vmatrix}
\cos x & -\sin x \\
-\cos x & \sin x
\end{vmatrix}
-
\begin{vmatrix}
\sin x & \cos x \\
-\cos x & \sin x
\end{vmatrix}
\right)
$$

$$
+ y'x
\begin{vmatrix}
-\sin x & -\cos x \\
-\cos x & \sin x
\end{vmatrix}
- y
\begin{vmatrix}
-\sin x & -\cos x \\
-\cos x & \sin x
\end{vmatrix}
$$

$$
= -xy''' + y'' - xy' + y = 0 .
$$

Hence, a third-order differential equation, for which functions x, $\sin x$, and $\cos x$ form a fundamental solution set, is $xy''' - y'' + xy' - y = 0$.

EXERCISES 6.2: Homogeneous Linear Equations with Constant Coefficients

1. The auxiliary equation

$$
r^3 + 2r^2 - 8r = 0 \qquad \Rightarrow \qquad r\left(r^2 + 2r - 8\right) = r(r - 2)(r + 4) = 0
$$

has roots $r = 0$, 2, and -4. Thus, a general solution has the form

$$
y = c_1 + c_2 e^{2x} + c_3 e^{-4x} .
$$

3. The auxiliary equation for this problem is $6r^3 + 7r^2 - r - 2 = 0$. By inspection we see that $r = -1$ is a root of this equation, and so with the aid of long division we can factor it as

$$
6r^3 + 7r^2 - r - 2 = (r + 1)(6r^2 + r - 2) = (r + 1)(3r + 2)(2r - 1) = 0.
$$

 Copyright © 2012 Pearson Education, Inc. Publishing as Addison-Wesley.

We see now that the roots to the auxiliary equation are $r = -1$, $-2/3$, and $1/2$. These roots are real and distinct. Therefore, a general solution to this problem is given by

$$z(x) = c_1 e^{-x} + c_2 e^{-2x/3} + c_3 e^{x/2}.$$

5. We can factor the auxiliary equation, $r^3 + 3r^2 + 28r + 26 = 0$, as follows.

$$\begin{aligned} r^3 + 3r^2 + 28r + 26 &= (r^3 + r^2) + (2r^2 + 2r) + (26r + 26) \\ &= r^2(r+1) + 2r(r+1) + 26(r+1) = (r+1)(r^2 + 2r + 26) = 0. \end{aligned}$$

Thus, either $r + 1 = 0 \Rightarrow r = -1$ or $r^2 + 2r + 26 = 0 \Rightarrow r = -1 \pm 5i$. Therefore, a general solution is given by

$$y(x) = c_1 e^{-x} + c_2 e^{-x} \cos 5x + c_3 e^{-x} \sin 5x.$$

7. Factoring the characteristic polynomial yields

$$\begin{aligned} 2r^3 - r^2 - 10r - 7 &= (2r^3 + 2r^2) + (-3r^2 - 3r) + (-7r - 7) \\ &= 2r^2(r+1) - 3r(r+1) - 7(r+1) = (r+1)(2r^2 - 3r - 7). \end{aligned}$$

Thus, the roots of the characteristic equation, $2r^3 - r^2 - 10r - 7 = 0$, are

$$r + 1 = 0 \quad \Rightarrow \quad r = -1,$$
$$2r^2 - 3r - 7 = 0 \quad \Rightarrow \quad r = \frac{3 \pm \sqrt{3^2 - 4(2)(-7)}}{4} = \frac{3 \pm \sqrt{65}}{4},$$

and a general solution is

$$y(x) = c_1 e^{-x} + c_2 e^{(3+\sqrt{65})x/4} + c_3 e^{(3-\sqrt{65})x/4}.$$

9. In the characteristic equation, we recognize a complete cube, namely,

$$r^3 - 9r^2 + 27r - 27 = 0 = (r-3)^3 = 0.$$

Thus, it has just one root, $r = 3$, of multiplicity three. Therefore, a general solution to the given differential equation is given by

$$u(x) = c_1 e^{3x} + c_2 x e^{3x} + c_3 x^2 e^{3x}.$$

Copyright © 2012 Pearson Education, Inc. Publishing as Addison-Wesley.

11. Since $r^4 + 4r^3 + 6r^2 + 4r + 1 = (r+1)^4$, the characteristic equation becomes $(r+1)^4 = 0$, and it has a root $r = -1$ of multiplicity four. Therefore, the functions e^{-x}, xe^{-x}, $x^2 e^{-x}$, and $x^3 e^{-x}$ form a fundamental solution set, and a general solution to the given differential equation is

$$y(x) = c_1 e^{-x} + c_2 x e^{-x} + c_3 x^2 e^{-x} + c_4 x^3 e^{-x} = \left(c_1 + c_2 x + c_3 x^2 + c_4 x^3 \right) e^{-x}.$$

13. The auxiliary equation in this problem is $r^4 + 4r^2 + 4 = \left(r^2 + 2 \right)^2 = 0$. Therefore, this equation has double roots $r = \pm\sqrt{2}\,i$ and, a general solution to this problem is given by

$$y(x) = c_1 \cos \sqrt{2}x + c_2 x \cos \sqrt{2}x + c_3 \sin \sqrt{2}x + c_4 x \sin \sqrt{2}x.$$

15. The roots the the auxiliary equation, $(r-1)^2 (r+3)(r^2 + 2r + 5)^2 = 0$, are

$$r = 1,\ 1,\ -3,\ -1 \pm 2i,\ -1 \pm 2i.$$

We note that 1 and $-1 \pm 2i$ are repeated roots. Therefore, a general solution to the differential equation, having given auxiliary equation, is

$$y(x) = c_1 e^x + c_2 x e^x + c_3 e^{-3x} + (c_4 + c_5 x)e^{-x}\cos 2x + (c_6 + c_7 x)e^{-x}\sin 2x.$$

17. From the differential operator, replacing D by r, we obtain the characteristic equation

$$(r+4)(r-3)(r+2)^3 (r^2 + 4r + 5)^2 r^5 = 0,$$

whose roots are

$$
\begin{aligned}
r + 4 = 0 &\quad\Rightarrow\quad r = -4, \\
r - 3 = 0 &\quad\Rightarrow\quad r = 3, \\
(r+2)^3 = 0 &\quad\Rightarrow\quad r = -2 \text{ of multiplicity 3}, \\
(r^2 + 4r + 5)^2 = 0 &\quad\Rightarrow\quad r = -2 \pm i \text{ of multiplicity 2}, \\
r^5 = 0 &\quad\Rightarrow\quad r = 0 \text{ of multiplicity 5}.
\end{aligned}
$$

Therefore, a general solution is given by

$$y(x) = c_1 e^{-4x} + c_2 e^{3x} + \left(c_3 + c_4 x + c_5 x^2 \right) e^{-2x} + (c_6 + c_7 x)\, e^{-2x} \cos x$$
$$+ (c_8 + c_9 x)\, e^{-2x} \sin x + c_{10} + c_{11} x + c_{12} x^2 + c_{13} x^3 + c_{14} x^4.$$

 Copyright © 2012 Pearson Education, Inc. Publishing as Addison-Wesley.

19. First, we find a general solution to the given equation. Solving the auxiliary equation,

$$
\begin{aligned}
r^3 - r^2 - 4r + 4 &= (r^3 - r^2) - (4r - 4) \\
&= (r - 1)(r^2 - 4) = (r - 1)(r + 2)(r - 2) = 0,
\end{aligned}
$$

yields three real distinct roots $r = 1$, -2, and 2. Thus, a general solution is

$$
y(x) = c_1 e^x + c_2 e^{-2x} + c_3 e^{2x} .
$$

Next, we find constants c_1, c_2, and c_3 such that the solution satisfies the initial conditions. Differentiating $y(x)$ twice and substituting the initial conditions, we obtain the system

$$
\begin{aligned}
y(0) &= \left(c_1 e^x + c_2 e^{-2x} + c_3 e^{2x} \right)\Big|_{x=0} = c_1 + c_2 + c_3 = -4, \\
y'(0) &= \left(c_1 e^x - 2c_2 e^{-2x} + 2c_3 e^{2x} \right)\Big|_{x=0} = c_1 - 2c_2 + 2c_3 = -1, \\
y''(0) &= \left(c_1 e^x + 4c_2 e^{-2x} + 4c_3 e^{2x} \right)\Big|_{x=0} = c_1 + 4c_2 + 4c_3 = -19.
\end{aligned}
$$

Solving yields

$$
c_1 = 1, \qquad c_2 = -2, \qquad c_3 = -3.
$$

With these coefficients, the solution to the given initial value problem becomes

$$
y(x) = e^x - 2e^{-2x} - 3e^{2x} .
$$

21. By inspection, $r = 2$ is a root of the characteristic equation, $r^3 - 4r^2 + 7r - 6 = 0$. Factoring yields

$$
r^3 - 4r^2 + 7r - 6 = (r - 2)(r^2 - 2r + 3) = 0.
$$

Therefore, the other two roots are the roots of $r^2 - 2r + 3 = 0$, which are $r = 1 \pm \sqrt{2}i$, and so a general solution to the given differential equation is given by

$$
y(x) = c_1 e^{2x} + \left(c_2 \cos \sqrt{2}x + c_3 \sin \sqrt{2}x \right) e^x .
$$

Differentiating, we obtain

$$
\begin{aligned}
y'(x) &= 2c_1 e^{2x} + \left[\left(c_2 + c_3\sqrt{2} \right) \cos \sqrt{2}x + \left(c_3 - c_2\sqrt{2} \right) \sin \sqrt{2}x \right] e^x , \\
y''(x) &= 4c_1 e^{2x} + \left[\left(2c_3\sqrt{2} - c_2 \right) \cos \sqrt{2}x - \left(2c_2\sqrt{2} + c_3 \right) \sin \sqrt{2}x \right] e^x .
\end{aligned}
$$

Hence, the initial conditions imply that

Copyright © 2012 Pearson Education, Inc. Publishing as Addison-Wesley.

$$y(0) = c_1 + c_2 = 1,$$
$$y'(0) = 2c_1 + c_2 + c_3\sqrt{2} = 0,$$
$$y''(0) = 4c_1 - c_2 + 2c_3\sqrt{2} = 0.$$

Solving yields $c_1 = 1$, $c_2 = 0$, and $c_3 = -\sqrt{2}$. Substituting these constants into the general solution, we get

$$y(x) = e^{2x} - \sqrt{2}e^x \sin\sqrt{2}x.$$

23. Rewriting the system in operator form yields

$$\left(D^3 - 1\right)[x] + (D+1)[y] = 0,$$
$$(D-1)[x] + y = 0.$$

Multiplying the second equation in this system by $(D+1)$ and subtracting the result from the first equation, we get

$$\left\{\left(D^3 - 1\right) - (D+1)(D-1)\right\}[x] = D^2(D-1)[x] = 0.$$

Since roots of the characteristic equation, $r^2(r-1) = 0$ are $r = 0$ of multiplicity two and $r = 1$, a general solution $x(t)$ is given by

$$x(t) = c_1 + c_2 t + c_3 e^t.$$

From the second equation in the original system, we obtain

$$y(t) = x(t) - x'(t) = \left(c_1 + c_2 t + c_3 e^t\right) - \left(c_1 + c_2 t + c_3 e^t\right)' = c_1 - c_2 + c_2 t.$$

25. A linear combination of the given functions

$$c_0 e^{rx} + c_1 x e^{rx} + c_2 x^2 e^{rx} + \cdots + c_{m-1} x^{m-1} e^{rx} = \left(c_0 + c_1 x + c_2 x^2 + \cdots + c_m x^m\right) e^{rx} \quad (6.3)$$

vanishes on an interval if and only if its polynomial factor, $c_0 + c_1 x + c_2 x^2 + \cdots + c_{m-1} x^{m-1}$, vanishes on this interval (since the exponential factor is never zero). But, as we have proved in Problem 27, Section 6.1, the system of monomials $\{1, x, \ldots, x^n\}$ is linearly independent on any interval. Thus, the linear combination (6.3) vanishes on an inteval if and only if it has all zero coefficients, i.e., $c_0 = c_1 = \ldots = c_{m-1} = 0$. Therefore, the system $\{e^{rx}, xe^{rx}, \ldots, x^{m-1}e^{rx}\}$ is linearly independent on any interval.

 Copyright © 2012 Pearson Education, Inc. Publishing as Addison-Wesley.

27. Solving the auxiliary equation

$$r^4 + 2r^3 - 3r^2 - r + \frac{1}{2} = 0$$

(using computer software) yields the roots

$$r_1 \approx 1.120, \; r_2 \approx 0.296, \; r_3 \approx -0.520, \; r_4 \approx -2.896.$$

Thus, all the roots are real and distinct. A general solution to the given equation is, therefore,

$$y(x) = c_1 e^{r_1 x} + c_2 e^{r_2 x} + c_3 e^{r_3 x} + c_4 e^{r_4 x} \approx c_1 e^{1.120x} + c_2 e^{0.296x} + c_3 e^{-0.520x} + c_4 e^{-2.896x}.$$

29. The auxiliary equation for this problem is

$$r^4 + 2r^3 + 4r^2 + 3r + 2 = 0.$$

Let

$$f(z) = z^4 + 2z^3 + 4z^2 + 3z + 2 \quad \Rightarrow \quad f'(z) = 4z^3 + 6z^2 + 8z + 3.$$

The Newton's recursion formula, given in the hint, becomes

$$z_{n+1} = z_n - \frac{z_n^4 + 2z_n^3 + 4z_n^2 + 3z_n + 2}{4z_n^3 + 6z_n^2 + 8z_n + 3}.$$

With an initial guess $z_0 = 1 + i$, this formula yields

$$z_1 = (1+i) - \frac{(1+i)^4 + 2(1+i)^3 + 4(1+i)^2 + 3(1+i) + 2}{4(1+i)^3 + 6(1+i)^2 + 8(1+i) + 3} \approx 0.4817 + 0.8373i,$$

$$z_2 = z_1 - \frac{z_1^4 + 2z_1^3 + 4z_1^2 + 3z_1 + 2}{4z_1^3 + 6z_1^2 + 8z_1 + 3} \approx 0.0528 + 0.7635i,$$

$$z_3 = z_2 - \frac{z_2^4 + 2z_2^3 + 4z_2^2 + 3z_2 + 2}{4z_2^3 + 6z_2^2 + 8z_2 + 3} \approx -0.2843 + 0.7899i,$$

$$\vdots$$

$$z_7 = z_6 - \frac{z_6^4 + 2z_6^3 + 4z_6^2 + 3z_6 + 2}{4z_6^3 + 6z_6^2 + 8z_6 + 3} \approx -0.5000 + 0.8660i,$$

$$z_8 = z_7 - \frac{z_7^4 + 2z_7^3 + 4z_7^2 + 3z_7 + 2}{4z_7^3 + 6z_7^2 + 8z_7 + 3} \approx -0.5000 + 0.8660i.$$

Therefore, two roots of the auxiliary equation are

$$r_1 = z_8 \approx -0.5 + 0.866i \quad \text{and} \quad r_2 = \overline{r_1} \approx -0.5 - 0.866i.$$

Copyright © 2012 Pearson Education, Inc. Publishing as Addison-Wesley.

Similarly, we find two other roots. With the initial guess $z_0 = -1 - 2i$, we find that

$$z_1 = (-1 - 2i) - \frac{(-1 - 2i)^4 + 2(-1 - 2i)^3 + 4(-1 - 2i)^2 + 3(-1 - 2i) + 2}{4(-1 - 2i)^3 + 6(-1 - 2i)^2 + 8(-1 - 2i) + 3}$$
$$\approx -0.8307 - 1.6528i\,,$$

$$\vdots$$

$$z_6 \approx -0.5000 - 1.3229i\,,$$

$$z_7 \approx -0.5000 - 1.3229i\,.$$

Therefore, the other two roots are

$$r_3 = z_7 \approx -0.5 - 1.323i \qquad \text{and} \qquad r_4 = \overline{r_3} = -0.5 + 1.323i\,.$$

Thus, the auxiliary equation has four complex roots, and a general solution to the given differential equation is given by

$$y \approx c_1 e^{-0.5x} \cos(0.866x) + c_2 e^{-0.5x} \sin(0.866x) + c_3 e^{-0.5x} \cos(1.323x) + c_4 e^{-0.5x} \sin(1.323x)\,.$$

31. **(a)** Letting $y(x) = x^r$, we see that

$$
\begin{aligned}
y' &= rx^{r-1}\,, \\
y'' &= r(r-1)x^{r-2} = (r^2 - r)x^{r-2}\,, \\
y''' &= r(r-1)(r-2)x^{r-3} = (r^3 - 3r^2 + 2r)x^{r-3}\,.
\end{aligned}
\tag{6.4}
$$

Thus, if $y = x^r$ is a solution to this third order Cauchy-Euler equation, then

$$x^3(r^3 - 3r^2 + 2r)x^{r-3} + x^2(r^2 - r)x^{r-2} - 2xrx^{r-1} + 2x^r = 0$$
$$\Rightarrow \quad (r^3 - 3r^2 + 2r)x^r + (r^2 - r)x^r - 2rx^r + 2x^r = 0$$
$$\Rightarrow \quad (r^3 - 2r^2 - r + 2)x^r = 0. \tag{6.5}$$

Therefore, we get $r^3 - 2r^2 - r + 2 = 0$. Factoring yields

$$r^3 - 2r^2 - r + 2 = (r^3 - 2r^2) - (r - 2) = (r - 2)(r^2 - 1) = (r - 2)(r + 1)(r - 1) = 0.$$

Thus, (6.5) is satisfied for $x > 0$ if and only if $r = \pm 1$ and $r = 2$. Hence, three solutions to the given differential equation are $y = x$, $y = x^{-1}$, and $y = x^2$. Since these functions are linearly independent, they form a fundamental solution set.

 Copyright © 2012 Pearson Education, Inc. Publishing as Addison-Wesley.

(b) Let $y(x) = x^r$. In addition to (6.4), we need the fourth derivative of $y(x)$.

$$y^{(4)} = (y''')' = r(r-1)(r-2)(r-3)x^{r-4} = (r^4 - 6r^3 + 11r^2 - 6r)x^{r-4}.$$

Thus, if $y = x^r$ is a solution to this fourth order Cauchy-Euler equation, then we have

$$x^4(r^4 - 6r^3 + 11r^2 - 6r)x^{r-4} + 6x^3(r^3 - 3r^2 + 2r)x^{r-3}$$

$$+2x^2(r^2 - r)x^{r-2} - 4xrx^{r-1} + 4x^r = 0$$

$$\Rightarrow \quad (r^4 - 6r^3 + 11r^2 - 6r)x^r + 6(r^3 - 3r^2 + 2r)x^r + 2(r^2 - r)x^r - 4rx^r + 4x^r = 0$$

or, after some algebra,

$$(r^4 - 5r^2 + 4)x^r = 0. \tag{6.6}$$

Therefore, $r^4 - 5r^2 + 4 = 0$. Factoring this equation yields

$$r^4 - 5r^2 + 4 = (r^2 - 4)(r^2 - 1) = (r-2)(r+2)(r-1)(r+1) = 0.$$

Thus, (6.6) is satisfied if and only if $r = \pm 1, \pm 2$, and so four solutions to the given differential equation are $y = x$, $y = x^{-1}$, $y = x^2$, and $y = x^{-2}$. These functions are linearly independent and form a fundamental solution set.

(c) Substituting $y = x^r$ into this differential equation yields

$$(r^3 - 3r^2 + 2r)x^r - 2(r^2 - r)x^r + 13rx^r - 13x^r = 0$$

$$\Rightarrow \quad (r^3 - 5r^2 + 17r - 13)x^r = 0.$$

Thus, in order for $y = x^r$ to be a solution to this differential equation for $x > 0$, we must have $r^3 - 5r^2 + 17r - 13 = 0$. By inspection, we find that $r = 1$ is a root to this equation. Factoring yields

$$(r-1)(r^2 - 4r + 13) = 0.$$

We find the remaining two roots by using the quadratic formula. Thus, we obtain $r = 1$ and $r = 2 \pm 3i$. For $r = 1$, we get a solution $y_1(x) = x$; for $r = 2 \pm 3i$ (using the hint), we see that another (complex-valued) solution is given by

$$y(x) = x^{2+3i} = x^2 \left[\cos(3\ln x) + i\sin(3\ln x)\right].$$

Therefore, by Lemma 2 in Section 4.3, we find that other two real-valued solutions to the given differential equation are

$$y_2(x) = x^2 \cos(3 \ln x) \qquad \text{and} \qquad y_3(x) = x^2 \sin(3 \ln x).$$

Since $y_1(x)$, $y_2(x)$, and $y_3(x)$ are linearly independent, we obtain a fundamental solution set

$$\left\{ x, x^2 \cos(3 \ln x), x^2 \sin(3 \ln x) \right\}.$$

33. With suggested values of $m_1 = m_2 = 1$, $k_1 = 3$, and $k_2 = 2$, the system (34)–(35) in the text becomes

$$\begin{aligned} x'' + 5x - 2y &= 0, \\ y'' - 2x + 2y &= 0. \end{aligned} \qquad (6.7)$$

(a) Expressing

$$y = \frac{x'' + 5x}{2}$$

from the first equation and substituting this formula into the second equation, we obtain (36).

$$\begin{aligned} \frac{1}{2} \left(x'' + 5x\right)'' - 2x + \left(x'' + 5x\right) &= 0 \\ \Rightarrow \qquad \left(x^{(4)} + 5x''\right) - 4x + 2\left(x'' + 5x\right) &= 0 \\ \Rightarrow \qquad x^{(4)} + 7x'' + 6x &= 0. \end{aligned} \qquad (6.8)$$

(b) The characteristic equation, corresponding to (6.8), is

$$r^4 + 7r^2 + 6 = 0.$$

This equation is of quadratic type. Substitution $s = r^2$ yields

$$s^2 + 7s + 6 = 0 \qquad \Rightarrow \qquad s = -1, \, -6.$$

Thus,

$$r = \pm\sqrt{-1} = \pm i \qquad \text{and} \qquad r = \pm\sqrt{-6} = \pm i\sqrt{6},$$

and a general solution to (6.8) is given by

$$x(t) = c_1 \cos t + c_2 \sin t + c_3 \cos \sqrt{6}t + c_4 \sin \sqrt{6}t.$$

 Copyright © 2012 Pearson Education, Inc. Publishing as Addison-Wesley.

(c) We have mentioned in (a), the first equation in (6.7) implies that $y = \left(x'' + 5x\right)/2$. Substituting $x(t)$ yields

$$y(t) = \frac{1}{2}\left[\left(c_1 \cos t + c_2 \sin t + c_3 \cos \sqrt{6}t + c_4 \sin \sqrt{6}t\right)''\right.$$

$$\left. + 5\left(c_1 \cos t + c_2 \sin t + c_3 \cos \sqrt{6}t + c_4 \sin \sqrt{6}t\right)\right]$$

$$= \frac{1}{2}\left[\left(-c_1 \cos t - c_2 \sin t - 6c_3 \cos \sqrt{6}t - 6c_4 \sin \sqrt{6}t\right)\right.$$

$$\left. + 5\left(c_1 \cos t + c_2 \sin t + c_3 \cos \sqrt{6}t + c_4 \sin \sqrt{6}t\right)\right]$$

$$= 2c_1 \cos t + 2c_2 \sin t - \frac{c_3}{2}\cos \sqrt{6}t - \frac{c_4}{2}\sin \sqrt{6}t\,.$$

(d) The initial conditions, $x(0) = y(0) = 1$ and $x'(0) = y'(0) = 0$, imply a system of linear equations for c_1, c_2, c_3, and c_4. Namely,

$$
\begin{aligned}
x(0) &= c_1 + c_3 = 1, & c_1 &= 3/5, \\
y(0) &= 2c_1 - (c_3/2) = 1, & c_3 &= 2/5, \\
x'(0) &= c_2 + c_4\sqrt{6} = 0, & \Rightarrow \quad c_2 &= 0, \\
y'(0) &= 2c_2 - (c_4\sqrt{6}/2) = 0 & c_4 &= 0.
\end{aligned}
$$

Thus, the solution to this initial value problem is

$$x(t) = \frac{3}{5}\cos t + \frac{2}{5}\cos \sqrt{6}t\,, \qquad y(t) = \frac{6}{5}\cos t - \frac{1}{5}\cos \sqrt{6}t\,.$$

35. Solving the characteristic equation yields

$$EIr^4 - k = 0 \qquad \Rightarrow \qquad r^4 = \frac{k}{EI}$$

$$\Rightarrow \qquad r^2 = \sqrt{\frac{k}{EI}} \qquad \text{or} \qquad r^2 = -\sqrt{\frac{k}{EI}}$$

$$\Rightarrow \qquad r = \pm\sqrt[4]{\frac{k}{EI}} \qquad \text{or} \qquad r = \pm\sqrt[4]{-\frac{k}{EI}} = \pm i\sqrt[4]{\frac{k}{EI}}\,.$$

The first two roots are real numbers, the other two are pure imaginary numbers. Therefore, a general solution to the vibrating beam equation is

$$y(x) = C_1 e^{\sqrt{k/(EI)}\,x} + C_2 e^{-\sqrt{k/(EI)}\,x} + C_3 \cos\left(\sqrt[4]{\frac{k}{EI}}\,x\right) + C_4 \sin\left(\sqrt[4]{\frac{k}{EI}}\,x\right).$$

Using the identities

$$e^{ax} = \cosh ax + \sinh ax, \qquad e^{-ax} = \cosh ax - \sinh ax,$$

Copyright © 2012 Pearson Education, Inc. Publishing as Addison-Wesley.

we can express the solution in terms of hyperbolic and trigonometric functions as follows.

$$y(x) = C_1 e^{\sqrt{k/(EI)}x} + C_2 e^{-\sqrt{k/(EI)}x} + C_3 \cos\left(\sqrt[4]{\frac{k}{EI}}\,x\right) + C_4 \sin\left(\sqrt[4]{\frac{k}{EI}}\,x\right)$$

$$= C_1\left[\cosh\left(\sqrt[4]{\frac{k}{EI}}\,x\right) + \sinh\left(\sqrt[4]{\frac{k}{EI}}\,x\right)\right] + C_2\left[\cosh\left(\sqrt[4]{\frac{k}{EI}}\,x\right) - \sinh\left(\sqrt[4]{\frac{k}{EI}}\,x\right)\right]$$

$$+ C_3 \cos\left(\sqrt[4]{\frac{k}{EI}}\,x\right) + C_4 \sin\left(\sqrt[4]{\frac{k}{EI}}\,x\right)$$

$$= c_1 \cosh\left(\sqrt[4]{\frac{k}{EI}}\,x\right) + c_2 \sinh\left(\sqrt[4]{\frac{k}{EI}}\,x\right) + c_3 \cos\left(\sqrt[4]{\frac{k}{EI}}\,x\right) + c_4 \sin\left(\sqrt[4]{\frac{k}{EI}}\,x\right),$$

where $c_1 := C_1 + C_2$, $c_2 := C_1 - C_2$, $c_3 := C_3$, and $c_4 := C_4$ are arbitrary constants.

EXERCISES 6.3:   Undetermined Coefficients and the Annihilator Method

1. The corresponding homogeneous equation for this problem is $y''' - 2y'' - 5y' + 6y = 0$, which has the associated auxiliary equation $r^3 - 2r^2 - 5r + 6 = 0$. By inspection we see that $r = 1$ is a root. Therefore, the polynomial can be factored as

$$r^3 - 2r^2 - 5r + 6 = (r - 1)(r^2 - r - 6) = (r - 1)(r - 3)(r + 2).$$

Thus, the roots of the auxiliary equation are $r = 1$, 3, and -2, and a general solution to the homogeneous equation is

$$y_h(x) = C_1 e^x + C_2 e^{3x} + C_3 e^{-2x}.$$

The nonhomogeneous term, $g(x) = e^x + x^2$, is the sum of an exponential term and a polynomial term. Therefore, according to the superposition principle (see Section 4.5), this equation has a particular solution of the form

$$y_p(x) = x^{s_1} c_1 e^x + x^{s_2}\left(c_2 + c_3 x + c_4 x^2\right).$$

Since e^x is a solution to the associated homogeneous equation and xe^x is not, we set $s_1 = 1$. Since none of the terms x^2, x, or 1 is a solution to the associated homogeneous equation, we set $s_2 = 0$. Thus, the form of a particular solution is

$$y_p(x) = c_1 xe^x + c_2 + c_3 x + c_4 x^2.$$

Copyright © 2012 Pearson Education, Inc. Publishing as Addison-Wesley.

3. The associated homogeneous equation for this equation is $y''' + 3y'' - 4y = 0$. The corresponding auxiliary equation, $y^3 + 3r^2 - 4 = 0$, by inspection, has a root $r = 1$. Thus, the auxiliary equation can be factored as

$$(r-1)(r^2 + 4r + 4) = (r-1)(r+2)^2 = 0,$$

and we see that the other two roots of the auxiliary equation are $r = -2, -2$. Therefore, a general solution to the homogeneous equation is

$$y_h(x) = C_1 e^x + C_2 e^{-2x} + C_3 x e^{-2x}.$$

The nonhomogeneous term here is $g(x) = e^{-2x}$. Therefore, a particular solution to the original differential equation has the form $y_p(x) = x^s c_1 e^{-2x}$. Since both e^{-2x} and $x e^{-2x}$ are solutions to the associated homogeneous equation, we set $s = 2$. Thus, the form of a particular solution is $y_p(x) = c_1 x^2 e^{-2x}$. (Note that this means that $r = -2$ is a root of multiplicity three of the auxiliary equation, associated with the operator equation $A[L[y]](x) = 0$, where A is an annihilator of the nonhomogeneous term, $g(x) = e^{-2x}$, and $L := D^3 + 3D^2 - 4$ is the given linear differential operator.)

5. In the solution to Problem 1, we determined that a general solution to the homogeneous differential equation, corresponding to this problem, is

$$y_h(x) = C_1 e^x + C_2 e^{3x} + C_3 e^{-2x},$$

and that a particular solution has the form

$$y_p(x) = c_1 x e^x + c_2 + c_3 x + c_4 x^2.$$

By differentiating $y_p(x)$, we find

$$y_p'(x) = c_1 x e^x + c_1 e^x + c_3 + 2c_4 x$$
$$\Rightarrow \quad y_p''(x) = c_1 x e^x + 2c_1 e^x + 2c_4$$
$$\Rightarrow \quad y_p'''(x) = c_1 x e^x + 3c_1 e^x.$$

Substituting these expressions into the original differential equation, we obtain

$$y_p'''(x) - 2y_p''(x) - 5y_p'(x) + 6y_p(x) = c_1 x e^x + 3c_1 e^x - 2c_1 x e^x - 4c_1 e^x - 4c_4$$
$$-5c_1 x e^x - 5c_1 e^x - 5c_3 - 10c_4 x + 6c_1 x e^x + 6c_2 + 6c_3 x + 6c_4 x^2 = e^x + x^2$$

Copyright © 2012 Pearson Education, Inc. Publishing as Addison-Wesley.

$$\Rightarrow \qquad -6c_1 e^x + (-4c_4 - 5c_3 + 6c_2) + (-10c_4 + 6c_3)x + 6c_4 x^2 = e^x + x^2 \,.$$

Equating coefficients yields

$$
\begin{aligned}
-6c_1 &= 1 & \Rightarrow \quad c_1 &= -1/6 \,, \\
6c_4 &= 1 & \Rightarrow \quad c_4 &= 1/6 \,, \\
-10c_4 + 6c_3 &= 0 & \Rightarrow \quad c_3 &= 10c_4/6 = 5/18 \,, \\
-4c_4 - 5c_3 + 6c_2 &= 0 & \Rightarrow \quad c_2 &= (4c_4 + 5c_3)/6 = 37/108 \,.
\end{aligned}
$$

Thus, a general solution to the given nonhomogeneous equation is

$$y(x) = y_h(x) + y_p(x) = C_1 e^x + C_2 e^{3x} + C_3 e^{-2x} - \frac{1}{6}\,xe^x + \frac{1}{6}\,x^2 + \frac{5}{18}\,x + \frac{37}{108}\,.$$

7. In Problem 3, we found a general solution to the associated homogeneous equation, that is,

$$y_h(x) = C_1 e^x + C_2 e^{-2x} + C_3 xe^{-2x} \,.$$

The form of a particular solution to the nonhomogeneous equation was $y_p(x) = c_1 x^2 e^{-2x}$. Differentiating $y_p(x)$ yields

$$
\begin{aligned}
y_p'(x) &= 2c_1 xe^{-2x} - 2c_1 x^2 e^{-2x} = 2c_1(x - x^2)e^{-2x} \\
\Rightarrow \quad y_p''(x) &= -4c_1(x - x^2)e^{-2x} + 2c_1(1 - 2x)e^{-2x} = 2c_1(2x^2 - 4x + 1)e^{-2x} \\
\Rightarrow \quad y_p'''(x) &= -4c_1(2x^2 - 4x + 1)e^{-2x} + 2c_1(4x - 4)e^{-2x} = 4c_1(-2x^2 + 6x - 3)e^{-2x} \,.
\end{aligned}
$$

Substituting these expressions into the original equation, we obtain

$$y_p'''(x) + 3y_p''(x) - 4y_p(x) = 4c_1(-2x^2 + 6x - 3)e^{-2x}$$

$$+6c_1(2x^2 - 4x + 1)e^{-2x} - 4c_1 x^2 e^{-2x} = e^{-2x}$$

$$\Rightarrow \qquad -6c_1 e^{-2x} = e^{-2x} \,.$$

Equating coefficients, we see that $c_1 = -1/6$. Thus, a general solution to the nonhomogeneous differential equation is given by

$$y(x) = y_h(x) + y_p(x) = C_1 e^x + C_2 e^{-2x} + C_3 xe^{-2x} - \frac{1}{6}\,x^2 e^{-2x} \,.$$

9. Solving the auxiliary equation, $r^3 - 3r^2 + 3r - 1 = (r - 1)^3 = 0$, we find that $r = 1$ is its root of multiplicity three. Therefore, a general solution to the associated homogeneous equation is

$$y_h(x) = c_1 e^x + c_2 xe^x + c_3 x^2 e^x \,.$$

 Copyright © 2012 Pearson Education, Inc. Publishing as Addison-Wesley.

The nonhomogeneous term, e^x, suggests a particular solution of the form $y_p(x) = Ax^s e^x$, where we choose $s = 3$, since $r = 1$ is a root of the auxiliary equation of multiplicity three. Thus, $y_p(x) = Ax^3 e^x$. Differentiating $y_p(x)$ yields

$$y_p'(x) = A\left(x^3 + 3x^2\right)e^x,$$
$$y_p''(x) = A\left(x^3 + 6x^2 + 6x\right)e^x,$$
$$y_p'''(x) = A\left(x^3 + 9x^2 + 18x + 6\right)e^x.$$

By substituting these expressions into the original equation, we obtain

$$y_p''' - 3y_p'' + 3y_p' - y = e^x$$
$$\Rightarrow \quad \left[A\left(x^3 + 9x^2 + 18x + 6\right)e^x\right] - 3\left[A\left(x^3 + 6x^2 + 6x\right)e^x\right]$$
$$+3\left[A\left(x^3 + 3x^2\right)e^x\right] - Ax^3 e^x = e^x$$
$$\Rightarrow \quad 6Ae^x = e^x \quad \Rightarrow \quad A = \frac{1}{6},$$

and so $y_p(x) = x^3 e^x / 6$. A general solution to the given equation then has the form

$$y(x) = y_h(x) + y_p(x) = c_1 e^x + c_2 x e^x + c_3 x^2 e^x + \frac{1}{6}x^3 e^x.$$

11. The operator D^5, that is, the fifth derivative operator, annihilates any polynomial of degree at most four. In particular, D^5 annihilates the polynomial $x^4 - x^2 + 11$.

13. According to (i) (see the text), the operator $[D - (-7)] = (D + 7)$ annihilates the exponential function e^{-7x}.

15. The operator $(D - 2)$ annihilates the function $f_1(x) := e^{2x}$ and the operator $(D - 1)$ annihilates the function $f_2(x) := e^x$. Thus, the product of these operators, namely, $(D - 2)(D - 1)$, annihilates both functions and so, by linearity, it annihilates their algebraic sum.

17. This function has the same form as the functions given in the case (iv) in the text. Here, we see that $\alpha = -1$, $\beta = 2$, and $m - 1 = 2$. Thus, the operator

$$\left[(D - \{-1\})^2 + 2^2\right]^3 = \left[(D + 1)^2 + 4\right]^3$$

annihilates this function.

Copyright © 2012 Pearson Education, Inc. Publishing as Addison-Wesley.

19. Given function as a sum of two functions. The first term, xe^{-2x}, is of the type (ii) with $m = 2$ and $r = -2$; so $[D - (-2)]^2 = (D + 2)^2$ annihilates this function. The second term, $xe^{-5x} \sin 3x$, is annihilated by

$$\left[(D - (-5))^2 + 3^2\right]^2 = \left[(D + 5)^2 + 9\right]^2,$$

according to (iv). Therefore, the product $(D+2)^2 \left[(D + 5)^2 + 9\right]^2$ annihilates the function $xe^{-2x} + xe^{-5x} \sin 3x$.

21. In operator form, the given equation can be written as

$$\left(D^2 - 5D + 6\right)[u] = \cos 2x + 1.$$

The function $g(x) = \cos 2x + 1$ is a sum of two functions: $\cos 2x$ is of type (iii) with $\beta = 2$, and so it is annihilated by $(D^2 + 4)$; 1, being a constant, is annihilated by D. Therefore, the operator $D(D^2 + 4)$ annihilates the right-hand side, $g(x)$. Applying this operator to both sides of the differential equation, given in this problem, yields

$$D\left(D^2 + 4\right)\left(D^2 - 5D + 6\right)[u] = D\left(D^2 + 4\right)[\cos 2x + 1] = 0.$$

Factoring, we obtain

$$D\left(D^2 + 4\right)(D - 3)(D - 2)[u] = 0.$$

This equation has the associated auxiliary equation $r\left(r^2 + 4\right)(r - 3)(r - 2) = 0$, with roots $r = 2, 3, 0, \pm 2i$. Thus, a general solution to the homogeneous differential equation, associated with this auxiliary equation, is

$$u(x) = c_1 e^{2x} + c_2 e^{3x} + c_3 \cos 2x + c_4 \sin 2x + c_5.$$

The homogeneous equation, $u'' - 5u' + 6u = 0$, corresponding to the original problem, has the auxiliary equation $r^2 - 5r + 6 = (r - 2)(r - 3) = 0$. Therefore, it has a general solution $u_h(x) = c_1 e^{2x} + c_2 e^{3x}$. Since a general solution to the original problem is given by

$$u(x) = u_h(x) + u_p(x) = c_1 e^{2x} + c_2 e^{3x} + u_p(x),$$

and since $u(x)$ must be of the form

$$u(x) = c_1 e^{2x} + c_2 e^{3x} + c_3 \cos 2x + c_4 \sin 2x + c_5,$$

we see that

$$u_p(x) = c_3 \cos 2x + c_4 \sin 2x + c_5.$$

 Copyright © 2012 Pearson Education, Inc. Publishing as Addison-Wesley.

23. The function $g(x) = e^{3x} - x^2$ is annihilated by the operator $A := D^3(D-3)$. Applying the operator A to both sides of the given differential equation yields

$$A\left[y'' - 5y' + 6y\right] = A\left[e^{3x} - x^2\right] = 0$$

$$\Rightarrow \quad D^3(D-3)(D^2 - 5D + 6)[y] = D^3(D-3)^2(D-2)[y] = 0.$$

This differential equation has the associated auxiliary equation

$$r^3(r-3)^2(r-2) = 0,$$

whose roots are $r = 0$ of multiplicity three, $r = 3$ of multiplicity two, and $r = 2$. Thus, a general solution to the corresponding homogeneous equation is

$$y(x) = c_1 e^{2x} + c_2 e^{3x} + c_3 x e^{3x} + c_4 x^2 + c_5 x + c_6.$$

The homogeneous equation, $y'' - 5y' + 6y = 0$, corresponding to the original problem, is the same as in Problem 21 (with u replaced by y). Therefore, its solution is

$$y_h(x) = c_1 e^{2x} + c_2 e^{3x}.$$

Since, by the superposition principle,

$$y(x) = y_h(x) + y_p(x) = c_1 e^{2x} + c_2 e^{3x} + y_p(x)$$

and since $y(x)$ must be of the form

$$y(x) = c_1 e^{2x} + c_2 e^{3x} + c_3 x e^{3x} + c_4 x^2 + c_5 x + c_6,$$

we see that

$$y_p(x) = c_3 x e^{3x} + c_4 x^2 + c_5 x + c_6.$$

25. First, we rewrite the equation in operator form, that is,

$$\left(D^2 - 6D + 9\right)[y] = \sin 2x + x \quad \Rightarrow \quad (D-3)^2[y] = \sin 2x + x.$$

In this problem, the right-hand side is a sum of two functions. The first function, $\sin 2x$, is annihilated by $(D^2 + 4)$, and the operator D^2 annihilates the term x. So, $A := D^2(D^2 + 4)$ annihilates the function $\sin 2x + x$. Applying this operator to the original equation (in operator form) yields

$$D^2(D^2 + 4)(D-3)^2[y] = D^2(D^2 + 4)[\sin 2x + x] = 0. \tag{6.9}$$

Copyright © 2012 Pearson Education, Inc. Publishing as Addison-Wesley.

This homogeneous equation has associated characteristic equation

$$r^2(r^2 + 4)(r - 3)^2 = 0$$

with simple roots $r = \pm 2i$, and double roots $r = 0$ and $r = 3$. Therefore, a general solution to (6.9) is given by

$$y(x) = c_1 e^{3x} + c_2 x e^{3x} + c_3 + c_4 x + c_5 \cos 2x + c_6 \sin 2x. \tag{6.10}$$

Since the homogeneous equation, $(D - 3)^2[y] = 0$, which corresponds to the original one, has a general solution $y_h(x) = c_1 e^{3x} + c_2 x e^{3x}$, the last four terms in (6.10) give the form of a particular solution.

27. Since

$$y'' + 2y' + 2y = \left(D^2 + 2D + 2\right)[y] = \left\{(D + 1)^2 + 1\right\}[y],$$

the auxiliary equation for this problem is $(r + 1)^2 + 1 = 0$, whose roots are $r = -1 \pm i$. Therefore, a general solution to the corresponding homogeneous equation is

$$y_h(x) = c_1 e^{-x} \cos x + c_2 e^{-x} \sin x. \tag{6.11}$$

Applying the operator $D^3\{(D + 1)^2 + 1\}$ to the given equation, which annihilates its right-hand side, yields

$$D^3\left\{(D + 1)^2 + 1\right\}\left\{(D + 1)^2 + 1\right\}[y] = D^3\left\{(D + 1)^2 + 1\right\}\left[e^{-x}\cos x + x^2\right] = 0$$
$$\Rightarrow \quad D^3\left[(D + 1)^2 + 1\right]^2[y] = 0. \tag{6.12}$$

The corresponding auxiliary equation, $r^3[(r+1)^2+1]^2 = 0$, has a root $r = 0$ of multiplicity three and double roots $r = -1 \pm i$. Therefore, a general solution to (6.12) is given by

$$y(x) = c_1 e^{-x} \cos x + c_2 e^{-x} \sin x + c_3 x e^{-x} \cos x + c_4 x e^{-x} \sin x + c_5 x^2 + c_6 x + c_7.$$

Since $y(x) = y_h(x) + y_p(x)$, from (6.11) we conclude that

$$y_p(x) = c_3 x e^{-x} \cos x + c_4 x e^{-x} \sin x + c_5 x^2 + c_6 x + c_7.$$

29. In operator form, the equation becomes

$$\left(D^3 - 2D^2 + D\right)[z] = D(D - 1)^2[z] = x - e^x. \tag{6.13}$$

 Copyright © 2012 Pearson Education, Inc. Publishing as Addison-Wesley.

Solving the corresponding auxiliary equation, $r(r-1)^2 = 0$, we find that $r = 0$ and $r = 1$ of multiplicity two. Therefore,

$$z_h(x) = C_1 + C_2 e^x + C_3 x e^x$$

is a general solution to the homogeneous equation, associated with the original equation. To annihilate the right-hand side in (6.13), we apply the operator $D^2(D-1)$ to this equation. Thus, we obtain

$$D^2(D-1)D(D-1)^2[z] = D^2(D-1)\left[x - e^x\right] \quad \Rightarrow \quad D^3(D-1)^3 = 0.$$

Solving the corresponding auxiliary equation, $r^3(r-1)^3 = 0$, we see that $r = 0$ and $r = 1$ are its roots of multiplicity three. Hence, a general solution is given by

$$z(x) = c_1 + c_2 x + c_3 x^2 + c_4 e^x + c_5 x e^x + c_6 x^2 e^x .$$

This general solution, when compared with $z_h(x)$, gives

$$z_p(x) = c_2 x + c_3 x^2 + c_6 x^2 e^x .$$

31. Writing this equation in operator form yields

$$\left(D^3 + 2D^2 - 9D - 18\right)[y] = -18x^2 - 18x + 22 . \tag{6.14}$$

Since,

$$D^3 + 2D^2 - 9D - 18 = D^2(D+2) - 9(D+2) = (D+2)\left(D^2 - 9\right) = (D+2)(D-3)(D+3),$$

(6.14) becomes

$$(D+2)(D-3)(D+3)[y] = -18x^2 - 18x + 22 .$$

The auxiliary equation in this problem is $(r+2)(r-3)(r+3) = 0$ with roots $r = -2, 3, -3$. Hence, a general solution to the corresponding homogeneous equation has the form

$$y_h(x) = c_1 e^{-2x} + c_2 e^{3x} + c_3 e^{-3x} .$$

Since the operator D^3 annihilates the nonhomogeneous term in the original equation and $r = 0$ is not a root of the auxiliary equation, we seek for a particular solution of the form

$$y_p(x) = C_0 x^2 + C_1 x + C_2 .$$

Copyright © 2012 Pearson Education, Inc. Publishing as Addison-Wesley.

Substituting y_p into the given equation (for convenience, in operator form) yileds

$$\left(D^3 + 2D^2 - 9D - 18\right)\left[C_0x^2 + C_1x + C_2\right] = -18x^2 - 18x + 22$$

$$\Rightarrow \quad 0 + 2\left(2C_0\right) - 9\left[2C_0x + C_1\right] - 18\left[C_0x^2 + C_1x + C_2\right] = -18x^2 - 18x + 22$$

$$\Rightarrow \quad -18C_0x^2 + (-18C_1 - 18C_0)x + (-18C_2 - 9C_1 + 4C_0) = -18x^2 - 18x + 22.$$

Equating coefficients, we obtain the system

$$\begin{aligned} -18C_0 &= -18, \\ -18C_1 - 18C_0 &= -18, \\ -18C_2 - 9C_1 + 4C_0 &= 22 \end{aligned} \qquad \Rightarrow \qquad \begin{aligned} C_0 &= 1, \\ C_1 &= 0, \\ C_2 &= -1. \end{aligned}$$

Thus, $y_p(x) = x^2 - 1$, and so

$$y(x) = y_h(x) + y_p(x) = c_1e^{-2x} + c_2e^{3x} + c_3e^{-3x} + x^2 - 1$$

is a general solution to the original nonhomogeneous equation. Next, we satisfy the initial conditions. Differentiation yields

$$\begin{aligned} y'(x) &= -2c_1e^{-2x} + 3c_2e^{3x} - 3c_3e^{-3x} + 2x, \\ y''(x) &= 4c_1e^{-2x} + 9c_2e^{3x} + 9c_3e^{-3x} + 2. \end{aligned}$$

Therefore,

$$\begin{aligned} -2 &= y(0) = c_1 + c_2 + c_3 - 1, \\ -8 &= y'(0) = -2c_1 + 3c_2 - 3c_3, \\ -12 &= y''(0) = 4c_1 + 9c_2 + 9c_3 + 2 \end{aligned} \qquad \Rightarrow \qquad \begin{aligned} c_1 + c_2 + c_3 &= -1, \\ -2c_1 + 3c_2 - 3c_3 &= -8, \\ 4c_1 + 9c_2 + 9c_3 &= -14. \end{aligned}$$

Solving this system, we find that $c_1 = 1$, $c_2 = -2$, and $c_3 = 0$, and so

$$y(x) = e^{-2x} - 2e^{3x} + x^2 - 1$$

gives the solution to the initial value problem.

33. Let us write given equation in operator form.

$$\left(D^3 - 2D^2 - 3D + 10\right)[y] = (34x - 16)e^{-2x} - 10x^2 + 6x + 34.$$

By inspection, $r = -2$ is a root of the characteristic equation, $r^3 - 2r^2 - 3r + 10 = 0$. Using, say, long division we find that

$$r^3 - 2r^2 - 3r + 10 = (r+2)\left(r^2 - 4r + 5\right) = (r+2)\left[(r-2)^2 + 1\right],$$

Copyright © 2012 Pearson Education, Inc. Publishing as Addison-Wesley.

and so the other two roots of the auxiliary equation are $r = 2 \pm i$. This gives a general solution to the corresponding homogeneous equation of the form

$$y_h(x) = c_1 e^{-2x} + (c_2 \cos x + c_3 \sin x) e^{2x}.$$

According to the nonhomogeneous term, we look for a particular solution to the original equation of the form

$$y_p(x) = x (C_0 x + C_1) e^{-2x} + C_2 x^2 + C_3 x + C_4,$$

where the factor x in the exponential term appears due to the fact that $r = -2$ is a simple root of the characteristic equation. Substituting $y_p(x)$ into the given equation and simplifying yields

$$\left(D^3 - 2D^2 - 3D + 10\right) [y_p(x)] = (34x - 16)e^{-2x} - 10x^2 + 6x + 34$$

$$\Rightarrow \quad (34C_0 x + 17C_1 - 16C_0) \, e^{-2x} + 10C_2 x^2 + (10C_3 - 6C_2)x$$

$$+ 10C_4 - 3C_3 - 4C_2 = (34x - 16)e^{-2x} - 10x^2 + 6x + 34.$$

Equating corresponding coefficients, we obtain the system

$$
\begin{array}{llll}
34C_0 = 34, & & & C_0 = 1, \\
17C_1 - 16C_0 = -16, & & & C_1 = 0, \\
10C_2 = -10, & & \Rightarrow & C_2 = -1, \\
10C_3 - 6C_2 = 6, & & & C_3 = 0, \\
10C_4 - 3C_3 - 4C_2 = 34 & & & C_4 = 3.
\end{array}
$$

Thus, $y_p(x) = x^2 e^{-2x} - x^2 + 3$ and

$$y(x) = y_h(x) + y_p(x) = c_1 e^{-2x} + (c_2 \cos x + c_3 \sin x) e^{2x} + x^2 e^{-2x} - x^2 + 3$$

is a general solution to the given nonhomogeneous equation. Next, we find constants c_1, c_2, and c_3 such that the initial conditions are satisfied. Differentiating yields

$$y'(x) = -2c_1 e^{-2x} + [(2c_2 + c_3) \cos x + (2c_3 - c_2) \sin x] e^{2x} + (2x - 2x^2)e^{-2x} - 2x,$$

$$y''(x) = 4c_1 e^{-2x} + [(3c_2 + 4c_3) \cos x + (3c_3 - 4c_2) \sin x] e^{2x} + (2 - 8x + 4x^2)e^{-2x} - 2.$$

Therefore,

$$
\begin{array}{llll}
3 = y(0) = c_1 + c_2 + 3, & & & c_1 + c_2 = 0, \\
0 = y'(0) = -2c_1 + 2c_2 + c_3, & & \Rightarrow & -2c_1 + 2c_2 + c_3 = 0, \\
0 = y''(0) = 4c_1 + 3c_2 + 4c_3 & & & 4c_1 + 3c_2 + 4c_3 = 0.
\end{array}
$$

Copyright © 2012 Pearson Education, Inc. Publishing as Addison-Wesley.

The solution of this homogeneous linear system is $c_1 = c_2 = c_3 = 0$. Hence, the answer is $y(x) = x^2 e^{-2x} - x^2 + 3$.

35. If $a_0 = 0$, then equation (4) becomes

$$a_n y^{(n)} + a_{n-1} y^{(n-1)} + \cdots + a_1 y' = f(x)$$

or, in operator form,

$$\left(a_n D^n + a_{n-1} D^{n-1} + \cdots + a_1 D \right) [y] = f(x)$$
$$\Rightarrow \qquad D \left(a_n D^{n-1} + a_{n-1} D^{n-2} + \cdots + a_1 \right) [y] = f(x). \qquad (6.15)$$

Since the operator D^{m+1} annihilates any polynomial $f(x) = b_m x^m + \cdots + b_0$, applying D^{m+1} to both sides in (6.15) yields

$$D^{m+1} D \left(a_n D^{n-1} + a_{n-1} D^{n-2} + \cdots + a_1 \right) [y] = D^{m+1}[f(x)] = 0$$
$$\Rightarrow \qquad D^{m+2} \left(a_n D^{n-1} + a_{n-1} D^{n-2} + \cdots + a_1 \right) [y] = 0. \qquad (6.16)$$

The auxiliary equation, corresponding to this homogeneous equation, is

$$r^{m+2} \left(a_n r^{n-1} + a_{n-1} r^{n-2} + \cdots + a_1 \right) = 0. \qquad (6.17)$$

Since $a_1 \neq 0$,

$$\left(a_n r^{n-1} + a_{n-1} r^{n-2} + \cdots + a_1 \right) \Big|_{r=0} = a_1 \neq 0,$$

which means that $r = 0$ is not a root of this polynomial. Thus, for the auxiliary equation (6.17), $r = 0$ is a root of exact multiplicity $m + 2$, and a general solution to (6.16) is given by

$$y(x) = c_0 + c_1 x + \cdots + c_{m+1} x^{m+1} + Y(x), \qquad (6.18)$$

where $Y(x)$, being associated with roots of $a_n r^{n-1} + a_{n-1} r^{n-2} + \cdots + a_1 = 0$, is a general solution to $(a_n D^{n-1} + a_{n-1} D^{n-2} + \cdots + a_1) [y] = 0$. (One can write down $Y(x)$ explicitly but there is no need in doing this.)

On the other hand, the auxiliary equation for the homogeneous equation, corresponding to (6.15), is $r(a_n r^{n-1} + a_{n-1} r^{n-2} + \cdots + a_1) = 0$, and $r = 0$ is its simple root. Hence, a general solution $y_h(x)$ to the homogeneous equation is given by

$$y_h(x) = c_0 + Y(x), \qquad (6.19)$$

 Copyright © 2012 Pearson Education, Inc. Publishing as Addison-Wesley.

where $Y(x)$ is the same as in (6.18). Since $y(x) = y_h(x) + y_p(x)$, it follows from (6.18) and (6.19) that

$$y_p(x) = c_1 x + \cdots + c_{m+1} x^{m+1} = x \left(c_1 + \cdots + c_{m+1} x^m \right).$$

37. Writing the equation (4) in operator form yields

$$\left(a_n D^n + a_{n-1} D^{n-1} + \cdots + a_0 \right) [y] = f(x). \tag{6.20}$$

The characteristic equation, corresponding to the associated homogeneous equation, is

$$a_n r^n + a_{n-1} r^{n-1} + \cdots + a_0 = 0. \tag{6.21}$$

Suppose that $r = \beta i$ is a root of (6.21) of multiplicity $s \geq 0$. ($s = 0$ means that $r = \beta i$ is *not* a root.) Then (6.21) can be factored as

$$a_n r^n + a_{n-1} r^{n-1} + \cdots + a_0 = \left(r^2 + \beta^2 \right)^s \left(a_n r^{n-2s} + \cdots + \frac{a_0}{\beta^{2s}} \right) = 0$$

and so a general solution to the homogeneous equation is given by

$$\begin{aligned} y_h(x) &= (c_1 \cos \beta x + c_2 \sin \beta x) + x(c_3 \cos \beta x + c_4 \sin \beta x) \\ &\quad + \cdots + x^{s-1}(c_{2s-1} \cos \beta x + c_{2s} \sin \beta x) + Y(x), \end{aligned} \tag{6.22}$$

where $Y(x)$ is the part of $y_h(x)$ corresponding to the roots of

$$a_n r^{n-2s} + \cdots + \frac{a_0}{\beta^{2s}} = 0.$$

Since the operator $(D^2 + \beta^2)$ annihilates $f(x) = a \cos \beta x + b \sin \beta x$, applying this operator to both sides in (6.20), we obtain

$$(D^2 + \beta^2) \left(a_n D^n + a_{n-1} D^{n-1} + \cdots + a_0 \right) [y] = (D^2 + \beta^2)[f(x)] = 0.$$

The corresponding auxiliary equation,

$$\left(r^2 + \beta^2 \right) \left(a_n r^n + a_{n-1} r^{n-1} + \cdots + a_0 \right) = \left(r^2 + \beta^2 \right)^{s+1} \left(a_n r^{n-2s} + \cdots + a_0/\beta^{2s} \right) = 0$$

has a root $r = \beta i$ of multiplicity $s + 1$. Hence, a general solution to this equation is given by

$$\begin{aligned} y(x) &= (c_1 \cos \beta x + c_2 \sin \beta x) + x(c_3 \cos \beta x + c_4 \sin \beta x) \\ &\quad + \cdots + x^{s-1}(c_{2s-1} \cos \beta x + c_{2s} \sin \beta x) + x^s(c_{2s+1} \cos \beta x + c_{2s+2} \sin \beta x) + Y(x). \end{aligned}$$

Copyright © 2012 Pearson Education, Inc. Publishing as Addison-Wesley.

Since, $y(x) = y_h(x) + y_p(x)$, comparing $y(x)$ with $y_h(x)$ given in (6.22), we conclude that

$$y_p(x) = x^s(c_{2s+1}\cos\beta x + c_{2s+2}\sin\beta x).$$

All that remains is to note that, for $m < s$, the functions $x^m\cos\beta x$ and $x^m\sin\beta x$ are present in (6.22), meaning that they are solutions to the homogeneous equation, corresponding to (6.20). Thus, s is the smallest number m such that $x^m\cos\beta x$ and $x^m\sin\beta x$ are not solutions to the corresponding homogeneous equation.

39. Writing the system in operator form yields

$$(D^2 - 1)[x] + y = 0,$$
$$x + (D^2 - 1)[y] = e^{3t}.$$

Subtracting the first equation from the second one, multiplied by $(D^2 - 1)$, we get

$$\left\{(D^2 - 1)[x] + (D^2 - 1)^2[y]\right\} - \left\{(D^2 - 1)[x] + y\right\} = (D^2 - 1)[e^{3t}] - 0 = 8e^{3t}$$
$$\Rightarrow \quad \left\{(D^2 - 1)^2 - 1\right\}[y] = 8e^{3t}$$
$$\Rightarrow \quad D^2(D^2 - 2)[y] = 8e^{3t}. \qquad (6.23)$$

The auxiliary equation, $r^2(r^2 - 2) = 0$, has simple roots $r = \pm\sqrt{2}$ and a double root $r = 0$. Hence, the function

$$y_h(t) = c_1 + c_2 t + c_3 e^{\sqrt{2}t} + c_4 e^{-\sqrt{2}t}$$

is a general solution to the homogeneous equation, corresponding to (6.23). A particular solution to (6.23) has the form

$$y_p(t) = Ae^{3t}.$$

Substituting yields

$$D^2(D^2 - 2)[Ae^{3t}] = (D^4 - 2D^2)[Ae^{3t}] = 81Ae^{3t} - (2)9Ae^{3t} = 63Ae^{3t} = 8e^{3t}$$
$$\Rightarrow \quad A = \frac{8}{63}$$
$$\Rightarrow \quad y_p(t) = Ae^{3t} = \frac{8e^{3t}}{63},$$

and so

$$y(t) = y_p(t) + y_h(t) = \frac{8e^{3t}}{63} + c_1 + c_2 t + c_3 e^{\sqrt{2}t} + c_4 e^{-\sqrt{2}t}$$

Copyright © 2012 Pearson Education, Inc. Publishing as Addison-Wesley.

is a general solution to (6.23). We find $x(t)$ from the second equation in the original system.

$$x(t) = e^{3t} + y(t) - y''(t)$$

$$= e^{3t} + \left(\frac{8e^{3t}}{63} + c_1 + c_2 t + c_3 e^{\sqrt{2}t} + c_4 e^{-\sqrt{2}t} \right) - \left(\frac{72e^{3t}}{63} + 2c_3 e^{\sqrt{2}t} + 2c_4 e^{-\sqrt{2}t} \right)$$

$$= -\frac{e^{3t}}{63} + c_1 + c_2 t - c_3 e^{\sqrt{2}t} - c_4 e^{-\sqrt{2}t}.$$

EXERCISES 6.4:   Method of Variation of Parameters

1. To apply the method of variation of parameters, we have to find a fundamental solution set for the corresponding homogeneous equation, which is

$$y''' - 3y'' + 4y = 0.$$

Factoring the auxiliary polynomial, $r^3 - 3r^2 + 4$, yields

$$r^3 - 3r^2 + 4 = \left(r^3 + r^2 \right) - \left(4r^2 - 4 \right)$$

$$= r^2(r+1) - 4(r-1)(r+1) = (r+1)(r-2)^2.$$

Therefore, $r = -1$, 2, and 2 are the roots of the auxiliary equation, and $y_1 = e^{-x}$, $y_2 = e^{2x}$, and $y_3 = xe^{2x}$ form a fundamental solution set. According to the variation of parameters method, we seek for a particular solution of the form

$$y_p(x) = v_1(x)y_1(x) + v_2(x)y_2(x) + v_3(x)y_3(x)$$

$$= v_1(x)e^{-x} + v_2(x)e^{2x} + v_3(x)xe^{2x}.$$

To find functions v_j's, we need four determinants: the Wronskian $W[y_1, y_2, y_3](x)$, $W_1(x)$, $W_2(x)$, and $W_3(x)$ given in the formula (10) in the text. Thus, we compute

$$W\left[e^{-x}, e^{2x}, xe^{2x}\right](x) = \begin{vmatrix} e^{-x} & e^{2x} & xe^{2x} \\ -e^{-x} & 2e^{2x} & (1+2x)e^{2x} \\ e^{-x} & 4e^{2x} & (4+4x)e^{2x} \end{vmatrix} = e^{-x}e^{2x}e^{2x} \begin{vmatrix} 1 & 1 & x \\ -1 & 2 & 1+2x \\ 1 & 4 & 4+4x \end{vmatrix} = 9e^{3x},$$

$$W_1(x) = (-1)^{3-1}W\left[e^{2x}, xe^{2x}\right](x) = \begin{vmatrix} e^{2x} & xe^{2x} \\ 2e^{2x} & (1+2x)e^{2x} \end{vmatrix} = e^{4x},$$

$$W_2(x) = (-1)^{3-2}W\left[e^{-x}, xe^{2x}\right](x) = -\begin{vmatrix} e^{-x} & xe^{2x} \\ -e^{-x} & (1+2x)e^{2x} \end{vmatrix} = -(1+3x)e^x,$$

Copyright © 2012 Pearson Education, Inc. Publishing as Addison-Wesley.

$$W_3(x) = (-1)^{3-3} W\left[e^{-x}, e^{2x}\right](x) = \begin{vmatrix} e^{-x} & e^{2x} \\ -e^{-x} & 2e^{2x} \end{vmatrix} = 3e^x.$$

Substituting these expressions into the formula (11), we obtain

$$v_1(x) = \int \frac{g(x)W_1(x)}{W[e^{-x}, e^{2x}, xe^{2x}]}\, dx = \int \frac{e^{2x}e^{4x}}{9e^{3x}}\, dx = \frac{1}{27}e^{3x},$$

$$v_2(x) = \int \frac{g(x)W_2(x)}{W[e^{-x}, e^{2x}, xe^{2x}]}\, dx = \int \frac{-e^{2x}(1+3x)e^{x}}{9e^{3x}}\, dx = -\frac{x}{9} - \frac{x^2}{6},$$

$$v_3(x) = \int \frac{g(x)W_3(x)}{W[e^{-x}, e^{2x}, xe^{2x}]}\, dx = \int \frac{e^{2x}3e^{x}}{9e^{3x}}\, dx = \frac{x}{3},$$

where we have chosen zero integration constants. Thus, the formula (12) in the text gives a particular solution

$$y_p(x) = \frac{1}{27}e^{3x}e^{-x} - \left(\frac{x}{9} + \frac{x^2}{6}\right)e^{2x} + \frac{x}{3}xe^{2x} = \frac{e^{2x}}{27} - \frac{xe^{2x}}{9} + \frac{x^2e^{2x}}{6}.$$

We note that the first two terms in $y_p(x)$ are solutions to the corresponding homogeneous equation. Thus, another (and simpler) answer is $y_p(x) = x^2 e^{2x}/6$.

3. Let us find a fundamental solution set for the corresponding homogeneous equation,

$$z''' + 3z'' - 4z = 0.$$

Factoring the auxiliary polynomial, $r^3 + 3r^2 - 4$, yields

$$r^3 + 3r^2 - 4 = \left(r^3 - r^2\right) + \left(4r^2 - 4\right) = r^2(r-1) + 4(r+1)(r-1) = (r-1)(r+2)^2.$$

Therefore, $r = 1, -2$, and -2 are the roots of the auxiliary equation, and so the functions $z_1 = e^x$, $z_2 = e^{-2x}$, and $z_3 = xe^{-2x}$ form a fundamental solution set. A particular solution then has the form

$$z_p(x) = v_1(x)z_1(x) + v_2(x)z_2(x) + v_3(x)z_3(x) = v_1(x)e^x + v_2(x)e^{-2x} + v_3(x)xe^{-2x}. \tag{6.24}$$

To find functions v_j's, we need four determinants: the Wronskian $W[z_1, z_2, z_3](x)$ and $W_1(x)$, $W_2(x)$, $W_3(x)$ given in the formula (10). Thus, we compute

$$W\left[e^x, e^{-2x}, xe^{-2x}\right](x) = \begin{vmatrix} e^x & e^{-2x} & xe^{-2x} \\ e^x & -2e^{-2x} & (1-2x)e^{-2x} \\ e^x & 4e^{-2x} & (4x-4)e^{-2x} \end{vmatrix} = e^{-3x}\begin{vmatrix} 1 & 1 & x \\ 1 & -2 & 1-2x \\ 1 & 4 & 4x-4 \end{vmatrix} = 9e^{-3x},$$

 Copyright © 2012 Pearson Education, Inc. Publishing as Addison-Wesley.

$$W_1(x) = (-1)^{3-1} W\left[e^{-2x}, xe^{-2x}\right](x) = \begin{vmatrix} e^{-2x} & xe^{-2x} \\ -2e^{-2x} & (1-2x)e^{-2x} \end{vmatrix} = e^{-4x},$$

$$W_2(x) = (-1)^{3-2} W\left[e^{x}, xe^{-2x}\right](x) = -\begin{vmatrix} e^{x} & xe^{-2x} \\ e^{x} & (1-2x)e^{-2x} \end{vmatrix} = (3x-1)e^{-x},$$

$$W_3(x) = (-1)^{3-3} W\left[e^{x}, e^{-2x}\right](x) = \begin{vmatrix} e^{x} & e^{-2x} \\ e^{x} & -2e^{-2x} \end{vmatrix} = -3e^{-x}.$$

Substituting these expressions into the formula (11), we obtain

$$v_1(x) = \int \frac{g(x)W_1(x)}{W\left[e^{x}, e^{-2x}, xe^{-2x}\right]}\, dx = \int \frac{e^{2x}e^{-4x}}{9e^{-3x}}\, dx = \frac{1}{9}\, e^{x},$$

$$v_2(x) = \int \frac{g(x)W_2(x)}{W\left[e^{x}, e^{-2x}, xe^{-2x}\right]}\, dx = \int \frac{e^{2x}(3x-1)e^{-x}}{9e^{-3x}}\, dx$$

$$= \frac{1}{9}\int (3x-1)e^{4x}\, dx = \left(\frac{x}{12} - \frac{7}{144}\right) e^{4x},$$

$$v_3(x) = \int \frac{g(x)W_3(x)}{W\left[e^{x}, e^{-2x}, xe^{-2x}\right]}\, dx = \int \frac{e^{2x}(-3e^{-x})}{9e^{-3x}}\, dx = -\frac{1}{12}\, e^{4x}.$$

Substituting these expressions into (6.24) yields

$$z_p(x) = \frac{1}{9}\, e^{x}e^{x} + \left(\frac{x}{12} - \frac{7}{144}\right) e^{4x}e^{-2x} - \frac{1}{12}\, e^{4x}xe^{-2x} = \frac{1}{16}\, e^{2x}.$$

5. Since the nonhomogeneous term, $g(x) = \tan x$, is not a solution to a homogeneous linear differential equation with constant coefficients, we find a particular solution using the method of variation of parameters. Thus, first we find a fundamental solution set for the corresponding homogeneous equation, $y''' + y' = 0$. Its auxiliary equation is $r^3 + r = 0$, which factors to $r^3 + r = r(r^2 + 1)$. Thus, the roots to this auxiliary equation are $r = 0, \pm i$. Therefore, a fundamental solution set to the homogeneous equation is $\{1, \cos x, \sin x\}$ and

$$y_p(x) = v_1(x) + v_2(x)\cos x + v_3(x)\sin x.$$

Next, we find the four determinants $W[1, \cos x, \sin x](x)$, $W_1(x)$, $W_2(x)$, and $W_3(x)$. That is, we calculate

$$W[1, \cos x, \sin x](x) = \begin{vmatrix} 1 & \cos x & \sin x \\ 0 & -\sin x & \cos x \\ 0 & -\cos x & -\sin x \end{vmatrix} = \sin^2 x + \cos^2 x = 1,$$

Copyright © 2012 Pearson Education, Inc. Publishing as Addison-Wesley.

$$W_1(x) = (-1)^{3-1} W[\cos x, \sin x](x) = \begin{vmatrix} \cos x & \sin x \\ -\sin x & \cos x \end{vmatrix} = \left(\cos^2 x + \sin^2 x\right) = 1,$$

$$W_2(x) = (-1)^{3-2} W[1, \sin x](x) = -\begin{vmatrix} 1 & \sin x \\ 0 & \cos x \end{vmatrix} = -\cos x,$$

$$W_3(x) = (-1)^{3-3} W[1, \cos x](x) = \begin{vmatrix} 1 & \cos x \\ 0 & -\sin x \end{vmatrix} = -\sin x.$$

By using the formula (11), we can now get $v_1(x)$, $v_2(x)$, and $v_3(x)$. Since $g(x) = \tan x$, we have (assuming that all constants of integration equal zero)

$$v_1(x) = \int \frac{g(x) W_1(x)}{W[1, \cos x, \sin x](x)} \, dx = \int \tan x \, dx = \ln(\sec x),$$

$$v_2(x) = \int \frac{g(x) W_2(x)}{W[1, \cos x, \sin x](x)} \, dx = \int \tan x (-\cos x) \, dx = -\int \sin x \, dx = \cos x,$$

$$v_3(x) = \int \frac{g(x) W_3(x)}{W[1, \cos x, \sin x](x)} \, dx = \int \tan x (-\sin x) \, dx = -\int \frac{\sin^2 x}{\cos x} \, dx$$

$$= -\int \frac{1 - \cos^2 x}{\cos x} \, dx = \int (\cos x - \sec x) \, dx = \sin x - \ln(\sec x + \tan x).$$

Therefore, we have

$$y_p(x) = v_1(x) + v_2(x) \cos x + v_3(x) \sin x$$

$$= \ln(\sec x) + \cos^2 x + \sin^2 x - \sin x \ln(\sec x + \tan x),$$

which simplifies to

$$y_p(x) = \ln(\sec x) + 1 - (\sin x) \ln(\sec x + \tan x).$$

Since $y(x) \equiv 1$ is a solution to the corresponding homogeneous equation, we may choose

$$y_p(x) = \ln(\sec x) - (\sin x) \ln(\sec x + \tan x).$$

Note: We left the absolute value signs off in $\ln(\sec x)$ and $\ln(\sec x + \tan x)$, because of the given domain, $0 < x < \pi/2$, where $\sec x$ and $\tan x$ are positive.

7. First, we divide the differential equation by x^3 to obtain its standard form,

$$y''' - 3x^{-1} y'' + 6x^{-2} y' - 6x^{-3} y = x^{-4}, \qquad x > 0,$$

 Copyright © 2012 Pearson Education, Inc. Publishing as Addison-Wesley.

from which we see that $g(x) = x^{-4}$. Given that $\{x, x^2, x^3\}$ is a fundamental solution set for the corresponding homogeneous equation, we are looking for a particular solution of the form

$$y_p(x) = v_1(x)x + v_3(x)x^2 + v_3(x)x^3 . \tag{6.25}$$

Evaluating determinants $W[x, x^2, x^3](x)$, $W_1(x)$, $W_2(x)$, and $W_3(x)$ yields

$$W[x, x^2, x^3](x) = \begin{vmatrix} x & x^2 & x^3 \\ 1 & 2x & 3x^2 \\ 0 & 2 & 6x \end{vmatrix} = x \begin{vmatrix} 2x & 3x^2 \\ 2 & 6x \end{vmatrix} - \begin{vmatrix} x^2 & x^3 \\ 2 & 6x \end{vmatrix} = 2x^3 ,$$

$$W_1(x) = (-1)^{3-1} W[x^2, x^3](x) = \begin{vmatrix} x^2 & x^3 \\ 2x & 3x^2 \end{vmatrix} = x^4 ,$$

$$W_2(x) = (-1)^{3-2} W[x, x^3](x) = - \begin{vmatrix} x & x^3 \\ 1 & 3x^2 \end{vmatrix} = -2x^3 ,$$

$$W_3(x) = (-1)^{3-3} W[x, x^2](x) = \begin{vmatrix} x & x^2 \\ 1 & 2x \end{vmatrix} = x^2 .$$

So,

$$v_1(x) = \int \frac{g(x)W_1(x)}{W[x, x^2, x^3](x)} \, dx = \int \frac{x^{-4}x^4}{2x^3} \, dx = -\frac{1}{4x^2} ,$$

$$v_2(x) = \int \frac{g(x)W_2(x)}{W[x, x^2, x^3](x)} \, dx = \int \frac{x^{-4}(-2x^3)}{2x^3} \, dx = \frac{1}{3x^3} ,$$

$$v_3(x) = \int \frac{g(x)W_3(x)}{W[x, x^2, x^3](x)} \, dx = \int \frac{x^{-4}(x^2)}{2x^3} \, dx = -\frac{1}{8x^4} .$$

Substitution back into (6.25) yields

$$y_p(x) = \left(-\frac{1}{4x^2} \right) x + \left(\frac{1}{3x^3} \right) x^2 + \left(-\frac{1}{8x^4} \right) x^3 = -\frac{1}{24x} .$$

Thus,

$$y(x) = y_h(x) + y_p(x) = c_1 x + c_2 x^2 + c_3 x^3 - \frac{1}{24x} .$$

9. To find a particular solution to the given nonhomogeneous equation, we use the method of variation of parameters. First, we calculate the determinants $W[e^x, e^{-x}, e^{2x}](x)$, $W_1(x)$, $W_2(x)$, and $W_3(x)$ (see (10)). Thus, we have

$$W[e^x, e^{-x}, e^{2x}](x) = \begin{vmatrix} e^x & e^{-x} & e^{2x} \\ e^x & -e^{-x} & 2e^{2x} \\ e^x & e^{-x} & 4e^{2x} \end{vmatrix} = (e^x)(e^{-x})(e^{2x}) \begin{vmatrix} 1 & 1 & 1 \\ 1 & -1 & 2 \\ 1 & 1 & 4 \end{vmatrix} = -6e^{2x},$$

Copyright © 2012 Pearson Education, Inc. Publishing as Addison-Wesley.

$$W_1(x) = \begin{vmatrix} 0 & e^{-x} & e^{2x} \\ 0 & -e^{-x} & 2e^{2x} \\ 1 & e^{-x} & 4e^{2x} \end{vmatrix} = (-1)^{3-1} \begin{vmatrix} e^{-x} & e^{2x} \\ -e^{-x} & 2e^{2x} \end{vmatrix} = 2e^{x} + e^{x} = 3e^{x},$$

$$W_2(x) = \begin{vmatrix} e^{x} & 0 & e^{2x} \\ e^{x} & 0 & 2e^{2x} \\ e^{x} & 1 & 4e^{2x} \end{vmatrix} = (-1)^{3-2} \begin{vmatrix} e^{x} & e^{2x} \\ e^{x} & 2e^{2x} \end{vmatrix} = -\left(2e^{3x} - e^{3x}\right) = -e^{3x},$$

$$W_3(x) = \begin{vmatrix} e^{x} & e^{-x} & 0 \\ e^{x} & -e^{-x} & 0 \\ e^{x} & e^{-x} & 1 \end{vmatrix} = (-1)^{3-3} \begin{vmatrix} e^{x} & e^{-x} \\ e^{x} & -e^{-x} \end{vmatrix} = -1 - 1 = -2.$$

Therefore, according to the formula (12), a particular solution $y_p(x)$ is given by

$$y_p(x) = e^{x} \int \frac{3e^{x} g(x)}{-6e^{2x}}\, dx + e^{-x} \int \frac{-e^{3x} g(x)}{-6e^{2x}}\, dx + e^{2x} \int \frac{-2g(x)}{-6e^{2x}}\, dx$$

$$= -\frac{e^{x}}{2} \int e^{-x} g(x)\, dx + \frac{e^{-x}}{6} \int e^{x} g(x)\, dx + \frac{e^{2x}}{3} \int e^{-2x} g(x)\, dx.$$

11. First, we find a fundamental solution set for the corresponding homogeneous equation,

$$x^3 y''' - 3xy' + 3y = 0. \tag{6.26}$$

We involve the procedure of solving Cauchy-Euler equations, discussed in Section 4.7. We look for solutions to (6.26) of the form $y = x^r$. Substituting into (6.26) and simplifying yields

$$x^3 r(r-1)(r-2)x^{r-3} - 3xrx^{r-1} + 3x^r = \left(r^3 - 3r^2 - r + 3\right) x^r = 0\,.$$

Factoring the auxiliary equation, we obtain

$$r^3 - 3r^2 - r + 3 = r^2(r-3) - (r-3) = (r-1)(r+1)(r-3) = 0.$$

Therefore, $r = \pm 1$ and $r = 3$ are the roots, and the functions

$$y_1(x) = x\,, \quad y_2(x) = x^{-1}\,, \quad y_3(x) = x^3$$

form a fundamental solution set for the homogeneous equation (6.26). Next, we apply the variation of parameters method to find a particular solution to the original equation of the form

$$y_p(x) = v_1(x)x + v_2(x)x^{-1} + v_3(x)x^3\,. \tag{6.27}$$

 Copyright © 2012 Pearson Education, Inc. Publishing as Addison-Wesley.

To find functions $v_1(x)$, $v_2(x)$, and $v_3(x)$, we use the formula (11). Thus, we compute

$$W[x, x^{-1}, x^3](x) = \begin{vmatrix} x & x^{-1} & x^3 \\ 1 & -x^{-2} & 3x^2 \\ 0 & 2x^{-3} & 6x \end{vmatrix} = x \begin{vmatrix} -x^{-2} & 3x^2 \\ 2x^{-3} & 6x \end{vmatrix} - \begin{vmatrix} x^{-1} & x^3 \\ 2x^{-3} & 6x \end{vmatrix} = -16,$$

$$W_1(x) = (-1)^{3-1} W[x^{-1}, x^3](x) = \begin{vmatrix} x^{-1} & x^3 \\ -x^{-2} & 3x^2 \end{vmatrix} = 4x,$$

$$W_2(x) = (-1)^{3-2} W[x, x^3](x) = - \begin{vmatrix} x & x^3 \\ 1 & 3x^2 \end{vmatrix} = -2x^3,$$

$$W_3(x) = (-1)^{3-3} W[x, x^{-1}](x) = \begin{vmatrix} x & x^{-1} \\ 1 & -x^{-2} \end{vmatrix} = -2x^{-1}.$$

Writing the given equation in standard form,

$$y''' - \frac{3}{x^2} y' + \frac{3}{x^3} y = x \cos x,$$

we see that the nonhomogeneous term is $g(x) = x \cos x$. By (11),

$$v_1(x) = \int \frac{(x \cos x)(4x)}{-16} \, dx = -\frac{1}{4} \int x^2 \cos x \, dx = -\frac{1}{4} \left(x^2 \sin x + 2x \cos x - 2 \sin x \right),$$

$$v_2(x) = \int \frac{(x \cos x)(-2x^3)}{-16} \, dx = \frac{1}{8} \int x^4 \cos x \, dx$$

$$= \frac{1}{8} \left(x^4 \sin x + 4x^3 \cos x - 12x^2 \sin x - 24x \cos x + 24 \sin x \right),$$

$$v_3(x) = \int \frac{x \cos x \left(2x^{-1} \right)}{-16} \, dx = \frac{1}{8} \int \cos x \, dx = \frac{1}{8} \sin x,$$

where we have used integration by parts to evaluate $v_1(x)$ and $v_2(x)$. Substituting these functions into (6.27) and simplifying yields

$$y_p(x) = -\frac{\left(x^2 \sin x + 2x \cos x - 2 \sin x \right) x}{4}$$

$$+ \frac{\left(x^4 \sin x + 4x^3 \cos x - 12x^2 \sin x - 24x \cos x + 24 \sin x \right) x^{-1}}{8} + \frac{x^3 \sin x}{8}$$

$$= -x \sin x - 3 \cos x + 3x^{-1} \sin x.$$

Thus, the answer is

$$y(x) = y_h(x) + y_p(x)$$

$$= c_1 x + c_2 x^{-1} + c_3 x^3 - x \sin x - 3 \cos x + 3x^{-1} \sin x.$$

Copyright © 2012 Pearson Education, Inc. Publishing as Addison-Wesley.

13. Since

$$W_k(x) = \begin{vmatrix} y_1 & \cdots & y_{k-1} & 0 & y_{k+1} & \cdots & y_n \\ y_1' & \cdots & y_{k-1}' & 0 & y_{k+1}' & \cdots & y_n' \\ y_1'' & \cdots & y_{k-1}'' & 0 & y_{k+1}'' & \cdots & y_n'' \\ \vdots & & \vdots & \vdots \vdots & & \vdots \\ y_1^{(n-2)} & \cdots & y_{k-1}^{(n-2)} & 0 & y_{k+1}^{(n-2)} & \cdots & y_n^{(n-2)} \\ y_1^{(n-1)} & \cdots & y_{k-1}^{(n-1)} & 1 & y_{k+1}^{(n-1)} & \cdots & y_n^{(n-1)} \end{vmatrix},$$

the kth column of this determinant consists of all zeros, except for the last entry. There-fore, expanding $W_k(x)$ by the cofactors in the kth column yields

$$W_k(x) = (0)C_{1,k} + (0)C_{2,k} + \cdots + (0)C_{n-1,k} + (1)C_{n,k}$$

$$= (-1)^{n+k} \begin{vmatrix} y_1 & \cdots & y_{k-1} & y_{k+1} & \cdots & y_n \\ y_1' & \cdots & y_{k-1}' & y_{k+1}' & \cdots & y_n' \\ \vdots & & \vdots & \vdots & & \vdots \\ y_1^{(n-2)} & \cdots & y_{k-1}^{(n-2)} & y_{k+1}^{(n-2)} & \cdots & y_n^{(n-2)} \end{vmatrix}$$

$$= (-1)^{n+k} W[y_1, \ldots, y_{k-1}, y_{k+1}, \ldots, y_n](x).$$

Finally,

$$(-1)^{n+k} = (-1)^{(n-k)+(2k)} = (-1)^{n-k}.$$

REVIEW PROBLEMS

1. (a) In notation of Theorem 1, we have

$$p_1(x) \equiv 0, \ p_2(x) = -\ln x, \ p_3(x) = x, \ p_4(x) \equiv 2, \ \text{and} \, g(x) = \cos 3x.$$

All these functions, except $p_2(x)$, are continuous on $(-\infty, \infty)$, and $p_2(x)$ is defined and continuous on $(0, \infty)$. Thus, Theorem 1 guarantees the existence of a unique solution on $(0, \infty)$.

(b) By dividing both sides of the given differential equation by $x^2 - 1$, we rewrite the equation in standard form, that is,

$$y''' + \frac{\sin x}{x^2 - 1} y'' + \frac{\sqrt{x+4}}{x^2 - 1} y' + \frac{e^x}{x^2 - 1} y = \frac{x^2 + 3}{x^2 - 1}.$$

Thus, we see that

$$p_1(x) = \frac{\sin x}{x^2 - 1}, \quad p_2(x) = \frac{\sqrt{x+4}}{x^2 - 1}, \quad p_3(x) = \frac{e^x}{x^2 - 1}, \quad \text{and} \quad g(x) = \frac{x^2 + 3}{x^2 - 1}.$$

 Copyright © 2012 Pearson Education, Inc. Publishing as Addison-Wesley.

Functions $p_1(x)$, $p_3(x)$, and $g(x)$ are defined and continuous on $(-\infty, \infty)$ except $x = \pm 1$; $p_2(x)$ is defined and continuous on $\{x \geq -4, x \neq \pm 1\}$. Thus, for $p_1(x)$, $p_2(x)$, $p_3(x)$, and $g(x)$, the common open domain of continuity is $\{x > 4, x \neq \pm 1\}$. This set consists of three intervals,

$$(-4, -1), \quad (-1, 1), \quad \text{and} \quad (1, \infty).$$

Theorem 1 guarantees the existence of a unique solution on each of these open intervals.

3. A linear combination,

$$c_1 \sin x + c_2 x \sin x + c_3 x^2 \sin x + c_4 x^3 \sin x = \left(c_1 + c_2 x + c_3 x^2 + c_4 x^3\right) \sin x \quad (6.28)$$

vanishes identically on $(-\infty, \infty)$ if and only if the polynomial $c_1 + c_2 x + c_3 x^2 + c_4 x^3$ vanishes identically on $(-\infty, \infty)$. Since the number of real zeros of a polynomial does not exceed its degree, unless it's the zero polynomial, we conclude that the linear combination (6.28) vanishes identically on $(-\infty, \infty)$ if and only if $c_1 = c_2 = c_3 = c_4 = 0$. This means that the given functions are linearly independent on $(-\infty, \infty)$.

5. (a) Solving the auxiliary equation yields

$$(r + 5)^2 (r - 2)^3 (r^2 + 1)^2 = 0 \quad \Rightarrow \quad \begin{aligned} (r + 5)^2 &= 0 \quad \text{or} \\ (r - 2)^3 &= 0 \quad \text{or} \\ (r^2 + 1)^2 &= 0. \end{aligned}$$

Thus, the roots of the auxiliary equation are

$$\begin{aligned} r &= -5 \quad \text{of multiplicity 2,} \\ r &= 2 \quad \text{of multiplicity 3,} \\ r &= \pm i \quad \text{of multiplicity 2.} \end{aligned}$$

According to the equations (22) and (28) of Section 6.2, the set of functions (assuming that x is the independent variable)

$$e^{-5x}, \ xe^{-5x}, \ e^{2x}, \ xe^{2x}, \ x^2 e^{2x}, \ \cos x, \ x \cos x, \ \sin x, \ x \sin x$$

forms a linearly independent solution set. Thus, a general solution is given by

$$e^{-5x}\left(c_1 + c_2 x\right) + e^{2x}\left(c_3 + c_4 x + c_5 x^2\right) + (\cos x)\left(c_6 + c_7 x\right) + (\sin x)\left(c_8 + c_9 x\right).$$

Copyright © 2012 Pearson Education, Inc. Publishing as Addison-Wesley.

(b) Solving the auxiliary equation yields

$$r^4(r-1)^2(r^2+2r+4)^2 = 0 \quad \Rightarrow \quad \begin{aligned} r^4 &= 0 \quad \text{or} \\ (r-1)^2 &= 0 \quad \text{or} \\ (r^2+2r+4)^2 &= 0. \end{aligned}$$

Thus, the roots of the auxiliary equation are

$$\begin{aligned} r &= 0 && \text{of multiplicity 4,} \\ r &= 1 && \text{of multiplicity 2,} \\ r &= -1 \pm \sqrt{3}i && \text{of multiplicity 2.} \end{aligned}$$

Using (22) and (28), Section 6.2, we conclude that the set of functions

$$1,\; x,\; x^2,\; x^3,\; e^x,\; xe^x,\; e^{-x}\cos\sqrt{3}x,\; xe^{-x}\cos\sqrt{3}x,\; \sin\sqrt{3}x,\; xe^{-x}\sin\sqrt{3}x$$

is an independent solution set. A general solution is then given by

$$c_1 + c_2x + c_3x^2 + c_4x^3 + c_5e^x + c_6xe^x + c_7e^{-x}\cos\sqrt{3}x + c_8xe^{-x}\cos\sqrt{3}x$$
$$+ c_9\sin\sqrt{3}x + c_{10}xe^{-x}\sin\sqrt{3}x$$
$$= c_1 + c_2x + c_3x^2 + c_4x^3 + e^x\left(c_5 + c_6x\right) + \left(e^{-x}\cos\sqrt{3}x\right)\left(c_7 + c_8x\right)$$
$$+ \left(e^{-x}\sin\sqrt{3}x\right)\left(c_9 + c_{10}x\right).$$

7. (a) D^3, since the third derivative of a quadratic polynomial is identically zero.

(b) The function $e^{3x}+x-1$ is the sum of e^{3x} and $x-1$. The function $x-1$ is annihilated by D^2, the second derivative operator, and, according to (i), Section 6.3, $(D-3)$ annihilates e^{3x}. Therefore, the composite operator

$$D^2(D-3) = (D-3)D^2$$

annihilates both functions and, hence, their sum.

(c) The function $x\sin 2x$ is of the form, given in (iv), Section 6.3 with $m=2$, $\alpha=0$, and $\beta=2$. Thus, the operator

$$\left[(D-0)^2 + 2^2\right]^2 = \left(D^2 + 4\right)^2$$

annihilates this function.

 Copyright © 2012 Pearson Education, Inc. Publishing as Addison-Wesley.

(d) We again use (iv), this time with $m = 3$, $\alpha = -2$, and $\beta = 3$, to conclude that the given function is annihilated by

$$\left\{ [D - (-2)]^2 + 3^2 \right\}^3 = \left[(D + 2)^2 + 9 \right]^3 .$$

(e) Representing the given function as a linear combination,

$$\left(x^2 - 2x \right) + \left(xe^{-x} \right) + (\sin 2x) - (\cos 3x),$$

we find an annihilator for each term. Thus, we have:

$$\begin{aligned}
x^2 - 2x \quad &\text{is annihilated by } D^3 , \\
xe^{-x} \quad &\text{is annihilated by } [D - (-1)]^2 = (D + 1)^2 \quad \text{(by (ii))}, \\
\sin 2x \quad &\text{is annihilated by } D^2 + 2^2 = D^2 + 4 \qquad\quad \text{(by (iii))}, \\
\cos 3x \quad &\text{is annihilated by } D^2 + 3^2 = D^2 + 9 \qquad\quad \text{(by (iii))}.
\end{aligned}$$

Therefore, the product,

$$D^3 (D + 1)^2 (D^2 + 4)(D^2 + 9),$$

annihilates the given function.

9. A general solution to the corresponding homogeneous equation,

$$x^3 y''' - 2x^2 y'' - 5xy' + 5y = 0,$$

is given by

$$y_h(x) = c_1 x + c_2 x^5 + c_3 x^{-1}$$

. We now apply the variation of parameters method, described in Section 6.4, and seek for a particular solution to the original nonhomogeneous equation of the form

$$y_p(x) = v_1(x)x + v_2(x)x^5 + v_3(x)x^{-1} .$$

Since

$$\begin{aligned}
(x)' &= 1, & (x)'' &= 0 , \\
(x^5)' &= 5x^4, & (x^5)'' &= 20x^3 , \\
(x^{-1})' &= -x^{-2}, & (x^{-1})'' &= 2x^{-3} ,
\end{aligned}$$

Copyright © 2012 Pearson Education, Inc. Publishing as Addison-Wesley.

the Wronskian $W[x, x^5, x^{-1}](x)$ and determinants $W_k(x)$, given in (10), Section 6.4, become

$$W[x, x^5, x^{-1}](x) = \begin{vmatrix} x & x^5 & x^{-1} \\ 1 & 5x^4 & -x^{-2} \\ 0 & 20x^3 & 2x^{-3} \end{vmatrix}$$

$$= (x)\begin{vmatrix} 5x^4 & -x^{-2} \\ 20x^3 & 2x^{-3} \end{vmatrix} - (1)\begin{vmatrix} x^5 & x^{-1} \\ 20x^3 & 2x^{-3} \end{vmatrix} = 48x^2 \,,$$

$$W_1(x) = (-1)^{3-1}\begin{vmatrix} x^5 & x^{-1} \\ 5x^4 & -x^{-2} \end{vmatrix} = -6x^3 \,,$$

$$W_2(x) = (-1)^{3-2}\begin{vmatrix} x & x^{-1} \\ 1 & -x^{-2} \end{vmatrix} = 2x^{-1} \,,$$

$$W_3(x) = (-1)^{3-3}\begin{vmatrix} x & x^5 \\ 1 & 5x^4 \end{vmatrix} = 4x^5 \,.$$

Now we divide both sides of the given equation by x^3 to obtain an equation in standard form, that is,

$$y''' - 2x^{-1}y'' - 5x^{-2}y' + 5x^{-3}y = x^{-5} \,.$$

Hence, the right-hand side, $g(x)$, in formula (1), Section 6.4, equals to x^{-5}. Applying the formula (11) yields

$$v_1(x) = \int \frac{x^{-5}(-6x^3)}{48x^2}\, dx = -\frac{1}{8}\int x^{-4}\, dx = \frac{1}{24}\, x^{-3} \,,$$

$$v_2(x) = \int \frac{x^{-5}(2x^{-1})}{48x^2}\, dx = \frac{1}{24}\int x^{-8}\, dx = -\frac{1}{168}\, x^{-7} \,,$$

$$v_3(x) = \int \frac{x^{-5}(4x^5)}{48x^2}\, dx = \frac{1}{12}\int x^{-2}\, dx = -\frac{1}{12}\, x^{-1} \,.$$

Therefore,

$$y_p(x) = \left(\frac{1}{24}\, x^{-3}\right)x + \left(-\frac{1}{168}\, x^{-7}\right)x^5 + \left(-\frac{1}{12}\, x^{-1}\right)x^{-1}$$

$$= \left(\frac{1}{24} - \frac{1}{168} - \frac{1}{12}\right)x^{-2} = -\frac{1}{21}\, x^{-2} \,,$$

and a general solution to the given equation is given by

$$y(x) = y_h(x) + y_p(x) = c_1 x + c_2 x^5 + c_3 x^{-1} - \frac{1}{21}\, x^{-2} \,.$$

 Copyright © 2012 Pearson Education, Inc. Publishing as Addison-Wesley.

CHAPTER 7: Laplace Transforms

EXERCISES 7.2:   **Definition of the Laplace Transform**

1. For $s > 0$, using Definition 1 and integration by parts, we compute

$$\mathcal{L}\{t\}(s) = \int_0^\infty e^{-st} t\, dt = \lim_{N\to\infty} \int_0^N e^{-st} t\, dt = \lim_{N\to\infty} \int_0^N t\, d\left(-\frac{e^{-st}}{s}\right)$$

$$= \lim_{N\to\infty}\left[-\frac{te^{-st}}{s}\Big|_0^N + \frac{1}{s}\int_0^N e^{-st}\, dt\right] = \lim_{N\to\infty}\left[-\frac{te^{-st}}{s}\Big|_0^N - \frac{e^{-st}}{s^2}\Big|_0^N\right]$$

$$= \lim_{N\to\infty}\left[-\frac{Ne^{-sN}}{s} + 0 - \frac{e^{-sN}}{s^2} + \frac{1}{s^2}\right] = \frac{1}{s^2}$$

because, for $s > 0$, $e^{-sN} \to 0$ and $Ne^{-sN} = N/e^{sN} \to 0$, as $N \to \infty$.

3. For $s > 6$, we have

$$\mathcal{L}\{t\}(s) = \int_0^\infty e^{-st} e^{6t}\, dt = \int_0^\infty e^{(6-s)t}\, dt = \lim_{N\to\infty} \int_0^N e^{(6-s)t}\, dt$$

$$= \lim_{N\to\infty}\left[\frac{e^{(6-s)t}}{6-s}\Big|_0^N\right] = \lim_{N\to\infty}\left[\frac{e^{(6-s)N}}{6-s} - \frac{1}{6-s}\right] = 0 - \frac{1}{6-s} = \frac{1}{s-6}.$$

5. For $s > 0$,

$$\mathcal{L}\{\cos 2t\}(s) = \int_0^\infty e^{-st} \cos 2t\, dt = \lim_{N\to\infty} \int_0^N e^{-st} \cos 2t\, dt$$

$$= \lim_{N\to\infty}\left[\frac{e^{-st}(-s\cos 2t + 2\sin 2t)}{s^2 + 4}\Big|_0^N\right]$$

$$= \lim_{N\to\infty}\left[\frac{e^{-sN}(-s\cos 2N + 2\sin 2N)}{s^2 + 4} - \frac{-s}{s^2 + 4}\right] = \frac{s}{s^2 + 4},$$

where we have used the integral table on the inside front cover to obtain an antiderivative of $e^{-st}\cos 2t$.

Copyright © 2012 Pearson Education, Inc. Publishing as Addison-Wesley.

Chapter 7

7. For $s > 2$,

$$
\begin{aligned}
\mathcal{L}\left\{e^{2t}\cos 3t\right\}(s) &= \int_0^\infty e^{-st}e^{2t}\cos 3t\, dt = \int_0^\infty e^{(2-s)t}\cos 3t\, dt \\[2mm]
&= \lim_{N\to\infty}\left[\frac{e^{(2-s)t}\left((2-s)\cos 3t + 3\sin 3t\right)}{(2-s)^2+9}\Big|_0^N\right] \\[2mm]
&= \lim_{N\to\infty}\frac{e^{(2-s)N}\left[(2-s)\cos 3N + 3\sin 3N\right]-(2-s)}{(2-s)^2+9} = \frac{s-2}{(s-2)^2+9}.
\end{aligned}
$$

9. As in Example 4 in the text, we first break the integral in Definition 1 into two parts. Thus,

$$
\mathcal{L}\{f(t)\}(s) = \int_0^\infty e^{-st}f(t)\, dt = \int_0^2 e^{-st}\cdot 0\, dt + \int_2^\infty te^{-st}\, dt = \int_2^\infty te^{-st}\, dt.
$$

An antiderivative of te^{-st} was, in fact, obtained in Problem 1. Thus, we have for $s > 0$,

$$
\begin{aligned}
\int_2^\infty te^{-st}\, dt &= \lim_{N\to\infty}\left[\left(-\frac{te^{-st}}{s}-\frac{e^{-st}}{s^2}\right)\Big|_2^N\right] = \lim_{N\to\infty}\left[-\frac{Ne^{-sN}}{s}-\frac{e^{-sN}}{s^2}+\frac{2e^{-2s}}{s}+\frac{e^{-2s}}{s^2}\right] \\[2mm]
&= \frac{2e^{-2s}}{s}+\frac{e^{-2s}}{s^2} = e^{-2s}\left(\frac{2}{s}+\frac{1}{s^2}\right) = e^{-2s}\left(\frac{2s+1}{s^2}\right).
\end{aligned}
$$

11. In this problem, $f(t)$ is also a piecewise defined function. So, we split the integral and obtain

$$
\begin{aligned}
\mathcal{L}\{f(t)\}(s) &= \int_0^\infty e^{-st}f(t)\, dt = \int_0^\pi e^{-st}\sin t\, dt + \int_\pi^\infty e^{-st}\cdot 0\, dt = \int_0^\pi e^{-st}\sin t\, dt \\[2mm]
&= \frac{e^{-st}(-s\sin t - \cos t)}{s^2+1}\Big|_0^\pi = \frac{e^{-\pi s}-(-1)}{s^2+1} = \frac{e^{-\pi s}+1}{s^2+1},
\end{aligned}
$$

which is valid for all s.

13. By linearity of the Laplace transform,

$$
\mathcal{L}\left\{6e^{-3t}-t^2+2t-8\right\}(s) = 6\mathcal{L}\left\{e^{-3t}\right\}(s) - \mathcal{L}\left\{t^2\right\}(s) + 2\mathcal{L}\{t\}(s) - 8\mathcal{L}\{1\}(s).
$$

From Table 7.1 in the text, we see that

$$
\mathcal{L}\left\{e^{-3t}\right\}(s) = \frac{1}{s-(-3)} = \frac{1}{s+3}, \qquad s > -3;
$$

$$
\mathcal{L}\left\{t^2\right\}(s) = \frac{2!}{s^{2+1}} = \frac{2}{s^3}, \quad \mathcal{L}\{t\}(s) = \frac{1!}{s^{1+1}} = \frac{1}{s^2}, \quad \mathcal{L}\{1\}(s) = \frac{1}{s}, \qquad s > 0.
$$

Copyright © 2012 Pearson Education, Inc. Publishing as Addison-Wesley.

Thus, the formula

$$\mathcal{L}\left\{6e^{-3t} - t^2 + 2t - 8\right\}(s) = 6\frac{1}{s+3} - \frac{2}{s^3} + 2\frac{1}{s^2} - 8\frac{1}{s} = \frac{6}{s+3} - \frac{2}{s^3} + \frac{2}{s^2} - \frac{8}{s},$$

is valid for s in the intersection of the sets $s > -3$ and $s > 0$, which is $s > 0$.

15. Using linearity of the Laplace transform and Table 7.1, we get

$$\mathcal{L}\left\{t^3 - te^t + e^{4t}\cos t\right\}(s) = \mathcal{L}\left\{t^3\right\}(s) - \mathcal{L}\left\{te^t\right\}(s) + \mathcal{L}\left\{e^{4t}\cos t\right\}(s)$$

$$= \frac{3!}{s^{3+1}} - \frac{1!}{(s-1)^{1+1}} + \frac{s-4}{(s-4)^2 + 1^2}$$

$$= \frac{6}{s^4} - \frac{1}{(s-1)^2} + \frac{s-4}{(s-4)^2 + 1},$$

which is valid for $s > 4$.

17. Using linearity of Laplace transform and Table 7.1, we get

$$\mathcal{L}\left\{e^{3t}\sin 6t - t^3 + e^t\right\}(s) = \mathcal{L}\left\{e^{3t}\sin 6t\right\}(s) - \mathcal{L}\left\{t^3\right\}(s) + \mathcal{L}\left\{e^t\right\}(s)$$

$$= \frac{6}{(s-3)^2 + 6^2} - \frac{3!}{s^{3+1}} + \frac{1}{s-1}$$

$$= \frac{6}{(s-3)^2 + 36} - \frac{6}{s^4} + \frac{1}{s-1},$$

valid for $s > 3$.

19. For $s > 5$, we have

$$\mathcal{L}\left\{t^4 e^{5t} - e^t\cos\sqrt{7}t\right\}(s) = \mathcal{L}\left\{t^4 e^{5t}\right\}(s) - \mathcal{L}\left\{e^t\cos\sqrt{7}t\right\}(s)$$

$$= \frac{4!}{(s-5)^{4+1}} - \frac{s-1}{(s-1)^2 + (\sqrt{7})^2} = \frac{24}{(s-5)^5} - \frac{s-1}{(s-1)^2 + 7}.$$

21. Since the function $g_1(t) \equiv 1$ is continuous on $(-\infty, \infty)$ and $f(t) = g_1(t)$ for t in $[0,1]$, we conclude that $f(t)$ is continuous on $[0,1)$ and continuous from the left at $t = 1$. The function $g_2(t) \equiv (t-2)^2$ is also continuous on $(-\infty, \infty)$, and so $f(t)$ (which is the same as $g_2(t)$ on $(1,10]$) is continuous on $(1,10]$. Moreover,

$$\lim_{t\to 1+} f(t) = \lim_{t\to 1+} g_2(t) = g_2(1) = (1-2)^2 = 1 = f(1),$$

which implies that $f(t)$ is continuous from the right at $t = 1$. Thus, $f(t)$ is continuous at $t = 1$ and, therefore, is continuous at any t in $[0,10]$.

Copyright © 2012 Pearson Education, Inc. Publishing as Addison-Wesley.

Chapter 7

23. All the functions involved in the definition of $f(t)$, that is, $g_1(t) \equiv 1$, $g_2(t) = t - 1$, and $g_3(t) = t^2 - 4$, are continuous on $(-\infty, \infty)$. So, $f(t)$, being a restriction of these functions, on $[0,1)$, $(1,3)$, and $(3,10]$, respectively, is continuous on these three intervals. At points $t = 1$ and $t = 3$, $f(t)$ is not defined and so is not continuous. But, one-sided limits

$$\lim_{t \to 1^-} f(t) = \lim_{t \to 1^-} g_1(t) = g_1(1) = 1\,,$$
$$\lim_{t \to 1^+} f(t) = \lim_{t \to 1^+} g_2(t) = g_2(1) = 0\,,$$
$$\lim_{t \to 3^-} f(t) = \lim_{t \to 3^-} g_2(t) = g_2(3) = 2\,,$$
$$\lim_{t \to 3^+} f(t) = \lim_{t \to 3^+} g_3(t) = g_3(3) = 5\,,$$

exist and pairwise different. Therefore, $f(t)$ has jump discontinuities at $t = 1$, $t = 3$ and, hence, is piecewise continuous on $[0,10]$.

25. Given function is a rational function and, therefore, continuous on its domain, which is all reals except zeros of the denominator. Solving $t^2 + 7t + 10 = 0$, we conclude that the points of discontinuity of $f(t)$ are $t = -2$ and $t = -5$. These points are not in $[0,10]$. So, $f(t)$ is continuous on $[0,10]$.

27. Since

$$\lim_{t \to 0^+} f(t) = \lim_{t \to 0^+} \frac{1}{t} = \infty,$$

$f(t)$ has an infinite discontinuity at $t = 0$, and so neither continuous nor piecewise continuous on $[0,10]$.

29. (a) First, observe that $|t^3 \sin t| \leq t^3$ for all $t > 0$. Next, applying L'Hospital's rule yields

$$\lim_{t \to \infty} \frac{t^3}{e^{\alpha t}} = \lim_{t \to \infty} \frac{3t^2}{\alpha e^{\alpha t}} = \lim_{t \to \infty} \frac{6t}{\alpha^2 e^{\alpha t}} = \lim_{t \to \infty} \frac{6}{\alpha^3 e^{\alpha t}} = 0$$

for all $\alpha > 0$. Thus, fixed $\alpha > 0$, for some $T = T(\alpha) > 0$, we have $t^3 \leq e^{\alpha t} \leq 1$ so that $t^3/e^{\alpha t} \leq 1$, for all $t > T$. Therefore,

$$|t^3 \sin t| \leq t^3 \leq e^{\alpha t}, \quad t > T,$$

and $t^3 \sin t$ is of exponential order α, for any $\alpha > 0$.

(b) Clearly, for any t, $|f(t)| = 100e^{49t}$, and so Definition 3 is satisfied with $M = 100$, $\alpha = 49$, and any T. Hence, $f(t)$ is of exponential order 49.

 Copyright © 2012 Pearson Education, Inc. Publishing as Addison-Wesley.

(c) Since

$$\lim_{t\to\infty}\frac{f(t)}{e^{\alpha t}}=\lim_{t\to\infty}e^{t^3-\alpha t}=\lim_{t\to\infty}e^{(t^2-\alpha)t}=\infty,$$

we see that $f(t)$ grows faster than $e^{\alpha t}$ for any α. Thus, $f(t)$ is *not* of exponential order.

(d) Similarly to (a), for any $\alpha>0$, we get

$$\lim_{t\to\infty}\frac{|t\ln t|}{e^{\alpha t}}=\lim_{t\to\infty}\frac{t\ln t}{e^{\alpha t}}=\lim_{t\to\infty}\frac{\ln t+1}{\alpha e^{\alpha t}}=\lim_{t\to\infty}\frac{1/t}{\alpha^2 e^{\alpha t}}=0,$$

and so $f(t)$ is of exponential order α for any positive α.

(e) Since,

$$f(t)=\cosh\left(t^2\right)=\frac{e^{t^2}+e^{-t^2}}{2}>\frac{1}{2}e^{t^2}$$

and e^{t^2} grows faster than $e^{\alpha t}$ for any fixed α (see the discussion after Definition 3 in the text), we conclude that $\cosh\left(t^2\right)$ is *not* of exponential order.

(f) This function is bounded on $(-\infty,\infty)$. Indeed,

$$|f(t)|=\left|\frac{1}{t^2+1}\right|=\frac{1}{t^2+1}\le\frac{1}{0+1}=1,$$

and so Definition 3 is satisfied with $M=1$ and $\alpha=0$. Hence, $f(t)$ is of exponential order 0.

(g) The function $\sin\left(t^2\right)$ is bounded, namely, $|\sin\left(t^2\right)|\le1$. For any fixed $\beta>0$, the limit of $t^4/e^{\beta t}$ as $t\to\infty$ is zero, which implies that $t^4\le e^{\beta t}$ for all $t>T=T(\beta)>0$. Thus,

$$\left|\sin\left(t^2\right)+t^4e^{6t}\right|\le1+e^{\beta t}e^{6t}\le2e^{(\beta+6)t},\quad t>T.$$

This means that $f(t)$ is of exponential order α for any $\alpha>6$.

(h) The function $3+\cos4t$ is bounded because

$$|3+\cos4t|\le3+|\cos4t|\le4.$$

Therefore, by the triangle inequality,

$$|f(t)|\ge\left|e^{t^2}\right|-|3+\cos4t|\ge e^{t^2}-4,$$

and, therefore, for any fixed α, $f(t)$ grows faster than $e^{\alpha t}$ (because e^{t^2} does, and the other term is bounded). So, $f(t)$ is *not* of exponential order.

Copyright © 2012 Pearson Education, Inc. Publishing as Addison-Wesley.

(i) Clearly, for any $t \geq 0$,

$$\frac{t^2}{t+1} = \frac{t}{t+1} \, t < (1)t = t \, .$$

Therefore,

$$e^{t^2/(t+1)} < e^t \, ,$$

and Definition 3 holds with $M = 1$, $\alpha = 1$, and $T = 0$.

(j) Since, for any x, $-1 \leq \sin x \leq 1$, the given function is bounded. Indeed,

$$\left| \sin \left(e^{t^2} \right) + e^{\sin t} \right| \leq \left| \sin \left(e^{t^2} \right) \right| + e^{\sin t} \leq 1 + e$$

Thus, it is of exponential order 0.

31. (a) We have

$$\mathcal{L} \left\{ e^{(a+ib)t} \right\} (s) := \int_0^\infty e^{-st} e^{(a+ib)t} \, dt = \int_0^\infty e^{(a+ib-s)t} \, dt = \lim_{N \to \infty} \int_0^N e^{(a+ib-s)t} \, dt$$

$$= \lim_{N \to \infty} \left[\frac{e^{(a+ib-s)t}}{a+ib-s} \Big|_0^N \right] = \frac{1}{a+ib-s} \lim_{N \to \infty} \left[e^{(a-s+ib)N} - 1 \right]. \quad (7.1)$$

Writing

$$e^{(a-s+ib)x} = e^{(a-s)x} e^{ibx} \, ,$$

we see that the first factor vanishes at ∞ if $a - s < 0$, while the second factor is bounded. (Indeed, $\left| e^{ibx} \right| \equiv 1$.) Thus, the limit in (7.1) exists if and only if $a - s < 0$. Assuming that $s > a$, we get

$$\frac{1}{a+ib-s} \lim_{N \to \infty} \left(e^{(a-s+ib)N} - 1 \right) = \frac{1}{a+ib-s} (0-1) = \frac{1}{s-(a+ib)} \, .$$

(b) Note that $s - (a + ib) = (s - a) - ib$. Multiplying the result in (a) by the complex conjugate of the denominator, that is, $(s - a) + bi$, we get

$$\frac{1}{s-(a+ib)} = \frac{(s-a)+ib}{[(s-a)-ib] \cdot [(s-a)+ib]} = \frac{(s-a)+ib}{(s-a)^2+b^2} \, ,$$

where we used the fact that, for any complex number z, $z\bar{z} = |z|^2$.

(c) From (a) and (b) we know that

$$\mathcal{L} \left\{ e^{(a+ib)t} \right\} (s) = \frac{(s-a)+ib}{(s-a)^2+b^2} \, .$$

Copyright © 2012 Pearson Education, Inc. Publishing as Addison-Wesley.

Writing

$$\frac{(s-a)+ib}{(s-a)^2+b^2} = \frac{s-a}{(s-a)^2+b^2} + \frac{b}{(s-a)^2+b^2}i,$$

we see that

$$\mathrm{Re}\left[\mathcal{L}\left\{e^{(a+ib)t}\right\}(s)\right] = \mathrm{Re}\left[\frac{s-a}{(s-a)^2+b^2} + \frac{b}{(s-a)^2+b^2}i\right] = \frac{s-a}{(s-a)^2+b^2}, \quad (7.2)$$

$$\mathrm{Im}\left[\mathcal{L}\left\{e^{(a+ib)t}\right\}(s)\right] = \mathrm{Im}\left[\frac{s-a}{(s-a)^2+b^2} + \frac{b}{(s-a)^2+b^2}i\right] = \frac{b}{(s-a)^2+b^2}. \quad (7.3)$$

On the other hand, by Euler's formulas,

$$\mathrm{Re}\left[e^{-st}e^{(a+ib)t}\right] = e^{-st}\mathrm{Re}\left[e^{at}(\cos bt + i\sin bt)\right] = e^{-st}e^{at}\cos bt\,,$$

and so

$$\mathrm{Re}\left[\mathcal{L}\left\{e^{(a+ib)t}\right\}(s)\right] = \mathrm{Re}\left[\int_0^\infty e^{-st}e^{(a+ib)t}\,dt\right] = \int_0^\infty \mathrm{Re}\left[e^{-st}e^{(a+ib)t}\right]dt$$

$$= \int_0^\infty e^{-st}e^{at}\cos bt\,dt = \mathcal{L}\left\{e^{at}\cos bt\right\}(s),$$

which together with (7.2) gives the last entry in Table 7.1. Similarly,

$$\mathrm{Im}\left[\mathcal{L}\left\{e^{(a+ib)t}\right\}(s)\right] = \mathcal{L}\left\{e^{at}\sin bt\right\}(s),$$

and so (7.3) gives the Laplace transform of $e^{at}\sin bt$.

33. Let $f(t)$ be a piecewise continuous function on $[a,b]$, and let a function $g(t)$ be continuous on $[a,b]$. At any point of continuity of $f(t)$, the function $(fg)(t)$ is continuous as the product of two continuous functions at this point. Suppose now that c is a point of discontinuity of $f(t)$. Then one-sided limits

$$\lim_{t\to c^-} f(t) = L_- \quad \text{and} \quad \lim_{t\to c^+} f(t) = L_+$$

exist. At the same time, continuity of $g(t)$ yields

$$\lim_{t\to c^-} g(t) = \lim_{t\to c^+} g(t) = \lim_{t\to c} g(t) = g(c).$$

Thus, the product rule implies that one-sided limits

$$\lim_{t\to c^-} (fg)(t) = \lim_{t\to c^-} f(t) \cdot \lim_{t\to c^-} g(t) = L_- g(c)$$

Copyright © 2012 Pearson Education, Inc. Publishing as Addison-Wesley.

$$\lim_{t \to c^+} (fg)(t) = \lim_{t \to c^+} f(t) \lim_{t \to c^+} g(t) = L_+ g(c)$$

exist. So, $(fg)(t)$ has a jump (even removable if $g(c) = 0$) discontinuity at $t = c$.

Therefore, the product $(fg)(t)$ is continuous at any point on $[a, b]$ except possibly a finite number of points (namely, points of discontinuity of $f(t)$).

EXERCISES 7.3: Properties of the Laplace Transform

1. Using linearity of the Laplace transform, we get

$$\mathcal{L} \left\{ t^2 + e^t \sin 2t \right\} (s) = \mathcal{L} \left\{ t^2 \right\} (s) + \mathcal{L} \left\{ e^t \sin 2t \right\} (s).$$

From Table 7.1 in Section 7.2, we know that

$$\mathcal{L} \left\{ t^2 \right\} (s) = \frac{2!}{s^3} = \frac{2}{s^3}, \quad \mathcal{L} \left\{ e^t \sin 2t \right\} (s) = \frac{2}{(s-1)^2 + 2^2} = \frac{2}{(s-1)^2 + 4}.$$

Thus,

$$\mathcal{L} \left\{ t^2 + e^t \sin 2t \right\} (s) = \frac{2}{s^3} + \frac{2}{(s-1)^2 + 4}.$$

3. By linearity of the Laplace transform,

$$\mathcal{L} \left\{ e^{-t} \cos 3t + e^{6t} - 1 \right\} (s) = \mathcal{L} \left\{ e^{-t} \cos 3t \right\} (s) + \mathcal{L} \left\{ e^{6t} \right\} (s) - \mathcal{L} \left\{ 1 \right\} (s).$$

From Table 7.1 of the text we see that

$$\mathcal{L} \left\{ e^{-t} \cos 3t \right\} (s) = \frac{s - (-1)}{[s - (-1)]^2 + 3^2} = \frac{s + 1}{(s+1)^2 + 9}, \quad s > -1; \tag{7.4}$$

$$\mathcal{L} \left\{ e^{6t} \right\} (s) = \frac{1}{s - 6}, \quad s > 6; \tag{7.5}$$

$$\mathcal{L} \left\{ 1 \right\} (s) = \frac{1}{s}, \quad s > 0. \tag{7.6}$$

Since all of (7.4), (7.5), and (7.6) hold for $s > 6$, we see that our answer,

$$\mathcal{L} \left\{ e^{-t} \cos 3t + e^{6t} - 1 \right\} (s) = \frac{s + 1}{(s+1)^2 + 9} + \frac{1}{s - 6} - \frac{1}{s},$$

is valid for $s > 6$. Note that (7.4) and (7.5) could also be obtained from the Laplace transforms for $f(t) = \cos 3t$ and $f(t) \equiv 1$, respectively, by applying Theorem 3.

 Copyright © 2012 Pearson Education, Inc. Publishing as Addison-Wesley.

5. We use the linearity of the Laplace transform and Table 7.1 to get

$$\mathcal{L}\left\{2t^2 e^{-t} - t + \cos 4t\right\}(s) = 2\mathcal{L}\left\{t^2 e^{-t}\right\}(s) - \mathcal{L}\left\{t\right\}(s) + \mathcal{L}\left\{\cos 4t\right\}(s)$$

$$= 2 \cdot \frac{2}{(s+1)^3} - \frac{1}{s^2} + \frac{s}{s^2 + 4^2} = \frac{4}{(s+1)^3} - \frac{1}{s^2} + \frac{s}{s^2 + 16},$$

which is valid for $s > 0$.

7. Since $(t-1)^4 = t^4 - 4t^3 + 6t^2 - 4t + 1$, we have

$$\mathcal{L}\left\{(t-1)^4\right\}(s) = \mathcal{L}\left\{t^4\right\}(s) - 4\mathcal{L}\left\{t^3\right\}(s) + 6\mathcal{L}\left\{t^2\right\}(s) - 4\mathcal{L}\left\{t\right\}(s) + \mathcal{L}\left\{1\right\}(s).$$

From Table 7.1 in the text, for $s > 0$ we get

$$\mathcal{L}\left\{t^4\right\}(s) = \frac{4!}{s^5} = \frac{24}{s^5},$$

$$\mathcal{L}\left\{t^3\right\}(s) = \frac{3!}{s^4} = \frac{6}{s^4},$$

$$\mathcal{L}\left\{t^2\right\}(s) = \frac{2!}{s^3} = \frac{2}{s^3},$$

$$\mathcal{L}\left\{t\right\}(s) = \frac{1!}{s^2} = \frac{1}{s^2},$$

$$\mathcal{L}\left\{1\right\}(s) = \frac{1}{s}.$$

Thus,

$$\mathcal{L}\left\{(t-1)^4\right\}(s) = \frac{24}{s^5} - \frac{24}{s^4} + \frac{12}{s^3} - \frac{4}{s^2} + \frac{1}{s}, \quad s > 0.$$

9. Since

$$\mathcal{L}\left\{e^{-t}\sin 2t\right\}(s) = \frac{2}{(s+1)^2 + 4},$$

we use Theorem 6 to conclude that

$$\mathcal{L}\left\{e^{-t}t\sin 2t\right\}(s) = \mathcal{L}\left\{t\left(e^{-t}\sin 2t\right)\right\}(s)$$

$$= = -\left[\mathcal{L}\left\{e^{-t}\sin 2t\right\}(s)\right]' = -\left[\frac{2}{(s+1)^2 + 4}\right]'$$

$$= -2(-1)\left[(s+1)^2 + 4\right]^{-2}\left[(s+1)^2 + 4\right]' = \frac{4(s+1)}{\left[(s+1)^2 + 4\right]^2}.$$

11. We use the definition of $\cosh x$ and the linear property of the Laplace transform.

$$\mathcal{L}\left\{\cosh bt\right\}(s) = \mathcal{L}\left\{\frac{e^{bt} + e^{-bt}}{2}\right\}(s)$$

$$= \frac{1}{2}\left[\mathcal{L}\left\{e^{bt}\right\}(s) + \mathcal{L}\left\{e^{-bt}\right\}(s)\right] = \frac{1}{2}\left(\frac{1}{s-b} + \frac{1}{s+b}\right) = \frac{s}{s^2 - b^2}.$$

Copyright © 2012 Pearson Education, Inc. Publishing as Addison-Wesley.

13. In this problem, we need the trigonometric identity $\sin^2 t = (1 - \cos 2t)/2$ and linearity of the Laplace transform.

$$\mathcal{L}\left\{\sin^2 t\right\}(s) = \mathcal{L}\left\{\frac{1 - \cos 2t}{2}\right\}(s)$$

$$= \frac{1}{2}[\mathcal{L}\left\{1\right\}(s) - \mathcal{L}\left\{\cos 2t\right\}(s)] = \frac{1}{2}\left(\frac{1}{s} - \frac{s}{s^2 + 4}\right) = \frac{2}{s(s^2 + 4)}.$$

15. From the trigonometric identity $\cos^2 t = (1 + \cos 2t)/2$, we find that

$$\cos^3 t = \cos t \cos^2 t = \frac{1}{2}\cos t + \frac{1}{2}\cos t \cos 2t.$$

Next, we write

$$\cos t \cos 2t = \frac{1}{2}[\cos(2t + t) + \cos(2t - t)] = \frac{1}{2}\cos 3t + \frac{1}{2}\cos t.$$

Thus,

$$\cos^3 t = \frac{1}{2}\cos t + \frac{1}{4}\cos 3t + \frac{1}{4}\cos t = \frac{3}{4}\cos t + \frac{1}{4}\cos 3t.$$

We now use linearity of the Laplace transform and Table 7.1 to find that

$$\mathcal{L}\left\{\cos^3 t\right\}(s) = \frac{3}{4}\mathcal{L}\left\{\cos t\right\}(s) + \frac{1}{4}\mathcal{L}\left\{\cos 3t\right\}(s) = \frac{3s}{4(s^2 + 1)} + \frac{s}{4(s^2 + 9)},$$

which holds for $s > 0$.

17. Since $\sin A \sin B = [\cos(A - B) - \cos(A + B)]/2$, we get

$$\mathcal{L}\left\{\sin 2t \sin 5t\right\}(s) = \mathcal{L}\left\{\frac{\cos 3t - \cos 7t}{2}\right\}(s) = \frac{1}{2}[\mathcal{L}\left\{\cos 3t\right\}(s) - \mathcal{L}\left\{\cos 7t\right\}(s)]$$

$$= \frac{1}{2}\left(\frac{s}{s^2 + 9} - \frac{s}{s^2 + 49}\right) = \frac{20s}{(s^2 + 9)(s^2 + 49)}.$$

19. Since $\sin A \cos B = [\sin(A + B) + \sin(A - B)]/2$, we get

$$\mathcal{L}\left\{\cos nt \sin mt\right\}(s) = \mathcal{L}\left\{\frac{\sin[(m + n)t] + \sin[(m - n)t]}{2}\right\}(s)$$

$$= \frac{m + n}{2[s^2 + (m + n)^2]} + \frac{m - n}{2[s^2 + (m - n)^2]}.$$

21. By the translation property of the Laplace transform (Theorem 3),

$$\mathcal{L}\left\{e^{at}\cos bt\right\}(s) = \mathcal{L}\left\{\cos bt\right\}(s - a) = \left.\frac{u}{u^2 + b^2}\right|_{u = s - a} = \frac{s - a}{(s - a)^2 + b^2}.$$

 Copyright © 2012 Pearson Education, Inc. Publishing as Addison-Wesley.

23. By the product rule,

$$(t \sin bt)' = (t)' \sin bt + t(\sin bt)' = \sin bt + bt \cos bt.$$

Therefore, using Theorem 4 and the entry 30 on the inside back cover of the text, that is, $\mathcal{L}\{t \sin bt\}(s) = (2bs)/[(s^2 + b^2)^2]$, we obtain

$$
\begin{aligned}
\mathcal{L}\{\sin bt + bt \cos bt\}(s) &= \mathcal{L}\{(t \sin bt)'\}(s) = s\mathcal{L}\{t \sin bt\}(s) - (t \sin bt)\big|_{t=0} \\
&= \frac{s(2bs)}{(s^2 + b^2)^2} - 0 = \frac{2bs^2}{(s^2 + b^2)^2}.
\end{aligned}
$$

25. (a) By Theorem 6,

$$\mathcal{L}\{t \cos bt\}(s) = -\left(\mathcal{L}\{\cos bt\}(s)\right)' = -\left(\frac{s}{s^2 + b^2}\right)' = \frac{s^2 - b^2}{(s^2 + b^2)^2}, \qquad s > 0.$$

(b) Using the same property, we get

$$
\begin{aligned}
\mathcal{L}\{t^2 \cos bt\}(s) &= \mathcal{L}\{t(t \cos bt)\}(s) = -\left[\mathcal{L}\{t \cos bt\}(s)\right]' \\
&= -\left[\frac{s^2 - b^2}{(s^2 + b^2)^2}\right]' = \frac{2s^3 - 6sb^2}{(s^2 + b^2)^3}, \qquad s > 0.
\end{aligned}
$$

27. First, we observe that, since $f(t)$ is piecewise continuous on $[0, \infty)$ and $f(t)/t$ has a finite limit, as $t \to 0^+$, $f(t)/t$ is also piecewise continuous on $[0, \infty)$. Next, for $t \geq 1$, $|f(t)/t| \leq |f(t)|$, and we see that $f(t)/t$ is of exponential order α because $f(t)$ is. These observations and Theorem 2, Section 7.2, show that $\mathcal{L}\{f(t)/t\}$ exists. When the result of Problem 26(b) is applied to $f(t)/t$, we see that

$$\lim_{N \to \infty} \mathcal{L}\left\{\frac{f(t)}{t}\right\}(N) = 0.$$

Theorem 6 yields

$$F(s) = \int_0^\infty e^{-st} f(t)\, dt = \int_0^\infty \frac{te^{-st} f(t)}{t}\, dt = -\frac{d}{ds}\mathcal{L}\left\{\frac{f(t)}{t}\right\}(s).$$

Thus,

$$
\begin{aligned}
\int_s^\infty F(u)\, du &= \int_s^\infty \left[-\frac{d}{du}\mathcal{L}\left\{\frac{f(t)}{t}\right\}(u)\right] du = \int_\infty^s \frac{d}{du}\mathcal{L}\left\{\frac{f(t)}{t}\right\}(u)\, du \\
&= \mathcal{L}\left\{\frac{f(t)}{t}\right\}(s) - \lim_{N \to \infty} \mathcal{L}\left\{\frac{f(t)}{t}\right\}(N) = \mathcal{L}\left\{\frac{f(t)}{t}\right\}(s).
\end{aligned}
$$

Copyright © 2012 Pearson Education, Inc. Publishing as Addison-Wesley.

Chapter 7

29. From linearity properties (2) and (3), Section 7.2, we have

$$\mathcal{L}\{g(t)\}(s) = \mathcal{L}\{y''(t) + 6y'(t) + 10y(t)\}(s)$$
$$= \mathcal{L}\{y''(t)\}(s) + 6\mathcal{L}\{y'(t)\}(s) + 10\mathcal{L}\{y(t)\}(s).$$

Next, applying properties (2) and (4) yields

$$\mathcal{L}\{g(t)\}(s) = \left[s^2\mathcal{L}\{y\}(s) - sy(0) - y'(0)\right] + 6\left[s\mathcal{L}\{y\}(s) - y(0)\right] + 10\mathcal{L}\{y\}(s).$$

Keeping in mind the fact that the initial conditions are zero, the equation becomes

$$G(s) = \left(s^2 + 6s + 10\right)Y(s), \quad \text{where} \quad Y(s) = \mathcal{L}\{y\}(s).$$

Therefore, the transfer function $H(s)$ is given by

$$H(s) = \frac{Y(s)}{G(s)} = \frac{1}{s^2 + 6s + 10}.$$

31. Using Definition 1 of the Laplace transform in Section 7.2, we obtain

$$\mathcal{L}\{g(t)\}(s) \;=\; \int_0^\infty e^{-st}g(t)\,dt = \int_0^c (0)\,dt + \int_c^\infty e^{-st}f(t-c)\,dt = \left(t - c \to u,\ dt \to du\right)$$
$$= \int_0^\infty e^{-s(u+c)}f(u)\,du = e^{-cs}\int_0^\infty e^{-su}f(u)\,du = e^{-cs}\mathcal{L}\{f(t)\}(s).$$

33. The graphs of the function $f(t) = t$ and its translation $g(t)$ to the right by $c = 1$ are shown in Fig. **7–A(a)** on page 455. We use the result of Problem 31 to find $\mathcal{L}\{g(t)\}$.

$$\mathcal{L}\{g(t)\}(s) = e^{-(1)s}\mathcal{L}\{t\}(s) = \frac{e^{-s}}{s^2}.$$

35. The graphs of the function $f(t) = \sin t$ and its translation $g(t)$ to the right by $c = \pi/2$ units are shown in Fig. **7–A(b)**, page 455. We use the formula in Problem 31 to find $\mathcal{L}\{g(t)\}$.

$$\mathcal{L}\{g(t)\}(s) = e^{-(\pi/2)s}\mathcal{L}\{\sin t\}(s) = \frac{e^{-(\pi/2)s}}{s^2 + 1}.$$

37. Since $f'(t)$ is of exponential order on $[0, \infty)$, for some α, $M > 0$, and $T > 0$,

$$|f'(t)| \le Me^{\alpha t} \quad \text{for all } t \ge T. \tag{7.7}$$

 Copyright © 2012 Pearson Education, Inc. Publishing as Addison-Wesley.

On the other hand, piecewise continuity of $f'(t)$ on $[0, \infty)$ implies that $f'(t)$ is bounded on any finite interval, in particular, on $[0, T]$. That is,

$$|f'(t)| \le C, \qquad \text{for all} \quad t \text{ in } [0, T]. \tag{7.8}$$

From (7.7) and (7.8) it follows that, for $s > \alpha$,

$$\int_0^\infty e^{-st}|f'(t)|\, dt = \int_0^T e^{-st}|f'(t)|\, dt + \int_T^\infty e^{-st}|f'(t)|\, dt \le C \int_0^T e^{-st}\, dt + M \int_T^\infty e^{-st}e^{\alpha t}\, dt$$

$$= \left. \frac{Ce^{-st}}{-s} \right|_0^T + \lim_{N \to \infty} \left[\left. \frac{Me^{(\alpha - s)t}}{\alpha - s} \right|_T^N \right] = \frac{C\left(1 - e^{-sT}\right)}{s} + \frac{Me^{(\alpha - s)T}}{s - \alpha} \to 0$$

as $s \to \infty$. Therefore, (7) yields

$$0 \le |s\mathcal{L}\{f\}(s) - f(0)| = |\mathcal{L}\{f'\}(s)| \le \int_0^\infty e^{-st}|f'(t)|\, dt \to 0 \quad \text{as } s \to \infty.$$

Hence, by the squeeze theorem, we get

$$\lim_{s \to \infty} [s\mathcal{L}\{f\}(s) - f(0)] = 0 \quad \Leftrightarrow \quad \lim_{s \to \infty} s\mathcal{L}\{f\}(s) = f(0).$$

EXERCISES 7.4:   Inverse Laplace Transform

1. From Table 7.1, Section 7.2, the function $6/(s-1)^4 = (3!)/(s-1)^4$ is the Laplace transform of $e^{\alpha t}t^n$ with $\alpha = 1$ and $n = 3$. Therefore,

$$\mathcal{L}^{-1}\left\{ \frac{6}{(s-1)^4} \right\}(t) = e^t t^3.$$

3. Writing

$$\frac{s+1}{s^2 + 2s + 10} = \frac{s+1}{(s^2 + 2s + 1) + 9} = \frac{s+1}{(s+1)^2 + 3^2},$$

we see that this function is the Laplace transform of $e^{-t}\cos 3t$ (the last entry in Table 7.1 with $\alpha = -1$ and $b = 3$). Hence,

$$\mathcal{L}^{-1}\left\{ \frac{s+1}{s^2 + 2s + 10} \right\}(t) = e^{-t}\cos 3t.$$

5. We complete the square in the denominator and use linearity of the inverse Laplace transform to get

$$\mathcal{L}^{-1}\left\{ \frac{1}{s^2 + 4s + 8} \right\}(t) = \mathcal{L}^{-1}\left\{ \frac{1}{(s+2)^2 + 2^2} \right\}(t)$$

Copyright © 2012 Pearson Education, Inc. Publishing as Addison-Wesley.

$$= \frac{1}{2} \mathcal{L}^{-1} \left\{ \frac{2}{(s+2)^2 + 2^2} \right\} (t) = \frac{e^{-2t} \sin 2t}{2}.$$

(See the Laplace transform of $e^{\alpha t} \sin bt$ in Table 7.1.)

7. By completing the square in the denominator, we can rewrite $(2s+16)/(s^2+4s+13)$ as

$$\frac{2s+16}{s^2+4s+4+9} = \frac{2s+16}{(s+2)^2+3^2} = \frac{2(s+2)}{(s+2)^2+3^2} + \frac{4(3)}{(s+2)^2+3^2}.$$

Thus, by linearity of the inverse Laplace transform,

$$\mathcal{L}^{-1} \left\{ \frac{2s+16}{s^2+4s+13} \right\} (t) = 2\mathcal{L}^{-1} \left\{ \frac{s+2}{(s+2)^2+3^2} \right\} (t) + 4\mathcal{L}^{-1} \left\{ \frac{3}{(s+2)^2+3^2} \right\} (t)$$

$$= 2e^{-2t} \cos 3t + 4e^{-2t} \sin 3t.$$

9. We complete the square in the denominator, rewrite the given function as a sum of two entries in Table 7.1, and use linearity of the inverse Laplace transform. This yields

$$\frac{3s-15}{2s^2-4s+10} = \frac{3}{2} \cdot \frac{s-5}{s^2-2s+5} = \frac{3}{2} \cdot \frac{(s-1)-4}{(s-1)^2+2^2} = \frac{(3/2)(s-1)}{(s-1)^2+2^2} - \frac{3(2)}{(s-1)^2+2^2}$$

$$\Rightarrow \quad \mathcal{L}^{-1} \left\{ \frac{3s-15}{2s^2-4s+10} \right\} = \frac{3}{2} \mathcal{L}^{-1} \left\{ \frac{s-1}{(s-1)^2+2^2} \right\} - 3\mathcal{L}^{-1} \left\{ \frac{2}{(s-1)^2+2^2} \right\}$$

$$= \frac{3}{2} e^t \cos 2t - 3e^t \sin 2t.$$

11. In this problem, we use the method of partial fractions decomposition. Since the denominator, $(s-1)(s+2)(s+5)$, is a product of three non-repeated linear factors, the expansion has the form

$$\frac{s^2-26s-47}{(s-1)(s+2)(s+5)} = \frac{A}{s-1} + \frac{B}{s+2} + \frac{C}{s+5}$$

$$= \frac{A(s+2)(s+5) + B(s-1)(s+5) + C(s-1)(s+2)}{(s-1)(s+2)(s+5)}.$$

Therefore,

$$s^2 - 26s - 47 = A(s+2)(s+5) + B(s-1)(s+5) + C(s-1)(s+2). \tag{7.9}$$

Evaluating both sides of (7.9) for $s = 1$, $s = -2$, and $s = -5$, we find constants A, B, and C.

$$s = 1: \quad (1)^2 - 26(1) - 47 = A(1+2)(1+5) \quad \Rightarrow \quad A = -4,$$
$$s = -2: \quad (-2)^2 - 26(-2) - 47 = B(-2-1)(-2+5) \quad \Rightarrow \quad B = -1,$$
$$s = -5: \quad (-5)^2 - 26(-5) - 47 = C(-5-1)(-5+2) \quad \Rightarrow \quad C = 6.$$

Copyright © 2012 Pearson Education, Inc. Publishing as Addison-Wesley.

Hence,

$$\frac{s^2 - 26s - 47}{(s-1)(s+2)(s+5)} = \frac{6}{s+5} - \frac{1}{s+2} - \frac{4}{s-1}.$$

13. The denominator has a simple linear factor s and a double linear factor $s+1$. Thus, we look for the decomposition of the form

$$\frac{-2s^2 - 3s - 2}{s(s+1)^2} = \frac{A}{s} + \frac{B}{s+1} + \frac{C}{(s+1)^2} = \frac{A(s+1)^2 + Bs(s+1) + Cs}{s(s+1)^2},$$

which yields

$$-2s^2 - 3s - 2 = A(s+1)^2 + Bs(s+1) + Cs. \tag{7.10}$$

Evaluating this identity for $s = 0$ and $s = -1$, we find A and C, respectively.

$$s = 0: \quad -2 = A(0+1)^2 \qquad\qquad \Rightarrow \qquad A = -2,$$
$$s = -1: \quad -2(-1)^2 - 3(-1) - 2 = C(-1) \quad \Rightarrow \qquad C = 1.$$

To find B, we compare the coefficients at s^2 in both sides of (7.10).

$$-2 = A + B \qquad \Rightarrow \qquad B = -2 - A = 0.$$

Hence,

$$\frac{-2s^2 - 3s - 2}{s(s+1)^2} = \frac{1}{(s+1)^2} - \frac{2}{s}.$$

15. First, we complete the square in the quadratic polynomial $s^2 - 2s + 5$ to make sure that this polynomial is irreducible and to find the form of the decomposition. Since

$$s^2 - 2s + 5 = (s^2 - 2s + 1) + 4 = (s-1)^2 + 2^2,$$

we have

$$\frac{-8s - 2s^2 - 14}{(s+1)(s^2 - 2s + 5)} = \frac{A}{s+1} + \frac{B(s-1) + C(2)}{(s-1)^2 + 2^2}$$
$$= \frac{A\left[(s-1)^2 + 4\right] + \left[B(s-1) + 2C\right](s+1)}{(s+1)\left[(s-1)^2 + 4\right]},$$

which yields

$$-8s - 2s^2 - 14 = A\left[(s-1)^2 + 4\right] + \left[B(s-1) + 2C\right](s+1).$$

Taking $s = -1$, $s = 1$, and $s = 0$, we find A, B, and C, respectively.

$$s = -1: \quad 8(-1) - 2(-1)^2 - 14 = A\left[(-1-1)^2 + 4\right] \qquad\qquad \Rightarrow A = -3,$$
$$s = 1: \quad 8(1) - 2(1)^2 - 14 = A\left[(1-1)^2 + 4\right] + 2C(1+1) \qquad\qquad \Rightarrow C = 1,$$
$$s = 0: \quad 8(0) - 2(0)^2 - 14 = A\left[(0-1)^2 + 4\right] + \left[B(0-1) + 2C\right](0+1) \Rightarrow B = 1,$$

Copyright © 2012 Pearson Education, Inc. Publishing as Addison-Wesley.

and so

$$\frac{-8s - 2s^2 - 14}{(s+1)(s^2 - 2s + 5)} = -\frac{3}{s+1} + \frac{(s-1)+2}{(s-1)^2 + 4}.$$

17. First, we need to factor the denominator. Since $s^2 + s - 6 = (s-2)(s+3)$, we have

$$\frac{3s+5}{s(s^2 + s - 6)} = \frac{3s+5}{s(s-2)(s+3)}.$$

Since the denominator has only non-repeated linear factors, we can write

$$\frac{3s+5}{s(s-2)(s+3)} = \frac{A}{s} + \frac{B}{s-2} + \frac{C}{s+3}.$$

Clearing fractions gives us

$$3s + 5 = A(s-2)(s+3) + Bs(s+3) + Cs(s-2).$$

With $s = 0$, this yields $5 = A(-2)(3)$ so that $A = -5/6$. Taking $s = 2$, we get $11 = B(2)(5)$ so that $B = 11/10$. Finally, substituting $s = -3$ yields $-4 = C(-3)(-5)$ so that $C = -4/15$. Thus,

$$\frac{3s+5}{s(s^2 + s - 6)} = -\frac{5}{6s} + \frac{11}{10(s-2)} - \frac{4}{15(s+3)}.$$

19. First, observe that the quadratic polynomial $s^2 + 2s + 2$ is irreducible, because its discriminant $2^2 - 4(1)(2) = -4$ is negative. Since the denominator has one non-repeated linear factor and one non-repeated quadratic factor, we can write

$$\frac{1}{(s-3)(s^2 + 2s + 2)} = \frac{1}{(s-3)[(s+1)^2 + 1]} = \frac{A}{s-3} + \frac{B(s+1)+C}{(s+1)^2 + 1}.$$

Clearing fractions, we get

$$1 = A\left[(s+1)^2 + 1\right] + [B(s+1) + C](s - 3). \tag{7.11}$$

With $s = 3$, this yields $1 = 17A$ so that $A = 1/17$. Substituting $s = -1$, we see that $1 = A(1) + C(-4)$, or $C = (A - 1)/4 = -4/17$. Finally, the coefficient $A + B$ at s^2 in the right-hand side of (7.11) must be the same as that in the left-hand side, that is, 0. Thus, $B = -A = -1/17$ and

$$\frac{1}{(s-3)(s^2 + 2s + 2)} = \frac{1}{17}\left[\frac{1}{s-3} - \frac{s+1}{(s+1)^2 + 1} - \frac{4}{(s+1)^2 + 1}\right].$$

Copyright © 2012 Pearson Education, Inc. Publishing as Addison-Wesley.

21. Since the denominator contains only non-repeated linear factors, the partial fractions decomposition has the form

$$\frac{6s^2 - 13s + 2}{s(s-1)(s-6)} = \frac{A}{s} + \frac{B}{s-1} + \frac{C}{s-6} = \frac{A(s-1)(s-6) + Bs(s-6) + Cs(s-1)}{s(s-1)(s-6)}.$$

Therefore,

$$6s^2 - 13s + 2 = A(s-1)(s-6) + Bs(s-6) + Cs(s-1).$$

Evaluating both sides of this equation for $s = 0$, $s = 1$, and $s = 6$, we find constants A, B, and C.

$$\begin{aligned}
s = 0: & \quad 2 = 6A & \Rightarrow & \quad A = 1/3, \\
s = 1: & \quad -5 = -5B & \Rightarrow & \quad B = 1, \\
s = 6: & \quad 140 = 30C & \Rightarrow & \quad C = 14/3.
\end{aligned}$$

Hence,

$$\frac{6s^2 - 13s + 2}{s(s-1)(s-6)} = \frac{1/3}{s} + \frac{1}{s-1} + \frac{14/3}{s-6},$$

and the linear property of the inverse Laplace transform yields

$$\mathcal{L}^{-1}\left\{\frac{6s^2 - 13s + 2}{s(s-1)(s-6)}\right\} = \frac{1}{3}\mathcal{L}^{-1}\left\{\frac{1}{s}\right\} + \mathcal{L}^{-1}\left\{\frac{1}{s-1}\right\} + \frac{14}{3}\mathcal{L}^{-1}\left\{\frac{1}{s-6}\right\} = \frac{1}{3} + e^t + \frac{14}{3}e^{6t}.$$

23. In this problem, the denominator of $F(s)$ has a simple linear factor $s+1$ and a double linear factor $s+3$. Thus, the decomposition is the form

$$\frac{5s^2 + 34s + 53}{(s+3)^2(s+1)} = \frac{A}{(s+3)^2} + \frac{B}{s+3} + \frac{C}{s+1} = \frac{A(s+1) + B(s+1)(s+3) + C(s+3)^2}{(s+3)^2(s+1)}.$$

Therefore, we have

$$5s^2 + 34s + 53 = A(s+1) + B(s+1)(s+3) + C(s+3)^2.$$

Substitutions $s = -3$ and $s = -1$ yields values of A and C, respectively.

$$\begin{aligned}
s = -3: & \quad -4 = -2A & \Rightarrow & \quad A = 2, \\
s = -1: & \quad 24 = 4C & \Rightarrow & \quad C = 6.
\end{aligned}$$

To find B, we take, say, $s = 0$ and get

$$53 = A + 3B + 9C \quad \Rightarrow \quad B = \frac{53 - A - 9C}{3} = -1.$$

Hence,

$$\mathcal{L}^{-1}\left\{\frac{5s^2 + 34s + 53}{(s+3)^2(s+1)}\right\}(t) = 2\mathcal{L}^{-1}\left\{\frac{1}{(s+3)^2}\right\}(t) - \mathcal{L}^{-1}\left\{\frac{1}{s+3}\right\}(t) + 6\mathcal{L}^{-1}\left\{\frac{1}{s+1}\right\}(t)$$

$$= 2te^{-3t} - e^{-3t} + 6e^{-t}.$$

Copyright © 2012 Pearson Education, Inc. Publishing as Addison-Wesley.

25. Observing that the quadratic $s^2 + 2s + 5 = (s+1)^2 + 2^2$ is irreducible, we conclude that the partial fractions decomposition for $F(s)$ has the form

$$\frac{7s^2 + 23s + 30}{(s-2)(s^2 + 2s + 5)} = \frac{A}{s-2} + \frac{B(s+1) + C(2)}{(s+1)^2 + 2^2}.$$

Clearing fractions gives us

$$7s^2 + 23s + 30 = A\left[(s+1)^2 + 4\right] + \left[B(s+1) + C(2)\right](s-2).$$

With $s = 2$, this yields $104 = 13A$ so that $A = 8$; $s = -1$ gives $14 = A(4) + C(-6)$, or $C = 3$. Finally, the coefficient $A + B$ at s^2 in the right-hand side must match that in the left-hand side, which is 7. So, $B = 7 - A = -1$. Therefore,

$$\frac{7s^2 + 23s + 30}{(s-2)(s^2 + 2s + 5)} = \frac{8}{s-2} + \frac{-(s+1) + 3(2)}{(s+1)^2 + 2^2},$$

which yields

$$\mathcal{L}^{-1}\left\{\frac{7s^2 + 23s + 30}{(s-2)(s^2 + 2s + 5)}\right\} = 8\mathcal{L}^{-1}\left\{\frac{1}{s-2}\right\} - \mathcal{L}^{-1}\left\{\frac{s+1}{(s+1)^2 + 2^2}\right\}$$

$$+ 3\mathcal{L}^{-1}\left\{\frac{2}{(s+1)^2 + 2^2}\right\} = 8e^{2t} - e^{-t}\cos 2t + 3e^{-t}\sin 2t.$$

27. First, we find $F(s)$.

$$F(s)\left(s^2 - 4\right) = \frac{5}{s+1} \quad \Rightarrow \quad F(s) = \frac{5}{(s+1)(s^2 - 4)} = \frac{5}{(s+1)(s-2)(s+2)}.$$

The partial fractions expansion yields

$$\frac{5}{(s+1)(s-2)(s+2)} = \frac{A}{s+1} + \frac{B}{s-2} + \frac{C}{s+2}.$$

Clearing fractions gives us

$$5 = A(s-2)(s+2) + B(s+1)(s+2) + C(s+1)(s-2).$$

With $s = -1$, $s = 2$, and $s = -2$ this yields $A = -5/3$, $B = 5/12$, and $C = 5/4$. So,

$$\mathcal{L}^{-1}\{F(s)\}(t) = -\frac{5}{3}\mathcal{L}^{-1}\left\{\frac{1}{s+1}\right\}(t) + \frac{5}{12}\mathcal{L}^{-1}\left\{\frac{1}{s-2}\right\}(t) + \frac{5}{4}\mathcal{L}^{-1}\left\{\frac{1}{s+2}\right\}(t)$$

$$= -\frac{5}{3}e^{-t} + \frac{5}{12}e^{2t} + \frac{5}{4}e^{-2t}.$$

 Copyright © 2012 Pearson Education, Inc. Publishing as Addison-Wesley.

29. Solving for $F(s)$ yields

$$F(s) = \frac{10s^2 + 12s + 14}{(s+2)(s^2 - 2s + 2)} = \frac{10s^2 + 12s + 14}{(s+2)[(s-1)^2 + 1]}.$$

Since, in the denominator, we have non-repeated linear and quadratic factors, we seek for the decomposition

$$\frac{10s^2 + 12s + 14}{(s+2)[(s-1)^2 + 1]} = \frac{A}{s+2} + \frac{B(s-1) + C(1)}{(s-1)^2 + 1}.$$

Clearing fractions, we conclude that

$$10s^2 + 12s + 14 = A[(s-1)^2 + 1] + [B(s-1) + C](s+2).$$

Substitution $s = -2$ into this equation yields $30 = 10A$ or $A = 3$. With $s = 1$, we get $36 = A + 3C$ and so $C = (36 - A)/3 = 11$. Finally, substitution $s = 0$ results $14 = 2A + 2(C - B)$ or $B = A + C - 7 = 7$. Now we apply the linearity of the inverse Laplace transform and obtain

$$\mathcal{L}^{-1}\{F(s)\}(t) = 3\mathcal{L}^{-1}\left\{\frac{1}{s+2}\right\}(t) + 7\mathcal{L}^{-1}\left\{\frac{s-1}{(s-1)^2 + 1}\right\}(t)$$

$$+ 11\mathcal{L}^{-1}\left\{\frac{1}{(s-1)^2 + 1}\right\}(t) = 3e^{-2t} + 7e^t \cos t + 11e^t \sin t.$$

31. Functions $f_1(t)$, $f_2(t)$, and $f_3(t)$ coincide for all t in $[0, \infty)$ except a finite number of points. Since the Laplace transform a function is a definite integral, it does not depend on values of the function at a finite number of points. Therefore, in (a), (b), and (c) we have one and the same answer, that is,

$$\mathcal{L}\{f_1(t)\}(s) = \mathcal{L}\{f_2(t)\}(s) = \mathcal{L}\{f_3(t)\}(s) = \mathcal{L}\{t\}(s) = \frac{1}{s^2}.$$

By Definition 4, the inverse Laplace transform is a continuous function on $[0, \infty)$. The function $f_3(t) = t$ clearly satisfies this condition, while $f_1(t)$ and $f_2(t)$ have removable discontinuities at $t = 2$ and $t = 1$, $t = 6$, respectively. Therefore,

$$\mathcal{L}^{-1}\left\{\frac{1}{s^2}\right\}(t) = f_3(t) = t.$$

33. We are looking for $\mathcal{L}^{-1}\{F(s)\}(t) = f(t)$. According to the formula given in the text (with $n = 1$),

$$f(t) = \frac{-1}{t}\mathcal{L}^{-1}\left\{\frac{dF}{ds}\right\}(t)$$

Copyright © 2012 Pearson Education, Inc. Publishing as Addison-Wesley.

Since

$$F(s) = \ln\left(\frac{s+2}{s-5}\right) = \ln(s+2) - \ln(s-5),$$

we have

$$\frac{dF(s)}{ds} = \frac{d}{ds}\left[\ln(s+2) - \ln(s-5)\right] = \frac{1}{s+2} - \frac{1}{s-5}$$

$$\Rightarrow \quad \mathcal{L}^{-1}\left\{\frac{dF}{ds}\right\}(t) = \mathcal{L}^{-1}\left\{\frac{1}{s+2} - \frac{1}{s-5}\right\}(t) = e^{-2t} - e^{5t}$$

$$\Rightarrow \quad \mathcal{L}^{-1}\left\{F(s)\right\}(t) = \frac{-1}{t}\left(e^{-2t} - e^{5t}\right) = \frac{e^{5t} - e^{-2t}}{t}.$$

35. Taking the derivative of $F(s)$, we get

$$\frac{dF(s)}{ds} = \frac{d}{ds}\left(\ln\frac{s^2+9}{s^2+1}\right) = \frac{d}{ds}\left[\ln(s^2+9) - \ln(s^2+1)\right] = \frac{2s}{s^2+9} - \frac{2s}{s^2+1}.$$

So, using the linear property of the inverse Laplace transform, we obtain

$$\mathcal{L}^{-1}\left\{\frac{dF(s)}{ds}\right\}(t) = 2\mathcal{L}^{-1}\left\{\frac{s}{s^2+9}\right\}(t) - 2\mathcal{L}^{-1}\left\{\frac{s}{s^2+1}\right\}(t) = 2(\cos 3t - \cos t).$$

Thus,

$$\mathcal{L}^{-1}\left\{F(s)\right\}(t) = \frac{-1}{t}\mathcal{L}^{-1}\left\{\frac{dF(s)}{ds}\right\}(t) = \frac{2(\cos t - \cos 3t)}{t}.$$

37. By the definition, $\mathcal{L}^{-1}\left\{F_1\right\}(t)$ and $\mathcal{L}^{-1}\left\{F_2\right\}(t)$ are continuous functions on $[0,\infty)$. Therefore, their sum, $\left(\mathcal{L}^{-1}\left\{F_1\right\} + \mathcal{L}^{-1}\left\{F_2\right\}\right)(t)$, is also continuous on $[0,\infty)$. Furthermore, linearity of the Laplace transform yields

$$\mathcal{L}\left\{\left(\mathcal{L}^{-1}\left\{F_1\right\} + \mathcal{L}^{-1}\left\{F_2\right\}\right)\right\}(s) = \mathcal{L}\left\{\mathcal{L}^{-1}\left\{F_1\right\}\right\}(s) + \mathcal{L}\left\{\mathcal{L}^{-1}\left\{F_2\right\}\right\}(s) = F_1(s) + F_2(s).$$

Therefore, $\mathcal{L}^{-1}\left\{F_1\right\} + \mathcal{L}^{-1}\left\{F_2\right\}$ is a continuous function on $[0,\infty)$, whose Laplace transform is $F_1 + F_2$. By the definition of the inverse Laplace transform, this function is the inverse Laplace transform of $F_1 + F_2$, that is,

$$\mathcal{L}^{-1}\left\{F_1\right\}(t) + \mathcal{L}^{-1}\left\{F_2\right\}(t) = \mathcal{L}^{-1}\left\{F_1 + F_2\right\}(t),$$

and (3) in Theorem 7 is proved.

To show (4), we use the continuity of $\mathcal{L}^{-1}\left\{F\right\}$ to conclude that $c\mathcal{L}^{-1}\left\{F\right\}$ is a continuous function. Since the linearity of the Laplace transform yields

$$\mathcal{L}\left\{c\mathcal{L}^{-1}\left\{F\right\}\right\}(s) = c\mathcal{L}\left\{\mathcal{L}^{-1}\left\{F\right\}\right\}(s) = cF(s),$$

we have $c\mathcal{L}^{-1}\left\{F\right\}(t) = \mathcal{L}^{-1}\left\{cF\right\}(t)$.

 Copyright © 2012 Pearson Education, Inc. Publishing as Addison-Wesley.

39. In this problem, the denominator $Q(s) := s(s-1)(s+2)$ has only non-repeated linear factors, and so the partial fractions decomposition has the form

$$F(s) := \frac{2s+1}{s(s-1)(s+2)} = \frac{A}{s} + \frac{B}{s-1} + \frac{C}{s+2}.$$

To find A, B, and C, we use the residue formula, given in Problem 38. This yields

$$A = \lim_{s\to 0} sF(s) = \lim_{s\to 0} \frac{2s+1}{(s-1)(s+2)} = \frac{2(0)+1}{(0-1)(0+2)} = -\frac{1}{2},$$

$$B = \lim_{s\to 1}(s-1)F(s) = \lim_{s\to 1}\frac{2s+1}{s(s+2)} = \frac{2(1)+1}{(1)(1+2)} = 1,$$

$$C = \lim_{s\to -2}(s+2)F(s) = \lim_{s\to 2}\frac{2s+1}{s(s-1)} = \frac{2(-2)+1}{(-2)(-2-1)} = -\frac{1}{2}.$$

Therefore,

$$\frac{2s+1}{s(s-1)(s+2)} = -\frac{1/2}{s} + \frac{1}{s-1} - \frac{1/2}{s+2}.$$

41. In notation of Problem 40,

$$P(s) = 3s^2 - 16s + 5, \qquad Q(s) = (s+1)(s-3)(s-2).$$

We can apply the Heaviside's expansion formula, because $Q(s)$ has only nonrepeated linear factors. We need the values of $P(s)$ and $Q'(s)$ at the points $r_1 = -1$, $r_2 = 3$, and $r_3 = 2$. Using the product rule, we find that

$$Q'(s) = (s-3)(s-2) + (s+1)(s-2) + (s+1)(s-3),$$

and so

$$Q'(-1) = (-1-3)(-1-2) = 12, \quad Q'(3) = (3+1)(3-2) = 4,$$

$$Q'(2) = (2+1)(2-3) = -3.$$

Also, we compute

$$P(-1) = 24, \quad P(3) = -16, \quad P(2) = -15.$$

Therefore,

$$\mathcal{L}^{-1}\left\{\frac{3s^2 - 16s + 5}{(s+1)(s-3)(s-2)}\right\}(t) = \frac{P(-1)}{Q'(-1)}e^{(-1)t} + \frac{P(3)}{Q'(3)}e^{(3)t} + \frac{P(2)}{Q'(2)}e^{(2)t}$$

$$= 2e^{-t} - 4e^{3t} + 5e^{2t}.$$

Copyright © 2012 Pearson Education, Inc. Publishing as Addison-Wesley.

43. Since $s^2 - 2s + 5 = (s-1)^2 + 2^2$, we see that the denominator of $F(s)$ has non-repeated linear factor $s + 2$ and nonrepeated irreducible quadratic factor $s^2 - 2s + 5$ with $\alpha = 1$ and $\beta = 2$ (in notation of Problem 40). Thus, the partial fractions decomposition has the form

$$F(s) = \frac{6s^2 + 28}{(s^2 - 2s + 5)(s+2)} = \frac{A(s-1) + 2B}{(s-1)^2 + 2^2} + \frac{C}{s+2}.$$

We find C by applying the real residue formula derived in Problem 38.

$$C = \lim_{s \to -2} \frac{(s+2)(6s^2 + 28)}{(s^2 - 2s + 5)(s+2)} = \lim_{s \to -2} \frac{6s^2 + 28}{s^2 - 2s + 5} = \frac{52}{13} = 4.$$

Next, we use the complex residue formula, given in Problem 42, to find A and B. Since $\alpha = 1$ and $\beta = 2$, the formula becomes

$$2B + 2Ai = \lim_{s \to 1 + 2i} \frac{(s^2 - 2s + 5)(6s^2 + 28)}{(s^2 - 2s + 5)(s+2)}$$

$$= \lim_{s \to 1 + 2i} \frac{6s^2 + 28}{s+2} = \frac{6(1 + 2i)^2 + 28}{(1 + 2i) + 2} = \frac{10 + 24i}{3 + 2i}.$$

Dividing, we get

$$2B + i2A = \frac{(10 + 24i)(3 - 2i)}{(3 + 2i)(3 - 2i)} = \frac{78 + 52i}{13} = 6 + 4i.$$

Taking the real and imaginary parts yields

$$\begin{array}{ccc} 2B = 6, & & B = 3, \\ 2A = 4 & \Rightarrow & A = 2. \end{array}$$

Therefore,

$$\frac{6s^2 + 28}{(s^2 - 2s + 5)(s+2)} = \frac{2(s-1) + 2(3)}{(s-1)^2 + 2^2} + \frac{4}{s+2}.$$

EXERCISES 7.5:   Solving Initial Value Problems

1. Let $Y(s) := \mathcal{L}\{y\}(s)$. Taking the Laplace transform of both sides of the given differential equation and using its linearity, we obtain

$$\mathcal{L}\{y''\}(s) - 2\mathcal{L}\{y'\}(s) + 5Y(s) = \mathcal{L}\{0\}(s) = 0. \tag{7.12}$$

We can express $\mathcal{L}\{y''\}(s)$ and $\mathcal{L}\{y'\}(s)$ in terms of $Y(s)$ using the initial conditions and Theorem 5 of Section 7.3.

$$\mathcal{L}\{y'\}(s) = sY(s) - y(0) = sY(s) - 2,$$

   Copyright © 2012 Pearson Education, Inc. Publishing as Addison-Wesley.

$$\mathcal{L}\{y''\}(s) = s^2 Y(s) - sy(0) - y'(0) = s^2 Y(s) - 2s - 4.$$

Substituting back into (7.12) and solving for $Y(s)$ yields

$$\left[s^2 Y(s) - 2s - 4\right] - 2\left[sY(s) - 2\right] + 5Y(s) = 0$$

$$\Rightarrow \quad Y(s)\left(s^2 - 2s + 5\right) = 2s\,.$$

Thus,

$$Y(s) = \frac{2s}{s^2 - 2s + 5} = \frac{2s}{(s-1)^2 + 2^2} = \frac{2(s-1)}{(s-1)^2 + 2^2} + \frac{2}{(s-1)^2 + 2^2}\,.$$

Applying now the inverse Laplace transform to both sides, we obtain

$$y(t) = 2\mathcal{L}^{-1}\left\{\frac{s-1}{(s-1)^2 + 2^2}\right\}(t) + \mathcal{L}^{-1}\left\{\frac{2}{(s-1)^2 + 2^2}\right\}(t) = 2e^t \cos 2t + e^t \sin 2t.$$

3. Let $Y(s) := \mathcal{L}\{y\}(s)$. Taking the Laplace transform of both sides of the given differential equation,

$$y'' + 6y' + 9y = 0\,,$$

and using linearity of the Laplace transform, we get

$$\mathcal{L}\{y''\}(s) + 6\mathcal{L}\{y'\}(s) + 9Y(s) = 0.$$

By formula (4), Section 7.3, we have

$$\mathcal{L}\{y'\}(s) = sY(s) - y(0) = sY(s) + 1,$$

$$\mathcal{L}\{y''\}(s) = s^2 Y(s) - sy(0) - y'(0) = s^2 Y(s) + s - 6.$$

Therefore,

$$\left[s^2 Y(s) + s - 6\right] + 6\left[sY(s) + 1\right] + 9Y(s) = 0$$

$$\Rightarrow \quad Y(s)\left(s^2 + 6s + 9\right) + s = 0$$

$$\Rightarrow \quad Y(s) = \frac{-s}{s^2 + 6s + 9} = \frac{-s}{(s+3)^2} = \frac{3}{(s+3)^2} - \frac{1}{s+3}\,,$$

where the last equality comes from the partial fraction expansion of $-s/(s+3)^2$. We apply the inverse Laplace transform to both sides and use Table 7.1 to obtain

$$y(t) = 3\mathcal{L}^{-1}\left\{\frac{1}{(s+3)^2}\right\}(t) - \mathcal{L}^{-1}\left\{\frac{1}{s+3}\right\}(t) = 3te^{-3t} - e^{-3t}\,.$$

Copyright © 2012 Pearson Education, Inc. Publishing as Addison-Wesley.

Chapter 7

5. Let $W(s) = \mathcal{L}\{w\}(s)$. Then taking the Laplace transform of both sides of the equation and using its linearity yields

$$\mathcal{L}\{w''\}(s) + W(s) = \mathcal{L}\{t^2 + 2\}(s) = \mathcal{L}\{t^2\}(s) + 2\mathcal{L}\{1\}(s) = \frac{2}{s^3} + \frac{2}{s}.$$

Since $\mathcal{L}\{w''\}(s) = s^2 W(s) - sw(0) - w'(0) = s^2 W(s) - s + 1$, we have

$$[s^2 W(s) - s + 1] + W(s) = \frac{2}{s^3} + \frac{2}{s}$$

$$\Rightarrow \quad (s^2 + 1) W(s) = s - 1 + \frac{2(s^2 + 1)}{s^3} \quad \Rightarrow \quad W(s) = \frac{s}{s^2 + 1} - \frac{1}{s^2 + 1} + \frac{2}{s^3}.$$

Now, taking the inverse Laplace transform, we obtain

$$w = \mathcal{L}^{-1}\left\{\frac{s}{s^2 + 1}\right\} - \mathcal{L}^{-1}\left\{\frac{1}{s^2 + 1}\right\} + \mathcal{L}^{-1}\left\{\frac{2}{s^3}\right\} = \cos t - \sin t + t^2.$$

7. Let $Y(s) := \mathcal{L}\{y\}(s)$. Using the initial conditions and Theorem 5 in Section 7.3 we can express $\mathcal{L}\{y''\}(s)$ and $\mathcal{L}\{y'\}(s)$ in terms of $Y(s)$, namely,

$$\mathcal{L}\{y'\}(s) = sY(s) - y(0) = sY(s) - 5,$$
$$\mathcal{L}\{y''\}(s) = s^2 Y(s) - sy(0) - y'(0) = s^2 Y(s) - 5s + 4.$$

Taking the Laplace transform of both sides of the given differential equation and using its linearity, we obtain

$$\mathcal{L}\{y'' - 7y' + 10y\}(s) = \mathcal{L}\{9\cos t + 7\sin t\}(s)$$

$$\Rightarrow \quad [s^2 Y(s) - 5s + 4] - 7[sY(s) - 5] + 10Y(s) = \frac{9s}{s^2 + 1} + \frac{7}{s^2 + 1}$$

$$\Rightarrow \quad (s^2 - 7s + 10) Y(s) = \frac{9s + 7}{s^2 + 1} + 5s - 39 = \frac{5s^3 - 39s^2 + 14s - 32}{s^2 + 1}$$

$$\Rightarrow \quad Y(s) = \frac{9s + 7}{s^2 + 1} + 5s - 39 = \frac{5s^3 - 39s^2 + 14s - 32}{(s^2 + 1)(s^2 - 7s + 10)} = \frac{5s^3 - 39s^2 + 14s - 32}{(s^2 + 1)(s - 5)(s - 2)}.$$

The partial fractions decomposition of $Y(s)$ has the form

$$\frac{5s^3 - 39s^2 + 14s - 32}{(s^2 + 1)(s - 5)(s - 2)} = \frac{As + B}{s^2 + 1} + \frac{C}{s - 5} + \frac{D}{s - 2}.$$

Clearing fractions yields

$$5s^3 - 39s^2 + 14s - 32 = (As + B)(s - 5)(s - 2) + C(s^2 + 1)(s - 2) + D(s^2 + 1)(s - 5).$$

Copyright © 2012 Pearson Education, Inc. Publishing as Addison-Wesley.

We substitute $s = 5$ and $s = 2$ to find C and D, respectively, and then $s = 0$ to find B.

$$s = 5: \quad -312 = 78C \qquad \Rightarrow \qquad C = -4,$$
$$s = 2: \quad -120 = -15D \qquad \Rightarrow \qquad D = 8,$$
$$s = 0: \quad -32 = 10B - 2C - 5D \qquad \Rightarrow \qquad B = 0.$$

Equating the coefficients at s^3, we also get $A + C + D = 5$, which implies that $A = 1$. Thus,

$$Y(s) = \frac{s}{s^2 + 1} - \frac{4}{s - 5} + \frac{8}{s - 2} \qquad \Rightarrow \qquad y(t) = \mathcal{L}^{-1}\left\{Y(s)\right\}(t) = \cos t - 4e^{5t} + 8e^{2t}.$$

9. First, note that the initial conditions are given at $t = 1$. Thus, to use the method of Laplace transform, we make a shift in t and move the initial conditions to $t = 0$.

$$z''(t) + 5z'(t) - 6z(t) = 21e^{t-1}$$
$$\Rightarrow \quad z''(t + 1) + 5z'(t + 1) - 6z(t + 1) = 21e^{(t+1)-1} = 21e^{t}. \qquad (7.13)$$

Now, let $y(t) := z(t + 1)$. Then the chain rule yields

$$y'(t) = z'(t + 1)(t + 1)' = z'(t + 1),$$
$$y''(t) = [y'(t)]' = z''(t + 1)(t + 1)' = z''(t + 1),$$

and (7.13) becomes

$$y''(t) + 5y'(t) - 6y(t) = 21e^{t} \qquad (7.14)$$

with initial conditions

$$y(0) = z(0 + 1) = z(1) = -1, \qquad y'(0) = z'(0 + 1) = z'(1) = 9.$$

With $Y(s) := \mathcal{L}\left\{y(t)\right\}(s)$, we apply the Laplace transform to both sides of (7.14) and obtain

$$\mathcal{L}\left\{y''\right\}(s) + 5\mathcal{L}\left\{y'\right\}(s) - 6Y(s) = \mathcal{L}\left\{21e^{t}\right\}(s) = \frac{21}{s - 1}. \qquad (7.15)$$

By Theorem 5, Section 7.3,

$$\mathcal{L}\left\{y'\right\}(s) = sY(s) - y(0) = sY(s) + 1,$$
$$\mathcal{L}\left\{y''\right\}(s) = s^2Y(s) - sy(0) - y'(0) = s^2Y(s) + s - 9.$$

Substituting these expressions back into (7.15) and solving for $Y(s)$ yields

$$\left[s^2Y(s) + s - 9\right] + 5\left[sY(s) + 1\right] - 6Y(s) = \frac{21}{s - 1}$$

Copyright © 2012 Pearson Education, Inc. Publishing as Addison-Wesley.

$$\Rightarrow \qquad \left(s^2 + 5s - 6\right) Y(s) = \frac{21}{s-1} - s + 4 = \frac{-s^2 + 5s + 17}{s - 1}$$

$$\Rightarrow \qquad Y(s) = \frac{-s^2 + 5s + 17}{(s-1)(s^2 + 5s - 6)} = \frac{-s^2 + 5s + 17}{(s-1)(s-1)(s+6)} = \frac{-s^2 + 5s + 17}{(s-1)^2(s+6)}.$$

The partial fractions decomposition for $Y(s)$ has the form

$$\frac{-s^2 + 5s + 17}{(s-1)^2(s+6)} = \frac{A}{(s-1)^2} + \frac{B}{s-1} + \frac{C}{s+6}.$$

Clearing fractions yields

$$-s^2 + 5s + 17 = A(s+6) + B(s-1)(s+6) + C(s-1)^2.$$

Substitutions $s = 1$ and $s = -6$ give $A = 3$ and $C = -1$. Also, with $s = 0$, we have $17 = 6A - 6B + C$ or $B = 0$. Therefore,

$$Y(s) = \frac{3}{(s-1)^2} - \frac{1}{s+6} \qquad \Rightarrow \qquad y(t) = \mathcal{L}^{-1}\left\{\frac{3}{(s-1)^2} - \frac{1}{s+6}\right\}(t) = 3te^t - e^{-6t}.$$

Finally, shifting the argument back, we obtain

$$z(t) = y(t-1) = 3(t-1)e^{t-1} - e^{-6(t-1)}.$$

11. As in the previous problem (and in Example 3 in the text), we first need to shift the initial conditions to $t = 0$. If we set $v(t) = y(t+2)$, the initial value problem for $v(t)$ becomes

$$v''(t) - v(t) = (t+2) - 2 = t, \qquad v(0) = y(2) = 3, \; v'(0) = y'(2) = 0.$$

Taking the Laplace transform of both sides of this new differential equation gives us

$$\mathcal{L}\left\{v''\right\}(s) - \mathcal{L}\left\{v\right\}(s) = \mathcal{L}\left\{t\right\}(s) = \frac{1}{s^2}.$$

If we denote $V(s) := \mathcal{L}\left\{v\right\}(s)$ and express $\mathcal{L}\left\{v''\right\}(s)$ in terms of $V(s)$ using the formula (4) in Section 7.3 (with $n = 2$), that is, $\mathcal{L}\left\{v''\right\}(s) = s^2 V(s) - 3s$, we obtain

$$\left[s^2 V(s) - 3s\right] - V(s) = \frac{1}{s^2}$$

$$\Rightarrow \qquad V(s) = \frac{3s^3 + 1}{s^2(s^2 - 1)} = \frac{3s^3 + 1}{s^2(s+1)(s-1)} = -\frac{1}{s^2} + \frac{1}{s+1} + \frac{2}{s-1}.$$

Hence,

$$v(t) = \mathcal{L}^{-1}\left\{V(s)\right\}(t) = \mathcal{L}^{-1}\left\{-\frac{1}{s^2} + \frac{1}{s+1} + \frac{2}{s-1}\right\}(t) = -t + e^{-t} + 2e^t.$$

Since $v(t) = y(t+2)$, we have $y(t) = v(t-2)$, and so

$$y(t) = -(t-2) + e^{-(t-2)} + 2e^{t-2} = 2 - t + e^{2-t} + 2e^{t-2}.$$

Copyright © 2012 Pearson Education, Inc. Publishing as Addison-Wesley.

13. To shift the initial conditions to $t = 0$, we make the substitution $x(t) := y(t + \pi/2)$ in the original equation and use the fact that

$$x'(t) = y'(t + \pi/2), \qquad x''(t) = y''(t + \pi/2).$$

This yields

$$y''(t) - y'(t) - 2y(t) = -8\cos t - 2\sin t$$
$$\Rightarrow \quad -8\cos\left(t + \frac{\pi}{2}\right) - 2\sin\left(t + \frac{\pi}{2}\right) = -8\cos\left(t + \frac{\pi}{2}\right) - 2\sin\left(t + \frac{\pi}{2}\right)$$
$$= 8\sin t - 2\cos t$$
$$\Rightarrow \quad x''(t) - x'(t) - 2x(t) = 8\sin t - 2\cos t, \qquad x(0) = 1, \ x'(0) = 0.$$

Taking the Laplace transform of both sides in this differential equation and using the fact that, with $X(s) := \mathcal{L}\{x\}(s)$,

$$\mathcal{L}\{x'\}(s) = sX(s) - 1 \quad \text{and} \quad \mathcal{L}\{x''\}(s) = s^2 X(s) - s$$

(which come from the initial conditions), we obtain

$$\left[s^2 X(s) - s\right] - \left[sX(s) - 1\right] - 2X(s) = \mathcal{L}\{8\sin t - 2\cos t\}(s) = \frac{8}{s^2 + 1} - \frac{2s}{s^2 + 1}$$
$$\Rightarrow \quad \left(s^2 - s - 2\right)X(s) = \frac{8 - 2s}{s^2 + 1} + s - 1 = \frac{s^3 - s^2 - s + 7}{s^2 + 1}$$
$$\Rightarrow \quad X(s) = \frac{s^3 - s^2 - s + 7}{(s^2 + 1)(s^2 - s - 2)} = \frac{s^3 - s^2 - s + 7}{(s^2 + 1)(s - 2)(s + 1)}.$$

We seek for a partial fractions decomposition of $X(s)$ in the form

$$\frac{s^3 - s^2 - s + 7}{(s^2 + 1)(s - 2)(s + 1)} = \frac{As + B}{s^2 + 1} + \frac{C}{(s - 2)} + \frac{D}{s + 1}.$$

Solving yields

$$A = \frac{7}{5}, \quad B = -\frac{11}{5}, \quad C = \frac{3}{5}, \quad D = -1.$$

Therefore,

$$X(s) = \frac{(7/5)s}{s^2 + 1} + \frac{(-11/5)}{s^2 + 1} + \frac{(3/5)}{(s - 2)} - \frac{1}{s + 1}$$
$$\Rightarrow \quad x(t) = \mathcal{L}^{-1}\{X(s)\}(t) = \frac{7}{5}\cos t - \frac{11}{5}\sin t + \frac{3}{5}e^{2t} - e^{-t}.$$

Finally, since $y(t) = x(t - \pi/2)$, we obtain the solution

$$\begin{aligned} y(t) &= \frac{7}{5}\cos\left(t - \frac{\pi}{2}\right) - \frac{11}{5}\sin\left(t - \frac{\pi}{2}\right) + \frac{3}{5}e^{2(t - \pi/2)} - e^{-(t - \pi/2)} \\ &= \frac{7}{5}\sin t + \frac{11}{5}\cos t + \frac{3}{5}e^{2t - \pi} - e^{(\pi/2) - t}. \end{aligned}$$

Copyright © 2012 Pearson Education, Inc. Publishing as Addison-Wesley.

15. Taking the Laplace transform of $y'' - 3y' + 2y = \cos t$ and applying linearity of the Laplace transform yields

$$\mathcal{L}\{y''\}(s) - 3\mathcal{L}\{y'\}(s) + 2\mathcal{L}\{y\}(s) = \mathcal{L}\{\cos t\}(s) = \frac{s}{s^2+1}. \tag{7.16}$$

If we denote $Y(s) = \mathcal{L}\{y\}(s)$ and apply the formula (4), Section 7.3, we get

$$\mathcal{L}\{y'\}(s) = sY(s), \qquad \mathcal{L}\{y''\}(s) = s^2Y(s) + 1.$$

Substitution back into (7.16) yields

$$\left[s^2Y(s) + 1\right] - 3\left[sY(s)\right] + 2Y(s) = \frac{s}{s^2+1}$$
$$\Rightarrow \quad \left(s^2 - 3s + 2\right)Y(s) = \frac{s}{s^2+1} - 1 = \frac{-s^2+s-1}{s^2+1}$$
$$\Rightarrow \quad Y(s) = \frac{-s^2+s-1}{(s^2+1)(s^2-3s+2)} = \frac{-s^2+s-1}{(s^2+1)(s-1)(s-2)}.$$

17. With $Y(s) := \mathcal{L}\{y\}(s)$, we find that

$$\mathcal{L}\{y'\}(s) = sY(s) - y(0) = sY(s) - 1, \quad \mathcal{L}\{y''\}(s) = s^2Y(s) - sy(0) - y'(0) = s^2Y(s) - s,$$

and so the Laplace transform of both sides of the original equation yields

$$\mathcal{L}\{y'' + y' - y\}(s) = \mathcal{L}\{t^3\}(s)$$
$$\Rightarrow \quad \left[s^2Y(s) - s\right] + \left[sY(s) - 1\right] - Y(s) = \frac{6}{s^4}$$
$$\Rightarrow \quad Y(s) = \frac{1}{s^2+s-1}\left(\frac{6}{s^4} + s + 1\right) = \frac{s^5 + s^4 + 6}{s^4(s^2+s-1)}.$$

19. Let us denote $Y(s) := \mathcal{L}\{y\}(s)$. From the initial conditions, we have

$$\mathcal{L}\{y'\}(s) = sY(s) - y(0) = sY(s) - 1, \quad \mathcal{L}\{y''\}(s) = s^2Y(s) - sy(0) - y'(0) = s^2Y(s) - s - 1.$$

The Laplace transform, applied to both sides of the given equation, yields

$$\left[s^2Y(s) - s - 1\right] + 5\left[sY(s) - 1\right] - Y(s) = \mathcal{L}\{e^t\}(s) - \mathcal{L}\{1\}(s)$$
$$= \frac{1}{s-1} - \frac{1}{s} = \frac{1}{s(s-1)}$$
$$\Rightarrow \quad \left(s^2 + 5s - 1\right)Y(s) = \frac{1}{s(s-1)} + s + 6 = \frac{s^3 + 5s^2 - 6s + 1}{s(s-1)}$$
$$\Rightarrow \quad Y(s) = \frac{s^3 + 5s^2 - 6s + 1}{s(s-1)(s^2+5s-1)}.$$

 Copyright © 2012 Pearson Education, Inc. Publishing as Addison-Wesley.

21. Applying the Laplace transform to both sides of the given equation yields

$$\mathcal{L}\left\{y''\right\}(s) - 2\mathcal{L}\left\{y'\right\}(s) + \mathcal{L}\left\{t\right\}(s) = \mathcal{L}\left\{\cos t\right\}(s) - \mathcal{L}\left\{\sin t\right\}(s) = \frac{s-1}{s^2+1}.$$

If $\mathcal{L}\left\{y\right\}(s) =: Y(s)$, then it follows from the initial conditions hat

$$\mathcal{L}\left\{y'\right\}(s) = sY(s) - 1, \quad \mathcal{L}\left\{y''\right\}(s) = s^2Y(s) - s - 3.$$

Therefore, $Y(s)$ satisfies

$$\left[s^2Y(s) - s - 3\right] - 2\left[sY(s) - 1\right] + Y(s) = \frac{s-1}{s^2+1}.$$

Solving for $Y(s)$ gives us

$$\left(s^2 - 2s + 1\right)Y(s) = \frac{s-1}{s^2+1} + s + 1 = \frac{s^3 + s^2 + 2s}{s^2+1}$$

$$\Rightarrow \quad Y(s) = \frac{s^3 + s^2 + 2s}{(s^2+1)(s^2 - 2s + 1)} = \frac{s^3 + s^2 + 2s}{(s^2+1)(s-1)^2}.$$

23. In this equation, the right-hand side is a piecewise defined function. Let us find its Laplace transform first.

$$\mathcal{L}\left\{g(t)\right\}(s) = \int_0^\infty e^{-st}g(t)\,dt = \int_0^2 e^{-st}(t)\,dt + \int_2^\infty e^{-st}(5)\,dt$$

$$= \left.\frac{te^{-st}}{-s}\right|_0^2 - \int_0^2 \frac{e^{-st}}{-s}\,dt + \lim_{N\to\infty}\left.\frac{5e^{-st}}{-s}\right|_2^N$$

$$= -\left(\frac{2e^{-2s}}{s}\right) - \left(\frac{e^{-2s}}{s^2} + \frac{1}{s^2}\right) + \frac{5e^{-2s}}{s} = \frac{1 + 3se^{-2s} - e^{-2s}}{s^2},$$

where we have applied integration by parts. Using this formula and taking the Laplace transform of the given equation yields

$$\mathcal{L}\left\{y''\right\}(s) + 4\mathcal{L}\left\{y\right\}(s) = \mathcal{L}\left\{g(t)\right\}(s)$$

$$\Rightarrow \quad s^2\mathcal{L}\left\{y\right\}(s) + s + 4\mathcal{L}\left\{y\right\}(s) = \mathcal{L}\left\{g(t)\right\}(s)$$

$$\Rightarrow \quad \left(s^2 + 4\right)\mathcal{L}\left\{y\right\}(s) = \mathcal{L}\left\{g(t)\right\}(s) - s = \frac{-s^3 + 1 + 3se^{-2s} - e^{-2s}}{s^2}$$

$$\Rightarrow \quad \mathcal{L}\left\{y\right\}(s) = \frac{-s^3 + 1 + 3se^{-2s} - e^{-2s}}{s^2(s^2 + 4)}.$$

Copyright © 2012 Pearson Education, Inc. Publishing as Addison-Wesley.

25. Taking the Laplace transform of $y''' - y'' + y' - y = 0$ and applying linearity of the Laplace transform yields

$$\mathcal{L}\{y'''\}(s) - \mathcal{L}\{y''\}(s) + \mathcal{L}\{y'\}(s) - \mathcal{L}\{y\}(s) = \mathcal{L}\{0\}(s) = 0. \qquad (7.17)$$

If we denote $Y(s) := \mathcal{L}\{y\}(s)$ and apply the formula (4), Section 7.3, we get

$$\mathcal{L}\{y'\}(s) = sY(s) - 1, \quad \mathcal{L}\{y''\}(s) = s^2 Y(s) - s - 1, \quad \mathcal{L}\{y'''\}(s) = s^3 Y(s) - s^2 - s - 3.$$

Combining these equations with (7.17) gives us

$$\left[s^3 Y(s) - s^2 - s - 3\right] - \left[s^2 Y(s) - s - 1\right] + \left[sY(s) - 1\right] - Y(s) = 0$$

$$\Rightarrow \qquad \left(s^3 - s^2 + s - 1\right) Y(s) = s^2 + 3$$

$$\Rightarrow \qquad Y(s) = \frac{s^2 + 3}{s^3 - s^2 + s - 1} = \frac{s^2 + 3}{(s-1)(s^2 + 1)}.$$

Expanding $Y(s)$ by partial fractions results

$$Y(s) = \frac{2}{s-1} - \frac{s+1}{s^2+1} = \frac{2}{s-1} - \frac{s}{s^2+1} - \frac{1}{s^2+1}.$$

From Table 7.1, Section 7.2, we see that

$$y(t) = \mathcal{L}^{-1}\{Y(s)\}(t) = 2e^t - \cos t - \sin t.$$

27. Let $Y(s) := \mathcal{L}\{y\}(s)$. Then, by Theorem 5 in Section 7.3,

$$\mathcal{L}\{y'\}(s) = sY(s) - y(0) = sY(s) + 4,$$

$$\mathcal{L}\{y''\}(s) = s^2 Y(s) - sy(0) - y'(0) = s^2 Y(s) + 4s - 4,$$

$$\mathcal{L}\{y'''\}(s) = s^3 Y(s) - s^2 y(0) - sy'(0) - y''(0) = s^3 Y(s) + 4s^2 - 4s + 2.$$

Using these formulas and applying the Laplace transform to both sides of the given differential equation, we get

$$\left[s^3 Y(s) + 4s^2 - 4s + 2\right] + 3\left[s^2 Y(s) + 4s - 4\right] + 3\left[sY(s) + 4\right] + Y(s) = 0$$

$$\Rightarrow \qquad \left(s^3 + 3s^2 + 3s + 1\right) Y(s) + \left(4s^2 + 8s + 2\right) = 0$$

$$\Rightarrow \qquad Y(s) = -\frac{4s^2 + 8s + 2}{s^3 + 3s^2 + 3s + 1} = -\frac{4s^2 + 8s + 2}{(s+1)^3}.$$

Therefore, the partial fractions decomposition of $Y(s)$ has the form

$$-\frac{4s^2 + 8s + 2}{(s+1)^3} = \frac{A}{(s+1)^3} + \frac{B}{(s+1)^2} + \frac{C}{s+1} = \frac{A + B(s+1) + C(s+1)^2}{(s+1)^3}$$

 Copyright © 2012 Pearson Education, Inc. Publishing as Addison-Wesley.

$$\Rightarrow \qquad -(4s^2 + 8s + 2) = A + B(s + 1) + C(s + 1)^2.$$

Substitution $s = -1$ yields $A = 2$. Equating coefficients at s^2, we get $C = -4$. Finally, substituting $s = 0$, we obtain

$$-2 = A + B + C \qquad \Rightarrow \qquad B = -2 - A - C = 0.$$

Therefore,

$$Y(s) = \frac{2}{(s + 1)^3} + \frac{-4}{s + 1} \qquad \Rightarrow \qquad y(t) = \mathcal{L}^{-1}\{Y\}(t) = t^2 e^{-t} - 4e^{-t} = \left(t^2 - 4\right)e^{-t}.$$

29. Using the initial conditions, $y(0) = a$ and $y'(0) = b$, and the formula (4) of Section 7.3, we conclude that

$$\mathcal{L}\{y'\}(s) = sY(s) - y(0) = sY(s) - a,$$
$$\mathcal{L}\{y''\}(s) = s^2 Y(s) - sy(0) - y'(0) = s^2 Y(s) - as - b,$$

where $Y(s) := \mathcal{L}\{y\}(s)$. Applying the Laplace transform to the original equation yields

$$\left[s^2 Y(s) - as - b\right] - 4\left[sY(s) - a\right] + 3Y(s) = \mathcal{L}\{0\}(s) = 0$$
$$\Rightarrow \qquad \left(s^2 - 4s + 3\right)Y(s) = as + b - 4a$$
$$\Rightarrow \qquad Y(s) = \frac{as + b - 4a}{s^2 - 4s + 3} = \frac{as + b - 4a}{(s - 1)(s - 3)} = \frac{A}{s - 1} + \frac{B}{s - 3}.$$

Solving for A and B, we find that $A = (3a - b)/2$, $B = (b - a)/2$. Hence,

$$Y(s) = \frac{(3a - b)/2}{s - 1} + \frac{(b - a)/2}{s - 3}$$
$$\Rightarrow \qquad y(t) = \mathcal{L}^{-1}\{Y\}(t) = \frac{3a - b}{2}\mathcal{L}^{-1}\left\{\frac{1}{s - 1}\right\}(t) + \frac{b - a}{2}\mathcal{L}^{-1}\left\{\frac{1}{s - 3}\right\}(t)$$
$$= \frac{3a - b}{2}e^t + \frac{b - a}{2}e^{3t}.$$

31. Similarly to Problem 29, we have

$$\mathcal{L}\{y'\}(s) = sY(s) - y(0) = sY(s) - a,$$
$$\mathcal{L}\{y''\}(s) = s^2 Y(s) - sy(0) - y'(0) = s^2 Y(s) - as - b,$$

with $Y(s) := \mathcal{L}\{y\}(s)$. Thus, the Laplace transform of both sides of the the given equation yields

$$\mathcal{L}\{y'' + 2y' + 2y\}(s) = \mathcal{L}\{5\}(s)$$

$$\Rightarrow \qquad \left[s^2 Y(s) - as - b\right] + 2\left[sY(s) - a\right] + 2Y(s) = \frac{5}{s}$$

$$\Rightarrow \qquad \left(s^2 + 2s + 2\right) Y(s) = \frac{5}{s} + as + 2a + b = \frac{as^2 + (2a+b)s + 5}{s}$$

$$\Rightarrow \qquad Y(s) = \frac{as^2 + (2a+b)s + 5}{s(s^2 + 2s + 2)} = \frac{as^2 + (2a+b)s + 5}{s[(s+1)^2 + 1]}.$$

We seek for a decomposition of $Y(s)$ of the form

$$\frac{as^2 + (2a+b)s + 5}{s[(s+1)^2 + 1]} = \frac{A}{s} + \frac{B(s+1) + C}{(s+1)^2 + 1(1)}.$$

Clearing fractions, we obtain

$$as^2 + (2a+b)s + 5 = A\left[(s+1)^2 + 1\right] + \left[B(s+1) + C\right]s.$$

Substitutions $s = 0$ and $s = -1$ give us

$$s = 0: \qquad\qquad 5 = 2A \qquad \Rightarrow \qquad A = 5/2,$$
$$s = -1: \quad 5 - a - b = A - C \quad \Rightarrow \qquad C = A + a + b - 5 = a + b - 5/2.$$

To find B, we compare coefficients at s^2.

$$a = A + B \qquad \Rightarrow \qquad B = a - A = a - 5/2.$$

So,

$$Y(s) = \frac{5/2}{s} + \frac{(a - 5/2)(s+1)}{(s+1)^2 + 1} + \frac{a + b - 5/2}{(s+1)^2 + 1}$$

$$\Rightarrow \qquad y(t) = \mathcal{L}^{-1}\{Y\}(t) = \frac{5}{2} + \left(a - \frac{5}{2}\right) e^{-t}\cos t + \left(a + b - \frac{5}{2}\right) e^{-t}\sin t.$$

33. By Theorem 6 in Section 7.3,

$$\mathcal{L}\left\{t^2 y'(t)\right\}(s) = (-1)^2 \frac{d^2}{ds^2}\left[\mathcal{L}\left\{y'(t)\right\}(s)\right] = \frac{d^2}{ds^2}\left[\mathcal{L}\left\{y'(t)\right\}(s)\right]. \qquad (7.18)$$

On the other hand, the equation (4) in Section 7.3 says that

$$\mathcal{L}\left\{y'(t)\right\}(s) = sY(s) - y(0), \qquad Y(s) := \mathcal{L}\left\{y\right\}(s).$$

Substitution back into (7.18) yields

$$\mathcal{L}\left\{t^2 y'(t)\right\}(s) = \frac{d^2}{ds^2}\left[sY(s) - y(0)\right] = \frac{d}{ds}\left\{\frac{d}{ds}\left[sY(s) - y(0)\right]\right\}$$

$$= \frac{d}{ds}\left[sY'(s) + Y(s)\right] = \left[sY''(s) + Y'(s)\right] + Y'(s) = sY''(s) + 2Y'(s).$$

 Copyright © 2012 Pearson Education, Inc. Publishing as Addison-Wesley.

35. Taking the Laplace transform of $y'' + 3ty' - 6y = 1$ and applying linearity of the Laplace transform, we conclude that

$$\mathcal{L}\{y''\}(s) + 3\mathcal{L}\{ty'\}(s) - 6\mathcal{L}\{y\}(s) = \mathcal{L}\{1\}(s) = \frac{1}{s}. \tag{7.19}$$

If we let $Y(s) = \mathcal{L}\{y\}(s)$ we get

$$\mathcal{L}\{y''\}(s) = s^2 Y(s) - sy(0) - y'(0) = s^2 Y(s). \tag{7.20}$$

Furthermore, as it was shown in Example 4, Section 4.5,

$$\mathcal{L}\{ty'\}(s) = -sY'(s) - Y(s). \tag{7.21}$$

Substituting (7.20) and (7.21) back into (7.19) yields

$$s^2 Y(s) + 3\left[-sY'(s) - Y(s)\right] - 6Y(s) = \frac{1}{s}$$

$$\Rightarrow \quad -3sY'(s) + \left(s^2 - 9\right)Y(s) = \frac{1}{s}$$

$$\Rightarrow \quad Y'(s) + \left(\frac{3}{s} - \frac{s}{3}\right)Y(s) = -\frac{1}{3s^2}.$$

This is a first order linear differential equation in $Y(s)$, which can be solved by methods of Section 2.3. Namely, it has an integrating factor

$$\mu(s) = \exp\left[\int\left(\frac{3}{s} - \frac{s}{3}\right)ds\right] = \exp\left(3\ln s - \frac{s^2}{6}\right) = s^3 e^{-s^2/6}.$$

Thus,

$$\begin{aligned} Y(s) &= \frac{1}{\mu(s)}\int \mu(s)\left(-\frac{1}{3s^2}\right)ds = \frac{1}{s^3 e^{-s^2/6}}\int \frac{-s}{3}e^{-s^2/6}\,ds \\ &= \frac{1}{s^3 e^{-s^2/6}}\left(e^{-s^2/6} + C\right) = \frac{1}{s^3}\left(1 + Ce^{s^2/6}\right). \end{aligned}$$

The constant C must be zero in order to ensure that $Y(s) \to 0$ as $s \to \infty$. Therefore, $Y(s) = 1/s^3$, and from Table 7.1 we get

$$y(t) = \mathcal{L}^{-1}\left\{\frac{1}{s^3}\right\}(t) = \frac{1}{2}\mathcal{L}^{-1}\left\{\frac{2}{s^3}\right\}(t) = \frac{t^2}{2}.$$

37. We apply the Laplace transform to the given equation and obtain

$$\mathcal{L}\{ty''\}(s) - 2\mathcal{L}\{y'\}(s) + \mathcal{L}\{ty\}(s) = 0. \tag{7.22}$$

Copyright © 2012 Pearson Education, Inc. Publishing as Addison-Wesley.

Using Theorem 5 (Section 7.3) and the initial conditions, we express $\mathcal{L}\{y''\}(s)$ and $\mathcal{L}\{y'\}(s)$ in terms of $Y(s) := \mathcal{L}\{y\}(s)$.

$$\mathcal{L}\{y'\}(s) = sY(s) - y(0) = sY(s) - 1, \tag{7.23}$$

$$\mathcal{L}\{y''\}(s) = s^2 Y(s) - sy(0) - y'(0) = s^2 Y(s) - s. \tag{7.24}$$

We now involve Theorem 6 in Section 7.3 to get

$$\mathcal{L}\{ty\}(s) = -\frac{d}{ds}\left[\mathcal{L}\{y\}(s)\right] = -Y'(s). \tag{7.25}$$

Theorem 6 and the equation (7.24) also give

$$\mathcal{L}\{ty''\}(s) = -\frac{d}{ds}\left[\mathcal{L}\{y''\}(s)\right] = -\frac{d}{ds}\left[s^2 Y(s) - s\right] = 1 - 2sY(s) - s^2 Y'(s). \tag{7.26}$$

Substituting (7.23), (7.25), and (7.26) into (7.22), we obtain

$$\left[1 - 2sY(s) - s^2 Y'(s)\right] - 2\left[sY(s) - 1\right] + \left[-Y'(s)\right] = 0$$

$$\Rightarrow \quad -\left(s^2 + 1\right)Y'(s) - 4sY(s) + 3 = 0$$

$$\Rightarrow \quad Y'(s) + \frac{4s}{s^2 + 1}Y(s) = \frac{3}{s^2 + 1}.$$

The integrating factor of this first order linear differential equation is

$$\mu(s) = \exp\left(\int \frac{4s}{s^2+1}\,ds\right) = e^{2\ln(s^2+1)} = \left(s^2 + 1\right)^2.$$

Hence,

$$Y(s) = \frac{1}{\mu(s)}\int \mu(s)\left(\frac{3}{s^2+1}\right)ds = \frac{1}{(s^2+1)^2}\int 3\left(s^2+1\right)ds$$

$$= \frac{1}{(s^2+1)^2}\left(s^3 + 3s + C\right) = \frac{(s^3 + s) + (2s + C)}{(s^2+1)^2}$$

$$= \frac{s}{s^2+1} + \frac{2s}{(s^2+1)^2} + \frac{C}{(s^2+1)^2},$$

where C is an arbitrary constant. Therefore,

$$y(t) = \mathcal{L}^{-1}\{Y\}(t) = \mathcal{L}^{-1}\left\{\frac{s}{s^2+1}\right\}(t) + \mathcal{L}^{-1}\left\{\frac{2s}{(s^2+1)^2}\right\}(t) + \frac{C}{2}\mathcal{L}^{-1}\left\{\frac{2}{(s^2+1)^2}\right\}(t).$$

Using formulas (24), (29) and (30) on the inside back cover of the text, we finally get

$$y(t) = \cos t + t\sin t + c(\sin t - t\cos t),$$

where $c := C/2$ is an arbitrary constant.

   Copyright © 2012 Pearson Education, Inc. Publishing as Addison-Wesley.

39. Similarly to Example 5, we have the initial value problem (18), namely,

$$Iy''(t) = -ke(t), \qquad y(0) = 0, \quad y'(0) = 0,$$

for a model of the mechanism. This equation leads to the equation (19) for the Laplace transforms $Y(s) := \mathcal{L}\{y(t)\}(s)$ and $E(s) := \mathcal{L}\{e(t)\}(s)$, that is,

$$s^2 I Y(s) = -kE(s). \tag{7.27}$$

But, this time, $e(t) = y(t) - a$, and so

$$E(s) = \mathcal{L}\{y(t) - a\}(s) = Y(s) - \frac{a}{s} \quad \Rightarrow \quad Y(s) = E(s) + \frac{a}{s}.$$

Substituting this relation into (7.27) yields

$$s^2 I E(s) + aIs = -kE(s) \quad \Rightarrow \quad E(s) = -\frac{-aIs}{s^2 I + k} = -\frac{as}{s^2 + (k/I)}.$$

Taking the inverse Laplace transform, we obtain

$$e(t) = \mathcal{L}^{-1}\{E(s)\}(t) = -a\mathcal{L}^{-1}\left\{\frac{s}{s^2 + (\sqrt{k/I})^2}\right\}(t) = -a\cos\left(\sqrt{k/I}\,t\right).$$

41. As in Problem 40, the differential equation, modeling the automatic pilot, is

$$Iy''(t) = -ke(t) - \mu e'(t), \tag{7.28}$$

but now the error is $e(t) = y(t) - at$. Let $Y(s) := \mathcal{L}\{y(t)\}(s)$, $E(s) := \mathcal{L}\{e(t)\}(s)$. Notice that (see Example 5) we have $y(0) = y'(0) = 0$, and so $e(0) = 0$. Using these initial conditions and Theorem 5 in Section 7.3, we obtain

$$\mathcal{L}\{y''(t)\}(s) = s^2 Y(s) \qquad \text{and} \qquad \mathcal{L}\{e'(t)\}(s) = sE(s).$$

Applying the Laplace transform to both sides of (7.28), we then conclude that

$$I\mathcal{L}\{y''(t)\}(s) = -k\mathcal{L}\{e(t)\}(s) - \mu\mathcal{L}\{e'(t)\}(s)$$
$$\Rightarrow \quad Is^2 Y(s) = -kE(s) - \mu sE(s) = -(k + \mu s)E(s). \tag{7.29}$$

Since $e(t) = y(t) - at$,

$$E(s) = \mathcal{L}\{e(t)\}(s) = \mathcal{L}\{y(t) - at\}(s) = Y(s) - a\mathcal{L}\{t\}(s) = Y(s) - \frac{a}{s^2}$$

Copyright © 2012 Pearson Education, Inc. Publishing as Addison-Wesley.

Chapter 7

or $Y(s) = E(s) + a/s^2$. Substituting $Y(s)$ back into (7.29) yields

$$Is^2\left(E(s) + \frac{a}{s^2}\right) = -(k + \mu s)E(s)$$

$$\Rightarrow \quad \left(Is^2 + \mu s + k\right)E(s) = -aI$$

$$\Rightarrow \quad E(s) = \frac{-aI}{Is^2 + \mu s + k} = \frac{-a}{s^2 + (\mu/I)s + (k/I)}.$$

Completing the square in the denominator, we write $E(s)$ in the form suitable for inverse Laplace transform.

$$E(s) = \frac{-a}{[s + \mu/(2I)]^2 + (k/I) - \mu^2/(4I^2)}$$

$$= \frac{-a}{[s + \mu/(2I)]^2 + (4kI - \mu^2)/(4I^2)}$$

$$= \frac{-2Ia}{\sqrt{4kI - \mu^2}} \frac{\sqrt{4kI - \mu^2}/(2I)}{[s + \mu/(2I)]^2 + (4kI - \mu^2)/(4I^2)}.$$

Thus, using Table 7.1, Section 7.2, we find that

$$e(t) = \mathcal{L}^{-1}\left\{E(s)\right\}(t) = \frac{-2Ia}{\sqrt{4kI - \mu^2}}e^{-\mu t/(2I)}\sin\left(\frac{\sqrt{4kI - \mu^2}\,t}{2I}\right).$$

Compare this with Example 5 of the text and observe how, for moderate damping with $\mu < 2\sqrt{kI}$, the oscillations die out exponentially.

EXERCISES 7.6:   Transforms of Discontinuous and Periodic Functions

1. To find the Laplace transform of $g(t) = (t-1)^2 u(t-1)$ we apply formula (8), with $a = 1$ and $f(t) = t^2$. This yields

$$\mathcal{L}\left\{(t-1)^2 u(t-1)\right\}(s) = e^{-s}\mathcal{L}\left\{t^2\right\}(s) = \frac{2e^{-s}}{s^3}.$$

The graph of $g(t) = (t-1)^2 u(t-1)$ is shown in Fig. **7–B(a)**, page 455.

3. The graph of the function $y = t^2 u(t-2)$ is shown in Fig. **7–B(b)**, page 455. For this function, formula (11) is more convenient. We observe that $g(t) = t^2$ and $a = 2$. Hence,

$$g(t + a) = g(t + 2) = (t + 2)^2 = t^2 + 4t + 4.$$

Now the Laplace transform of $g(t + 2)$ is

$$\mathcal{L}\left\{t^2 + 4t + 4\right\}(s) = \mathcal{L}\left\{t^2\right\}(s) + 4\mathcal{L}\left\{t\right\}(s) + 4\mathcal{L}\left\{1\right\}(s) = \frac{2}{s^3} + \frac{4}{s^2} + \frac{4}{s}.$$

Copyright © 2012 Pearson Education, Inc. Publishing as Addison-Wesley.

Hence, by formula (11), we have

$$\mathcal{L}\left\{t^2 u(t-2)\right\}(s) = e^{-2s}\mathcal{L}\left\{g(t+2)\right\}(s) = e^{-2s}\left(\frac{2}{s^3}+\frac{4}{s^2}+\frac{4}{s}\right) = \frac{e^{-2s}(4s^2+4s+2)}{s^3}.$$

5. Using the rectangular window function on the intervals $(0,1)$, $(1,2)$, and $(2,3)$ and the unit step function on $(3,\infty)$ we write

$$g(t) = 0\cdot\Pi_{0,1}(t) + 2\cdot\Pi_{1,2}(t) + 1\cdot\Pi_{2,3}(t) + 3u(t-3) = 2\Pi_{1,2}(t) + \Pi_{2,3}(t) + 3u(t-3).$$

From formulas (6) and (7) we find

$$\mathcal{L}\{g\}(s) = 2\frac{e^{-s}-e^{-2s}}{s} + \frac{e^{-2s}-e^{-3s}}{s} + \frac{3e^{-3s}}{s} = \frac{2e^{-s}-e^{-2s}+2e^{-3s}}{s}.$$

7. We observe from the graph that

$$g(t) = \begin{cases} 0, & t<1, \\ t, & 1<t<2, \\ 1, & t>2. \end{cases}$$

Using the rectangular window function on the intervals $(0,1)$ and $(1,2)$ and the unit step function on $(2,\infty)$, we write

$$\begin{aligned} g(t) &= 0\cdot\Pi_{0,1}(t) + t\cdot\Pi_{1,2}(t) + u(t-2) \\ &= t[u(t-1)-u(t-2)] + u(t-2) = tu(t-1) - (t-1)u(t-2). \end{aligned}$$

Taking the Laplace transform of both sides and using (11), we find that the Laplace transform of the function $g(t)$ is given by

$$\begin{aligned} \mathcal{L}\{g(t)\}(s) &= \mathcal{L}\{tu(t-1)\}(s) - \mathcal{L}\{(t-1)u(t-2)\}(s) \\ &= e^{-s}\mathcal{L}\{(t+1)\}(s) - e^{-2s}\mathcal{L}\{(t-1)+2\}(s) \\ &= \left(e^{-s}-e^{-2s}\right)\mathcal{L}\{t+1\}(s) = \left(e^{-s}-e^{-2s}\right)\left(\frac{1}{s^2}+\frac{1}{s}\right) \\ &= \frac{(e^{-s}-e^{-2s})(s+1)}{s^2}. \end{aligned}$$

9. First, we find the formula for $g(t)$ from the picture given.

$$g(t) = \begin{cases} 0, & t<1, \\ t-1, & 1<t<2, \\ 3-t, & 2<t<3, \\ 0, & 3<t. \end{cases}$$

Copyright © 2012 Pearson Education, Inc. Publishing as Addison-Wesley.

Using the rectangular window function on the intervals $(0, 1)$, $(1, 2)$, and $(2, 3)$ and the unit step function on $(3, \infty)$ we write

$$
\begin{aligned}
g(t) &= 0 \cdot \Pi_{0,1}(t) + (t - 1) \cdot \Pi_{1,2}(t) + (3 - t) \cdot \Pi_{2,3}(t) + 0 \cdot u(t - 3) \\
&= (t - 1)[u(t - 1) - u(t - 2)] + (3 - t)[u(t - 2) - u(t - 3)] \\
&= (t - 1)u(t - 1) + (4 - 2t)u(t - 2) + (t - 3)u(t - 3).
\end{aligned}
$$

Therefore,

$$
\begin{aligned}
\mathcal{L}\{g(t)\}(s) &= \mathcal{L}\{(t - 1)u(t - 1)\}(s) + \mathcal{L}\{(4 - 2t)u(t - 2)\}(s) + \mathcal{L}\{(t - 3)u(t - 3)\}(s) \\
&= e^{-s}\mathcal{L}\{(t + 1) - 1\}(s) + e^{-2s}\mathcal{L}\{4 - 2(t + 2)\}(s) + e^{-3s}\mathcal{L}\{(t + 3) - 3\}(s) \\
&= e^{-s}\mathcal{L}\{t\}(s) - 2e^{-2s}\mathcal{L}\{t\}(s) + e^{-3s}\mathcal{L}\{t\}(s) = \frac{e^{-s} - 2e^{-2s} + e^{-3s}}{s^2}.
\end{aligned}
$$

11. Here, we use formula (6) with $a = 2$ and $F(s) = 1/(s - 1)$. Since

$$
f(t) = \mathcal{L}^{-1}\{F(s)\}(t) = \mathcal{L}^{-1}\left\{\frac{1}{s - 1}\right\}(t) = e^t \qquad \Rightarrow \qquad f(t - 2) = e^{t-2},
$$

we get

$$
\mathcal{L}^{-1}\left\{\frac{e^{-2s}}{s - 1}\right\}(t) = f(t - 2)u(t - 2) = e^{t-2}u(t - 2).
$$

13. Using the linear property of the inverse Laplace transform, we obtain

$$
\mathcal{L}^{-1}\left\{\frac{e^{-2s} - 3e^{-4s}}{s + 2}\right\}(t) = \mathcal{L}^{-1}\left\{\frac{e^{-2s}}{s + 2}\right\}(t) - 3\mathcal{L}^{-1}\left\{\frac{e^{-4s}}{s + 2}\right\}(t).
$$

To each term in this equation, we apply now formula (9) with $F(s) = 1/(s + 2)$, $a = 2$ and $a = 4$, respectively. Since

$$
f(t) := \mathcal{L}^{-1}\{F(s)\}(t) = \mathcal{L}^{-1}\{1/(s + 2)\}(t) = e^{-2t},
$$

we get

$$
\begin{aligned}
\mathcal{L}^{-1}\left\{\frac{e^{-2s}}{s + 2}\right\}(t) - 3\mathcal{L}^{-1}\left\{\frac{e^{-4s}}{s + 2}\right\}(t) &= f(t - 2)u(t - 2) - 3f(t - 4)u(t - 4) \\
&= e^{-2(t-2)}u(t - 2) - 3e^{-2(t-4)}u(t - 4).
\end{aligned}
$$

15. Since

$$
F(s) := \frac{s}{s^2 + 4s + 5} = \frac{s}{(s + 2)^2 + 1^2} = \frac{s + 2}{(s + 2)^2 + 1^2} - 2\frac{1}{(s + 2)^2 + 1^2}
$$

 Copyright © 2012 Pearson Education, Inc. Publishing as Addison-Wesley.

$$\Rightarrow \qquad f(t) := \mathcal{L}^{-1}\{F(s)\}(t) = e^{-2t}(\cos t - 2\sin t),$$

applying Theorem 8, we get

$$\mathcal{L}^{-1}\left\{\frac{se^{-3s}}{s^2+4s+5}\right\}(t) = f(t-3)u(t-3) = e^{-2(t-3)}[\cos(t-3) - 2\sin(t-3)]u(t-3).$$

17. By partial fractions decomposition,

$$\frac{s-5}{(s+1)(s+2)} = -\frac{6}{s+1} + \frac{7}{s+2},$$

so that

$$\mathcal{L}^{-1}\left\{\frac{e^{-3s}(s-5)}{(s+1)(s+2)}\right\}(t) = -6\mathcal{L}^{-1}\left\{\frac{e^{-3s}}{s+1}\right\}(t) + 7\mathcal{L}^{-1}\left\{\frac{e^{-3s}}{s+2}\right\}(t)$$

$$= \left[-6\mathcal{L}^{-1}\left\{\frac{1}{s+1}\right\} + 7\mathcal{L}^{-1}\left\{\frac{1}{s+2}\right\}\right](t-3)u(t-3)$$

$$= \left[-6e^{-(t-3)} + 7e^{-2(t-3)}\right]u(t-3) = \left(7e^{6-2t} - 6e^{3-t}\right)u(t-3).$$

19. In this problem, we apply methods of Section 7.5 of solving initial value problems using the Laplace transform. Taking the Laplace transform of both sides of the given equation and using the linear property of the Laplace transform, we get

$$\mathcal{L}\{I''\}(s) + 2\mathcal{L}\{I'\}(s) + 2\mathcal{L}\{I\}(s) = \mathcal{L}\{g(t)\}(s). \tag{7.30}$$

Let us denote $\mathbf{I}(s) := \mathcal{L}\{I\}(s)$. By Theorem 5, Section 7.3,

$$\begin{aligned}\mathcal{L}\{I'\}(s) &= s\mathbf{I}(s) - I(0) = s\mathbf{I}(s) - 10,\\ \mathcal{L}\{I''\}(s) &= s^2\mathbf{I}(s) - sI(0) - I'(0) = s^2\mathbf{I}(s) - 10s.\end{aligned} \tag{7.31}$$

Using the rectangular window function on the intervals $(0, 3\pi)$ and $(3\pi, 4\pi)$ and the unit step function on $(4\pi, \infty)$ we write

$$g(t) = 20\cdot\Pi_{0,3\pi}(t) + 0\cdot\Pi_{3\pi,4\pi}(t) + 20u(t-4\pi) = 20 - 20u(t-3\pi) + 20u(t-4\pi).$$

Therefore,

$$\mathcal{L}\{g(t)\}(s) = \mathcal{L}\{20 - 20u(t-3\pi) + 20u(t-4\pi)\}(s)$$

$$= 20\mathcal{L}\{1 - u(t-3\pi) + u(t-4\pi)\}(s) = 20\left(\frac{1}{s} - e^{-3\pi s} + e^{-4\pi s}\right).$$

Substituting this expression and (7.31) into (7.30) yields

$$\left[s^2\mathbf{I}(s) - 10s\right] + 2\left[s\mathbf{I}(s) - 10\right] + 2\mathbf{I}(s) = 20\left(\frac{1}{s} - \frac{e^{-3\pi s}}{s} + \frac{e^{-4\pi s}}{s}\right)$$

Copyright © 2012 Pearson Education, Inc. Publishing as Addison-Wesley.

$$\Rightarrow \quad \mathbf{I}(s) = 10\frac{1}{s} + 20\frac{-e^{-3\pi s} + e^{-4\pi s}}{s[(s+1)^2 + 1]}. \qquad (7.32)$$

Since $\mathcal{L}^{-1}\{1/s\}(t) = 1$ and

$$\mathcal{L}^{-1}\left\{\frac{1}{s[(s+1)^2 + 1]}\right\}(t) = \mathcal{L}^{-1}\left\{\frac{1}{2}\left[1s - \frac{s+1}{(s+1)^2 + 1} - \frac{1}{(s+1)^2 + 1}\right]\right\}(t)$$

$$= \frac{1}{2}\left[1 - e^{-t}(\cos t + \sin t)\right],$$

applying the inverse Laplace transform to both sides of (7.32) yields

$$I(t) = \mathcal{L}^{-1}\left\{10\frac{1}{s} + 20\frac{-e^{-3\pi s}}{s[(s+1)^2 + 1]} + 20\frac{e^{-4\pi s}}{s[(s+1)^2 + 1]}\right\}(t)$$

$$= 10 - 10u(t - 3\pi)\left[1 - e^{-(t-3\pi)}\left(\cos(t - 3\pi) + \sin(t - 3\pi)\right)\right]$$

$$+ 10u(t - 4\pi)\left[1 - e^{-(t-4\pi)}\left(\cos(t - 4\pi) + \sin(t - 4\pi)\right)\right]$$

$$= 10 - 10u(t - 3\pi)\left[1 + e^{-(t-3\pi)}\left(\cos t + \sin t\right)\right]$$

$$+ 10u(t - 4\pi)\left[1 - e^{-(t-4\pi)}\left(\cos t + \sin t\right)\right].$$

The graph of the solution, $y = I(t)$, $0 < t < 8\pi$, is depicted in Fig. **7–C**, page 456.

21. In the windowed version (14) of $f(t)$, $T = 2$ and

$$f_T(t) = t\Pi_{0,2}(t) = t - tu(t - 2).$$

Thus,

$$F_T(s) = \frac{1}{s^2} - e^{-2s}\mathcal{L}\{t + 2\}(s) = -\frac{2e^{-2s}}{s} - \frac{e^{-2s}}{s^2} + \frac{1}{s^2} = \frac{1 - 2se^{-2s} - e^{-2s}}{s^2}.$$

From Theorem 9, we obtain

$$\mathcal{L}\{f(t)\}(s) = \frac{F_T(s)}{1 - e^{-2s}} = \frac{1 - 2se^{-2s} - e^{-2s}}{s^2(1 - e^{-2s})}.$$

The graph of the function $y = f(t)$ is given in Fig. B.45 in the answers of the text.

23. Here, we use the formula (15). With the period $T = 2$, the windowed version $f_T(t)$ of $f(t)$ is

$$f_T(t) = \begin{cases} f(t), & 0 < t < 2, \\ 0, & \text{otherwise} \end{cases} = \begin{cases} e^{-t}, & 0 < t < 1, \\ 1, & 1 < t < 2, \\ 0, & \text{otherwise.} \end{cases}$$

 Copyright © 2012 Pearson Education, Inc. Publishing as Addison-Wesley.

Using the function $\Pi_{a,b}(t)$, we can write

$$
\begin{aligned}
f_T(t) &= e^{-t}\Pi_{0,1}(t) + \Pi_{1,2}(t) \\
&= e^{-t}[1 - u(t-1)] + [u(t-1) - u(t-2)] = e^{-t} + (1 - e^{-t})u(t-1) - u(t-2).
\end{aligned}
$$

Therefore,

$$
\begin{aligned}
F_T(s) &= \frac{1}{s+1} + e^{-s}\mathcal{L}\left\{1 - e^{-(t+1)}\right\}(s) - \frac{e^{-2s}}{s} \\
&= \frac{1}{s+1} + \frac{e^{-s}}{s} - \frac{e^{-s-1}}{s+1} - \frac{e^{-2s}}{s}
\end{aligned}
$$

and, by (15),

$$
\mathcal{L}\left\{f(t)\right\}(s) = \frac{1}{1 - e^{-2s}}\left(\frac{1 - e^{-s-1}}{s+1} + \frac{e^{-s} - e^{-2s}}{s}\right).
$$

The graph of $f(t)$ is shown in Fig. B.46 in the answers of the text.

25. Similarly to Example 6, $f(t)$ is a periodic function with period $T = 2a$, whose windowed version has the form

$$
f_{2a}(t) = 1 - u(t - a), \qquad 0 < t < 2a.
$$

Thus, using linearity of the Laplace transform and formula (6) for the Laplace transform of the unit step function, we get

$$
F_{2a}(s) = \mathcal{L}\left\{f_{2a}(t)\right\}(s) = \mathcal{L}\left\{1\right\}(s) - \mathcal{L}\left\{u(t-a)\right\}(s) = \frac{1}{s} - \frac{e^{-as}}{s} = \frac{1 - e^{-as}}{s}.
$$

Applying now Theorem 9 yields

$$
\mathcal{L}\left\{f(t)\right\}(s) = \frac{1}{1 - e^{-2as}}\frac{1 - e^{-as}}{s} = \frac{1}{(1 - e^{-as})(1 + e^{-as})}\frac{1 - e^{-as}}{s} = \frac{1}{s(1 + e^{-as})}.
$$

27. Observe that if we let

$$
f_{2a}(t) = \begin{cases} f(t), & 0 < t < 2a, \\ 0, & \text{otherwise} \end{cases}
$$

denote the windowed version of $f(t)$, then from formula (15) we have

$$
\mathcal{L}\left\{f(t)\right\}(s) = \frac{\mathcal{L}\left\{f_{2a}(t)\right\}(s)}{1 - e^{-2as}} = \frac{\mathcal{L}\left\{f_{2a}(t)\right\}(s)}{(1 - e^{-as})(1 + e^{-as})}.
$$

Now,

$$
f_{2a}(t) = \frac{t}{a}\Pi_{0,a}(t) + \left(2 - \frac{t}{a}\right)\Pi_{a,2a}(t) = \frac{t}{a}[1 - u(t-a)] + \left(2 - \frac{t}{a}\right)[u(t-a) - u(t-2a)].
$$

Copyright © 2012 Pearson Education, Inc. Publishing as Addison-Wesley.

Hence,

$$\mathcal{L}\left\{f_{2a}(t)\right\}(s) = \frac{1}{a}\mathcal{L}\left\{t\right\}(s) - \frac{2}{a}\mathcal{L}\left\{(t-a)u(t-a)\right\}(s) + \frac{1}{a}\mathcal{L}\left\{(t-2a)u(t-2a)\right\}(s)$$

$$= \frac{1}{a}\frac{1}{s^2} - \frac{2}{a}\frac{e^{-as}}{s^2} + \frac{1}{a}\frac{e^{-as}}{s^2} = \frac{1}{as^2}\left(1 - 2e^{-as} + e^{-2as}\right) = \frac{(1-e^{-as})^2}{as^2}$$

and

$$\mathcal{L}\left\{f(t)\right\}(s) = \frac{(1-e^{-as})^2/(as^2)}{(1-e^{-as})(1+e^{-as})} = \frac{1-e^{-as}}{as^2(1+e^{-as})}.$$

29. Applying the Laplace transform to both sides of the given differential equation, we obtain

$$\mathcal{L}\left\{y''\right\}(s) + \mathcal{L}\left\{y\right\}(s) = \mathcal{L}\left\{u(t-3)\right\}(s) = \frac{e^{-3s}}{s}.$$

Since

$$\mathcal{L}\left\{y''\right\}(s) = s^2\mathcal{L}\left\{y\right\}(s) - sy(0) - y'(0) = s^2\mathcal{L}\left\{y\right\}(s) - 1,$$

substituting yields

$$\left[s^2\mathcal{L}\left\{y\right\}(s) - 1\right] + \mathcal{L}\left\{y\right\}(s) = \frac{e^{-3s}}{s}$$

$$\Rightarrow \quad \mathcal{L}\left\{y\right\}(s) = \frac{1}{s^2+1} + \frac{e^{-3s}}{s(s^2+1)} = \frac{1}{s^2+1} + e^{-3s}\left(\frac{1}{s} - \frac{s}{s^2+1}\right).$$

By the formula (9),

$$\mathcal{L}^{-1}\left\{e^{-3s}\left(\frac{1}{s} - \frac{s}{s^2+1}\right)\right\}(t) = \mathcal{L}^{-1}\left\{\frac{1}{s} - \frac{s}{s^2+1}\right\}(t-3)u(t-3) = [1 - \cos(t-3)]u(t-3).$$

Hence,

$$y(t) = \mathcal{L}^{-1}\left\{\frac{1}{s^2+1} + e^{-3s}\left(\frac{1}{s} - \frac{s}{s^2+1}\right)\right\}(t) = \sin t + [1 - \cos(t-3)]u(t-3)$$

The graph of the solution is shown in Fig. B.47 in the answers of the text.

31. We apply the Laplace transform to both sides of the given differential equation and get

$$\mathcal{L}\left\{y''\right\}(s) + \mathcal{L}\left\{y\right\}(s) = \mathcal{L}\left\{t - (t-4)u(t-2)\right\}(s)$$

$$= \frac{1}{s^2} - \mathcal{L}\left\{(t-4)u(t-2)\right\}(s). \tag{7.33}$$

Since $(t-4)u(t-2) = [(t-2) - 2]u(t-2)$, we can use the formula (8) from Theorem 8 to find its Laplace transform. With $f(t) = t - 2$ and $a = 2$, formula (8) yields

$$\mathcal{L}\left\{(t-4)u(t-2)\right\}(s) = e^{-2s}\mathcal{L}\left\{t-2\right\}(s) = e^{-2s}\left(\frac{1}{s^2} - \frac{2}{s}\right).$$

 Copyright © 2012 Pearson Education, Inc. Publishing as Addison-Wesley.

Also,

$$\mathcal{L}\{y''\}(s) = s^2\mathcal{L}\{y\}(s) - sy(0) - y'(0) = s^2\mathcal{L}\{y\}(s) - 1.$$

Substitution these expressions back into (7.33) gives

$$\left[s^2\mathcal{L}\{y\}(s) - 1\right] + \mathcal{L}\{y\}(s) = \frac{1}{s^2} - e^{-2s}\left(\frac{1}{s^2} - \frac{2}{s}\right)$$

$$\Rightarrow \qquad \mathcal{L}\{y\}(s) = \frac{1}{s^2} - e^{-2s}\frac{1 - 2s}{s^2(s^2 + 1)} = \frac{1}{s^2} - e^{-2s}\left(\frac{1}{s^2} - \frac{2}{s} + \frac{2s}{s^2 + 1} - \frac{1}{s^2 + 1}\right).$$

Applying now the inverse Laplace transform and using the formula (9), we obtain

$$y(t) = \mathcal{L}^{-1}\left\{\frac{1}{s^2} - e^{-2s}\left(\frac{1}{s^2} - \frac{2}{s} + \frac{2s}{s^2 + 1} - \frac{1}{s^2 + 1}\right)\right\}(t)$$

$$= t - \mathcal{L}^{-1}\left\{\frac{1}{s^2} - \frac{2}{s} + \frac{2s}{s^2 + 1} - \frac{1}{s^2 + 1}\right\}(t - 2)u(t - 2)$$

$$= t - \left[(t - 2) - 2 + 2\cos(t - 2) - \sin(t - 2)\right]u(t - 2)$$

$$= t + \left[4 - t + \sin(t - 2) - 2\cos(t - 2)\right]u(t - 2).$$

See Fig. B.48 in the answers of the text.

33. By the formula (6),

$$\mathcal{L}\{u(t - 2\pi) - u(t - 4\pi)\}(s) = \frac{e^{-2\pi s}}{s} - \frac{e^{-4\pi s}}{s}.$$

Thus, taking the Laplace transform of $y'' + 2y' + 2y = u(t - 2\pi) - u(t - 4\pi)$ and applying the initial conditions, $y(0) = y'(0) = 1$, gives us

$$\left[s^2Y(s) - s - 1\right] + 2\left[sY(s) - 1\right] + 2Y(s) = \frac{e^{-2\pi s} - e^{-4\pi s}}{s},$$

where $Y(s)$ is the Laplace transform of $y(t)$. Solving for $Y(s)$ yields

$$Y(s) = \frac{s + 3}{s^2 + 2s + 2} + \frac{e^{-2\pi s} - e^{-4\pi s}}{s(s^2 + 2s + 2)}$$

$$= \frac{s + 1}{(s + 1)^2 + 1^2} + \frac{2(1)}{(s + 1)^2 + 1^2} + \frac{e^{-2\pi s}}{s[(s + 1)^2 + 1^2]} - \frac{e^{-4\pi s}}{s[(s + 1)^2 + 1^2]}. \qquad (7.34)$$

Since

$$\frac{1}{s[(s + 1)^2 + 1^2]} = \frac{1}{2}\frac{(s^2 + 2s + 2) - (s^2 + 2s)}{s[(s + 1)^2 + 1^2]} = \frac{1}{2}\left[\frac{1}{s} - \frac{s + 1}{(s + 1)^2 + 1^2} - \frac{1}{(s + 1)^2 + 1^2}\right],$$

we have

$$\mathcal{L}^{-1}\left\{\frac{1}{s[(s + 1)^2 + 1^2]}\right\}(t) = \mathcal{L}^{-1}\left\{\frac{1}{2}\left[\frac{1}{s} - \frac{s + 1}{(s + 1)^2 + 1^2} - \frac{1}{(s + 1)^2 + 1^2}\right]\right\}(t)$$

Copyright © 2012 Pearson Education, Inc. Publishing as Addison-Wesley.

$$= \frac{1}{2} \left[1 - e^{-t} \cos t - e^{-t} \sin t \right]$$

and, by (6),

$$\mathcal{L}^{-1} \left\{ \frac{e^{-2\pi s}}{s[(s+1)^2 + 1^2]} \right\}(t) = \frac{1}{2} \left[1 - e^{-(t-2\pi)} \cos(t - 2\pi) - e^{-(t-2\pi)} \sin(t - 2\pi) \right] u(t - 2\pi)$$

$$= \frac{1}{2} \left[1 - e^{2\pi - t}(\cos t + \sin t) \right] u(t - 2\pi)$$

$$\mathcal{L}^{-1} \left\{ \frac{e^{-4\pi s}}{s[(s+1)^2 + 1^2]} \right\}(t) = \frac{1}{2} \left[1 - e^{-(t-4\pi)} \cos(t - 4\pi) - e^{-(t-4\pi)} \sin(t - 4\pi) \right] u(t - 4\pi)$$

$$= \frac{1}{2} \left[1 - e^{4\pi - t}(\cos t + \sin t) \right] u(t - 4\pi).$$

Finally, taking the inverse Laplace transform in (7.34) yields

$$y(t) = e^{-t} \cos t + 2e^{-t} \sin t + \frac{1}{2} \left[1 - e^{2\pi - t}(\cos t + \sin t) \right] u(t - 2\pi)$$

$$- \frac{1}{2} \left[1 - e^{4\pi - t}(\cos t + \sin t) \right] u(t - 4\pi).$$

35. We take the Laplace transform of the both sides of the given equation and obtain

$$\mathcal{L} \{z''\}(s) + 3\mathcal{L} \{z'\}(s) + 2\mathcal{L} \{z\}(s) = \mathcal{L} \{e^{-3t} u(t - 2)\}(s). \tag{7.35}$$

We use the initial conditions, $z(0) = 2$ and $z'(0) = -3$, and the formula (4) in Section 7.3 to express $\mathcal{L} \{z'\}(s)$ and $\mathcal{L} \{z''\}(s)$ in terms of $Z(s) := \mathcal{L} \{z\}(s)$. That is,

$$\mathcal{L} \{z'\}(s) = sZ(s) - z(0) = sZ(s) - 2,$$

$$\mathcal{L} \{z''\}(s) = s^2 Z(s) - sz(0) - z'(0) = s^2 Z(s) - 2s + 3.$$

In the right-hand side of (7.35), we can use, say, the translation property of the Laplace transform (Theorem 3, Section 7.3) and the Laplace transform of the unit step function (the formula (6), Section 7.6).

$$\mathcal{L} \{e^{-3t} u(t - 2)\}(s) = \mathcal{L} \{u(t - 2)\}(s + 3) = \frac{e^{-2(s+3)}}{s + 3}.$$

Therefore, (7.35) becomes

$$\left[s^2 Z(s) - 2s + 3 \right] + 3 \left[sZ(s) - 2 \right] + 2Z(s) = \frac{e^{-2(s+3)}}{s + 3}$$

$$\Rightarrow \quad \left(s^2 + 3s + 2 \right) Z(s) = 2s + 3 + \frac{e^{-2(s+3)}}{s + 3}$$

 Copyright © 2012 Pearson Education, Inc. Publishing as Addison-Wesley.

$$\Rightarrow \quad Z(s) = \frac{2s+3}{s^2+3s+2} + e^{-2s-6}\frac{1}{(s+3)(s^2+3s+2)}$$

$$= \frac{1}{s+1} + \frac{1}{s+2} + e^{-2s-6}\left(\frac{1/2}{s+3} - \frac{1}{s+2} + \frac{1/2}{s+1}\right).$$

Hence,

$$z(t) = \mathcal{L}^{-1}\left\{\frac{1}{s+1} + \frac{1}{s+2} + e^{-6}e^{-2s}\left(\frac{1/2}{s+3} - \frac{1}{s+2} + \frac{1/2}{s+1}\right)\right\}(t)$$

$$= \mathcal{L}^{-1}\left\{\frac{1}{s+1}\right\}(t) + \mathcal{L}^{-1}\left\{\frac{1}{s+2}\right\}(t)$$

$$+ \frac{e^{-6}}{2}\left[\mathcal{L}^{-1}\left\{\frac{1}{s+3}\right\} - 2\mathcal{L}^{-1}\left\{\frac{1}{s+2}\right\} + \mathcal{L}^{-1}\left\{\frac{1}{s+1}\right\}\right](t-2)u(t-2)$$

$$= e^{-t} + e^{-2t} + \frac{e^{-6}}{2}\left[e^{-3(t-2)} - 2e^{-2(t-2)} + e^{-(t-2)}\right]u(t-2)$$

$$= e^{-t} + e^{-2t} + \frac{1}{2}\left[e^{-3t} - 2e^{-2(t+1)} + e^{-(t+4)}\right]u(t-2).$$

37. Since

$$\mathcal{L}\{g(t)\}(s) = \mathcal{L}\{\sin(t)\Pi_{0,2\pi}(t)\}(s) = \mathcal{L}\{\sin(t) - \sin(t)u(t-2\pi)(t)\}(s)$$

$$= \frac{1}{s^2+1} - e^{-2\pi s}\mathcal{L}\{\sin(t+2\pi)\}(s) = \frac{1-e^{-2\pi s}}{s^2+1},$$

applying the Laplace transform to the original equation yields

$$\mathcal{L}\{y''\}(s) + 4\mathcal{L}\{y\}(s) = \mathcal{L}\{g(t)\}(s)$$

$$\Rightarrow \quad [s^2\mathcal{L}\{y\}(s) - s - 3] + 4\mathcal{L}\{y\}(s) = \frac{1-e^{-2\pi s}}{s^2+1}$$

$$\Rightarrow \quad \mathcal{L}\{y\}(s) = \frac{s+3}{s^2+4} + \frac{1}{(s^2+1)(s^2+4)} - \frac{e^{-2\pi s}}{(s^2+1)(s^2+4)}.$$

Using partial fractions decomposition,

$$\frac{1}{(s^2+1)(s^2+4)} = \frac{1}{3}\frac{(s^2+4)-(s^2+1)}{(s^2+1)(s^2+4)} = \frac{1}{3}\left(\frac{1}{s^2+1} - \frac{1}{6}\frac{2}{s^2+4}\right),$$

we conclude that

$$\mathcal{L}\{y\}(s) = \frac{s}{s^2+4} + \frac{4}{3}\frac{2}{s^2+4} + \frac{1}{3}\frac{1}{s^2+1} - e^{-2\pi s}\left(\frac{1}{3}\frac{1}{s^2+1} - \frac{1}{6}\frac{2}{s^2+4}\right)$$

and so

$$y(t) = \mathcal{L}^{-1}\left\{\frac{s}{s^2+4}\right\}(t) + \frac{4}{3}\mathcal{L}^{-1}\left\{\frac{2}{s^2+4}\right\}(t) + \mathcal{L}^{-1}\left\{\frac{1}{3}\frac{1}{s^2+1}\right\}(t)$$

Copyright © 2012 Pearson Education, Inc. Publishing as Addison-Wesley.

$$-\mathcal{L}^{-1}\left\{\frac{1}{3}\frac{1}{s^2+1}-\frac{1}{6}\frac{2}{s^2+4}\right\}(t-2\pi)u(t-2\pi)$$

$$=\cos 2t+\frac{4}{3}\sin 2t+\frac{1}{3}\sin t-\left[\frac{1}{3}\sin(t-2\pi)-\frac{1}{6}\sin 2(t-2\pi)\right]u(t-2\pi)$$

$$=\cos 2t+\frac{4}{3}\sin 2t+\frac{1}{3}\sin t-\left(\frac{1}{3}\sin t-\frac{1}{6}\sin 2t\right)u(t-2\pi)$$

$$=\cos 2t+\frac{1}{3}\left[1-u(t-2\pi)\right]\sin t+\frac{1}{6}\left[8+u(t-2\pi)\right]\sin 2t\,.$$

39. Using the rectangular window function on the intervals $(0,1)$ and $(1,5)$ and the unit step function on $(5,\infty)$ we write

$$
\begin{aligned}
g(t) &= 0\cdot\Pi_{0,1}(t)+t\cdot\Pi_{1,5}(t)+u(t-5)\\
&= t[u(t-1)-u(t-5)]+u(t-5)=tu(t-1)+(1-t)u(t-5).
\end{aligned}
$$

Thus, the formula (11) yields

$$\mathcal{L}\{g(t)\}(s)=e^{-s}\mathcal{L}\{t+1\}(s)-e^{-5s}\mathcal{L}\{t+4\}(s)=e^{-s}\left(\frac{1}{s^2}+\frac{1}{s}\right)-e^{-5s}\left(\frac{1}{s^2}+\frac{4}{s}\right).$$

Let $Y(s):=\mathcal{L}\{y\}(s)$. Applying the Laplace transform to the given equation and using the initial conditions, we obtain

$$
\begin{aligned}
\mathcal{L}\{y''\}(s)+5\mathcal{L}\{y'\}(s)+6Y(s) &= \mathcal{L}\{g(t)\}(s)\\
\Rightarrow\quad \left[s^2Y(s)-2\right]+5\left[sY(s)\right]+6Y(s) &= \mathcal{L}\{g(t)\}(s)\\
\Rightarrow\quad \left(s^2+5s+6\right)Y(s) &= 2+e^{-s}\left(\frac{1}{s^2}+\frac{1}{s}\right)-e^{-5s}\left(\frac{1}{s^2}+\frac{4}{s}\right)\\
\Rightarrow\quad Y(s) &= \frac{2}{s^2+5s+6}+e^{-s}\frac{s+1}{s^2(s^2+5s+6)}-e^{-5s}\frac{4s+1}{s^2(s^2+5s+6)}\,.\quad(7.36)
\end{aligned}
$$

Using partial fractions decomposition, we can write

$$
\begin{aligned}
\frac{2}{s^2+5s+6} &= \frac{2}{s+2}-\frac{2}{s+3},\\
\frac{s+1}{s^2(s^2+5s+6)} &= \frac{1/36}{s}+\frac{1/6}{s^2}-\frac{1/4}{s+2}+\frac{2/9}{s+3},\\
\frac{4s+1}{s^2(s^2+5s+6)} &= \frac{1/6}{s^2}+\frac{19/36}{s}-\frac{7/4}{s+2}+\frac{11/9}{s+3}\,.
\end{aligned}
$$

Therefore,

$$\mathcal{L}^{-1}\left\{\frac{2}{s^2+5s+6}\right\}(t)=2e^{-2t}-2e^{-3t}\,,$$

     Copyright © 2012 Pearson Education, Inc. Publishing as Addison-Wesley.

$$\mathcal{L}^{-1}\left\{\frac{s+1}{s^2(s^2+5s+6)}\right\}(t) = \frac{1}{36} + \frac{t}{6} - \frac{e^{-2t}}{4} + \frac{2e^{-3t}}{9},$$

$$\mathcal{L}^{-1}\left\{\frac{4s+1}{s^2(s^2+5s+6)}\right\}(t) = \frac{19}{36} + \frac{t}{6} - \frac{7e^{-2t}}{4} + \frac{11e^{-3t}}{9}.$$

Using these equations and taking the inverse Laplace transform in (7.36), we finally get

$$y(t) = 2e^{-2t} - 2e^{-3t} + \left[\frac{1}{36} + \frac{t-1}{6} - \frac{e^{-2(t-1)}}{4} + \frac{2e^{-3(t-1)}}{9}\right]u(t-1)$$

$$- \left[\frac{19}{36} + \frac{t-5}{6} - \frac{7e^{-2(t-5)}}{4} + \frac{11e^{-3(t-5)}}{9}\right]u(t-5).$$

41. First, observe that, for $s > 0$ and $T > 0$, we have $0 < e^{-Ts} < 1$ so that

$$\frac{1}{1 - e^{-Ts}} = 1 + e^{-Ts} + e^{-2Ts} + e^{-3Ts} + \cdots \tag{7.37}$$

and the series converges for all $s > 0$. Thus,

$$\frac{1}{(s+\alpha)(1-e^{-Ts})} = \frac{1}{s+\alpha}\frac{1}{1-e^{-Ts}} = \frac{1}{s+\alpha}\left(1 + e^{-Ts} + e^{-2Ts} + e^{-3Ts} + \cdots\right)$$

$$= \frac{1}{s+\alpha} + \frac{e^{-Ts}}{s+\alpha} + \frac{e^{-2Ts}}{s+\alpha} + \cdots.$$

Hence,

$$\mathcal{L}^{-1}\left\{\frac{1}{(s+\alpha)(1-e^{-Ts})}\right\}(t) = \mathcal{L}^{-1}\left\{\frac{1}{s+\alpha} + \frac{e^{-Ts}}{s+\alpha} + \frac{e^{-2Ts}}{s+\alpha} + \cdots\right\}(t). \tag{7.38}$$

Taking for granted that linearity of the inverse Laplace transform extends to the infinite sum in (7.38) and ignoring convergence questions, we obtain

$$\mathcal{L}^{-1}\left\{\frac{1}{(s+\alpha)(1-e^{-Ts})}\right\} = \mathcal{L}^{-1}\left\{\frac{1}{s+\alpha}\right\} + \mathcal{L}^{-1}\left\{\frac{e^{-Ts}}{s+\alpha}\right\} + \mathcal{L}^{-1}\left\{\frac{e^{-2Ts}}{s+\alpha}\right\} + \cdots$$

$$= e^{-\alpha t} + e^{-\alpha(t-T)}u(t-T) + e^{-\alpha(t-2T)}u(t-2T) + \cdots.$$

43. Using the expansion (7.37), obtained in Problem 41, we can represent $\mathcal{L}\{g\}(s)$ as

$$\mathcal{L}\{g\}(s) = \frac{\beta}{s^2+\beta^2}\frac{1}{1-e^{-Ts}} = \frac{\beta}{s^2+\beta^2}\left(1 + e^{-Ts} + e^{-2Ts} + e^{-3Ts} + \cdots\right)$$

$$= \frac{\beta}{s^2+\beta^2} + e^{-Ts}\frac{\beta}{s^2+\beta^2} + e^{-2Ts}\frac{\beta}{s^2+\beta^2} + \cdots.$$

Since $\mathcal{L}^{-1}\{\beta/(s^2+\beta^2)\}(t) = \sin\beta t$, using linearity of the inverse Laplace transform (extended to infinite series) and formula (6) in Theorem 8, we obtain

$$g(t) = \mathcal{L}^{-1}\left\{\frac{\beta}{s^2+\beta^2}\right\}(t) + \mathcal{L}^{-1}\left\{\frac{\beta}{s^2+\beta^2}\right\}(t-T)u(t-T)$$

Copyright © 2012 Pearson Education, Inc. Publishing as Addison-Wesley.

$$+\mathcal{L}^{-1}\left\{\frac{\beta}{s^2+\beta^2}\right\}(t-2T)u(t-2T)+\cdots$$

$$=\sin\beta t+[\sin\beta(t-T)]u(t-T)+[\sin\beta(t-2T)]u(t-2T)+\cdots.$$

45. To apply the Laplace transform method to the given initial value problem, we find $\mathcal{L}\{f\}(s)$ first. Since the period of $f(t)$ is $T=1$ and $f(t)=e^t$ on $(0,1)$, the windowed version of $f(t)$ is

$$f_1(t)=\begin{cases}e^t,&0<t<1,\\0,&\text{otherwise,}\end{cases}$$

and so $f_1(t)=e^t\Pi_{0,1}(t)=e^t[1-u(t-1)]$. Thus

$$F_1(s)=\frac{1}{s-1}-e^{-s}\mathcal{L}\{e^{t+1}\}(s)=\frac{1-e^{1-s}}{s-1}.$$

Hence, Theorem 9 yields

$$\mathcal{L}\{f\}(s)=\frac{1-e^{1-s}}{(s-1)(1-e^{-s})}.$$

We now apply the Laplace transform to the given differential equation and obtain

$$\mathcal{L}\{y''\}(s)+3\mathcal{L}\{y'\}(s)+2\mathcal{L}\{y\}(s)=\frac{1-e^{1-s}}{(s-1)(1-e^{-s})}$$

$$\Rightarrow\quad[s^2\mathcal{L}\{y\}(s)]+3[s\mathcal{L}\{y\}(s)]+2\mathcal{L}\{y\}(s)=\frac{1-e^{1-s}}{(s-1)(1-e^{-s})}$$

$$\Rightarrow\quad\mathcal{L}\{y\}(s)=\frac{1-e^{1-s}}{(s-1)(s^2+3s+2)(1-e^{-s})}=\frac{1-e^{1-s}}{(s-1)(s+1)(s+2)(1-e^{-s})}$$

$$\Rightarrow\quad\mathcal{L}\{y\}(s)=\frac{e}{(s-1)(s+1)(s+2)}+\frac{1-e}{1-e^{-s}}\frac{1}{(s-1)(s+1)(s+2)}.$$

Using the partial fractions decomposition

$$\frac{1}{(s-1)(s+1)(s+2)}=\frac{1/6}{s-1}-\frac{1/2}{s+1}+\frac{1/3}{s+2},$$

we find that

$$\mathcal{L}\{y\}(s)=\frac{e/6}{s-1}-\frac{e/2}{s+1}+\frac{e/3}{s+2}+\frac{1-e}{6}\frac{1}{(s-1)(1-e^{-s})}$$

$$-\frac{1-e}{2}\frac{1}{(s+1)(1-e^{-s})}+\frac{1-e}{3}\frac{1}{(s+2)(1-e^{-s})}$$

$$\Rightarrow\quad y(t)=\frac{e}{6}e^t-\frac{e}{2}e^{-t}+\frac{e}{3}e^{-2t}+\frac{1-e}{6}\mathcal{L}^{-1}\left\{\frac{1}{(s-1)(1-e^{-s})}\right\}(t)$$

$$-\frac{1-e}{2}\mathcal{L}^{-1}\left\{\frac{1}{(s+1)(1-e^{-s})}\right\}(t)+\frac{1-e}{3}\mathcal{L}^{-1}\left\{\frac{1}{(s+2)(1-e^{-s})}\right\}(t).\quad(7.39)$$

 Copyright © 2012 Pearson Education, Inc. Publishing as Addison-Wesley.

To each of the inverse Laplace transforms in (7.39) we can apply results of Problem 42(a) with $T = 1$ and $\alpha = -1, 1$, and 2, respectively. Thus, for $n < t < n + 1$, we have

$$\mathcal{L}^{-1}\left\{\frac{1}{(s-1)(1-e^{-s})}\right\}(t) = e^t \frac{e^{-(n+1)}-1}{e^{-1}-1},$$

$$\mathcal{L}^{-1}\left\{\frac{1}{(s+1)(1-e^{-s})}\right\}(t) = e^{-t} \frac{e^{n+1}-1}{e-1},$$

$$\mathcal{L}^{-1}\left\{\frac{1}{(s+2)(1-e^{-s})}\right\}(t) = e^{-2t} \frac{e^{2(n+1)}-1}{e^2-1}.$$

Finally, substitution back into (7.39) yields

$$y(t) = \frac{e}{6}e^t - \frac{e}{2}e^{-t} + \frac{e}{3}e^{-2t} + \frac{1-e}{6}e^t \frac{e^{-(n+1)}-1}{e^{-1}-1}$$

$$-\frac{1-e}{2}e^{-t}\frac{e^{n+1}-1}{e-1} + \frac{1-e}{3}e^{-2t}\frac{e^{2(n+1)}-1}{e^2-1}$$

$$= \frac{e^{t-n}}{6} - \frac{e^{-t}(1+e-e^{n+1})}{2} + \frac{e^{-2t}(1+e+e^2-e^{2n+2})}{3(e+1)}.$$

47. Since

$$e^t = \sum_{k=0}^{\infty} \frac{t^k}{k!} \tag{7.40}$$

and

$$\mathcal{L}\left\{t^k\right\}(s) = \frac{k!}{s^{k+1}},$$

using linearity of the Laplace transform, we have

$$\mathcal{L}\left\{e^t\right\}(s) = \mathcal{L}\left\{\sum_{k=0}^{\infty}\frac{t^k}{k!}\right\}(s) = \sum_{k=0}^{\infty}\frac{\mathcal{L}\left\{t^k\right\}(s)}{k!} = \sum_{k=0}^{\infty}\frac{k!/s^{k+1}}{k!} = \frac{1}{s}\sum_{k=0}^{\infty}\left(\frac{1}{s}\right)^k. \tag{7.41}$$

We apply now the summation formula for geometric series, that is,

$$1 + x + x^2 + \cdots = \frac{1}{1-x},$$

which is valid for $|x| < 1$. With $x = 1/s$, $s > 1$, (7.41) yields

$$\mathcal{L}\left\{e^t\right\}(s) = \frac{1}{s}\frac{1}{1-(1/s)} = \frac{1}{s-1}.$$

49. Recall that the Taylor's series for $\cos t$ about $t = 0$ is

$$\cos t = 1 - \frac{t^2}{2!} + \frac{t^4}{4!} - \frac{t^6}{6!} + \cdots + (-1)^n\frac{t^{2n}}{(2n)!} + \cdots,$$

so that

$$\frac{1 - \cos t}{t} = \frac{t}{2!} - \frac{t^3}{4!} + \frac{t^5}{6!} + \cdots + (-1)^{n+1}\frac{t^{2n-1}}{(2n)!} + \cdots .$$

Thus,

$$\mathcal{L}\left\{\frac{1 - \cos t}{t}\right\}(s) = \frac{1}{2!}\mathcal{L}\{t\}(s) - \frac{1}{4!}\mathcal{L}\{t^3\}(s) + \cdots + \frac{(-1)^{n+1}}{(2n)!}\mathcal{L}\{t^{2n-1}\}(s) + \cdots$$

$$= \frac{1}{2}\frac{1}{s^2} - \frac{1}{4}\frac{1}{s^4} + \cdots + \frac{(-1)^{n+1}}{2n}\frac{1}{s^{2n}} + \cdots$$

$$= \sum_{n=1}^{\infty}\frac{(-1)^{n+1}}{2n}\frac{1}{s^{2n}} = \sum_{n=1}^{\infty}\frac{(-1)^{n+1}}{2ns^{2n}} .$$

To sum this series, recall that

$$\ln(1 - x) = -\sum_{n=1}^{\infty}\frac{x^n}{n} .$$

Hence,

$$\ln\left(1 + \frac{1}{s^2}\right) = -\sum_{n=1}^{\infty}\frac{(-1)^n}{ns^{2n}} = \sum_{n=1}^{\infty}\frac{(-1)^{n+1}}{ns^{2n}} . \tag{7.42}$$

Thus, we have

$$\frac{1}{2}\ln\left(1 + \frac{1}{s^2}\right) = \sum_{n=1}^{\infty}\frac{(-1)^{n+1}}{2ns^{2n}} = \mathcal{L}\left\{\frac{1 - \cos t}{t}\right\}(s) .$$

This formula can also be obtained by using the result of Problem 27 in Section 7.3.

51. Here, we use the formula (20) from the text.

(a) With $r = -1/2$, formula (20) yields

$$\mathcal{L}\{t^{-1/2}\}(s) = \frac{\Gamma[(-1/2) + 1]}{s^{(-1/2)+1}} = \frac{\Gamma(1/2)}{s^{1/2}} = \frac{\sqrt{\pi}}{\sqrt{s}} = \sqrt{\frac{\pi}{s}} .$$

(b) This time, $r = 7/2$, and (20) becomes

$$\mathcal{L}\{t^{7/2}\}(s) = \frac{\Gamma[(7/2) + 1]}{s^{(7/2)+1}} = \frac{\Gamma(9/2)}{s^{9/2}} .$$

From the recursive formula (19) we find that

$$\Gamma\left(\frac{9}{2}\right) = \Gamma\left(\frac{7}{2} + 1\right) = \frac{7}{2}\Gamma\left(\frac{7}{2}\right) = \cdots = \frac{7}{2}\frac{5}{2}\frac{3}{2}\frac{1}{2}\Gamma\left(\frac{1}{2}\right) = \frac{105\sqrt{\pi}}{16} .$$

Therefore,

$$\mathcal{L}\{t^{7/2}\}(s) = \frac{105\sqrt{\pi}}{16s^{9/2}} .$$

 Copyright © 2012 Pearson Education, Inc. Publishing as Addison-Wesley.

53. The Laplace transform of $f(t) = \sin t$ is listed in Table 7.1, Section 7.2, that is

$$F(s) = \frac{1}{s^2 + 1}, \quad s > 0.$$

More generally, observe that $\sin t$ is periodic with period $T = 2\pi k$, for any integer $k > 0$. For this choice of T we have

$$f_T(t) = (\sin t)\Pi_{0,2\pi k}(t) = (\sin t)[1 - u(t - 2\pi k)] = \sin t - (\sin t)u(t - 2\pi k).$$

Thus using formula (11) in the text we get

$$F_T(s) = \frac{1}{s^2 + 1} - e^{-2\pi ks}\mathcal{L}\{\sin(t + 2\pi k)\}(s) = F(s) - e^{-2\pi ks}F(s) = F(s)\left(1 - e^{-sT}\right).$$

55. Substituting $-1/s$ for t into the power series expansion (7.40) yields

$$e^{-1/s} = 1 - \frac{1}{s} + \frac{1}{2!s^2} - \frac{1}{3!s^3} + \cdots + \frac{(-1)^n}{n!s^n} + \cdots.$$

Thus, we have

$$s^{-1/2}e^{-1/s} = \frac{1}{s^{1/2}} - \frac{1}{s^{3/2}} + \frac{1}{2!s^{5/2}} + \cdots + \frac{(-1)^n}{n!s^{n+(1/2)}} + \cdots = \sum_{n=0}^{\infty} \frac{(-1)^n}{n!s^{n+(1/2)}}.$$

By Problem 52,

$$\mathcal{L}^{-1}\left\{\frac{1}{s^{n+(1/2)}}\right\}(t) = \frac{2^n t^{n-(1/2)}}{1 \cdot 3 \cdot 5 \cdots (2n - 1)\sqrt{\pi}},$$

so that

$$\mathcal{L}^{-1}\left\{s^{-1/2}e^{-1/s}\right\}(t) = \mathcal{L}^{-1}\left\{\sum_{n=0}^{\infty} \frac{(-1)^n}{n!s^{n+(1/2)}}\right\}(t)$$

$$= \sum_{n=0}^{\infty} \frac{(-1)^n}{n!}\mathcal{L}^{-1}\left\{\frac{1}{s^{n+(1/2)}}\right\}(t) = \sum_{n=0}^{\infty} \frac{(-1)^n}{n!}\frac{2^n t^{n-(1/2)}}{1 \cdot 3 \cdots (2n - 1)\sqrt{\pi}}.$$

Multiplying the nth term by $2^n[1 \cdot 2 \cdots n]/[2 \cdot 4 \cdots (2n)] = 1$, we obtain

$$\mathcal{L}^{-1}\left\{s^{-1/2}e^{-1/s}\right\}(t) = \sum_{n=0}^{\infty} \frac{(-1)^n(2^n)^2 t^{n-(1/2)}}{(2n)!\sqrt{\pi}} = \sum_{n=0}^{\infty} \frac{(-1)^n(2^n)^2 t^n}{(2n)!\sqrt{\pi t}}$$

$$= \frac{1}{\sqrt{\pi t}}\sum_{n=0}^{\infty} \frac{(-1)^n(2\sqrt{t})^{2n}}{(2n)!} = \frac{\cos(2\sqrt{t})}{\sqrt{\pi t}}.$$

Copyright © 2012 Pearson Education, Inc. Publishing as Addison-Wesley.

57. Using (7.42) and assuming that the inverse Laplace transform of the series can be computed term-wise, we get

$$\mathcal{L}^{-1}\left\{\ln\left(1+\frac{1}{s^2}\right)\right\}(t) = \mathcal{L}^{-1}\left\{\sum_{n=1}^{\infty}\frac{(-1)^{n+1}}{ns^{2n}}\right\}(t) = \sum_{n=1}^{\infty}\frac{(-1)^{n+1}}{n}\mathcal{L}^{-1}\left\{\frac{1}{s^{2n}}\right\}(t).$$

From Table 7.1 in Section 7.2, $\mathcal{L}^{-1}\left\{1/s^{k+1}\right\} = t^k/k!$, $k = 1, 2, \ldots$. With $k = 2n - 1$, this yields

$$\mathcal{L}^{-1}\left\{\frac{1}{s^{2n}}\right\}(t) = \frac{t^{2n-1}}{(2n-1)!}, \qquad n = 1, 2, \ldots$$

and, therefore,

$$\mathcal{L}^{-1}\left\{\ln\left(1+\frac{1}{s^2}\right)\right\}(t) = \sum_{n=1}^{\infty}\frac{(-1)^{n+1}}{n}\frac{t^{2n-1}}{(2n-1)!} = -\frac{2}{t}\sum_{n=1}^{\infty}\frac{(-1)^n}{(2n)!}t^{2n}. \qquad (7.43)$$

Since

$$\cos t = \sum_{n=0}^{\infty}\frac{(-1)^n}{(2n)!}t^{2n} = 1 + \sum_{n=1}^{\infty}\frac{(-1)^n}{(2n)!}t^{2n},$$

(7.43) implies that

$$\mathcal{L}^{-1}\left\{\ln\left(1+\frac{1}{s^2}\right)\right\}(t) = -\frac{2}{t}\left(\cos t - 1\right) = \frac{2(1-\cos t)}{t}.$$

59. In this problem, we use the method of solving "mixing problems" discussed in Section 3.2. So, let $x(t)$ denote the mass of salt in the tank at time t with $t = 0$ denoting the moment, when the process started. Thus, using the formula

$$\text{mass} = \text{volume} \times \text{concentration},$$

we have the initial condition

$$x(0) = 500\,(\text{L}) \times 0.2\,(\text{kg/L}) = 100\,(\text{kg}).$$

For the rate of change of $x(t)$, that is, $x'(t)$, we use the relation

$$x'(t) = \text{input rate} - \text{output rate}. \qquad (7.44)$$

While the output rate (through the exit valve C) can be computed as

$$\text{output rate} = \frac{x(t)}{500}\,(\text{kg/L}) \times 12\,(\text{L/min}) = \frac{3x(t)}{125}\,(\text{kg/min})$$

 Copyright © 2012 Pearson Education, Inc. Publishing as Addison-Wesley.

for all t, the input rate has different formulas for the first 10 minutes and after that. Namely,

$$0 < t < 10 \text{ (valve A)}: \quad \text{input rate} = 12\,(\text{L/min}) \times 0.4\,(\text{kg/L}) = 4.8\,(\text{kg/min});$$

$$t > 10 \text{ (valve B)}: \quad \text{input rate} = 12\,(\text{L/min}) \times 0.6\,(\text{kg/L}) = 7.2\,(\text{kg/min}).$$

In other words, the input rate is a function of t, which can be written as

$$\text{input rate} = g(t) = \begin{cases} 4.8, & 0 < t < 10, \\ 7.2, & t > 10. \end{cases}$$

Using the unit step function, we can express $g(t) = 4.8 + 2.4u(t-10)$. Thus, (7.44) gives

$$x'(t) = g(t) - \frac{3x(t)}{125} \quad \Rightarrow \quad x'(t) + \frac{3}{125}x(t) = 4.8 + 2.4u(t-10) \qquad (7.45)$$

with the initial condition $x(0) = 100$. Taking the Laplace transform of both sides yields

$$\mathcal{L}\{x'\}(s) + \frac{3}{125}\mathcal{L}\{x\}(s) = \mathcal{L}\{4.8 + 2.4u(t-10)\}(s) = \frac{4.8}{s} + \frac{2.4e^{-10s}}{s}$$

$$\Rightarrow \quad [s\mathcal{L}\{x\}(s) - 100] + \frac{3}{125}\mathcal{L}\{x\}(s) = \frac{4.8}{s} + \frac{2.4e^{-10s}}{s}$$

$$\Rightarrow \quad \mathcal{L}\{x\}(s) = \frac{100s + 4.8}{s[s + (3/125)]} + \frac{2.4}{s[s + (3/125)]}e^{-10s}. \qquad (7.46)$$

Since

$$\frac{2.4}{s[s + (3/125)]} = 100\left(\frac{1}{s} - \frac{1}{s + (3/125)}\right),$$

$$\frac{100s + 4.8}{s[s + (3/125)]} = 100\left(\frac{2}{s} - \frac{1}{s + (3/125)}\right),$$

applying the inverse Laplace transform in (7.46), we get

$$x(t) = 100\left(2 - e^{-3t/125}\right) + 100\left(1 - e^{-3(t-10)/125}\right)u(t-10).$$

Finally, dividing by the volume of the solution in the tank, which constantly equals to 500 L, we conclude that

$$\text{concentration} = 0.4 - 0.2e^{-3t/125} + 0.2\left(1 - e^{-3(t-10)/125}\right)u(t-10).$$

61. In this problem, the solution still enters the tank at a rate 12 L/min, but leaves the tank at a rate 6 L/min. Thus, every minute, the volume of the solution in the tank increases by $12 - 6 = 6$ (L). Therefore, the volume, as a function of t, is given by $500 + 6t$, and so

$$\text{output rate} = \frac{x(t)}{500 + 6t}\,(\text{kg/L}) \times 6\,(\text{L/min}) = \frac{3x(t)}{250 + 3t}\,(\text{kg/min}).$$

Copyright © 2012 Pearson Education, Inc. Publishing as Addison-Wesley.

Instead of equation (7.45) in Problem 61, we now have

$$x'(t) = g(t) - \frac{3x(t)}{250 + 3t} \qquad \Rightarrow \qquad (250 + 3t)x'(t) + 3x(t) = (250 + 3t)[4.8 + 2.4u(t - 10)].$$

(Note that this equation has polynomial coefficients and so it can be solved using the Laplace transform method. See the discussion and Example 4 in Section 7.5.)

EXERCISES 7.7: Convolution

1. Let $Y(s) := \mathcal{L}\{y\}(s)$, $G(s) := \mathcal{L}\{g\}(s)$. Taking the Laplace transform of both sides of the given differential equation and using the linear property of the Laplace transform, we obtain

$$\mathcal{L}\{y''\}(s) - 2\mathcal{L}\{y'\}(s) + Y(s) = G(s).$$

The initial conditions and Theorem 5, Section 7.3, imply that

$$\mathcal{L}\{y'\}(s) = sY(s) + 1,$$
$$\mathcal{L}\{y''\}(s) = s^2 Y(s) + s - 1.$$

Thus, substituting yields

$$\left[s^2 Y(s) + s - 1\right] - 2\left[sY(s) + 1\right] + Y(s) = G(s)$$
$$\Rightarrow \qquad \left(s^2 - 2s + 1\right) Y(s) = 3 - s + G(s)$$
$$\Rightarrow \qquad Y(s) = \frac{3 - s}{s^2 - 2s + 1} + \frac{G(s)}{s^2 - 2s + 1} = \frac{2}{(s-1)^2} - \frac{1}{s-1} + \frac{G(s)}{(s-1)^2}.$$

Taking now the inverse Laplace transform, we obtain

$$y(t) = 2\mathcal{L}^{-1}\left\{\frac{1}{(s-1)^2}\right\}(t) - \mathcal{L}^{-1}\left\{\frac{1}{s-1}\right\}(t) + \mathcal{L}^{-1}\left\{\frac{G(s)}{(s-1)^2}\right\}(t).$$

Using Table 7.1, we find

$$\mathcal{L}^{-1}\left\{\frac{1}{s-1}\right\}(t) = e^t, \qquad \mathcal{L}^{-1}\left\{\frac{1}{(s-1)^2}\right\}(t) = te^t,$$

and, by the convolution theorem,

$$\mathcal{L}^{-1}\left\{\frac{G(s)}{(s-1)^2}\right\}(t) = \mathcal{L}^{-1}\left\{\frac{1}{(s-1)^2} G(s)\right\}(t) = \left(te^t\right) * g(t) = \int_0^t (t - v)e^{t-v} g(v)\, dv.$$

Thus,

$$y(t) = 2te^t - e^t + \int_0^t (t - v)e^{t-v} g(v)\, dv.$$

     Copyright © 2012 Pearson Education, Inc. Publishing as Addison-Wesley.

3. Taking the Laplace transform of $y'' + 4y' + 5y = g(t)$ and applying the initial conditions $y(0) = y'(0) = 1$, give us

$$\left[s^2 Y(s) - s - 1\right] + 4\left[sY(s) - 1\right] + 5Y(s) = G(s),$$

where $Y(s) := \mathcal{L}\{y\}(s)$, $G(s) := \mathcal{L}\{g\}(s)$. Thus,

$$Y(s) = \frac{s+5}{s^2 + 4s + 5} + \frac{G(s)}{s^2 + 4s + 5} = \frac{s+2}{(s+2)^2 + 1} + \frac{3}{(s+2)^2 + 1} + \frac{G(s)}{(s+2)^2 + 1}.$$

Taking the inverse Laplace transform of $Y(s)$, with help of the convolution theorem, yields

$$y(t) = e^{-2t}\cos t + 3e^{-2t}\sin t + \int_0^t e^{-2(t-v)}\sin(t-v)g(v)\,dv.$$

5. Since $\mathcal{L}^{-1}\{1/s\}(t) = 1$ and $\mathcal{L}^{-1}\{1/(s^2 + 1)\}(t) = \sin t$, writing

$$\frac{1}{s(s^2 + 1)} = \frac{1}{s} \cdot \frac{1}{s^2 + 1}.$$

and using the convolution theorem, we obtain

$$\mathcal{L}^{-1}\left\{\frac{1}{s(s^2 + 1)}\right\}(t) = 1 * \sin t = \int_0^t \sin v\,dv = -\cos v\,\Big|_{v=0}^{v=t} = 1 - \cos t.$$

7. From Table 7.1, $\mathcal{L}^{-1}\{1/(s-a)\}(t) = e^{at}$. Therefore, using linearity of the inverse Laplace transform and the convolution theorem, we get

$$\mathcal{L}^{-1}\left\{\frac{14}{(s+2)(s-5)}\right\}(t) = 14\mathcal{L}^{-1}\left\{\frac{1}{s+2} \cdot \frac{1}{s-5}\right\}(t)$$

$$= 14e^{-2t} * e^{5t} = 14\int_0^t e^{-2(t-v)}e^{5v}\,dv$$

$$= 14e^{-2t}\int_0^t e^{7v}\,dv = 2e^{-2t}\left(e^{7t} - 1\right) = 2\left(e^{5t} - e^{-2t}\right).$$

9. Since $s/(s^2 + 1)^2 = [s/(s^2 + 1)] \cdot [1/(s^2 + 1)]$, the convolution theorem tells us that

$$\mathcal{L}^{-1}\left\{\frac{s}{(s^2 + 1)^2}\right\}(t) = \mathcal{L}^{-1}\left\{\frac{s}{s^2 + 1} \cdot \frac{1}{s^2 + 1}\right\}(t) = \cos t * \sin t = \int_0^t \cos(t - v)\sin v\,dv.$$

Copyright © 2012 Pearson Education, Inc. Publishing as Addison-Wesley.

Using the identity $\sin\alpha\cos\beta = [\sin(\alpha + \beta) + \sin(\alpha - \beta)]/2$, we get

$$\mathcal{L}^{-1}\left\{\frac{s}{(s^2+1)^2}\right\}(t) = \frac{1}{2}\int_0^t [\sin t + \sin(t - 2v)]\,dv$$

$$= \frac{1}{2}\left(v\sin t + \frac{\cos(t-2v)}{2}\right)\Bigg|_{v=0}^{v=t} = \frac{t\sin t}{2}.$$

11. Using the hint, we can write

$$\frac{s}{(s-1)(s+2)} = \frac{1}{s+2} + \frac{1}{(s-1)(s+2)},$$

so that, by Theorem 11,

$$\mathcal{L}^{-1}\left\{\frac{s}{(s-1)(s+2)}\right\}(t) = \mathcal{L}^{-1}\left\{\frac{1}{s+2}\right\}(t) + \mathcal{L}^{-1}\left\{\frac{1}{(s-1)(s+2)}\right\}(t)$$

$$= e^{-2t} + e^t * e^{-2t} = e^{-2t} + \int_0^t e^{t-v}e^{-2v}\,dv$$

$$= e^{-2t} + e^t\int_0^t e^{-3v}\,dv = e^{-2t} - \frac{e^t}{3}\left(e^{-3t} - 1\right) = \frac{2e^{-2t}}{3} + \frac{e^t}{3}.$$

13. Note that $f(t) = t * e^{3t}$. Hence, by the convolution theorem,

$$\mathcal{L}\{f(t)\}(s) = \mathcal{L}\{t\}(s)\mathcal{L}\{e^{3t}\}(s) = \frac{1}{s^2}\cdot\frac{1}{s-3} = \frac{1}{s^2(s-3)}.$$

15. Since

$$\int_0^t y(v)\sin(t-v)\,dv = \sin t * y(t),$$

denoting $Y(s) := \mathcal{L}\{y\}(s)$ and taking the Laplace transform of both sides of the original equation, we obtain

$$Y(s) + 3\mathcal{L}\{\sin t * y(t)\}(s) = \mathcal{L}\{t\}(s)$$

$$\Rightarrow \quad Y(s) + 3\mathcal{L}\{\sin t\}(s)Y(s) = Y(s) + \frac{3}{s^2+1}Y(s) = \frac{1}{s^2}$$

$$\Rightarrow \quad Y(s) = \frac{s^2+1}{s^2(s^2+4)} = \frac{(1/4)}{s^2} + \frac{(3/8)(2)}{s^2+2^2}$$

$$\Rightarrow \quad y(t) = \mathcal{L}^{-1}\left\{\frac{(1/4)}{s^2} + \frac{(3/8)(2)}{s^2+2^2}\right\}(t) = \frac{t}{4} + \frac{3\sin 2t}{8}.$$

 Copyright © 2012 Pearson Education, Inc. Publishing as Addison-Wesley.

17. We use the convolution Theorem 11 to find the Laplace transform of the integral term.

$$\mathcal{L}\left\{\int_0^t (t-v)y(v)\,dv\right\}(s) = \mathcal{L}\{t * y(t)\}(s) = \mathcal{L}\{t\}(s)\mathcal{L}\{y(t)\}(s) = \frac{Y(s)}{s^2},$$

where $Y(s)$ denotes the Laplace transform of $y(t)$. Thus, taking the Laplace transform of both sides of the given equation yields

$$Y(s) + \frac{Y(s)}{s^2} = \frac{1}{s} \quad \Rightarrow \quad Y(s) = \frac{s}{s^2+1} \quad \Rightarrow \quad y(t) = \mathcal{L}^{-1}\left\{\frac{s}{s^2+1}\right\}(t) = \cos t.$$

19. By the convolution theorem,

$$\mathcal{L}\left\{\int_0^t (t-v)^2 y(v)\,dv\right\}(s) = \mathcal{L}\{t^2 * y(t)\}(s) = \mathcal{L}\{t^2\}(s)\mathcal{L}\{y(t)\}(s) = \frac{2Y(s)}{s^3}.$$

Hence, applying the Laplace transform to the original equation gives us

$$Y(s) + \frac{2Y(s)}{s^3} = \mathcal{L}\{t^3 + 3\}(s) = \frac{6}{s^4} + \frac{3}{s}$$

$$\Rightarrow \quad Y(s) = \frac{s^3}{s^3 + 2} \cdot \frac{6 + 3s^3}{s^4} = \frac{3}{s}$$

$$\Rightarrow \quad y(t) = \mathcal{L}^{-1}\left\{\frac{3}{s}\right\}(t) = 3.$$

21. As in Example 3, we first rewrite the given integro-differential equation as

$$y'(t) + y(t) - y(t) * \sin t = -\sin t, \qquad y(0) = 1. \tag{7.47}$$

We now take the Laplace transform of (7.47) to obtain

$$[sY(s) - 1] + Y(s) - \frac{1}{s^2+1}Y(s) = -\frac{1}{s^2+1},$$

where $Y(s) := \mathcal{L}\{y\}(s)$. Thus,

$$Y(s) = \frac{s^2}{s^3 + s^2 + s} = \frac{s}{s^2 + s + 1} = \frac{s}{(s+1/2)^2 + 3/4}$$

$$= \frac{s+1/2}{(s+1/2)^2 + 3/4} - \frac{(1/\sqrt{3})(\sqrt{3}/2)}{(s+1/2)^2 + 3/4}.$$

Taking the inverse Laplace transform yields

$$y(t) = e^{-t/2}\cos\left(\frac{\sqrt{3}t}{2}\right) - \frac{1}{\sqrt{3}}e^{-t/2}\sin\left(\frac{\sqrt{3}t}{2}\right).$$

Copyright © 2012 Pearson Education, Inc. Publishing as Addison-Wesley.

23. Taking the Laplace transform of the differential equation and assuming zero initial conditions, we obtain

$$s^2 Y(s) + 9Y(s) = G(s),$$

where $Y(s) := \mathcal{L}\{y\}(s)$, $G(s) := \mathcal{L}\{g\}(s)$. Thus,

$$H(s) = \frac{Y(s)}{G(s)} = \frac{1}{s^2 + 9}.$$

The impulse response function is then

$$h(t) = \mathcal{L}^{-1}\{H(s)\}(t) = \mathcal{L}^{-1}\left\{\frac{1}{s^2 + 9}\right\}(t) = \frac{1}{3}\mathcal{L}^{-1}\left\{\frac{3}{s^2 + 3^2}\right\}(t) = \frac{\sin 3t}{3}.$$

To solve the initial value problem, we need the solution to the corresponding homogeneous problem. Its auxiliary equation, $r^2 + 9 = 0$, has roots, $r = \pm 3i$. Thus, a general solution to the homogeneous equation is $y_h(t) = C_1 \cos 3t + C_2 \sin 3t$. Applying the initial conditions $y(0) = 2$ and $y'(0) = -3$, we obtain

$$\begin{aligned} 2 &= y(0) = (C_1 \cos 3t + C_2 \sin 3t)\big|_{t=0} = C_1, \\ -3 &= y'(0) = (-3C_1 \sin 3t + 3C_2 \cos 3t)\big|_{t=0} = 3C_2 \end{aligned} \qquad \Rightarrow \qquad \begin{aligned} C_1 &= 2, \\ C_2 &= -1. \end{aligned}$$

So, $y_k(t) = 2\cos 3t - \sin 3t$, and the formula for the solution to the original initial value problem is

$$y = (h * g)(t) + y_k(t) = \frac{1}{3}\int_0^t g(v)\sin 3(t - v)\, dv + 2\cos 3t - \sin 3t.$$

25. Taking the Laplace transform of both sides of the given equation and assuming zero initial conditions, we get

$$\mathcal{L}\{y'' - y' - 6y\}(s) = \mathcal{L}\{g(t)\}(s) \qquad \Rightarrow \qquad s^2 Y(s) - sY(s) - 6Y(s) = G(s).$$

Thus,

$$H(s) = \frac{Y(s)}{G(s)} = \frac{1}{s^2 - s - 6} = \frac{1}{(s - 3)(s + 2)}$$

is the transfer function. The impulse response function $h(t)$ is then given by

$$h(t) = \mathcal{L}^{-1}\left\{\frac{1}{(s - 3)(s + 2)}\right\}(t) = e^{3t} * e^{-2t}$$

$$= \int_0^t e^{3(t-v)} e^{-2v}\, dv = e^{3t}\, \frac{e^{-5v}}{-5}\bigg|_{v=0}^{v=t} = \frac{e^{3t} - e^{-2t}}{5}.$$

   Copyright © 2012 Pearson Education, Inc. Publishing as Addison-Wesley.

To solve the given initial value problem, we use Theorem 12. To this end, we need the solution $y_k(t)$ to the corresponding initial value problem for the homogeneous equation. That is,

$$y'' - y' - 6y = 0, \qquad y(0) = 1, \quad y'(0) = 8$$

(see (19) in the text). Applying the Laplace transform yields

$$\left[s^2 Y_k(s) - s - 8\right] - \left[s Y_k(s) - 1\right] - 6 Y_k(s) = 0$$

$$\Rightarrow \qquad Y_k(s) = \frac{s+7}{s^2 - s - 6} = \frac{s+7}{(s-3)(s+2)} = \frac{2}{s-3} - \frac{1}{s+2}$$

$$\Rightarrow \qquad y_k(t) = \mathcal{L}^{-1}\left\{Y_k(s)\right\}(t) = \mathcal{L}^{-1}\left\{\frac{2}{s-3} - \frac{1}{s+2}\right\}(t) = 2e^{3t} - e^{-2t}.$$

Therefore,

$$y(t) = (h * g)(t) + y_k(t) = \frac{1}{5}\int_0^t \left[e^{3(t-v)} - e^{-2(t-v)}\right] g(v)\, dv + 2e^{3t} - e^{-2t}.$$

27. Taking the Laplace transform and assuming zero initial conditions, we compute the transfer function $H(s)$.

$$s^2 Y(s) - 2sY(s) + 5Y(s) = G(s) \qquad \Rightarrow \qquad H(s) = \frac{Y(s)}{G(s)} = \frac{1}{s^2 - 2s + 5}.$$

Therefore, the impulse response function is

$$h(t) = \mathcal{L}^{-1}\left\{H(s)\right\}(t) = \mathcal{L}^{-1}\left\{\frac{1}{(s-1)^2 + 2^2}\right\}(t) = \frac{1}{2} e^t \sin 2t.$$

Next, we find the solution $y_k(t)$ to the corresponding initial value problem for the homogeneous equation,

$$y'' - 2y' + 5y = 0, \qquad y(0) = 0, \quad y'(0) = 2.$$

Since the associated equation, $r^2 - 2r + 5 = 0$, has roots $r = 1 \pm 2i$, a general solution to the homogeneous equations is

$$y_h(t) = e^t \left(C_1 \cos 2t + C_2 \sin 2t\right).$$

We satisfy the initial conditions by solving

$$\begin{aligned} 0 &= y(0) = C_1 \\ 2 &= y'(0) = C_1 + 2C_2 \end{aligned} \qquad \Rightarrow \qquad \begin{aligned} C_1 &= 0, \\ C_2 &= 1. \end{aligned}$$

Copyright © 2012 Pearson Education, Inc. Publishing as Addison-Wesley.

Hence, $y_k(t) = e^t \sin 2t$ and

$$y(t) = (h * g)(t) + y_k(t) = \frac{1}{2} \int_0^t e^{t-v} \sin 2(t-v) g(v) \, dv + e^t \sin 2t$$

is the desired solution.

29. With the given data, the initial value problem becomes

$$5I''(t) + 20I'(t) + \frac{1}{0.005} I(t) = e(t), \qquad I(0) = -1, \quad I'(0) = 8.$$

Using the formula (15), we find the transfer function

$$H(s) = \frac{1}{5s^2 + 20s + 200} = \frac{1}{5} \frac{1}{(s+2)^2 + 6^2}.$$

Therefore,

$$h(t) = \mathcal{L}^{-1} \left\{ \frac{1}{5} \frac{1}{(s+2)^2 + 6^2} \right\} (t) = \frac{1}{30} e^{-2t} \sin 6t.$$

Next, we consider the initial value problem

$$5I''(t) + 20I'(t) + 200I(t) = 0, \qquad I(0) = -1, \quad I'(0) = 8$$

for the corresponding homogeneous equation. Its auxiliary equation, $5r^2 + 20r + 200 = 0$, has roots $r = -2 \pm 6i$, which imply a general solution

$$I_h(t) = e^{-2t} \left(C_1 \cos 6t + C_2 \sin 6t \right).$$

Next, we find constants C_1 and C_2 such that the solution satisfies the initial conditions.

$$\begin{aligned} -1 &= I(0) = C_1, \\ 8 &= I'(0) = -2C_1 + 6C_2 \end{aligned} \qquad \Rightarrow \qquad \begin{aligned} C_1 &= -1, \\ C_2 &= 1, \end{aligned}$$

and so $I_k(t) = e^{-2t} (\sin 6t - \cos 6t)$. Finally,

$$I(t) = h(t) * e(t) + I_k(t) = \frac{1}{30} \int_0^t e(v) e^{-2(t-v)} \sin 6(t-v) \, dv + e^{-2t} (\sin 6t - \cos 6t).$$

31. By the convolution theorem, we get

$$\mathcal{L} \{1 * 1 * 1\} (s) = \mathcal{L} \{1\} (s) \mathcal{L} \{1 * 1\} (s) = \mathcal{L} \{1\} (s) \mathcal{L} \{1\} (s) \mathcal{L} \{1\} (s) = \left(\frac{1}{s} \right)^3 = \frac{1}{s^3}.$$

Applying the inverse Laplace transform yields

$$1 * 1 * 1 = \mathcal{L}^{-1} \left\{ \frac{1}{s^3} \right\} (t) = \frac{1}{2} \mathcal{L}^{-1} \left\{ \frac{2}{s^3} \right\} (t) = \frac{1}{2} t^2.$$

Copyright © 2012 Pearson Education, Inc. Publishing as Addison-Wesley.

33. Using the linear property of integrals, we have

$$f * (g + h) = \int_0^t f(t - v)(g + h)(v)\, dv = \int_0^t f(t - v)[g(v) + h(v)]\, dv$$

$$= \int_0^t f(t - v)g(v)\, dv + \int_0^t f(t - v)h(v)\, dv = f * g + f * h.$$

35. Since

$$\int_0^t f(v)\, dv = \int_0^t 1 \cdot f(v)\, dv = 1 * f(t),$$

we conclude that

$$\mathcal{L}\left\{\int_0^t f(v)\, dv\right\}(s) = \mathcal{L}\left\{1 * f(t)\right\}(s) = \mathcal{L}\left\{1\right\}(s)\mathcal{L}\left\{f(t)\right\}(s) = \frac{1}{s}F(s).$$

Hence, applying the inverse Laplace transform, we get

$$\int_0^t f(v)\, dv = \mathcal{L}^{-1}\left\{\frac{1}{s}F(s)\right\}(t).$$

(Note that the integral in the left-hand side is a continuous function.)

37. Actually, this statement holds for any continuously differentiable function $h(t)$ on $[0, \infty)$ satisfying $h(0) = 0$. Indeed, first of all,

$$(h * g)(0) = \int_0^t h(t - v)g(v)\, dv\, \bigg|_{t=0} = \int_0^0 h(-v)g(v)\, dv = 0\,,$$

since the interval of integration has zero length. Next, we apply the Leibniz's rule to find the derivative of $(h * g)(t)$.

$$(h * g)'(t) = \left(\int_0^t h(t - v)g(v)\, dv\right)'$$

$$= \int_0^t \frac{\partial h(t - v)g(v)}{\partial t}\, dv + h(t - v)g(v)\, \bigg|_{v=t}$$

$$= \int_0^t h'(t - v)g(v)\, dv + h(0)g(t) = \int_0^t h'(t - v)g(v)\, dv\,,$$

Copyright © 2012 Pearson Education, Inc. Publishing as Addison-Wesley.

since $h(0) = 0$. Therefore,

$$(h * g)'(0) = \int_0^0 h'(-v)g(v)\,dv = 0\,.$$

EXERCISES 7.8: Impulses and the Dirac Delta Function

1. Applying the equation (3), we get

$$\int_{-\infty}^{\infty} (t^2 - 1)\delta(t)\,dt = (t^2 - 1)\big|_{t=0} = -1.$$

3. By the equation (3),

$$\int_{-\infty}^{\infty} (\sin 3t)\delta\left(t - \frac{\pi}{2}\right)dt = \sin\left(3 \cdot \frac{\pi}{2}\right) = -1.$$

5. The formula (6) for the Laplace transform of the Dirac delta function yields

$$\int_0^{\infty} e^{-2t}\delta(t - 1)\,dt = \mathcal{L}\left\{\delta(t - 1)\right\}(2) = e^{-s}\big|_{s=2} = e^{-2}\,.$$

7. Using linearity of the Laplace transform and (6), we get

$$\mathcal{L}\left\{\delta(t - 1) - \delta(t - 3)\right\}(s) = \mathcal{L}\left\{\delta(t - 1)\right\}(s) - \mathcal{L}\left\{\delta(t - 3)\right\}(s) = e^{-s} - e^{-3s}\,.$$

9. Since $\delta(t - 1) = 0$ for $t < 1$,

$$\mathcal{L}\left\{t\delta(t - 1)\right\}(s) = \int_0^{\infty} e^{-st}t\delta(t - 1)\,dt = \int_{-\infty}^{\infty} e^{-st}t\delta(t - 1)\,dt = e^{-st}t\big|_{t=1} = e^{-s}$$

by the equation (3).

Another way to solve this problem is to use Theorem 6 of Section 7.3.

$$\mathcal{L}\left\{t\delta(t - 1)\right\}(s) = -\frac{d}{ds}\mathcal{L}\left\{\delta(t - 1)\right\}(s) = -\frac{d\left(e^{-s}\right)}{ds} = e^{-s}\,.$$

Copyright © 2012 Pearson Education, Inc. Publishing as Addison-Wesley.

11. Since $\delta(t - \pi) = 0$ for $t < \pi$, we use the definition of the Laplace transform and the formula (3) to conclude that

$$\mathcal{L}\{(\sin t)\delta(t - \pi)\}(s) = \int_0^\infty e^{-st}(\sin t)\delta(t - \pi)\,dt$$

$$= \int_{-\infty}^\infty e^{-st}(\sin t)\delta(t - \pi)\,dt = e^{-\pi t}\sin\pi = 0.$$

13. Let $W(s) := \mathcal{L}\{w\}(s)$. Using the initial conditions and Theorem 5 in Section 7.3, we find that

$$\mathcal{L}\{w''\}(s) = s^2 W(s) - sw(0) - w'(0) = s^2 W(s).$$

Thus, applying the Laplace transform to both sides of the given equation yields

$$s^2 W(s) + W(s) = \mathcal{L}\{\delta(t - \pi)\}(s) = e^{-\pi s} \quad\Rightarrow\quad W(s) = \frac{e^{-\pi s}}{s^2 + 1}.$$

Taking now the inverse Laplace transform of both sides and using Theorem 8 of Section 7.6, we get

$$w(t) = \mathcal{L}^{-1}\left\{\frac{e^{-\pi s}}{s^2 + 1}\right\}(t) = \mathcal{L}^{-1}\left\{\frac{1}{s^2 + 1}\right\}(t - \pi)u(t - \pi)$$

$$= \sin(t - \pi)u(t - \pi) = -(\sin t)u(t - \pi).$$

15. Let $Y := \mathcal{L}\{y\}$. Taking the Laplace transform of $y'' + 2y' - 3y = \delta(t - 1) - \delta(t - 2)$ and applying the initial conditions $y(0) = 2$, $y'(0) = -2$, we obtain

$$\left[s^2 Y(s) - 2s + 2\right] + 2\left[sY(s) - 2\right] - 3Y(s) = \mathcal{L}\{\delta(t - 1) - \delta(t - 2)\}(s) = e^{-s} - e^{-2s}$$

$$\Rightarrow\quad Y(s) = \frac{2s + 2 + e^{-s} - e^{-2s}}{s^2 + 2s - 3} = \frac{2s + 2}{(s + 3)(s - 1)} + \frac{e^{-s}}{(s + 3)(s - 1)} - \frac{e^{-2s}}{(s + 3)(s - 1)}$$

$$= \frac{1}{s - 1} + \frac{1}{s + 3} + \frac{e^{-s}}{4}\left(\frac{1}{s - 1} - \frac{1}{s + 3}\right) - \frac{e^{-2s}}{4}\left(\frac{1}{s - 1} - \frac{1}{s + 3}\right),$$

so that by Theorem 8, Section 7.6, we get

$$y(t) = e^t + e^{-3t} + \frac{1}{4}\left(e^{t-1} - e^{-3(t-1)}\right)u(t - 1) - \frac{1}{4}\left(e^{t-2} - e^{-3(t-2)}\right)u(t - 2).$$

17. Let $Y := \mathcal{L}\{y\}$. We use the initial conditions to find that

$$\mathcal{L}\{y''\}(s) = s^2 Y(s) - sy(0) - y'(0) = s^2 Y(s) - 2.$$

Copyright © 2012 Pearson Education, Inc. Publishing as Addison-Wesley.

Chapter 7

Thus, taking the Laplace transform of both sides and using the formula (6), we get

$$\left[s^2 Y(s) - 2\right] - Y(s) = 4\mathcal{L}\left\{\delta(t-2)\right\}(s) + \mathcal{L}\left\{t^2\right\}(s) = 4e^{-2s} + \frac{2}{s^3}$$

$$\Rightarrow \quad Y(s) = \frac{4e^{-2s}}{s^2 - 1} + \frac{2(s^3 + 1)}{s^3(s^2 - 1)} = 2e^{-2s}\left(\frac{1}{s-1} - \frac{1}{s+1}\right) + \frac{2}{s-1} - \frac{2}{s^3} - \frac{2}{s}.$$

Now we can apply the inverse Laplace transform.

$$y(t) = \mathcal{L}^{-1}\left\{2e^{-2s}\left(\frac{1}{s-1} - \frac{1}{s+1}\right) + \frac{2}{s-1} - \frac{2}{s^3} - \frac{2}{s}\right\}(t)$$

$$= 2\left(\mathcal{L}^{-1}\left\{\frac{1}{s-1}\right\} - \mathcal{L}^{-1}\left\{\frac{1}{s+1}\right\}\right)(t-2)u(t-2)$$

$$+ 2\mathcal{L}^{-1}\left\{\frac{1}{s-1}\right\}(t) - \mathcal{L}^{-1}\left\{\frac{2}{s^3}\right\}(t) - 2\mathcal{L}^{-1}\left\{\frac{1}{s}\right\}(t)$$

$$= 2\left(e^{t-2} - e^{2-t}\right)u(t-2) + 2e^t - t^2 - 2.$$

19. Let $W(s) := \mathcal{L}\left\{w\right\}(s)$. We apply the Laplace transform to the given equation and obtain

$$\mathcal{L}\left\{w''\right\}(s) + 6\mathcal{L}\left\{w'\right\}(s) + 5W(s) = \mathcal{L}\left\{e^t \delta(t-1)\right\}(s). \tag{7.48}$$

From the formula (4), Section 7.3, we see that

$$\begin{aligned}\mathcal{L}\left\{w'\right\}(s) &= sW(s) - w(0) = sW(s),\\ \mathcal{L}\left\{w''\right\}(s) &= s^2 W(s) - sw(0) - w'(0) = s^2 W(s) - 4.\end{aligned} \tag{7.49}$$

In addition, the translation property (1) of the Laplace transform (see Section 7.3) yields

$$\mathcal{L}\left\{e^t \delta(t-1)\right\}(s) = \mathcal{L}\left\{\delta(t-1)\right\}(s-1) = e^{-(s-1)} = e^{1-s}. \tag{7.50}$$

Substituting (7.49) and (7.50) back into (7.48), we obtain

$$\left[s^2 W(s) - 4\right] + 6\left[sW(s)\right] + 5W(s) = e^{1-s}$$

$$\Rightarrow \quad W(s) = \frac{4 + e^{1-s}}{s^2 + 6s + 5} = \frac{4 + e^{1-s}}{(s+1)(s+5)} = \frac{1}{s+1} - \frac{1}{s+5} + \frac{e}{4}e^{-s}\left(\frac{1}{s+1} - \frac{1}{s+5}\right).$$

Finally, the inverse Laplace transform of both sides of this equation yields

$$w(t) = e^{-t} - e^{-5t} + \frac{e}{4}\left(e^{-(t-1)} - e^{-5(t-1)}\right)u(t-1).$$

21. We apply the Laplace transform to the given equation, solve the resulting equation for $\mathcal{L}\left\{y\right\}(s)$, and then use the inverse Laplace transform. This gives us

$$\mathcal{L}\left\{y''\right\}(s) + \mathcal{L}\left\{y\right\}(s) = \mathcal{L}\left\{\delta(t-2\pi)\right\}(s)$$

 Copyright © 2012 Pearson Education, Inc. Publishing as Addison-Wesley.

$$\Rightarrow \quad \left[s^2 \mathcal{L}\{y\}(s) - 1\right] + \mathcal{L}\{y\}(s) = e^{-2\pi s} \quad \Rightarrow \quad \mathcal{L}\{y\}(s) = \frac{1 + e^{-2\pi s}}{s^2 + 1}$$

$$\Rightarrow \quad y(t) = \mathcal{L}^{-1}\left\{\frac{1}{s^2 + 1}\right\}(t) + \mathcal{L}^{-1}\left\{\frac{1}{s^2 + 1}\right\}(t - 2\pi)u(t - 2\pi)$$

$$= \sin t + [\sin(t - 2\pi)]u(t - 2\pi) = [1 + u(t - 2\pi)]\sin t.$$

The graph of the solution is shown in Fig. B.49 in the answers of the text.

23. The solution to the initial value problem

$$y'' + y = \delta(t - 2\pi), \qquad y(0) = 0, \quad y'(0) = 1$$

is given in Problem 21, that is

$$y_1(t) = [1 + u(t - 2\pi)]\sin t.$$

Thus, if $y_2(t)$ is the solution to the initial value problem

$$y'' + y = -\delta(t - \pi), \qquad y(0) = 0, \quad y'(0) = 0, \tag{7.51}$$

then, by the superposition principle (see Section 4.5), $y(t) = y_1(t) + y_2(t)$ is the desired solution. The Laplace transform of both sides in (7.51) yields

$$s^2 \mathcal{L}\{y\}(s) + \mathcal{L}\{y\}(s) = -e^{-\pi s} \quad \Rightarrow \quad \mathcal{L}\{y\}(s) = -\frac{e^{-\pi s}}{s^2 + 1}$$

$$\Rightarrow \quad y_2(t) = -\mathcal{L}^{-1}\left\{\frac{1}{s^2 + 1}\right\}(t - \pi)u(t - \pi) = -\sin(t - \pi)u(t - \pi) = u(t - \pi)\sin t.$$

(We have used zero initial conditions to express $\mathcal{L}\{y''\}$ in terms of $\mathcal{L}\{y\}$.) Therefore, the answer is

$$y(t) = y_1(t) + y_2(t) = [1 + u(t - 2\pi)]\sin t + u(t - \pi)\sin t = [1 + u(t - \pi) + u(t - 2\pi)]\sin t.$$

The sketch of this curve is given in Fig. B.50.

25. Taking the Laplace transform of $y'' + 4y' + 8y = \delta(t)$ with zero initial conditions yields

$$s^2 Y(s) + 4sY(s) + 8Y(s) = \mathcal{L}\{\delta(t)\}(s) = 1.$$

Solving for $Y(s)$, we obtain

$$Y(s) = \frac{1}{s^2 + 4s + 8} = \frac{1}{(s + 2)^2 + 4} = \frac{1}{2}\frac{2}{(s + 2)^2 + 2^2}$$

Copyright © 2012 Pearson Education, Inc. Publishing as Addison-Wesley.

so that

$$h(t) = \frac{1}{2} \mathcal{L}^{-1} \left\{ \frac{2}{(s+2)^2 + 2^2} \right\} (t) = \frac{1}{2} e^{-2t} \sin 2t.$$

Notice that the transfer function $H(s)$ for $y'' + 4y' + 8y = g(t)$ with $y(0) = y'(0) = 0$ is given by $H(s) = 1/(s^2 + 4s + 8)$, so that again

$$h(t) = \mathcal{L}^{-1} \left\{ H(s) \right\} (t) = \frac{1}{2} e^{-2t} \sin 2t.$$

27. The Laplace transform of both sides of the given equation, with zero initial conditions and $g(t) = \delta(t)$, gives us

$$s^2 \mathcal{L} \{y\} (s) - 2s \mathcal{L} \{y\} (s) + 5 \mathcal{L} \{y\} (s) = \mathcal{L} \{\delta(t)\} (s) = 1$$

$$\Rightarrow \quad \mathcal{L} \{y\} (s) = \frac{1}{s^2 - 2s + 5} = \frac{1}{(s-1)^2 + 2^2}.$$

Applying the inverse Laplace transform now yields

$$h(t) = \mathcal{L}^{-1} \left\{ \frac{1}{(s-1)^2 + 2^2} \right\} (t) = \frac{1}{2} e^t \sin 2t.$$

29. We solve the given initial value problem for the displacement $x(t)$. Let $X(s) := \mathcal{L} \{x\} (s)$. Applying the Laplace transform to the differential equation, we obtain

$$\mathcal{L} \{x''\} (s) + 9X(s) = \mathcal{L} \left\{ -3\delta \left(t - \frac{\pi}{2} \right) \right\} (s) = -3e^{-\pi s/2}.$$

Since

$$\mathcal{L} \{x''\} (s) = s^2 X(s) - sx(0) - x'(0) = s^2 X(s) - s,$$

the above equation becomes

$$\left[s^2 X(s) - s \right] + 9X(s) = -3e^{-\pi s/2} \quad \Rightarrow \quad X(s) = \frac{s - 3e^{-\pi s/2}}{s^2 + 9} = \frac{s}{s^2 + 3^2} - e^{-\pi s/2} \frac{3}{s^2 + 3^2}.$$

Therefore,

$$x(t) = \mathcal{L}^{-1} \left\{ \frac{s}{s^2 + 3^2} - e^{-\pi s/2} \frac{3}{s^2 + 3^2} \right\} (t)$$

$$= \cos 3t - \left[\sin 3 \left(t - \frac{\pi}{2} \right) \right] u \left(t - \frac{\pi}{2} \right) = \left[1 - u \left(t - \frac{\pi}{2} \right) \right] \cos 3t.$$

Since, for $t > \pi/2$, $u(t - \pi/2) \equiv 1$, we conclude that

$$x(t) \equiv 0 \quad \text{for} \quad t > \frac{\pi}{2}.$$

This means that the mass stops after the hit and remains in the equilibrium position thereafter.

 Copyright © 2012 Pearson Education, Inc. Publishing as Addison-Wesley.

31. By taking the Laplace transform of

$$ay'' + by' + cy = \delta(t), \qquad y(0) = y'(0) = 0,$$

and solving for $Y := \mathcal{L}\{y\}$, we find that the transfer function $H(s)$ is given by

$$H(s) = \frac{1}{as^2 + bs + c}.$$

If the roots of the polynomial $as^2 + bs + c$ are real and distinct, say r_1 and r_2, then

$$H(s) = \frac{1}{(s - r_1)(s - r_2)} = \frac{1/(r_1 - r_2)}{s - r_1} - \frac{1/(r_1 - r_2)}{s - r_2}.$$

Thus,

$$h(t) = \frac{1}{r_1 - r_2}\left(e^{r_1 t} - e^{r_2 t}\right)$$

and, clearly, $h(t)$ is bounded if and only if r_1 and r_2 are non-positive. If the roots of $as^2 + bs + c$ are complex, then (by the quadratic formula) they are given by

$$-\frac{b}{2a} \pm \frac{\sqrt{4ac - b^2}}{2a}\, i$$

so that their real part is $-b/(2a)$. Now,

$$H(s) = \frac{1}{as^2 + bs + c} = \frac{1}{a} \cdot \frac{1}{s^2 + (b/a)s + (c/a)} = \frac{1}{a} \cdot \frac{1}{[s + b/(2a)]^2 + (4ac - b^2)/(4a^2)}$$

$$= \frac{2}{\sqrt{4ac - b^2}} \cdot \frac{\sqrt{4ac - b^2}/(2a)}{[s + b/(2a)]^2 + [\sqrt{4ac - b^2}/(2a)]^2}$$

and we conclude that

$$h(t) = \frac{2}{\sqrt{4ac - b^2}} e^{-(b/2a)t} \sin\left(\frac{\sqrt{4ac - b^2}}{2a}\, t\right).$$

Thus, $h(t)$ is bounded if and only if $b/(2a)$ is nonnegative.

33. Let a function $f(t)$ be defined on $(-\infty, \infty)$ and continuous in a neighborhood of the origin. Since $\delta(t) = 0$ for any $t \neq 0$, so does the product $f(t)\delta(t)$. Therefore,

$$\int_{-\infty}^{\infty} f(t)\delta(t)\, dt = \int_{-\varepsilon}^{\varepsilon} f(t)\delta(t)\, dt \qquad \text{for any } \varepsilon > 0. \tag{7.52}$$

By the mean value theorem, for any $\varepsilon > 0$ small enough (so that $f(t)$ is continuous on $(-\varepsilon, \varepsilon)$) there exists a point ζ_ε in $(-\varepsilon, \varepsilon)$ such that

$$\int_{-\varepsilon}^{\varepsilon} f(t)\delta(t)\, dt = f(\zeta_\varepsilon)\int_{-\varepsilon}^{\varepsilon} \delta(t)\, dt = f(\zeta_\varepsilon)\int_{-\infty}^{\infty} \delta(t)\, dt = f(\zeta_\varepsilon).$$

Copyright © 2012 Pearson Education, Inc. Publishing as Addison-Wesley.

Together with (7.52), this yields

$$\int_{-\infty}^{\infty} f(t)\delta(t)\, dt = f(\zeta_\varepsilon) \quad \text{for any } \varepsilon > 0.$$

We now take limit, as $\varepsilon \to 0$, of both sides.

$$\lim_{\varepsilon \to 0} \left[\int_{-\infty}^{\infty} f(t)\delta(t)\, dt \right] = \lim_{\varepsilon \to 0} f(\zeta_\varepsilon).$$

Note that the integral in the left-hand side does not depend on ε, and so the limit equals to the integral itself. In the right-hand side, since ζ_ε belongs to $(-\varepsilon, \varepsilon)$, $\zeta_\varepsilon \to 0$ as $\varepsilon \to 0$, and the continuity of $f(t)$ implies that $f(\zeta_\varepsilon)$ converges to $f(0)$, as $\varepsilon \to 0$. Combining these observations, we get the required.

35. Following the hint, we solve the initial value problem

$$EIy^{(4)}(x) = L\delta(x - \lambda), \qquad y(0) = y'(0) = 0, \ y''(0) = A, \ y'''(0) = B.$$

Using these initial conditions and Theorem 5 of Section 7.3 with $n = 4$, we obtain

$$\mathcal{L}\left\{ y^{(4)}(x) \right\}(s) = s^4 \mathcal{L}\left\{ y(x) \right\}(s) - sA - B,$$

and so the Laplace transform of the given equation yields

$$EI\left[s^4 \mathcal{L}\left\{ y(x) \right\}(s) - sA - B \right] = L\mathcal{L}\left\{ \delta(x - \lambda) \right\}(s) = Le^{-\lambda s}.$$

Therefore,

$$\mathcal{L}\left\{ y(x) \right\}(s) = \frac{L}{EI}\frac{e^{-\lambda s}}{s^4} + \frac{A}{s^3} + \frac{B}{s^4}$$

$$\Rightarrow \quad y(x) = \mathcal{L}^{-1}\left\{ \frac{L}{EI}\frac{e^{-\lambda s}}{s^4} + \frac{A}{s^3} + \frac{B}{s^4} \right\}(x)$$

$$= \frac{L}{EI3!}\mathcal{L}^{-1}\left\{ \frac{3!}{s^4} \right\}(x - \lambda)u(x - \lambda) + \frac{A}{2!}\mathcal{L}^{-1}\left\{ \frac{2!}{s^3} \right\}(x) + \frac{B}{3!}\mathcal{L}^{-1}\left\{ \frac{3!}{s^4} \right\}(x)$$

$$= \frac{L}{6EI}(x - \lambda)^3 u(x - \lambda) + \frac{A}{2}x^2 + \frac{B}{6}x^3. \tag{7.53}$$

Next, we are looking for A and B such that $y''(2\lambda) = y'''(2\lambda) = 0$. Note that, for $x > \lambda$, $u(x - \lambda) \equiv 1$ and so (7.53) becomes

$$y(x) = \frac{L}{6EI}(x - \lambda)^3 + \frac{A}{2}x^2 + \frac{B}{6}x^3.$$

 Copyright © 2012 Pearson Education, Inc. Publishing as Addison-Wesley.

Differentiating, we get

$$y''(x) = \frac{L}{EI}(x - \lambda) + A + Bx \qquad \text{and} \qquad y'''(x) = \frac{L}{EI} + B.$$

Hence, A and B must satisfy

$$
\begin{aligned}
0 &= y''(2\lambda) = [L/(EI)](2\lambda - \lambda) + A + 2B\lambda, \\
0 &= y'''(2\lambda) = L/(EI) + B
\end{aligned}
\qquad \Rightarrow \qquad
\begin{aligned}
A &= \lambda L/(EI), \\
B &= -L/(EI).
\end{aligned}
$$

Substituting A and B back into (7.53) yields the solution

$$y(x) = \frac{L}{6EI}\left[(x - \lambda)^3 u(x - \lambda) + 3\lambda x^2 - x^3\right].$$

EXERCISES 7.9: Solving Linear Systems with Laplace Transforms

1. Let $X(s) = \mathcal{L}\{x\}(s)$, $Y(s) = \mathcal{L}\{y\}(s)$. Applying the Laplace transform to both sides of the given equations yields

$$
\begin{aligned}
\mathcal{L}\{x'\}(s) &= 3X(s) - 2Y(s), \\
\mathcal{L}\{y'\}(s) &= 3Y(s) - 2X(s).
\end{aligned}
\tag{7.54}
$$

Since

$$
\begin{aligned}
\mathcal{L}\{x'\}(s) &= sX(s) - x(0) = sX(s) - 1, \\
\mathcal{L}\{y'\}(s) &= sY(s) - y(0) = sY(s) - 1,
\end{aligned}
$$

the system (7.54) becomes

$$
\begin{aligned}
sX(s) - 1 &= 3X(s) - 2Y(s), \\
sY(s) - 1 &= 3Y(s) - 2X(s)
\end{aligned}
\qquad \Rightarrow \qquad
\begin{aligned}
(s - 3)X(s) + 2Y(s) &= 1, \\
2X(s) + (s - 3)Y(s) &= 1.
\end{aligned}
\tag{7.55}
$$

Subtracting the second equation from the first one yields

$$(s - 5)X(s) + (5 - s)Y(s) = 0 \qquad \Rightarrow \qquad X(s) = Y(s).$$

So, from the first equation in (7.55) we get

$$(s - 3)X(s) + 2X(s) = 1 \quad \Rightarrow \quad X(s) = \frac{1}{s - 1} \quad \Rightarrow \quad x(t) = \mathcal{L}^{-1}\left\{\frac{1}{s - 1}\right\}(t) = e^t.$$

Since $Y(s) = X(s)$, $y(t) = x(t) = e^t$.

3. Let $Z(s) = \mathcal{L}\{z\}(s)$, $W(s) = \mathcal{L}\{w\}(s)$. Using the initial conditions, we conclude that

$$\mathcal{L}\{z'\}(s) = sZ(s) - z(0) = sZ(s) - 1, \quad \mathcal{L}\{w'\}(s) = sW(s) - w(0) = sW(s).$$

Copyright © 2012 Pearson Education, Inc. Publishing as Addison-Wesley.

Using these equations and taking the Laplace transform of the equations in the given system, we obtain

$$[sZ(s) - 1] + [sW(s)] = Z(s) - W(s), \quad \Rightarrow \quad (s-1)Z(s) + (s+1)W(s) = 1,$$
$$[sZ(s) - 1] - [sW(s)] = Z(s) - W(s) \quad\quad\quad (s-1)Z(s) - (s-1)W(s) = 1. \quad (7.56)$$

Subtracting equations yields

$$2sW(s) = 0 \quad \Rightarrow \quad W(s) = 0 \quad \Rightarrow \quad w(t) = \mathcal{L}^{-1}\{0\}(t) \equiv 0.$$

Substituting $W(s)$ into either equation in (7.56), we obtain

$$(s-1)Z(s) = 1 \quad \Rightarrow \quad Z(s) = \frac{1}{s-1} \quad \Rightarrow \quad z(t) = \mathcal{L}^{-1}\left\{\frac{1}{s-1}\right\}(t) = e^t.$$

5. Denote $X(s) = \mathcal{L}\{x\}(s)$, $Y(s) = \mathcal{L}\{y\}(s)$. The Laplace transform of the given equations leads us to a system

$$\mathcal{L}\{x'\}(s) = Y(s) + \mathcal{L}\{\sin t\}(s),$$
$$\mathcal{L}\{y'\}(s) = X(s) + 2\mathcal{L}\{\cos t\}(s),$$

which becomes

$$sX(s) - 2 = Y(s) + 1/(s^2 + 1), \quad \Rightarrow \quad sX(s) - Y(s) = (2s^2 + 3)/(s^2 + 1),$$
$$sY(s) = X(s) + 2s/(s^2 + 1) \quad\quad\quad -X(s) + sY(s) = 2s/(s^2 + 1)$$

after expressing $\mathcal{L}\{x'\}$ and $\mathcal{L}\{y'\}$ in terms of $X(s)$ and $Y(s)$. Multiplying the second equation by s and adding the result to the first equation, we get

$$(s^2 - 1)Y(s) = \frac{4s^2 + 3}{s^2 + 1} \quad \Rightarrow \quad Y(s) = \frac{4s^2 + 3}{(s-1)(s+1)(s^2 + 1)}.$$

Since the partial fractions decomposition for $Y(s)$ is

$$\frac{4s^2 + 3}{(s-1)(s+1)(s^2 + 1)} = \frac{7/4}{s-1} - \frac{7/4}{s+1} + \frac{1/2}{s^2 + 1},$$

taking the inverse Laplace transform yields the solution

$$y(t) = \mathcal{L}^{-1}\left\{\frac{7/4}{s-1} - \frac{7/4}{s+1} + \frac{1/2}{s^2 + 1}\right\}(t) = \frac{7}{4}e^t - \frac{7}{4}e^{-t} + \frac{1}{2}\sin t.$$

From the second equation in the original system,

$$x(t) = y' - 2\cos t = \frac{7}{4}e^t + \frac{7}{4}e^{-t} - \frac{3}{2}\cos t.$$

     Copyright © 2012 Pearson Education, Inc. Publishing as Addison-Wesley.

7. First, we write this system in the form

$$x' - 4x + 6y = 9e^{-3t},$$
$$x - y' + y = 5e^{-3t}. \tag{7.57}$$

By taking the Laplace transform of both sides of these differential equations and using its linearity, we obtain

$$\mathcal{L}\{x'\}(s) - 4X(s) + 6Y(s) = 9/(s+3),$$
$$X(s) - \mathcal{L}\{y'\}(s) + Y(s) = 5/(s+3), \tag{7.58}$$

where $X(s)$ and $Y(s)$ are the Laplace transforms of $x(t)$ and $y(t)$, respectively. Using the initial conditions $x(0) = -9$ and $y(0) = 4$, we can express

$$\mathcal{L}\{x'\}(s) = sX(s) - x(0) = sX(s) + 9,$$
$$\mathcal{L}\{y'\}(s) = sY(s) - y(0) = sY(s) - 4.$$

Substituting these expressions into the system (7.58) and simplifying yields

$$(s-4)X(s) + 6Y(s) = -9 + \frac{9}{s+3} = \frac{-9s - 18}{s+3},$$
$$X(s) + (1-s)Y(s) = -4 + \frac{5}{s+3} = \frac{-4s - 7}{s+3}.$$

Multiplying the second equation by $(4-s)$, adding the resulting equations, and simplifying, we obtain

$$\left(s^2 - 5s + 10\right)Y(s) = \frac{(4s + 7)(s - 4)}{s+3} + \frac{-9s - 18}{s+3} = \frac{4s^2 - 18s - 46}{s+3}$$
$$\Rightarrow \quad Y(s) = \frac{4s^2 - 18s - 46}{(s+3)(s^2 - 5s + 10)}.$$

Note that the quadratic $s^2 - 5s + 10 = (s - 5/2)^2 + 15/4$ is irreducible. The partial fractions decomposition yields

$$Y(s) = \frac{1}{17}\left[\frac{46s - 334}{(s - 5/2)^2 + 15/4} + \frac{22}{s+3}\right]$$
$$= \frac{1}{17}\left[46\left(\frac{s - 5/2}{(s - 5/2)^2 + 15/4}\right) - \frac{146\sqrt{15}}{5}\left(\frac{\sqrt{15}/2}{(s - 5/2)^2 + 15/4}\right) + 22\frac{1}{s+3}\right],$$

and so

$$y(t) = \mathcal{L}^{-1}\{Y(s)\}(t) = \frac{46}{17}e^{5t/2}\cos\left(\frac{\sqrt{15}t}{2}\right) - \frac{146\sqrt{15}}{85}e^{5t/2}\sin\left(\frac{\sqrt{15}t}{2}\right) + \frac{22}{17}e^{-3t}.$$

Copyright © 2012 Pearson Education, Inc. Publishing as Addison-Wesley.

From the second equation in the system (7.57), we find that

$$x(t) = 5e^{-3t} + y'(t) - y(t) = 5e^{-3t} + \frac{115}{17} e^{5t/2} \cos\left(\frac{\sqrt{15}t}{2}\right)$$

$$- \left(\frac{23\sqrt{15}}{17} + \frac{73\sqrt{15}}{17}\right) e^{5t/2} \sin\left(\frac{\sqrt{15}t}{2}\right) - \frac{219}{17} e^{5t/2} \cos\left(\frac{\sqrt{15}t}{2}\right) - \frac{66}{17} e^{-3t}$$

$$= -\frac{150}{17} e^{5t/2} \cos\left(\frac{\sqrt{15}t}{2}\right) - \frac{334\sqrt{15}}{85} e^{5t/2} \sin\left(\frac{\sqrt{15}t}{2}\right) - \frac{3}{17} e^{-3t}.$$

9. Taking the Laplace transform of both sides of these differential equations yields

$$\mathcal{L}\{x''\}(s) + X(s) + 2\mathcal{L}\{y'\}(s) = 0,$$
$$-3\mathcal{L}\{x''\}(s) - 3X(s) + 2\mathcal{L}\{y''\}(s) + 4Y(s) = 0,$$

where $X(s) = \mathcal{L}\{x\}(s)$, $Y(s) = \mathcal{L}\{y\}(s)$. Using the initial conditions $x(0) = 2$, $x'(0) = -7$ and $y(0) = 4$, $y'(0) = -9$, we see that

$$\mathcal{L}\{x''\}(s) = s^2 X(s) - sx(0) - x'(0) = s^2 X(s) - 2s + 7,$$
$$\mathcal{L}\{y'\}(s) = sY(s) - y(0) = sY(s) - 4,$$
$$\mathcal{L}\{y''\}(s) = s^2 Y(s) - sy(0) - y'(0) = s^2 Y(s) - 4s + 9.$$

Substituting these expressions into the system given above yields

$$[s^2 X(s) - 2s + 7] + X(s) + 2[sY(s) - 4] = 0,$$
$$-3[s^2 X(s) - 2s + 7] - 3X(s) + 2[s^2 Y(s) - 4s + 9] + 4Y(s) = 0,$$

which simplifies to

$$(s^2 + 1)X(s) + 2sY(s) = 2s + 1,$$
$$-3(s^2 + 1)X(s) + 2(s^2 + 2)Y(s) = 2s + 3. \tag{7.59}$$

Multiplying the first equation by 3 and adding the result to the second equation, we eliminate $X(s)$. Thus, we obtain

$$(2s^2 + 6s + 4)Y(s) = 8s + 6 \quad \Rightarrow \quad Y(s) = \frac{4s + 3}{(s+2)(s+1)} = \frac{5}{s+2} - \frac{1}{s+1},$$

where we have factored the expression $2s^2 + 6s + 4$ and used the partial fractions expansion. Taking the inverse Laplace transform, we get

$$y(t) = \mathcal{L}^{-1}\{Y(s)\}(t) = 5\mathcal{L}^{-1}\left\{\frac{1}{s+2}\right\}(t) - \mathcal{L}^{-1}\left\{\frac{1}{s+1}\right\}(t) = 5e^{-2t} - e^{-t}.$$

 Copyright © 2012 Pearson Education, Inc. Publishing as Addison-Wesley.

To find the solution $x(t)$, we examine the system (7.59) again. This time, we eliminate the function $Y(s)$ by multiplying the first equation by (s^2+2), the second one – by $(-s)$, and adding the results. Thus, we have

$$\left(s^2+3s+2\right)\left(s^2+1\right)X(s)=2s^3-s^2+s+2$$
$$\Rightarrow\quad X(s)=\frac{2s^3-s^2+s+2}{(s+2)(s+1)(s^2+1)}.$$

Expressing $X(s)$ in a partial fractions decomposition, we find that

$$X(s)=\frac{4}{s+2}-\frac{1}{s+1}-\frac{s}{s^2+1},$$

and so

$$x(t)=\mathcal{L}^{-1}\left\{\frac{4}{s+2}-\frac{1}{s+1}-\frac{s}{s^2+1}\right\}(t)=4e^{-2t}-e^{-t}-\cos t.$$

Hence, the solution to this initial value problem is

$$x(t)=4e^{-2t}-e^{-t}-\cos t\quad\text{and}\quad y(t)=5e^{-2t}-e^{-t}.$$

11. Since

$$\mathcal{L}\left\{x'\right\}(s)=sX(s)-x(0)=sX(s),$$
$$\mathcal{L}\left\{y'\right\}(s)=sY(s)-y(0)=sY(s),$$

applying the Laplace transform to the given equations yields

$$sX(s)+Y(s)=\mathcal{L}\left\{1-u(t-2)\right\}(s)=\frac{1}{s}-\frac{e^{-2s}}{s}=\frac{1-e^{-2s}}{s},$$
$$X(s)+sY(s)=\mathcal{L}\left\{0\right\}(s)=0.$$

From the second equation, $X(s)=-sY(s)$. Substituting this into the first equation, we eliminate $X(s)$ and obtain

$$-s^2Y(s)+Y(s)=\frac{1-e^{-2s}}{s}$$
$$\Rightarrow\quad Y(s)=\frac{1-e^{-2s}}{s(1-s^2)}=\left(1-e^{-2s}\right)\left(\frac{1}{s}-\frac{1/2}{s-1}-\frac{1/2}{s+1}\right).$$

Using now the linear property of the inverse Laplace transform and the formula (6), Section 7.6, we get

$$y(t)=\mathcal{L}^{-1}\left\{\frac{1}{s}-\frac{1/2}{s-1}-\frac{1/2}{s+1}\right\}(t)-\mathcal{L}^{-1}\left\{\frac{1}{s}-\frac{1/2}{s-1}-\frac{1/2}{s+1}\right\}(t-2)u(t-2)$$

Copyright © 2012 Pearson Education, Inc. Publishing as Addison-Wesley.

$$= 1 - \frac{e^t + e^{-t}}{2} - \left(1 - \frac{e^{t-2} + e^{-(t-2)}}{2}\right)u(t-2).$$

Since, from the second equation in the original system, $x = -y'$, we have

$$x(t) = -\left[1 - \frac{e^t + e^{-t}}{2} - \left(1 - \frac{e^{t-2} + e^{-(t-2)}}{2}\right)u(t-2)\right]$$

$$= \frac{e^t - e^{-t}}{2} - \left(\frac{e^{t-2} - e^{-(t-2)}}{2}\right)u(t-2).$$

13. Since, by formula (8), Section 7.6,

$$\mathcal{L}\left\{(\sin t)u(t-\pi)\right\}(s) = e^{-\pi s}\mathcal{L}\left\{\sin(t+\pi)\right\}(s) = e^{-\pi s}\mathcal{L}\left\{-\sin t\right\}(s) = -\frac{e^{-\pi s}}{s^2+1},$$

applying the Laplace transform to the given system yields

$$\mathcal{L}\left\{x'\right\}(s) - \mathcal{L}\left\{y'\right\}(s) = \mathcal{L}\left\{(\sin t)u(t-\pi)\right\}(s),$$
$$\mathcal{L}\left\{x\right\}(s) + \mathcal{L}\left\{y'\right\}(s) = \mathcal{L}\left\{0\right\}(s)$$

$$\Rightarrow \qquad [sX(s) - 1] - [sY(s) - 1] = -\frac{e^{-\pi s}}{s^2+1},$$
$$X(s) + [sY(s) - 1] = 0,$$

where we have used the initial conditions $x(0) = 1$, $y(0) = 1$, and Theorem 4, Section 7.3, to express $\mathcal{L}\left\{x'\right\}(s)$ and $\mathcal{L}\left\{y'\right\}(s)$ in terms of $X(s) := \mathcal{L}\left\{x\right\}(s)$ and $Y(s) := \mathcal{L}\left\{y\right\}(s)$. This system simplifies to

$$X(s) - Y(s) = -\frac{e^{-\pi s}}{s(s^2+1)},$$
$$X(s) + sY(s) = 1.$$

From the second equation, $X(s) = 1 - sY(s)$ and, after this substitution, the first equation becomes

$$1 - sY(s) - Y(s) = -\frac{e^{-\pi s}}{s(s^2+1)}.$$

Solving for $Y(s)$ yields

$$Y(s) = \left[1 + \frac{e^{-\pi s}}{s(s^2+1)}\right]\frac{1}{s+1} \doteq \frac{1}{s+1} + \frac{e^{-\pi s}}{s(s+1)(s^2+1)}.$$

Using partial fractions, we express

$$Y(s) = \frac{1}{s+1} + e^{-\pi s}\left(\frac{1}{s} - \frac{1/2}{s+1} - \frac{(1/2)s}{s^2+1} - \frac{1/2}{s^2+1}\right).$$

Copyright © 2012 Pearson Education, Inc. Publishing as Addison-Wesley.

Thus,

$$y(t) = e^{-t} + \left[1 - \frac{1}{2} e^{-(t-\pi)} - \frac{1}{2} \cos(t - \pi) - \frac{1}{2} \sin(t - \pi) \right] u(t - \pi)$$

$$= e^{-t} + \left(1 - \frac{1}{2} e^{-(t-\pi)} + \frac{1}{2} \cos t + \frac{1}{2} \sin t \right) u(t - \pi).$$

Finally,

$$x(t) = -y'(t) = e^{-t} - \left(\frac{1}{2} e^{-(t-\pi)} - \frac{1}{2} \sin t + \frac{1}{2} \cos t \right) u(t - \pi).$$

15. First, note that the initial conditions are given at the point $t = 1$. Thus, for the Laplace transform method, we have to shift the argument to get zero initial point. This can be done by denoting $u(t) := x(t + 1)$ and $v(t) := y(t + 1)$. The chain rule yields

$$u'(t) = x'(t + 1)(t + 1)' = x'(t + 1), \quad v'(t) = y'(t + 1)(t + 1)' = y'(t + 1).$$

In the original system, we substitute $t + 1$ for t to get

$$x'(t + 1) - 2y(t + 1) = 2,$$
$$x'(t + 1) + x(t + 1) - y'(t + 1) = (t + 1)^2 + 2(t + 1) - 1,$$

and make u and v substitution.

$$u'(t) - 2v(t) = 2,$$
$$u'(t) + u(t) - v'(t) = (t + 1)^2 + 2(t + 1) - 1 = t^2 + 4t + 2$$

with initial conditions $u(0) = 1$, $v(0) = 0$. Taking the Laplace transform and using the formula (2) of Section 7.3, we obtain the system

$$[sU(s) - 1] - 2V(s) = \frac{2}{s},$$

$$[sU(s) - 1] + U(s) - sV(s) = \frac{2}{s^3} + \frac{4}{s^2} + \frac{2}{s},$$

where $U(s) = \mathcal{L}\{u\}(s)$, $V(s) = \mathcal{L}\{v\}(s)$. Expressing

$$U(s) = \frac{2V(s)}{s} + \frac{2}{s^2} + \frac{1}{s}$$

from the first equation and substituting this into the second equation, we obtain

$$\left[\frac{2}{s} + 2V(s) \right] + \left[\frac{2V(s)}{s} + \frac{2}{s^2} + \frac{1}{s} \right] - sV(s) = \frac{2}{s^3} + \frac{4}{s^2} + \frac{2}{s},$$

which gives us

$$V(s) = \frac{1}{s^2} \quad \Rightarrow \quad U(s) = \frac{2}{s^3} + \frac{2}{s^2} + \frac{1}{s}.$$

Copyright © 2012 Pearson Education, Inc. Publishing as Addison-Wesley.

Applying now the inverse Laplace transform yields

$$u(t) = t^2 + 2t + 1 = (t+1)^2, \qquad v(t) = \mathcal{L}^{-1}\left\{\frac{1}{s^2}\right\}(t) = t.$$

Finally,

$$x(t) = u(t-1) = t^2 \quad \text{and} \quad y(t) = v(t-1) = t-1.$$

17. First, we make a shift in t to move the initial conditions to $t = 0$. Let

$$u(t) := x(t+2) \qquad \text{and} \qquad v(t) := y(t+2).$$

With t replaced by $t+2$, the original system becomes

$$x'(t+2) + x(t+2) - y'(t+2) = 2te^t,$$
$$x''(t+2) - x'(t+2) - 2y(t+2) = -e^t$$

or

$$u'(t) + u(t) - v'(t) = 2te^t, \qquad \text{with} \qquad \begin{aligned} u(0) &= 0, \\ u'(0) &= 1, \\ v(0) &= 1. \end{aligned}$$
$$u''(t) - u'(t) - 2v(t) = -e^t$$

Applying the Laplace transform to these equations and expressing $\mathcal{L}\{u''\}$, $\mathcal{L}\{u'\}$, and $\mathcal{L}\{v'\}$ in terms of $U = \mathcal{L}\{u\}$ and $V = \mathcal{L}\{v\}$ (see the formula (4), Section 7.3), we obtain

$$[sU(s)] + U(s) - [sV(s) - 1] = 2\mathcal{L}\{te^t\}(s) = \frac{2}{(s-1)^2},$$

$$\left[s^2 U(s) - 1\right] - [sU(s)] - 2V(s) = -\frac{1}{s-1}.$$

We multiply the first equation by two, the second equation – by s, and subtract the resulting equations in order to eliminate $V(s)$. Thus, we get

$$\left[s(s^2 - s) - 2(s+1)\right] U(s) = s - \frac{s}{s-1} - \frac{4}{(s-1)^2} + 2$$

$$\Rightarrow \qquad \left(s^3 - s^2 - 2s - 2\right) U(s) = \frac{s^3 - s^2 - 2s - 2}{(s-1)^2} \qquad \Rightarrow \qquad U(s) = \frac{1}{(s-1)^2}.$$

The inverse Laplace transform then yields

$$u(t) = \mathcal{L}^{-1}\left\{\frac{1}{(s-1)^2}\right\}(t) = te^t \qquad \Rightarrow \qquad x(t) = u(t-2) = (t-2)e^{t-2}.$$

We find $y(t)$ from the second equation in the original system.

$$y(t) = \frac{x''(t) - x'(t) + e^{t-2}}{2} = \frac{te^{t-2} - (t-1)e^{t-2} + e^{t-2}}{2} = e^{t-2}.$$

 Copyright © 2012 Pearson Education, Inc. Publishing as Addison-Wesley.

19. We take the Laplace transform of the given equations and use the initial conditions to obtain a system for the Laplace transforms.

$$sX(s) + 6 = 3X(s) + Y(s) - 2Z(s),$$
$$sY(s) - 2 = -X(s) + 2Y(s) + Z(s),$$
$$sZ(s) + 12 = 4X(s) + Y(s) - 3Z(s).$$

Simplifying yields

$$(s - 3)X(s) - Y(s) + 2Z(s) = -6,$$
$$X(s) + (s - 2)Y(s) - Z(s) = 2, \tag{7.60}$$
$$-4X(s) - Y(s) + (s + 3)Z(s) = -12.$$

To solve this system, we use a substitution to eliminate the function $Y(s)$. Thus, we solve the first equation in (7.60) for $Y(s)$ and obtain

$$Y(s) = (s - 3)X(s) + 2Z(s) + 6.$$

Substituting this expression into the two remaining equations in (7.60) and simplifying yields

$$(s^2 - 5s + 7)X(s) + (2s - 5)Z(s) = -6s + 14,$$
$$-(s + 1)X(s) + (s + 1)Z(s) = -6. \tag{7.61}$$

Next, we eliminate the function $X(s)$ from the system (7.61). To do this, we can either multiply the first equation by $(s + 1)$, the second one – by $(s^2 - 5s + 7)$, and add the results, or solve the last equation in (7.61) for $X(s)$ to obtain

$$X(s) = Z(s) + \frac{6}{s + 1} \tag{7.62}$$

and substitute this expression into the first equation in (7.61). Either way yields

$$Z(s) = \frac{-12s^2 + 38s - 28}{(s + 1)(s^2 - 3s + 2)} = \frac{-12s^2 + 38s - 28}{(s + 1)(s - 2)(s - 1)}.$$

Now, $Z(s)$ has a partial fractions decomposition

$$Z(s) = \frac{-13}{s + 1} + \frac{1}{s - 1}.$$

Therefore, taking the inverse Laplace transform of both sides of this equation, we obtain

$$z(t) = \mathcal{L}^{-1}\{Z(s)\}(t) = \mathcal{L}^{-1}\left\{\frac{-13}{s + 1} + \frac{1}{s - 1}\right\}(t) = -13e^{-t} + e^t.$$

Copyright © 2012 Pearson Education, Inc. Publishing as Addison-Wesley.

To find $X(s)$, we use (7.62) and the expression found for $Z(s)$. Thus, we have

$$X(s) = Z(s) + \frac{6}{s+1} = \frac{-13}{s+1} + \frac{1}{s-1} + \frac{6}{s+1} = \frac{-7}{s+1} + \frac{1}{s-1}$$

$$\Rightarrow \quad x(t) = \mathcal{L}^{-1}\{X(s)\}(t) = \mathcal{L}^{-1}\left\{\frac{-7}{s+1} + \frac{1}{s-1}\right\}(t) = -7e^{-t} + e^t.$$

To find $y(t)$, we can substitute the expressions, that we have already found for $X(s)$ and $Z(s)$, into $Y(s) = (s-3)X(s) + 2Z(s) + 6$ or we can return to the original system of differential equations and use $x(t)$ and $z(t)$ to solve for $y(t)$. For the latter approach, we solve the first equation in the original system for $y(t)$ to obtain

$$y(t) = x'(t) - 3x(t) + 2z(t)$$

$$= 7e^{-t} + e^t + 21e^{-t} - 3e^t - 26e^{-t} + 2e^t = 2e^{-t}.$$

Therefore, the solution to the given initial value problem is

$$x(t) = -7e^{-t} + e^t, \qquad y(t) = 2e^{-t}, \qquad z(t) = -13e^{-t} + e^t.$$

21. We refer the reader to Section 5.1 for details on systems, governing interconnected tanks. All the arguments provided remain in force except for the one affected by the new "valve condition", which is the formula for the input rate for the tank A. In Section 5.1, just fresh water was pumped into the tank A, and so there was no salt coming from outside of the system into the tank A . Now we have more complicated rule: the incoming liquid is fresh water for the first $5\,\text{min}$, but then it changes to a solution, having a concentration of $2\,\text{kg/L}$. This solution contributes additional

$$2\,(\text{kg/L}) \times 6\,(\text{L/min}) = 12\,(\text{kg/min})$$

to the input rate into the tank A. Thus, from the valve, we have

$$\begin{cases} 0, & t < 5, \\ 12, & t > 5 \end{cases} = 12u(t-5)\,(\text{kg/min})$$

of salt coming to the tank A. With this change, the system (1), Section 5.1, becomes

$$\begin{aligned} x' &= -x/3 + y/12 + 12u(t-5), \\ y' &= x/3 - y/3. \end{aligned} \tag{7.63}$$

 Copyright © 2012 Pearson Education, Inc. Publishing as Addison-Wesley.

Also, we have initial conditions $x(0) = x_0 = 0$, $y(0) = y_0 = 4$. Let $X := \mathcal{L}\{x\}$ and $Y := \mathcal{L}\{y\}$. Taking the Laplace transform of both equations in (7.63), we get

$$\mathcal{L}\{x'\}(s) = -\frac{1}{3}X(s) + \frac{1}{12}Y(s) + 12\mathcal{L}\{u(t-5)\}(s),$$

$$\mathcal{L}\{y'\}(s) = \frac{1}{3}X(s) - \frac{1}{3}Y(s).$$

Since $\mathcal{L}\{u(t-5)\}(s) = e^{-5s}/s$ and

$$\mathcal{L}\{x'\}(s) = sX(s) - x(0) = sX(s),$$

$$\mathcal{L}\{y'\}(s) = sY(s) - y(0) = sY(s) - 4,$$

we obtain

$$sX(s) = -\frac{1}{3}X(s) + \frac{1}{12}Y(s) + \frac{12e^{-5s}}{s},$$

$$sY(s) - 4 = \frac{1}{3}X(s) - \frac{1}{3}Y(s),$$

which simplifies to

$$4(3s+1)X(s) - Y(s) = \frac{144e^{-5s}}{s},$$

$$-X(s) + (3s+1)Y(s) = 12.$$

From the second equation in this system, we have $X(s) = (3s+1)Y(s) - 12$. Substituting into the first equation yields

$$4(3s+1)\left[(3s+1)Y(s) - 12\right] - Y(s) = \frac{144e^{-5s}}{s}$$

$$\Rightarrow \quad \left[4(3s+1)^2 - 1\right]Y(s) - 48(3s+1) + \frac{144e^{-5s}}{s}.$$

Note that

$$4(3s+1)^2 - 1 = \left[2(3s+1)+1\right] \cdot \left[2(3s+1)-1\right] = (6s+3)(6s+1) = 36\left(s+\frac{1}{2}\right)\left(s+\frac{1}{6}\right).$$

Therefore,

$$Y(s) = \frac{4(3s+1)}{3(s+1/2)(s+1/6)} + \frac{4e^{-5s}}{s(s+1/2)(s+1/6)}$$

$$= \frac{2}{(s+1/2)} + \frac{2}{(s+1/6)} + e^{-5s}\left(\frac{48}{s} + \frac{24}{s+1/2} - \frac{72}{s+1/6}\right)$$

by partial fractions decomposition. Taking the inverse Laplace transform of both sides and applying Theorem 8, Section 7.6, for the inverse Laplace transform of the term having the exponential factor, we get

$$y(t) = 2\mathcal{L}^{-1}\left\{\frac{1}{(s+1/2)}\right\}(t) + 2\mathcal{L}^{-1}\left\{\frac{1}{(s+1/6)}\right\}(t)$$

Copyright © 2012 Pearson Education, Inc. Publishing as Addison-Wesley.

$$+ \left[48\mathcal{L}^{-1}\left\{\frac{1}{s}\right\} + 24\mathcal{L}^{-1}\left\{\frac{1}{s+1/2}\right\} - 72\mathcal{L}^{-1}\left\{\frac{1}{s+1/6}\right\} \right] (t-5)u(t-5)$$

$$= 2e^{-t/2} + 2e^{-t/6} + \left[48 + 24e^{-(t-5)/2} - 72e^{-(t-5)/6} \right] u(t-5).$$

From the second equation in (7.63), after some algebra, we find $x(t)$.

$$x(t) = 3y'(t) + y = -e^{-t/2} + e^{-t/6} + \left[48 - 12e^{-(t-5)/2} - 36e^{-(t-5)/6} \right] u(t-5).$$

23. Recall that Kirchhoff's voltage law says that, in an electrical circuit consisting of an inductor of $L(\mathrm{H})$, a resistor of $R(\Omega)$, a capacitor of $C(\mathrm{F})$, and a voltage source of $E(\mathrm{V})$,

$$E_L + E_R + E_C = E, \tag{7.64}$$

where E_L, E_R, and E_C denote the voltage drops across the inductor, resistor, and capacitor, respectively. These voltage drops are given by

$$E_L = L\frac{dI}{dt}, \qquad E_R := RI, \qquad E_C := \frac{q}{C}, \tag{7.65}$$

where I denotes the current passing through the correspondent element.

Kirchhoff's current law also states that the algebraic sum of currents passing through any point in an electrical network equals to zero.

The electrical network shown in Fig. 7.30 consists of three closed circuits: loop 1 through the battery, $R_1 = 2\,\Omega$ resistor, $L_1 = 0.1\,\mathrm{H}$ inductor, and $L_2 = 0.2\,\mathrm{H}$ inductor; loop 2 through the inductor L_1 and $R_2 = 1\,\Omega$ resistor; loop 3 through the battery, resistors R_1 and R_2, and inductor L_2. We apply Kirchhoff's voltage law (7.64) to two of these loops, say, the loop 1 and the loop 2 (since the equation obtained from Kirchhoff's voltage law for the third loop is a linear combination of the other two) and Kirchhoff's current law to one of the junction points, say, the upper one. Thus, choosing the clockwise direction in the loops and using formulas (7.65), we obtain

Loop 1:
$$E_{R_1} + E_{L_1} + E_{L_2} = E \qquad \Rightarrow \qquad 2I_1 + 0.1I_3' + 0.2I_1' = 6;$$

Loop 2:
$$E_{L_1} + E_{R_2} = 0 \qquad \Rightarrow \qquad 0.1I_3' - I_2 = 0$$

with the negative sign due to the counterclockwise direction of the current I_2 in this loop;

 Copyright © 2012 Pearson Education, Inc. Publishing as Addison-Wesley.

Upper junction point:

$$I_1 - I_2 - I_3 = 0.$$

Therefore, we have the following system for the currents I_1, I_2, and I_3.

$$\begin{aligned}
2I_1 + 0.1I_3' + 0.2I_1' &= 6, \\
0.1I_3' - I_2 &= 0, \\
I_1 - I_2 - I_3 &= 0
\end{aligned}$$ (7.66)

with initial conditions $I_1(0) = I_2(0) = I_3(0) = 0$.

Let $\mathbf{I}_1(s) := \mathcal{L}\{I_1\}(s)$, $\mathbf{I}_2(s) := \mathcal{L}\{I_2\}(s)$, and $\mathbf{I}_3(s) := \mathcal{L}\{I_3\}(s)$. Using the initial conditions, we conclude that

$$\begin{aligned}
\mathcal{L}\{I_1'\}(s) &= s\mathbf{I}_1(s) - I_1(0) = s\mathbf{I}_1(s), \\
\mathcal{L}\{I_3'\}(s) &= s\mathbf{I}_3(s) - I_3(0) = s\mathbf{I}_3(s).
\end{aligned}$$

Using these equations and taking the Laplace transform of the equations in (7.66), we come up with

$$\begin{aligned}
(0.2s + 2)\mathbf{I}_1(s) + 0.1s\mathbf{I}_3(s) &= \frac{6}{s}, \\
0.1s\mathbf{I}_3(s) - \mathbf{I}_2(s) &= 0, \\
\mathbf{I}_1(s) - \mathbf{I}_2(s) - \mathbf{I}_3(s) &= 0.
\end{aligned}$$

Expressing $\mathbf{I}_2(s) = 0.1s\mathbf{I}_3(s)$ from the second equation and substituting this into the third equation, we get

$$\mathbf{I}_1(s) - 0.1s\mathbf{I}_3(s) - \mathbf{I}_3(s) = 0 \quad \Rightarrow \quad \mathbf{I}_1(s) = (0.1s + 1)\mathbf{I}_3(s).$$

The latter, when substituted into the first equation, yields

$$(0.2s + 2)(0.1s + 1)\mathbf{I}_3(s) + 0.1s\mathbf{I}_3(s) = \frac{6}{s}$$

$$\Rightarrow \qquad \left[2(0.1s + 1)^2 + 0.1s\right]\mathbf{I}_3(s) = \frac{6}{s}$$

$$\Rightarrow \qquad \mathbf{I}_3(s) = \frac{6}{s[2(0.1s + 1)^2 + 0.1s]} = \frac{300}{s(s + 20)(s + 5)}.$$

We use the partial fractions decomposition to find that

$$\mathbf{I}_3(s) = \frac{3}{s} + \frac{1}{s + 20} - \frac{4}{s + 5}$$

and so

$$I_3(t) = \mathcal{L}^{-1}\left\{\frac{3}{s} + \frac{1}{s + 20} - \frac{4}{s + 5}\right\}(t) = 3 + e^{-20t} - 4e^{-5t}.$$

Copyright © 2012 Pearson Education, Inc. Publishing as Addison-Wesley.

Now we can find $I_2(t)$ using the second equation in (7.66).

$$I_2(t) = 0.1 I_3'(t) = 0.1 \left(3 + e^{-20t} - 4e^{-5t}\right)' = -2e^{-20t} + 2e^{-5t}.$$

Finally, the third equation in (7.66) yields

$$I_1(t) = I_2(t) + I_3(t) = 3 - e^{-20t} - 2e^{-5t}.$$

REVIEW PROBLEMS

1. By the definition of Laplace transform,

$$\mathcal{L}\{f\}(s) = \int_0^\infty e^{-st} f(t)\, dt = \int_0^2 e^{-st}(3)\, dt + \int_2^\infty e^{-st}(6-t)\, dt.$$

For the first integral, we have

$$\int_0^2 e^{-st}(3)\, dt = \frac{3e^{-st}}{-s}\Bigg|_{t=0}^{t=2} = \frac{3(1-e^{-2s})}{s}.$$

The second integral is an improper integral. Using integration by parts, we obtain

$$\int_2^\infty e^{-st}(6-t)\, dt = \lim_{M\to\infty}\int_2^M e^{-st}(6-t)\, dt = \lim_{M\to\infty}\left[(6-t)\frac{e^{-st}}{-s}\Bigg|_{t=2}^{t=M} - \int_2^M \frac{e^{-st}}{-s}(-1)dt\right]$$

$$= \lim_{M\to\infty}\left[\frac{4e^{-2s}}{s} - \frac{(6-M)e^{-sM}}{s} + \frac{e^{-st}}{s^2}\Bigg|_{t=2}^{t=M}\right]$$

$$= \lim_{M\to\infty}\left[\frac{4e^{-2s}}{s} - \frac{(6-M)e^{-sM}}{s} + \frac{e^{-sM}}{s^2} - \frac{e^{-2s}}{s^2}\right] = \frac{4e^{-2s}}{s} - \frac{e^{-2s}}{s^2}.$$

Thus,

$$\mathcal{L}\{f\}(s) = \frac{3(1-e^{-2s})}{s} + \frac{4e^{-2s}}{s} - \frac{e^{-2s}}{s^2} = \frac{3}{s} + e^{-2s}\left(\frac{1}{s} - \frac{1}{s^2}\right).$$

3. From Table 7.1, using the formula for the Laplace transform of $e^{at}t^n$ with $n = 2$ and $a = -9$, we get

$$\mathcal{L}\{t^2 e^{-9t}\}(s) = \frac{2!}{[s-(-9)]^3} = \frac{2}{(s+9)^3}.$$

5. We use linearity of the Laplace transform and Table 7.1 to obtain

$$\mathcal{L}\{e^{2t} - t^3 + t^2 - \sin 5t\}(s) = \mathcal{L}\{e^{2t}\}(s) - \mathcal{L}\{t^3\}(s) + \mathcal{L}\{t^2\}(s) - \mathcal{L}\{\sin 5t\}(s)$$

$$= \frac{1}{s-2} - \frac{3!}{s^4} + \frac{2!}{s^3} - \frac{5}{s^2+5^2} = \frac{1}{s-2} - \frac{6}{s^4} + \frac{2}{s^3} - \frac{5}{s^2+25}.$$

 Copyright © 2012 Pearson Education, Inc. Publishing as Addison-Wesley.

7. We apply Theorem 6 of Section 7.3 to obtain

$$\mathcal{L}\left\{t\cos 6t\right\}(s) = -\frac{d}{ds}\mathcal{L}\left\{\cos 6t\right\}(s) = -\frac{d}{ds}\left(\frac{s}{s^2+6^2}\right)$$
$$= -\frac{(s^2+36)-s(2s)}{(s^2+36)^2} = \frac{s^2-36}{(s^2+36)^2}.$$

9. We apply the formula (11), Section 7.6, and the linearity of the Laplace transform to get

$$\mathcal{L}\left\{t^2 u(t-4)\right\}(s) = e^{-4s}\mathcal{L}\left\{(t+4)^2\right\}(s) = e^{-4s}\mathcal{L}\left\{t^2+8s+16\right\}(s)$$
$$= e^{-4s}\left(\frac{2}{s^3}+\frac{8}{s^2}+\frac{16}{s}\right) = 2e^{-4s}\left(\frac{1}{s^3}+\frac{4}{s^2}+\frac{8}{s}\right).$$

11. Using linearity of the inverse Laplace transform and Table 7.1, we find

$$\mathcal{L}^{-1}\left\{\frac{7}{(s+3)^3}\right\}(t) = \frac{7}{2!}\mathcal{L}^{-1}\left\{\frac{2!}{[s-(-3)]^3}\right\}(t) = \frac{7}{2}t^2 e^{-3t}.$$

13. We apply partial fractions decomposition to find the inverse Laplace transform. Since the quadratic polynomial $s^2+4s+13 = (s+2)^2+3^2$ is irreducible, the partial fractions decomposition for the given function has the form

$$\frac{4s^2+13s+19}{(s-1)(s^2+4s+13)} = \frac{A}{s-1} + \frac{B(s+2)+C(3)}{(s+2)^2+3^2}.$$

Clearing fractions yields

$$4s^2+13s+19 = A[(s+2)^2+3^2] + [B(s+2)+C(3)](s-1).$$

With $s = 1$, this gives $36 = 18A$ or $A = 2$. Substituting $s = -2$, we get

$$9 = 9A - 9C \qquad \Rightarrow \qquad C = A - 1 = 1.$$

Finally, with $s = 0$, we compute

$$19 = 13A + (2B+3C)(-1) \qquad \Rightarrow \qquad B = 2.$$

Thus,

$$\frac{4s^2+13s+19}{(s-1)(s^2+4s+13)} = \frac{2}{s-1} + \frac{2(s+2)+(1)(3)}{(s+2)^2+3^2},$$

and so

$$\mathcal{L}^{-1}\left\{\frac{4s^2+13s+19}{(s-1)(s^2+4s+13)}\right\}(t) = 2\mathcal{L}^{-1}\left\{\frac{1}{s-1}\right\}(t) + 2\mathcal{L}^{-1}\left\{\frac{s+2}{(s+2)^2+3^2}\right\}(t)$$
$$+ \mathcal{L}^{-1}\left\{\frac{3}{(s+2)^2+3^2}\right\}(t)$$
$$= 2e^t + 2e^{-2t}\cos 3t + e^{-2t}\sin 3t.$$

Copyright © 2012 Pearson Education, Inc. Publishing as Addison-Wesley.

15. The partial fractions decomposition for the given function has the form

$$\frac{2s^2 + 3s - 1}{(s+1)^2(s+2)} = \frac{A}{(s+1)^2} + \frac{B}{s+1} + \frac{C}{s+2} = \frac{A(s+2) + B(s+1)(s+2) + C(s+1)^2}{(s+1)^2(s+2)}.$$

Thus,

$$2s^2 + 3s - 1 = A(s+2) + B(s+1)(s+2) + C(s+1)^2.$$

We evaluate both sides of this equation at $s = -2, -1$, and 0. This yields

$$s = -2: \quad 2(-2)^2 + 3(-2) - 1 = C(-2+1)^2 \quad \Rightarrow \quad C = 1,$$
$$s = -1: \quad 2(-1)^2 + 3(-1) - 1 = A(-1+2) \quad \Rightarrow \quad A = -2,$$
$$s = 0: \quad -1 = 2A + 2B + C \quad \Rightarrow \quad B = (-1 - 2A - C)/2 = 1.$$

Therefore,

$$\mathcal{L}^{-1}\left\{\frac{2s^2 + 3s - 1}{(s+1)^2(s+2)}\right\}(t) = \mathcal{L}^{-1}\left\{\frac{-2}{(s+1)^2} + \frac{1}{s+1} + \frac{1}{s+2}\right\}(t) = -2te^{-t} + e^{-t} + e^{-2t}.$$

17. First, we apply Theorem 8 of Section 7.6 to get

$$\mathcal{L}^{-1}\left\{\frac{e^{-2s}(4s+2)}{(s-1)(s+2)}\right\}(t) = \mathcal{L}^{-1}\left\{\frac{4s+2}{(s-1)(s+2)}\right\}(t-2)u(t-2). \tag{7.67}$$

Using partial fractions decomposition yields

$$\frac{4s+2}{(s-1)(s+2)} = \frac{2}{s-1} + \frac{2}{s+2} \quad \Rightarrow \quad \mathcal{L}^{-1}\left\{\frac{4s+2}{(s-1)(s+2)}\right\}(t) = 2e^t + 2e^{-2t}.$$

Therefore, it follows from (7.67) that

$$\mathcal{L}^{-1}\left\{\frac{e^{-2s}(4s+2)}{(s-1)(s+2)}\right\}(t) = \left[2e^{t-2} + 2e^{-2(t-2)}\right]u(t-2) = \left(2e^{t-2} + 2e^{4-2t}\right)u(t-2).$$

19. Applying the Laplace transform to both sides of the given equation and using linearity of the Laplace transform yields

$$\mathcal{L}\{y'' - 7y' + 10y\}(s) = \mathcal{L}\{y''\}(s) - 7\mathcal{L}\{y'\}(s) + 10\mathcal{L}\{y\}(s) = 0. \tag{7.68}$$

By Theorem 5 of Section 7.3,

$$\mathcal{L}\{y'\}(s) = s\mathcal{L}\{y\}(s) - y(0) = s\mathcal{L}\{y\}(s),$$
$$\mathcal{L}\{y''\}(s) = s^2\mathcal{L}\{y\}(s) - sy(0) - y'(0) = s^2\mathcal{L}\{y\}(s) + 3,$$

where we have used the initial conditions, $y(0) = 0$ and $y'(0) = -3$. Substituting these expressions into (7.68), we get

$$\left[s^2\mathcal{L}\{y\}(s) + 3\right] - 7\left[s\mathcal{L}\{y\}(s)\right] + 10\mathcal{L}\{y\}(s) = 0$$

 Copyright © 2012 Pearson Education, Inc. Publishing as Addison-Wesley.

$$\Rightarrow \qquad (s^2 - 7s + 10)\mathcal{L}\{y\}(s) + 3 = 0$$

$$\Rightarrow \qquad \mathcal{L}\{y\}(s) = \frac{-3}{s^2 - 7s + 10} = \frac{-3}{(s-2)(s-5)} = \frac{1}{s-2} - \frac{1}{s-5}.$$

Thus,

$$y(t) = \mathcal{L}^{-1}\left\{\frac{1}{s-2} - \frac{1}{s-5}\right\}(t) = \mathcal{L}^{-1}\left\{\frac{1}{s-2}\right\}(t) - \mathcal{L}^{-1}\left\{\frac{1}{s-5}\right\}(t) = e^{2t} - e^{5t}.$$

21. Let $Y(s) := \mathcal{L}\{y\}(s)$. Taking the Laplace transform of both sides of the given equation and using properties of the Laplace transform, we obtain

$$\mathcal{L}\{y'' + 2y' + 2y\}(s) = \mathcal{L}\{t^2 + 4t\}(s) = \frac{2}{s^3} + \frac{4}{s^2} = \frac{2 + 4s}{s^3}.$$

Since

$$\mathcal{L}\{y'\}(s) = sY(s) - y(0) = sY(s), \qquad \mathcal{L}\{y''\}(s) = s^2 Y(s) - sy(0) - y'(0) = s^2 Y(s) + 1,$$

we have

$$\left[s^2 Y(s) + 1\right] + 2\left[sY(s)\right] + 2Y(s) = \frac{2 + 4s}{s^3}$$

$$\Rightarrow \qquad (s^2 + 2s + 2)Y(s) = \frac{2 + 4s}{s^3} - 1 = \frac{2 + 4s - s^3}{s^3}$$

$$\Rightarrow \qquad Y(s) = \frac{2 + 4s - s^3}{s^3(s^2 + 2s + 2)} = \frac{2 + 4s - s^3}{s^3[(s+1)^2 + 1^2]}.$$

The partial fractions decomposition for $Y(s)$ has the form

$$\frac{2 + 4s - s^3}{s^3[(s+1)^2 + 1^2]} = \frac{A}{s^3} + \frac{B}{s^2} + \frac{C}{s} + \frac{D(s+1) + E(1)}{(s+1)^2 + 1^2}.$$

Clearing fractions, we obtain

$$2 + 4s - s^3 = A[(s+1)^2 + 1] + Bs[(s+1)^2 + 1] + Cs^2[(s+1)^2 + 1] + [D(s+1) + E]s^3.$$

Comparing coefficients at the corresponding power of s in both sides of this equation yields

$$
\begin{array}{llll}
s^0: & 2 = 2A & \Rightarrow & A = 1, \\
s^1: & 4 = 2A + 2B & \Rightarrow & B = (4 - 2A)/2 = 1, \\
s^2: & 0 = A + 2B + 2C & \Rightarrow & C = -(A + 2B)/2 = -3/2, \\
s^4: & 0 = C + D & \Rightarrow & D = -C = 3/2, \\
s^3: & -1 = B + 2C + D + E & \Rightarrow & E = -1 - B - 2C - D = -1/2.
\end{array}
$$

Copyright © 2012 Pearson Education, Inc. Publishing as Addison-Wesley.

Therefore,

$$Y(s) = \frac{1}{s^3} + \frac{1}{s^2} - \frac{3/2}{s} + \frac{(3/2)(s+1)}{(s+1)^2 + 1^2} - \frac{(1/2)(1)}{(s+1)^2 + 1^2}$$

$$\Rightarrow \quad y(t) = \mathcal{L}^{-1}\{Y(s)\}(t) = \frac{t^2}{2} + t - \frac{3}{2} + \frac{3}{2} e^{-t}\cos t - \frac{1}{2} e^{-t}\sin t.$$

23. By the formula (6) in Section 7.6,

$$\mathcal{L}\{u(t-1)\}(s) = \frac{e^{-s}}{s}.$$

Thus, applying the Laplace transform to both sides of the given equation and using the initial conditions, we get

$$\mathcal{L}\{y'' + 3y' + 4y\}(s) = \frac{e^{-s}}{s}$$

$$\Rightarrow \quad [s^2 Y(s) - 1] + 3[sY(s)] + 4Y(s) = \frac{e^{-s}}{s}$$

$$\Rightarrow \quad Y(s) = \frac{1}{s^2 + 3s + 4} + \frac{e^{-s}}{s(s^2 + 3s + 4)}$$

$$\Rightarrow \quad Y(s) = \frac{1}{(s+3/2)^2 + (\sqrt{7}/2)^2} + e^{-s}\frac{1}{s[(s+3/2)^2 + (\sqrt{7}/2)^2]},$$

where $Y(s) := \mathcal{L}\{y\}(s)$. To apply the inverse Laplace transform, we need the partial fractions decomposition of the last fraction.

$$\frac{1}{s[(s+3/2)^2 + (\sqrt{7}/2)^2]} = \frac{A}{s} + \frac{B(s+3/2) + C(\sqrt{7}/2)}{(s+3/2)^2 + (\sqrt{7}/2)^2}.$$

Solving for A, B, and C yields

$$A = \frac{1}{4}, \qquad B = -\frac{1}{4}, \qquad C = -\frac{3}{4\sqrt{7}}.$$

Therefore,

$$Y(s) = \frac{1}{(s+3/2)^2 + (\sqrt{7}/2)^2}$$

$$+ e^{-s}\left[\frac{1/4}{s} - \frac{(1/4)(s+3/2)}{(s+3/2)^2 + (\sqrt{7}/2)^2} - \frac{(3/4\sqrt{7})(\sqrt{7}/2)}{(s+3/2)^2 + (\sqrt{7}/2)^2}\right],$$

and the inverse Laplace transform gives

$$y(t) = \mathcal{L}^{-1}\left\{\frac{1}{(s+3/2)^2 + (\sqrt{7}/2)^2}\right\}(t)$$

 Copyright © 2012 Pearson Education, Inc. Publishing as Addison-Wesley.

$$+\mathcal{L}^{-1}\left\{\frac{1/4}{s}-\frac{(1/4)(s+3/2)}{(s+3/2)^2+(7/4)}-\frac{(3/4\sqrt{7})(\sqrt{7}/2)}{(s+3/2)^2+(7/4)}\right\}(t-1)u(t-1)$$

$$=\frac{2}{\sqrt{7}}e^{-3t/2}\sin\left(\frac{\sqrt{7}t}{2}\right)$$

$$+\left[\frac{1}{4}-\frac{1}{4}e^{-3(t-1)/2}\cos\left(\frac{\sqrt{7}(t-1)}{2}\right)-\frac{3}{4\sqrt{7}}e^{-3(t-1)/2}\sin\left(\frac{\sqrt{7}(t-1)}{2}\right)\right]u(t-1).$$

25. Let $Y(s):=\mathcal{L}\{y\}(s)$. Then, from the initial conditions, we have

$$\mathcal{L}\{y'\}(s)=sY(s)-y(0)=sY(s),$$

$$\mathcal{L}\{y''\}(s)=s^2Y(s)-sy(0)-y'(0)=s^2Y(s).$$

Moreover, Theorem 6 of Section 7.3 yields

$$\mathcal{L}\{ty'\}(s)=-\frac{d}{ds}\mathcal{L}\{y'\}(s)=-\frac{d}{ds}[sY(s)]=-sY'(s)-Y(s),$$

$$\mathcal{L}\{ty''\}(s)=-\frac{d}{ds}\mathcal{L}\{y''\}(s)=-\frac{d}{ds}[s^2Y(s)]=-s^2Y'(s)-2sY(s).$$

Hence, applying the Laplace transform to the given equation and using its linearity, we obtain

$$\mathcal{L}\{ty''+2(t-1)y'-2y\}(s)=\mathcal{L}\{ty''\}(s)+2\mathcal{L}\{ty'\}(s)-2\mathcal{L}\{y'\}(s)-2\mathcal{L}\{y\}(s)=0$$

$$\Rightarrow\quad\left[-s^2Y'(s)-2sY(s)\right]+2\left[-sY'(s)-Y(s)\right]-2\left[sY(s)\right]-2Y(s)=0$$

$$\Rightarrow\quad-s(s+2)Y'(s)-4(s+1)Y(s)=0$$

$$\Rightarrow\quad Y'(s)+\frac{4(s+1)}{s(s+2)}Y(s)=0.$$

Separating variables and integrating yields

$$\frac{dY}{Y}=-\frac{4(s+1)}{s(s+2)}ds=-2\left(\frac{1}{s}+\frac{1}{s+2}\right)ds$$

$$\Rightarrow\quad\ln|Y|=-2(\ln|s|+\ln|s+2|)+C$$

$$\Rightarrow\quad Y(s)=\pm\frac{e^C}{s^2(s+2)^2}=\frac{c_1}{s^2(s+2)^2},$$

where $c_1\neq0$ is an arbitrary constant. Allowing $c_1=0$, we also get the solution $Y(s)\equiv0$, which was lost in separation of variables. Thus,

$$Y(s)=\frac{c_1}{s^2(s+2)^2}=\frac{c_1}{4}\left[\frac{1}{s^2}-\frac{1}{s}+\frac{1}{(s+2)^2}+\frac{1}{s+2}\right]$$

Copyright © 2012 Pearson Education, Inc. Publishing as Addison-Wesley.

and so

$$y(t) = \mathcal{L}^{-1}\{Y(s)\}(t)$$
$$= \frac{c_1}{4}\left(t - 1 + te^{-2t} + e^{-2t}\right) = c\left(t - 1 + te^{-2t} + e^{-2t}\right),$$

where $c = c_1/4$ is an arbitrary constant.

27. Note that the original equation can be written in the form

$$y(t) + t * y(t) = e^{-3t}.$$

Let $Y(s) := \mathcal{L}\{y\}(s)$. Applying the Laplace transform to both sides of this equation and using Theorem 11 of Section 7.7, we obtain

$$\mathcal{L}\{y(t) + t * y(t)\}(s) = Y(s) + \mathcal{L}\{t\}(s)Y(s) = \mathcal{L}\{e^{-3t}\}(s)$$
$$\Rightarrow \qquad Y(s) + \frac{1}{s^2}Y(s) = \frac{1}{s + 3}$$
$$\Rightarrow \qquad Y(s) = \frac{s^2}{(s + 3)(s^2 + 1)}.$$

The partial fractions decomposition for $Y(s)$ has the form

$$\frac{s^2}{(s+3)(s^2+1)} = \frac{A}{s+3} + \frac{Bs+C}{s^2+1} = \frac{A(s^2+1) + (Bs+C)(s+3)}{(s+3)(s^2+1)}.$$

Thus,

$$s^2 = A(s^2 + 1) + (Bs + C)(s + 3).$$

Evaluating both sides of this equation at $s = -3$, 0, and -2 yields

$$
\begin{array}{llll}
s = -3: & \Rightarrow & 9 = A(10) & \Rightarrow & A = 9/10, \\
s = 0: & \Rightarrow & 0 = A + 3C & \Rightarrow & C = -A/3 = -3/10, \\
s = -2: & \Rightarrow & 4 = 5A - 2B + C & \Rightarrow & B = (5A + C - 4)/2 = 1/10.
\end{array}
$$

Therefore,

$$Y(s) = \frac{9/10}{s+3} + \frac{(1/10)s}{s^2+1} - \frac{3/10}{s^2+1}$$
$$\Rightarrow \qquad y(t) = \mathcal{L}^{-1}\{Y(s)\}(t) = \frac{9}{10}e^{-3t} + \frac{1}{10}\cos t - \frac{3}{10}\sin t.$$

29. To find the transfer function, we use results of Section 7.7. Comparing the given equation with (14), we find that $a = 1$, $b = -5$, and $c = 6$. Thus, (15) yields

$$H(s) = \frac{1}{as^2 + bs + c} = \frac{1}{s^2 - 5s + 6}.$$

 Copyright © 2012 Pearson Education, Inc. Publishing as Addison-Wesley.

The impulse response function $h(t) = \mathcal{L}^{-1}\{H(s)\}(t)$. Using partial fractions, we see that

$$H(s) = \frac{1}{s^2 - 5s + 6} = \frac{1}{(s-3)(s-2)} = \frac{1}{s-3} - \frac{1}{s-2}$$

$$\Rightarrow \quad h(t) = \mathcal{L}^{-1}\left\{\frac{1}{s-3} - \frac{1}{s-2}\right\}(t) = e^{3t} - e^{2t}.$$

31. Let $X(s) := \mathcal{L}\{x\}(s)$, $Y(s) := \mathcal{L}\{y\}(s)$. Using initial conditions, we obtain

$$\mathcal{L}\{x'\}(s) = sX(s) - x(0) = sX(s),$$

$$\mathcal{L}\{y'\}(s) = sY(s) - y(0) = sY(s).$$

Therefore, applying the Laplace transform to both sides of the given equations yields

$$sX(s) + Y(s) = \mathcal{L}\{0\}(s) = 0,$$

$$X(s) + sY(s) = \mathcal{L}\{1 - u(t-2)\}(s) = \frac{1}{s} - \frac{e^{-2s}}{s} = \frac{1 - e^{-2s}}{s}.$$

Expressing

$$Y(s) = -sX(s)$$

from the first equation and substituting this into the second equation, we eliminate $Y(s)$.

$$X(s) - s^2 X(s) = \frac{1 - e^{-2s}}{s}$$

$$\Rightarrow \quad X(s) = -\frac{1 - e^{-2s}}{s(s^2 - 1)} = -\frac{1 - e^{-2s}}{s(s-1)(s+1)}.$$

We now need to find the partial fractions decomposition for

$$\frac{1}{s(s-1)(s+1)} = \frac{A}{s} + \frac{B}{s-1} + \frac{C}{s+1}$$

$$= \frac{A(s-1)(s+1) + Bs(s+1) + Cs(s-1)}{s(s-1)(s+1)}$$

$$= \frac{(A+B+C)s^2 + (B-C)s + (-A)}{s(s-1)(s+1)}.$$

This leads to the system

$$\begin{cases} A + B + C = 0, \\ B - C = 0, \\ -A = 1. \end{cases}$$

Copyright © 2012 Pearson Education, Inc. Publishing as Addison-Wesley.

Solving for A, B, and C yields

$$A = -1,$$
$$B = \frac{1}{2},$$
$$C = \frac{1}{2}.$$

Thus,

$$-\frac{1}{s(s-1)(s+1)} = \frac{1}{s} - \frac{1/2}{s-1} - \frac{1/2}{s+1},$$

the inverse Laplace transform yields

$$x(t) = \mathcal{L}^{-1}\left\{\left(1 - e^{-2s}\right)\left(\frac{1}{s} - \frac{1/2}{s-1} - \frac{1/2}{s+1}\right)\right\}(t)$$

$$= \mathcal{L}^{-1}\left\{\frac{1}{s} - \frac{1/2}{s-1} - \frac{1/2}{s+1}\right\}(t) - \mathcal{L}^{-1}\left\{\frac{1}{s} - \frac{1/2}{s-1} - \frac{1/2}{s+1}\right\}(t-2)u(t-2)$$

$$= 1 - \frac{e^{t} + e^{-t}}{2} - \left[1 - \frac{e^{t-2} + e^{-(t-2)}}{2}\right]u(t-2).$$

We now find $y(t)$ from the first equation in the original system.

$$y(t) = -x'(t) = \frac{e^{t} - e^{-t}}{2} - \frac{e^{t-2} - e^{-(t-2)}}{2}u(t-2).$$

 Copyright © 2012 Pearson Education, Inc. Publishing as Addison-Wesley.

FIGURES

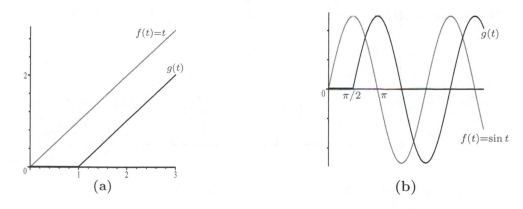

(a)

(b)

Figure 7–A: Graphs of functions in Problems 33 and 35.

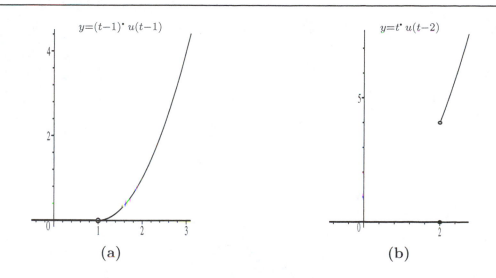

(a)

(b)

Figure 7–B: Graphs of functions in Problems 1 and 3.

Copyright © 2012 Pearson Education, Inc. Publishing as Addison-Wesley.

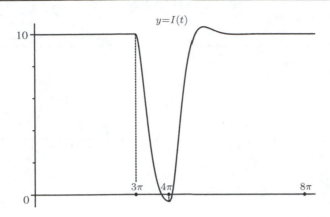

Figure 7–C: The graph of the function $y = I(t)$ in Problem 19.

 Copyright © 2012 Pearson Education, Inc. Publishing as Addison-Wesley.

CHAPTER 8: Series Solutions of Differential Equations

EXERCISES 8.1:   Introduction: The Taylor Polynomial Approximation

1. To find Taylor approximations

$$y(0) + \frac{y'(0)}{1!}\, x + \frac{y''(0)}{2!}\, x^2 + \frac{y''(0)}{3!}\, x^3 + \cdots ,$$

we need the values of $y(0)$, $y'(0)$, $y''(0)$, etc. $y(0)$ is provided by the initial condition, $y(0) = 1$. Substituting $x = 0$ into the given differential equation,

$$y'(x) = x^2 + y(x)^2 , \tag{8.1}$$

we obtain

$$y'(0) = 0^2 + y(0)^2 = 0 + 1^2 = 1.$$

Differentiating both sides of (8.1) yields

$$y''(x) = 2x + 2y(x)y'(x),$$

and so

$$y''(0) = 2(0) + 2y(0)y'(0) = 0 + 2(1)(1) = 2.$$

Hence,

$$y(x) = 1 + \frac{1}{1!}\, x + \frac{2}{2!}\, x^2 + \cdots = 1 + x + x^2 + \cdots .$$

3. Using the initial condition, $y(0) = 0$, we substitute $x = 0$ and $y = 0$ into the given equation and find $y'(0)$.

$$y'(0) = \sin(0) + e^0 = 1.$$

To determine $y''(0)$, we differentiate the given equation with respect to x and substitute $x = 0$, $y = 0$, and $y' = 1$.

$$y''(x) = (\sin y + e^x)' = (\sin y)' + (e^x)' = y' \cos y + e^x,$$

Copyright © 2012 Pearson Education, Inc. Publishing as Addison-Wesley.   457

$$y''(0) = 1 \cdot \cos 0 + e^0 = 2.$$

Similarly, differentiating $y''(x)$ and substituting, we obtain

$$y'''(x) = (y' \cos y + e^x)' = (y' \cos y)' + (e^x)' = y'' \cos y - (y')^2 \sin y + e^x,$$

$$y'''(0) = y''(0) \cos 0 - (y'(0))^2 \sin y(0) + e^0 = 2\cos 0 - (1)^2 \sin 0 + 1 = 3.$$

Thus, the first three nonzero terms in the Taylor polynomial approximations to the solution of the given initial value problem are

$$
\begin{aligned}
y(x) &= y(0) + \frac{y'(0)}{1!}x + \frac{y''(0)}{2!}x^2 + \frac{y'''(0)}{3!}x^3 + \cdots \\
&= 0 + \frac{1}{1}x + \frac{2}{2}x^2 + \frac{3}{6}x^3 + \cdots = x + x^2 + \frac{1}{2}x^3 + \cdots.
\end{aligned}
$$

5. We need the values of $x(0)$, $x'(0)$, $x''(0)$, etc. The first two are given by the initial conditions, $x(0) = 1$ and $x'(0) = 0$. Writing the given equation in the form

$$x''(t) = -tx(t) \tag{8.2}$$

we find that

$$x''(0) = -0 \cdot x(0) = -0 \cdot 1 = 0.$$

Differentiating (8.2) and substituting $t = 0$, we conclude that

$$
\begin{aligned}
x'''(t) &= -\left[tx'(t) + x(t)\right] &\Rightarrow&& x'''(0) &= -\left[0 \cdot x'(0) + x(0)\right] = -1, \\
x^{(4)}(t) &= -\left[tx''(t) + 2x'(t)\right] &\Rightarrow&& x^{(4)}(0) &= -\left[0 \cdot x''(0) + 2x'(0)\right] = 0, \\
x^{(5)}(t) &= -\left[tx'''(t) + 3x''(t)\right] &\Rightarrow&& x^{(5)}(0) &= -\left[0 \cdot x'''(0) + 3x''(0)\right] = 0, \\
x^{(6)}(t) &= -\left[tx^{(4)}(t) + 4x'''(t)\right] &\Rightarrow&& x^{(6)}(0) &= -\left[0 \cdot x^{(4)}(0) + 4x'''(0)\right] = 4.
\end{aligned}
$$

Therefore,

$$x(t) = 1 - \frac{1}{3!}t^3 + \frac{4}{6!}t^6 + \cdots = 1 - \frac{t^3}{6} + \frac{t^6}{180} + \cdots.$$

7. We use the initial conditions to find $y''(0)$. Writing the given equation in the form

$$y''(\theta) = -y(\theta)^3 + \sin \theta$$

and substituting $\theta = 0$, $y(0) = 0$, we get

$$y''(0) = -y(0)^3 + \sin 0 = 0.$$

 Copyright © 2012 Pearson Education, Inc. Publishing as Addison-Wesley.

Differentiating the given equation yields

$$y''' = (y'')' = -\left(y^3\right)' + (\sin\theta)' = -3y^2y' + \cos\theta$$

$$\Rightarrow \quad y'''(0) = -3y(0)^2y'(0) + \cos 0 = -3(0)^2(0) + 1 = 1.$$

Similarly, we get

$$y^{(4)} = (y''')' = -3y^2y'' - 6y\left(y'\right)^2 - \sin\theta$$

$$\Rightarrow \quad y^{(4)}(0) = -3y(0)^2y''(0) - 6y(0)\left(y'(0)\right)^2 - \sin 0 = 0.$$

To simplify further computations, we observe that the Taylor series for $y(\theta)$ has the form

$$y(\theta) = \frac{1}{3!}\theta^3 + \cdots .$$

Hence, the Taylor series expansion for $y(\theta)^3$ begins with the term $(1/3!)^3\theta^9$, so that

$$\left[y(\theta)^3\right]^{(k)}\Big|_{\theta=0} = 0 \quad \text{for} \quad k = 0, 1, \ldots, 8\,.$$

Thus,

$$y^{(5)} = -\left(y^3\right)''' - \cos\theta \quad \Rightarrow \quad y^{(5)}(0) = -\left(y^3\right)'''\Big|_{\theta=0} - \cos 0 = -1,$$

$$y^{(6)} = -\left(y^3\right)^{(4)} + \sin\theta \quad \Rightarrow \quad y^{(6)}(0) = -\left(y^3\right)^{(4)}\Big|_{\theta=0} - \sin 0 = 0,$$

$$y^{(7)} = -\left(y^3\right)^{(5)} + \cos\theta \quad \Rightarrow \quad y^{(7)}(0) = -\left(y^3\right)^{(5)}\Big|_{\theta=0} + \cos 0 = 1\,,$$

and the first three nonzero terms of the Taylor approximations are

$$y(\theta) = \frac{1}{3!}\theta^3 - \frac{1}{5!}\theta^5 + \frac{1}{7!}\theta^7 + \cdots = \frac{1}{6}\theta^3 - \frac{1}{120}\theta^5 + \frac{1}{5040}\theta^7 + \cdots .$$

9. (a) To construct $p_3(x)$, we need $f(1)$, $f'(1)$, $f''(1)$, and $f'''(1)$. Thus, we compute

$$f(x) = \ln x \quad \Rightarrow \quad f(1) = \ln 1 = 0,$$
$$f'(x) = x^{-1} \quad \Rightarrow \quad f'(1) = (1)^{-1} = 1,$$
$$f''(x) = -x^{-2} \quad \Rightarrow \quad f''(1) = -(1)^{-2} = -1,$$
$$f'''(x) = 2x^{-3} \quad \Rightarrow \quad f'''(1) = 2(1)^{-3} = 2,$$

and so

$$p_3(x) = 0 + \frac{1}{1!}(x-1) + \frac{-1}{2!}(x-1)^2 + \frac{2}{3!}(x-1)^3$$

$$= x - 1 - \frac{(x-1)^2}{2} + \frac{(x-1)^3}{3}\,.$$

Copyright © 2012 Pearson Education, Inc. Publishing as Addison-Wesley.

(b) To apply the error formula (6), we first compute

$$f^{(4)}(x) = [f'''(x)]' = (2x^{-3})' = -6x^{-4}.$$

Thus, (6) yields

$$\epsilon_3(x) := \ln x - p_3(x) = \frac{f^{(4)}(\xi)}{4!}(x - x_0)^4 = \frac{-6\xi^{-4}}{24}(x - 1)^4 = -\frac{(x - 1)^4}{4\xi^4}$$

$$\Rightarrow \quad |\ln(1.5) - p_3(1.5)| = \left|-\frac{(1.5 - 1)^4}{4\xi^4}\right| = \frac{(0.5)^4}{4\xi^4} \le \frac{(0.5)^4}{4} = 0.015625,$$

where we have used the fact that $\xi > 1$.

(c) Direct calculations give us

$$|\ln(1.5) - p_3(1.5)| \approx \left|0.405465 - \left(0.5 - \frac{(0.5)^2}{2} + \frac{(0.5)^3}{3}\right)\right| \approx 0.011202.$$

(d) See Figure B.51 in the answers of the text.

11. First, we rewrite the given equation in the form $y'' = -py' - qy + g$. On the right-hand side of this equation, the function y' is differentiable (since y'' exists) and the functions y, p, q, and g are differentiable (even twice). Thus, we conclude that its left-hand side, y'', is differentiable as a product, sum, and difference of differentiable functions. Therefore, y''' exists, and it is given by

$$y''' = (-py' - qy + g)' = -p'y' - py'' - q'y - qy' + g'.$$

Similarly, we conclude that the right-hand side of this equation is a differentiable function, since all the functions involved are differentiable. Hence, y''' (the left-hand side) is also differentiable, i.e., $y^{(4)}$ does exist.

13. Given $k = r = A = 1$ and $\omega = 10$, the Duffing's equation becomes

$$y'' + y + y^3 = \cos 10t \quad \Rightarrow \quad y'' = -y - y^3 + \cos 10t. \tag{8.3}$$

Substituting the initial conditions, $y(0) = 0$ and $y'(0) = 1$, into (8.3) yields

$$y''(0) = -y(0) - y(0)^3 + \cos(10 \cdot 0) = -0 - (0)^3 + \cos 0 = 1.$$

Differentiating (8.3), we conclude that

$$y''' = \left(-y - y^3 + \cos 10t\right)' = -y' - 3y^2y' - 10\sin 10t,$$

which, when evaluated at $t = 0$, gives

$$y'''(0) = -y'(0) - 3y(0)^2 y'(0) - 10\sin(10 \cdot 0) = -1 - 3(0)^2(1) - 10\sin 0 = -1.$$

Thus, the Taylor approximations to the solution of the given initial value problem are

$$y(t) = y(0) + \frac{y'(0)}{1!} t + \frac{y''(0)}{2!} t^2 + \frac{y'''(0)}{3!} t^3 + \cdots = t + \frac{1}{2} t^2 - \frac{1}{6} t^3 + \cdots .$$

15. For the Taylor polynomial $p_2(x)$, we need $y(0)$, $y'(0)$, and $y''(0)$. We already know $y(0)$ and $y'(0)$ from the initial conditions, $y(0) = 1$ and $y'(0) = 0$. Expressing $y''(x)$ from the given equation yields

$$y''(x) = -\frac{2y'(x) + xy(x)}{x}. \tag{8.4}$$

The formal substitution of $x = 0$ into (8.4) gives "0/0" – an indeterminate form. On the other hand, since the differentiability of a function implies its continuity, and we are given that $y(x)$ has derivatives of all orders at $x = 0$, we conclude that all the derivatives of $y(x)$ are continuous at $x = 0$. Therefore,

$$y''(0) = \lim_{x \to 0} y''(x),$$

and we can find the above limit by applying L'Hôpital's rule. Namely,

$$
\begin{aligned}
y''(0) &= \lim_{x \to 0} \left[-\frac{2y'(x) + xy(x)}{x} \right] \\
&= -\lim_{x \to 0} \frac{[2y'(x) + xy(x)]'}{(x)'} = -\lim_{x \to 0} [2y''(x) + xy'(x) + y(x)].
\end{aligned}
$$

This limit can be found by substitution due to the continuity of $y(x)$ and its derivatives at $x = 0$. Hence,

$$y''(0) = -[2y''(0) + 0 \cdot y'(0) + y(0)] = -2y''(0) - 1.$$

Solving for $y''(0)$ yields $y''(0) = -1/3$, and so

$$p_2(x) = y(0) + \frac{y'(0)}{1!} x + \frac{y''(0)}{2!} x^2 = 1 - \frac{x^2}{6}.$$

EXERCISES 8.2:   Power Series and Analytic Functions

1. Since $a_n = 2^{-n}/(n+1)$, the ratio test yields

$$\lim_{n \to \infty} \left| \frac{a_n}{a_{n+1}} \right| = \lim_{n \to \infty} \frac{2^{-n}/(n+1)}{2^{-(n+1)}/(n+2)} = \lim_{n \to \infty} \frac{n+2}{2^{-1}(n+1)} = 2.$$

Copyright © 2012 Pearson Education, Inc. Publishing as Addison-Wesley.

So, the radius of convergence is $\rho = 2$. In this power series, $x_0 = 1$. Hence, the endpoints of the interval of convergence are

$$x_1 = x_0 + \rho = 1 + 2 = 3,$$
$$x_2 = x_0 - \rho = 1 - 2 = -1.$$

At the point x_1, the series becomes

$$\sum_{n=0}^{\infty} \frac{2^{-n}}{n+1} (3-1)^n = \sum_{n=0}^{\infty} \frac{1}{n+1} = \infty$$

(a harmonic series); at the point x_2 we have

$$\sum_{n=0}^{\infty} \frac{2^{-n}}{n+1} (-1-1)^n = \sum_{n=0}^{\infty} \frac{(-1)^n}{n+1} < \infty$$

(by the alternating series test). Therefore, the set of convergence is $[-1, 3)$.

3. We use the ratio test given in Theorem 2 to find the radius of convergence for this power series. Since $a_n = n^2/2^n$, we see that

$$\frac{a_n}{a_{n+1}} = \frac{n^2/2^n}{(n+1)^2/2^{n+1}} = \frac{2n^2}{(n+1)^2}.$$

Therefore, we have

$$\lim_{n\to\infty} \left| \frac{a_n}{a_{n+1}} \right| = \lim_{n\to\infty} \left| \frac{2n^2}{(n+1)^2} \right| = 2 \lim_{n\to\infty} \frac{n^2}{(n+1)^2} = 2.$$

Thus, the radius of convergence is $\rho = 2$, and this power series converges absolutely for $|x + 2| < 2$, that is, for $-4 < x < 0$. We must now check the endpoints of this interval. For $x = -4$, $x + 2 = -2$, and the series

$$\sum_{n=0}^{\infty} \frac{n^2(-2)^n}{2^n} = \sum_{n=0}^{\infty} (-1)^n n^2$$

diverges, since the nth term, $a_n = (-1)^n n^2$, does not approach zero, as n approaches infinity. (Recall that it is necessary for the nth term of a convergent series to approach zero, as n approaches infinity.) Next, we check the endpoint $x = 0$, where $x + 2 = 2$. At this point, the series

$$\sum_{n=0}^{\infty} \frac{n^2 2^n}{2^n} = \sum_{n=0}^{\infty} n^2$$

is also divergent. Therefore, the given power series converges in the open interval $(-4, 0)$ and diverges outside of this interval.

     Copyright © 2012 Pearson Education, Inc. Publishing as Addison-Wesley.

5. With $a_n = 3/n^3$, the ratio test gives

$$L = \lim_{n \to \infty} \frac{3/n^3}{3/(n+1)^3} = \lim_{n \to \infty} \left(\frac{n+1}{n} \right)^3 = \left(\lim_{n \to \infty} \frac{n+1}{n} \right)^3 = 1.$$

Therefore, the radius of convergence is $\rho = 1$. At the points $x_0 \pm \rho = 2 \pm 1$, that is, $x = 3$ and $x = 1$, we have convergent series

$$\sum_{n=0}^{\infty} \frac{3}{n^3} \quad \text{and} \quad \sum_{n=0}^{\infty} \frac{3(-1)^n}{n^3}.$$

Therefore, the set of convergence of the given series is the closed interval $[1, 3]$.

7. By writing

$$\sum_{k=0}^{\infty} a_{2k} x^{2k} = \sum_{k=0}^{\infty} a_{2k} \left(x^2 \right)^k = \sum_{k=0}^{\infty} b_k z^k,$$

where $b_k := a_{2k}$ and $z := x^2$, we obtain a power series centered at the origin. The ratio test then yields the radius of convergence $\rho = L$, where

$$L = \lim_{k \to \infty} \left| \frac{b_k}{b_{k+1}} \right| = \lim_{k \to \infty} \left| \frac{a_{2k}}{a_{2k+2}} \right|.$$

So, the series $\sum_{k=0}^{\infty} b_k z^k$ converges for $|z| < L$ and diverges for $|z| > L$. Since $z = x^2$,

$$|z| < L \quad \Leftrightarrow \quad |x| < \sqrt{L}.$$

Hence, the original series converges for $|x| < \sqrt{L}$ and diverges for $|x| > \sqrt{L}$. By the definition, $\sqrt{L}$ is its radius of convergence.

The second statement can be proved in a similar way, since

$$\sum_{k=0}^{\infty} a_{2k+1} x^{2k+1} = x \sum_{k=0}^{\infty} a_{2k+1} \left(x^2 \right)^k = x \sum_{k=0}^{\infty} b_k z^k,$$

where $b_k := a_{2k+1}$ and $z := x^2$.

9. Since the addition of power series reduces to the addition of their coefficients, we make the following changes in indices of summation.

$$f(x): \quad n \to k \quad \Rightarrow \quad f(x) = \sum_{k=0}^{\infty} \frac{1}{k+1} x^k,$$

$$g(x): \quad n - 1 \to k \quad \Rightarrow \quad g(x) = \sum_{k=0}^{\infty} 2^{-(k+1)} x^k.$$

Copyright © 2012 Pearson Education, Inc. Publishing as Addison-Wesley.

Chapter 8

Therefore,

$$f(x) + g(x) = \sum_{k=0}^{\infty} \frac{1}{k+1} x^k + \sum_{k=0}^{\infty} 2^{-(k+1)} x^k = \sum_{k=0}^{\infty} \left[\frac{1}{k+1} + 2^{-k-1} \right] x^k.$$

11. We want to find a power series expansion of the product $f(x)g(x)$, where

$$f(x) = \sum_{n=0}^{\infty} \frac{x^n}{n!} = 1 + x + \frac{x^2}{2} + \frac{x^3}{6} + \frac{x^4}{24} + \cdots,$$

$$g(x) = \sin x = \sum_{k=0}^{\infty} \frac{(-1)^k}{(2k+1)!} x^{2k+1} = x - \frac{x^3}{6} + \frac{x^5}{120} - \frac{x^7}{7!} + \cdots.$$

Therefore, we have

$$
\begin{aligned}
f(x)g(x) &= \left(1 + x + \frac{x^2}{2} + \frac{x^3}{6} + \frac{x^4}{24} + \cdots \right) \left(x - \frac{x^3}{6} + \frac{x^5}{120} - \frac{x^7}{7!} + \cdots \right) \\
&= x + x^2 + \left(\frac{1}{2} - \frac{1}{6} \right) x^3 + \left(\frac{1}{6} - \frac{1}{6} \right) x^4 + \left(\frac{1}{24} - \frac{1}{12} + \frac{1}{120} \right) x^5 + \cdots \\
&= x + x^2 + \frac{1}{3} x^3 + \cdots.
\end{aligned}
$$

Note that, since the radius of convergence for both series is $\rho = \infty$, the expansion of the product $f(x)g(x)$ also converges for all values of x.

13. Using formula (6), we obtain

$$
\begin{aligned}
f(x)g(x) &= \left[\sum_{n=0}^{\infty} \frac{(-1)^n}{n!} x^n \right] \left[\sum_{n=0}^{\infty} (-1)^n x^n \right] \\
&= \left(1 - x + \frac{1}{2} x^2 - \frac{1}{6} x^3 + \cdots \right) \left(1 - x + x^2 - x^3 + \cdots \right) \\
&= (1)(1) + [(1)(-1) + (-1)(1)] x + \left[(1)(1) + (-1)(-1) + \left(\frac{1}{2} \right) (1) \right] x^2 + \cdots \\
&= 1 - 2x + \frac{5}{2} x^2 + \cdots.
\end{aligned}
$$

15. (a) Let

$$q(x) = \sum_{n=0}^{\infty} a_n x^n.$$

Multiplying both sides of the given equation by $\sum_{n=0}^{\infty} x^n/n!$, we obtain

$$\left(\sum_{n=0}^{\infty} a_n x^n \right) \left(\sum_{n=0}^{\infty} \frac{1}{n!} x^n \right) = \sum_{n=0}^{\infty} \frac{1}{2^n} x^n.$$

Thus, the right-hand side is the Cauchy product of $q(x)$ and $\sum_{n=0}^{\infty} x^n/n!$.

 Copyright © 2012 Pearson Education, Inc. Publishing as Addison-Wesley.

(b) With $c_n = 1/2^n$ and $b_n = 1/n!$, the formula (6) yields

$$n = 0 : \quad \frac{1}{2^0} = c_0 = a_0 b_0 = a_0 \frac{1}{0!} = a_0 \,;$$

$$n = 1 : \quad \frac{1}{2^1} = c_1 = a_0 b_1 + a_1 b_0 = a_0 \frac{1}{1!} + a_1 \frac{1}{0!} = a_0 + a_1 \,;$$

$$n = 2 : \quad \frac{1}{2^2} = c_2 = a_0 b_2 + a_1 b_1 + a_2 b_0 = a_0 \frac{1}{2!} + a_1 \frac{1}{1!} + a_2 \frac{1}{0!} = \frac{a_0}{2} + a_1 + a_2 \,;$$

$$n = 3 : \quad \frac{1}{2^3} = c_3 = a_0 b_3 + a_1 b_2 + a_2 b_1 + a_3 b_0 = \frac{a_0}{6} + \frac{a_1}{2} + a_2 + a_3 \,;$$

$$\vdots$$

(c) The system in (b) simplifies to

$$
\begin{aligned}
1 &= a_0 \,, & a_0 &= 1 \,, \\
1/2 &= a_0 + a_1 \,, & a_1 &= 1/2 - a_0 = -1/2 \,, \\
1/4 &= a_0/2 + a_1 + a_2 \,, \quad \Rightarrow & a_2 &= 1/4 - a_0/2 - a_1 = 1/4 \,, \\
1/8 &= a_0/6 + a_1/2 + a_2 + a_3 \,, & a_3 &= 1/8 - a_0/6 - a_1/2 - a_2 = -1/24 \,, \\
\vdots & & \vdots &
\end{aligned}
$$

Thus,

$$q(x) = 1 - \frac{1}{2} x + \frac{1}{4} x^2 - \frac{1}{24} x^3 + \cdots .$$

17. Since

$$\lim_{n \to \infty} \left| \frac{a_n}{a_{n+1}} \right| = \lim_{n \to \infty} \left| \frac{(-1)^n}{(-1)^{n+1}} \right| = \lim_{n \to \infty} 1 = 1,$$

by the ratio test, we find the radius of convergence of the given series to be $\rho = 1 > 0$. Therefore, Theorem 4 can be applied. This yields

$$\left[(1 + x)^{-1} \right]' = \sum_{n=1}^{\infty} (-1)^n n x^{n-1} \quad \Rightarrow \quad -(1 + x)^{-2} = \sum_{n=1}^{\infty} (-1)^n n x^{n-1} \,;$$

the radius of convergence of this series is also equal to 1.

19. Here we assume that this series has a positive radius of convergence. Thus, since we have

$$f(x) = \sum_{k=0}^{\infty} a_k x^{2k} = a_0 + a_1 x^2 + a_2 x^4 + a_3 x^6 + \cdots + a_k x^{2k} + \cdots ,$$

we can differentiate term by term to obtain

$$f'(x) = 0 + 2a_1 x + a_2 4 x^3 + a_3 6 x^5 + \cdots + a_k 2k x^{2k-1} + \cdots = \sum_{k=1}^{\infty} 2k a_k x^{2k-1} .$$

Copyright © 2012 Pearson Education, Inc. Publishing as Addison-Wesley.

Note that the summation for $f(x)$ starts at $k = 0$, while the summation for $f'(x)$ starts at $k = 1$.

21. Using the ratio test, we find that the radius ρ of convergence of the given series is

$$\rho = \lim_{n \to \infty} \left| \frac{(-1)^n}{(-1)^{n+1}} \right| = 1.$$

Thus, by Theorem 4,

$$g(x) = \int_0^x f(t)\, dt = \int_0^x \left[\sum_{n=0}^{\infty} (-1)^n t^n \right] dt$$

$$= \sum_{n=0}^{\infty} (-1)^n \int_0^x t^n\, dt = \sum_{n=0}^{\infty} (-1)^n \frac{1}{n+1} t^{n+1} \Big|_0^x = \sum_{n=0}^{\infty} \frac{(-1)^n}{n+1} x^{n+1}.$$

On the other hand,

$$g(x) = \int_0^x \frac{dt}{1+t} = \ln(1+t) \Big|_0^x = \ln(1+x), \quad x \in (-1, 1).$$

23. Setting $k = n - 1$, we get $n = k + 1$. Note that $k = 0$, when $n = 1$. Hence, substituting into the given series yields

$$\sum_{n=1}^{\infty} n a_n x^{n-1} = \sum_{k=0}^{\infty} (k+1) a_{k+1} x^k.$$

25. We let $n + 1 = k$, so that $n = k - 1$; for $n = 0$, $k = 1$. Thus,

$$\sum_{n=0}^{\infty} a_n x^{n+1} = \sum_{k=1}^{\infty} a_{k-1} x^k.$$

27. Termwise multiplication yields

$$x^2 \sum_{n=0}^{\infty} n(n+1) a_n x^n = \sum_{n=0}^{\infty} n(n+1) a_n x^n x^2 = \sum_{n=0}^{\infty} n(n+1) a_n x^{n+2}.$$

Now we can shift the summation index by letting $k = n + 2$. Thus, we have $n = k - 2$, $n + 1 = k - 1$, $k = 2$, when $n = 0$, and so

$$\sum_{n=0}^{\infty} n(n+1) a_n x^{n+2} = \sum_{k=2}^{\infty} (k-2)(k-1) a_{k-2} x^k.$$

By replacing k by n, we obtain the desired form.

    Copyright © 2012 Pearson Education, Inc. Publishing as Addison-Wesley.

29. We need to determine the nth derivative of $f(x)$ at $x = \pi$. Thus, we observe that

$$
\begin{aligned}
f(x) &= \cos x & \Rightarrow & \quad f(\pi) = \cos \pi = -1, \\
f'(x) &= \sin x & \Rightarrow & \quad f'(\pi) = -\sin \pi = 0, \\
f''(x) &= -\cos x & \Rightarrow & \quad f''(\pi) = -\cos \pi = 1, \\
f'''(x) &= \sin x & \Rightarrow & \quad f'''(\pi) = \sin \pi = 0, \\
f^{(4)}(x) &= \cos x & \Rightarrow & \quad f^{(4)}(\pi) = \cos \pi = -1.
\end{aligned}
$$

Since $f^{(4)}(x) = \cos x = f(x)$, the four derivatives, given above, will be repeated in a cycle. Thus, we see that $f^{(n)}(\pi) = 0$ if n is odd and $f^{(n)}(\pi) = (-1)^{(n/2)+1}$ if n is even. Therefore, the Taylor series for f about $x_0 = \pi$ is given by

$$
f(x) = -1 + 0 + \frac{1}{2!}(x-\pi)^2 + 0 - \frac{1}{4!}(x-\pi)^4 + \cdots = \sum_{n=0}^{\infty} \frac{(-1)^{n+1}(x-\pi)^{2n}}{(2n)!}.
$$

31. Writing

$$
f(x) = \frac{1+x}{1-x} = \frac{(1-x)+2x}{1-x} = 1 + \frac{2x}{1-x},
$$

we can use the power series expansion (3) (i.e., the geometric series) to obtain the desired Taylor series. Thus, we have

$$
f(x) = 1 + \frac{2x}{1-x} = 1 + 2x \sum_{k=0}^{\infty} x^k = 1 + 2 \sum_{k=0}^{\infty} x^{k+1}.
$$

Shifting the summation index, that is, letting $k+1 = n$, yields

$$
f(x) = 1 + 2 \sum_{k=0}^{\infty} x^{k+1} = 1 + 2 \sum_{n=1}^{\infty} x^n.
$$

33. Using the formula

$$
c_j = \frac{f^{(j)}(x_0)}{j!}
$$

for the coefficients of the Taylor series expansion for $f(x)$ about x_0, we find that

$$
\begin{aligned}
f(x_0) &= x^3 + 3x - 4 \big|_{x=1} = 0 & \Rightarrow & \quad c_0 = 0, \\
f'(x_0) &= 3x^2 + 3 \big|_{x=1} = 6 & \Rightarrow & \quad c_1 = 6/1! = 6, \\
f''(x_0) &= 6x \big|_{x=1} = 6 & \Rightarrow & \quad c_2 = 6/2! = 3, \\
f'''(x) &\equiv 6 & \Rightarrow & \quad c_3 = 6/3! = 1, \\
f^{(j)}(x) &\equiv 0 & \Rightarrow & \quad c_j = 0 \quad \text{for } j \geq 4.
\end{aligned}
$$

Therefore,

$$
x^3 + 3x - 4 = 6(x-1) + 3(x-1)^2 + (x-1)^3.
$$

Copyright © 2012 Pearson Education, Inc. Publishing as Addison-Wesley.

35. (a) We have

$$\frac{1}{x} = \frac{1}{1+(x-1)} = \frac{1}{1-s}, \qquad \text{where} \quad s = -(x-1).$$

Since $1/(1-s) = \sum_{n=0}^{\infty} s^n$, the substitution $s = -(x-1)$ yields

$$\frac{1}{x} = \frac{1}{1-s} = \sum_{n=0}^{\infty} s^n = \sum_{n=0}^{\infty} [-(x-1)]^n = \sum_{n=0}^{\infty} (-1)^n (x-1)^n,$$

which is valid for

$$|s| = |x-1| < 1 \qquad \Rightarrow \qquad 0 < x < 2.$$

(b) Since the above series has positive radius of convergence ($\rho = 1$), Theorem 4 can be applied. Hence, for $0 < x < 2$,

$$\ln x = \int_1^x \frac{1}{t}\, dt = \int_1^x \left[\sum_{n=0}^{\infty} (-1)^n (t-1)^n \right] dt = \sum_{n=0}^{\infty} (-1)^n \int_1^x (t-1)^n\, dt$$

$$= \sum_{n=0}^{\infty} (-1)^n \frac{1}{n+1} (t-1)^{n+1} \Big|_1^x = \sum_{n=0}^{\infty} \frac{(-1)^n}{n+1} (x-1)^{n+1} = \sum_{k=1}^{\infty} \frac{(-1)^{k-1}}{k} (x-1)^k.$$

37. For $n = 0$, $f^{(0)}(0) := f(0) = 0$ by the definition of $f(x)$.

To find $f'(0)$, we use the definition of the derivative.

$$f'(0) = \lim_{x \to 0} \frac{f(x) - f(0)}{x - 0} = \lim_{x \to 0} \frac{e^{-1/x^2}}{x}. \tag{8.5}$$

We compute the left-hand side and the right-hand side limits by making the substitution $t = 1/x$. Note that $t \to +\infty$, as $x \to 0^+$, and $t \to -\infty$, as $x \to 0^-$. Thus, we have

$$\lim_{x \to 0^{\pm}} \frac{e^{-1/x^2}}{x} = \lim_{t \to \pm\infty} t e^{-t^2} = \lim_{t \to \pm\infty} \frac{t}{e^{t^2}} = \lim_{t \to \pm\infty} \frac{1}{2te^{t^2}} = 0,$$

where we applied L'Hospital's rule to the indeterminate form ∞/∞. Therefore, the limit in (8.5) exists and equals to 0. For any $x \neq 0$,

$$f'(x) = \left(e^{-1/x^2} \right)' = e^{-1/x^2} \left(-\frac{1}{x^2} \right)' = \frac{2}{x^3} e^{-1/x^2}.$$

Next, we proceed by induction. Assuming that, for some $n \geq 1$,

$$f^{(n)}(0) = 0 \quad \text{and} \quad f^{(n)}(x) = p\left(\frac{1}{x} \right) e^{-1/x^2}, \qquad x \neq 0,$$

 Copyright © 2012 Pearson Education, Inc. Publishing as Addison-Wesley.

where $p(t)$ is a polynomial in t, we show that

$$f^{(n+1)}(0) = 0 \quad \text{and} \quad f^{(n+1)}(x) = q\left(\frac{1}{x}\right)e^{-1/x^2}, \quad x \neq 0,$$

where $q(t)$ is a polynomial in t. This will imply that $f^{(n)}(0) = 0$ for all $n \geq 0$.

Indeed, substituting $t = 1/x$ in the one-sided limits yields

$$\lim_{x \to 0^\pm} \frac{f^{(n)}(x) - f^{(n)}(0)}{x - 0} = \lim_{x \to 0^\pm} \frac{p(1/x)e^{-1/x^2}}{x} = \lim_{t \to \pm\infty} \frac{tp(t)}{e^{t^2}} = \lim_{t \to \pm\infty} \frac{r(t)}{e^{t^2}},$$

where $r(t) = a_0 t^k + \cdots + a_k$ is a polynomial. Applying the L'Hospital's rule k times, we obtain

$$\lim_{t \to \pm\infty} \frac{r(t)}{e^{t^2}} = \lim_{t \to \pm\infty} \frac{r'(t)}{(e^{t^2})'} = \lim_{t \to \pm\infty} \frac{r'(t)}{2te^{t^2}}$$

$$= \lim_{t \to \pm\infty} \frac{r''(t)}{(4t^2 + 2)e^{t^2}} = \cdots = \lim_{t \to \pm\infty} \frac{a_0 k!}{(2^k t^k + \cdots)e^{t^2}} = 0.$$

Since both one-sided limits exist and are equal, the regular limit exists and equals to the same number. That is,

$$f^{(n+1)}(0) = \lim_{x \to 0} \frac{f^{(n)}(x) - f^{(n)}(0)}{x - 0} = 0.$$

For any $x \neq 0$,

$$f^{(n+1)}(x) = \left[p\left(\frac{1}{x}\right)e^{-1/x^2}\right]' = \left[p'\left(\frac{1}{x}\right)\left(\frac{1}{x}\right)'\right]e^{-1/x^2} + p\left(\frac{1}{x}\right)\left[e^{-1/x^2}\left(-\frac{1}{x^2}\right)'\right]$$

$$= \left[-p'\left(\frac{1}{x}\right)\frac{1}{x^2} + p\left(\frac{1}{x}\right)\frac{2}{x^3}\right]e^{-1/x^2} = q\left(\frac{1}{x}\right)e^{-1/x^2},$$

where $q(t) = -p'(t)t^2 + p(t)2t^3$.

EXERCISES 8.3: Power Series Solutions to Linear Differential Equations

1. Dividing the given equation by $(x + 1)$ yields

$$y'' - \frac{x^2}{x + 1}y' + \frac{3}{x + 1}y = 0.$$

Thus, we see that

$$p(x) = -\frac{x^2}{x + 1}, \qquad q(x) = \frac{3}{x + 1}.$$

These functions are rational functions, and so they are analytic everywhere except, perhaps, zeros of their denominators. Solving $x + 1 = 0$, we find that $x = -1$, which is a point of infinite discontinuity for both functions. Consequently, $x = -1$ is the only singular point of the given equation.

Copyright © 2012 Pearson Education, Inc. Publishing as Addison-Wesley.

Chapter 8

3. Writing the equation in standard form yields

$$y'' + \frac{2}{\theta^2 - 2}\, y' + \frac{\sin\theta}{\theta^2 - 2}\, y = 0.$$

The coefficients,

$$p(\theta) = \frac{2}{\theta^2 - 2} \qquad \text{and} \qquad q(\theta) = \frac{\sin\theta}{\theta^2 - 2},$$

are quotients of analytic functions, and so they are analytic everywhere except zeros $\theta = \pm\sqrt{2}$ of the denominator, where they have infinite discontinuities. Hence, the given equation has two singular points, $\theta = \pm\sqrt{2}$.

5. In standard form, the equation becomes

$$x'' + \frac{t+1}{t^2 - t - 2}\, x' - \frac{t-2}{t^2 - t - 2}\, x = 0.$$

Hence,

$$p(t) = \frac{t+1}{t^2 - t - 2} = \frac{t+1}{(t+1)(t-2)}, \qquad q(t) = -\frac{t-2}{t^2 - t - 2} = -\frac{t-2}{(t+1)(t-2)}.$$

The point $t = -1$ is a removable singularity for $p(t)$ since, for $t \neq -1$, we can cancel $(t+1)$-term in the numerator and denominator, and so $p(t)$ becomes analytic at $t = -1$ if we set

$$p(-1) := \lim_{t \to -1} p(t) = \lim_{t \to -1} \frac{1}{t-2} = -\frac{1}{3}.$$

At the point $t = 2$, $p(t)$ has infinite discontinuity. Thus, $p(t)$ is analytic everywhere except $t = 2$. Similarly, $q(t)$ is analytic everywhere except $t = -1$. Therefore, the given equation has two singular points, $t = -1$ and $t = 2$.

7. Writing the equation in standard form, we obtain

$$y'' + \left(\frac{\cos x}{\sin x}\right) y = 0.$$

Thus, $p(x) \equiv 0$. We also see that

$$q(x) = \frac{\cos x}{\sin x} = \cot x.$$

Note that $q(x)$ is the quotient of two analytic functions ($\cos x$ and $\sin x$), which have power series expansions about any real number x with infinite radius of convergence. Thus, $q(x)$ also has a power series expansion with positive radius of convergence about every real number x as long as the denominator, $\sin x$, is not zero. Since $\sin x = 0$ at $x = n\pi$, $n = 0, \pm 1, \ldots$, the differential equation is singular only at these points.

 Copyright © 2012 Pearson Education, Inc. Publishing as Addison-Wesley.

9. Dividing the differential equation by $\sin\theta$, we get

$$y'' - \frac{\ln\theta}{\sin\theta}y = 0.$$

Thus, $p(\theta) \equiv 0$ and $q(\theta) = -\ln\theta/\sin\theta$. The function $q(\theta)$ is not defined for $\theta \le 0$ because of the logarithmic term, and it has infinite discontinuities at positive zeros of the denominator.

$$\sin\theta = 0 \quad \Rightarrow \quad \theta = k\pi, \quad k = 1, 2, 3, \dots.$$

At any other point θ, $q(\theta)$ is analytic as a quotient of two analytic functions. Hence, the singular points of the given equation are $\theta \le 0$ and $\theta = k\pi$, $k = 1, 2, 3, \dots$.

11. The coefficient, $x+2$, is a polynomial, and so it is analytic everywhere. Therefore, $x = 0$ is an ordinary point for the given equation. We seek a power series solution of the form

$$y(x) = \sum_{n=0}^{\infty} a_n x^n \quad \Rightarrow \quad y'(x) = \sum_{n=1}^{\infty} n a_n x^{n-1},$$

where we have applied Theorem 4 to find the power series expansion of $y'(x)$. We now substitute the power series for y and y' into the given differential equation and obtain

$$\sum_{n=1}^{\infty} n a_n x^{n-1} + (x+2)\sum_{n=0}^{\infty} a_n x^n = 0$$

$$\Rightarrow \quad \sum_{n=1}^{\infty} n a_n x^{n-1} + \sum_{n=0}^{\infty} 2a_n x^n + \sum_{n=0}^{\infty} a_n x^{n+1} = 0. \tag{8.6}$$

To sum these series, we make shifts in indices of summation so that they sum over the same power of x. In the first sum, we set $k = n-1$ so that $n = k+1$, where k runs from 0 to ∞; in the second sum, we just replace n by k; in the third sum, we let $k = n+1$ so that $n = k-1$, and the summation starts from $k = 1$. Thus, the equation (8.6) becomes

$$\sum_{k=0}^{\infty}(k+1)a_{k+1}x^k + \sum_{k=0}^{\infty} 2a_k x^k + \sum_{k=1}^{\infty} a_{k-1}x^k = 0$$

$$\Rightarrow \quad \left[a_1 + \sum_{k=1}^{\infty}(k+1)a_{k+1}x^k\right] + \left[2a_0 + \sum_{k=1}^{\infty} 2a_k x^k\right] + \sum_{k=1}^{\infty} a_{k-1}x^k = 0$$

$$\Rightarrow \quad (a_1 + 2a_0) + \sum_{k=1}^{\infty}\left[(k+1)a_{k+1} + 2a_k + a_{k-1}\right]x^k = 0.$$

Copyright © 2012 Pearson Education, Inc. Publishing as Addison-Wesley.

Chapter 8

For the power series in the left-hand side to vanish, we need all zero coefficients. Hence,

$$a_1 + 2a_0 = 0,$$

$$(k+1)a_{k+1} + 2a_k + a_{k-1} = 0 \quad \text{for all} \quad k \geq 1.$$

This yields

$$
\begin{aligned}
a_1 + 2a_0 &= 0 & \Rightarrow \quad a_1 &= -2a_0, \\
k = 1: \quad 2a_2 + 2a_1 + a_0 &= 0 & \Rightarrow \quad a_2 &= (-2a_1 - a_0)/2 = (4a_0 - a_0)/2 = 3a_0/2, \\
k = 2: \quad 3a_3 + 2a_2 + a_1 &= 0 & \Rightarrow \quad a_3 &= (-2a_2 - a_1)/3 = (-3a_0 + 2a_0)/3 = -a_0/3, \\
&\vdots
\end{aligned}
$$

Therefore,

$$y(x) = a_0 - 2a_0 x + \frac{3a_0}{2} x^2 - \frac{a_0}{3} x^3 + \cdots = a_0 \left(1 - 2x + \frac{3x^2}{2} - \frac{x^3}{3} + \cdots \right),$$

where a_0 is an arbitrary constant (which is, actually, $y(0)$).

13. This equation has no singular points since the coefficients $p(x) \equiv 0$ and $q(x) = -x^2$ are analytic everywhere. So, let

$$z(x) = \sum_{k=0}^{\infty} a_k x^k \quad \Rightarrow \quad z'(x) = \sum_{k=1}^{\infty} k a_k x^{k-1} \quad \Rightarrow \quad z''(x) = \sum_{k=2}^{\infty} k(k-1) a_k x^{k-2},$$

where we have used Theorem 4 differentiating the series termwise. Substituting z and z'' into the given equation yields

$$
\begin{aligned}
z'' - x^2 z &= \sum_{k=2}^{\infty} k(k-1) a_k x^{k-2} - x^2 \sum_{k=0}^{\infty} a_k x^k \\
&= \sum_{k=2}^{\infty} k(k-1) a_k x^{k-2} - \sum_{k=0}^{\infty} a_k x^{k+2}.
\end{aligned}
$$

We now shift indices of summation so that they sum over the same power of x. For the first sum, we substitute $n = k - 2$ so that $k = n + 2$, $k - 1 = n + 1$, and the summation starts from $n = 0$. In the second sum, we let $n = k + 2$, which yields $k = n - 2$ and $n = 2$ as the starting index. Thus, we obtain

$$z'' - x^2 z = \sum_{n=0}^{\infty} (n+2)(n+1) a_{n+2} x^n - \sum_{n=2}^{\infty} a_{n-2} x^n.$$

 Copyright © 2012 Pearson Education, Inc. Publishing as Addison-Wesley.

Next step in writing the right-hand side as a single power series is to start both summations at the same index. To do this, we observe that

$$\sum_{n=0}^{\infty}(n+2)(n+1)a_{n+2}x^n - \sum_{n=2}^{\infty}a_{n-2}x^n$$

$$= 2a_2 + 6a_3x + \sum_{n=2}^{\infty}(n+2)(n+1)a_{n+2}x^n - \sum_{n=2}^{\infty}a_{n-2}x^n$$

$$= 2a_2 + 6a_3x + \sum_{n=2}^{\infty}\left[(n+2)(n+1)a_{n+2} - a_{n-2}\right]x^n\,.$$

In order for this power series to be zero, all coefficients must be zero. Therefore, we obtain

$$2a_2 = 0, \qquad 6a_3 = 0 \qquad \text{and} \qquad (n+2)(n+1)a_{n+2} - a_{n-2} = 0, \quad n \geq 2\,.$$

From the first two equations, we find that $a_2 = 0$ and $a_3 = 0$. Next, we take $n = 2$ and $n = 3$ in the above recurrence relation and get

$$n = 2: \quad (4)(3)a_4 - a_0 = 0 \qquad \Rightarrow \qquad a_4 = a_0/12\,,$$
$$n = 3: \quad (5)(4)a_5 - a_1 = 0 \qquad \Rightarrow \qquad a_5 = a_1/20\,.$$

Hence,

$$z(x) = \sum_{k=0}^{\infty}a_kx^k = a_0 + a_1x + (0)x^2 + (0)x^3 + \frac{a_0}{12}x^4 + \frac{a_1}{20}x^5 + \cdots$$

$$= a_0\left(1 + \frac{x^4}{12} + \cdots\right) + a_1\left(x + \frac{x^5}{20} + \cdots\right)\,.$$

15. $x = 0$ is an ordinary point for this equation, since the functions $p(x) = x - 1$ and $q(x) = 1$ are both analytic everywhere. Thus, we can assume that the solution to this linear differential equation has a power series expansion with positive radius of convergence about the point $x = 0$. That is, we assume that

$$y(x) = a_0 + a_1x + a_2x^2 + a_3x^3 + \cdots = \sum_{n=0}^{\infty}a_nx^n. \tag{8.7}$$

We have to find the coefficients a_n. So, we substitute $y(x)$ and its derivatives into the given differential equation. To accomplish this, first we find $y'(x)$ and $y''(x)$. Since $y(x)$ has a power series expansion with positive radius of convergence about the point $x = 0$, we can find its derivatives by differentiating the series (8.7) termwise. Thus, we have

$$y'(x) = 0 + a_1 + 2a_2x + 3a_3x^2 + \cdots = \sum_{n=1}^{\infty}na_nx^{n-1}\,,$$

Copyright © 2012 Pearson Education, Inc. Publishing as Addison-Wesley.

$$y''(x) = 2a_2 + 6a_3x + \cdots = \sum_{n=2}^{\infty} n(n-1)a_n x^{n-2}.$$

Substituting these expressions into the given differential equation, we obtain

$$y'' + (x-1)y' + y = \sum_{n=2}^{\infty} n(n-1)a_n x^{n-2} + (x-1)\sum_{n=1}^{\infty} na_n x^{n-1} + \sum_{n=0}^{\infty} a_n x^n = 0.$$

Simplifying yields

$$\sum_{n=2}^{\infty} n(n-1)a_n x^{n-2} + \sum_{n=1}^{\infty} na_n x^n - \sum_{n=1}^{\infty} na_n x^{n-1} + \sum_{n=0}^{\infty} a_n x^n = 0. \tag{8.8}$$

We now want to rewrite the left-hand side of this equation as a single power series. This will allow us to find expressions for the coefficients. Therefore, we need to shift the indices in each power series in (8.8) so that they sum over same powers of x. Thus, we let $k = n - 2$ in the first sum and note that this means that $n = k + 2$ and that $k = 0$ for $n = 2$.

$$\sum_{n=2}^{\infty} n(n-1)a_n x^{n-2} = \sum_{k=0}^{\infty} (k+2)(k+1)a_{k+2} x^k.$$

In the third power series, we let $k = n - 1$, which implies that $n = k + 1$ and that $k = 0$ if $n = 1$. Thus, we see that

$$\sum_{n=1}^{\infty} na_n x^{n-1} = \sum_{k=0}^{\infty} (k+1)a_{k+1} x^k.$$

For the second and the last power series, we just replace n by k. Substituting all these expressions into (8.8) yields

$$\sum_{k=0}^{\infty} (k+2)(k+1)a_{k+2} x^k + \sum_{k=1}^{\infty} ka_k x^k - \sum_{k=0}^{\infty} (k+1)a_{k+1} x^k + \sum_{k=0}^{\infty} a_k x^k = 0.$$

Our next step in writing the left-hand side as a single power series is to start all of the summations at the same index. To do this, we observe that

$$\sum_{k=0}^{\infty} (k+2)(k+1)a_{k+2} x^k = (2)(1)a_2 x^0 + \sum_{k=1}^{\infty} (k+2)(k+1)a_{k+2} x^k,$$

$$\sum_{k=0}^{\infty} (k+1)a_{k+1} x^k = (1)a_1 x^0 + \sum_{k=1}^{\infty} (k+1)a_{k+1} x^k,$$

$$\sum_{k=0}^{\infty} a_k x^k = a_0 x^0 + \sum_{k=1}^{\infty} a_k x^k.$$

 Copyright © 2012 Pearson Education, Inc. Publishing as Addison-Wesley.

Thus, all of the summations now start at $k = 1$. Therefore, we have

$$(2)(1)a_2x^0 + \sum_{k=1}^{\infty}(k+2)(k+1)a_{k+2}x^k + \sum_{k=1}^{\infty} ka_kx^k$$

$$-(1)a_1x^0 - \sum_{k=1}^{\infty}(k+1)a_{k+1}x^k + a_0x^0 + \sum_{k=1}^{\infty} a_kx^k = 0$$

$$\Rightarrow \quad 2a_2 - a_1 + a_0 + \sum_{k=1}^{\infty}\left((k+2)(k+1)a_{k+2}x^k + ka_kx^k - (k+1)a_{k+1}x^k + a_kx^k\right) = 0$$

$$\Rightarrow \quad 2a_2 - a_1 + a_0 + \sum_{k=1}^{\infty}\left((k+2)(k+1)a_{k+2} + (k+1)a_k - (k+1)a_{k+1}\right)x^k = 0.$$

In order for this power series to vanish, all coefficients must be zero. Therefore, we obtain

$$2a_2 - a_1 + a_0 = 0 \quad \Rightarrow \quad a_2 = \frac{a_1 - a_0}{2},$$

and

$$(k+2)(k+1)a_{k+2} + (k+1)a_k - (k+1)a_{k+1} = 0$$

$$\Rightarrow \quad a_{k+2} = \frac{a_{k+1} - a_k}{k+2}, \quad k \geq 1,$$

where we have canceled out the factor $(k+1)$ from the recurrence relation. *Note that we have solved the recurrence relation for a_{k+2}, the coefficient with the largest subscript. Also, note that the first value for k in the recurrence formula is the same as that used in the summation notation.* Substituting $k = 1$ into the recurrence relation, we find that

$$a_3 = \frac{a_2 - a_1}{3} = \frac{\dfrac{a_1 - a_0}{2} - a_1}{3} = \frac{-(a_1 + a_0)}{6}.$$

By letting $k = 2$, we obtain

$$a_4 = \frac{a_3 - a_2}{4} = \frac{\dfrac{-(a_1 + a_0)}{6} - \dfrac{a_1 - a_0}{2}}{4} = \frac{-2a_1 + a_0)}{12},$$

where we have plugged in the values for a_2 and a_3 found above. Continuing this process allows us to find as many coefficients for the power series expansion of the solution as we want. Notice that the coefficients, we just found, involve only the variables a_0 and a_1. From the recurrence equation, we see that this will be the case for all coefficients of the power series solution. Thus, a_0 and a_1 are arbitrary constants, and they will be our

Copyright © 2012 Pearson Education, Inc. Publishing as Addison-Wesley.

arbitrary constants in a general solution. Substituting the formulas for the coefficients into

$$y(x) = \sum_{n=0}^{\infty} a_n x^n = a_0 + a_1 x + a_2 x^2 + a_3 x^3 + a_4 x^4 + \cdots,$$

yields

$$
\begin{aligned}
y(x) &= a_0 + a_1 x + \frac{a_1 - a_0}{2} x^2 + \frac{-(a_1 + a_0)}{6} x^3 + \frac{-2a_1 + a_0}{12} x^4 + \cdots \\
&= a_0 \left(1 - \frac{x^2}{2} - \frac{x^3}{6} + \frac{x^4}{12} + \cdots \right) + a_1 \left(x + \frac{x^2}{2} - \frac{x^3}{6} - \frac{x^4}{6} + \cdots \right).
\end{aligned}
$$

19. Since $x = 0$ is an ordinary point for the given equation, we seek for a power series expansion of a general solution of the form

$$y(x) = \sum_{n=0}^{\infty} a_n x^n \qquad \Rightarrow \qquad y'(x) = \sum_{n=1}^{\infty} n a_n x^{n-1}.$$

Substituting $y(x)$ and $y'(x)$ into the given equation, we obtain

$$\sum_{n=1}^{\infty} n a_n x^{n-1} - 2x \sum_{n=0}^{\infty} a_n x^n = \sum_{n=1}^{\infty} n a_n x^{n-1} - \sum_{n=0}^{\infty} 2 a_n x^{n+1} = 0.$$

We shift the indices of summations so that they sum over the same powers of x. In the first sum, we let $k = n - 1$. Then $n = k + 1$ and the summation starts from $k = 0$. In the second sum, we let $k = n + 1$. Then $n = k - 1$, and $k = 1$ for $n = 0$. Thus, we have

$$\sum_{k=0}^{\infty} (k+1) a_{k+1} x^k - \sum_{k=1}^{\infty} 2 a_{k-1} x^k = a_1 + \sum_{k=1}^{\infty} [(k+1) a_{k+1} - 2 a_{k-1}] x^k = 0.$$

In order for this power series to be zero, each coefficient must be zero. That is,

$$
\begin{aligned}
&a_1 = 0, \\
&(k+1) a_{k+1} - 2 a_{k-1} = 0, \quad k \geq 1
\end{aligned}
\quad \Rightarrow \quad
\begin{aligned}
&a_1 = 0, \\
&a_{k+1} = 2 a_{k-1}/(k+1), \quad k \geq 1.
\end{aligned}
$$

Since $a_1 = 0$, it follows from this recurrence relation that all odd coefficients are zeros. Indeed,

$$a_3 = \frac{2a_1}{3} = 0, \qquad a_5 = \frac{2a_3}{5} = 0, \qquad \text{etc.}$$

For even coefficients, we have

$$
\begin{aligned}
k = 1: \quad & a_2 = 2a_0/2, \\
k = 3: \quad & a_4 = 2a_2/4 = 2[2a_0/2]/4 = 2^2 a_0/(2 \cdot 4), \\
k = 5: \quad & a_6 = 2a_4/6 = 2[2^2 a_0/(2 \cdot 4)]/6 = 2^3 a_0/(2 \cdot 4 \cdot 6), \\
& \vdots
\end{aligned}
$$

    Copyright © 2012 Pearson Education, Inc. Publishing as Addison-Wesley.

The pattern for the even coefficients is now apparent. Namely,

$$a_{2k} = \frac{2^k a_0}{2 \cdot 4 \cdots (2k)} = \frac{2^k a_0}{2^k (1 \cdot 2 \cdots k)} = \frac{a_0}{k!}, \qquad k = 1, 2, \ldots .$$

This formula remains correct for $k = 0$ as well (since $0! := 1$). Thus,

$$y(x) = \sum_{k=0}^{\infty} \frac{a_0}{k!} x^{2k} = a_0 \sum_{k=0}^{\infty} \frac{x^{2k}}{k!},$$

where a_0 is an arbitrary constant.

21. Since $x = 0$ is an ordinary point for this differential equation, we assume that the solution has a power series expansion with a positive radius of convergence about $x = 0$. Thus, we have

$$y(x) = \sum_{n=0}^{\infty} a_n x^n \quad \Rightarrow \quad y'(x) = \sum_{n=1}^{\infty} n a_n x^{n-1} \quad \Rightarrow \quad y''(x) = \sum_{n=2}^{\infty} n(n-1) a_n x^{n-2}.$$

Plugging in these expressions into the differential equation, we obtain

$$y'' - xy' + 4y = \sum_{n=2}^{\infty} n(n-1) a_n x^{n-2} - x \sum_{n=1}^{\infty} n a_n x^{n-1} + 4 \sum_{n=0}^{\infty} a_n x^n = 0$$

$$\Rightarrow \quad \sum_{n=2}^{\infty} n(n-1) a_n x^{n-2} - \sum_{n=1}^{\infty} n a_n x^n + \sum_{n=0}^{\infty} 4 a_n x^n = 0.$$

In order for each power series to sum over the same powers of x, we shift the index in the first summation by letting $k = n - 2$, and we let $k = n$ in the other two power series. Thus, we have

$$\sum_{k=0}^{\infty} (k+2)(k+1) a_{k+2} x^k - \sum_{k=1}^{\infty} k a_k x^k + \sum_{k=0}^{\infty} 4 a_k x^k = 0.$$

Next, we want that all the summations start at the same index. Therefore, we take the first terms in the first and the last power series out of the summation sign. This yields

$$(2)(1) a_2 x^0 + \sum_{k=1}^{\infty} (k+2)(k+1) a_{k+2} x^k - \sum_{k=1}^{\infty} k a_k x^k + 4 a_0 x^0 + \sum_{k=1}^{\infty} 4 a_k x^k = 0$$

$$\Rightarrow \quad 2a_2 + 4a_0 + \sum_{k=1}^{\infty} (k+2)(k+1) a_{k+2} x^k - \sum_{k=1}^{\infty} k a_k x^k + \sum_{k=1}^{\infty} 4 a_k x^k = 0$$

$$\Rightarrow \quad 2a_2 + 4a_0 + \sum_{k=1}^{\infty} \left[(k+2)(k+1) a_{k+2} + (-k+4) a_k \right] x^k = 0.$$

Copyright © 2012 Pearson Education, Inc. Publishing as Addison-Wesley.

By setting each coefficient of this power series to be zero, we see that

$$2a_2 + 4a_0 = 0 \quad \Rightarrow \quad a_2 = \frac{-4a_0}{2} = -2a_0\,,$$

$$(k+2)(k+1)a_{k+2} + (-k+4)a_k = 0 \quad \Rightarrow \quad a_{k+2} = \frac{(k-4)a_k}{(k+2)(k+1)}\,, \quad k \ge 1,$$

where we have solved the recurrence equation, the last equation for a_{k+2}, the coefficient with the largest subscript. Thus, we have

$$k = 1 \quad \Rightarrow \quad a_3 = \frac{-3a_1}{3 \cdot 2} = \frac{-a_1}{2}\,,$$

$$k = 2 \quad \Rightarrow \quad a_4 = \frac{-2a_2}{4 \cdot 3} = \frac{(-2)(-4)a_0}{4 \cdot 3 \cdot 2} = \frac{a_0}{3}\,,$$

$$k = 3 \quad \Rightarrow \quad a_5 = \frac{-a_3}{5 \cdot 4} = \frac{(-3)(-1)a_1}{5 \cdot 4 \cdot 3 \cdot 2} = \frac{a_1}{40}\,,$$

$$k = 4 \quad \Rightarrow \quad a_6 = 0,$$

$$k = 5 \quad \Rightarrow \quad a_7 = \frac{a_5}{7 \cdot 6} = \frac{(-3)(-1)(1)a_1}{7 \cdot 6 \cdot 5 \cdot 4 \cdot 3 \cdot 2} = \frac{a_1}{560}\,,$$

$$k = 6 \quad \Rightarrow \quad a_8 = \frac{2a_6}{8 \cdot 7} = 0,$$

$$k = 7 \quad \Rightarrow \quad a_9 = \frac{3a_7}{9 \cdot 8} = \frac{(-3)(-1)(1)(3)a_1}{9!}\,,$$

$$k = 8 \quad \Rightarrow \quad a_{10} = \frac{4a_8}{10 \cdot 9} = 0,$$

$$k = 9 \quad \Rightarrow \quad a_{11} = \frac{5a_9}{11 \cdot 10} = \frac{(-3)(-1)(1)(3)(5)a_1}{11!}\,.$$

Now we can see a pattern. (*Note that it is easier to determine such a pattern if we consider specific coefficients that have not been multiplied out.*)

First, we notice that a_0 and a_1 can be chosen arbitrarily.

Next, we see that

$$a_{2n} = 0 \quad \text{for} \quad n > 2$$

and that a general formula for odd coefficients is given by

$$a_{2n+1} = \frac{(-3)(-1)(1) \cdots (2n-5)a_1}{(2n+1)!}\,.$$

Notice that this formula is also valid for a_3 and a_5. Substituting these expressions for the coefficients into the solution

$$y(x) = \sum_{n=0}^{\infty} a_n x^n = a_0 + a_1 x + a_2 x^2 + a_3 x^3 + a_4 x^4 + \cdots$$

 Copyright © 2012 Pearson Education, Inc. Publishing as Addison-Wesley.

yields

$$y(x) = a_0 + a_1 x - 2a_0 x^2 - \frac{a_1}{2} x^3 + \frac{a_0}{3} x^4 + \frac{a_1}{40} x^5 + \cdots$$

$$+ \frac{(-3)(-1)(1)\cdots(2n-5)a_1}{(2n+1)!} x^{2n+1} + \cdots$$

$$= a_0 \left[1 - 2x^2 + \frac{x^4}{3} \right] + a_1 \left[x - \frac{x^3}{2} + \frac{x^5}{40} + \cdots + \frac{(-3)(-1)(1)\cdots(2n-5)}{(2n+1)!} x^{2n+1} + \cdots \right]$$

$$= a_0 \left[1 - 2x^2 + \frac{x^4}{3} \right] + a_1 \left[x + \sum_{k=1}^{\infty} \frac{(-3)(-1)(1)\cdots(2k-5)}{(2k+1)!} x^{2k+1} \right].$$

29. Since $x = 0$ is an ordinary point for this differential equation, we can assume that a solution to this problem is given by

$$y(x) = \sum_{n=0}^{\infty} a_n x^n \quad \Rightarrow \quad y'(x) = \sum_{n=1}^{\infty} n a_n x^{n-1} \quad \Rightarrow \quad y''(x) = \sum_{n=2}^{\infty} n(n-1) a_n x^{n-2}.$$

By substituting the initial conditions, $y(0) = 1$ and $y'(0) = -2$, into the first two equations, we see that

$$y(0) = a_0 = 1 \quad \text{and} \quad y'(0) = a_1 = -2.$$

Next, we substitute the expressions found above for $y(x)$, $y'(x)$, and $y''(x)$ into the differential equation to obtain

$$y'' + y' - xy = \sum_{n=2}^{\infty} n(n-1) a_n x^{n-2} + \sum_{n=1}^{\infty} n a_n x^{n-1} - x \sum_{n=0}^{\infty} a_n x^n = 0$$

$$\Rightarrow \quad \sum_{n=2}^{\infty} n(n-1) a_n x^{n-2} + \sum_{n=1}^{\infty} n a_n x^{n-1} - \sum_{n=0}^{\infty} a_n x^{n+1} = 0.$$

By setting $k = n - 2$ in the first power series, $k = n - 1$ in the second power series, and $k = n + 1$ in the last power series, we shift the indices, so that x is raised to the power k in each of them. Thus, we obtain

$$\sum_{k=0}^{\infty} (k+2)(k+1) a_{k+2} x^k + \sum_{k=0}^{\infty} (k+1) a_{k+1} x^k - \sum_{k=1}^{\infty} a_{k-1} x^k = 0.$$

We can start all the summations at the same index if we remove the first term from each of the first two power series above. Therefore, we have

$$(2)(1)a_2 + \sum_{k=1}^{\infty} (k+2)(k+1) a_{k+2} x^k + (1)a_1 + \sum_{k=1}^{\infty} (k+1) a_{k+1} x^k - \sum_{k=1}^{\infty} a_{k-1} x^k = 0$$

Copyright © 2012 Pearson Education, Inc. Publishing as Addison-Wesley.

$$\Rightarrow \qquad 2a_2 + a_1 + \sum_{k=1}^{\infty} \left[(k+2)(k+1)a_{k+2} + (k+1)a_{k+1} - a_{k-1} \right] x^k = 0.$$

Thus, all the coefficients must be zero.

$$2a_2 + a_1 = 0 \qquad \Rightarrow \qquad a_2 = \frac{-a_1}{2},$$

$$(k+2)(k+1)a_{k+2} + (k+1)a_{k+1} - a_{k-1} = 0$$

$$\Rightarrow \qquad a_{k+1} = \frac{a_{k-1} - (k+1)a_{k+1}}{(k+2)(k+1)}, \qquad k \geq 1.$$

Therefore,

$$k = 1 \qquad \Rightarrow \qquad a_3 = \frac{a_0 - 2a_2}{3 \cdot 2} = \frac{a_0}{6} + \frac{a_1}{6}.$$

Using the fact that $a_0 = 1$ and $a_1 = -2$, which we found from the initial conditions, we compute

$$a_2 = \frac{-(-2)}{2} = 1, \quad a_3 = \frac{1}{6} + \frac{-2}{6} = -\frac{1}{6}.$$

By substituting these coefficients, we obtain a cubic polynomial approximation

$$y(x) = 1 - 2x + x^2 - \frac{x^3}{6}.$$

The graphs of the linear, quadratic and cubic polynomial approximations can be easily generated by using a standard software utility.

31. The point $x_0 = 0$ is an ordinary point for the given equation, since $p(x) = 2x/(x^2 + 2)$ and $q(x) = 3/(x^2 + 2)$ are analytic at zero. Hence, we can express a general solution as

$$y(x) = \sum_{n=0}^{\infty} a_n x^n.$$

Substituting this expansion into the given differential equation yields

$$(x^2 + 2) \sum_{n=2}^{\infty} n(n-1)a_n x^{n-2} + 2x \sum_{n=1}^{\infty} n a_n x^{n-1} + 3 \sum_{n=0}^{\infty} a_n x^n = 0$$

$$\Rightarrow \qquad \sum_{n=2}^{\infty} n(n-1)a_n x^n + \sum_{n=2}^{\infty} 2n(n-1)a_n x^{n-2} + \sum_{n=1}^{\infty} 2n a_n x^n + \sum_{n=0}^{\infty} 3a_n x^n = 0.$$

To sum over like powers x^k, we put $k = n - 2$ into the second sum and $k = n$ into the others.

$$\sum_{k=2}^{\infty} k(k-1)a_k x^k + \sum_{k=0}^{\infty} 2(k+2)(k+1)a_{k+2} x^k + \sum_{k=1}^{\infty} 2k a_k x^k + \sum_{k=0}^{\infty} 3a_k x^k = 0.$$

     Copyright © 2012 Pearson Education, Inc. Publishing as Addison-Wesley.

Next, we separate the terms corresponding to $k = 0$ and $k = 1$ and combine the rest under one summation.

$$(4a_2 + 3a_0) + (12a_3 + 5a_1)x + \sum_{k=2}^{\infty} \left[k(k-1)a_k + 2(k+2)(k+1)a_{k+2} + 2ka_k + 3a_k \right] x^k = 0.$$

Setting the coefficients equal to zero and simplifying, we get

$$4a_2 + 3a_0 = 0,$$
$$12a_3 + 5a_1 = 0,$$
$$(k^2 + k + 3)a_k + 2(k+2)(k+1)a_{k+2} = 0, \qquad k \geq 2$$

$$\Rightarrow \qquad \begin{aligned} a_2 &= -3a_0/4, \\ a_3 &= -5a_1/12, \\ a_{k+2} &= -(k^2 + k + 3)a_k/[2(k+2)(k+1)], \qquad k \geq 2. \end{aligned}$$

From the initial conditions, we have

$$a_0 = y(0) = 1 \qquad \text{and} \qquad a_1 = y'(0) = 2.$$

Therefore,

$$a_2 = -3(1)/4 = -3/4,$$
$$a_3 = -5(2)/12 = -5/6,$$

and the cubic polynomial approximation to the solution is

$$y(x) = a_0 + a_1 x + x_2 x^2 + a_3 x^3 = 1 + 2x - \frac{3x^2}{4} - \frac{5x^3}{6}.$$

33. In Problem 7, Exercises 8.2, we have shown that the radius of convergence of a power series $\sum_{n=0}^{\infty} a_{2n} x^{2n}$ is $\rho = \sqrt{L}$, where

$$L = \lim_{n \to \infty} \left| \frac{a_{2n}}{a_{2n+2}} \right|.$$

In the series (13), $a_{2n} = (-1)^n/n!$, and so

$$L = \lim_{n \to \infty} \left| \frac{(-1)^n/n!}{(-1)^{n+1}/(n+1)!} \right| = \lim_{n \to \infty} (n+1) = \infty.$$

Therefore, $\rho = \infty$.

35. With the given values of parameters, we have an initial value problem

$$\frac{1}{10} q''(t) + \left(1 + \frac{t}{10} \right) q'(t) + \frac{1}{2} q(t) = 0, \qquad q(0) = 10, \quad q'(0) = 0.$$

Copyright © 2012 Pearson Education, Inc. Publishing as Addison-Wesley.

Simplifying yields

$$q''(t) + (10 + t)\, q'(t) + 5q(t) = 0, \qquad q(0) = 10, \quad q'(0) = 0.$$

The point $t = 0$ is an ordinary point for this equation. Let $q(t) = \sum_{n=0}^{\infty} a_n t^n$ be the power series expansion of $q(t)$ about $t = 0$. Substituting this series into the above differential equation, we obtain

$$\sum_{n=2}^{\infty} n(n-1)a_n t^{n-2} + (10 + t)\sum_{n=1}^{\infty} na_n t^{n-1} + 5\sum_{n=0}^{\infty} a_n t^n = 0$$

$$\Rightarrow \quad \sum_{n=2}^{\infty} n(n-1)a_n t^{n-2} + \sum_{n=1}^{\infty} 10na_n t^{n-1} + \sum_{n=1}^{\infty} na_n t^n + \sum_{n=0}^{\infty} 5a_n t^n = 0.$$

Setting $k = n - 2$ in the first summation, $k = n - 1$ in the second summation, and $k = n$ in the last two summations, we obtain

$$\sum_{k=0}^{\infty}(k+2)(k+1)a_{k+2}t^k + \sum_{k=0}^{\infty} 10(k+1)a_{k+1}t^k + \sum_{k=1}^{\infty} ka_k t^k + \sum_{k=0}^{\infty} 5a_k t^k = 0.$$

Separating the terms, corresponding to $k = 0$, and combining the rest under one sum yields

$$(2a_2 + 10a_1 + 5a_0) + \sum_{k=1}^{\infty} \left[(k+2)(k+1)a_{k+2} + 10(k+1)a_{k+1} + (k+5)a_k\right]t^k = 0.$$

Setting the coefficients equal to zero, we obtain the recurrence relations

$$\begin{aligned} 2a_2 + 10a_1 + 5a_0 &= 0, \\ (k+2)(k+1)a_{k+2} + 10(k+1)a_{k+1} + (k+5)a_k &= 0, \qquad k \geq 1. \end{aligned} \qquad (8.9)$$

Next, we use the initial conditions to find a_0 and a_1.

$$a_0 = q(0) = 10, \qquad a_1 = q'(0) = 0.$$

From the first equation in (8.9), we have

$$a_2 = \frac{-10a_1 - 5a_0}{2} = -25.$$

Taking $k = 1$ and $k = 2$ in the second equation in (8.9), we find a_3 and a_4.

$$\begin{aligned} k = 1: \quad & 6a_3 + 20a_2 + 6a_1 = 0 \quad \Rightarrow \quad a_3 = -(20a_2 + 6a_1)/6 = 250/3, \\ k = 2: \quad & 12a_4 + 30a_3 + 7a_2 = 0 \quad \Rightarrow \quad a_4 = -(30a_3 + 7a_2)/12 = -775/4. \end{aligned}$$

Hence,

$$q(t) = 10 + (0)t - 25t^2 + \frac{250t^3}{3} - \frac{775t^4}{4} + \cdots = 10 - 25t^2 + \frac{250t^3}{3} - \frac{775t^4}{4} + \cdots .$$

 Copyright © 2012 Pearson Education, Inc. Publishing as Addison-Wesley.

EXERCISES 8.4: Equations with Analytic Coefficients

3. In this equation, $p(x) = 0$ and $q(x) = -3/(1 + x + x^2)$. Therefore, singular points occur when

$$1 + x + x^2 = 0 \quad \Rightarrow \quad x = -\frac{1}{2} \pm \frac{\sqrt{3}}{2} i.$$

Thus, $x = 1$ is an ordinary point for this equation, and we can find a power series solution with radius of convergence of at least the minimum of the distances from the point $x = 1$ to $x = (-1/2) \pm (\sqrt{3}/2)i$, which equal $\sqrt{3}$. Therefore, the radius of convergence for the power series solution to this differential equation about $x = 1$ is at least $\sqrt{3}$.

9. We see that $x = 0$ and $x = 2$ are the only singular points for this differential equation and, so $x = 1$ is an ordinary point. Therefore, according to Theorem 5, there exists a power series solution to this equation about the point $x = 1$ with radius of convergence of at least 1, which is the smaller of the distances from $x = 1$ to either $x = 0$ or $x = 2$. That is, we have a general solution to the given differential equation of the form

$$y(x) = \sum_{n=0}^{\infty} a_n (x - 1)^n \,,$$

which is convergent for all x in the interval $(0, 2)$, the interval on which the inequality $|x - 1| < 1$ is satisfied. To find this solution we proceed as in Example 3. Thus, we make a substitution $t = x - 1$. (Note that $dx/dt = 1$.) We then define a new function

$$
\begin{aligned}
Y(t) &:= y(t + 1) = y(x) \\
\Rightarrow \quad \frac{dY}{dt} &= \left(\frac{dy}{dx}\right)\left(\frac{dx}{dt}\right) = \left(\frac{dy}{dx}\right) \cdot 1 = \frac{dy}{dx} \\
\Rightarrow \quad \frac{d^2Y}{dt^2} &= \frac{d}{dt}\left(\frac{dY}{dt}\right) = \frac{d}{dt}\left(\frac{dy}{dx}\right) = \left(\frac{d^2y}{dx^2}\right)\left(\frac{dx}{dt}\right) = \frac{d^2y}{dx^2}\,.
\end{aligned}
$$

Hence, with substitutions $t = x - 1$ and $Y(t) = y(t + 1)$, we transform the given differential equation into

$$
\begin{aligned}
\left[(t + 1)^2 - 2(t + 1)\right] y''(t + 1) + 2y(t + 1) &= 0 \\
\Rightarrow \quad \left[(t + 1)^2 - 2(t + 1)\right] Y''(t) + 2Y(t) &= 0 \\
\Rightarrow \quad \left(t^2 - 1\right) Y''(t) + 2Y(t) &= 0. \quad (8.10)
\end{aligned}
$$

To find a general solution to (8.10), we note that $t = 0$ is its ordinary point. Thus, we

Copyright © 2012 Pearson Education, Inc. Publishing as Addison-Wesley.

can assume that there is a power series solution of the form

$$Y(t) = \sum_{n=0}^{\infty} a_n t^n,$$

which converges for all t in $(-1, 1)$. Substituting $Y(t)$ into the equation (8.10) yields

$$(t^2 - 1) \sum_{n=2}^{\infty} n(n-1)a_n t^{n-2} + 2 \sum_{n=0}^{\infty} a_n t^n = 0$$

$$\Rightarrow \quad \sum_{n=2}^{\infty} n(n-1)a_n t^n - \sum_{n=2}^{\infty} n(n-1)a_n t^{n-2} + \sum_{n=0}^{\infty} 2a_n t^n = 0.$$

Making a shift in the index, $k = n - 2$, in the second power series above and replacing n by k in the other two power series gives us

$$\sum_{k=2}^{\infty} k(k-1)a_k t^k - \sum_{k=0}^{\infty} (k+2)(k+1)a_{k+2} t^k + \sum_{k=0}^{\infty} 2a_k t^k = 0.$$

Thus,

$$\sum_{k=2}^{\infty} k(k-1)a_k t^k - (2)(1)a_2 - (3)(2)a_3 t - \sum_{k=2}^{\infty} (k+2)(k+1)a_{k+2} t^k$$

$$+ 2a_0 + 2a_1 t + \sum_{k=2}^{\infty} 2a_k t^k = 0$$

$$\Rightarrow \quad 2a_0 - 2a_2 + (2a_1 - 6a_3)t + \sum_{k=2}^{\infty} [k(k-1)a_k - (k+2)(k+1)a_{k+2} + 2a_k] t^k = 0.$$

For this power series to vanish, its coefficients must be zero. Hence, we have

$$2a_0 - 2a_2 = 0 \quad \Rightarrow \quad a_2 = a_0, \qquad 2a_1 - 6a_3 = 0 \quad \Rightarrow \quad a_3 = \frac{a_1}{3},$$

$$k(k-1)a_k - (k+2)(k+1)a_{k+2} + 2a_k = 0, \qquad k \geq 2$$

$$\Rightarrow \quad a_{k+2} = \frac{k(k-1)a_k + 2a_k}{(k+2)(k+1)} \quad \Rightarrow \quad a_{k+2} = \frac{(k^2 - k + 2)a_k}{(k+2)(k+1)}, \quad k \geq 2.$$

Therefore, we see that

$$k = 2 \quad \Rightarrow \quad a_4 = \frac{4a_2}{4 \cdot 3} = \frac{a_2}{3} = \frac{a_0}{3},$$

$$k = 3 \quad \Rightarrow \quad a_5 = \frac{8a_3}{5 \cdot 4} = \frac{2a_1}{15}, \quad \text{etc.}$$

Plugging in these values for the coefficients into the power series solution,

$$Y(t) = \sum_{n=0}^{\infty} a_n t^n = a_0 + a_1 t + a_2 t^2 + a_3 t^3 + a_4 t^4 + \cdots,$$

 Copyright © 2012 Pearson Education, Inc. Publishing as Addison-Wesley.

yields

$$Y(t) = a_0 + a_1 t + a_0 t^2 + \frac{a_1 t^3}{3} + \frac{a_0 t^4}{3} + \frac{2a_1 t^5}{15} + \cdots$$

$$\Rightarrow \quad Y(t) = a_0 \left(1 + t^2 + \frac{t^4}{3} + \cdots \right) + a_1 \left(t + \frac{t^3}{3} + \frac{2t^5}{15} + \cdots \right). \quad (8.11)$$

Finally, we change back to the independent variable x. Recall that $Y(t) = y(t+1)$. Thus, $t = x - 1$, and so

$$Y(t) = Y(x-1) = y([x-1]+1) = y(x).$$

Thus, we replace t by $x - 1$ in the solution (8.11) and obtain a power series expansion of a general solution in the independent variable x.

$$y(x) = a_0 \left[1 + (x-1)^2 + \frac{1}{3}(x-1)^4 + \cdots \right]$$

$$+ a_1 \left[(x-1) + \frac{1}{3}(x-1)^3 + \frac{2}{15}(x-1)^5 + \cdots \right].$$

17. Here, $p(x) = 0$ and $q(x) = -\sin x$ are both analytic everywhere. Thus, $x = \pi$ is an ordinary point for this differential equation, and there are no singular points. Therefore, by Theorem 5, we can assume that this equation has a general power series solution about $x = \pi$ with an infinite radius of convergence (i.e., $\rho = \infty$). That is, we assume that we have a solution to this differential equation given by

$$y(x) = \sum_{n=0}^{\infty} a_n (x - \pi)^n \quad \Rightarrow \quad y'(x) = \sum_{n=1}^{\infty} n a_n (x - \pi)^{n-1},$$

which converges for all x. If we apply the initial conditions, $y(\pi) = 1$ and $y'(\pi) = 0$, we see that $a_0 = 1$ and $a_1 = 0$. To find a general solution, we combine the methods of Example 3 and Example 4. Thus, we first define a new function, $Y(t)$, using the substitution $t = x - \pi$. Thus, we denote

$$Y(t) := y(t + \pi) = y(x).$$

Hence, by the chain rule, we derive that $dY/dt = (dy/dx)(dx/dt) = dy/dx$ and, similarly, $d^2Y/dt^2 = d^2y/dx^2$. We now solve the transformed differential equation

$$\frac{d^2Y}{dt^2} - [\sin(t+\pi)] Y(t) = 0 \quad \Rightarrow \quad \frac{d^2Y}{dt^2} + (\sin t)Y(t) = 0. \quad (8.12)$$

Copyright © 2012 Pearson Education, Inc. Publishing as Addison-Wesley.

We seek a power series solution to the equation (8.12) of the form

$$Y(t) = \sum_{n=0}^{\infty} a_n t^n \quad \Rightarrow \quad Y'(t) = \sum_{n=1}^{\infty} n a_n t^{n-1} \quad \Rightarrow \quad Y''(t) = \sum_{n=2}^{\infty} n(n-1) a_n t^{n-2}.$$

Since the initial conditions, $y(\pi) = 1$ and $y'(\pi) = 0$, transform into $Y(0) = 1$ and $Y'(0) = 0$, we have

$$Y(0) = a_0 = 1 \quad \text{and} \quad Y'(0) = a_1 = 0.$$

Next we note that $q(t) = \sin t$ is an analytic function with a Maclaurin series given by

$$\sin t = \sum_{n=0}^{\infty} \frac{(-1)^n t^{2n+1}}{(2n+1)!} = t - \frac{t^3}{6} + \frac{t^5}{120} - \frac{t^7}{5040} + \cdots.$$

Substituting the expressions that we found for $Y(t)$, $Y''(t)$, and $\sin t$ into (8.12) yields

$$\sum_{n=2}^{\infty} n(n-1) a_n t^{n-2} + \left(t - \frac{t^3}{6} + \frac{t^5}{120} - \frac{t^7}{5040} + \cdots \right) \sum_{n=0}^{\infty} a_n t^n = 0.$$

Therefore,

$$\left(2a_2 + 6a_3 t + 12a_4 t^2 + 20a_5 t^3 + 30a_6 t^4 + \cdots \right) + t \left(a_0 + a_1 t + a_2 t^2 + a_3 t^3 + \cdots \right)$$
$$- \frac{t^3}{6} \left(a_0 + a_1 t + \cdots \right) + \cdots = 0$$
$$\Rightarrow \quad \left(2a_2 + 6a_3 t + 12a_4 t^2 + 20a_5 t^3 + 30a_6 t^4 + \cdots \right) + t \left(a_0 + a_1 t + a_2 t^2 + a_3 t^3 + \cdots \right)$$
$$+ \left(-\frac{a_0 t^3}{6} - \frac{a_1 t^4}{6} - \cdots \right) + \cdots = 0.$$

Collecting similar terms, we obtain

$$2a_2 + (6a_3 + a_0) t + (12a_4 + a_1) t^2 + \left(20a_5 + a_2 - \frac{a_0}{6} \right) t^3 + \left(30a_6 + a_3 - \frac{a_1}{6} \right) t^4 + \cdots = 0.$$

Setting the coefficients to zero and recalling that $a_0 = 1$ and $a_1 = 0$ yields

$$2a_2 = 0 \qquad\qquad \Rightarrow \qquad a_2 = 0,$$
$$6a_3 + a_0 = 0 \qquad\qquad \Rightarrow \qquad a_3 = -a_0/6 = -1/6,$$
$$12a_4 + a_1 = 0 \qquad\qquad \Rightarrow \qquad a_4 = -a_1/12 = 0,$$
$$20a_5 + a_2 - (a_0/6) = 0 \qquad \Rightarrow \qquad a_5 = [(a_0/6) - a_2]/20 = (1/6)/20 = 1/120,$$
$$30a_6 + a_3 - (a_1/6) = 0 \qquad \Rightarrow \qquad a_6 = [(a_1/6) - a_3]/30 = [0 + (1/6)]/30 = 1/180.$$

Plugging in these coefficients into the power series solution,

$$Y(t) = \sum_{n=0}^{\infty} a_n t^n = a_0 + a_1 t + a_2 t^2 + \cdots,$$

 Copyright © 2012 Pearson Education, Inc. Publishing as Addison-Wesley.

yields the solution to (8.12).

$$Y(t) = 1 + 0 + 0 - \frac{t^3}{6} + 0 + \frac{t^5}{120} + \frac{t^6}{180} + \cdots = 1 - \frac{t^3}{6} + \frac{t^5}{120} + \frac{t^6}{180} + \cdots .$$

Finally, we to find the solution to the original equation with the independent variable x. To this end, we recall that $t = x - \pi$, and so $Y(x - \pi) = y(x)$. Therefore, substituting these values into the equation above, we obtain

$$y(x) = 1 - \frac{1}{6}(x - \pi)^3 + \frac{1}{120}(x - \pi)^5 + \frac{1}{180}(x - \pi)^6 + \cdots .$$

21. We assume that this differential equation has a power series solution with a positive radius of convergence about the point $x = 0$. This is reasonable because all of the coefficients and the forcing function $g(x) = \sin x$ are analytic everywhere. Thus, we assume that

$$y(x) = \sum_{n=0}^{\infty} a_n x^n \qquad \Rightarrow \qquad y'(x) = \sum_{n=1}^{\infty} n a_n x^{n-1}.$$

By substituting these expressions and the Maclaurin expansion for $\sin x$ into the differential equation, $y'(x) - xy(x) = \sin x$, we obtain

$$\sum_{n=1}^{\infty} n a_n x^{n-1} - x \sum_{n=0}^{\infty} a_n x^n = \sum_{n=0}^{\infty} (-1)^n \frac{x^{2n+1}}{(2n+1)!} .$$

In the first power series on the left, we make the shift $k = n - 1$. In the second power series on the left, we make the shift $k = n + 1$. Thus, we obtain

$$\sum_{k=0}^{\infty} (k+1) a_{k+1} x^k - \sum_{k=1}^{\infty} a_{k-1} x^k = \sum_{n=0}^{\infty} (-1)^n \frac{x^{2n+1}}{(2n+1)!} .$$

Separating out the first term of the first power series on the left yields

$$a_1 + \sum_{k=1}^{\infty} (k+1) a_{k+1} x^k - \sum_{k=1}^{\infty} a_{k-1} x^k = \sum_{n=0}^{\infty} (-1)^n \frac{x^{2n+1}}{(2n+1)!}$$

$$\Rightarrow \qquad a_1 + \sum_{k=1}^{\infty} \left[(k+1) a_{k+1} - a_{k-1} \right] x^k = \sum_{n=0}^{\infty} (-1)^n \frac{x^{2n+1}}{(2n+1)!} .$$

Therefore, by expanding both of the power series, we have

$$a_1 + (2a_2 - a_0)x + (3a_3 - a_1)x^2 + (4a_4 - a_2)x^3 + (5a_5 - a_3)x^4$$
$$+ (6a_6 - a_4)x^5 + (7a_7 - a_5)x^6 + \cdots = x - \frac{x^3}{6} + \frac{x^5}{120} - \frac{x^7}{5040} + \cdots .$$

Copyright © 2012 Pearson Education, Inc. Publishing as Addison-Wesley.

By equating the coefficients of like powers of x, we obtain

$$a_1 = 0,$$

$$2a_2 - a_0 = 1 \qquad \Rightarrow \qquad a_2 = (a_0 + 1)/2\,,$$

$$3a_3 - a_1 = 0 \qquad \Rightarrow \qquad a_3 = a_1/3 = 0,$$

$$4a_4 - a_2 = -1/6 \qquad \Rightarrow \qquad a_4 = [a_2 - (1/6)]/4 = (a_0/8) + (1/12)\,,$$

$$5a_5 - a_3 = 0 \qquad \Rightarrow \qquad a_5 = a_3/5 = 0,$$

$$6a_6 - a_4 = 1/120 \qquad \Rightarrow \qquad a_6 = [a_4 - (1/120)]/6 = (a_0/48) + (11/720)\,.$$

Substituting these coefficients into the power series solution and noting that a_0 is an arbitrary constant, yields

$$
\begin{aligned}
y(x) &= \sum_{n=0}^{\infty} a_n x^n \\
&= a_0 + 0 + \left(\frac{a_0}{2} + \frac{1}{2} \right) x^2 + 0 + \left(\frac{a_0}{8} + \frac{1}{12} \right) x^4 + 0 + \left(\frac{a_0}{48} + \frac{11}{720} \right) x^6 + \cdots \\
&= a_0 \left[1 + \frac{1}{2} x^2 + \frac{1}{8} x^4 + \frac{1}{48} x^6 + \cdots \right] + \left[\frac{1}{2} x^2 + \frac{1}{12} x^4 + \frac{11}{720} x^6 + \cdots \right].
\end{aligned}
$$

27. Observe that $x = 0$ is an ordinary point for this differential equation. Therefore, we can assume that this equation has a power series solution about the point $x = 0$ with a positive radius of convergence. Thus, we have

$$y(x) = \sum_{n=0}^{\infty} a_n x^n \quad \Rightarrow \quad y'(x) = \sum_{n=1}^{\infty} n a_n x^{n-1} \quad \Rightarrow \quad y''(x) = \sum_{n=2}^{\infty} n(n-1) a_n x^{n-2}\,.$$

The Maclaurin series for $\tan x$ is

$$\tan x = x + \frac{x^3}{3} + \frac{2x^5}{15} + \cdots,$$

which is given in the table on the inside front cover of the text. Substituting the expressions for $y(x)$, $y'(x)$, $y''(x)$, and the Maclaurin series of $\tan x$ into the given differential equation yields

$$
\left(1 - x^2 \right) \sum_{n=2}^{\infty} n(n-1) a_n x^{n-2} - \sum_{n=1}^{\infty} n a_n x^{n-1} + \sum_{n=0}^{\infty} a_n x^n = x + \frac{x^3}{3} + \frac{2x^5}{15} + \cdots
$$

$$
\Rightarrow \quad \sum_{n=2}^{\infty} n(n-1) a_n x^{n-2} - \sum_{n=2}^{\infty} n(n-1) a_n x^n - \sum_{n=1}^{\infty} n a_n x^{n-1} + \sum_{n=0}^{\infty} a_n x^n
$$

$$
= x + \frac{x^3}{3} + \frac{2x^5}{15} + \cdots.
$$

 Copyright © 2012 Pearson Education, Inc. Publishing as Addison-Wesley.

Shifting the indices of the power series, we obtain

$$\sum_{k=0}^{\infty}(k+2)(k+1)a_{k+2}x^k - \sum_{k=2}^{\infty}k(k-1)a_kx^k - \sum_{k=0}^{\infty}(k+1)a_{k+1}x^k + \sum_{k=0}^{\infty}a_kx^k = x + \frac{x^3}{3} + \frac{2x^5}{15} + \cdots$$

Removing the first two terms from the summation notation in the first, third and fourth power series yields

$$(2)(1)a_2 + (3)(2)a_3x + \sum_{k=2}^{\infty}(k+2)(k+1)a_{k+2}x^k - \sum_{k=2}^{\infty}k(k-1)a_kx^k - (1)a_1 - (2)a_2x$$

$$- \sum_{k=2}^{\infty}(k+1)a_{k+1}x^k + a_0 + a_1x + \sum_{k=2}^{\infty}a_kx^k = x + \frac{x^3}{3} + \frac{2x^5}{15} + \cdots$$

$$\Rightarrow \quad (2a_2 - a_1 + a_0) + (6a_3 - 2a_2 + a_1)x$$

$$+ \sum_{k=2}^{\infty}\left[(k+2)(k+1)a_{k+2} - k(k-1)a_k - (k+1)a_{k+1} + a_k\right]x^k = x + \frac{x^3}{3} + \frac{2x^5}{15} + \cdots.$$

By equating the coefficients, we see that

$$2a_2 - a_1 + a_0 = 0 \qquad \Rightarrow \qquad a_2 = (a_1 - a_0)/2\,,$$
$$6a_3 - 2a_2 + a_1 = 1 \qquad \Rightarrow \qquad a_3 = (2a_2 - a_1 + 1)/6 = (1 - a_0)/6\,,$$
$$4 \cdot 3a_4 - 2a_2 - 3a_3 + a_2 = 0 \qquad \Rightarrow \qquad a_4 = (a_2 + 3a_3)/12 = (a_1 - 2a_0 + 1)/24\,.$$

Therefore, noting that a_0 and a_1 are arbitrary, we can substitute these coefficients into the power series solution $y(x) = \sum_{n=0}^{\infty}a_nx^n = a_0 + a_1x + a_2x^2 + a_3x^3 + a_4x^4 + \cdots$ to obtain

$$y(x) = a_0 + a_1x + \left(\frac{a_1}{2} - \frac{a_0}{2}\right)x^2 + \left(\frac{1}{6} - \frac{a_0}{6}\right)x^3 + \left(\frac{a_1}{24} - \frac{a_0}{12} + \frac{1}{24}\right)x^4 + \cdots$$

$$= a_0\left(1 - \frac{1}{2}x^2 - \frac{1}{6}x^3 - \frac{1}{12}x^4 + \cdots\right) + a_1\left(x + \frac{1}{2}x^2 + \frac{1}{24}x^4 + \cdots\right)$$

$$+ \left(\frac{1}{6}x^3 + \frac{1}{24}x^4 + \cdots\right).$$

EXERCISES 8.5: Cauchy-Euler (Equidimensional) Equations Revisited

5. Notice that, since $x > 0$, we can multiply this differential equation by x^2 to obtain

$$x^2\frac{d^2y}{dx^2} - 5x\frac{dy}{dx} + 13y = 0.$$

We see that this is a Cauchy-Euler equation. Thus, we assume that a solution has the form

$$y(x) = x^r \qquad \Rightarrow \qquad y'(x) = rx^{r-1} \qquad \Rightarrow \qquad y''(x) = r(r-1)x^{r-2}\,.$$

Copyright © 2012 Pearson Education, Inc. Publishing as Addison-Wesley.

Substituting these expressions into the differential equation yields

$$r(r-1)x^r - 5rx^r + 13x^r = 0 \quad \Rightarrow \quad \left(r^2 - 6r + 13\right)x^r = 0 \quad \Rightarrow \quad r^2 - 6r + 13 = 0.$$

Using the quadratic formula, we see that the roots to this equation are

$$r = \frac{6 \pm \sqrt{36 - 52}}{2} = 3 \pm 2i.$$

Therefore, using formulas (5) and (6) and Euler's formula, we get two linearly independent solutions

$$y_1(x) = x^3 \cos(2\ln x), \qquad y_2(x) = x^3 \sin(2\ln x).$$

Hence, a general solution to this equation is given by

$$y(x) = c_1 x^3 \cos(2\ln x) + c_2 x^3 \sin(2\ln x).$$

7. This equation is a third order Cauchy-Euler equation, and so we assume that a solution has the form $y(x) = x^r$. This implies that

$$y'(x) = rx^{r-1} \quad \Rightarrow \quad y''(x) = r(r-1)x^{r-2} \quad \Rightarrow \quad y'''(x) = r(r-1)(r-2)x^{r-3}.$$

Substituting these expressions into the differential equation, we obtain

$$[r(r-1)(r-2) + 4r(r-1) + 10r - 10]\,x^r = 0$$
$$\Rightarrow \quad \left[r^3 + r^2 + 8r - 10\right]x^r = 0 \quad \Rightarrow \quad r^3 + r^2 + 8r - 10 = 0.$$

By inspection, we see that $r = 1$ is a root of this equation. Thus, one solution to the differential equation is $y_1(x) = x$. We can factor the indicial equation as follows.

$$(r-1)(r^2 + 2r + 10) = 0.$$

Therefore, using the quadratic formula, we see that the other two roots of this equation are $r = -1 \pm 3i$. Thus, we can find two more linearly independent solutions to the given equation by using Euler's formula. Thus, three linearly independent solutions to this problem are given by

$$y_1(x) = x, \qquad y_2(x) = x^{-1}\cos(3\ln x), \qquad y_3(x) = x^{-1}\sin(3\ln x),$$

and a general solution is

$$y(x) = c_1 x + c_2 x^{-1}\cos(3\ln x) + c_3 x^{-1}\sin(3\ln x).$$

   Copyright © 2012 Pearson Education, Inc. Publishing as Addison-Wesley.

13. First, we find two linearly independent solutions to the associated homogeneous equation. Since this is a Cauchy-Euler equation, we assume that there are solutions of the form

$$y(x) = x^r \quad \Rightarrow \quad y'(x) = rx^{r-1} \quad \Rightarrow \quad y''(x) = r(r-1)x^{r-2}.$$

Substituting these expressions into the associated homogeneous equation yields

$$[r(r-1) - 2r + 2]\, x^r = 0 \quad \Rightarrow \quad r^2 - 3r + 2 = 0 \quad \Rightarrow \quad (r-1)(r-2) = 0.$$

Thus, the roots to this indicial equation are $r = 1, 2$. Therefore, a general solution to the associated homogeneous equation is

$$y_h(x) = c_1 x + c_2 x^2.$$

For the variation of parameters method, we let $y_1(x) = x$ and $y_2(x) = x^2$, and then assume that a particular solution has the form

$$y_p(x) = v_1(x)y_1(x) + v_2(x)y_2(x) = v_1(x)x + v_2(x)x^2.$$

In order to find $v_1(x)$ and $v_2(x)$, we use the formula (10). The Wronskian of y_1 and y_2 is

$$W[y_1, y_2](x) = y_1(x)y_2'(x) - y_2(x)y_1'(x) = 2x^2 - x^2 = x^2.$$

Next we write the differential equation, given in this problem, in standard form and find that $g(x) = x^{-5/2}$. Therefore, by (10), we have

$$v_1(x) = \int \frac{-x^{-5/2}x^2}{x^2}\, dx = \int (-x^{-5/2})dx = \frac{2}{3} x^{-3/2},$$

$$v_2(x) = \int \frac{x^{-5/2}x}{x^2}\, dx = \int x^{-7/2} dx = \frac{-2}{5} x^{-5/2}.$$

Thus, a particular solution is given by

$$y_p(x) = \left(\frac{2}{3} x^{-3/2}\right) x + \left(\frac{-2}{5} x^{-5/2}\right) x^2 = \frac{4}{15} x^{-1/2}.$$

Therefore, a general solution of the given nonhomogeneous differential equation is

$$y(x) = y_h(x) + y_p(x) = c_1 x + c_2 x^2 + \frac{4}{15} x^{-1/2}.$$

19. **(a)** For this linear differential operator L, we have

$$\begin{aligned} L[x^r](x) &= x^3 \left[r(r-1)(r-2)x^{r-3}\right] + x\left[rx^{r-1}\right] - x^r \\ &= r(r-1)(r-2)x^r + rx^r - x^r \\ &= \left(r^3 - 3r^2 + 3r - 1\right)x^r = (r-1)^3 x^r. \end{aligned}$$

Copyright © 2012 Pearson Education, Inc. Publishing as Addison-Wesley.

(b) From part (a), we see that $r = 1$ is a root of multiplicity three of the indicial equation. Thus, we have one solution given by

$$y_1(x) = x. \qquad (8.13)$$

To find two other linearly independent solutions, we use a method similar to that used in the text. Taking the partial derivative of $L[x^r](x) = (r-1)^3 x^r$ with respect to r, we get

$$\frac{\partial}{\partial r}\{L[x^r](x)\} = \frac{\partial}{\partial r}\{(r-1)^3 x^r\} = 3(r-1)^2 x^r + (r-1)^3 x^r \ln x$$

$$\Rightarrow \quad \frac{\partial^2}{\partial r^2}\{L[x^r](x)\} = \frac{\partial}{\partial r}\{3(r-1)^2 x^r + (r-1)^3 x^r \ln x\}$$

$$= 6(r-1)x^r + 6(r-1)^2 x^r \ln x + (r-1)^3 x^r (\ln x)^2.$$

Since $(r-1)$ is a factor of every term in $\partial\{L[x^r](x)\}/\partial r$ and $\partial^2\{L[x^r](x)\}/\partial r^2$, we see that

$$\frac{\partial}{\partial r}\{L[x^r](x)\}\Big|_{r=1} = 0, \qquad (8.14)$$

and

$$\frac{\partial^2}{\partial r^2}\{L[x^r](x)\}\Big|_{r=1} = 0. \qquad (8.15)$$

In order to find a second linearly independent solution, we find an alternative form of of the partial derivative in (8.14). Using the fact that

$$L[y](x) = x^3 y'''(x) + xy'(x) - y(x)$$

and proceeding as in the equation (9) in the text with $w(r, x) = x^r$, we obtain

$$\frac{\partial}{\partial r}\{L[x^r](x)\} = \frac{\partial}{\partial r}\{L[w](x)\} = \frac{\partial}{\partial r}\left\{x^3\frac{\partial^3 w}{\partial x^3} + x\frac{\partial w}{\partial x} - w\right\}$$

$$= x^3\frac{\partial^4 w}{\partial r\partial x^3} + x\frac{\partial^2 w}{\partial r\partial x} - \frac{\partial w}{\partial r} = x^3\frac{\partial^4 w}{\partial x^3 \partial r} + x\frac{\partial^2 w}{\partial x\partial r} - \frac{\partial w}{\partial r}$$

$$= x^3\frac{\partial^3}{\partial x^3}\left(\frac{\partial w}{\partial r}\right) + x\frac{\partial}{\partial x}\left(\frac{\partial w}{\partial r}\right) - \frac{\partial w}{\partial r} = L\left[\frac{\partial w}{\partial r}\right](x).$$

Combining this with (8.14) yields

$$\frac{\partial}{\partial r}\{L[x^r](x)\}\Big|_{r=1} L\left[\frac{\partial x^r}{\partial r}\Big|_{r=1}\right] = L[x^r \ln x\,|_{r=1}] = L[x\ln x] = 0.$$

Thus, a second linearly independent solution is given by

$$y_2(x) = x\ln x.$$

 Copyright © 2012 Pearson Education, Inc. Publishing as Addison-Wesley.

To find a third solution, we use the equation (8.15). We would like to find an alternative form for $\partial^2\{L[x^r](x)\}/\partial r^2$. To do this, we use the fact that

$$\frac{\partial}{\partial r}\{L[x^r](x)\} = x^3\frac{\partial^4 w}{\partial r\partial x^3} + x\frac{\partial^2 w}{\partial r\partial x} - \frac{\partial w}{\partial r}$$

and that mixed partial derivatives of $w(r,x)$ are equal. Thus, we have

$$\frac{\partial^2}{\partial r^2}\{L[x^r](x)\} = \frac{\partial}{\partial r}\left[\frac{\partial}{\partial r}\{L[x^r](x)\}\right] = \frac{\partial}{\partial r}\left\{x^3\frac{\partial^4 w}{\partial r\partial x^3} + x\frac{\partial^2 w}{\partial r\partial x} - \frac{\partial w}{\partial r}\right\}$$

$$= x^3\frac{\partial^5 w}{\partial r^2\partial x^3} + x\frac{\partial^3 w}{\partial r^2\partial x} - \frac{\partial^2 w}{\partial r^2} = x^3\frac{\partial^5 w}{\partial x^3\partial r^2} + x\frac{\partial^3 w}{\partial x\partial r^2} - \frac{\partial^2 w}{\partial r^2}$$

$$= x^3\frac{\partial^3}{\partial x^3}\left(\frac{\partial^2 w}{\partial r^2}\right) + x\frac{\partial}{\partial x}\left(\frac{\partial^2 w}{\partial r^2}\right) - \frac{\partial^2 w}{\partial r^2} = L\left[\frac{\partial^2 w}{\partial r^2}\right](x) = 0.$$

Therefore, combining this with the equation (8.15) yields

$$\frac{\partial^2}{\partial r^2}\{L[x^r](x)\}\bigg|_{r=1} = L\left[\frac{\partial^2(x^r)}{\partial r^2}\bigg|_{r=1}\right] = L\left[x(\ln x)^2\right] = 0,$$

where we have used the fact that $\partial^2 x^r/\partial r^2 = x^r(\ln x)^2$. Thus, we see that

$$y_3(x) = x(\ln x)^2,$$

which, by inspection, is linearly independent from y_1 and y_2. Hence, a general solution to the given differential equation is $y(x) = C_1 x + C_2 x\ln x + C_3 x(\ln x)^2$.

EXERCISES 8.6: Method of Frobenius

5. Putting this equation in standard form, we see that

$$p(x) == -\frac{x-1}{(x^2-1)^2} = -\frac{x-1}{(x-1)^2(x+1)^2} = -\frac{1}{(x-1)(x+1)^2},$$

$$q(x) = \frac{3}{(x^2-1)^2} = \frac{3}{(x-1)^2(x+1)^2}.$$

Thus, $x = \pm 1$ are singular points of this equation. To check if $x = 1$ is regular, we note that

$$(x-1)p(x) = -\frac{1}{(x+1)^2} \qquad \text{and} \qquad (x-1)^2 q(x) = \frac{3}{(x+1)^2}.$$

These functions are analytic at $x = 1$. Therefore, $x = 1$ is a regular singular point for this differential equation. Next, we check the singular point $x = -1$. Here,

$$(x+1)p(x) = -\frac{1}{(x-1)(x+1)}$$

is not analytic at $x = -1$. Therefore, $x = -1$ is an irregular singular point.

Copyright © 2012 Pearson Education, Inc. Publishing as Addison-Wesley.

13. By putting this equation in standard form, we see that

$$p(x) = \frac{x^2 - 4}{(x^2 - x - 2)^2} = \frac{(x-2)(x+2)}{(x-2)^2(x+1)^2} = \frac{x+2}{(x-2)(x+1)^2},$$

$$q(x) = \frac{-6x}{(x-2)^2(x+1)^2)}.$$

Thus, we have

$$(x-2)p(x) = \frac{x+2}{(x+1)^2} \qquad \text{and} \qquad (x-2)^2 q(x) = \frac{-6x}{(x+1)^2}.$$

Therefore, $x = 2$ is a regular singular point of this differential equation. We also observe that

$$\lim_{x \to 2}(x-2)p(x) = \lim_{x \to 2}\frac{x+2}{(x+1)^2} = \frac{4}{9} = p_0,$$

$$\lim_{x \to 2}(x-2)^2 q(x) = -\lim_{x \to 2}\frac{6x}{(x+1)^2} = -\frac{12}{9} = -\frac{4}{3} = q_0.$$

Thus, we can use the equation (16) in the text to obtain the indicial equation

$$r(r-1) + \frac{4r}{9} - \frac{4}{3} = 0 \qquad \Rightarrow \qquad r^2 - \frac{5r}{9} - \frac{4}{3} = 0.$$

By the quadratic formula, we see that the roots to this equation and, therefore, the exponents of the singularity $x = 2$, are given by

$$r_{1,2} = \frac{5 \pm \sqrt{25 + 432}}{18} = \frac{5 \pm \sqrt{457}}{18}.$$

21. Here $p(x) = x^{-1}$ and $q(x) = 1$. This implies that $xp(x) = 1$ and $x^2 q(x) = x^2$. Therefore, we see that $x = 0$ is a regular singular point for this differential equation, and so we can use the method of Frobenius to find a solution to this problem. We also note that $x = 0$ is the only singular point for this equation.) Thus, we assume that a solution has the form

$$w(r, x) = x^r \sum_{n=0}^{\infty} a_n x^n = \sum_{n=0}^{\infty} a_n x^{n+r}.$$

Notice that

$$p_0 = \lim_{x \to 0} xp(x) = \lim_{x \to 0} 1 = 1,$$

$$q_0 = \lim_{x \to 0} x^2 q(x) = \lim_{x \to 0} x^2 = 0.$$

Hence, the indicial equation is given by

$$r(r-1) + r = r^2 = 0.$$

    Copyright © 2012 Pearson Education, Inc. Publishing as Addison-Wesley.

This means that $r_1 = r_2 = 0$. Since $x = 0$ is the only singular point for this differential equation, we observe that the series solution $w(0, x)$, which we will find by the method of Frobenius, converges for all $x > 0$. To find the solution, we note that

$$w(r, x) = \sum_{n=0}^{\infty} a_n x^{n+r}$$

$$\Rightarrow \quad w'(r, x) = \sum_{n=0}^{\infty} (n+r) a_n x^{n+r-1}$$

$$\Rightarrow \quad w''(r, x) = \sum_{n=0}^{\infty} (n+r)(n+r-1) a_n x^{n+r-2}.$$

Substituting these expressions into the differential equation and simplifying yields

$$\sum_{n=0}^{\infty} (n+r)(n+r-1) a_n x^{n+r} + \sum_{n=0}^{\infty} (n+r) a_n x^{n+r} + \sum_{n=0}^{\infty} a_n x^{n+r+2} = 0.$$

Next, we want each power series to sum over x^{k+r}. Thus, we let $k = n$ in the first and second power series and shift the index in the last power series by letting $k = n + 2$.

$$\sum_{k=0}^{\infty} (k+r)(k+r-1) a_k x^{k+r} + \sum_{k=0}^{\infty} (k+r) a_k x^{k+r} + \sum_{k=2}^{\infty} a_{k-2} x^{k+r} = 0.$$

We separate out the first two terms from the first two power series so that all power series start at the same index. Thus, we have

$$(r-1) r a_0 x^r + r(1+r) a_1 x^{1+r} + \sum_{k=2}^{\infty} (k+r)(k+r-1) a_k x^{k+r}$$

$$+ r a_0 x^r + (1+r) a_1 x^{1+r} + \sum_{k=2}^{\infty} (k+r) a_k x^{k+r} + \sum_{k=2}^{\infty} a_{k-2} x^{k+r} = 0$$

$$\Rightarrow \quad [r(r-1) + r] a_0 x^r + [r(r+1) + (r+1)] a_1 x^{1+r}$$

$$+ \sum_{k=2}^{\infty} [(k+r)(k+r-1) a_k + (k+r) a_k + a_{k-2}] x^{k+r} = 0.$$

By equating coefficients and assuming that $a_0 \neq 0$, we obtain

$$r(r-1) + r = 0 \qquad \text{(the indicial equation)},$$

$$[r(r+1) + (r+1)] a_1 = 0 \qquad \Rightarrow \qquad (r+1)^2 a_1 = 0,$$

and, for $k \geq 2$, the recurrence relation

$$(k+r)(k+r-1) a_k + (k+r) a_k + a_{k-2} = 0 \qquad \Rightarrow \qquad a_k = \frac{-a_{k-2}}{(k+r)^2}, \qquad k \geq 2.$$

Copyright © 2012 Pearson Education, Inc. Publishing as Addison-Wesley.

Using the fact (found from the indicial equation) that $r_1 = 0$, we observe that $a_1 = 0$. Next, using the recurrence relation (and the fact that $r_1 = 0$), we see that

$$a_k = \frac{-a_{k-2}}{k^2}, \qquad k \geq 2.$$

Hence,

$$
\begin{aligned}
k = 2 &\quad\Rightarrow\quad a_2 = \frac{-a_0}{4}, \\
k = 3 &\quad\Rightarrow\quad a_3 = \frac{-a_1}{9} = 0, \\
k = 4 &\quad\Rightarrow\quad a_4 = \frac{-a_2}{16} = \frac{-(-a_0/4)}{16} = \frac{a_0}{64}, \\
k = 5 &\quad\Rightarrow\quad a_5 = \frac{-a_3}{25} = 0, \\
k = 6 &\quad\Rightarrow\quad a_6 = \frac{-a_4}{36} = \frac{-(a_0/64)}{36} = -\frac{a_0}{2304}.
\end{aligned}
$$

Substituting these coefficients into the solution

$$w(0, x) = \sum_{n=0}^{\infty} a_n x^n = a_0 + a_1 x + a_2 x^2 + a_3 x^3 + a_4 x^4 + a_5 x^5 + a_6 x^6 + \cdots,$$

we obtain the series solution given by

$$w(0, x) = a_0 \left[1 - \frac{1}{4} x^2 + \frac{1}{64} x^4 - \frac{1}{2304} x^6 + \cdots \right], \qquad x > 0.$$

25. For this equation, we see that

$$xp(x) = \frac{x}{2}, \qquad x^2 q(x) = -\frac{x+3}{4}.$$

Thus, $x = 0$ is a regular singular point for this equation, and we can use the method of Frobenius to find a solution. To this end, we compute

$$
\lim_{x \to 0} xp(x) = \lim_{x \to 0} \frac{x}{2} = 0,
$$

$$
\lim_{x \to 0} x^2 q(x) = \lim_{x \to 0} \left(-\frac{x+3}{4} \right) = -\frac{3}{4}.
$$

Therefore, by the equation (16) in the text, the indicial equation is

$$r(r-1) - \frac{3}{4} = 0 \quad\Rightarrow\quad 4r^2 - 4r - 3 = (2r+1)(2r-3) = 0.$$

 Copyright © 2012 Pearson Education, Inc. Publishing as Addison-Wesley.

This indicial equation has roots $r_1 = 3/2$ and $r_2 = -1/2$. By the method of Frobenius, we can assume that a solution to this differential equation has the form

$$w(r, x) = \sum_{n=0}^{\infty} a_n x^{n+r}$$

$$\Rightarrow \quad w'(r, x) = \sum_{n=0}^{\infty} (n + r) a_n x^{n+r-1}$$

$$\Rightarrow \quad w''(r, x) = \sum_{n=0}^{\infty} (n + r - 1)(n + r) a_n x^{n+r-2},$$

where $r = r_1 = 3/2$. Since $x = 0$ is the only singular point for this equation, we see that the solution, $w(3/2, x)$, converges for all $x > 0$. The first step in finding this solution is to plug in $w(r, x)$ and its first and second derivatives into the differential equation. Thus, we obtain

$$\sum_{n=0}^{\infty} 4(n + r - 1)(n + r) a_n x^{n+r} + \sum_{n=0}^{\infty} 2(n + r) a_n x^{n+r+1} - \sum_{n=0}^{\infty} a_n x^{n+r+1} - \sum_{n=0}^{\infty} 3 a_n x^{n+r} = 0.$$

By shifting indices, we can sum each power series over x^{k+r}. Thus, with the substitution $k = n$ in the first and the last power series and the substitution $k = n + 1$ in the two remaining power series, we obtain

$$\sum_{k=0}^{\infty} 4(k + r - 1)(k + r) a_k x^{k+r} + \sum_{k=1}^{\infty} 2(k + r - 1) a_{k-1} x^{k+r} - \sum_{k=1}^{\infty} a_{k-1} x^{k+r} - \sum_{k=0}^{\infty} 3 a_k x^{k+r} = 0.$$

Next, removing the $k = 0$ term from the first and the last power series, and writing the result as a single power series yields

$$4(r - 1) r a_0 x^r + \sum_{k=1}^{\infty} 4(k + r - 1)(k + r) a_k x^{k+r} + \sum_{k=1}^{\infty} 2(k + r - 1) a_{k-1} x^{k+r}$$

$$- \sum_{k=1}^{\infty} a_{k-1} x^{k+r} - 3 a_0 x^r - \sum_{k=1}^{\infty} 3 a_k x^{k+r} = 0$$

$$\Rightarrow \quad [4(r - 1)r - 3] a_0 x^r$$

$$+ \sum_{k=1}^{\infty} [4(k + r - 1)(k + r) a_k + 2(k + r - 1) a_{k-1} - a_{k-1} - 3 a_k] x^{k+r} = 0.$$

Equating coefficients, we see that each coefficient in the power series must be zero. Also, we are assuming that $a_0 \neq 0$. Therefore, we have

$$4(r - 1)r - 3 = 0 \qquad \text{(the indicial equation)},$$

Copyright © 2012 Pearson Education, Inc. Publishing as Addison-Wesley.

$$4(k+r-1)(k+r)a_k + 2(k+r-1)a_{k-1} - a_{k-1} - 3a_k = 0, \qquad k \geq 1.$$

Thus, the recurrence relation is given by

$$a_k = \frac{(3-2k-2r)a_{k-1}}{4(k+r-1)(k+r)-3}, \qquad k \geq 1.$$

For $r = r_1 = 3/2$, we have

$$a_k = \frac{-2ka_{k-1}}{4(k+1/2)(k+3/2)-3}, \qquad \Rightarrow \qquad a_k = \frac{-a_{k-1}}{2(k+2)}, \qquad k \geq 1.$$

Thus, we see that

$$
\begin{aligned}
k = 1 &\quad \Rightarrow \quad a_1 = \frac{-a_0}{2 \cdot 3} = \frac{-a_0}{2^0 \cdot 3!}, \\
k = 2 &\quad \Rightarrow \quad a_2 = \frac{-a_1}{2 \cdot 4} = \frac{a_0}{2 \cdot 2 \cdot 3 \cdot 4} = \frac{a_0}{2^1 \cdot 4!}, \\
k = 3 &\quad \Rightarrow \quad a_3 = \frac{-a_2}{2 \cdot 5} = \frac{-a_0}{2^2 \cdot 5!}, \\
k = 4 &\quad \Rightarrow \quad a_4 = \frac{-a_3}{2 \cdot 6} = \frac{a_0}{2^3 \cdot 6!}.
\end{aligned}
$$

Inspection of this sequence shows that we can write a_n, for $n \geq 1$, as

$$a_n = \frac{(-1)^n a_0}{2^{n-1}(n+2)!}.$$

Substituting these coefficients into the solution

$$w\left(\frac{3}{2}, x\right) = \sum_{n=0}^{\infty} a_n x^{n+(3/2)}$$

yields a power series solution

$$w\left(\frac{3}{2}, x\right) = a_0 x^{3/2} + a_0 \sum_{n=1}^{\infty} \frac{(-1)^n x^{n+(3/2)}}{2^{n-1}(n+2)!}, \qquad x > 0.$$

But, since substituting $n = 0$ into a_n yields $(-1)^0 a_0 / (2^{-1} 2!) = a_0$, the solution can be written as

$$w\left(\frac{3}{2}, x\right) = a_0 \sum_{n=0}^{\infty} \frac{(-1)^n x^{n+(3/2)}}{2^{n-1}(n+2)!}.$$

27. In this equation, we see that $p(x) = -1/x$ and $q(x) = -1$. Thus, the only singular point is $x = 0$. Since $xp(x) = -1$ and $x^2 q(x) = -x^2$, we see that $x = 0$ is a regular singular point for this equation, and so we can use the method of Frobenius to find a solution to

    Copyright © 2012 Pearson Education, Inc. Publishing as Addison-Wesley.

this equation. We also note that the solution that we will find, converges for all $x > 0$. We observe that

$$p_0 = \lim_{x \to 0} xp(x) = \lim_{x \to 0}(-1) = -1 \qquad \text{and} \qquad q_0 = \lim_{x \to 0} x^2 q(x) = \lim_{x \to 0}(-x^2) = 0.$$

Thus, according to the equation (16) in the text, the indicial equation for the point $x = 0$ is

$$r(r-1) - r = 0 \qquad \Rightarrow \qquad r(r-2) = 0.$$

Therefore, the roots to the indicial equation are $r_1 = 2$ and $r_2 = 0$. Hence, we use the method of Frobenius to find the solution $w(2, x)$. Letting

$$w(r, x) = \sum_{n=0}^{\infty} a_n x^{n+r},$$

we find that

$$w'(r, x) = \sum_{n=0}^{\infty} (n+r) a_n x^{n+r-1}, \qquad w''(r, x) = \sum_{n=0}^{\infty} (n+r-1)(n+r) a_n x^{n+r-2}.$$

Substituting these expressions into the differential equation and simplifying, we obtain

$$\sum_{n=0}^{\infty} (n+r-1)(n+r) a_n x^{n+r-1} - \sum_{n=0}^{\infty} (n+r) a_n x^{n+r-1} - \sum_{n=0}^{\infty} a_n x^{n+r+1} = 0.$$

Next, we shift the indices by letting $k = n-1$ in the first two power series and $k = n+1$ in the last power series. Thus, we have

$$\sum_{k=-1}^{\infty} (k+r)(k+r+1) a_{k+1} x^{k+r} - \sum_{k=-1}^{\infty} (k+r+1) a_{k+1} x^{k+r} - \sum_{k=1}^{\infty} a_{k-1} x^{k+r} = 0.$$

We can start all these summations at the same term if we separate out the first two terms (corresponding to $k = -1$ and $k = 0$) from the first two power series. Thus, we have

$$(r-1)r a_0 x^{r-1} + r(r+1) a_1 x^r + \sum_{k=1}^{\infty} (k+r)(k+r+1) a_{k+1} x^{k+r}$$

$$-r a_0 x^{r-1} - (r+1) a_1 x^r - \sum_{k=1}^{\infty} (k+r+1) a_{k+1} x^{k+r} - \sum_{k=1}^{\infty} a_{k-1} x^{k+r} = 0$$

$$\Rightarrow \qquad [(r-1)r - r] a_0 x^{r-1} + [r(r+1) - (r+1)] a_1 x^r$$

$$+ \sum_{k=1}^{\infty} [(k+r)(k+r+1) a_{k+1} - (k+r+1) a_{k+1} - a_{k-1}] x^{k+r} = 0.$$

Copyright © 2012 Pearson Education, Inc. Publishing as Addison-Wesley.

By equating coefficients and assuming that $a_0 \neq 0$, we obtain

$$r(r-1) - r = 0 \qquad \text{(the indicial equation)},$$

$$(r+1)(r-1)a_1 = 0, \tag{8.16}$$

$$(k+r)(k+r+1)a_{k+1} - (k+r+1)a_{k+1} - a_{k-1} = 0, \qquad k \geq 1,$$

where the last equation is the recurrence relation. Simplifying yields

$$a_{k+1} = \frac{a_{k-1}}{(k+r+1)(k+r-1)}, \qquad k \geq 1. \tag{8.17}$$

Next, we let $r = r_1 = 2$ in the equations (8.16) and (8.17) to obtain

$$3a_1 = 0 \qquad \Rightarrow \qquad a_1 = 0,$$

$$a_{k+1} = \frac{a_{k-1}}{(k+3)(k+1)}, \qquad k \geq 1.$$

Thus, we have

$$k = 1 \quad \Rightarrow \quad a_2 = \frac{a_0}{4 \cdot 2},$$

$$k = 2 \quad \Rightarrow \quad a_3 = \frac{a_1}{5 \cdot 3} = 0,$$

$$k = 3 \quad \Rightarrow \quad a_4 = \frac{a_2}{6 \cdot 4} = \frac{a_0}{6 \cdot 4 \cdot 4 \cdot 2} = \frac{a_0}{2^4 \cdot 3 \cdot 2 \cdot 2 \cdot 1 \cdot 1} = \frac{a_0}{2^4 \cdot 3! \cdot 2!},$$

$$k = 4 \quad \Rightarrow \quad a_5 = \frac{a_3}{7 \cdot 5} = 0,$$

$$k = 5 \quad \Rightarrow \quad a_6 = \frac{a_4}{8 \cdot 6} = \frac{a_0}{8 \cdot 6 \cdot 2^4 \cdot 3! \cdot 2!} = \frac{a_0}{2^6 \cdot 4! \cdot 3!}.$$

By inspection, we see that the coefficients of the power series solution $w(2, x)$ are

$$a_{2n-1} = 0, \qquad a_{2n} = \frac{a_0}{2^{2n} \cdot (n+1)! n!}, \qquad n \geq 1.$$

Thus, substituting these coefficients into the power series solution yields

$$w(2, x) = a_0 \sum_{n=0}^{\infty} \frac{x^{2n+2}}{2^{2n}(n+1)! \, n!}.$$

35. In applying the method of Frobenius, we seek for a solution of the form

$$w(r, x) = \sum_{n=0}^{\infty} a_n x^{n+r} \qquad \Rightarrow \qquad w'(r, x) = \sum_{n=0}^{\infty} (n+r)a_n x^{n+r-1}$$

$$\Rightarrow \qquad w''(r, x) = \sum_{n=0}^{\infty} (n+r-1)(n+r)a_n x^{n+r-2}$$

 Copyright © 2012 Pearson Education, Inc. Publishing as Addison-Wesley.

$$\Rightarrow \quad w'''(r,x) = \sum_{n=0}^{\infty} (n+r-2)(n+r-1)(n+r)a_n x^{n+r-3},$$

where we have differentiated the series term-wise. Substituting these expressions into the given differential equation and simplifying yields

$$\sum_{n=0}^{\infty} 6(n+r-2)(n+r-1)(n+r)a_n x^{n+r} + \sum_{n=0}^{\infty} 13(n+r-1)(n+r)a_n x^{n+r}$$

$$+ \sum_{n=0}^{\infty} (n+r)a_n x^{n+r} + \sum_{n=0}^{\infty} (n+r)a_n x^{n+r+1} + \sum_{n=0}^{\infty} a_n x^{n+r+1} = 0.$$

Shifting the index to $k = n + 1$ in the last two power series and letting $k = n$ in the other power series, we obtain

$$\sum_{k=0}^{\infty} 6(k+r-2)(k+r-1)(k+r)a_k x^{k+r} + \sum_{k=0}^{\infty} 13(k+r-1)(k+r)a_k x^{k+r}$$

$$+ \sum_{k=0}^{\infty} (k+r)a_k x^{k+r} + \sum_{k=1}^{\infty} (k-1+r)a_{k-1} x^{k+r} + \sum_{k=1}^{\infty} a_{k-1} x^{k+r} = 0.$$

Next, we remove the first term from each of the first three power series above, so that all summations start at $k = 1$. Thus, we have

$$6(r-2)(r-1)ra_0 x^r + \sum_{k=1}^{\infty} 6(k+r-2)(k+r-1)(k+r)a_k x^{k+r}$$

$$+13(r-1)ra_0 x^r + \sum_{k=1}^{\infty} 13(k+r-1)(k+r)a_k x^{k+r} + ra_0 x^r + \sum_{k=1}^{\infty} (k+r)a_k x^{k+r}$$

$$+ \sum_{k=1}^{\infty} (k-1+r)a_{k-1} x^{k+r} + \sum_{k=1}^{\infty} a_{k-1} x^{k+r} = 0$$

$$\Rightarrow \quad [6(r-2)(r-1)r + 13(r-1)r + r] a_0 x^r$$

$$+ \sum_{k=1}^{\infty} [6(k+r-2)(k+r-1)(k+r)a_k + 13(k+r-1)(k+r)a_k$$

$$+(k+r)a_k + (k-1+r)a_{k-1} + a_{k-1}] x^{k+r} = 0. \qquad (8.18)$$

If we assume that $a_0 \neq 0$ and set the coefficient of x^r equal to zero, we find that the indicial equation is

$$6(r-2)(r-1)r + 13(r-1)r + r = 0 \qquad \Rightarrow \qquad r^2(6r-5) = 0.$$

Hence, the roots to the indicial equation are 0, 0, and 5/6. We will find a solution, associated with the largest of these roots. That is, we will find $w(5/6, x)$. From (8.18),

Copyright © 2012 Pearson Education, Inc. Publishing as Addison-Wesley.

we have the recurrence relation

$$6(k+r-2)(k+r-1)(k+r)a_k + 13(k+r-1)(k+r)a_k$$
$$+(k+r)a_k + (k-1+r)a_{k-1} + a_{k-1} = 0, \qquad k \geq 1$$
$$\Rightarrow \qquad a_k = \frac{-a_{k-1}}{6(k+r-2)(k+r-1) + 13(k+r-1) + 1}, \qquad k \geq 1.$$

If we assume that $r = 5/6$, then this recurrence relation simplifies to

$$a_k = \frac{-a_{k-1}}{k(6k+5)}, \qquad k \geq 1.$$

Therefore, we have

$$k = 1 \qquad \Rightarrow \qquad a_1 = \frac{-a_0}{11},$$
$$k = 2 \qquad \Rightarrow \qquad a_2 = \frac{-a_1}{34} = \frac{a_0}{374},$$
$$k = 3 \qquad \Rightarrow \qquad a_3 = \frac{-a_2}{69} = \frac{-a_0}{25806}.$$

Substituting these coefficients into the solution $w(5/6, x) = \sum_{n=0}^{\infty} a_n x^{n+(5/6)}$, we obtain

$$w\left(\frac{5}{6}, x\right) = a_0\left(x^{5/6} - \frac{x^{11/6}}{11} + \frac{x^{17/6}}{374} - \frac{x^{23/6}}{25806} + \cdots\right).$$

41. If we let $z = 1/x$ (so that $dz/dx = -1/x^2$), then we can define a new function $Y(z)$ as

$$Y(z) := y\left(\frac{1}{z}\right) = y(x).$$

Thus, by the chain rule, we have

$$\frac{dy}{dx} = \left(\frac{dY}{dz}\right)\left(\frac{dz}{dx}\right) = \left(\frac{dY}{dz}\right)\left(-\frac{1}{x^2}\right) \qquad \Rightarrow \qquad -x^2\frac{dy}{dx} = \frac{dY}{dz}. \qquad (8.19)$$

Differentiating one more time yields

$$\begin{aligned}
\frac{d^2y}{dx^2} &= \frac{d}{dx}\left(\frac{dy}{dx}\right) = \frac{d}{dx}\left[\left(\frac{dY}{dz}\right)\left(-\frac{1}{x^2}\right)\right] \\
&= \left(\frac{dY}{dz}\right)\frac{d}{dx}\left(-\frac{1}{x^2}\right) + \frac{d}{dx}\left(\frac{dY}{dz}\right)\left(-\frac{1}{x^2}\right) \\
&= \left(\frac{dY}{dz}\right)\left(\frac{2}{x^3}\right) + \left[\left(\frac{d^2Y}{dz^2}\right)\left(\frac{dz}{dx}\right)\right]\left(-\frac{1}{x^2}\right) \\
&= \left(\frac{dY}{dz}\right)\left(\frac{2}{x^3}\right) + \left(\frac{d^2Y}{dz^2}\right)\left(-\frac{1}{x^2}\right)^2 = \frac{2}{x^3}\frac{dY}{dz} + \frac{1}{x^4}\frac{d^2Y}{dz^2}.
\end{aligned}$$

 Copyright © 2012 Pearson Education, Inc. Publishing as Addison-Wesley.

Hence, we have

$$x^3 \frac{d^2y}{dx^2} = 2\frac{dY}{dz} + \frac{1}{x}\frac{d^2Y}{dz^2} = 2\frac{dY}{dz} + z\frac{d^2Y}{dz^2}. \qquad (8.20)$$

By using the fact that $Y(z) = y(x)$ and the equations (8.19) and (8.20), we can now transform the original differential equation into

$$2\frac{dY}{dz} + z\frac{d^2Y}{dz^2} + \frac{dY}{dz} - Y = 0$$
$$\Rightarrow \qquad zY'' + 3Y' - Y = 0$$
$$\Rightarrow \qquad Y'' + \frac{3}{z}Y' - \frac{1}{z}Y = 0. \qquad (8.21)$$

To solve this linear differential equation, we note that

$$p(z) = \frac{3}{z} \qquad \Rightarrow \qquad zp(z) = 3,$$
$$q(z) = \frac{-1}{z} \qquad \Rightarrow \qquad z^2 q(z) = -z.$$

Therefore, $z = 0$ is a regular singular point for this equation, so that $x = \infty$ is a regular singular point for the original equation. To find a solution to (8.21), we compute

$$p_0 = \lim_{z\to 0} zp(z) = 3 \qquad \text{and} \qquad q_0 = \lim_{z\to 0} z^2 q(z) = 0.$$

Thus, the indicial equation for (8.21) is $r(r-1) + 3r = r(r+2) = 0$, which has roots $r_1 = 0$ and $r_2 = -2$. We seek for a solution of the form

$$w(r, z) = \sum_{n=0}^{\infty} a_n z^{n+r}.$$

Substituting this expression into (8.21) yields

$$z\sum_{n=0}^{\infty}(n+r-1)(n+r)a_n z^{n+r-2} + 3\sum_{n=0}^{\infty}(n+r)a_n z^{n+r-1} - \sum_{n=0}^{\infty} a_n z^{n+r} = 0.$$

Simplifying yields

$$\sum_{n=0}^{\infty}(n+r-1)(n+r)a_n z^{n+r-1} + \sum_{n=0}^{\infty} 3(n+r)a_n z^{n+r-1} - \sum_{n=0}^{\infty} a_n z^{n+r} = 0.$$

Making the shift of index $k = n - 1$ in the first two power series and denoting $k = n$ in the last power series allow us to sum each power series over the same powers of z.

$$\sum_{k=-1}^{\infty}(k+r)(k+r+1)a_{k+1} z^{k+r} + \sum_{k=-1}^{\infty} 3(k+r+1)a_{k+1} z^{n+r} - \sum_{k=0}^{\infty} a_k z^{k+r} = 0.$$

Copyright © 2012 Pearson Education, Inc. Publishing as Addison-Wesley.

Removing the first term from the first two power series, we can rewrite these three sums as a single power series. Therefore, we have

$$(r-1)ra_0 z^{r-1} + \sum_{k=0}^{\infty}(k+r)(k+r+1)a_{k+1}z^{k+r}$$

$$+3ra_0 z^{r-1} + \sum_{k=0}^{\infty}3(k+r+1)a_{k+1}z^{n+r} - \sum_{k=0}^{\infty}a_k z^{k+r} = 0$$

$$\Rightarrow \quad [(r-1)r + 3r]\,a_0 z^{r-1} + \sum_{k=0}^{\infty}[(k+r)(k+r+1)a_{k+1}$$

$$+3(k+r+1)a_{k+1} - a_k]\,z^{k+r} = 0.$$

Equating the coefficients and assuming that $a_0 \neq 0$, we obtain the indicial equation $(r-1)r + 3r = 0$ and the recurrence relation

$$(k+r)(k+r+1)a_{k+1} + 3(k+r+1)a_{k+1} - a_k = 0, \qquad k \geq 0$$

$$\Rightarrow \quad a_{k+1} = \frac{a_k}{(k+r+1)(k+r+3)}, \qquad k \geq 0.$$

Thus, for $r = r_1 = 0$, we obtain

$$a_{k+1} = \frac{a_k}{(k+1)(k+3)}, \qquad k \geq 3.$$

Since a_0 is an arbitrary number, we see from this recurrence equation that the next three coefficients are given by

$$k = 0 \qquad \Rightarrow \qquad a_1 = \frac{a_0}{3},$$

$$k = 1 \qquad \Rightarrow \qquad a_2 = \frac{a_1}{8} = \frac{a_0}{24},$$

$$k = 2 \qquad \Rightarrow \qquad a_3 = \frac{a_2}{15} = \frac{a_0}{360}.$$

Thus, from the method of Frobenius, we obtain a power series solution to (8.21) given by

$$Y(z) = w(0, z) = \sum_{n=0}^{\infty}a_n z^n = a_0\left(1 + \frac{1}{3}z + \frac{1}{24}z^2 + \frac{1}{360}z^3 + \cdots\right).$$

In order to find the solution to the original differential equation, we make the back substitution $z = 1/x$. Thus,

$$y(x) = Y\left(\frac{1}{x}\right) = a_0\left(1 + \frac{1}{3}x^{-1} + \frac{1}{24}x^{-2} + \frac{1}{360}x^{-3} + \cdots\right).$$

 Copyright © 2012 Pearson Education, Inc. Publishing as Addison-Wesley.

EXERCISES 8.7: Finding a Second Linearly Independent Solution

3. In Problem 21, Exercises 8.6, we found that one power series solution for this differential equation about the point $x = 0$ is given by

$$y_1(x) = 1 - \frac{1}{4}x^2 + \frac{1}{64}x^4 - \frac{1}{2304}x^6 + \cdots ,$$

where we let $a_0 = 1$. We also found that the roots to the indicial equation are $r_1 = r_2 = 0$. Thus, to find a second linearly independent solution about the regular singular point $x = 0$, we use part (b) of Theorem 7. This second solution has the form

$$y_2(x) = y_1(x) \ln x + \sum_{n=1}^{\infty} b_n x^n$$

$$\Rightarrow \qquad y_2'(x) = y_1'(x) \ln x + x^{-1}y_1(x) + \sum_{n=1}^{\infty} n b_n x^{n-1}$$

$$\Rightarrow \qquad y_2''(x) = y_1''(x) \ln x + 2x^{-1}y_1'(x) - x^{-2}y_1(x) + \sum_{n=1}^{\infty} n(n-1) b_n x^{n-2} .$$

Substituting these expressions into the differential equation yields

$$x^2 \left\{ y_1''(x) \ln x + 2x^{-1}y_1'(x) - x^{-2}y_1(x) + \sum_{n=1}^{\infty} n(n-1) b_n x^{n-2} \right\}$$

$$+ x \left\{ y_1'(x) \ln x + x^{-1}y_1(x) + \sum_{n=1}^{\infty} n b_n x^{n-1} \right\} + x^2 \left\{ y_1(x) \ln x + \sum_{n=1}^{\infty} b_n x^n \right\} = 0,$$

which simplifies to

$$x^2 y_1''(x) \ln x + 2xy_1'(x) - y_1(x) + \sum_{n=1}^{\infty} n(n-1) b_n x^n$$

$$+ xy_1'(x) \ln x + y_1(x) + \sum_{n=1}^{\infty} n b_n x^n + x^2 y_1(x) \ln x + \sum_{n=1}^{\infty} b_n x^{n+2} = 0,$$

$$\Rightarrow \qquad \left[x^2 y_1''(x) + xy_1'(x) + x^2 y_1(x) \right] \ln x + 2xy_1'(x)$$

$$+ \sum_{n=1}^{\infty} n(n-1) b_n x^n + \sum_{n=1}^{\infty} n b_n x^n + \sum_{n=1}^{\infty} b_n x^{n+2} = 0.$$

Therefore, since $y_1(x)$ is a solution to the given differential equation, the term in brackets equals zero, and the above equation reduces to

$$2xy_1'(x) + \sum_{n=1}^{\infty} n(n-1) b_n x^n + \sum_{n=1}^{\infty} n b_n x^n + \sum_{n=1}^{\infty} b_n x^{n+2} = 0.$$

Next, we make the substitution $k = n + 2$ in the last power series and the substitution $k = n$ in the other two power series so that we can sum all three series over the same power of x, namely x^k. Thus, we have

$$2xy_1'(x) + \sum_{k=1}^{\infty} k(k-1)b_k x^k + \sum_{k=1}^{\infty} kb_k x^k + \sum_{k=3}^{\infty} b_{k-2}x^k = 0.$$

Separating out the first two terms in the first two series and simplifying, we obtain

$$2xy_1'(x) + 0 + 2b_2 x^2 + \sum_{k=3}^{\infty} k(k-1)b_k x^k + b_1 x + 2b_2 x^2 + \sum_{k=3}^{\infty} kb_k x^k + \sum_{k=3}^{\infty} b_{k-2}x^k = 0$$

$$\Rightarrow \quad 2xy_1'(x) + b_1 x + 4b_2 x^2 + \sum_{k=3}^{\infty} \left(k^2 b_k + b_{k-2}\right) x^k = 0. \tag{8.22}$$

By differentiating the series for $y_1(x)$ termwise, we get

$$y_1'(x) = -\frac{1}{2}x + \frac{1}{16}x^3 - \frac{1}{384}x^5 + \cdots.$$

Thus, substituting this expression for $y_1'(x)$ into (8.22) and simplifying yields

$$\left(-x^2 + \frac{1}{8}x^4 - \frac{1}{192}x^6 + \cdots\right) + b_1 x + 4b_2 x^2 + \sum_{k=3}^{\infty} \left(k^2 b_k + b_{k-2}\right) x^k = 0.$$

Equating coefficients, we see that

$$b_1 = 0;$$
$$4b_2 - 1 = 0 \qquad \Rightarrow \qquad b_2 = \frac{1}{4};$$
$$9b_3 + b_1 = 0 \qquad \Rightarrow \qquad b_3 = 0;$$
$$\frac{1}{8} + 16b_4 + b_2 = 0 \qquad \Rightarrow \qquad b_4 = \frac{-3}{128};$$
$$25b_5 + b_3 = 0 \qquad \Rightarrow \qquad b_5 = 0;$$
$$\frac{-1}{192} + 36b_6 + b_4 = 0 \qquad \Rightarrow \qquad b_6 = \frac{11}{13824}.$$

Substituting these coefficients into the solution

$$y_2(x) = y_1(x)\ln x + \sum_{n=1}^{\infty} b_n x^n$$

yields

$$y_2(x) = y_1(x)\ln x + \frac{1}{4}x^2 - \frac{3}{128}x^4 + \frac{11}{13824}x^6 + \cdots.$$

Thus, a general solution to the given differential equation is

$$y(x) = c_1 y_1(x) + c_2 y_2(x),$$

 Copyright © 2012 Pearson Education, Inc. Publishing as Addison-Wesley.

where

$$y_1(x) = 1 - \frac{1}{4}x^2 + \frac{1}{64}x^4 - \frac{1}{2304}x^6 + \cdots,$$

$$y_2(x) = y_1(x)\ln x + \frac{1}{4}x^2 - \frac{3}{128}x^4 + \frac{11}{13,824}x^6 + \cdots.$$

7. In Problem 25 of Section 8.6, we found a solution to this differential equation about the regular singular point $x = 0$ given by

$$y_1(x) = \sum_{n=0}^{\infty} \frac{(-1)^n x^{n+(3/2)}}{2^{n-1}(n+2)!} = x^{3/2} - \frac{1}{6}x^{5/2} + \frac{1}{48}x^{7/2} + \cdots,$$

where we let $a_0 = 1$. We also found that the roots to the indicial equation for this problem are $r_1 = 3/2$ and $r_2 = -1/2$, and so $r_1 - r_2 = 2$. Thus, in order to find a second linearly independent solution about $x = 0$, we use part (c) of Theorem 7.

$$y_2(x) = Cy_1(x)\ln x + \sum_{n=0}^{\infty} b_n x^{n-(1/2)}, \qquad b_0 \neq 0$$

$$\Rightarrow \quad y_2'(x) = Cy_1'(x)\ln x + \frac{Cy_1(x)}{x} + \sum_{n=0}^{\infty}\left(n - \frac{1}{2}\right)b_n x^{n-(3/2)}$$

$$\Rightarrow \quad y_2''(x) = Cy_1''(x)\ln x + \frac{2Cy_1'(x)}{x} - \frac{Cy_1(x)}{x^2} + \sum_{n=0}^{\infty}\left(n - \frac{3}{2}\right)\left(n - \frac{1}{2}\right)b_n x^{n-(5/2)}.$$

Substituting these expressions into the differential equation yields

$$4x^2\left[Cy_1''(x)\ln x + \frac{2Cy_1'(x)}{x} - \frac{Cy_1(x)}{x^2} + \sum_{n=0}^{\infty}\left(n - \frac{3}{2}\right)\left(n - \frac{1}{2}\right)b_n x^{n-(5/2)}\right]$$

$$+ 2x^2\left[Cy_1'(x)\ln x + \frac{Cy_1(x)}{x} + \sum_{n=0}^{\infty}\left(n - \frac{1}{2}\right)b_n x^{n-(3/2)}\right]$$

$$- (x+3)\left[Cy_1(x)\ln x + \sum_{n=0}^{\infty}b_n x^{n-(1/2)}\right] = 0.$$

Multiplying through, we get

$$\left[4x^2 Cy_1''(x)\ln x + 8Cxy_1'(x) - 4Cy_1(x) + \sum_{n=0}^{\infty}4\left(n - \frac{3}{2}\right)\left(n - \frac{1}{2}\right)b_n x^{n-(1/2)}\right]$$

$$+ \left[2x^2 Cy_1'(x)\ln x + 2Cxy_1(x) + \sum_{n=0}^{\infty}2\left(n - \frac{1}{2}\right)b_n x^{n+(1/2)}\right]$$

$$- \left[Cxy_1(x)\ln x + \sum_{n=0}^{\infty}b_n x^{n+(1/2)} + 3Cy_1(x)\ln x + \sum_{n=0}^{\infty}3b_n x^{n-(1/2)}\right] = 0,$$

Copyright © 2012 Pearson Education, Inc. Publishing as Addison-Wesley.

which simplifies to

$$C\left[4x^2y_1''(x) + 2x^2y_1'(x) - xy_1(x) - 3y_1(x)\right]\ln x + 8Cxy_1'(x) + 2C(x-2)y_1(x)$$

$$+ \sum_{n=0}^{\infty}(2n-3)(2n-1)\,b_n x^{n-(1/2)} + \sum_{n=0}^{\infty}(2n-1)\,b_n x^{n+(1/2)}$$

$$- \sum_{n=0}^{\infty}b_n x^{n+(1/2)} - \sum_{n=0}^{\infty}3b_n x^{n-(1/2)} = 0.$$

Since $y_1(x)$ is a solution to the given differential equation, the term in brackets equals zero. By shifting indices so that each power series is summed over the same power of x, we have

$$8Cxy_1'(x) + 2C(x-2)y_1(x) + \sum_{k=0}^{\infty}(2k-3)(2k-1)\,b_k x^{k-(1/2)}$$

$$+ \sum_{k=1}^{\infty}(2k-3)\,b_{k-1}x^{k-(1/2)} - \sum_{k=1}^{\infty}b_{k-1}x^{k-(1/2)} - \sum_{k=0}^{\infty}3b_k x^{k-(1/2)} = 0.$$

Hence

$$8Cxy_1'(x) + 2C(x-2)y_1(x)$$

$$+ \sum_{k=1}^{\infty}\left[(2k-3)(2k-1)\,b_k + (2k-3)\,b_{k-1} - b_{k-1} - 3b_k\right]x^{k-(1/2)} = 0.$$

Substituting into this equation the expressions for $y_1(x)$ and $y_1'(x)$ given by

$$y_1(x) = \sum_{n=0}^{\infty}\frac{(-1)^n x^{n+(3/2)}}{2^{n-1}(n+2)!}, \qquad y_1'(x) = \sum_{n=0}^{\infty}\frac{(-1)^n[n+(3/2)]x^{n+(1/2)}}{2^{n-1}(n+2)!},$$

yields

$$\sum_{n=0}^{\infty}\frac{8C(-1)^n[n+(3/2)]x^{n+(3/2)}}{2^{n-1}(n+2)!} + \sum_{n=0}^{\infty}\frac{2C(-1)^n x^{n+(5/2)}}{2^{n-1}(n+2)!}$$

$$- \sum_{n=0}^{\infty}\frac{4C(-1)^n x^{n+(3/2)}}{2^{n-1}(n+2)!} + \sum_{k=1}^{\infty}\left[4k(k-2)b_k + 2(k-2)b_{k-1}\right]x^{k-(1/2)} = 0,$$

where we have simplified the expression in the last summation. Combining the first and third power series yields

$$\sum_{n=0}^{\infty}\frac{8C(-1)^n(n+1)x^{n+(3/2)}}{2^{n-1}(n+2)!} + \sum_{n=0}^{\infty}\frac{2C(-1)^n x^{n+(5/2)}}{2^{n-1}(n+2)!}$$

$$+ \sum_{k=1}^{\infty}\left[4k(k-2)b_k + 2(k-2)b_{k-1}\right]x^{k-(1/2)} = 0. \qquad (8.23)$$

 Copyright © 2012 Pearson Education, Inc. Publishing as Addison-Wesley.

By writing out the terms up to $x^{7/2}$, we obtain

$$8C\left[x^{3/2} - \frac{1}{3}x^{5/2} + \frac{3}{16}x^{7/2} + \cdots\right] + 2C\left[x^{5/2} - \frac{1}{6}x^{7/2} + \cdots\right]$$
$$+ \left[(-4b_1 - 2b_0)x^{1/2} + (12b_3 + 2b_2)x^{5/2} + (32b_4 + 4b_3)x^{7/2} + \cdots\right] = 0.$$

Setting the coefficients equal to zero yields

$$\begin{aligned}
-4b_1 - 2b_0 = 0 &\qquad\Rightarrow\qquad b_1 = -b_0/2; \\
8C = 0 &\qquad\Rightarrow\qquad C = 0; \\
-(8/3)C + 2C + 12b_3 + 2b_2 = 0 &\qquad\Rightarrow\qquad b_3 = -b_2/6; \\
(2/3)C - (1/3)C + 32b_4 + 4b_3 = 0 &\qquad\Rightarrow\qquad b_4 = -b_3/8 = b_2/48.
\end{aligned}$$

From this we see that b_0 and b_2 are arbitrary constants and that $C = 0$. Also, since $C = 0$, we can use the last power series in (8.23) to obtain the recurrence relation $b_k = b_{k-1}/(2k)$. Thus, every coefficient after b_4 will depend only on b_2 (not on b_0). Substituting these coefficients into the solution,

$$y_2(x) = Cy_1(x)\ln x + \sum_{n=0}^{\infty} b_n x^{n-(1/2)},$$

yields

$$y_2(x) = b_0\left(x^{-1/2} - \frac{1}{2}x^{1/2}\right) + b_2\left(x^{3/2} - \frac{1}{6}x^{5/2} + \frac{1}{48}x^{7/2} + \cdots\right).$$

The expression in the parentheses, following b_2, is just the series expansion of $y_1(x)$. Hence, in order to obtain a second linearly independent solution, we must choose $b_0 \neq 0$. Taking $b_0 = 1$ and $b_2 = 0$ gives

$$y_2(x) = x^{-1/2} - \frac{1}{2}x^{1/2}.$$

Therefore, a general solution is

$$y(x) = c_1 y_1(x) + c_2 y_2(x),$$

where

$$y_1(x) = x^{3/2} - \frac{1}{6}x^{5/2} + \frac{1}{48}x^{7/2} + \cdots \qquad\text{and}\qquad y_2(x) = x^{-1/2} - \frac{1}{2}x^{1/2}.$$

17. In Problem 35 of Section 8.6, we assumed that there exists a power series solution to this problem of the form $w(r,x) = \sum_{n=0}^{\infty} a_n x^{n+r}$. This assumption led us to the equation (see (8.18))

$$r^2(6r - 5)a_0 x^r + \sum_{k=1}^{\infty}\left\{(k+r)^2[6(k+r) - 5]a_k + (k+r)a_{k-1}\right\}x^{k+r} = 0. \qquad (8.24)$$

Copyright © 2012 Pearson Education, Inc. Publishing as Addison-Wesley.

From (8.24), we found the indicial equation, $r^2(6r - 5) = 0$, having roots $r = 0, 0, 5/6$. By using the root $r = 5/6$, we found the solution

$$y_1(x) = w(5/6, x) = x^{5/6} - \frac{x^{11/6}}{11} + \frac{x^{17/6}}{374} - \frac{x^{23/6}}{25806} + \cdots,$$

where we have chosen $a_0 = 1$ in $w(5/6, x)$. We now seek for two more linearly independent solutions to this differential equation. To find a second linearly independent solution, we will use the root $r = 0$ and set the coefficients (8.24) to zero to get a recurrence relation

$$k^2(6k - 5)a_k + ka_{k-1} = 0, \qquad k \geq 1.$$

Solving for a_k in terms of a_{k-1} gives

$$a_k = \frac{-a_{k-1}}{k(6k - 5)}, \qquad k \geq 1.$$

Thus, we have

$$
\begin{aligned}
k = 1 &\quad \Rightarrow \quad a_1 = -a_0, \\
k = 2 &\quad \Rightarrow \quad a_2 = \frac{-a_1}{14} = \frac{a_0}{14}, \\
k = 3 &\quad \Rightarrow \quad a_3 = \frac{-a_2}{39} = \frac{-a_0}{546}, \\
k = 4 &\quad \Rightarrow \quad a_4 = \frac{-a_3}{76} = \frac{a_0}{41496}, \\
k = 5 &\quad \Rightarrow \quad a_5 = \frac{-a_4}{125} = \frac{-a_0}{5187000}.
\end{aligned}
$$

Plugging in these coefficients into the solution $w(0, x)$ and setting $a_0 = 1$ yields a second linearly independent solution

$$y_2(x) = 1 - x + \frac{1}{14}x^2 - \frac{1}{546}x^3 + \frac{1}{41496}x^4 - \frac{1}{5187000}x^5 + \cdots.$$

To find a third linearly independent solution, we will use the repeated root $r = 0$ and assume that (similarly to the case of second order equations with repeated roots) the solution has the form

$$y_3(x) = y_2(x)\ln x + \sum_{n=1}^{\infty} c_n x^n.$$

Since the first three derivatives of $y_3(x)$ are given by

$$y_3'(x) = y_2'(x)\ln x + x^{-1}y_2(x) + \sum_{n=1}^{\infty} nc_n x^{n-1},$$

     Copyright © 2012 Pearson Education, Inc. Publishing as Addison-Wesley.

$$y_3''(x) = y_2''(x)\ln x + 2x^{-1}y_2'(x) - x^{-2}y_2(x) + \sum_{n=1}^{\infty}(n-1)nc_n x^{n-2},$$

$$y_3'''(x) = y_2'''(x)\ln x + 3x^{-1}y_2''(x) - 3x^{-2}y_2'(x) + 2x^{-3}y_2(x) + \sum_{n=1}^{\infty}(n-2)(n-1)nc_n x^{n-3},$$

substituting $y_3(x)$ into the given differential equation yields

$$6x^3 y'''(x) + 13x^2 y''(x) + \left(x + x^2\right)y'(x) + xy(x)$$

$$= 6x^3\left[y_2'''(x)\ln x + 3x^{-1}y_2''(x) - 3x^{-2}y_2'(x) + 2x^{-3}y_2(x) + \sum_{n=1}^{\infty}(n-2)(n-1)nc_n x^{n-3}\right]$$

$$+ 13x^2\left[y_2''(x)\ln x + 2x^{-1}y_2'(x) - x^{-2}y_2(x) + \sum_{n=1}^{\infty}(n-1)nc_n x^{n-2}\right]$$

$$+ \left(x + x^2\right)\left[y_2'(x)\ln x + x^{-1}y_2(x) + \sum_{n=1}^{\infty}nc_n x^{n-1}\right] + x\left[y_2(x)\ln x + \sum_{n=1}^{\infty}c_n x^n\right] = 0.$$

Since $y_2(x)$ is a solution to the given equation, this simplifies to

$$18x^2 y_2''(x) + 8xy_2'(x) + xy_2(x) + \sum_{n=1}^{\infty}6(n-2)(n-1)nc_n x^n$$

$$+ \sum_{n=1}^{\infty}13(n-1)nc_n x^n + \sum_{n=1}^{\infty}nc_n x^n + \sum_{n=1}^{\infty}nc_n x^{n+1} + \sum_{n=1}^{\infty}c_n x^{n+1} = 0.$$

By shifting indices and then starting all of the resulting power series at the same index, we can combine all of the summations above into a single power series. Thus, we have

$$18x^2 y_2''(x) + 8xy_2'(x) + xy_2(x)$$

$$+ c_1 x + \sum_{k=2}^{\infty}\left[6(k-2)(k-1)kc_k + 13(k-1)kc_k + kc_k + kc_{k-1}\right]x^k = 0. \quad (8.25)$$

Computing $y_2'(x)$ and $y_2''(x)$, we obtain

$$y_2'(x) = -1 + \frac{1}{7}x - \frac{1}{182}x^2 + \frac{1}{10374}x^3 + \cdots,$$

$$y_2''(x) = \frac{1}{7} - \frac{1}{91}x + \frac{1}{3458}x^2 + \cdots.$$

By substituting these expressions into (8.25), we get

$$18x^2\left(\frac{1}{7} - \frac{x}{91} + \frac{x^2}{3458} + \cdots\right) + 8x\left(-1 + \frac{x}{7} - \frac{x^2}{182} + \frac{x^3}{10374} + \cdots\right)$$

$$+ x\left(1 - x + \frac{x^2}{14} - \frac{x^3}{546} + \frac{x^4}{41496} + \cdots\right) + c_1 x + \sum_{k=2}^{\infty}\left[(6k^3 - 5k^2)c_k + kc_{k-1}\right]x^k = 0.$$

Copyright © 2012 Pearson Education, Inc. Publishing as Addison-Wesley.

Writing out the terms up to x^3 we find

$$(-7 + c_1)x + \left(\frac{19}{7} + 28c_2 + 2c_1\right)x^2 + \left(-\frac{31}{182} + 117c_3 + 3c_2\right)x^3 + \cdots = 0.$$

We now set all coefficients to zero and obtain

$$
\begin{aligned}
-7 + c_1 &= 0 & \Rightarrow & \quad c_1 = 7; \\
19/7 + 28c_2 + 2c_1 &= 0 & \Rightarrow & \quad c_2 = -117/196; \\
-31/182 + 117c_3 + 3c_2 &= 0 & \Rightarrow & \quad c_3 = 4997/298116.
\end{aligned}
$$

Therefore, plugging in these coefficients into the expansion

$$y_3(x) = y_2(x) \ln x + \sum_{n=1}^{\infty} c_n x^n$$

yields a third linearly independent solution

$$y_3(x) = y_2(x) \ln x + 7x - \frac{117}{196}x^2 + \frac{4997}{298116}x^3 + \cdots.$$

Thus, a general solution is $y(x) = c_1 y_1(x) + c_2 y_2(x) + c_3 y_3(x)$, where

$$
\begin{aligned}
y_1(x) &= x^{5/6} - \frac{x^{11/6}}{11} + \frac{x^{17/6}}{374} - \frac{x^{23/6}}{25806} + \cdots, \\
y_2(x) &= 1 - x + \frac{x^2}{14} - \frac{x^3}{546} + \frac{x^4}{41496} - \frac{x^5}{5187000} + \cdots, \\
y_3(x) &= y_2(x) \ln x + 7x - \frac{117x^2}{196} + \frac{4997x^3}{298116} + \cdots.
\end{aligned}
$$

23. We will try to find a solution of the form

$$y(x) = \sum_{n=0}^{\infty} a_n x^{n+r}$$

$$\Rightarrow \quad y'(x) = \sum_{n=0}^{\infty} (n+r)a_n x^{n+r-1}$$

$$\Rightarrow \quad y''(x) = \sum_{n=0}^{\infty} (n+r)(n+r-1)a_n x^{n+r-2}.$$

We substitute these expressions into the differential equation to obtain

$$x^2 y'' + y' - 2y = x^2 \sum_{n=0}^{\infty} (n+r)(n+r-1)a_n x^{n+r-2}$$

$$+ \sum_{n=0}^{\infty} (n+r)a_n x^{n+r-1} - 2\sum_{n=0}^{\infty} a_n x^{n+r} = 0$$

 Copyright © 2012 Pearson Education, Inc. Publishing as Addison-Wesley.

$$\Rightarrow \qquad \sum_{k=0}^{\infty}(k+r)(k+r-1)a_k x^{k+r} + \sum_{k=-1}^{\infty}(k+r+1)a_{k+1}x^{k+r} - \sum_{k=0}^{\infty}2a_k x^{k+r} = 0$$

$$\Rightarrow \qquad ra_0 x^{r-1} + \sum_{k=0}^{\infty}\left[(k+r)(k+r-1)a_k + (k+r+1)a_{k+1} - 2a_k\right]x^{k+r} = 0.$$

Assuming that $a_0 \neq 0$, we see that $ra_0 x^{r-1} = 0$ results $r = 0$. Plugging in $r = 0$ into the coefficients and noting that each of these coefficients must be zero yields a recurrence relation

$$k(k-1)a_k + (k+1)a_{k+1} - 2a_k = 0 \qquad \Rightarrow \qquad a_{k+1} = (2-k)a_k, \quad k \geq 0.$$

Thus, we see that the coefficients of the solution are given by

$$k = 0 \;\Rightarrow\; a_1 = 2a_0; \quad k = 1 \;\Rightarrow\; a_2 = a_1 = 2a_0;$$
$$k = 2 \;\Rightarrow\; a_3 = 0; \quad\;\; k = 3 \;\Rightarrow\; a_4 = -a_3 = 0.$$

Since each coefficient is a multiple of the previous coefficient, we see that $a_n = 0$ for $n \geq 3$. If we take $a_0 = 1$, we obtain a solution

$$y_1(x) = 1 + 2x + 2x^2.$$

We will now use the reduction of order procedure described in Problem 31, Section 6.1, to find a second linearly independent solution. Thus we look for a solution of the form

$$y(x) = y_1(x)v(x)$$
$$\Rightarrow \qquad y'(x) = y_1'(x)v(x) + y_1(x)v'(x)$$
$$\Rightarrow \qquad y''(x) = y_1''(x)v(x) + 2y_1'(x)v'(x) + y_1(x)v''(x).$$

Substituting $y(x)$, $y'(x)$, and $y''(x)$ into the given equation yields

$$\begin{aligned}
x^2 y'' + y' - 2y &= x^2\left(y_1''v + 2y_1'v' + y_1 v''\right) + \left(y_1'v + y_1 v'\right) - 2\left(y_1 v\right)\\
&= \left(x^2 y_1\right)v'' + \left(2x^2 y_1' + y_1\right)v' + \left(x^2 y_1'' + y_1' - 2y_1\right)v\\
&= \left(x^2 y_1\right)v'' + \left(2x^2 y_1' + y_1\right)v' = 0
\end{aligned}$$

(since y_1 is a solution, the coefficient at v equals to zero). With $w = v'$, the last equation becomes a first order separable equation, which can be solved by methods of Section 2.2.

$$\left[x^2 y_1(x)\right]w'(x) + \left[2x^2 y_1'(x) + y_1(x)\right]w(x) = 0$$

Copyright © 2012 Pearson Education, Inc. Publishing as Addison-Wesley.

$$\Rightarrow \qquad \frac{dw}{w} = -\frac{2x^2 y_1'(x) + y_1(x)}{x^2 y_1(x)} \, dx = -\left(\frac{2y'(x)}{y_1(x)} + \frac{1}{x^2} \right) dx$$

$$\Rightarrow \qquad \ln|w| = -\int \frac{2y'(x)dx}{y_1(x)} - \int \frac{dx}{x^2} = -2\ln|y_1(x)| + \frac{1}{x}$$

$$\Rightarrow \qquad w(x) = \exp\left[-2\ln|y_1(x)| + \frac{1}{x} \right] = \frac{e^{1/x}}{[y_1(x)]^2} , \qquad\qquad (8.26)$$

where we have taken zero integration constant and positive function w. Since

$$[y_1(x)]^2 = 4x^4 + 8x^3 + 8x^2 + 4x + 1 \qquad \text{and}$$

$$e^{1/x} = 1 + x^{-1} + \frac{x^{-2}}{2} + \frac{x^{-3}}{6} + \frac{x^{-4}}{24} + \cdots$$

(we have used the Maclaurin expansion for e^z with $z = 1/x$), performing long division with descending powers of x in each polynomial, we see that

$$\frac{e^{1/x}}{[y_1(x)]^2} = \frac{1 + x^{-1} + \dfrac{x^{-2}}{2} + \dfrac{x^{-3}}{6} + \dfrac{x^{-4}}{24} + \cdots}{4x^4 + 8x^3 + 8x^2 + 4x + 1} = \frac{1}{4}x^{-4} - \frac{1}{4}x^{-5} + \frac{1}{8}x^{-6} + \cdots .$$

Therefore, (8.26) yields

$$v'(x) = w(x) = \frac{1}{4}x^{-4} - \frac{1}{4}x^{-5} + \frac{1}{8}x^{-6} + \cdots$$

$$\Rightarrow \quad v(x) = \int \left(\frac{1}{4}x^{-4} - \frac{1}{4}x^{-5} + \frac{1}{8}x^{-6} + \cdots \right) dx = -\frac{1}{12}x^{-3} + \frac{1}{16}x^{-4} - \frac{1}{40}x^{-5} + \cdots$$

and so

$$y(x) = y_1(x)v(x) = \left(1 + 2x + 2x^2\right)\left(-\frac{1}{12}x^{-3} + \frac{1}{16}x^{-4} - \frac{1}{40}x^{-5} + \cdots \right)$$

$$= -\frac{1}{6}x^{-1} - \frac{1}{24}x^{-2} - \frac{1}{120}x^{-3} + \cdots$$

is a second linearly independent solution. Thus, a general solution to this differential equation is given by $y(x) = c_1 y_1(x) + c_2 y_2(x)$, where

$$y_1(x) = 1 + 2x + 2x^2 \qquad \text{and} \qquad y_2(x) = -\frac{1}{6}x^{-1} - \frac{1}{24}x^{-2} - \frac{1}{120}x^{-3} + \cdots .$$

EXERCISES 8.8: Special Functions

1. In this problem, we see that $\gamma = 1/2$, $\alpha + \beta + 1 = 4$, and $\alpha\beta = 2$. First we note that γ is not an integer. Next, by solving in the last two equations above simultaneously for

 Copyright © 2012 Pearson Education, Inc. Publishing as Addison-Wesley.

α and β, we see that either $\alpha = 1$ and $\beta = 1$ or $\alpha = 2$ and $\beta = 1$. Therefore, assuming that $\alpha = 1$ and $\beta = 2$, equations (10) and (17) in the text give two solutions

$$y_1(x) = F\left(1, 2; \frac{1}{2}; x\right) \qquad \text{and} \qquad y_2(x) = x^{1/2} F\left(\frac{3}{2}, \frac{5}{2}; \frac{3}{2}; x\right).$$

Therefore, a general solution for this differential equation is given by

$$y(x) = c_1 F\left(1, 2; \frac{1}{2}; x\right) + c_2 x^{1/2} F\left(\frac{3}{2}, \frac{5}{2}; \frac{3}{2}; x\right).$$

Notice that

$$F(\alpha, \beta; \gamma; x) = 1 + \sum_{n=0}^{\infty} \frac{(\alpha)_n (\beta)_n}{n!(\gamma)_n} x^n = 1 + \sum_{n=0}^{\infty} \frac{(\beta)_n (\alpha)_n}{n!(\gamma)_n} x^n = F(\beta, \alpha; \gamma; x).$$

Therefore, letting $\alpha = 2$ and $\beta = 1$ yields an equivalent form of the same solution given by

$$y(x) = c_1 F\left(2, 1; \frac{1}{2}; x\right) + c_2 x^{1/2} F\left(\frac{5}{2}, \frac{3}{2}; \frac{3}{2}; x\right).$$

13. This equation can be written as

$$x^2 y'' + xy' + \left(x^2 - \frac{1}{4}\right) y = 0.$$

Thus, $\nu^2 = 1/4$ which implies that $\nu = 1/2$. Since this is not an integer (even though 2ν is an integer), two linearly independent solutions to this problem are given by equations (25) and (26), that is

$$y_1(x) = J_{1/2}(x) = \sum_{n=0}^{\infty} \frac{(-1)^n}{n!\,\Gamma\,(3/2 + n)} \left(\frac{x}{2}\right)^{2n + (1/2)},$$

$$y_2(x) = J_{-1/2}(x) = \sum_{n=0}^{\infty} \frac{(-1)^n}{n!\,\Gamma\,(1/2 + n)} \left(\frac{x}{2}\right)^{2n - (1/2)}.$$

Therefore, a general solution to this differential equation is given by

$$y(x) = c_1 J_{1/2}(x) + c_2 J_{-1/2}(x).$$

15. In this problem, $\nu = 1$. Thus, one solution to this differential equation is given by

$$y_1(x) = J_1(x) = \sum_{n=0}^{\infty} \frac{(-1)^n}{n!\,\Gamma\,(2 + n)} \left(\frac{x}{2}\right)^{2n+1}.$$

Copyright © 2012 Pearson Education, Inc. Publishing as Addison-Wesley.

Since $J_{-1}(x)$ and $J_1(x)$ are linearly dependent, $J_{-1}(x)$ is not a second linearly independent solution to this problem. A second linearly independent solution can be found using the equation (30) in the text with $m = 1$. That is, we have

$$y_2(x) = Y_1(x) = \lim_{\nu \to 1} \frac{\cos(\nu\pi)J_\nu(x) - J_{-\nu}(x)}{\sin(\nu\pi)}.$$

Therefore, a general solution to this differential equation is given by

$$y(x) = c_1 J_1(x) + c_2 Y_1(x).$$

21. Let $y(x) = x^\nu J_\nu(x)$. Then, by the equation (31), we have

$$y'(x) = x^\nu J_{\nu-1}(x).$$

Therefore, we see that

$$
\begin{aligned}
y''(x) &= D_x\left[y'(x)\right] = D_x\left[x^\nu J_{\nu-1}(x)\right] = D_x\left\{x\left[x^{\nu-1}J_{\nu-1}(x)\right]\right\} \\
&= x^{\nu-1}J_{\nu-1}(x) + xD_x\left[x^{\nu-1}J_{\nu-1}(x)\right] = x^{\nu-1}J_{\nu-1}(x) + x^\nu J_{\nu-2}(x).
\end{aligned}
$$

Notice that, taking the last derivative, we have used the equation (31). By substituting these expressions into the left-hand side of the first differential equation given in the problem, we obtain

$$xy'' + (1 - 2\nu)y' + xy = x\left[x^{\nu-1}J_{\nu-1}(x) + x^\nu J_{\nu-2}(x)\right]$$
$$+ (1 - 2\nu)\left[x^\nu J_{\nu-1}(x)\right] + x\left[x^\nu J_\nu(x)\right]$$
$$= x^\nu J_{\nu-1}(x) + x^{\nu+1}J_{\nu-2}(x) + x^\nu J_{\nu-1}(x) - 2\nu x^\nu J_{\nu-1}(x) + x^{\nu+1}J_\nu(x). \quad (8.27)$$

Notice that, by the equation (33) in the text, we have

$$J_\nu(x) = \frac{2(\nu - 1)}{x}J_{\nu-1}(x) - J_{\nu-2}(x)$$
$$\Rightarrow \quad x^{\nu+1}J_\nu(x) = 2(\nu - 1)x^\nu J_{\nu-1}(x) - x^{\nu+1}J_{\nu-2}(x).$$

Replacing $x^{\nu+1}J_\nu(x)$ in (8.27) by the above expression and simplifying yields

$$xy'' + (1 - 2\nu)y' + xy = x^\nu J_{\nu-1}(x) + x^{\nu+1}J_{\nu-2}(x) + x^\nu J_{\nu-1}(x)$$
$$- 2\nu x^\nu J_{\nu-1}(x) + 2(\nu - 1)x^\nu J_{\nu-1}(x) - x^{\nu+1}J_{\nu-2}(x) = 0.$$

Therefore, $y(x) = x^\nu J_\nu(x)$ is a solution to this type of differential equation.

 Copyright © 2012 Pearson Education, Inc. Publishing as Addison-Wesley.

In order to find a solution to the differential equation $xy'' - 2y' + xy = 0$, we observe that this equation is of the same type as the equation given above with

$$1 - 2\nu = -2 \quad \Rightarrow \quad \nu = \frac{3}{2}.$$

Thus, a solution to the given equation is

$$y(x) = x^{3/2} J_{3/2}(x) = x^{3/2} \sum_{n=0}^{\infty} \frac{(-1)^n}{n!\Gamma(5/2 + n)} \left(\frac{x}{2}\right)^{2n+(3/2)}.$$

29. In Legendre polynomials, n is a fixed nonnegative integer. Thus, in the first polynomial, $n = 0$, and so $[n/2] = [0/2] = 0$. By the equation (43) in the text, we have

$$P_0(x) = 2^{-0} \frac{(-1)^0 0!}{0!0!0!} x^0 = 1.$$

Similarly,

$$n = 1 \quad \Rightarrow \quad \left[\frac{1}{2}\right] = 0 \quad \Rightarrow \quad P_1(x) = 2^{-1} \frac{(-1)^0 2!}{1!0!1!} x^1 = x,$$

$$n = 2 \quad \Rightarrow \quad \left[\frac{2}{2}\right] = 1 \quad \Rightarrow \quad P_2(x) = 2^{-2} \left[\frac{(-1)^0 4!}{2!0!2!} x^2 + \frac{(-1)^1 2!}{1!1!0!} x^0\right] = \frac{3x^2 - 1}{2},$$

$$n = 3 \quad \Rightarrow \quad \left[\frac{3}{2}\right] = 1 \quad \Rightarrow \quad P_3(x) = 2^{-3} \left[\frac{(-1)^0 6!}{3!0!3!} x^3 + \frac{(-1)^1 4!}{2!1!1!} x^1\right] = \frac{5x^3 - 3x}{2},$$

$$n = 4 \quad \Rightarrow \quad \left[\frac{4}{2}\right] = 2 \quad \Rightarrow \quad P_4(x) = 2^{-4} \left[\frac{(-1)^0 8!}{4!0!4!} x^4 + \frac{(-1)^1 6!}{3!1!2!} x^2 + \frac{(-1)^2 4!}{2!2!0!} x^0\right]$$

$$= \frac{35x^4 - 30x^2 + 3}{8}.$$

37. Since the Taylor series expansion of an analytic function $f(t)$ about $t = 0$ is given by

$$f(t) = \sum_{n=0}^{\infty} \frac{f^{(n)}(0)}{n!} t^n,$$

we see that $H(x)$ is just the nth derivative of $y(t) = e^{2tx-t^2}$ with respect to t evaluated at the point $t = 0$ (with x as a fixed parameter). Therefore, we have

$$
\begin{aligned}
y(t) &= e^{2tx-t^2} & \Rightarrow \quad & H_0(x) = y(0) = e^0 = 1, \\
y'(t) &= (2x - 2t)e^{2tx-t^2} & \Rightarrow \quad & H_1(x) = y'(0) = 2xe^0 = 2x, \\
y''(t) &= [-2 + (2x - 2t)^2] e^{2tx-t^2} & \Rightarrow \quad & H_2(x) = y''(0) = 4x^2 - 2, \\
y'''(t) &= [-6(2x - 2t) + (2x - 2t)^3] e^{2tx-t^2} & \Rightarrow \quad & H_3(x) = y'''(0) = 8x^3 - 12x.
\end{aligned}
$$

Copyright © 2012 Pearson Education, Inc. Publishing as Addison-Wesley.

39. To find the first four Laguerre polynomials, we need to find the first four derivatives of the function $y(x) = x^n e^{-x}$.

$$y^{(0)}(x) = x^n e^{-x},$$

$$y'(x) = \left(nx^{n-1} - x^n\right)e^{-x},$$

$$y''(x) = \left[n(n-1)x^{n-2} - 2nx^{n-1} + x^n\right]e^{-x},$$

$$y'''(x) = \left[n(n-1)(n-2)x^{n-3} - 3n(n-1)x^{n-2} + 3nx^{n-1} - x^n\right]e^{-x}.$$

Substituting these expressions into Rodrigues formula and plugging in the appropriate values of n yields

$$L_0(x) = \frac{e^x}{0!}x^0 e^{-x} = 1,$$

$$L_1(x) = \frac{e^x}{1!}\left[(1)x^{1-1} - x^1\right]e^{-x} = 1 - x,$$

$$L_2(x) = \frac{e^x}{2!}\left[2(2-1)x^{2-2} - 2 \cdot 2x^{2-1} + x^2\right]e^{-x} = \frac{2 - 4x + x^2}{2},$$

$$L_3(x) = \frac{e^x}{3!}\left[3(3-1)(3-2)x^{3-3} - 3 \cdot 3(3-1)x^{3-2} + 3 \cdot 3x^{3-1} - x^3\right]e^{-x}$$

$$= \frac{6 - 18x + 9x^2 - x^3}{6}.$$

REVIEW PROBLEMS

1. (a) To construct the Taylor polynomials

$$p_n(x) = y(0) + \frac{y'(0)}{1!}x + \frac{y''(0)}{2!}x^2 + \cdots + \frac{y^{(n)}(0)}{n!}x^n$$

approximating the solution to the given initial value problem, we need $y(0)$, $y'(0)$, etc. $y(0)$ is provided by the initial condition, $y(0) = 1$. The value of $y'(0)$ can be deduced from the given differential equation. We have

$$y'(0) = (0)y(0) - y(0)^2 = (0)(1) - (1)^2 = -1.$$

Differentiating both sides of the given equation, $y' = xy - y^2$, and substituting $x = 0$ into the result, we get

$$y'' = y + xy' - 2yy'$$

$$\Rightarrow \quad y''(0) = y(0) + (0)y'(0) - 2y(0)y'(0) = (1) + (0)(-1) - 2(1)(-1) = 3.$$

Differentiating once again yields

$$y''' = y' + y' + xy'' - 2y'y' - 2yy''$$

$$\Rightarrow \quad y'''(0) = (-1) + (-1) + (0)(3) - 2(-1)(-1) - 2(1)(3) = -10.$$

Thus,

$$p_3(x) = 1 + \frac{-1}{1!}x + \frac{3}{2!}x^2 + \frac{-10}{3!}x^3 = 1 - x + \frac{3x^2}{2} - \frac{5x^3}{3}.$$

(b) The values of $z(0)$ and $z'(0)$ are given. Namely, $z(0) = -1$ and $z'(0) = 1$. Substituting $x = 0$ into the differential equation yields

$$z''(0) - (0)^3 z'(0) + (0)z(0)^2 = 0 \quad \Rightarrow \quad z''(0) = 0.$$

We now differentiate the given equation and evaluate the result at $x = 0$.

$$z''' - 3x^2 z' - x^3 z'' + z^2 + 2xzz' = 0$$

$$\Rightarrow \quad z'''(0) = 3(0)^2 z'(0) + (0)^3 z''(0) - z(0)^2 - 2(0)z(0)z'(0) = -1.$$

One more differentiation yields

$$z^{(4)} - 6xz' - 3x^2 z'' - 3x^2 z'' - x^3 z''' + 2zz' + 2zz' + 2xz'z' + 2xzz'' = 0$$

$$\Rightarrow \quad z^{(4)}(0) = -4z(0)z'(0) = 4.$$

Hence,

$$p_4(x) = -1 + \frac{1}{1!}x + \frac{0}{2!}x^2 + \frac{-1}{3!}x^3 + \frac{4}{4!}x^4 = -1 + x - \frac{x^3}{6} + \frac{x^4}{6}.$$

3. (a) Since both $p(x) = x^2$ and $q(x) = -2$ are analytic at $x = 0$, a general solution to the given equation is also analytic at this point. Thus, it has an expansion

$$y = \sum_{k=0}^{\infty} a_k x^k \quad \Rightarrow \quad y' = \sum_{k=1}^{\infty} k a_k x^{k-1} \quad \Rightarrow \quad y'' = \sum_{k=2}^{\infty} k(k-1) a_k x^{k-2}.$$

Substituting these expansions into the original equation yields

$$\sum_{k=2}^{\infty} k(k-1) a_k x^{k-2} + x^2 \sum_{k=1}^{\infty} k a_k x^{k-1} - 2 \sum_{k=0}^{\infty} a_k x^k = 0$$

$$\Rightarrow \quad \sum_{k=2}^{\infty} k(k-1) a_k x^{k-2} + \sum_{k=1}^{\infty} k a_k x^{k+1} - \sum_{k=0}^{\infty} 2 a_k x^k = 0.$$

Copyright © 2012 Pearson Education, Inc. Publishing as Addison-Wesley.

We now shift the indices of summation so that all three sums contain like powers x^n. In the first sum, we let $k - 2 = n$; in the second sum, we let $k + 1 = n$; and we let $k = n$ in the third sum. This yields

$$\sum_{n=0}^{\infty}(n+2)(n+1)a_{n+2}x^n + \sum_{n=2}^{\infty}(n-1)a_{n-1}x^n - \sum_{n=0}^{\infty}2a_n x^n = 0.$$

Separating the terms, corresponding to $n = 0$ and $n = 1$, and combining the remaining series, we obtain

$$(2a_2 - 2a_0) + (6a_3 - 2a_1)x + \sum_{n=2}^{\infty}\left[(n+2)(n+1)a_{n+2} + (n-1)a_{n-1} - 2a_n\right]x^n = 0.$$

Therefore,

$$2a_2 - 2a_0 = 0,$$
$$6a_3 - 2a_1 = 0,$$
$$(n+2)(n+1)a_{n+2} + (n-1)a_{n-1} - 2a_n = 0, \quad n \geq 2.$$

Hence,

$$a_2 = a_0, \qquad a_3 = \frac{a_1}{3}, \qquad \text{and} \qquad a_{n+2} = \frac{2a_n - (n-1)a_{n-1}}{(n+2)(n+1)}, \quad n \geq 2,$$

and

$$
\begin{aligned}
y(x) &= a_0 + a_1 x + a_2 x^2 + a_3 x^3 + \cdots = a_0 + a_1 x + a_0 x^2 + \frac{a_1}{3}x^3 + \cdots \\
&= a_0\left(1 + x^2 + \cdots\right) + a_1\left(x + \frac{x^3}{3} + \cdots\right).
\end{aligned}
$$

5. Clearly, $x = 2$ is an ordinary point for the given equation because $p(x) = x - 2$ and $q(x) = -1$ are analytic everywhere. Thus, we seek for a solution of the form

$$w(x) = \sum_{k=0}^{\infty} a_k (x-2)^k.$$

Differentiating this power series yields

$$w'(x) = \sum_{k=1}^{\infty} k a_k (x-2)^{k-1},$$

$$w''(x) = \sum_{k=2}^{\infty} k(k-1) a_k (x-2)^{k-2}.$$

 Copyright © 2012 Pearson Education, Inc. Publishing as Addison-Wesley.

Therefore,

$$w'' + (x-2)w' - w = \sum_{k=2}^{\infty} k(k-1)a_k(x-2)^{k-2} + (x-2)\sum_{k=1}^{\infty} ka_k(x-2)^{k-1}$$

$$- \sum_{k=0}^{\infty} a_k(x-2)^k = 0$$

$$\Rightarrow \quad \sum_{k=2}^{\infty} k(k-1)a_k(x-2)^{k-2} + \sum_{k=1}^{\infty} ka_k(x-2)^k - \sum_{k=0}^{\infty} a_k(x-2)^k = 0.$$

Shifting the index of summation in the first sum yields

$$\sum_{n=0}^{\infty} (n+2)(n+1)a_{n+2}(x-2)^n + \sum_{n=1}^{\infty} na_n(x-2)^n - \sum_{n=0}^{\infty} a_n(x-2)^n = 0$$

$$\Rightarrow \quad \left[2a_2 + \sum_{n=1}^{\infty} (n+2)(n+1)a_{n+2}(x-2)^n \right]$$

$$+ \sum_{n=1}^{\infty} na_n(x-2)^n - \left[a_0 + \sum_{n=1}^{\infty} a_n(x-2)^n \right] = 0$$

$$\Rightarrow \quad (2a_2 - a_0) + \sum_{n=1}^{\infty} \left[(n+2)(n+1)a_{n+2} + (n-1)a_n \right](x-2)^n = 0,$$

where we have separated the terms corresponding to $n = 0$ and collected the rest under one summation. In order that the above power series equals zero, it must have all zero coefficients. Thus,

$$2a_2 - a_0 = 0,$$
$$(n+2)(n+1)a_{n+2} + (n-1)a_n = 0, \quad n \geq 1$$

$$\Rightarrow \quad a_2 = a_0/2,$$
$$a_{n+2} = (1-n)a_n/[(n+2)(n+1)], \quad n \geq 1.$$

For $n = 1$ and $n = 2$, the last equation gives $a_3 = 0$ and $a_4 = -a_2/12 = -a_0/24$. Therefore,

$$
\begin{aligned}
w(x) &= a_0 + a_1(x-2) + a_2(x-2)^2 + a_3(x-2)^3 + a_4(x-2)^4 + \cdots \\
&= a_0 + a_1(x-2) + \frac{a_0}{2}(x-2)^2 + (0)(x-2)^3 - \frac{1}{24}a_0(x-2)^4 + \cdots \\
&= a_0 \left[1 + \frac{(x-2)^2}{2} - \frac{(x-2)^4}{24} + \cdots \right] + a_1(x-2).
\end{aligned}
$$

7. **(a)** The point $x = 0$ is a regular singular point for the given equation because

$$p(x) = \frac{-5x}{x^2} = -\frac{5}{x}, \qquad q(x) = \frac{9-x}{x^2},$$

Copyright © 2012 Pearson Education, Inc. Publishing as Addison-Wesley.

and the limits

$$p_0 = \lim_{x \to 0} xp(x) = \lim_{x \to 0}(-5) = -5,$$

$$q_0 = \lim_{x \to 0} x^2 q(x) = \lim_{x \to 0}(9 - x) = 9$$

exist. The indicial equation (3), Section 8.6, becomes

$$r(r - 1) + (-5)r + 9 = 0 \quad \Rightarrow \quad r^2 - 6r + 9 = 0 \quad \Rightarrow \quad (r - 3)^2 = 0.$$

Hence, $r = 3$ is the exponent of the singularity $x = 0$, and a solution to the given differential equation has the form

$$y = x^3 \sum_{k=0}^{\infty} a_k x^k = \sum_{k=0}^{\infty} a_k x^{k+3}.$$

Substituting this power series into the given equation yields

$$x^2 \left(\sum_{k=0}^{\infty} a_k x^{k+3} \right)'' - 5x \left(\sum_{k=0}^{\infty} a_k x^{k+3} \right)' + (9 - x) \left(\sum_{k=0}^{\infty} a_k x^{k+3} \right) = 0$$

$$\Rightarrow \quad \sum_{k=0}^{\infty} (k+3)(k+2) a_k x^{k+3} - \sum_{k=0}^{\infty} 5(k+3) a_k x^{k+3} + (9 - x) \sum_{k=0}^{\infty} a_k x^{k+3} = 0$$

$$\Rightarrow \quad \sum_{k=0}^{\infty} \left[(k+3)(k+2) - 5(k+3) + 9 \right] a_k x^{k+3} - \sum_{k=0}^{\infty} a_k x^{k+4} = 0$$

$$\Rightarrow \quad \sum_{k=0}^{\infty} k^2 a_k x^{k+3} - \sum_{k=0}^{\infty} a_k x^{k+4} = 0$$

$$\Rightarrow \quad \sum_{n=1}^{\infty} n^2 a_n x^{n+3} - \sum_{n=1}^{\infty} a_{n-1} x^{n+3} = 0 \quad \Rightarrow \quad \sum_{n=1}^{\infty} \left(n^2 a_n - a_{n-1} \right) x^{n+3} = 0.$$

Thus,

$$n^2 a_n - a_{n-1} = 0 \quad \text{or} \quad a_n = \frac{a_{n-1}}{n^2}, \quad n \geq 1.$$

This recurrence relation yields

$$n = 1 : \quad a_1 = a_0 / (1)^2 = a_0,$$

$$n = 2 : \quad a_2 = a_1 / (2)^2 = a_0 / 4,$$

$$n = 3 : \quad a_3 = a_2 / (3)^2 = (a_0 / 4) / 9 = a_0 / 36.$$

Therefore,

$$\begin{aligned} y(x) &= x^3 \left(a_0 + a_1 x + a_2 x^2 + a_3 x^3 + \cdots \right) \\ &= x^3 \left(a_0 + a_0 x + \frac{a_0}{4} x^2 + \frac{a_0}{36} x^3 + \cdots \right) = a_0 \left(x^3 + x^4 + \frac{x^5}{4} + \frac{x^6}{36} + \cdots \right). \end{aligned}$$

Copyright © 2012 Pearson Education, Inc. Publishing as Addison-Wesley.

CHAPTER 9: Matrix Methods for Linear Systems

EXERCISES 9.1: Introduction

3. We start by expressing right-hand sides of all equations as dot products.

$$x + y + z = [1, 1, 1] \cdot [x, y, z], \quad 2z - x = [-1, 0, 2] \cdot [x, y, z], \quad 4y = [0, 4, 0] \cdot [x, y, z].$$

Thus, by the definition of the product of a matrix and a column vector, the matrix form of the given system is

$$\begin{bmatrix} x \\ y \\ z \end{bmatrix}' = \begin{bmatrix} 1 & 1 & 1 \\ -1 & 0 & 2 \\ 0 & 4 & 0 \end{bmatrix} \begin{bmatrix} x \\ y \\ z \end{bmatrix}.$$

7. First, we have to express the second derivative y'' in terms of the first derivative in order to rewrite the given equation as a first order system. Denoting y' by v, we get $y'' = v'$, and so

$$\begin{aligned} y' &= v, \\ mv' + bv + ky &= 0 \end{aligned} \quad \text{or} \quad \begin{aligned} y' &= v, \\ v' &= -\frac{k}{m}y - \frac{b}{m}v. \end{aligned}$$

Expressing the right-hand side of each equation as a dot product, we obtain

$$v = [0, 1] \cdot [y, v], \qquad -\frac{k}{m}y - \frac{b}{m}v = \left[-\frac{k}{m}, -\frac{b}{m}\right] \cdot [y, v].$$

Thus, the matrix form of the system is

$$\begin{bmatrix} y \\ v \end{bmatrix}' = \begin{bmatrix} 0 & 1 \\ -k/m & -b/m \end{bmatrix} \begin{bmatrix} y \\ v \end{bmatrix}.$$

11. Introducing auxiliary variables

$$x_1 = x, \quad x_2 = x', \quad x_3 = y, \quad x_4 = y',$$

we rewrite the given system in normal form.

$$\begin{aligned} x_1' &= x_2, \\ x_2' + 3x_1 + 2x_3 &= 0, \\ x_3' &= x_4, \\ x_4' - 2x_1 &= 0 \end{aligned} \quad \text{or} \quad \begin{aligned} x_1' &= x_2, \\ x_2' &= -3x_1 - 2x_3, \\ x_3' &= x_4, \\ x_4' &= 2x_1. \end{aligned}$$

Copyright © 2012 Pearson Education, Inc. Publishing as Addison-Wesley.

Since

$$x_2 = [0, 1, 0, 0] \cdot [x_1, x_2, x_3, x_4], \qquad -3x_1 - 2x_3 = [-3, 0, -2, 0] \cdot [x_1, x_2, x_3, x_4],$$
$$x_4 = [0, 0, 0, 1] \cdot [x_1, x_2, x_3, x_4], \qquad 2x_1 = [2, 0, 0, 0] \cdot [x_1, x_2, x_3, x_4],$$

the matrix is given by

$$
\begin{bmatrix} x_1 \\ x_2 \\ x_3 \\ x_4 \end{bmatrix}' =
\begin{bmatrix} 0 & 1 & 0 & 0 \\ -3 & 0 & -2 & 0 \\ 0 & 0 & 0 & 1 \\ 2 & 0 & 0 & 0 \end{bmatrix}
\begin{bmatrix} x_1 \\ x_2 \\ x_3 \\ x_4 \end{bmatrix}.
$$

EXERCISES 9.2: Review 1: Linear Algebraic Equations

3. Adding the first equation to the second one and the subtracting the first equation from the third equation, we see that the last to equations become $0 = 0$. Thus, what remains is

$$x_1 + x_2 - x_3 = 0.$$

Letting now $x_2 = s$, $x_3 = t$, we conclude that $x_1 = -s + t$, $(-\infty < s, t < \infty)$.

7. Subtracting the first equation, multiplied by 3, from the second equation yields

$$-x_1 + 3x_2 = 0,$$
$$0 = 0.$$

The last equation is trivially satisfied, so we ignore it. Thus, just one equation remains.

$$-x_1 + 3x_2 = 0 \qquad \Rightarrow \qquad x_1 = 3x_2.$$

Choosing $x_2 = s$ as a free variable, we get $x_1 = 3s$, where s is any number.

9. We eliminate x_1 from the first equation by adding $(1 - i)$ times the second equation to it.

$$[2 - (1 + i)(1 - i)]x_2 = 0,$$
$$-x_1 - (1 + i)x_2 = 0.$$

Since $(1 - i)(1 + i) = 1^2 - i^2 = 1 - (-1) = 2$, we obtain

$$
\begin{aligned}
0 &= 0, \\
-x_1 - (1 + i)x_2 &= 0
\end{aligned}
\qquad \Rightarrow \qquad
x_2 = -\frac{1}{1 + i} x_1 = \frac{-1 + i}{2} x_1.
$$

 Copyright © 2012 Pearson Education, Inc. Publishing as Addison-Wesley.

Assigning an arbitrary complex value to x_1, say $2s$, we see that the system has infinitely many solutions given by

$$x_1 = 2s, \qquad x_2 = (-1+i)s, \qquad \text{where } s \text{ is any complex number.}$$

11. It is slightly more convenient to switch the order of the equations.

$$
\begin{aligned}
-x_1 + x_2 + 5x_3 &= 0, \\
2x_1 + x_3 &= -1, \\
-3x_1 + x_2 + 4x_3 &= 1.
\end{aligned}
$$

We now eliminate x_1 from the second equation by adding twice the first one to it; and by subtracting the first equation, multiplied by 3, from the third one, we eliminate x_1.

$$
\begin{aligned}
-x_1 + x_2 + 5x_3 &= 0, \\
-2x_2 - 11x_3 &= 1, \\
2x_2 + 11x_3 &= -1.
\end{aligned}
$$

To make our computations more convenient, we multiply the first equation by 2.

$$
\begin{aligned}
-2x_1 + 2x_2 + 10x_3 &= 0, \\
-2x_2 - 11x_3 &= 1, \\
2x_2 + 11x_3 &= -1.
\end{aligned}
$$

Now we add the second equation to the others and obtain

$$
\begin{aligned}
-2x_1 - x_3 &= 1, \\
-2x_2 - 11x_3 &= 1, \qquad \text{or} \qquad
\begin{aligned}
-2x_1 - x_3 &= 1, \\
-2x_2 - 11x_3 &= 1.
\end{aligned} \\
0 &= 0
\end{aligned}
$$

Choosing x_3 as a free variable, i.e., $x_3 = s$, yields $x_1 = -(s+1)/2$, $x_2 = -(11s+1)/2$, $-\infty < s < \infty$.

13. The given system can be written in an equivalent form

$$
\begin{aligned}
(2-r)x_1 - 3x_2 &= 0, \\
x_1 - (2+r)x_2 &= 0.
\end{aligned}
$$

The variable x_1 can be eliminated from the first equation by subtracting $(2-r)$ times the second equation.

$$
\begin{aligned}
[-3 + (2-r)(2+r)]x_2 &= 0, \\
x_1 - (2+r)x_2 &= 0
\end{aligned}
\qquad \text{or} \qquad
\begin{aligned}
(1-r^2)\,x_2 &= 0, \\
x_1 - (2+r)x_2 &= 0.
\end{aligned}
\qquad (9.1)
$$

Copyright © 2012 Pearson Education, Inc. Publishing as Addison-Wesley.

If $1 - r^2 \neq 0$, i.e., $r \neq \pm 1$, then the first equation implies $x_2 = 0$. Substituting this into the second equation, we get x_1. Thus, the given system has a unique (zero) solution for any $r \neq \pm 1$, in particular, for $r = 2$.

If $r = 1$ or $r = -1$, then the first equation in (9.1) becomes trivial, $0 = 0$, and the system degenerates to

$$x_1 - (2 + r)x_2 = 0 \qquad \Rightarrow \qquad x_1 = (2 + r)x_2 \,.$$

Therefore, there are infinitely many solutions to the given system of the form

$$x_1 = (2 + r)s, \qquad x_2 = s, \qquad -\infty < s < \infty, \qquad r = \pm 1 \,.$$

In particular, for $r = 1$ we obtain

$$x_1 = 3s, \qquad x_2 = s, \qquad -\infty < s < \infty.$$

EXERCISES 9.3: Review 2: Matrices and Vectors

5. (a) $\mathbf{AB} = \begin{bmatrix} 1 & -2 \\ 2 & -3 \end{bmatrix} \begin{bmatrix} 1 & 0 \\ 1 & 1 \end{bmatrix} = \begin{bmatrix} 1-2 & 0-2 \\ 2-3 & 0-3 \end{bmatrix} = \begin{bmatrix} -1 & -2 \\ -1 & -3 \end{bmatrix}.$

   (b) $\mathbf{AC} = \begin{bmatrix} 1 & -2 \\ 2 & -3 \end{bmatrix} \begin{bmatrix} -1 & 1 \\ 2 & 1 \end{bmatrix} = \begin{bmatrix} -1-4 & 1-2 \\ -2-6 & 2-3 \end{bmatrix} = \begin{bmatrix} -5 & -1 \\ -8 & -1 \end{bmatrix}.$

   (c) By the distributive property of matrix multiplication, we have

   $$\mathbf{A\,(B + C)} = \mathbf{AB} + \mathbf{AC} = \begin{bmatrix} -1 & -2 \\ -1 & -3 \end{bmatrix} + \begin{bmatrix} -5 & -1 \\ -8 & -1 \end{bmatrix} = \begin{bmatrix} -6 & -3 \\ -9 & -4 \end{bmatrix}.$$

13. Author's note: We will use the abbreviation $R_i + cR_j \rightarrow R_k$ to denote the following row operation: "*add the row i to c times the row j and place the result into the row k.*" Similarly, $cR_j \rightarrow R_k$ will mean: "*multiply the row j by c and place the result into the row k.*"

    As in Example 1, we perform the row-reduction procedure on the matrix $[\mathbf{A}|\mathbf{I}]$. Thus, we have

    $$[\mathbf{A}|\mathbf{I}] = \begin{bmatrix} -2 & -1 & 1 & | & 1 & 0 & 0 \\ 2 & 1 & 0 & | & 0 & 1 & 0 \\ 3 & 1 & -1 & | & 0 & 0 & 1 \end{bmatrix}$$

 Copyright © 2012 Pearson Education, Inc. Publishing as Addison-Wesley.

$$
\begin{array}{c}
R_2 + R_1 \to R_2 \\
2R_3 + 3R_1 \to R_3
\end{array}
\qquad
\left[\begin{array}{ccc|ccc}
-2 & -1 & 1 & 1 & 0 & 0 \\
0 & 0 & 1 & 1 & 1 & 0 \\
0 & -1 & 1 & 3 & 0 & 2
\end{array}\right]
$$

$$
R_1 - R_3 \to R_1
\qquad
\left[\begin{array}{ccc|ccc}
-2 & 0 & 0 & -2 & 0 & -2 \\
0 & 0 & 1 & 1 & 1 & 0 \\
0 & -1 & 1 & 3 & 0 & 2
\end{array}\right]
$$

$$
\begin{array}{c}
-R_1/2 \to R_1 \\
R_3 \to R_2 \\
R_2 \to R_3
\end{array}
\qquad
\left[\begin{array}{ccc|ccc}
1 & 0 & 0 & 1 & 0 & 1 \\
0 & -1 & 1 & 3 & 0 & 2 \\
0 & 0 & 1 & 1 & 1 & 0
\end{array}\right]
$$

$$
-R_2 + R_3 \to R_2
\qquad
\left[\begin{array}{ccc|ccc}
1 & 0 & 0 & 1 & 0 & 1 \\
0 & 1 & 0 & -2 & 1 & -2 \\
0 & 0 & 1 & 1 & 1 & 0
\end{array}\right].
$$

Therefore, the inverse matrix is

$$
\mathbf{A}^{-1} = \left[\begin{array}{ccc}
1 & 0 & 1 \\
-2 & 1 & -2 \\
1 & 1 & 0
\end{array}\right].
$$

19. To find the inverse matrix $\mathbf{X}^{-1}(t)$, we use the method of Example 1 again. Thus, we start with

$$
[\mathbf{X}(t)|\mathbf{I}] = \left[\begin{array}{ccc|ccc}
e^t & e^{-t} & e^{2t} & 1 & 0 & 0 \\
e^t & -e^{-t} & 2e^{2t} & 0 & 1 & 0 \\
e^t & e^{-t} & 4e^{2t} & 0 & 0 & 1
\end{array}\right]
$$

$$
\begin{array}{c}
R_2 - R_1 \to R_2 \\
R_3 - R_1 \to R_3
\end{array}
\qquad
\left[\begin{array}{ccc|ccc}
e^t & e^{-t} & e^{2t} & 1 & 0 & 0 \\
0 & -2e^{-t} & e^{2t} & -1 & 1 & 0 \\
0 & 0 & 3e^{2t} & -1 & 0 & 1
\end{array}\right]
$$

$$
\begin{array}{c}
-R_2/2 \to R_2 \\
R_3/3 \to R_3
\end{array}
\qquad
\left[\begin{array}{ccc|ccc}
e^t & e^{-t} & e^{2t} & 1 & 0 & 0 \\
0 & e^{-t} & -e^{2t}/2 & 1/2 & -1/2 & 0 \\
0 & 0 & e^{2t} & -1/3 & 0 & 1/3
\end{array}\right]
$$

$$
\begin{array}{c}
R_1 - R_3 \to R_1 \\
R_2 - R_3/2 \to R_2
\end{array}
\qquad
\left[\begin{array}{ccc|ccc}
e^t & e^{-t} & 0 & 4/3 & 0 & -1/3 \\
0 & e^{-t} & 0 & 1/3 & -1/2 & 1/6 \\
0 & 0 & e^{2t} & -1/3 & 0 & 1/3
\end{array}\right]
$$

Copyright © 2012 Pearson Education, Inc. Publishing as Addison-Wesley.

$$R_1 - R_2 \to R_1 \qquad \left[\begin{array}{ccc|ccc} e^t & 0 & 0 & 1 & 1/2 & -1/2 \\ 0 & e^{-t} & 0 & 1/3 & -1/2 & 1/6 \\ 0 & 0 & e^{2t} & -1/3 & 0 & 1/3 \end{array}\right]$$

$$\begin{array}{c} e^{-t}R_1 \to R_1 \\ e^{t}R_2 \to R_2 \\ e^{-2t}R_3 \to R_3 \end{array} \qquad \left[\begin{array}{ccc|ccc} 1 & 0 & 0 & e^{-t} & e^{-t}/2 & -e^{-t}/2 \\ 0 & 1 & 0 & e^{t}/3 & -e^{t}/2 & e^{t}/6 \\ 0 & 0 & 1 & -e^{-2t}/3 & 0 & e^{-2t}/3 \end{array}\right].$$

Therefore, the inverse matrix $\mathbf{X}^{-1}(t)$ is given by

$$\mathbf{X}^{-1}(t) = \begin{bmatrix} e^{-t} & (1/2)e^{-t} & -(1/2)e^{-t} \\ (1/3)e^{t} & -(1/2)e^{t} & (1/6)e^{t} \\ -(1/3)e^{-2t} & 0 & (1/3)e^{-2t} \end{bmatrix}.$$

23. We calculate this determinant by finding its cofactor expansion about the first row. Therefore, we have

$$\begin{vmatrix} 1 & 0 & 0 \\ 3 & 1 & 2 \\ 1 & 5 & -2 \end{vmatrix} = (1)\begin{vmatrix} 1 & 2 \\ 5 & -2 \end{vmatrix} - 0 + 0 = -2 - 10 = -12.$$

37. We first calculate $\mathbf{X}'(t)$ by differentiating each entry of $\mathbf{X}(t)$. Thus, we obtain

$$\mathbf{X}'(t) = \begin{bmatrix} 2e^{2t} & 3e^{3t} \\ -2e^{2t} & -6e^{3t} \end{bmatrix}.$$

Substituting the matrix $\mathbf{X}(t)$ into the given differential equation and performing matrix multiplication yields

$$\begin{bmatrix} 1 & -1 \\ 2 & 4 \end{bmatrix}\mathbf{X} = \begin{bmatrix} 1 & -1 \\ 2 & 4 \end{bmatrix}\begin{bmatrix} e^{2t} & e^{3t} \\ -e^{2t} & -2e^{3t} \end{bmatrix}$$

$$= \begin{bmatrix} e^{2t}+e^{2t} & e^{3t}+2e^{3t} \\ 2e^{2t}-4e^{2t} & 2e^{3t}-8e^{3t} \end{bmatrix} = \begin{bmatrix} 2e^{2t} & 3e^{3t} \\ -2e^{2t} & -6e^{3t} \end{bmatrix} = \mathbf{X}'.$$

So, we see that $\mathbf{X}(t)$ does satisfy the given differential equation.

39. (a) To calculate $\int \mathbf{A}(t)\,dt$, we integrate each entry of $\mathbf{A}(t)$ to obtain

$$\int \mathbf{A}(t)\,dt = \begin{bmatrix} \int t\,dt & \int e^t\,dt \\ \int 1\,dt & \int e^t\,dt \end{bmatrix} = \begin{bmatrix} t^2/2 + c_1 & e^t + c_2 \\ t + c_3 & e^t + c_4 \end{bmatrix}.$$

 Copyright © 2012 Pearson Education, Inc. Publishing as Addison-Wesley.

(b) Taking the definite integral of each entry in $\mathbf{B}(t)$ yields

$$\int_0^1 \mathbf{B}(t)\,dt = \begin{bmatrix} \int_0^1 \cos t\,dt & -\int_0^1 \sin t\,dt \\ \int_0^1 \sin t\,dt & \int_0^1 \cos t\,dt \end{bmatrix}$$

$$= \begin{bmatrix} \sin t\,\big|_0^1 & \cos t\,\big|_0^1 \\ -\cos t\,\big|_0^1 & \sin t\,\big|_0^1 \end{bmatrix} = \begin{bmatrix} \sin 1 & \cos 1 - 1 \\ 1 - \cos 1 & \sin 1 \end{bmatrix}.$$

(c) By the product rule, we see that

$$\frac{d}{dt}\,[\mathbf{A}(t)\mathbf{B}(t)] = \mathbf{A}(t)\mathbf{B}'(t) + \mathbf{A}'(t)\mathbf{B}(t).$$

Therefore, we first calculate $\mathbf{A}'(t)$ and $\mathbf{B}'(t)$ by differentiating each entry of $\mathbf{A}(t)$ and $\mathbf{B}(t)$, respectively, to obtain

$$\mathbf{A}'(t) = \begin{bmatrix} 1 & e^t \\ 0 & e^t \end{bmatrix} \quad \text{and} \quad \mathbf{B}'(t) = \begin{bmatrix} -\sin t & -\cos t \\ \cos t & -\sin t \end{bmatrix}.$$

Hence, by matrix multiplication we have

$$\frac{d}{dt}\,[\mathbf{A}(t)\mathbf{B}(t)] = \mathbf{A}(t)\mathbf{B}'(t) + \mathbf{A}'(t)\mathbf{B}(t)$$

$$= \begin{bmatrix} t & e^t \\ 1 & e^t \end{bmatrix} \begin{bmatrix} -\sin t & -\cos t \\ \cos t & -\sin t \end{bmatrix} + \begin{bmatrix} 1 & e^t \\ 0 & e^t \end{bmatrix} \begin{bmatrix} \cos t & -\sin t \\ \sin t & \cos t \end{bmatrix}$$

$$= \begin{bmatrix} e^t\cos t - t\sin t & -t\cos t - e^t\sin t \\ e^t\cos t - \sin t & -\cos t - e^t\sin t \end{bmatrix} + \begin{bmatrix} \cos t + e^t\sin t & e^t\cos t - \sin t \\ e^t\sin t & e^t\cos t \end{bmatrix}$$

$$= \begin{bmatrix} (1+e^t)\cos t + (e^t - t)\sin t & (e^t - t)\cos t - (e^t + 1)\sin t \\ e^t\cos t + (e^t - 1)\sin t & (e^t - 1)\cos t - e^t\sin t \end{bmatrix}.$$

Thus,

$$\frac{d}{dt}\,[\mathbf{A}(t)\mathbf{B}(t)] = \begin{bmatrix} (1+e^t)\cos t + (e^t - t)\sin t & (e^t - t)\cos t - (e^t + 1)\sin t \\ e^t\cos t + (e^t - 1)\sin t & (e^t - 1)\cos t - e^t\sin t \end{bmatrix}.$$

One can also get this answer by differentiating term-wise the product

$$\mathbf{A}(t)\mathbf{B}(t) = \begin{bmatrix} t\cos t + e^t\sin t & e^t\cos t - t\sin t \\ \cos t + e^t\sin t & e^t\cos t - \sin t \end{bmatrix}.$$

Copyright © 2012 Pearson Education, Inc. Publishing as Addison-Wesley.

Chapter 9

EXERCISES 9.4: Linear Systems in Normal Form

1. To rewrite this system in matrix form, we define the vectors $\mathbf{x}(t) = \text{col}[x(t), y(t)]$ (which gives $\mathbf{x}'(t) = \text{col}[x'(t), y'(t)]$), $\mathbf{f}(t) = \text{col}[t^2, e^t]$, and the matrix

$$\mathbf{A}(t) = \begin{bmatrix} 3 & -1 \\ -1 & 2 \end{bmatrix}.$$

Thus, this system becomes a matrix differential equation

$$\begin{bmatrix} x(t) \\ y(t) \end{bmatrix}' = \begin{bmatrix} 3 & -1 \\ -1 & 2 \end{bmatrix} \begin{bmatrix} x(t) \\ y(t) \end{bmatrix} + \begin{bmatrix} t^2 \\ e^t \end{bmatrix}.$$

We can verify that this equation is equivalent to the original system by performing matrix multiplication and addition to obtain

$$\begin{bmatrix} x'(t) \\ y'(t) \end{bmatrix} = \begin{bmatrix} 3x(t) - y(t) \\ -x(t) + 2y(t) \end{bmatrix} + \begin{bmatrix} t^2 \\ e^t \end{bmatrix} = \begin{bmatrix} 3x(t) - y(t) + t^2 \\ -x(t) + 2y(t) + e^t \end{bmatrix}.$$

Since two vectors are equal if and only if their corresponding components are equal, we see that this vector equation is equivalent to

$$x'(t) = 3x(t) - y(t) + t^2,$$
$$y'(t) = -x(t) + 2y(t) + e^t,$$

which is the original system.

5. This equation can be written as a first order system in normal form by using the substitutions $x_1(t) = y(t)$ and $x_2(t) = y'(t)$. With these substitutions, we have

$$\begin{aligned} x_1'(t) &= 0 \cdot x_1(t) + x_2(t), \\ x_2'(t) &= 10x_1(t) + 3x_2(t) + \sin t. \end{aligned} \tag{9.2}$$

Let $\mathbf{x}(t) = \text{col}[x_1(t), x_2(t)]$ (which means that $\mathbf{x}'(t) = \text{col}[x_1'(t), x_2'(t)]$). We now write (9.2) as a matrix differential equation using the vector $\mathbf{f}(t) = \text{col}[0, \sin t]$ and the matrix

$$\mathbf{A} = \begin{bmatrix} 0 & 1 \\ 10 & 3 \end{bmatrix}.$$

Hence, the system (9.2) in normal form yields a matrix differential equation

$$\begin{bmatrix} x_1'(t) \\ x_2'(t) \end{bmatrix} = \begin{bmatrix} 0 & 1 \\ 10 & 3 \end{bmatrix} \begin{bmatrix} x_1(t) \\ x_2(t) \end{bmatrix} + \begin{bmatrix} 0 \\ \sin t \end{bmatrix}.$$

 Copyright © 2012 Pearson Education, Inc. Publishing as Addison-Wesley.

(As in Problem 1 above, we can see that this equation in matrix form is equivalent to the given system by performing matrix multiplication and addition and then noting that corresponding entries of equal vectors must be equal.)

7. This equation can be written as a first order system in normal form by using the substitutions $x_1(t) = w(t)$, $x_2(t) = w'(t)$, $x_3(t) = w''(t)$, and $x_4(t) = w'''(t)$. These substitutions give

$$x_1'(t) = 0 \cdot x_1(t) + x_2(t) + 0 \cdot x_3(t) + 0 \cdot x_4(t),$$
$$x_2'(t) = 0 \cdot x_1(t) + 0 \cdot x_2(t) + x_3(t) + 0 \cdot x_4(t),$$
$$x_3'(t) = 0 \cdot x_1(t) + 0 \cdot x_2(t) + 0 \cdot x_3(t) + x_4(t),$$
$$x_4'(t) = -x_1(t) + 0 \cdot x_2(t) + 0 \cdot x_3(t) + 0 \cdot x_4(t) + t^2 \,.$$

We now rewrite this system as a matrix differential equation, i.e., $\mathbf{x}' = \mathbf{A}\mathbf{x}$, by defining vectors $\mathbf{x}(t) = \mathrm{col}[x_1(t), x_2(t), x_3(t), x_4(t)]$ (so that $\mathbf{x}'(t) = \mathrm{col}[x_1'(t), x_2'(t), x_3'(t), x_4'(t)]$) and $\mathbf{f}(t) = \mathrm{col}[0, 0, 0, t^2]$, and the matrix

$$\mathbf{A} = \begin{bmatrix} 0 & 1 & 0 & 0 \\ 0 & 0 & 1 & 0 \\ 0 & 0 & 0 & 1 \\ -1 & 0 & 0 & 0 \end{bmatrix}.$$

Thus, the given fourth order differential equation is equivalent to the matrix system

$$\begin{bmatrix} x_1'(t) \\ x_2'(t) \\ x_3'(t) \\ x_4'(t) \end{bmatrix} = \begin{bmatrix} 0 & 1 & 0 & 0 \\ 0 & 0 & 1 & 0 \\ 0 & 0 & 0 & 1 \\ -1 & 0 & 0 & 0 \end{bmatrix} \begin{bmatrix} x_1(t) \\ x_2(t) \\ x_3(t) \\ x_4(t) \end{bmatrix} + \begin{bmatrix} 0 \\ 0 \\ 0 \\ t^2 \end{bmatrix}.$$

17. Notice that, using scalar multiplication, given vector functions can be written as

$$\begin{bmatrix} e^{2t} \\ 0 \\ 5e^{2t} \end{bmatrix}, \quad \begin{bmatrix} e^{2t} \\ e^{2t} \\ -e^{2t} \end{bmatrix}, \quad \begin{bmatrix} 0 \\ e^{3t} \\ 0 \end{bmatrix}.$$

Thus, as in Example 2 in the text, we will prove that these vectors are linearly independent by showing that the only way to get

$$c_1 \begin{bmatrix} e^{2t} \\ 0 \\ 5e^{2t} \end{bmatrix} + c_2 \begin{bmatrix} e^{2t} \\ e^{2t} \\ -e^{2t} \end{bmatrix} + c_3 \begin{bmatrix} 0 \\ e^{3t} \\ 0 \end{bmatrix} = \mathbf{0}$$

Copyright © 2012 Pearson Education, Inc. Publishing as Addison-Wesley.

for all t in $(-\infty, \infty)$ is to have $c_1 = c_2 = c_3 = 0$. Since the equation above must be true for all t, it must be true for $t = 0$. Thus, c_1, c_2, and c_3 must satisfy

$$c_1 \begin{bmatrix} 1 \\ 0 \\ 5 \end{bmatrix} + c_2 \begin{bmatrix} 1 \\ 1 \\ -1 \end{bmatrix} + c_3 \begin{bmatrix} 0 \\ 1 \\ 0 \end{bmatrix} = \mathbf{0},$$

which is equivalent to the system

$$c_1 + c_2 = 0,$$
$$c_2 + c_3 = 0,$$
$$5c_1 - c_2 = 0.$$

By solving the first and the last equations simultaneously, we see that $c_1 = c_2 = 0$. Substituting these values into the second equation yields $c_3 = 0$. Therefore, the original set of vectors is linearly independent on the interval $(-\infty, \infty)$.

19. Linearly independent. Say, for the first row of their linear combination one has $c_1 + c_2 t + c_3 t^2 = 0$ on $(-\infty, \infty)$ so that it must be a zero polynomial.

23. Since it is given that these vectors are solutions to the system $\mathbf{x}'(t) = \mathbf{A}\mathbf{x}(t)$, in order to determine whether they are linearly independent, we need to calculate their Wronskian. If their Wronskian is never zero, then these vectors are linearly independent, and so form a fundamental solution set. If the Wronskian is identically zero, then the vectors are linearly dependent, and so they do not form a fundamental solution set. Thus, we compute

$$W\left[\mathbf{x}_1, \mathbf{x}_2, \mathbf{x}_3\right](t) = \begin{vmatrix} e^{-t} & e^t & e^{3t} \\ 2e^{-t} & 0 & -e^{3t} \\ e^{-t} & e^t & 2e^{3t} \end{vmatrix}$$

$$= e^{-t} \begin{vmatrix} 0 & -e^{3t} \\ e^t & 2e^{3t} \end{vmatrix} - e^t \begin{vmatrix} 2e^{-t} & -e^{3t} \\ e^{-t} & 2e^{3t} \end{vmatrix} + e^{3t} \begin{vmatrix} 2e^{-t} & 0 \\ e^{-t} & e^t \end{vmatrix}$$

$$= e^{-t}\left(0 + e^{4t}\right) - e^t\left(4e^{2t} + e^{2t}\right) + e^{3t}(2 - 0) = -2e^{3t} \neq 0,$$

where we have used cofactors to calculate the determinant. Therefore, this set of vectors is linearly independent and so forms a fundamental solution set for the given system. A

    Copyright © 2012 Pearson Education, Inc. Publishing as Addison-Wesley.

fundamental matrix is then given by

$$\mathbf{X}(t) = \begin{bmatrix} e^{-t} & e^t & e^{3t} \\ 2e^{-t} & 0 & -e^{3t} \\ e^{-t} & e^t & 2e^{3t} \end{bmatrix},$$

and a general solution of the system is

$$\mathbf{x}(t) = \mathbf{X}(t)\mathbf{c} = c_1 \begin{bmatrix} e^{-t} \\ 2e^{-t} \\ e^{-t} \end{bmatrix} + c_2 \begin{bmatrix} e^t \\ 0 \\ e^t \end{bmatrix} + c_3 \begin{bmatrix} e^{3t} \\ -e^{3t} \\ 2e^{3t} \end{bmatrix}.$$

29. In order to show that $\mathbf{X}(t)$ is a fundamental matrix for the system, we must show first that each of its column vectors is a solution. Thus, we substitute each of the vectors

$$\mathbf{x}_1(t) = \begin{bmatrix} 6e^{-t} \\ -e^{-t} \\ -5e^{-t} \end{bmatrix}, \qquad \mathbf{x}_2(t) = \begin{bmatrix} -3e^{-2t} \\ e^{-2t} \\ e^{-2t} \end{bmatrix}, \qquad \mathbf{x}_3(t) = \begin{bmatrix} 2e^{3t} \\ e^{3t} \\ e^{3t} \end{bmatrix}$$

into the given system and obtain

$$\mathbf{A}\mathbf{x}_1(t) = \begin{bmatrix} 0 & 6 & 0 \\ 1 & 0 & 1 \\ 1 & 1 & 0 \end{bmatrix} \begin{bmatrix} 6e^{-t} \\ -e^{-t} \\ -5e^{-t} \end{bmatrix} = \begin{bmatrix} -6e^{-t} \\ e^{-t} \\ 5e^{-t} \end{bmatrix} = \mathbf{x}_1'(t),$$

$$\mathbf{A}\mathbf{x}_2(t) = \begin{bmatrix} 0 & 6 & 0 \\ 1 & 0 & 1 \\ 1 & 1 & 0 \end{bmatrix} \begin{bmatrix} -3e^{-2t} \\ e^{-2t} \\ e^{-2t} \end{bmatrix} = \begin{bmatrix} 6e^{-2t} \\ -2e^{-2t} \\ -2e^{-2t} \end{bmatrix} = \mathbf{x}_2'(t),$$

$$\mathbf{A}\mathbf{x}_3(t) = \begin{bmatrix} 0 & 6 & 0 \\ 1 & 0 & 1 \\ 1 & 1 & 0 \end{bmatrix} \begin{bmatrix} 2e^{3t} \\ e^{3t} \\ e^{3t} \end{bmatrix} = \begin{bmatrix} 6e^{3t} \\ 3e^{3t} \\ 3e^{3t} \end{bmatrix} = \mathbf{x}_3'(t).$$

Therefore, each column vector of $\mathbf{X}(t)$ is a solution to the given system on $(-\infty, \infty)$.

Next, we show that these vectors are linearly independent. Since they are solutions to a matrix differential equation, it is enough to show that their Wronskian is never zero. We find

$$W(t) = \begin{vmatrix} 6e^{-t} & -3e^{-2t} & 2e^{3t} \\ -e^{-t} & e^{-2t} & e^{3t} \\ -5e^{-t} & e^{-2t} & e^{3t} \end{vmatrix}$$

Copyright © 2012 Pearson Education, Inc. Publishing as Addison-Wesley.

$$= 6e^{-t} \begin{vmatrix} e^{-2t} & e^{3t} \\ e^{-2t} & e^{3t} \end{vmatrix} + 3e^{-2t} \begin{vmatrix} -e^{-t} & e^{3t} \\ -5e^{-t} & e^{3t} \end{vmatrix} + 2e^{3t} \begin{vmatrix} -e^{-t} & e^{-2t} \\ -5e^{-t} & e^{-2t} \end{vmatrix}$$

$$= 6e^{-t} \left(e^{t} - e^{t} \right) + 3e^{-2t} \left(-e^{2t} + 5e^{2t} \right) + 2e^{3t} \left(-e^{-3t} + 5e^{-3t} \right) = 20 \neq 0,$$

where we have used cofactors to calculate the determinant. Hence, these three vectors are linearly independent. Therefore, $\mathbf{X}(t)$ is a fundamental matrix for this system.

We now find the inverse of the matrix $\mathbf{X}(t)$ by performing a row-reduction on $[\mathbf{X}(t)|\mathbf{I}]$ to get $[\mathbf{I}|\mathbf{X}^{-1}(t)]$. (Alternatively, one can use other methods, such as the formula from linear algebra). Thus, we have

$$[\mathbf{X}(t)|\mathbf{I}] = \left[\begin{array}{ccc|ccc} 6e^{-t} & -3e^{-2t} & 2e^{3t} & 1 & 0 & 0 \\ -e^{-t} & e^{-2t} & e^{3t} & 0 & 1 & 0 \\ -5e^{-t} & e^{-2t} & e^{3t} & 0 & 0 & 1 \end{array} \right]$$

$$\begin{array}{c} -R_2 \to R_1 \\ R_1 \to R_2 \end{array} \left[\begin{array}{ccc|ccc} e^{-t} & -e^{-2t} & -e^{3t} & 0 & -1 & 0 \\ 6e^{-t} & -3e^{-2t} & 2e^{3t} & 1 & 0 & 0 \\ -5e^{-t} & e^{-2t} & e^{3t} & 0 & 0 & 1 \end{array} \right]$$

$$\begin{array}{c} R_2 - 6R_1 \to R_2 \\ R_3 + 5R_1 \to R_3 \end{array} \left[\begin{array}{ccc|ccc} e^{-t} & -e^{-2t} & -e^{3t} & 0 & -1 & 0 \\ 0 & 3e^{-2t} & 8e^{3t} & 1 & 6 & 0 \\ 0 & -4e^{-2t} & -4e^{3t} & 0 & -5 & 1 \end{array} \right]$$

$$\begin{array}{c} -R_3/4 \to R_2 \\ R_2 \to R_3 \end{array} \left[\begin{array}{ccc|ccc} e^{-t} & -e^{-2t} & -e^{3t} & 0 & -1 & 0 \\ 0 & e^{-2t} & e^{3t} & 0 & 5/4 & -1/4 \\ 0 & 3e^{-2t} & 8e^{3t} & 1 & 6 & 0 \end{array} \right]$$

$$\begin{array}{c} R_1 + R_2 \to R_1 \\ R_3 - 3R_2 \to R_3 \end{array} \left[\begin{array}{ccc|ccc} e^{-t} & 0 & 0 & 0 & 1/4 & -1/4 \\ 0 & e^{-2t} & e^{3t} & 0 & 5/4 & -1/4 \\ 0 & 0 & 5e^{3t} & 1 & 9/4 & 3/4 \end{array} \right]$$

$$\frac{1}{5} R_3 \to R_3 \left[\begin{array}{ccc|ccc} e^{-t} & 0 & 0 & 0 & 1/4 & -1/4 \\ 0 & e^{-2t} & e^{3t} & 0 & 5/4 & -1/4 \\ 0 & 0 & e^{3t} & 1/5 & 9/20 & 3/20 \end{array} \right]$$

$$R_2 - R_3 \to R_2 \left[\begin{array}{ccc|ccc} e^{-t} & 0 & 0 & 0 & 1/4 & -1/4 \\ 0 & e^{-2t} & 0 & -1/5 & 4/5 & -2/5 \\ 0 & 0 & e^{3t} & 1/5 & 9/20 & 3/20 \end{array} \right]$$

 Copyright © 2012 Pearson Education, Inc. Publishing as Addison-Wesley.

$$
\begin{array}{c}
e^{t}R_1 \to R_1 \\
e^{2t}R_2 \to R_2 \\
e^{-3t}R_3 \to R_3
\end{array}
\left[\begin{array}{ccc|ccc}
1 & 0 & 0 & 0 & e^{t}/4 & -e^{t}/4 \\
0 & 1 & 0 & -e^{2t}/5 & 4e^{2t}/5 & -2e^{2t}/5 \\
0 & 0 & 1 & e^{-3t}/5 & 9e^{-3t}/20 & 3e^{-3t}/20
\end{array}\right].
$$

Therefore, we see that

$$
\mathbf{X}^{-1}(t) = \left[\begin{array}{ccc}
0 & (1/4)e^{t} & -(1/4)e^{t} \\
-(1/5)e^{2t} & (4/5)e^{2t} & -(2/5)e^{2t} \\
(1/5)e^{-3t} & (9/20)e^{-3t} & (3/20)e^{-3t}
\end{array}\right].
$$

We can now use Problem 26 to find the solution to this differential equation for *any* initial value. In the given initial value problem $t_0 = 0$. Thus, substituting $t = 0$ into the matrix $\mathbf{X}^{-1}(t)$ yields

$$
\mathbf{X}^{-1}(0) = \left[\begin{array}{ccc}
0 & 1/4 & -1/4 \\
-1/5 & 4/5 & -2/5 \\
1/5 & 9/20 & 3/20
\end{array}\right].
$$

Hence, we see that the solution to this initial value problem is given by

$$
\begin{aligned}
\mathbf{x}(t) &= \mathbf{X}(t)\mathbf{X}^{-1}(0)\mathbf{x}(0) \\
&= \left[\begin{array}{ccc}
6e^{-t} & -3e^{-2t} & 2e^{3t} \\
-e^{-t} & e^{-2t} & e^{3t} \\
-5e^{-t} & e^{-2t} & e^{3t}
\end{array}\right]
\left[\begin{array}{ccc}
0 & 1/4 & -1/4 \\
-1/5 & 4/5 & -2/5 \\
1/5 & 9/20 & 3/20
\end{array}\right]
\left[\begin{array}{c}
-1 \\ 0 \\ 1
\end{array}\right] \\
&= \left[\begin{array}{ccc}
6e^{-t} & -3e^{-2t} & 2e^{3t} \\
-e^{-t} & e^{-2t} & e^{3t} \\
-5e^{-t} & e^{-2t} & e^{3t}
\end{array}\right]
\left[\begin{array}{c}
-1/4 \\ -1/5 \\ -1/20
\end{array}\right] \\
&= \left[\begin{array}{c}
-(3/2)e^{-t} + (3/5)e^{-2t} - (1/10)e^{3t} \\
(1/4)e^{-t} - (1/5)e^{-2t} - (1/20)e^{3t} \\
(5/4)e^{-t} - (1/5)e^{-2t} - (1/20)e^{3t}
\end{array}\right].
\end{aligned}
$$

There are two short cuts that can be used to solve the given problem. First, since we only need $\mathbf{X}^{-1}(0)$, it suffices to compute the inverse of $\mathbf{X}(0)$, not $\mathbf{X}(t)$. Second, upon finding $\mathbf{X}^{-1}(t)$, we could immediately conclude that $\det \mathbf{X}(0) \neq 0$ and, hence, $\mathbf{X}(t)$ is a fundamental matrix. Thus, it was not really necessary to compute the Wronskian.

35. Let $\phi(t)$ be an arbitrary solution to the system $\mathbf{x}'(t) = \mathbf{A}(t)\mathbf{x}(t)$ on an interval I. We want to find $\mathbf{c} = \mathrm{col}(c_1, c_2, \ldots, c_n)$ so that

$$
\phi(t) = c_1 \mathbf{x}_1(t) + c_2 \mathbf{x}_2(t) + \cdots + c_n \mathbf{x}_n(t),
$$

Copyright © 2012 Pearson Education, Inc. Publishing as Addison-Wesley.

where $x_1, x_2, \ldots, x_n$ are n linearly independent solutions for this system. Since

$$c_1 x_1(t) + c_2 x_2(t) + \cdots + c_n x_n(t) = X(t)c,$$

where $X(t)$ is the fundamental matrix, whose columns are the vectors $x_1, x_2, \ldots, x_n$, this equation can be written as

$$\phi(t) = X(t)c \qquad (9.3)$$

Since $x_1, x_2, \ldots, x_n$ are linearly independent solutions, their Wronskian is never zero. Therefore, as it was discussed in the text, $X(t)$ has the inverse at each point in I. Thus, $X^{-1}(t_0)$ exists for any t_0 in I, and (9.3) becomes

$$\phi(t_0) = X(t_0)c \qquad \Rightarrow \qquad X^{-1}(t_0)\phi(t_0) = X^{-1}(t_0)X(t_0)c = c.$$

Hence, if we define c_0 to be the vector $c_0 = X^{-1}(t_0)\phi(t_0)$, then the equation (9.3) is satisfied at t_0 (i.e., $\phi(t_0) = X(t_0)X^{-1}(t_0)\phi(t_0)$). To see that, with this c_0, (9.3) is true for all t in I (and so this is the vector that we need), notice that $\phi(t)$ and $X(t)c_0$ are both solutions to same initial value problem (with the initial value given at the point t_0). Therefore, by the uniqueness theorem (Theorem 2), these solutions must coincide on I, which means that $\phi(t) = X(t)c_0$ for all t in I.

EXERCISES 9.5: Homogeneous Linear Systems with Constant Coefficients

5. The characteristic equation for this matrix is given by

$$|A - rI| = \begin{vmatrix} 1-r & 0 & 0 \\ 0 & -r & 2 \\ 0 & 2 & -r \end{vmatrix} = (1-r) \begin{vmatrix} -r & 2 \\ 2 & -r \end{vmatrix}$$

$$= (1-r)\left(r^2 - 4\right) = (1-r)(r-2)(r+2) = 0.$$

Thus, the eigenvalues are $r = 1, 2, -2$. Substituting the eigenvalue $r = 1$, into the equation $(A - rI)u = 0$ yields

$$(A - I)u = \begin{bmatrix} 0 & 0 & 0 \\ 0 & -1 & 2 \\ 0 & 2 & -1 \end{bmatrix} \begin{bmatrix} u_1 \\ u_2 \\ u_3 \end{bmatrix} = \begin{bmatrix} 0 \\ 0 \\ 0 \end{bmatrix}, \qquad (9.4)$$

which is equivalent to the system

$$-u_2 + 2u_3 = 0,$$
$$2u_2 - u_3 = 0.$$

 Copyright © 2012 Pearson Education, Inc. Publishing as Addison-Wesley.

This system reduces to $u_2 = 0$, $u_3 = 0$, which does not assign any value to u_1. Thus, arbitrary $u_1 = s$ and $u_2 = 0$, $u_3 = 0$ satisfy the system (9.4). So, we see that the eigenvectors, associated with the eigenvalue $r = 1$, are given by

$$\mathbf{u}_1 = \text{col}\,(u_1, u_2, u_3) = \text{col}(s, 0, 0) = s\,\text{col}(1, 0, 0).$$

For $r = 2$, we observe that the equation $(\mathbf{A} - r\mathbf{I})\mathbf{u} = \mathbf{0}$ becomes

$$(\mathbf{A} - 2\mathbf{I})\mathbf{u} = \begin{bmatrix} -1 & 0 & 0 \\ 0 & -2 & 2 \\ 0 & 2 & -2 \end{bmatrix} \begin{bmatrix} u_1 \\ u_2 \\ u_3 \end{bmatrix} = \begin{bmatrix} 0 \\ 0 \\ 0 \end{bmatrix},$$

whose corresponding system of equations reduces to $u_1 = 0$, $u_2 = u_3$. Therefore, we can take arbitrary $u_2 = s$ (which gives $u_3 = s$) and find that the eigenvectors for this matrix, associated with the eigenvalue $r = 2$, are given by

$$\mathbf{u}_2 = \text{col}\,(u_1, u_2, u_3) = \text{col}(0, s, s) = s\,\text{col}(0, 1, 1).$$

For $r = -2$, we solve the equation

$$(\mathbf{A} + 2\mathbf{I})\mathbf{u} = \begin{bmatrix} 3 & 0 & 0 \\ 0 & 2 & 2 \\ 0 & 2 & 2 \end{bmatrix} \begin{bmatrix} u_1 \\ u_2 \\ u_3 \end{bmatrix} = \begin{bmatrix} 0 \\ 0 \\ 0 \end{bmatrix},$$

which reduces to $u_1 = 0$, $u_2 = -u_3$. Hence, u_3 is arbitrary, and so we will let $u_3 = s$ (which means that $u_2 = -s$). Thus, solutions to this system and, therefore, eigenvectors, associated with the eigenvalue $r = -2$, are

$$\mathbf{u}_3 = \text{col}\,(u_1, u_2, u_3) = \text{col}(0, -s, s) = s\,\text{col}(0, -1, 1).$$

13. First, we must find the eigenvalues and eigenvectors associated with the given matrix $\mathbf{A}$. Thus, we note that the characteristic equation for this matrix is given by

$$|\mathbf{A} - r\mathbf{I}| = \begin{vmatrix} 1-r & 2 & 2 \\ 2 & -r & 3 \\ 2 & 3 & -r \end{vmatrix} = 0$$

$$\Rightarrow \quad (1-r) \begin{vmatrix} -r & 3 \\ 3 & -r \end{vmatrix} - 2 \begin{vmatrix} 2 & 3 \\ 2 & -r \end{vmatrix} + 2 \begin{vmatrix} 2 & -r \\ 2 & 3 \end{vmatrix} = 0$$

Copyright © 2012 Pearson Education, Inc. Publishing as Addison-Wesley.

$$\Rightarrow \quad (1-r)\left(r^2 - 9\right) - 2(-2r - 6) + 2(6 + 2r) = (1-r)(r-2)(r+2) = 0$$

$$\Rightarrow \quad (r+3)[(1-r)(r-3) + 8] = 0 \quad \Rightarrow \quad (r+3)(r-5)(r+1) = 0.$$

Therefore, the eigenvalues are $r = -3, -1, 5$. To find an eigenvector, associated with the eigenvalue $r = -3$, we look for a vector $\mathbf{u} = \mathrm{col}(u_1, u_2, u_3)$, which satisfies the equation $(\mathbf{A} + 3\mathbf{I})\mathbf{u} = \mathbf{0}$. Thus, we have

$$(\mathbf{A} + 3\mathbf{I})\mathbf{u} = \begin{bmatrix} 4 & 2 & 2 \\ 2 & 3 & 3 \\ 2 & 3 & 3 \end{bmatrix} \begin{bmatrix} u_1 \\ u_2 \\ u_3 \end{bmatrix} = \begin{bmatrix} 0 \\ 0 \\ 0 \end{bmatrix} \quad \Rightarrow \quad \begin{bmatrix} 2 & 0 & 0 \\ 0 & 1 & 1 \\ 0 & 0 & 0 \end{bmatrix} \begin{bmatrix} u_1 \\ u_2 \\ u_3 \end{bmatrix} = \begin{bmatrix} 0 \\ 0 \\ 0 \end{bmatrix},$$

where we have obtained the last equation above by using elementary row operations. This matrix equation is equivalent to the system $u_1 = 0$, $u_2 = -u_3$. Hence, if we let $u_3 = s_1$ be arbitrary, then the eigenvectors, associated with the eigenvalue $r = -3$, are given by

$$\mathbf{u} = \mathrm{col}\,(u_1, u_2, u_3) = \mathrm{col}\,(0, -s_1, s_1) = s_1\mathrm{col}(0, -1, 1).$$

Thus, if we choose $s_1 = 1$, then vector $\mathbf{u}_1 = \mathrm{col}(0, -1, 1)$ is an eigenvector, associated with this eigenvalue. For the eigenvalue $r = -1$, we must find a vector $\mathbf{u}$, which satisfies the equation $(\mathbf{A} + \mathbf{I})\mathbf{u} = \mathbf{0}$.

$$(\mathbf{A} + \mathbf{I})\mathbf{u} = \begin{bmatrix} 2 & 2 & 2 \\ 2 & 1 & 3 \\ 2 & 3 & 1 \end{bmatrix} \begin{bmatrix} u_1 \\ u_2 \\ u_3 \end{bmatrix} = \begin{bmatrix} 0 \\ 0 \\ 0 \end{bmatrix} \quad \Rightarrow \quad \begin{bmatrix} 1 & 2 & 0 \\ 0 & 1 & -1 \\ 0 & 0 & 0 \end{bmatrix} \begin{bmatrix} u_1 \\ u_2 \\ u_3 \end{bmatrix} = \begin{bmatrix} 0 \\ 0 \\ 0 \end{bmatrix},$$

which is equivalent to $u_1 = -2u_2$, $u_3 = u_2$. Therefore, letting $u_2 = s_2$, we see that vectors, satisfying the equation $(\mathbf{A} + \mathbf{I})\mathbf{u} = \mathbf{0}$ and, hence, eigenvectors for the matrix $\mathbf{A}$, associated with the eigenvalue $r = -1$, are given by

$$\mathbf{u} = \mathrm{col}\,(u_1, u_2, u_3) = \mathrm{col}\,(-2s_2, s_2, s_2) = s_2\mathrm{col}(-2, 1, 1).$$

By taking $s_2 = 1$, we find that $\mathbf{u}_2 = \mathrm{col}(-2, 1, 1)$. In order to find an eigenvector, associated with the eigenvalue $r = 5$, we solve the equation $(\mathbf{A} - 5\mathbf{I})\mathbf{u} = \mathbf{0}$. Thus, we have

$$(\mathbf{A} - 5\mathbf{I})\mathbf{u} = \begin{bmatrix} -4 & 2 & 2 \\ 2 & -5 & 3 \\ 2 & 3 & -5 \end{bmatrix} \begin{bmatrix} u_1 \\ u_2 \\ u_3 \end{bmatrix} = \begin{bmatrix} 0 \\ 0 \\ 0 \end{bmatrix} \quad \Rightarrow \quad \begin{bmatrix} 1 & 0 & -1 \\ 0 & 1 & -1 \\ 0 & 0 & 0 \end{bmatrix} \begin{bmatrix} u_1 \\ u_2 \\ u_3 \end{bmatrix} = \begin{bmatrix} 0 \\ 0 \\ 0 \end{bmatrix};$$

 Copyright © 2012 Pearson Education, Inc. Publishing as Addison-Wesley.

which gives $u_1 = u_3$, $u_2 = u_3$. Thus, if we let $u_3 = s_3$, then the eigenvectors, associated with the eigenvalue $r = 5$, are

$$\mathbf{u} = \text{col}\,(u_1, u_2, u_3) = \text{col}\,(s_3, s_3, s_3) = s_3\text{col}(1, 1, 1).$$

Hence, by letting $s_3 = 1$, we get $\mathbf{u}_3 = \text{col}(1, 1, 1)$. Therefore, by Corollary 1 in the text, a fundamental solution set for this equation is given by

$$\left\{ e^{-3t}\mathbf{u}_1 \,, e^{-t}\mathbf{u}_2 \,, e^{5t}\mathbf{u}_3 \right\}.$$

Thus, a general solution to this system is

$$\mathbf{x}(t) = c_1 e^{-3t}\mathbf{u}_1 + c_2 e^{-t}\mathbf{u}_2 + c_3 e^{5t}\mathbf{u}_3 = c_1 e^{-3t}\begin{bmatrix} 0 \\ -1 \\ 1 \end{bmatrix} + c_2 e^{-t}\begin{bmatrix} -2 \\ 1 \\ 1 \end{bmatrix} + c_3 e^{5t}\begin{bmatrix} 1 \\ 1 \\ 1 \end{bmatrix}.$$

21. A fundamental matrix for this system has three columns, which are linearly independent solutions to the given system. Thus, we have to find them. To this end, we first find the eigenvalues for the matrix $\mathbf{A}$ by solving the characteristic equation

$$|\mathbf{A} - r\mathbf{I}| = \begin{vmatrix} -r & 1 & 0 \\ 0 & -r & 1 \\ 8 & -14 & 7 - r \end{vmatrix} = 0$$

$$\Rightarrow \quad r\begin{vmatrix} -r & 1 \\ -14 & 7 - r \end{vmatrix} - \begin{vmatrix} 0 & 1 \\ 8 & 7 - r \end{vmatrix} = 0$$

$$\Rightarrow \quad r^3 - 7r^2 + 14r - 8 = 0 \quad \Rightarrow \quad (r - 1)(r - 2)(r - 4) = 0.$$

Hence, $\mathbf{A}$ has three distinct eigenvalues $r = 1, 2, 4$ and, according to Theorem 6, the eigenvectors, associated with these eigenvalues, are linearly independent. Thus, these eigenvectors can used in finding three linearly independent solutions. To find an eigenvector, $\mathbf{u} = \text{col}(u_1, u_2, u_3)$, associated with the eigenvalue $r = 1$, we solve the equation $(\mathbf{A} - \mathbf{I})\mathbf{u} = 0$.

$$(\mathbf{A} - \mathbf{I})\mathbf{u} = \begin{bmatrix} -1 & 1 & 0 \\ 0 & -1 & 1 \\ 8 & -14 & 6 \end{bmatrix}\begin{bmatrix} u_1 \\ u_2 \\ u_3 \end{bmatrix} = \begin{bmatrix} 0 \\ 0 \\ 0 \end{bmatrix} \quad \Rightarrow \quad \begin{bmatrix} -1 & 0 & 1 \\ 0 & -1 & 1 \\ 0 & 0 & 0 \end{bmatrix}\begin{bmatrix} u_1 \\ u_2 \\ u_3 \end{bmatrix} = \begin{bmatrix} 0 \\ 0 \\ 0 \end{bmatrix},$$

which yields $u_1 = u_3$, $u_2 = u_3$. Thus, by letting $u_3 = 1$ (which implies that $u_1 = u_2 = 1$), we find that an eigenvector, which is associated with the eigenvalue $r = 1$, is given by

$\mathbf{u}_1 = \text{col}(1,1,1)$. To find an eigenvector, associated with the eigenvalue $r = 2$, we solve the equation

$$(\mathbf{A} - 2\mathbf{I})\mathbf{u} = \begin{bmatrix} -2 & 1 & 0 \\ 0 & -2 & 1 \\ 8 & -14 & 5 \end{bmatrix} \begin{bmatrix} u_1 \\ u_2 \\ u_3 \end{bmatrix} = \begin{bmatrix} 0 \\ 0 \\ 0 \end{bmatrix} \Rightarrow \begin{bmatrix} 4 & 0 & -1 \\ 0 & 2 & -1 \\ 0 & 0 & 0 \end{bmatrix} \begin{bmatrix} u_1 \\ u_2 \\ u_3 \end{bmatrix} = \begin{bmatrix} 0 \\ 0 \\ 0 \end{bmatrix},$$

which is equivalent to the system $4u_1 = u_3$, $2u_2 = u_3$. Hence, $u_3 = 4$ implies that $u_1 = 1$ and $u_2 = 2$. Therefore, an eigenvector, associated with the eigenvalue $r = 2$, is $\mathbf{u}_2 = \text{col}(1,2,4)$. In order to find an eigenvector, corresponding to the eigenvalue $r = 4$, we solve the equation

$$(\mathbf{A} - 4\mathbf{I})\mathbf{u} = \begin{bmatrix} -4 & 1 & 0 \\ 0 & -4 & 1 \\ 8 & -14 & 3 \end{bmatrix} \begin{bmatrix} u_1 \\ u_2 \\ u_3 \end{bmatrix} = \begin{bmatrix} 0 \\ 0 \\ 0 \end{bmatrix} \Rightarrow \begin{bmatrix} 16 & 0 & -1 \\ 0 & 4 & -1 \\ 0 & 0 & 0 \end{bmatrix} \begin{bmatrix} u_1 \\ u_2 \\ u_3 \end{bmatrix} = \begin{bmatrix} 0 \\ 0 \\ 0 \end{bmatrix},$$

which is equivalent to $16u_1 = u_3$, $4u_2 = u_3$. Therefore, $u_3 = 16$ implies that $u_1 = 1$ and $u_2 = 4$. Thus, an eigenvector, associated with $r = 4$, is the vector $\mathbf{u}_3 = \text{col}(1,4,16)$. Therefore, by Theorem 5 (or by Corollary 1), we see that three linearly independent solutions to this system are given by $e^t\mathbf{u}_1$, $e^{2t}\mathbf{u}_2$, and $e^{4t}\mathbf{u}_3$. Thus, a fundamental matrix for this system is

$$\begin{bmatrix} e^t & e^{2t} & e^{4t} \\ e^t & 2e^{2t} & 4e^{4t} \\ e^t & 4e^{2t} & 16e^{4t} \end{bmatrix}.$$

33. Since the coefficient matrix for this system is a 3×3 real symmetric matrix, by the discussion in the text, we know that we can find three linearly independent eigenvectors for this matrix. Therefore, to solve the given initial value problem, we first find three such eigenvectors. To do this, we compute the eigenvalues for this matrix. Solving the characteristic equation

$$|\mathbf{A} - r\mathbf{I}| = \begin{vmatrix} 1-r & -2 & 2 \\ -2 & 1-r & -2 \\ 2 & -2 & 1-r \end{vmatrix} = 0$$

$$\Rightarrow \quad (1-r)\begin{vmatrix} 1-r & -2 \\ -2 & 1-r \end{vmatrix} + 2\begin{vmatrix} -2 & -2 \\ 2 & 1-r \end{vmatrix} + 2\begin{vmatrix} -2 & 1-r \\ 2 & -2 \end{vmatrix} = 0$$

$$\Rightarrow \quad (1-r)\left[(1-r)^2 - 4\right] + 2\left[-2(1-r) + 4\right] + 2\left[4 - 2(1-r)\right] = 0$$

Copyright © 2012 Pearson Education, Inc. Publishing as Addison-Wesley.

$$\Rightarrow \qquad (1-r)(r-3)(r+1) + 8(r+1) = -(r+1)(r-5)(r+1) = 0$$

yields the eigenvalues $r = -1$ and $r = 5$, where $r = -1$ is of multiplicity two. In order to find an eigenvector, associated with the eigenvalue $r = 5$, we solve the equation

$$(\mathbf{A} - 5\mathbf{I})\mathbf{u} = \begin{bmatrix} -4 & -2 & 2 \\ -2 & -4 & -2 \\ 2 & -2 & -4 \end{bmatrix} \begin{bmatrix} u_1 \\ u_2 \\ u_3 \end{bmatrix} = \begin{bmatrix} 0 \\ 0 \\ 0 \end{bmatrix} \Rightarrow \begin{bmatrix} 1 & 0 & -1 \\ 0 & 1 & 1 \\ 0 & 0 & 0 \end{bmatrix} \begin{bmatrix} u_1 \\ u_2 \\ u_3 \end{bmatrix} = \begin{bmatrix} 0 \\ 0 \\ 0 \end{bmatrix}.$$

This results $u_1 = u_3$, $u_2 = -u_3$. Thus, if we let $u_3 = 1$, we see that an eigenvector, associated with the eigenvalue $r = 5$, is given by $\mathbf{u}_1 = \text{col}(u_1, u_2, u_3) = \text{col}(1, -1, 1)$. We must now find two more linearly independent eigenvectors for this coefficient matrix. These eigenvectors are associated with the eigenvalue $r = -1$. Thus, we solve the equation

$$(\mathbf{A} + \mathbf{I})\mathbf{u} = \begin{bmatrix} 2 & -2 & 2 \\ -2 & 2 & -2 \\ 2 & -2 & 2 \end{bmatrix} \begin{bmatrix} u_1 \\ u_2 \\ u_3 \end{bmatrix} = \begin{bmatrix} 0 \\ 0 \\ 0 \end{bmatrix} \Rightarrow \begin{bmatrix} 1 & -1 & 1 \\ 0 & 0 & 0 \\ 0 & 0 & 0 \end{bmatrix} \begin{bmatrix} u_1 \\ u_2 \\ u_3 \end{bmatrix} = \begin{bmatrix} 0 \\ 0 \\ 0 \end{bmatrix}, \quad (9.5)$$

which reduces to $u_1 - u_2 + u_3 = 0$. Therefore, if we assign arbitrarily values s to u_2 and v to u_3, we see that $u_1 = s - v$, and solutions to the equation (9.5) are

$$\mathbf{u} = \begin{bmatrix} s - v \\ s \\ v \end{bmatrix} = s \begin{bmatrix} 1 \\ 1 \\ 0 \end{bmatrix} + v \begin{bmatrix} -1 \\ 0 \\ 1 \end{bmatrix}.$$

By taking $s = 1$ and $v = 0$, we see that one solution to (9.5) is $\mathbf{u}_2 = \text{col}(1, 1, 0)$. Similarly, by letting $s = 0$ and $v = 1$, we find another eigenvector $\mathbf{u}_3 = \text{col}(-1, 0, 1)$. Since the eigenvectors $\mathbf{u}_1$, $\mathbf{u}_2$, and $\mathbf{u}_3$ are linearly independent, by Theorem 5, we see that a general solution to this system is given by

$$\mathbf{x}(t) = c_1 e^{5t} \begin{bmatrix} 1 \\ -1 \\ 1 \end{bmatrix} + c_2 e^{-t} \begin{bmatrix} 1 \\ 1 \\ 0 \end{bmatrix} + c_3 e^{-t} \begin{bmatrix} -1 \\ 0 \\ 1 \end{bmatrix}.$$

To find a solution, which satisfies the initial condition, we must solve the equation

$$\mathbf{x}(0) = c_1 \begin{bmatrix} 1 \\ -1 \\ 1 \end{bmatrix} + c_2 \begin{bmatrix} 1 \\ 1 \\ 0 \end{bmatrix} + c_3 \begin{bmatrix} -1 \\ 0 \\ 1 \end{bmatrix} = \begin{bmatrix} 1 & 1 & -1 \\ -1 & 1 & 0 \\ 1 & 0 & 1 \end{bmatrix} \begin{bmatrix} c_1 \\ c_2 \\ c_3 \end{bmatrix} = \begin{bmatrix} -2 \\ -3 \\ 2 \end{bmatrix}.$$

Copyright © 2012 Pearson Education, Inc. Publishing as Addison-Wesley.

This equation can be solved either by using elementary row operations on the augmented matrix, associated with this equation, or by solving the system

$$c_1 + c_2 - c_3 = -2,$$
$$-c_1 + c_2 = -3,$$
$$c_1 + c_3 = 2.$$

By either method, we find that $c_1 = 1$, $c_2 = -2$, and $c_3 = 1$. Therefore, the solution to this initial value problem is given by

$$\mathbf{x}(t) = e^{5t} \begin{bmatrix} 1 \\ -1 \\ 1 \end{bmatrix} - 2e^{-t} \begin{bmatrix} 1 \\ 1 \\ 0 \end{bmatrix} + e^{-t} \begin{bmatrix} -1 \\ 0 \\ 1 \end{bmatrix}$$

$$= \begin{bmatrix} e^{5t} - 2e^{-t} - e^{-t} \\ -e^{5t} - 2e^{-t} + 0 \\ e^{5t} + 0 + e^{-t} \end{bmatrix} = \begin{bmatrix} -3e^{-t} + e^{5t} \\ -2e^{-t} - e^{5t} \\ e^{-t} + e^{5t} \end{bmatrix}.$$

37. (a) In order to find the eigenvalues for the matrix $\mathbf{A}$, we solve the characteristic equation

$$|\mathbf{A} - r\mathbf{I}| = \begin{vmatrix} 2 - r & 1 & 6 \\ 0 & 2 - r & 5 \\ 0 & 0 & 2 - r \end{vmatrix} = 0 \quad \Rightarrow \quad (2 - r)^3 = 0.$$

Thus, $r = 2$ is an eigenvalue of multiplicity three. To find eigenvectors for the matrix $\mathbf{A}$, associated with this eigenvalue, we solve the equation

$$(\mathbf{A} - 2\mathbf{I})\mathbf{u} = \begin{bmatrix} 0 & 1 & 6 \\ 0 & 0 & 5 \\ 0 & 0 & 0 \end{bmatrix} \begin{bmatrix} u_1 \\ u_2 \\ u_3 \end{bmatrix} = \begin{bmatrix} 0 \\ 0 \\ 0 \end{bmatrix}. \tag{9.6}$$

This equation is equivalent to $u_2 = 0$, $u_3 = 0$. Therefore, we can assign arbitrary $u_1 = s$ and find that the vector

$$\mathbf{u} = \text{col}(u_1, u_2, u_3) = \text{col}(s, 0, 0) = s\,\text{col}(1, 0, 0)$$

solves this equation and, thus, is an eigenvector for the matrix $\mathbf{A}$. We also notice that the vectors $\mathbf{u} = s\,\text{col}(1, 0, 0)$ are the only vectors that solve (9.6). Hence, they are the only eigenvectors for the matrix $\mathbf{A}$.

 Copyright © 2012 Pearson Education, Inc. Publishing as Addison-Wesley.

(b) By taking $s = 1$, we find that, for the matrix $\mathbf{A}$, one eigenvector, associated with the eigenvalue $r = 2$, is $\mathbf{u}_1 = \text{col}(1, 0, 0)$. Therefore, by the way eigenvalues and eigenvectors were defined (as it was discussed in the text), we see that one solution to the system $\mathbf{x}' = \mathbf{A}\mathbf{x}$ is given by

$$\mathbf{x}_1(t) = e^{2t}\mathbf{u}_1 = e^{2t}\begin{bmatrix} 1 \\ 0 \\ 0 \end{bmatrix}.$$

(c) We know from part (b) that $\mathbf{u}_1 = \text{col}(1, 0, 0)$ is an eigenvector for $\mathbf{A}$, associated with the eigenvalue $r = 2$. Thus, $\mathbf{u}_1$ satisfies the equation

$$(\mathbf{A} - 2\mathbf{I})\mathbf{u}_1 = \mathbf{0} \qquad \Rightarrow \qquad \mathbf{A}\mathbf{u}_1 = 2\mathbf{u}_1. \qquad (9.7)$$

We want to find a constant vector $\mathbf{u}_2 = \text{col}(v_1, v_2, v_3)$ such that

$$\mathbf{x}_2(t) = te^{2t}\mathbf{u}_1 + e^{2t}\mathbf{u}_2$$

is a second solution to the system $\mathbf{x}' = \mathbf{A}\mathbf{x}$. To do this, we first show that $\mathbf{x}_2$ satisfies the equation $\mathbf{x}' = \mathbf{A}\mathbf{x}$ if and only if the vector $\mathbf{u}_2$ satisfies the equation $(\mathbf{A} - 2\mathbf{I})\mathbf{u}_2 = \mathbf{u}_1$. To this end, we find that

$$\mathbf{x}_2'(t) = 2te^{2t}\mathbf{u}_1 + e^{2t}\mathbf{u}_1 + 2e^{2t}\mathbf{u}_2 = 2te^{2t}\mathbf{u}_1 + e^{2t}\left(\mathbf{u}_1 + 2\mathbf{u}_2\right),$$

where we have used the fact that $\mathbf{u}_1$ and $\mathbf{u}_2$ are constant vectors. We also have

$$\begin{aligned}
\mathbf{A}\mathbf{x}_2(t) &= \mathbf{A}\left(te^{2t}\mathbf{u}_1 + e^{2t}\mathbf{u}_2\right) \\
&= \mathbf{A}\left(te^{2t}\mathbf{u}_1\right) + \mathbf{A}\left(e^{2t}\mathbf{u}_2\right) && \text{(distributivity of matrix multiplication)} \\
&= te^{2t}\left(\mathbf{A}\mathbf{u}_1\right) + e^{2t}\left(\mathbf{A}\mathbf{u}_2\right) && \text{(associativity of matrix multiplication)} \\
&= 2te^{2t}\mathbf{u}_1 + e^{2t}\mathbf{A}\mathbf{u}_2 && \text{(the equation (9.7))}.
\end{aligned}$$

Thus, if $\mathbf{x}_2(t)$ is a solution to the given system, we must have

$$\begin{aligned}
\mathbf{x}_2'(t) &= \mathbf{A}\mathbf{x}_2(t) \\
\Rightarrow \qquad 2te^{2t}\mathbf{u}_1 + e^{2t}\left(\mathbf{u}_1 + 2\mathbf{u}_2\right) &= 2te^{2t}\mathbf{u}_1 + e^{2t}\mathbf{A}\mathbf{u}_2 \\
\Rightarrow \qquad e^{2t}\left(\mathbf{u}_1 + 2\mathbf{u}_2\right) &= e^{2t}\mathbf{A}\mathbf{u}_2.
\end{aligned}$$

Dividing both sides of this equation by the nonzero term e^{2t} yields

$$\mathbf{u}_1 + 2\mathbf{u}_2 = \mathbf{A}\mathbf{u}_2 \qquad \Rightarrow \qquad (\mathbf{A} - 2\mathbf{I})\mathbf{u}_2 = \mathbf{u}_1.$$

Copyright © 2012 Pearson Education, Inc. Publishing as Addison-Wesley.

Since all of these steps are reversible, if a vector $\mathbf{u}_2$ satisfies this last equation, then $\mathbf{x}_2(t) = te^{2t}\mathbf{u}_1 + e^{2t}\mathbf{u}_2$ is a solution to the system $\mathbf{x}' = \mathbf{Ax}$. Now we can use the formula $(\mathbf{A} - 2\mathbf{I})\mathbf{u}_2 = \mathbf{u}_1$ to find the vector $\mathbf{u}_2 = \text{col}(v_1, v_2, v_3)$. Thus, we solve

$$(\mathbf{A} - 2\mathbf{I})\mathbf{u}_2 = \begin{bmatrix} 0 & 1 & 6 \\ 0 & 0 & 5 \\ 0 & 0 & 0 \end{bmatrix} \begin{bmatrix} v_1 \\ v_2 \\ v_3 \end{bmatrix} = \begin{bmatrix} 1 \\ 0 \\ 0 \end{bmatrix}.$$

This equation is equivalent to $v_2 + 6v_3 = 1$, $5v_3 = 0$, which implies that $v_2 = 1$, $v_3 = 0$. Therefore, the vector $\mathbf{u}_2 = \text{col}(0, 1, 0)$ satisfies the equation $(\mathbf{A} - 2\mathbf{I})\mathbf{u}_2 = \mathbf{u}_1$, and so

$$\mathbf{x}_2(t) = te^{2t} \begin{bmatrix} 1 \\ 0 \\ 0 \end{bmatrix} + e^{2t} \begin{bmatrix} 0 \\ 1 \\ 0 \end{bmatrix}$$

is a second solution to the given system. We can see, by inspection, that $\mathbf{x}_2(t)$ and $\mathbf{x}_1(t)$ are linearly independent.

(d) To find a third linearly independent solution to this system, we will try to find a solution of the form $\mathbf{x}_3(t) = (t^2 e^{2t}/2)\,\mathbf{u}_1 + te^{2t}\mathbf{u}_2 + e^{2t}\mathbf{u}_3$, where $\mathbf{u}_3$ is a constant vector and $\mathbf{u}_1$, $\mathbf{u}_2$ are the vectors that we found in parts (b) and (c).

To find $\mathbf{u}_3$, we proceed similarly to the part (c). We show first that $\mathbf{x}_3(t)$ is a solution to the given system if and only if the vector $\mathbf{u}_3$ satisfies the equation $(\mathbf{A} - 2\mathbf{I})\mathbf{u}_3 = \mathbf{u}_2$. To do this, we observe that

$$\mathbf{x}_3'(t) = te^{2t}\mathbf{u}_1 + t^2 e^{2t}\mathbf{u}_1 + e^{2t}\mathbf{u}_2 + 2te^{2t}\mathbf{u}_2 + 2e^{2t}\mathbf{u}_3\,.$$

Using the facts that

$$(\mathbf{A} - 2\mathbf{I})\mathbf{u}_1 = \mathbf{0} \qquad \Rightarrow \qquad \mathbf{A}\mathbf{u}_1 = 2\mathbf{u}_1 \tag{9.8}$$

and

$$(\mathbf{A} - 2\mathbf{I})\mathbf{u}_2 = \mathbf{u}_1 \qquad \Rightarrow \qquad \mathbf{A}\mathbf{u}_2 = \mathbf{u}_1 + 2\mathbf{u}_2\,, \tag{9.9}$$

 Copyright © 2012 Pearson Education, Inc. Publishing as Addison-Wesley.

we get

$$\mathbf{Ax_3}(t) = \mathbf{A}\left(\frac{t^2 e^{2t}}{2}\mathbf{u_1} + te^{2t}\mathbf{u_2} + e^{2t}\mathbf{u_3}\right)$$

$$= \mathbf{A}\left(\frac{t^2 e^{2t}}{2}\mathbf{u_1}\right) + \mathbf{A}\left(te^{2t}\mathbf{u_2}\right) + +\mathbf{A}\left(e^{2t}\mathbf{u_3}\right) \qquad \text{(distributive property)}$$

$$= \frac{t^2 e^{2t}}{2}\left(\mathbf{Au_1}\right) + te^{2t}\left(\mathbf{Au_2}\right) + e^{2t}\left(\mathbf{Au_3}\right) \qquad \text{(associative property)}$$

$$= \frac{t^2 e^{2t}}{2}\left(2\mathbf{u_1}\right) + te^{2t}\left(\mathbf{u_1} + 2\mathbf{u_2}\right) + e^{2t}\mathbf{Au_3} \qquad \text{(see (9.8) and (9.9))}$$

$$= t^2 e^{2t}\mathbf{u_1} + te^{2t}\mathbf{u_1} + 2te^{2t}\mathbf{u_2} + e^{2t}\mathbf{Au_3}.$$

Therefore, for $\mathbf{x_3}(t)$ to satisfy the given system, we must have

$$\mathbf{x_3'}(t) = \mathbf{Ax_3}(t)$$
$$\Rightarrow \quad te^{2t}\mathbf{u_1} + t^2 e^{2t}\mathbf{u_1} + e^{2t}\mathbf{u_2} + 2te^{2t}\mathbf{u_2} + 2e^{2t}\mathbf{u_3}$$
$$= t^2 e^{2t}\mathbf{u_1} + te^{2t}\mathbf{u_1} + 2te^{2t}\mathbf{u_2} + e^{2t}\mathbf{Au_3}$$
$$\Rightarrow \quad e^{2t}\mathbf{u_2} + 2e^{2t}\mathbf{u_3} = e^{2t}\mathbf{Au_3}$$
$$\Rightarrow \quad \mathbf{u_2} + 2\mathbf{u_3} = \mathbf{Au_3}$$
$$\Rightarrow \quad (\mathbf{A} - 2\mathbf{I})\mathbf{u_3} = \mathbf{u_2}.$$

Again, since these steps are reversible, we see that, if a vector $\mathbf{u_3}$ satisfies the equation $(\mathbf{A} - 2\mathbf{I})\mathbf{u_3} = \mathbf{u_2}$, then the vector

$$\mathbf{x_3}(t) = \frac{t^2 e^{2t}}{2}\mathbf{u_1} + te^{2t}\mathbf{u_2} + e^{2t}\mathbf{u_3}$$

gives a third linearly independent solution to the given system. Thus, we can use this equation to find $\mathbf{u_3} = \text{col}(w_1, w_2, w_3)$. Hence, we solve

$$(\mathbf{A} - 2\mathbf{I})\mathbf{u_3} = \begin{bmatrix} 0 & 1 & 6 \\ 0 & 0 & 5 \\ 0 & 0 & 0 \end{bmatrix}\begin{bmatrix} w_1 \\ w_2 \\ w_3 \end{bmatrix} = \begin{bmatrix} 0 \\ 1 \\ 0 \end{bmatrix}.$$

This equation is equivalent to the system $w_2 + 6w_3 = 0$, $5w_3 = 1$, which implies

Copyright © 2012 Pearson Education, Inc. Publishing as Addison-Wesley.

that $w_3 = 1/5$, $w_2 = -6/5$. Therefore, if we let $\mathbf{u}_3 = \text{col}(0, -6/5, 1/5)$, then

$$\mathbf{x}_3(t) = \frac{t^2}{2}e^{2t}\begin{bmatrix} 1 \\ 0 \\ 0 \end{bmatrix} + te^{2t}\begin{bmatrix} 0 \\ 1 \\ 0 \end{bmatrix} + e^{2t}\begin{bmatrix} 0 \\ -6/5 \\ 1/5 \end{bmatrix}$$

will be a third solution to the given system and, by inspection, we see that this solution is linearly independent from the solutions $\mathbf{x}_1(t)$ and $\mathbf{x}_2(t)$.

(e) Notice that

$$\begin{aligned}
(\mathbf{A} - 2\mathbf{I})^3\mathbf{u}_3 &= (\mathbf{A} - 2\mathbf{I})^2\left[(\mathbf{A} - 2\mathbf{I})\mathbf{u}_3\right] \\
&= (\mathbf{A} - 2\mathbf{I})^2\mathbf{u}_2 = (\mathbf{A} - 2\mathbf{I})\left[(\mathbf{A} - 2\mathbf{I})\mathbf{u}_2\right] = (\mathbf{A} - 2\mathbf{I})\mathbf{u}_1 = \mathbf{0}.
\end{aligned}$$

43. According to Problem 42, we will look for solutions of the form $\mathbf{x}(t) = t^r\mathbf{u}$, where r is an eigenvalue for the coefficient matrix and $\mathbf{u}$ is an associated eigenvector. To find the eigenvalues, we solve the equation

$$|\mathbf{A} - r\mathbf{I}| = \begin{vmatrix} 1 - r & 3 \\ -1 & 5 - r \end{vmatrix} = 0$$

$$\Rightarrow \quad (1 - r)(5 - r) + 3 = 0$$

$$\Rightarrow \quad r^2 - 6r + 8 = 0 \quad \Rightarrow \quad (r - 2)(r - 4) = 0.$$

Therefore, the coefficient matrix has eigenvalues $r = 2, 4$. Since these numbers real and distinct, Theorem 6 assures that their associated eigenvectors are linearly independent. To find an eigenvector $\mathbf{u} = \text{col}(u_1, u_2)$, associated with the eigenvalue $r = 2$, we solve the system

$$(\mathbf{A} - 2\mathbf{I})\mathbf{u} = \begin{bmatrix} -1 & 3 \\ -1 & 3 \end{bmatrix}\begin{bmatrix} u_1 \\ u_2 \end{bmatrix} = \begin{bmatrix} 0 \\ 0 \end{bmatrix},$$

which is equivalent to $-u_1 + 3u_2 = 0$. Thus, letting $u_2 = 1$ implies $u_1 = 3$. Hence, we the vector $\mathbf{u}_1 = \text{col}(3, 1)$ is an eigenvector for the coefficient matrix of the given system, associated with the eigenvalue $r = 2$. According to Problem 42, one solution to this system is given by

$$\mathbf{x}_1(t) = t^2\mathbf{u}_1 = t^2\begin{bmatrix} 3 \\ 1 \end{bmatrix}.$$

To find an eigenvector, associated with the eigenvalue $r = 4$, we solve the equation

$$(\mathbf{A} - 4\mathbf{I})\mathbf{u} = \begin{bmatrix} -3 & 3 \\ -1 & 1 \end{bmatrix}\begin{bmatrix} u_1 \\ u_2 \end{bmatrix} = \begin{bmatrix} 0 \\ 0 \end{bmatrix},$$

 Copyright © 2012 Pearson Education, Inc. Publishing as Addison-Wesley.

which yields $u_1 = u_2$. Thus, we let $u_1 = u_2 = 1$, and so an eigenvector, associated with the eigenvalue $r = 4$, is given by $\mathbf{u}_2 = \operatorname{col}(1, 1)$. Therefore, another solution to the given system is

$$\mathbf{x}_2(t) = t^4 \mathbf{u}_2 = t^4 \begin{bmatrix} 1 \\ 1 \end{bmatrix}.$$

Clearly the solutions $\mathbf{x}_1(t)$ and $\mathbf{x}_2(t)$ are linearly independent. So, a general solution to the given system is

$$\mathbf{x}(t) = c_1 t^2 \begin{bmatrix} 3 \\ 1 \end{bmatrix} + c_2 t^4 \begin{bmatrix} 1 \\ 1 \end{bmatrix} = c_1 \begin{bmatrix} 3t^2 \\ t^2 \end{bmatrix} + c_2 \begin{bmatrix} t^4 \\ t^4 \end{bmatrix}, \quad t > 0.$$

EXERCISES 9.6: Complex Eigenvalues

3. To find the eigenvalues for the matrix $\mathbf{A}$, we solve the characteristic equation given by

$$|\mathbf{A} - r\mathbf{I}| = \begin{vmatrix} 1 - r & 2 & -1 \\ 0 & 1 - r & 1 \\ 0 & -1 & 1 - r \end{vmatrix} = 0$$

$$\Rightarrow \quad (1 - r) \begin{vmatrix} 1 - r & 1 \\ -1 & 1 - r \end{vmatrix} - 0 + 0 = 0$$

$$\Rightarrow \quad (1 - r) \left[(1 - r)^2 + 1 \right] = (1 - r) \left(r^2 - 2r + 2 \right) = 0.$$

From this equation and the quadratic formula, we see that the roots to the characteristic equation and, therefore, the eigenvalues for the matrix $\mathbf{A}$ are $r = 1$ and $r = 1 \pm i$. To find an eigenvector $\mathbf{u} = \operatorname{col}(u_1, u_2, u_3)$, associated with the real eigenvalue $r = 1$, we solve the system

$$(\mathbf{A} - \mathbf{I})\mathbf{u} = \begin{bmatrix} 0 & 2 & -1 \\ 0 & 0 & 1 \\ 0 & -1 & 0 \end{bmatrix} \begin{bmatrix} u_1 \\ u_2 \\ u_3 \end{bmatrix} = \begin{bmatrix} 0 \\ 0 \\ 0 \end{bmatrix},$$

which implies that $u_2 = 0$, $u_3 = 0$. Therefore, setting $u_1 = s$ yields eigenvectors

$$\mathbf{u} = \operatorname{col}(u_1, u_2, u_3) = \operatorname{col}(s, 0, 0) = s \operatorname{col}(1, 0, 0).$$

Hence, if we let $s = 1$, we see that one eigenvector, associated with the eigenvalue $r = 1$, is $\mathbf{u}_1 = \operatorname{col}(1, 0, 0)$. Thus,

$$\mathbf{x}_1(t) = e^t \mathbf{u}_1 = e^t \begin{bmatrix} 1 \\ 0 \\ 0 \end{bmatrix}$$

Copyright © 2012 Pearson Education, Inc. Publishing as Addison-Wesley.

is a solution to the given system. In order to find an eigenvector $\mathbf{z} = \text{col}(z_1, z_2, z_3)$, associated with the complex eigenvalue $r = 1 + i$, we solve the equation

$$[\mathbf{A} - (1+i)\mathbf{I}]\mathbf{z} = \begin{bmatrix} -i & 2 & -1 \\ 0 & -i & 1 \\ 0 & -1 & i \end{bmatrix} \begin{bmatrix} z_1 \\ z_2 \\ z_3 \end{bmatrix} = \begin{bmatrix} 0 \\ 0 \\ 0 \end{bmatrix}.$$

This equation is equivalent to the system

$$-iz_1 + 2z_2 - z_3 = 0 \qquad \text{and} \qquad -iz_2 + z_3 = 0.$$

Thus, if we let $z_2 = s$, then we see that $z_3 = is$ and

$$-iz_1 = -2z_2 + z_3 = -2s + is \qquad \Rightarrow \qquad (i)(-iz_1) = (i)(-2s + is)$$

$$\Rightarrow \qquad z_1 = -2is - s = -s(1 + 2i),$$

where we have used the fact that $i^2 = -1$. Hence, eigenvectors, associated with the eigenvalue $r = 1 + i$, are $\mathbf{z} = s\,\text{col}(-1 - 2i, 1, i)$. By taking $s = 1$, we find an eigenvector

$$\mathbf{z}_1 = \begin{bmatrix} -1 - 2i \\ 1 \\ i \end{bmatrix} = \begin{bmatrix} -1 \\ 1 \\ 0 \end{bmatrix} + i \begin{bmatrix} -2 \\ 0 \\ 1 \end{bmatrix} = \mathbf{a} + i\mathbf{b},$$

where $\mathbf{a} = \text{col}(-1, 1, 0)$ and $\mathbf{b} = \text{col}(-2, 0, 1)$. Since, for $r = 1 + i$, $\alpha = 1$, $\beta = 1$, according to the formulas (6) and (7) in the text, two other linearly independent real-valued solutions to the given system are given by

$$\mathbf{x}_2(t) = (e^t \cos t)\mathbf{a} - (e^t \sin t)\mathbf{b} \qquad \text{and} \qquad \mathbf{x}_3(t) = (e^t \sin t)\mathbf{a} + (e^t \cos t)\mathbf{b}.$$

Hence, a general solution to the system in this problem is

$$\mathbf{x}(t) = c_1\mathbf{x}_2(t) + c_2\mathbf{x}_3(t) + c_3\mathbf{x}_1(t)$$
$$= c_1 e^t \cos t \begin{bmatrix} -1 \\ 1 \\ 0 \end{bmatrix} - c_1 e^t \sin t \begin{bmatrix} -2 \\ 0 \\ 1 \end{bmatrix} + c_2 e^t \sin t \begin{bmatrix} -1 \\ 1 \\ 0 \end{bmatrix} + c_2 e^t \cos t \begin{bmatrix} -2 \\ 0 \\ 1 \end{bmatrix}$$
$$+ c_3 e^t \begin{bmatrix} 1 \\ 0 \\ 0 \end{bmatrix}.$$

 Copyright © 2012 Pearson Education, Inc. Publishing as Addison-Wesley.

7. In order to find a fundamental matrix for this system, we need three linearly independent solutions. Thus, we seek for the eigenvalues of the matrix $\mathbf{A}$. Solving the characteristic equation

$$|\mathbf{A} - r\mathbf{I}| = \begin{vmatrix} -r & 0 & 1 \\ 0 & -r & -1 \\ 0 & 1 & -r \end{vmatrix} = 0$$

$$\Rightarrow \quad -r\begin{vmatrix} -r & -1 \\ 1 & -r \end{vmatrix} - 0 + 0 = 0 \quad \Rightarrow \quad -r\left(r^2 + 1\right) = 0$$

yields $r = 0$ and $r = \pm i$. To find an eigenvector $\mathbf{u} = \mathrm{col}(u_1, u_2, u_3)$, associated with the real eigenvalue $r = 0$, we solve the equation

$$(\mathbf{A} - 0\mathbf{I})\mathbf{u} = \begin{bmatrix} 0 & 0 & 1 \\ 0 & 0 & -1 \\ 0 & 1 & 0 \end{bmatrix}\begin{bmatrix} u_1 \\ u_2 \\ u_3 \end{bmatrix} = \begin{bmatrix} 0 \\ 0 \\ 0 \end{bmatrix},$$

which is equivalent to the system $u_3 = 0$, $u_2 = 0$. Thus, if we let u_1 to have an arbitrary value s, then the vectors

$$\mathbf{u} = \mathrm{col}(u_1, u_2, u_3) = \mathrm{col}(s, 0, 0) = s\,\mathrm{col}(1, 0, 0)$$

satisfy this equation and, therefore, are eigenvectors for the matrix $\mathbf{A}$, associated with the eigenvalue $r = 0$. Letting $s = 1$, we find that one of these eigenvectors is $\mathbf{u} = \mathrm{col}(1, 0, 0)$. Thus, one solution to the given system is given by

$$\mathbf{x}_1(t) = e^0\mathbf{u} = \begin{bmatrix} 1 \\ 0 \\ 0 \end{bmatrix}.$$

To find two other linearly independent real-valued solutions, we first look for a complex eigenvector $\mathbf{z} = \mathrm{col}(z_1, z_2, z_3)$, associated with the complex eigenvalue $r = i$.

$$(\mathbf{A} - i\mathbf{I})\mathbf{z} = \begin{bmatrix} -i & 0 & 1 \\ 0 & -i & -1 \\ 0 & 1 & -i \end{bmatrix}\begin{bmatrix} z_1 \\ z_2 \\ z_3 \end{bmatrix} = \begin{bmatrix} 0 \\ 0 \\ 0 \end{bmatrix},$$

which is equivalent to the system $iz_1 = z_3$ and $iz_2 = -z_3$. Thus, if we let $z_3 = is$ (which means that we must have $z_1 = s$ and $z_2 = -s$), then we see that the vectors

$$\mathbf{z} = \mathrm{col}(z_1, z_2, z_3) = \mathrm{col}(s, -s, is) = s\,\mathrm{col}(1, -1, i),$$

Copyright © 2012 Pearson Education, Inc. Publishing as Addison-Wesley.

will be eigenvectors for the matrix $\mathbf{A}$, associated with the eigenvalue $r = i$. Therefore, by letting $s = 1$, we find one of these eigenvectors.

$$\mathbf{z} = \begin{bmatrix} 1 \\ -1 \\ i \end{bmatrix} = \begin{bmatrix} 1 \\ -1 \\ 0 \end{bmatrix} + i \begin{bmatrix} 0 \\ 0 \\ 1 \end{bmatrix} = \mathbf{a} + i\mathbf{b},$$

where $\mathbf{a} = \mathrm{col}(1, -1, 0)$ and $\mathbf{b} = \mathrm{col}(0, 0, 1)$. Since, for $r = i$, $\alpha = 0$, $\beta = 1$, by (6) and (7) in the text, two other linearly independent solutions to our system are

$$\mathbf{x}_2(t) = (\cos t)\mathbf{a} - (\sin t)\mathbf{b} = \begin{bmatrix} \cos t \\ -\cos t \\ 0 \end{bmatrix} - \begin{bmatrix} 0 \\ 0 \\ \sin t \end{bmatrix} = \begin{bmatrix} \cos t \\ -\cos t \\ \sin t \end{bmatrix}$$

and

$$\mathbf{x}_3(t) = (\sin t)\mathbf{a} + (\cos t)\mathbf{b} = \begin{bmatrix} \sin t \\ -\sin t \\ 0 \end{bmatrix} + \begin{bmatrix} 0 \\ 0 \\ \cos t \end{bmatrix} = \begin{bmatrix} \sin t \\ -\sin t \\ \cos t \end{bmatrix}.$$

Finally, since a fundamental matrix for our system is formed by three columns, which are linearly independent solutions to the system, we see that

$$\mathbf{X}(t) = \begin{bmatrix} 1 & \cos t & \sin t \\ 0 & -\cos t & -\sin t \\ 0 & -\sin t & \cos t \end{bmatrix}.$$

17. We will assume that $t > 0$. According to Problem 42 in Exercises 9.5, a solution to this Cauchy-Euler system has the form $\mathbf{x}(t) = t^r \mathbf{u}$, where r is an eigenvalue for the coefficient matrix of the system and $\mathbf{u}$ is an eigenvector associated with this eigenvalue. Therefore, we first must find the eigenvalues for this matrix by solving the characteristic equation

$$|\mathbf{A} - r\mathbf{I}| = \begin{vmatrix} -1 - r & -1 & 0 \\ 2 & -1 - r & 1 \\ 0 & 1 & -1 - r \end{vmatrix} = 0$$

$$\Rightarrow \quad (-1 - r)\begin{vmatrix} -1 - r & 1 \\ 1 & -1 - r \end{vmatrix} + \begin{vmatrix} 2 & 1 \\ 0 & -1 - r \end{vmatrix} = 0$$

$$\Rightarrow \quad (-1 - r)\left[(-1 - r)^2 - 1\right] + 2(-1 - r) = -(1 + r)\left(r^2 + 2r + 2\right) = 0.$$

From this equation and by using the quadratic formula, we see that the eigenvalues for this coefficient matrix are $r = -1, -1 \pm i$. The eigenvectors, associated with the real

    Copyright © 2012 Pearson Education, Inc. Publishing as Addison-Wesley.

eigenvalue $r = -1$, are the vectors $\mathbf{u} = \mathrm{col}(u_1, u_2, u_3)$, which satisfy the equation

$$(\mathbf{A} + \mathbf{I})\mathbf{u} = \begin{bmatrix} 0 & -1 & 0 \\ 2 & 0 & 1 \\ 0 & 1 & 0 \end{bmatrix} \begin{bmatrix} u_1 \\ u_2 \\ u_3 \end{bmatrix} = \begin{bmatrix} 0 \\ 0 \\ 0 \end{bmatrix},$$

which is equivalent to the system $u_2 = 0$, $2u_1 + u_3 = 0$. Thus, by letting $u_1 = 1$ (which means that $u_3 = -2$), we see that the vector

$$\mathbf{u} = \mathrm{col}(u_1, u_2, u_3) = \mathrm{col}(1, 0, -2)$$

is an eigenvector of the coefficient matrix, associated with the eigenvalue $r = -1$. Hence, according to Problem 42, Exercises 9.5, we see that a solution to this Cauchy-Euler system is given by

$$\mathbf{x}_1(t) = t^{-1}\mathbf{u} = t^{-1} \begin{bmatrix} 1 \\ 0 \\ -2 \end{bmatrix} = \begin{bmatrix} t^{-1} \\ 0 \\ -2t^{-1} \end{bmatrix}.$$

To find eigenvectors $\mathbf{z} = \mathrm{col}(z_1, z_2, z_3)$, associated with the complex eigenvalue $r = -1 + i$, we solve the equation

$$(\mathbf{A} - (-1 + i)\mathbf{I})\mathbf{z} = \begin{bmatrix} -i & -1 & 0 \\ 2 & -i & 1 \\ 0 & 1 & -i \end{bmatrix} \begin{bmatrix} z_1 \\ z_2 \\ z_3 \end{bmatrix} = \begin{bmatrix} 0 \\ 0 \\ 0 \end{bmatrix}$$

$$\Rightarrow \begin{bmatrix} -i & -1 & 0 \\ 0 & i & 1 \\ 0 & 0 & 0 \end{bmatrix} \begin{bmatrix} z_1 \\ z_2 \\ z_3 \end{bmatrix} = \begin{bmatrix} 0 \\ 0 \\ 0 \end{bmatrix},$$

which is equivalent to the system $-iz_1 - z_2 = 0$, $iz_2 + z_3 = 0$. Thus, if we let $z_1 = 1$, we get $z_2 = -i$ and $z_3 = -1$. Therefore, one eigenvector for the coefficient matrix, associated with $r = -1 + i$, is the vector $\mathbf{z} = \mathrm{col}(1, -i, -1)$, and another solution to our system is $\mathbf{x}(t) = t^{-1+i}\,\mathbf{z}$. Since we need real-valued solutions to this problem, applying Euler's formula, we obtain

$$t^{-1+i} = t^{-1}t^i = t^{-1}e^{i\ln t} = t^{-1}[\cos(\ln t) + i\sin(\ln t)], \qquad t > 0.$$

Hence, the solution $\mathbf{x}(t)$ becomes

$$\mathbf{x}(t) \;=\; t^{-1+i}\mathbf{z} = t^{-1}[\cos(\ln t) + i\sin(\ln t)]\mathbf{z}$$

Copyright © 2012 Pearson Education, Inc. Publishing as Addison-Wesley.

$$= t^{-1}[\cos(\ln t) + i \sin(\ln t)] \begin{bmatrix} 1 \\ -i \\ -1 \end{bmatrix} = \begin{bmatrix} t^{-1}\cos(\ln t) \\ t^{-1}\sin(\ln t) \\ -t^{-1}\cos(\ln t) \end{bmatrix} + i \begin{bmatrix} t^{-1}\sin(\ln t) \\ -t^{-1}\cos(\ln t) \\ -t^{-1}\sin(\ln t) \end{bmatrix}.$$

Thus, by Lemma 2 in Section 4.3 (adapted to systems), we see that two other linearly independent solutions to this Cauchy-Euler system are

$$\mathbf{x}_2(t) = \begin{bmatrix} t^{-1}\cos(\ln t) \\ t^{-1}\sin(\ln t) \\ -t^{-1}\cos(\ln t) \end{bmatrix} \quad \text{and} \quad \mathbf{x}_3(t) = \begin{bmatrix} t^{-1}\sin(\ln t) \\ -t^{-1}\cos(\ln t) \\ -t^{-1}\sin(\ln t) \end{bmatrix},$$

and, hence, a general solution is given by

$$\mathbf{x}(t) = c_1 \begin{bmatrix} t^{-1} \\ 0 \\ -2t^{-1} \end{bmatrix} + c_2 \begin{bmatrix} t^{-1}\cos(\ln t) \\ t^{-1}\sin(\ln t) \\ -t^{-1}\cos(\ln t) \end{bmatrix} + c_3 \begin{bmatrix} t^{-1}\sin(\ln t) \\ -t^{-1}\cos(\ln t) \\ -t^{-1}\sin(\ln t) \end{bmatrix}.$$

EXERCISES 9.7: Nonhomogeneous Linear Systems

3. We must first find a general solution to the corresponding homogeneous system. Therefore, we first find the eigenvalues for the coefficient matrix $\mathbf{A}$ by solving the characteristic equation

$$|\mathbf{A} - r\mathbf{I}| = \begin{vmatrix} 1-r & -2 & 2 \\ -2 & 1-r & 2 \\ 2 & 2 & 1-r \end{vmatrix} = 0$$

$$\Rightarrow \quad (1-r)\begin{vmatrix} 1-r & 2 \\ 2 & 1-r \end{vmatrix} + 2\begin{vmatrix} -2 & 2 \\ 2 & 1-r \end{vmatrix} + 2\begin{vmatrix} -2 & 1-r \\ 2 & 2 \end{vmatrix} = 0$$

$$\Rightarrow \quad (1-r)\left[(1-r)^2 - 4\right] + 2\left[-2(1-r) - 4\right] + 2\left[-4 - 2(1-r)\right] = 0$$

$$\Rightarrow \quad (1-r)(r^2 - 2r - 3) + 4(2r - 6) = 0$$

$$\Rightarrow \quad (1-r)(r+1)(r-3) + 8(r-3) = (r-3)(r^2 - 9) = (r-3)^2(r+3) = 0.$$

Thus, the eigenvalues for the matrix $\mathbf{A}$ are $r = 3, -3$, where $r = 3$ is an eigenvalue of multiplicity two. Notice that, even though the matrix $\mathbf{A}$ has only two distinct eigenvalues, we are still guaranteed three linearly independent eigenvectors, because $\mathbf{A}$ is a 3×3 real symmetric matrix. To find an eigenvector, associated with the eigenvalue $r = -3$,

 Copyright © 2012 Pearson Education, Inc. Publishing as Addison-Wesley.

we must find a vector $\mathbf{u} = \text{col}(u_1, u_2, u_3)$, which satisfies the system

$$(\mathbf{A} + 3\mathbf{I})\mathbf{u} = \begin{bmatrix} 4 & -2 & 2 \\ -2 & 4 & 2 \\ 2 & 2 & 4 \end{bmatrix} \begin{bmatrix} u_1 \\ u_2 \\ u_3 \end{bmatrix} = \begin{bmatrix} 0 \\ 0 \\ 0 \end{bmatrix} \Rightarrow \begin{bmatrix} 1 & 0 & 1 \\ 0 & 1 & 1 \\ 0 & 0 & 0 \end{bmatrix} \begin{bmatrix} u_1 \\ u_2 \\ u_3 \end{bmatrix} = \begin{bmatrix} 0 \\ 0 \\ 0 \end{bmatrix},$$

equivalent to $u_1 + u_3 = 0$, $u_2 + u_3 = 0$. Hence, by letting $u_3 = -1$, we have $u_1 = u_2 = 1$, and so the vector $\mathbf{u}_1 = \text{col}(1, 1, -1)$ satisfies our system. Therefore, this vector is an eigenvector for the matrix $\mathbf{A}$, associated with the eigenvalue $r = -3$. Thus, one solution to the corresponding homogeneous system is given by

$$\mathbf{x}_1(t) = e^{-3t}\mathbf{u}_1 = e^{-3t} \begin{bmatrix} 1 \\ 1 \\ -1 \end{bmatrix}.$$

To find eigenvectors $\mathbf{u} = \text{col}(u_1, u_2, u_3)$, associated with the eigenvalue $r = 3$, we solve the equation

$$(\mathbf{A} - 3\mathbf{I})\mathbf{u} = \begin{bmatrix} -2 & -2 & 2 \\ -2 & -2 & 2 \\ 2 & 2 & -2 \end{bmatrix} \begin{bmatrix} u_1 \\ u_2 \\ u_3 \end{bmatrix} = \begin{bmatrix} 0 \\ 0 \\ 0 \end{bmatrix},$$

which is equivalent to the equation $u_1 + u_2 - u_3 = 0$. Thus, if we let $u_3 = s$ and $u_2 = v$, then we get $u_1 = s - v$. Hence, solutions to this equation and, therefore, eigenvectors for $\mathbf{A}$, associated with the eigenvalue $r = 3$, are the vectors

$$\mathbf{u} = \begin{bmatrix} s - v \\ v \\ s \end{bmatrix} = s \begin{bmatrix} 1 \\ 0 \\ 1 \end{bmatrix} + v \begin{bmatrix} -1 \\ 1 \\ 0 \end{bmatrix},$$

where s and v are arbitrary scalars. Letting $s = 1$ and $v = 0$, we obtain an eigenvector $\mathbf{u}_2 = \text{col}(1, 0, 1)$. Similarly, by letting $s = 0$ and $v = 1$, we obtain another eigenvector, $\mathbf{u}_3 = \text{col}(-1, 1, 0)$, which is linearly independent from $\mathbf{u}_2$, as we can see by inspection. Hence, two more solutions to the corresponding homogeneous system, which are linearly independent from each other and from $\mathbf{x}_1(t)$, are given by

$$\mathbf{x}_2(t) = e^{3t}\mathbf{u}_2 = e^{3t} \begin{bmatrix} 1 \\ 0 \\ 1 \end{bmatrix} \quad \text{and} \quad \mathbf{x}_3(t) = e^{3t}\mathbf{u}_3 = e^{3t} \begin{bmatrix} -1 \\ 1 \\ 0 \end{bmatrix}.$$

Copyright © 2012 Pearson Education, Inc. Publishing as Addison-Wesley.

Thus, we obtain a general solution to the corresponding homogeneous system given by

$$\mathbf{x}_h(t) = c_1 e^{-3t} \begin{bmatrix} 1 \\ 1 \\ -1 \end{bmatrix} + c_2 e^{3t} \begin{bmatrix} 1 \\ 0 \\ 1 \end{bmatrix} + c_3 e^{3t} \begin{bmatrix} -1 \\ 1 \\ 0 \end{bmatrix}.$$

To find a particular solution to the nonhomogeneous system, we note that

$$\mathbf{f}(t) = \begin{bmatrix} 2e^t \\ 4e^t \\ -2e^t \end{bmatrix} = e^t \begin{bmatrix} 2 \\ 4 \\ -2 \end{bmatrix} = e^t \mathbf{g},$$

where $\mathbf{g} = \text{col}(2, 4, -2)$. Therefore, we will assume that a particular solution to the nonhomogeneous system has the form $\mathbf{x}_p(t) = e^t \mathbf{a}$, where $\mathbf{a} = \text{col}(a_1, a_2, a_3)$ is a constant vector, which must be determined. Hence, we see that $\mathbf{x}_p'(t) = e^t \mathbf{a}$. By substituting $\mathbf{x}_p(t)$ into the given system, we obtain

$$e^t \mathbf{a} = \mathbf{A}\mathbf{x}_p(t) + \mathbf{f}(t) = \mathbf{A}e^t \mathbf{a} + e^t \mathbf{g} = e^t \mathbf{A}\mathbf{a} + e^t \mathbf{g}.$$

Therefore, we have

$$e^t \mathbf{a} = e^t \mathbf{A}\mathbf{a} + e^t \mathbf{g}$$

$$\Rightarrow \quad \mathbf{a} = \mathbf{A}\mathbf{a} + \mathbf{g} \quad \Rightarrow \quad (\mathbf{I} - \mathbf{A})\mathbf{a} = \mathbf{g}$$

$$\Rightarrow \quad \begin{bmatrix} 0 & 2 & -2 \\ 2 & 0 & -2 \\ -2 & -2 & 0 \end{bmatrix} \begin{bmatrix} a_1 \\ a_2 \\ a_3 \end{bmatrix} = \begin{bmatrix} 2 \\ 4 \\ -2 \end{bmatrix}.$$

The last equation above can be solved by either performing elementary row operations on the augmented matrix or by solving the system

$$\begin{aligned} 2a_2 - 2a_3 &= 2, \\ 2a_1 - 2a_3 &= 4, \\ -2a_1 - 2a_2 &= -2. \end{aligned}$$

Either way, we obtain $a_1 = 1$, $a_2 = 0$, and $a_3 = -1$. Thus, a particular solution to the nonhomogeneous system is given by

$$\mathbf{x}_p(t) = e^t \mathbf{a} = e^t \begin{bmatrix} 1 \\ 0 \\ -1 \end{bmatrix},$$

 Copyright © 2012 Pearson Education, Inc. Publishing as Addison-Wesley.

and so a general solution to the nonhomogeneous system is

$$\mathbf{x}(t) = \mathbf{x}_h(t) + \mathbf{x}_p(t) = c_1 e^{-3t} \begin{bmatrix} 1 \\ 1 \\ -1 \end{bmatrix} + c_2 e^{3t} \begin{bmatrix} 1 \\ 0 \\ 1 \end{bmatrix} + c_3 e^{3t} \begin{bmatrix} -1 \\ 1 \\ 0 \end{bmatrix} + e^t \begin{bmatrix} 1 \\ 0 \\ -1 \end{bmatrix}.$$

13. First, we must find a fundamental matrix for the corresponding homogeneous system $\mathbf{x}' = \mathbf{A}\mathbf{x}$. To this end, we find the eigenvalues of the matrix $\mathbf{A}$ by solving the characteristic equation

$$|\mathbf{A} - r\mathbf{I}| = \begin{vmatrix} 2 - r & 1 \\ -3 & -2 - r \end{vmatrix} = 0 \quad \Rightarrow \quad (2 - r)(-2 - r) + 3 = 0 \quad \Rightarrow \quad r^2 - 1 = 0.$$

Thus, the eigenvalues of the coefficient matrix $\mathbf{A}$ are $r = \pm 1$. The eigenvectors, associated with the eigenvalue $r = 1$, are the vectors $\mathbf{u} = \mathrm{col}(u_1, u_2)$ which satisfy the equation

$$(\mathbf{A} - \mathbf{I})\mathbf{u} = \begin{bmatrix} 1 & 1 \\ -3 & -3 \end{bmatrix} \begin{bmatrix} u_1 \\ u_2 \end{bmatrix} = \begin{bmatrix} 0 \\ 0 \end{bmatrix}.$$

This equation reduces to $u_1 + u_2 = 0$. Therefore, if we let $u_1 = 1$, then we have $u_2 = -1$, so that an eigenvector of the matrix $\mathbf{A}$, associated with the eigenvalue $r = 1$, is $\mathbf{u}_1 = \mathrm{col}(1, -1)$. Hence, one solution to the corresponding homogeneous system is given by

$$\mathbf{x}_1(t) = e^t \mathbf{u}_1 = e^t \begin{bmatrix} 1 \\ -1 \end{bmatrix} = \begin{bmatrix} e^t \\ -e^t \end{bmatrix}.$$

To find an eigenvector, associated with the eigenvalue $r = -1$, we solve the equation

$$(\mathbf{A} + \mathbf{I})\mathbf{u} = \begin{bmatrix} 3 & 1 \\ -3 & -1 \end{bmatrix} \begin{bmatrix} u_1 \\ u_2 \end{bmatrix} = \begin{bmatrix} 0 \\ 0 \end{bmatrix},$$

which is equivalent to the equation $3u_1 + u_2 = 0$. Since $u_1 = 1$ and $u_2 = -3$ satisfy this equation, an eigenvector for the matrix $\mathbf{A}$, associated with $r = -1$, is $\mathbf{u}_2 = \mathrm{col}(1, -3)$. Thus, another linearly independent solution to the corresponding homogeneous system is

$$\mathbf{x}_2(t) = e^{-t} \mathbf{u}_2 = e^{-t} \begin{bmatrix} 1 \\ -3 \end{bmatrix} = \begin{bmatrix} e^{-t} \\ -3e^{-t} \end{bmatrix}.$$

Hence, a general solution to the homogeneous system is given by

$$\mathbf{x}_h(t) = c_1 \begin{bmatrix} e^t \\ -e^t \end{bmatrix} + c_2 \begin{bmatrix} e^{-t} \\ -3e^{-t} \end{bmatrix},$$

Copyright © 2012 Pearson Education, Inc. Publishing as Addison-Wesley.

and a fundamental matrix is

$$\mathbf{X}(t) = \begin{bmatrix} e^t & e^{-t} \\ -e^t & -3e^{-t} \end{bmatrix}.$$

To find the inverse matrix $\mathbf{X}^{-1}(t)$, we perform a row-reduction on the matrix $[\mathbf{X}(t)|\mathbf{I}]$. Thus,

$$[\mathbf{X}(t)|\mathbf{I}] = \left[\begin{array}{cc|cc} e^t & e^{-t} & 1 & 0 \\ -e^t & -3e^{-t} & 0 & 1 \end{array} \right] \longrightarrow \left[\begin{array}{cc|cc} e^t & e^{-t} & 1 & 0 \\ 0 & -2e^{-t} & 1 & 1 \end{array} \right]$$

$$\longrightarrow \left[\begin{array}{cc|cc} e^t & 0 & 3/2 & 1/2 \\ 0 & e^{-t} & -1/2 & -1/2 \end{array} \right] \longrightarrow \left[\begin{array}{cc|cc} 1 & 0 & (3/2)e^{-t} & (1/2)e^{-t} \\ 0 & 1 & -(1/2)e^{-t} & -(1/2)e^{-t} \end{array} \right].$$

Therefore, we see that

$$\mathbf{X}^{-1}(t) = \begin{bmatrix} (3/2)e^{-t} & (1/2)e^{-t} \\ -(1/2)e^{-t} & -(1/2)e^{-t} \end{bmatrix}$$

and

$$\mathbf{X}^{-1}(t)\mathbf{f}(t) = \begin{bmatrix} (3/2)e^{-t} & (1/2)e^{-t} \\ -(1/2)e^{-t} & -(1/2)e^{-t} \end{bmatrix} \begin{bmatrix} 2e^t \\ 4e^t \end{bmatrix} = \begin{bmatrix} 5 \\ -3e^{2t} \end{bmatrix},$$

and so we have

$$\int \mathbf{X}^{-1}(t)\mathbf{f}(t)\, dt = \begin{bmatrix} \int (5)dt \\ -3\int e^{2t}dt \end{bmatrix} = \begin{bmatrix} 5t \\ -(3/2)e^{2t} \end{bmatrix},$$

where we have taken zero constants of integration. Thus, by the equation (8) in the text, we see that

$$\mathbf{x}_p(t) = \begin{bmatrix} e^t & e^{-t} \\ -e^t & -3e^{-t} \end{bmatrix} \begin{bmatrix} 5t \\ -(3/2)e^{2t} \end{bmatrix} = \begin{bmatrix} 5te^t - (3/2)e^t \\ -5te^t + (9/2)e^t \end{bmatrix}.$$

Therefore, by adding $\mathbf{x}_h(t)$ and $\mathbf{x}_p(t)$ we obtain

$$\mathbf{x}(t) = c_1 \begin{bmatrix} e^t \\ -e^t \end{bmatrix} + c_2 \begin{bmatrix} e^{-t} \\ -3e^{-t} \end{bmatrix} + \begin{bmatrix} 5te^t - (3/2)e^t \\ -5te^t + (9/2)e^t \end{bmatrix}.$$

We remark that this answer is the same as that in the text, as can be seen by replacing c_1 by $c_1 + 9/4$.

 Copyright © 2012 Pearson Education, Inc. Publishing as Addison-Wesley.

15. We must first find a fundamental matrix for the associated homogeneous system. We will do this by finding the solutions derived from the eigenvalues and the associated eigenvectors for the coefficient matrix $\mathbf{A}$. We find the eigenvalues by solving the characteristic equation

$$|\mathbf{A} - r\mathbf{I}| = \begin{vmatrix} -4 - r & 2 \\ 2 & -1 - r \end{vmatrix} = 0$$

$$\Rightarrow \quad (-4 - r)(-1 - r) - 4 = 0 \quad \Rightarrow \quad r^2 + 5r = 0.$$

Thus, the eigenvalues of the matrix $\mathbf{A}$ are $r = -5, 0$. An eigenvector for this matrix, associated with the eigenvalue $r = 0$, is $\mathbf{u} = \mathrm{col}(u_1, u_2)$, which satisfies the equation

$$(\mathbf{A} - 0 \cdot \mathbf{I})\,\mathbf{u} = \mathbf{A}\mathbf{u} = \begin{bmatrix} -4 & 2 \\ 2 & -1 \end{bmatrix} \begin{bmatrix} u_1 \\ u_2 \end{bmatrix} = \begin{bmatrix} 0 \\ 0 \end{bmatrix}.$$

This equation is equivalent to $2u_1 = u_2$. Therefore, if we let $u_1 = 1$ and $u_2 = 2$, then the vector $\mathbf{u}_1 = \mathrm{col}(1, 2)$ satisfies this equation and is, therefore, an eigenvector for the matrix $\mathbf{A}$, associated with the eigenvalue $r = 0$. Hence, one solution to the corresponding homogeneous system is given by

$$\mathbf{x}_1(t) = e^{(0)t}\mathbf{u}_1 = \begin{bmatrix} 1 \\ 2 \end{bmatrix}.$$

To find an eigenvector, associated with the eigenvalue $r = -5$, we solve the equation

$$(\mathbf{A} + 5\mathbf{I})\mathbf{u} = \begin{bmatrix} 1 & 2 \\ 2 & 4 \end{bmatrix} \begin{bmatrix} u_1 \\ u_2 \end{bmatrix} = \begin{bmatrix} 0 \\ 0 \end{bmatrix},$$

which is equivalent to the equation $u_1 + 2u_2 = 0$. Thus, by letting $u_2 = 1$ and $u_1 = -2$, we see that the vector $\mathbf{u}_2 = \mathrm{col}(u_1, u_2) = \mathrm{col}(-2, 1)$ satisfies this equation and is, therefore, an eigenvector for $\mathbf{A}$, associated with the eigenvalue $r = -5$. Hence, since the two eigenvalues of $\mathbf{A}$ are distinct, we see that another linearly independent solution to the corresponding homogeneous system is given by

$$\mathbf{x}_2(t) = e^{-5t}\mathbf{u}_2 = e^{-5t} \begin{bmatrix} -2 \\ 1 \end{bmatrix}.$$

By combining these two solutions, we find a general solution to the homogeneous system,

$$\mathbf{x}_h(t) = c_1 \begin{bmatrix} 1 \\ 2 \end{bmatrix} + c_2 e^{-5t} \begin{bmatrix} -2 \\ 1 \end{bmatrix}.$$

Copyright © 2012 Pearson Education, Inc. Publishing as Addison-Wesley.

So, a fundamental matrix for this system is

$$\mathbf{X}(t) = \begin{bmatrix} 1 & -2e^{-5t} \\ 2 & e^{-5t} \end{bmatrix}.$$

We now use the equation (10) in the text to find a particular solution to the given nonhomogeneous system. Thus, we need to find the inverse matrix $\mathbf{X}^{-1}(t)$. This can be done, for example, by performing row-reduction on the matrix $[\mathbf{X}(t)|\mathbf{I}]$ to obtain the matrix $[\mathbf{I}|\mathbf{X}^{-1}(t)]$. On this way, we find that the required inverse matrix is given by

$$\mathbf{X}^{-1}(t) = \begin{bmatrix} 1/5 & 2/5 \\ -(2/5)e^{5t} & (1/5)e^{5t} \end{bmatrix}.$$

Therefore, we have

$$\mathbf{X}^{-1}(t)\mathbf{f}(t) = \begin{bmatrix} 1/5 & 2/5 \\ -(2/5)e^{5t} & (1/5)e^{5t} \end{bmatrix} \begin{bmatrix} t^{-1} \\ 4 + 2t^{-1} \end{bmatrix} = \begin{bmatrix} t^{-1} + (8/5) \\ (4/5)e^{5t} \end{bmatrix}.$$

From this equation, we see that

$$\int \mathbf{X}^{-1}(t)\mathbf{f}(t)\,dt = \begin{bmatrix} \int [t^{-1} + (8/5)]\,dt \\ \int (4/5)e^{5t}\,dt \end{bmatrix} = \begin{bmatrix} \ln|t| + (8/5)t \\ (4/25)e^{5t} \end{bmatrix},$$

where we have taken zero integration constants. Hence, by the equation (10) in the text, we obtain

$$\begin{aligned} \mathbf{x}_p(t) &= \mathbf{X}(t) \int \mathbf{X}^{-1}(t)\mathbf{f}(t)\,dt \\ &= \begin{bmatrix} 1 & -2e^{-5t} \\ 2 & e^{-5t} \end{bmatrix} \begin{bmatrix} \ln|t| + (8/5)t \\ (4/25)e^{5t} \end{bmatrix} = \begin{bmatrix} \ln|t| + (8/5)t - (8/25) \\ 2\ln|t| + (16/5)t + (4/25) \end{bmatrix}. \end{aligned}$$

Adding $\mathbf{x}_h(t)$ and $\mathbf{x}_p(t)$ yields a general solution to the given nonhomogeneous system,

$$\mathbf{x}(t) = c_1 \begin{bmatrix} 1 \\ 2 \end{bmatrix} + c_2 e^{-5t} \begin{bmatrix} -2 \\ 1 \end{bmatrix} + \begin{bmatrix} \ln|t| + (8/5)t - (8/25) \\ 2\ln|t| + (16/5)t + (4/25) \end{bmatrix}.$$

21. We will find the solution to this initial value problem by using the equation (13) in the text. Therefore, we have to find a fundamental matrix for the associated homogeneous system. This means that we need the eigenvalues and the corresponding eigenvectors for the coefficient matrix of this system. Solving the characteristic equation,

$$|\mathbf{A} - r\mathbf{I}| = \begin{vmatrix} -r & 2 \\ -1 & 3 - r \end{vmatrix} = 0$$

 Copyright © 2012 Pearson Education, Inc. Publishing as Addison-Wesley.

$$\Rightarrow \quad -r(3-r)+2=0 \quad \Rightarrow \quad r^2-3r+2=0 \quad \Rightarrow \quad (r-2)(r-1)=0\,,$$

yields $r=1,2$. To find an eigenvector $\mathbf{u}=\operatorname{col}(u_1,u_2)$, associated with the eigenvalue $r=1$, we solve the system

$$(\mathbf{A}-\mathbf{I})\mathbf{u} = \begin{bmatrix} -1 & 2 \\ -1 & 2 \end{bmatrix}\begin{bmatrix} u_1 \\ u_2 \end{bmatrix} = \begin{bmatrix} 0 \\ 0 \end{bmatrix}.$$

This system is equivalent to the equation $u_1 = 2u_2$. Thus, $u_1 = 2$ and $u_2 = 1$ is a set of values, which satisfies this equation and, therefore, the vector $\mathbf{u}_1 = \operatorname{col}(2,1)$ is an eigenvector for the coefficient matrix, corresponding to the eigenvalue $r=1$. Hence, one solution to the associated homogeneous system is given by

$$\mathbf{x}_1(t) = e^t\mathbf{u}_1 = e^t\begin{bmatrix} 2 \\ 1 \end{bmatrix} = \begin{bmatrix} 2e^t \\ e^t \end{bmatrix}.$$

Similarly, by solving the equation

$$(\mathbf{A}-2\mathbf{I})\mathbf{u} = \begin{bmatrix} -2 & 2 \\ -1 & 1 \end{bmatrix}\begin{bmatrix} u_1 \\ u_2 \end{bmatrix} = \begin{bmatrix} 0 \\ 0 \end{bmatrix},$$

we find that an eigenvector for the coefficient matrix, associated with the eigenvalue $r=2$, is $\mathbf{u}_2 = \operatorname{col}(1,1)$. Thus, another linearly independent solution to the associated homogeneous problem is

$$\mathbf{x}_2(t) = e^{2t}\mathbf{u}_2 = e^{2t}\begin{bmatrix} 1 \\ 1 \end{bmatrix} = \begin{bmatrix} e^{2t} \\ e^{2t} \end{bmatrix}.$$

By combining these two solutions, we obtain a general solution to the homogeneous system

$$\mathbf{x}_h(t) = c_1\begin{bmatrix} 2e^t \\ e^t \end{bmatrix} + c_2\begin{bmatrix} e^{2t} \\ e^{2t} \end{bmatrix}$$

and the fundamental matrix

$$\mathbf{X}(t) = \begin{bmatrix} 2e^t & e^{2t} \\ e^t & e^{2t} \end{bmatrix}.$$

In order to use the equation (13) in the text, we must also find the inverse of the fundamental matrix. One way of doing this is to perform a row-reduction on the matrix $[\mathbf{X}(t)|\mathbf{I}]$ to obtain the matrix $[\mathbf{I}|\mathbf{X}^{-1}(t)]$. Thus, we find that

$$\mathbf{X}^{-1}(t) = \begin{bmatrix} e^{-t} & -e^{-t} \\ -e^{-2t} & 2e^{-2t} \end{bmatrix}.$$

Copyright © 2012 Pearson Education, Inc. Publishing as Addison-Wesley.

From this we see that

$$\mathbf{X}^{-1}(s)\mathbf{f}(s) = \begin{bmatrix} e^{-s} & -e^{-s} \\ -e^{-2s} & 2e^{-2s} \end{bmatrix}\begin{bmatrix} e^s \\ -e^s \end{bmatrix} = \begin{bmatrix} 2 \\ -3e^{-s} \end{bmatrix}.$$

(a) Using the initial conditions $t_0 = 0$ and $\mathbf{x}(0) = \text{col}(5,4)$, we obtain

$$\mathbf{X}^{-1}(0) = \begin{bmatrix} 1 & -1 \\ -1 & 2 \end{bmatrix}.$$

Therefore,

$$\int_{t_0}^{t} \mathbf{X}^{-1}(s)\mathbf{f}(s)\,ds = \int_{0}^{t} \mathbf{X}^{-1}(s)\mathbf{f}(s)\,ds = \begin{bmatrix} \int_0^t (2)\,ds \\ \int_0^t (-3e^{-s})\,ds \end{bmatrix} = \begin{bmatrix} 2t \\ 3e^{-t} - 3 \end{bmatrix},$$

from which it follows that

$$\mathbf{X}(t)\int_{t_0}^{t} \mathbf{X}^{-1}(s)\mathbf{f}(s)\,ds = \begin{bmatrix} 2e^t & e^{2t} \\ e^t & e^{2t} \end{bmatrix}\begin{bmatrix} 2t \\ 3e^{-t} - 3 \end{bmatrix} = \begin{bmatrix} 4te^t + 3e^t - 3e^{2t} \\ 2te^t + 3e^t - 3e^{2t} \end{bmatrix}.$$

We also find that

$$\mathbf{X}(t)\mathbf{X}^{-1}(t_0)\mathbf{x}_0 = \begin{bmatrix} 2e^t & e^{2t} \\ e^t & e^{2t} \end{bmatrix}\begin{bmatrix} 1 & -1 \\ -1 & 2 \end{bmatrix}\begin{bmatrix} 5 \\ 4 \end{bmatrix}$$

$$= \begin{bmatrix} 2e^t & e^{2t} \\ e^t & e^{2t} \end{bmatrix}\begin{bmatrix} 1 \\ 3 \end{bmatrix} = \begin{bmatrix} 2e^t + 3e^{2t} \\ e^t + 3e^{2t} \end{bmatrix}.$$

Hence, substituting these expressions into the equation (13) in the text, we obtain the solution to the given initial value problem

$$\mathbf{x}(t) = \mathbf{X}(t)\mathbf{X}^{-1}(t_0)\mathbf{x}_0 + \mathbf{X}(t)\int_{t_0}^{t} \mathbf{X}^{-1}(s)\mathbf{f}(s)\,ds$$

$$= \begin{bmatrix} 2e^t + 3e^{2t} \\ e^t + 3e^{2t} \end{bmatrix} + \begin{bmatrix} 4te^t + 3e^t - 3e^{2t} \\ 2te^t + 3e^t - 3e^{2t} \end{bmatrix} = \begin{bmatrix} 4te^t + 5e^t \\ 2te^t + 4e^t \end{bmatrix}.$$

(b) Using the initial conditions $t_0 = 1$ and $\mathbf{x}(1) = \text{col}(0,1)$, we get

$$\mathbf{X}^{-1}(1) = \begin{bmatrix} e^{-1} & -e^{-1} \\ -e^{-2} & 2e^{-2} \end{bmatrix}.$$

Therefore,

$$\int_{t_0}^{t} \mathbf{X}^{-1}(s)\mathbf{f}(s)\,ds = \int_{1}^{t} \mathbf{X}^{-1}(s)\mathbf{f}(s)\,ds$$

Copyright © 2012 Pearson Education, Inc. Publishing as Addison-Wesley.

$$= \begin{bmatrix} \int_1^t (2)\,ds \\ \int_1^t (-3e^{-s})\,ds \end{bmatrix} = \begin{bmatrix} 2t - 2 \\ 3e^{-t} - 3e^{-1} \end{bmatrix},$$

from which it follows that

$$\mathbf{X}(t) \int_{t_0}^t \mathbf{X}^{-1}(s)\mathbf{f}(s)\,ds = \begin{bmatrix} 2e^t & e^{2t} \\ e^t & e^{2t} \end{bmatrix} \begin{bmatrix} 2t - 2 \\ 3e^{-t} - 3e^{-1} \end{bmatrix}$$

$$= \begin{bmatrix} 4te^t - 4e^t + 3e^t - 3e^{2t-1} \\ 2te^t - 2e^t + 3e^t - 3e^{2t-1} \end{bmatrix} = \begin{bmatrix} 4te^t - e^t - 3e^{2t-1} \\ 2te^t + e^t - 3e^{2t-1} \end{bmatrix}.$$

We also find that

$$\mathbf{X}(t)\mathbf{X}^{-1}(t_0)\mathbf{x}_0 = \begin{bmatrix} 2e^t & e^{2t} \\ e^t & e^{2t} \end{bmatrix} \begin{bmatrix} e^{-1} & -e^{-1} \\ -e^{-2} & 2e^{-2} \end{bmatrix} \begin{bmatrix} 0 \\ 1 \end{bmatrix}$$

$$= \begin{bmatrix} 2e^t & e^{2t} \\ e^t & e^{2t} \end{bmatrix} \begin{bmatrix} -e^{-1} \\ 2e^{-2} \end{bmatrix} = \begin{bmatrix} -2e^{t-1} + 2e^{2t-2} \\ -e^{t-1} + 2e^{2t-2} \end{bmatrix}.$$

Hence, by substituting these expressions into the equation (13), we obtain the solution

$$\mathbf{x}(t) = \mathbf{X}(t)\mathbf{X}^{-1}(t_0)\mathbf{x}_0 + \mathbf{X}(t) \int_{t_0}^t \mathbf{X}^{-1}(s)\mathbf{f}(s)\,ds$$

$$= \begin{bmatrix} -2e^{t-1} + 2e^{2t-2} \\ -e^{t-1} + 2e^{2t-2} \end{bmatrix} + \begin{bmatrix} 4te^t - e^t - 3e^{2t-1} \\ 2te^t + e^t - 3e^{2t-1} \end{bmatrix}$$

$$= \begin{bmatrix} -2e^{t-1} + 2e^{2(t-1)} - 3e^{2t-1} + (4t - 1)e^t \\ -e^{t-1} + 2e^{2(t-1)} - 3e^{2t-1} + (2t + 1)e^t \end{bmatrix}.$$

to the given initial value problem.

(c) Using the initial conditions $\mathbf{x}(5) = \mathrm{col}(1,0)$ and $t_0 = 5$, we have

$$\mathbf{X}^{-1}(5) = \begin{bmatrix} e^{-5} & -e^{-5} \\ -e^{-10} & 2e^{-10} \end{bmatrix}.$$

Therefore,

$$\int_{t_0}^t \mathbf{X}^{-1}(s)\mathbf{f}(s)\,ds = \int_5^t \mathbf{X}^{-1}(s)\mathbf{f}(s)\,ds = \begin{bmatrix} \int_5^t (2)\,ds \\ \int_5^t (-3e^{-s})\,ds \end{bmatrix} = \begin{bmatrix} 2t - 10 \\ 3e^{-t} - 3e^{-5} \end{bmatrix},$$

Copyright © 2012 Pearson Education, Inc. Publishing as Addison-Wesley.

from which it follows that

$$\mathbf{X}(t) \int_{t_0}^{t} \mathbf{X}^{-1}(s)\mathbf{f}(s)\,ds = \begin{bmatrix} 2e^t & e^{2t} \\ e^t & e^{2t} \end{bmatrix} \begin{bmatrix} 2t-10 \\ 3e^{-t} - 3e^{-5} \end{bmatrix}$$

$$= \begin{bmatrix} 4te^t - 20e^t + 3e^t - 3e^{2t-5} \\ 2te^t - 10e^t + 3e^t - 3e^{2t-5} \end{bmatrix} = \begin{bmatrix} 4te^t - 17e^t - 3e^{2t-5} \\ 2te^t - 7e^t - 3e^{2t-5} \end{bmatrix}.$$

We also find that

$$\mathbf{X}(t)\mathbf{X}^{-1}(t_0)\mathbf{x}_0 = \begin{bmatrix} 2e^t & e^{2t} \\ e^t & e^{2t} \end{bmatrix} \begin{bmatrix} e^{-5} & -e^{-5} \\ -e^{-10} & 2e^{-10} \end{bmatrix} \begin{bmatrix} 1 \\ 0 \end{bmatrix}$$

$$= \begin{bmatrix} 2e^t & e^{2t} \\ e^t & e^{2t} \end{bmatrix} \begin{bmatrix} e^{-5} \\ -e^{-10} \end{bmatrix} = \begin{bmatrix} 2e^{t-5} - e^{2t-10} \\ e^{t-5} - e^{2t-10} \end{bmatrix}.$$

Hence, substituting these expressions into (13) in the text, we obtain the solution to the given initial value problem

$$\mathbf{x}(t) = \mathbf{X}(t)\mathbf{X}^{-1}(t_0)\mathbf{x}_0 + \mathbf{X}(t) \int_{t_0}^{t} \mathbf{X}^{-1}(s)\mathbf{f}(s)\,ds$$

$$= \begin{bmatrix} 2e^{t-5} - e^{2t-10} \\ e^{t-5} - e^{2t-10} \end{bmatrix} + \begin{bmatrix} 4te^t - 17e^t - 3e^{2t-5} \\ 2te^t - 7e^t - 3e^{2t-5} \end{bmatrix}$$

$$= \begin{bmatrix} 2e^{t-5} - e^{2t-10} + 4te^t - 17e^t - 3e^{2t-5} \\ e^{t-5} - e^{2t-10} + 2te^t - 7e^t - 3e^{2t-5} \end{bmatrix},$$

which is equivalent to the answer in the text.

25. (a) We will find a fundamental solution set to the corresponding homogeneous system by using the eigenvalues and associated eigenvectors for the coefficient matrix. First, we solve the characteristic equation

$$|\mathbf{A} - r\mathbf{I}| = \begin{vmatrix} -r & 1 \\ -2 & 3-r \end{vmatrix} = 0 \quad \Rightarrow \quad -r(3-r) + 2 = 0$$

$$\Rightarrow \quad r^2 - 3r + 2 = 0 \quad \Rightarrow \quad (r-2)(r-1) = 0.$$

Thus, we see that the eigenvalues for the coefficient matrix are $r = 1, 2$. Since these eigenvalues are real and distinct, the associated eigenvectors are linearly independent, and so the solutions, derived from these eigenvectors, are also linearly

Copyright © 2012 Pearson Education, Inc. Publishing as Addison-Wesley.

independent. We find an eigenvector, associated with the eigenvalue $r = 1$, by solving the equation

$$(\mathbf{A} - \mathbf{I})\mathbf{u} = \begin{bmatrix} -1 & 1 \\ -2 & 2 \end{bmatrix} \begin{bmatrix} u_1 \\ u_2 \end{bmatrix} = \begin{bmatrix} -u_1 + u_2 \\ -2u_1 + 2u_2 \end{bmatrix} = \mathbf{0}.$$

Since the vector $\mathbf{u}_1 = \text{col}(u_1, u_2) = \text{col}(1, 1)$ satisfies this equation, we see that $\mathbf{u}_1$ is an eigenvector, and so one solution to the homogeneous problem is given by

$$\mathbf{x}_1(t) = e^t \mathbf{u}_1 = e^t \begin{bmatrix} 1 \\ 1 \end{bmatrix}.$$

To find an eigenvector, associated with the eigenvalue $r = 2$, we solve the equation

$$(\mathbf{A} - 2\mathbf{I})\mathbf{u} = \begin{bmatrix} -2 & 1 \\ -2 & 1 \end{bmatrix} \begin{bmatrix} u_1 \\ u_2 \end{bmatrix} = \begin{bmatrix} -2u_1 + u_2 \\ -2u_1 + u_2 \end{bmatrix} = \mathbf{0}.$$

The vector $\mathbf{u}_2 = \text{col}(u_1, u_2) = \text{col}(1, 2)$ is a vector which satisfies this equation, and so it is an eigenvector of the coefficient matrix, associated with the eigenvalue $r = 2$. Thus, another linearly independent solution to the corresponding homogeneous problem is given by

$$\mathbf{x}_2(t) = e^{2t} \mathbf{u}_2 = e^{2t} \begin{bmatrix} 1 \\ 2 \end{bmatrix},$$

and a fundamental solution set for the homogeneous system is

$$\left\{ e^t \begin{bmatrix} 1 \\ 1 \end{bmatrix}, e^{2t} \begin{bmatrix} 1 \\ 2 \end{bmatrix} \right\}.$$

(b) If we assume that $\mathbf{x}_p(t) = te^t \mathbf{a}$ for some constant vector $\mathbf{a} = \text{col}(a_1, a_2)$, then we have

$$\mathbf{x}_p'(t) = te^t \mathbf{a} + e^t \mathbf{a} = \begin{bmatrix} te^t a_1 \\ te^t a_2 \end{bmatrix} + \begin{bmatrix} e^t a_1 \\ e^t a_2 \end{bmatrix} = \begin{bmatrix} te^t a_1 + e^t a_1 \\ te^t a_2 + e^t a_2 \end{bmatrix}.$$

We also have

$$\begin{bmatrix} 0 & 1 \\ -2 & 3 \end{bmatrix} \mathbf{x}_p(t) + \mathbf{f}(t) = \begin{bmatrix} 0 & 1 \\ -2 & 3 \end{bmatrix} \begin{bmatrix} te^t a_1 \\ te^t a_2 \end{bmatrix} + \begin{bmatrix} e^t \\ 0 \end{bmatrix} = \begin{bmatrix} te^t a_2 + e^t \\ -2te^t a_1 + 3te^t a_2 \end{bmatrix}.$$

Thus, if $\mathbf{x}_p(t) = te^t \mathbf{a}$ is to satisfy this system, we must have

$$\begin{bmatrix} te^t a_1 + e^t a_1 \\ te^t a_2 + e^t a_2 \end{bmatrix} = \begin{bmatrix} te^t a_2 + e^t \\ -2te^t a_1 + 3te^t a_2 \end{bmatrix},$$

Copyright © 2012 Pearson Education, Inc. Publishing as Addison-Wesley.

which means that

$$te^t a_1 + e^t a_1 = te^t a_2 + e^t,$$
$$te^t a_2 + e^t a_2 = -2te^t a_1 + 3te^t a_2.$$

Dividing by e^t and equating coefficients yields

$$a_1 = a_2, \qquad a_1 = 1,$$
$$a_2 = -2a_1 + 3a_2, \quad a_2 = 0.$$

Since this set of equations implies that $1 = a_1 = a_2 = 0$, which is inconsistent, we conclude that this system has no solutions. Therefore, one cannot find a vector $\mathbf{a}$, for which $\mathbf{x}_p(t) = te^t \mathbf{a}$ is a particular solution to this problem.

(c) Assuming that

$$\mathbf{x}_p(t) = te^t \mathbf{a} + e^t \mathbf{b} = \begin{bmatrix} te^t a_1 \\ te^t a_2 \end{bmatrix} + \begin{bmatrix} e^t b_1 \\ e^t b_2 \end{bmatrix} = \begin{bmatrix} te^t a_1 + e^t b_1 \\ te^t a_2 + e^t b_2 \end{bmatrix},$$

where $\mathbf{a} = \mathrm{col}(a_1, a_2)$ and $\mathbf{b} = \mathrm{col}(b_1, b_2)$ are two constant vectors, yields

$$\mathbf{x}_p'(t) = te^t \mathbf{a} + e^t \mathbf{a} + e^t \mathbf{b} = \begin{bmatrix} te^t a_1 + e^t a_1 + e^t b_1 \\ te^t a_2 + e^t a_2 + e^t a_2 \end{bmatrix}.$$

We also see that

$$\begin{bmatrix} 0 & 1 \\ -2 & 3 \end{bmatrix} \mathbf{x}_p(t) + \mathbf{f}(t) = \begin{bmatrix} 0 & 1 \\ -2 & 3 \end{bmatrix} \begin{bmatrix} te^t a_1 + e^t b_1 \\ te^t a_2 + e^t b_2 \end{bmatrix} + \begin{bmatrix} e^t \\ 0 \end{bmatrix}$$
$$= \begin{bmatrix} te^t a_2 + e^t b_2 + e^t \\ -2te^t a_1 - 2e^t b_1 + 3te^t a_2 + 3e^t b_2 \end{bmatrix}.$$

Thus, if $\mathbf{x}_p(t)$ satisfies this system, we must have

$$\begin{bmatrix} te^t a_1 + e^t a_1 + e^t b_1 \\ te^t a_2 + e^t a_2 + e^t b_2 \end{bmatrix} = \begin{bmatrix} te^t a_2 + e^t b_2 + e^t \\ -2te^t a_1 - 2e^t b_1 + 3te^t a_2 + 3e^t b_2 \end{bmatrix}, \tag{9.10}$$

which implies that

$$te^t a_1 + e^t a_1 + e^t b_1 = te^t a_2 + e^t b_2 + e^t,$$
$$te^t a_2 + e^t a_2 + e^t b_2 = -2te^t a_1 - 2e^t b_1 + 3te^t a_2 + 3e^t b_2.$$

Dividing each equation by e^t and equating the coefficients yields

$$a_1 = a_2, \qquad a_1 + b_1 = b_2 + 1,$$
$$a_2 = -2a_1 + 3a_2, \quad a_2 + b_2 = -2b_1 + 3b_2. \tag{9.11}$$

 Copyright © 2012 Pearson Education, Inc. Publishing as Addison-Wesley.

The second and the last equations give us

$$b_1 - b_2 = 1 - a_1 ,$$
$$2b_1 - 2b_2 = -a_2 .$$
(9.12)

By multiplying the first equation by 2, we obtain the system

$$2b_1 - 2b_2 = 2 - 2a_1 ,$$
$$2b_1 - 2b_2 = -a_2 ,$$

which results $2 - 2a_1 + a_2 = 0$. Applying the first equation in (9.11) (i.e., $a_1 = a_2$) yields $a_1 = a_2 = 2$. By substituting these values for a_1 and a_2 into the first equation in (9.12), we see that both equations reduce to $b_2 = b_1 + 1$. Thus, b_1 can have any value, say $b_1 = s$, and so the set $a_1 = a_2 = 2$, $b_1 = s$, $b_2 = s + 1$, satisfies all of the equations given in (9.11) and, hence, the system given in (9.10). Therefore, particular solutions to the nonhomogeneous equation in this problem are

$$\mathbf{x}_p(t) = te^t \begin{bmatrix} 2 \\ 2 \end{bmatrix} + e^t \begin{bmatrix} s \\ s+1 \end{bmatrix} = te^t \begin{bmatrix} 2 \\ 2 \end{bmatrix} + e^t \begin{bmatrix} 0 \\ 1 \end{bmatrix} + se^t \begin{bmatrix} 1 \\ 1 \end{bmatrix} .$$

But, since the vector $\mathbf{u} = e^t \mathrm{col}(1, 1)$ is a solution to the corresponding homogeneous system, the last term can be incorporated into the solution $\mathbf{x}_h(t)$, and so we obtain

$$\mathbf{x}_p(t) = te^t \begin{bmatrix} 2 \\ 2 \end{bmatrix} + e^t \begin{bmatrix} 0 \\ 1 \end{bmatrix} .$$

(d) To find a general solution to the nonhomogeneous system, we first form the solution to the corresponding homogeneous system using the fundamental solution set found in part (a).

$$\mathbf{x}_h(t) = c_1 e^t \begin{bmatrix} 1 \\ 1 \end{bmatrix} + c_2 e^{2t} \begin{bmatrix} 1 \\ 2 \end{bmatrix} .$$

By adding a particular solution found in part (c), we obtain a general solution

$$\mathbf{x}(t) = c_1 e^t \begin{bmatrix} 1 \\ 1 \end{bmatrix} + c_2 e^{2t} \begin{bmatrix} 1 \\ 2 \end{bmatrix} + te^t \begin{bmatrix} 2 \\ 2 \end{bmatrix} + e^t \begin{bmatrix} 0 \\ 1 \end{bmatrix} .$$

EXERCISES 9.8: The Matrix Exponential Function

3. (a) From the characteristic equation, $|\mathbf{A} - r\mathbf{I}| = 0$, we obtain

$$|\mathbf{A} - r\mathbf{I}| = \begin{vmatrix} 2 - r & 1 & -1 \\ -3 & -1 - r & 1 \\ 9 & 3 & -4 - r \end{vmatrix} = 0$$

Copyright © 2012 Pearson Education, Inc. Publishing as Addison-Wesley.

$$\Rightarrow \quad (2-r)\begin{vmatrix} -1-r & 1 \\ 3 & -4-r \end{vmatrix} - \begin{vmatrix} -3 & 1 \\ 9 & -4-r \end{vmatrix} + (-1)\begin{vmatrix} -3 & -1-r \\ 9 & 3 \end{vmatrix} = 0$$

$$\Rightarrow \quad (2-r)[(-1-r)(-4-r)-3] - [-3(-4-r)-9] - [-9-9(-1-r)] = 0$$

$$\Rightarrow \quad r^3 + 3r^2 + 3r + 1 = (r+1)^3 = 0.$$

Therefore, for the matrix $\mathbf{A}$, $r = -1$ is an eigenvalue of multiplicity three. Thus, by the Cayley-Hamilton theorem (with $r = -1$ and $k = 3$), we have

$$(\mathbf{A} + \mathbf{I})^3 = \mathbf{0}.$$

(b) In order to find $e^{\mathbf{A}t}$, we first notice that

$$\begin{aligned}
e^{\mathbf{A}t} &= e^{[-\mathbf{I}+(\mathbf{A}+\mathbf{I})]t} \quad \text{(commutative and associative properties of matrix addition)} \\
&= e^{-\mathbf{I}t}e^{(\mathbf{A}+\mathbf{I})t} \quad \text{(property (d) in the text, since } (\mathbf{A}+\mathbf{I})\mathbf{I} = \mathbf{I}(\mathbf{A}+\mathbf{I})) \\
&= e^{-t}\mathbf{I}e^{(\mathbf{A}+\mathbf{I})t} \quad \text{(property (e) in the text)} \\
&= e^{-t}e^{(\mathbf{A}+\mathbf{I})t}.
\end{aligned}$$

Therefore, to find $e^{\mathbf{A}t}$, we need only to find $e^{(\mathbf{A}+\mathbf{I})t}$ and then multiply it by e^{-t}. By the formula (2), since $(\mathbf{A}+\mathbf{I})^3 = \mathbf{0}$ (which implies that $(\mathbf{A}+\mathbf{I})^n = \mathbf{0}$ for $n \geq 3$), we have

$$\begin{aligned}
e^{(\mathbf{A}+\mathbf{I})t} &= \mathbf{I} + (\mathbf{A}+\mathbf{I})t + (\mathbf{A}+\mathbf{I})^2\left(\frac{t^2}{2}\right) + \cdots + (\mathbf{A}+\mathbf{I})^n\left(\frac{t^n}{n!}\right) + \cdots \\
&= \mathbf{I} + (\mathbf{A}+\mathbf{I})t + (\mathbf{A}+\mathbf{I})^2\left(\frac{t^2}{2}\right).
\end{aligned} \tag{9.13}$$

Since

$$(\mathbf{A}+\mathbf{I})^2 = \begin{bmatrix} 3 & 1 & -1 \\ -3 & 0 & 1 \\ 9 & 3 & -3 \end{bmatrix}\begin{bmatrix} 3 & 1 & -1 \\ -3 & 0 & 1 \\ 9 & 3 & -3 \end{bmatrix} = \begin{bmatrix} -3 & 0 & 1 \\ 0 & 0 & 0 \\ -9 & 0 & 3 \end{bmatrix},$$

the equation (9.13) becomes

$$\begin{aligned}
e^{(\mathbf{A}+\mathbf{I})t} &= \begin{bmatrix} 1 & 0 & 0 \\ 0 & 1 & 0 \\ 0 & 0 & 1 \end{bmatrix} + \begin{bmatrix} 3t & t & -t \\ -3t & 0 & t \\ 9t & 3t & -3t \end{bmatrix} + \begin{bmatrix} -(3/2)t^2 & 0 & (1/2)t^2 \\ 0 & 0 & 0 \\ -(9/2)t^2 & 0 & (3/2)t^2 \end{bmatrix} \\
&= \begin{bmatrix} 1+3t-(3/2)t^2 & t & -t+(1/2)t^2 \\ -3t & 1 & t \\ 9t-(9/2)t^2 & 3t & 1-3t+(3/2)t^2 \end{bmatrix}.
\end{aligned}$$

Copyright © 2012 Pearson Education, Inc. Publishing as Addison-Wesley.

Hence, we have

$$e^{\mathbf{A}t} = e^{-t}\begin{bmatrix} 1 + 3t - (3/2)t^2 & t & -t + (1/2)t^2 \\ -3t & 1 & t \\ 9t - (9/2)t^2 & 3t & 1 - 3t + (3/2)t^2 \end{bmatrix}.$$

9. We have $e^{\mathbf{A}t} = \mathbf{X}(t)\mathbf{X}^{-1}(0)$, where $\mathbf{X}(t)$ is a fundamental matrix for the system $\mathbf{x}' = \mathbf{A}\mathbf{x}$. We construct this fundamental matrix from three linearly independent solutions derived from the eigenvalues and associated eigenvectors for the matrix $\mathbf{A}$. Thus, we solve the characteristic equation

$$|\mathbf{A} - r\mathbf{I}| = \begin{vmatrix} -r & 1 & 0 \\ 0 & -r & 1 \\ 1 & -1 & 1 - r \end{vmatrix} = 0$$

$$\Rightarrow \quad (-r)\begin{vmatrix} -r & 1 \\ -1 & 1-r \end{vmatrix} - \begin{vmatrix} 0 & 1 \\ 1 & 1-r \end{vmatrix} = 0$$

$$\Rightarrow \quad -r[-r(1-r) + 1] + 1 = -r^3 + r^2 - r + 1 = -(r-1)(r^2+1) = 0.$$

Therefore, the eigenvalues of the matrix $\mathbf{A}$ are $r = 1$ and $r = \pm i$. To find an eigenvector $\mathbf{u} = \mathrm{col}(u_1, u_2, u_3)$, associated with the eigenvalue $r = 1$, we solve the system

$$(\mathbf{A} - \mathbf{I})\mathbf{u} = \mathbf{0} \qquad \Rightarrow \qquad \begin{bmatrix} -1 & 1 & 0 \\ 0 & -1 & 1 \\ 1 & -1 & 0 \end{bmatrix}\begin{bmatrix} u_1 \\ u_2 \\ u_3 \end{bmatrix} = \begin{bmatrix} 0 \\ 0 \\ 0 \end{bmatrix}$$

$$\Rightarrow \qquad \begin{bmatrix} -1 & 0 & 1 \\ 0 & -1 & 1 \\ 0 & 0 & 0 \end{bmatrix}\begin{bmatrix} u_1 \\ u_2 \\ u_3 \end{bmatrix} = \begin{bmatrix} 0 \\ 0 \\ 0 \end{bmatrix}.$$

This system is equivalent to $u_1 = u_3$, $u_2 = u_3$. Hence, u_3 is free to be arbitrary, say $u_3 = 1$. Then $u_1 = u_2 = 1$, and so the vector $\mathbf{u} = \mathrm{col}(1, 1, 1)$ is an eigenvector, associated with $r = 1$. Hence, one solution to the system $\mathbf{x}' = \mathbf{A}\mathbf{x}$ is given by

$$\mathbf{x}_1(t) = e^t\mathbf{u} = e^t\begin{bmatrix} 1 \\ 1 \\ 1 \end{bmatrix} = \begin{bmatrix} e^t \\ e^t \\ e^t \end{bmatrix}.$$

Since the eigenvalue $r = i$ is complex, we first find a complex-valued eigenvector, associated with this eigenvalue. Such an eigenvector, $\mathbf{z} = \mathrm{col}(z_1, z_2, z_3)$, must satisfy the

Copyright © 2012 Pearson Education, Inc. Publishing as Addison-Wesley.

equation

$$(A - iI)z = \begin{bmatrix} -i & 1 & 0 \\ 0 & -i & 1 \\ 1 & -1 & 1-i \end{bmatrix} \begin{bmatrix} z_1 \\ z_2 \\ z_3 \end{bmatrix} = \begin{bmatrix} 0 \\ 0 \\ 0 \end{bmatrix},$$

which is equivalent to the system $z_1 = -z_3$, $z_2 = -iz_3$. One solution to this system is $z_3 = 1$, $z_1 = -1$, and $z_2 = -i$, and so an eigenvector for A, associated with the eigenvalue $r = i$, is given by

$$z = \begin{bmatrix} z_1 \\ z_2 \\ z_3 \end{bmatrix} = \begin{bmatrix} -1 \\ -i \\ 1 \end{bmatrix} = \begin{bmatrix} -1 \\ 0 \\ 1 \end{bmatrix} + i \begin{bmatrix} 0 \\ -1 \\ 0 \end{bmatrix} = a + ib,$$

where $a = \text{col}(-1, 0, 1)$ and $b = \text{col}(0, -1, 0)$. By the equations (6) and (7) in the text (with $\alpha = 0$, $\beta = 1$), we see that two other linearly independent solutions to the system $x' = Ax$ are given by

$$x_2(t) = e^{(0)t}(\cos t)a - e^{(0)t}(\sin t)b = \begin{bmatrix} -\cos t \\ 0 \\ \cos t \end{bmatrix} - \begin{bmatrix} 0 \\ -\sin t \\ 0 \end{bmatrix} = \begin{bmatrix} -\cos t \\ \sin t \\ \cos t \end{bmatrix},$$

$$x_3(t) = e^{(0)t}(\sin t)a + e^{(0)t}(\cos t)b = \begin{bmatrix} -\sin t \\ 0 \\ \sin t \end{bmatrix} + \begin{bmatrix} 0 \\ -\cos t \\ 0 \end{bmatrix} = \begin{bmatrix} -\sin t \\ -\cos t \\ \sin t \end{bmatrix}.$$

Thus, a fundamental matrix for this system is

$$X(t) = \begin{bmatrix} e^t & -\cos t & -\sin t \\ e^t & \sin t & -\cos t \\ e^t & \cos t & \sin t \end{bmatrix} \quad\Rightarrow\quad X(0) = \begin{bmatrix} 1 & -1 & 0 \\ 1 & 0 & -1 \\ 1 & 1 & 0 \end{bmatrix}.$$

To find the inverse of $X(0)$ we can, for example, perform a row-reduction on the augmented matrix $[X(0)|I]$ to obtain the matrix $[I|X^{-1}(0)]$. Thus, we see that

$$X^{-1}(0) = \begin{bmatrix} 1/2 & 0 & 1/2 \\ -1/2 & 0 & 1/2 \\ 1/2 & -1 & 1/2 \end{bmatrix}.$$

Hence,

$$e^{At} = X(t)X^{-1}(0) = \begin{bmatrix} e^t & -\cos t & -\sin t \\ e^t & \sin t & -\cos t \\ e^t & \cos t & \sin t \end{bmatrix} \begin{bmatrix} 1/2 & 0 & 1/2 \\ -1/2 & 0 & 1/2 \\ 1/2 & -1 & 1/2 \end{bmatrix}$$

Copyright © 2012 Pearson Education, Inc. Publishing as Addison-Wesley.

$$= \frac{1}{2} \begin{bmatrix} e^t + \cos t - \sin t & 2\sin t & e^t - \cos t - \sin t \\ e^t - \cos t - \sin t & 2\cos t & e^t - \cos t + \sin t \\ e^t - \cos t + \sin t & -2\sin t & e^t + \cos t + \sin t \end{bmatrix}.$$

11. The first step in finding $e^{\mathbf{A}t}$, using a fundamental matrix for the system $\mathbf{x}' = \mathbf{A}\mathbf{x}$, is to find the eigenvalues for the matrix $\mathbf{A}$. Thus, we solve the characteristic equation

$$|\mathbf{A} - r\mathbf{I}| = \begin{vmatrix} 5-r & -4 & 0 \\ 1 & -r & 2 \\ 0 & 2 & 5-r \end{vmatrix} = 0$$

$$\Rightarrow \quad (5-r)\begin{vmatrix} -r & 2 \\ 2 & 5-r \end{vmatrix} + 4\begin{vmatrix} 1 & 2 \\ 0 & 5-r \end{vmatrix} = 0$$

$$\Rightarrow \quad (5-r)[-r(5-r) - 4] + 4(5-r) = -r(r-5)^2 = 0.$$

Therefore, the eigenvalues of $\mathbf{A}$ are $r = 0$ and $r = 5$ of multiplicity two. Next, we find the eigenvectors and generalized eigenvectors for the matrix $\mathbf{A}$ and, from these vectors, derive three linearly independent solutions of the system $\mathbf{x}' = \mathbf{A}\mathbf{x}$. To find the eigenvector, associated with the eigenvalue $r = 0$, we solve the equation

$$(\mathbf{A} - 0 \cdot \mathbf{I})\mathbf{u} = \mathbf{A}\mathbf{u} = \begin{bmatrix} 5 & -4 & 0 \\ 1 & 0 & 2 \\ 0 & 2 & 5 \end{bmatrix}\begin{bmatrix} u_1 \\ u_2 \\ u_3 \end{bmatrix} = \begin{bmatrix} 0 \\ 0 \\ 0 \end{bmatrix} \Rightarrow \begin{bmatrix} 1 & 0 & 2 \\ 0 & 2 & 5 \\ 0 & 0 & 0 \end{bmatrix}\begin{bmatrix} u_1 \\ u_2 \\ u_3 \end{bmatrix} = \begin{bmatrix} 0 \\ 0 \\ 0 \end{bmatrix}.$$

This equation is equivalent to the system $u_1 = -2u_3$, $2u_2 = -5u_3$, and a solution to this system is $u_3 = 2$, $u_1 = -4$, $u_2 = -5$. Therefore, an eigenvector of the matrix $\mathbf{A}$, associated with the eigenvalue $r = 0$, is given by

$$\mathbf{u}_1 = \mathrm{col}\,(u_1\,u_2\,u_3) = \mathrm{col}(-4, -5, 2),$$

and so

$$\mathbf{x}_1(t) = e^0\mathbf{u}_1 = \begin{bmatrix} -4 \\ -5 \\ 2 \end{bmatrix}$$

is a solution to the system $\mathbf{x}' = \mathbf{A}\mathbf{x}$. To find an eigenvector, associated with the eigenvalue $r = 5$, we solve the equation

$$(\mathbf{A} - 5\mathbf{I})\mathbf{u} = \begin{bmatrix} 0 & -4 & 0 \\ 1 & -5 & 2 \\ 0 & 2 & 0 \end{bmatrix}\begin{bmatrix} u_1 \\ u_2 \\ u_3 \end{bmatrix} = \begin{bmatrix} 0 \\ 0 \\ 0 \end{bmatrix},$$

Copyright © 2012 Pearson Education, Inc. Publishing as Addison-Wesley.

which is equivalent to the system $u_2 = 0$, $u_1 = -2u_3$. One solution to this system is $u_1 = -2$, $u_2 = 0$, and $u_3 = 1$. Thus, an eigenvector of the matrix $\mathbf{A}$, associated with the eigenvalue $r = 5$, is

$$\mathbf{u}_2 = \text{col}\,(u_1\,u_2\,u_3) = \text{col}(-2, 0, 1),$$

and so another linearly independent solution to the system $\mathbf{x}' = \mathbf{A}\mathbf{x}$ is given by

$$\mathbf{x}_2(t) = e^{5t}\mathbf{u}_2 = e^{5t}\begin{bmatrix} -2 \\ 0 \\ 1 \end{bmatrix} = \begin{bmatrix} -2e^{5t} \\ 0 \\ e^{5t} \end{bmatrix}.$$

Since $r = 5$ is an eigenvalue of multiplicity two, we can find a generalized eigenvector (with $k = 2$), associated with the eigenvalue $r = 5$ and linearly independent from the vector $\mathbf{u}_2$. Thus, we solve the equation

$$(\mathbf{A} - 5\mathbf{I})^2\mathbf{u} = \mathbf{0}. \tag{9.14}$$

Because

$$(\mathbf{A} - 5\mathbf{I})^2 = \begin{bmatrix} 0 & -4 & 0 \\ 1 & -5 & 2 \\ 0 & 2 & 0 \end{bmatrix}\begin{bmatrix} 0 & -4 & 0 \\ 1 & -5 & 2 \\ 0 & 2 & 0 \end{bmatrix} = \begin{bmatrix} -4 & 20 & -8 \\ -5 & 25 & -10 \\ 2 & -10 & 4 \end{bmatrix},$$

we see that the equation (9.14) becomes

$$\begin{bmatrix} -4 & 20 & -8 \\ -5 & 25 & -10 \\ 2 & -10 & 4 \end{bmatrix}\begin{bmatrix} u_1 \\ u_2 \\ u_3 \end{bmatrix} = \begin{bmatrix} 0 \\ 0 \\ 0 \end{bmatrix} \Rightarrow \begin{bmatrix} -1 & 5 & -2 \\ 0 & 0 & 0 \\ 0 & 0 & 0 \end{bmatrix}\begin{bmatrix} u_1 \\ u_2 \\ u_3 \end{bmatrix} = \begin{bmatrix} 0 \\ 0 \\ 0 \end{bmatrix}.$$

This equation is equivalent to $-u_1 + 5u_2 - 2u_3 = 0$ and, therefore, is satisfied if we let $u_2 = s$, $u_3 = v$, and $u_1 = 5s - 2v$ for any values of s and v. Hence, solutions to the equation (9.14) are given by

$$\mathbf{u} = \begin{bmatrix} u_1 \\ u_2 \\ u_3 \end{bmatrix} = \begin{bmatrix} 5s - 2v \\ s \\ v \end{bmatrix} = s\begin{bmatrix} 5 \\ 1 \\ 0 \end{bmatrix} + v\begin{bmatrix} -2 \\ 0 \\ 1 \end{bmatrix}.$$

Notice that the vectors $v\,\text{col}(-2, 0, 1) = v\mathbf{u}_2$ are the eigenvectors, associated with the eigenvalue $r = 5$, that we found earlier. Since we are looking for a vector, which satisfies (9.14) and is linearly independent from $\mathbf{u}_2$, we choose $s = 1$ and $v = 0$. Thus, a

Copyright © 2012 Pearson Education, Inc. Publishing as Addison-Wesley.

generalized eigenvector for the matrix $\mathbf{A}$, associated with the eigenvalue $r = 5$ and linearly independent from the eigenvector $\mathbf{u}_2$ is given by $\mathbf{u}_3 = \text{col}(5, 1, 0)$. Hence, by the formula (8) in the text, we see that another linearly independent solution to the system $\mathbf{x}' = \mathbf{A}\mathbf{x}$ is given by

$$\mathbf{x}_3(t) = e^{\mathbf{A}t}\mathbf{u}_3 = e^{5t}\left[\mathbf{u}_3 + t(\mathbf{A} - 5\mathbf{I})\mathbf{u}_3\right]$$

$$= e^{5t}\begin{bmatrix} 5 \\ 1 \\ 0 \end{bmatrix} + te^{5t}\begin{bmatrix} 0 & -4 & 0 \\ 1 & -5 & 2 \\ 0 & 2 & 0 \end{bmatrix}\begin{bmatrix} 5 \\ 1 \\ 0 \end{bmatrix}$$

$$= e^{5t}\begin{bmatrix} 5 \\ 1 \\ 0 \end{bmatrix} + te^{5t}\begin{bmatrix} -4 \\ 0 \\ 2 \end{bmatrix} = \begin{bmatrix} 5e^{5t} - 4te^{5t} \\ e^{5t} \\ 2te^{5t} \end{bmatrix},$$

where we have used the fact that, by our choice of $\mathbf{u}_3$, $(\mathbf{A} - 5\mathbf{I})^2\mathbf{u}_3 = \mathbf{0}$ and so the product $(\mathbf{A} - 5\mathbf{I})^n\mathbf{u}_3 = \mathbf{0}$ for $n \geq 2$. (This is the reason why we used the generalized eigenvector to calculate $\mathbf{x}_3(t)$. The Cayley-Hamilton theorem states that $\mathbf{A}$ satisfies its characteristic equation, which is $\mathbf{A}(\mathbf{A} - 5\mathbf{I})^2 = \mathbf{0}$. However, we cannot conclude from this fact that $(\mathbf{A} - 5\mathbf{I})^2 = \mathbf{0}$ because, in matrix multiplication, it is possible for two nonzero matrices to have zero product.)

Our last step is to find a fundamental matrix for the system $\mathbf{x}' = \mathbf{A}\mathbf{x}$, using the linearly independent solutions found above, and then employ this fundamental matrix to calculate $e^{\mathbf{A}t}$. Thus, from $\mathbf{x}_1(t)$, $\mathbf{x}_2(t)$, and $\mathbf{x}_3(t)$ we get a fundamental matrix

$$\mathbf{X}(t) = \begin{bmatrix} -4 & -2e^{5t} & 5e^{5t} - 4te^{5t} \\ -5 & 0 & e^{5t} \\ 2 & e^{5t} & 2te^{5t} \end{bmatrix} \quad \Rightarrow \quad \mathbf{X}(0) = \begin{bmatrix} -4 & -2 & 5 \\ -5 & 0 & 1 \\ 2 & 1 & 0 \end{bmatrix}.$$

We can find the inverse matrix $\mathbf{X}^{-1}(0)$ by (for example) performing a row-reduction on the matrix $[\mathbf{X}(0)|\mathbf{I}]$ to obtain the matrix $[\mathbf{I}|\mathbf{X}^{-1}(0)]$. Thus, we find

$$\mathbf{X}^{-1}(0) = \frac{1}{25}\begin{bmatrix} 1 & -5 & 2 \\ -2 & 10 & 21 \\ 5 & 0 & 10 \end{bmatrix}.$$

Therefore, by the formula (3) in the text, we see that

$$e^{\mathbf{A}t} = \mathbf{X}(t)\mathbf{X}^{-1}(0) = \frac{1}{25}\begin{bmatrix} -4 & -2e^{5t} & 5e^{5t} - 4te^{5t} \\ -5 & 0 & e^{5t} \\ 2 & e^{5t} & 2te^{5t} \end{bmatrix}\begin{bmatrix} 1 & -5 & 2 \\ -2 & 10 & 21 \\ 5 & 0 & 10 \end{bmatrix}$$

Copyright © 2012 Pearson Education, Inc. Publishing as Addison-Wesley.

$$= \frac{1}{25} \begin{bmatrix} -4 + 29e^{5t} - 20te^{5t} & 20 - 20e^{5t} & -8 + 8e^{5t} - 40te^{5t} \\ -5 + 5e^{5t} & 25 & -10 + 10e^{5t} \\ 2 - 2e^{5t} + 10te^{5t} & -10 + 10e^{5t} & 4 + 21e^{5t} + 20te^{5t} \end{bmatrix}.$$

17. We first calculate the eigenvalues for the matrix $\mathbf{A}$ by solving the characteristic equation

$$|\mathbf{A} - r\mathbf{I}| = \begin{vmatrix} -r & 1 & 0 \\ 0 & -r & 1 \\ -2 & -5 & -4 - r \end{vmatrix} = 0$$

$$\Rightarrow \quad (-r) \begin{vmatrix} -r & 1 \\ -5 & -4 - r \end{vmatrix} - \begin{vmatrix} 0 & 1 \\ -2 & -4 - r \end{vmatrix} = 0$$

$$\Rightarrow \quad -r[-r(-4 - r) + 5] - 2 = -\left(r^3 + 4r^2 + 5r + 2\right) = -(r + 1)^2(r + 2) = 0.$$

Thus, the eigenvalues of $\mathbf{A}$ are $r = -1$ of multiplicity two and $r = -2$. To find an eigenvector $\mathbf{u} = \mathrm{col}(u_1, u_2, u_3)$, associated with the eigenvalue $r = -1$, we solve the equation

$$(\mathbf{A} + \mathbf{I})\mathbf{u} = \begin{bmatrix} 1 & 1 & 0 \\ 0 & 1 & 1 \\ -2 & -5 & -3 \end{bmatrix} \begin{bmatrix} u_1 \\ u_2 \\ u_3 \end{bmatrix} = \begin{bmatrix} 0 \\ 0 \\ 0 \end{bmatrix},$$

which is equivalent to $u_1 = u_3$, $u_2 = -u_3$. Letting $u_3 = 1$ (so that $u_1 = 1$ and $u_2 = -1$), we see that an eigenvector for the matrix $\mathbf{A}$, associated with the eigenvalue $r = -1$, is

$$\mathbf{u}_1 = \mathrm{col}(u_1, u_2, u_3) = \mathrm{col}(1, -1, 1).$$

Hence, one solution to the system $\mathbf{x}' = \mathbf{A}\mathbf{x}$ is given by

$$\mathbf{x}_1(t) = e^{-t}\mathbf{u}_1 = e^{-t} \begin{bmatrix} 1 \\ -1 \\ 1 \end{bmatrix}.$$

Since $r = -1$ is an eigenvalue of multiplicity two, we can find a generalized eigenvector, associated with this eigenvalue (with $k = 2$), which is linearly independent from $\mathbf{u}_1$. To do this, we solve the equation

$$(\mathbf{A} + \mathbf{I})^2\mathbf{u} = \mathbf{0}$$

$$\Rightarrow \quad \begin{bmatrix} 1 & 1 & 0 \\ 0 & 1 & 1 \\ -2 & -5 & -3 \end{bmatrix} \begin{bmatrix} 1 & 1 & 0 \\ 0 & 1 & 1 \\ -2 & -5 & -3 \end{bmatrix} \begin{bmatrix} u_1 \\ u_2 \\ u_3 \end{bmatrix} = \begin{bmatrix} 0 \\ 0 \\ 0 \end{bmatrix}$$

Copyright © 2012 Pearson Education, Inc. Publishing as Addison-Wesley.

$$\Rightarrow \quad \begin{bmatrix} 1 & 2 & 1 \\ -2 & -4 & -2 \\ 4 & 8 & 4 \end{bmatrix} \begin{bmatrix} u_1 \\ u_2 \\ u_3 \end{bmatrix} = \begin{bmatrix} 0 \\ 0 \\ 0 \end{bmatrix} \quad \Rightarrow \quad \begin{bmatrix} 1 & 2 & 1 \\ 0 & 0 & 0 \\ 0 & 0 & 0 \end{bmatrix} \begin{bmatrix} u_1 \\ u_2 \\ u_3 \end{bmatrix} = \begin{bmatrix} 0 \\ 0 \\ 0 \end{bmatrix},$$

which is equivalent to $u_1 + 2u_2 + u_3 = 0$. This equation is satisfied if we let $u_3 = s$, $u_2 = v$, and $u_1 = -2v - s$, s and v are arbitrary. Thus, generalized eigenvectors, associated with the eigenvalue $r = -1$, are given by

$$\mathbf{u} = \begin{bmatrix} u_1 \\ u_2 \\ u_3 \end{bmatrix} = \begin{bmatrix} -2v - s \\ v \\ s \end{bmatrix} = s \begin{bmatrix} -1 \\ 0 \\ 1 \end{bmatrix} + v \begin{bmatrix} -2 \\ 1 \\ 0 \end{bmatrix}.$$

Hence, letting $s = 2$ and $v = -1$, we find one of such generalized eigenvectors,

$$\mathbf{u}_2 = \text{col}(0, -1, 2),$$

which, by inspection, is linearly independent from $\mathbf{u}_1$. Therefore, a second linearly independent solution to the system $\mathbf{x}' = \mathbf{A}\mathbf{x}$ is

$$\begin{aligned}
\mathbf{x}_2(t) &= e^{\mathbf{A}t}\mathbf{u}_2 = e^{-t}\left[\mathbf{u}_2 + t(\mathbf{A} + \mathbf{I})\mathbf{u}_2\right] \\
&= e^{-t}\begin{bmatrix} 0 \\ -1 \\ 2 \end{bmatrix} + te^{-t}\begin{bmatrix} 1 & 1 & 0 \\ 0 & 1 & 1 \\ -2 & -5 & -3 \end{bmatrix}\begin{bmatrix} 0 \\ -1 \\ 2 \end{bmatrix} \\
&= e^{-t}\begin{bmatrix} 0 \\ -1 \\ 2 \end{bmatrix} + te^{-t}\begin{bmatrix} -1 \\ 1 \\ -1 \end{bmatrix} = e^{-t}\begin{bmatrix} -t \\ -1 + t \\ 2 - t \end{bmatrix}.
\end{aligned}$$

In order to obtain a third linearly independent solution to this system, we find an eigenvector, associated with the eigenvalue $r = -2$, by solving the equation

$$(\mathbf{A} + 2\mathbf{I})\mathbf{u} = \begin{bmatrix} 2 & 1 & 0 \\ 0 & 2 & 1 \\ -2 & -5 & -2 \end{bmatrix} \begin{bmatrix} u_1 \\ u_2 \\ u_3 \end{bmatrix} = \begin{bmatrix} 0 \\ 0 \\ 0 \end{bmatrix}.$$

This equation is equivalent to the system $2u_1 + u_2 = 0$, $2u_2 + u_3 = 0$. One solution to this system is given by $u_1 = 1$, $u_2 = -2$, and $u_3 = 4$. Thus, an eigenvector, associated with the eigenvalue $r = -2$, is

$$\mathbf{u}_3 = \text{col}(u_1, u_2, u_3) = \text{col}(1, -2, 4),$$

Copyright © 2012 Pearson Education, Inc. Publishing as Addison-Wesley.

Chapter 9

and so a third linearly independent solution to the given system is

$$\mathbf{x}_3(t) = e^{-2t}\mathbf{u}_3 = e^{-2t}\begin{bmatrix} 1 \\ -2 \\ 4 \end{bmatrix}.$$

Hence, combining all three linearly independent solutions that we have found, we see that a general solution to our system is

$$\mathbf{x}(t) = c_1 e^{-t}\begin{bmatrix} 1 \\ -1 \\ 1 \end{bmatrix} + c_2 e^{-t}\begin{bmatrix} -t \\ -1+t \\ 2-t \end{bmatrix} + c_3 e^{-2t}\begin{bmatrix} 1 \\ -2 \\ 4 \end{bmatrix}.$$

23. In Problem 3, we found that

$$e^{\mathbf{A}t} = e^{-t}\begin{bmatrix} 1+3t-(3/2)t^2 & t & -t+(1/2)t^2 \\ -3t & 1 & t \\ 9t-(9/2)t^2 & 3t & 1-3t+(3/2)t^2 \end{bmatrix}.$$

In order to use the variation of parameters formula (the equation (16) in the text), we need to find expressions for

$$e^{\mathbf{A}t}\mathbf{x}_0 \qquad \text{and} \qquad \int_0^t e^{\mathbf{A}(t-s)}\mathbf{f}(s)\,ds\,,$$

where we have used the fact that $t_0 = 0$.

First, we notice that

$$\int_0^t e^{\mathbf{A}(t-s)}\mathbf{f}(s)\,ds = \int_0^t e^{\mathbf{A}t-\mathbf{A}s}\mathbf{f}(s)\,ds = e^{\mathbf{A}t}\int_0^t e^{-\mathbf{A}s}\mathbf{f}(s)\,ds\,.$$

Since $\mathbf{f}(s) = \mathrm{col}(0, s, 0)$, we obtain

$$e^{-\mathbf{A}s}\mathbf{f}(s) = e^s\begin{bmatrix} 1-3s-(3/2)s^2 & -s & s+(1/2)s^2 \\ 3s & 1 & - \\ -9s-(9/2)s^2 & -3s & 1+3s+(3/2)s^2 \end{bmatrix}\begin{bmatrix} 0 \\ s \\ 0 \end{bmatrix}$$

$$= e^s\begin{bmatrix} -s^2 \\ s \\ -3s^2 \end{bmatrix} = \begin{bmatrix} -s^2 e^s \\ s e^s \\ -3s^2 e^s \end{bmatrix}.$$

    Copyright © 2012 Pearson Education, Inc. Publishing as Addison-Wesley.

Therefore,

$$\int_0^t e^{\mathbf{A}(t-s)}\mathbf{f}(s)\,ds = e^{\mathbf{A}t}\int_0^t c^{-\mathbf{A}s}\mathbf{f}(s)\,ds = e^{\mathbf{A}t}\begin{bmatrix} \int_0^t(-s^2e^s)ds \\ \int_0^t(se^s)ds \\ \int_0^t(-3s^2e^s)ds \end{bmatrix}.$$

We use integration by parts to evaluate integrals.

$$\int_0^t (se^s)\,ds = se^s\,\big|_0^t - \int_0^t e^s\,ds = 1 + e^t\,(t-1)\ ,$$

$$\int_0^t \left(-s^2e^s\right)ds = -s^2e^s\,\big|_0^t + 2\int_0^t (se^s)\,ds$$

$$= -t^2e^t + 2\left[1 + e^t\,(t-1)\right] = 2 - e^t\left(t^2 - 2t + 2\right)\ ,$$

$$\int_0^t \left(-3s^2e^s\right)ds = 3\int_0^t \left(-s^2e^s\right)ds = 6 - 3e^t\left(t^2 - 2t + 2\right)\ .$$

Thus,

$$\int_0^t e^{\mathbf{A}(t-s)}\mathbf{f}(s)\,ds = e^{\mathbf{A}t}\begin{bmatrix} 2 - e^t(t^2 - 2t + 2) \\ 1 + e^t(t-1) \\ 6 - 3e^t(t^2 - 2t + 2) \end{bmatrix}.$$

Since

$$\mathbf{x}_0 = \begin{bmatrix} 0 \\ 3 \\ 0 \end{bmatrix},$$

we get

$$e^{\mathbf{A}t}\mathbf{x}_0 = e^{-t}\begin{bmatrix} 1 + 3t - (3/2)t^2 & t & -t + (1/2)t^2 \\ -3t & 1 & t \\ 9t - (9/2)t^2 & 3t & 1 - 3t + (3/2)t^2 \end{bmatrix}\begin{bmatrix} 0 \\ 3 \\ 0 \end{bmatrix} = e^{-t}\begin{bmatrix} 3t \\ 3 \\ 9t \end{bmatrix}.$$

Finally, substituting these expressions into the variation of parameters formula (16) yields

$$\mathbf{x}(t) = e^{\mathbf{A}t}\mathbf{x}_0 + \int_0^t e^{\mathbf{A}(t-s)}\mathbf{f}(s)\,ds$$

$$= e^{-t}\begin{bmatrix} 3t \\ 3 \\ 9t \end{bmatrix} + e^{\mathbf{A}t}\begin{bmatrix} 2 - e^t(t^2 - 2t + 2) \\ 1 + e^t(t-1) \\ 6 - 3e^t(t^2 - 2t + 2) \end{bmatrix},$$

Copyright © 2012 Pearson Education, Inc. Publishing as Addison-Wesley.

where

$$e^{\mathbf{A}t} = e^{-t} \begin{bmatrix} 1 + 3t - (3/2)t^2 & t & -t + (1/2)t^2 \\ -3t & 1 & t \\ 9t - (9/2)t^2 & 3t & 1 - 3t + (3/2)t^2 \end{bmatrix}.$$

 Copyright © 2012 Pearson Education, Inc. Publishing as Addison-Wesley.

CHAPTER 10: Partial Differential Equations

EXERCISES 10.2:   Method of Separation of Variables

5. To find a general solution to this equation, we first observe that the auxiliary equation, associated with the corresponding homogeneous equation, is given by $r^2 - 1 = 0$. This equation has roots $r = \pm 1$. Thus, the solution to the corresponding homogeneous equation is given by

$$y_h(x) = C_1 e^x + C_2 e^{-x}.$$

By the method of undetermined coefficients, we see that the form of a particular solution to the given nonhomogeneous equation is

$$y_p(x) = Ax + B, \tag{10.1}$$

where we have used the fact that neither $y = 1$ nor $y = x$ is a solution to the corresponding homogeneous equation. To find A and B, we note that

$$y_p'(x) = A \qquad \text{and} \qquad y_p''(x) = 0.$$

By substituting these expressions into the original differential equation, we obtain

$$y_p''(x) - y_p(x) = -Ax - B = 1 - 2x.$$

Equating coefficients yields $A = 2$ and $B = -1$. Substituting these values into (10.1) yields

$$y_p(x) = 2x - 1.$$

Thus,

$$y(x) = y_h(x) + y_p(x) = C_1 e^x + C_2 e^{-x} + 2x - 1.$$

Next, we find C_1 and C_2 so that the solution $y(x)$ satisfies the boundary conditions. That is, we are looking for C_1 and C_2 satisfying

$$y(0) = C_1 + C_2 - 1 = 0 \qquad \text{and} \qquad y(1) = C_1 e + C_2 e^{-1} + 1 = 1 + e.$$

Copyright © 2012 Pearson Education, Inc. Publishing as Addison-Wesley.

From the first equation, $C_2 = 1 - C_1$. Substituting this expression for C_2 into the second equation and simplifying yields

$$e - e^{-1} = C_1 \left(e - e^{-1} \right).$$

Thus, $C_1 = 1$ and $C_2 = 0$. Therefore,

$$y(x) = e^x + 2x - 1$$

is the only solution to the given boundary value problem.

13. First note that the auxiliary equation in this problem is $r^2 + \lambda = 0$. To find eigenvalues, which yields nontrivial solutions, we consider three cases: $\lambda < 0$, $\lambda = 0$, and $\lambda > 0$.

<u>Case 1: $\lambda < 0$.</u> In this case, the roots to the auxiliary equation are $r = \pm\sqrt{-\lambda}$. Therefore, a general solution to the differential equation $y'' + \lambda y = 0$ is given by

$$y(x) = C_1 e^{\sqrt{-\lambda}x} + C_2 e^{-\sqrt{-\lambda}x}.$$

In order to apply the boundary conditions, we need to find $y'(x)$. Thus, we have

$$y'(x) = \sqrt{-\lambda}C_1 e^{\sqrt{-\lambda}x} - \sqrt{-\lambda}C_2 e^{-\sqrt{-\lambda}x}.$$

The boundary conditions give us

$$
\begin{aligned}
y(0) - y'(0) &= C_1 + C_2 - \sqrt{-\lambda}C_1 + \sqrt{-\lambda}C_2 = 0 \\
\Rightarrow \quad & \left(1 - \sqrt{-\lambda}\right)C_1 + \left(1 + \sqrt{-\lambda}\right)C_2 = 0, \\
y(\pi) &= C_1 e^{\sqrt{-\lambda}\pi} + C_2 e^{-\sqrt{-\lambda}\pi} = 0 \\
\Rightarrow \quad & C_2 = -C_1 e^{2\sqrt{-\lambda}\pi}.
\end{aligned}
$$

Combining these expressions, we find that

$$
\begin{aligned}
&\left(1 - \sqrt{-\lambda}\right)C_1 - \left(1 + \sqrt{-\lambda}\right)C_1 e^{2\sqrt{-\lambda}\pi} = 0 \\
\Rightarrow \quad & C_1 \left[\left(1 - \sqrt{-\lambda}\right) - \left(1 + \sqrt{-\lambda}\right) e^{2\sqrt{-\lambda}\pi}\right] = 0.
\end{aligned}
\tag{10.2}
$$

(10.2) is satisfied if $C_1 = 0$ or

$$e^{2\sqrt{-\lambda}\pi} = \frac{1 - \sqrt{-\lambda}}{1 + \sqrt{-\lambda}}.$$

   Copyright © 2012 Pearson Education, Inc. Publishing as Addison-Wesley.

But, since $\sqrt{-\lambda} > 0$, we see that $e^{2\sqrt{-\lambda}\pi} > 1$ while $(1-\sqrt{-\lambda})/(1+\sqrt{-\lambda}) < 1$. Therefore, the only way for the equation (10.2) to be true is having $C_1 = 0$. This means that $C_2 = 0$ as well and so, for $\lambda < 0$, we have only the trivial solution.

Case 2: $\lambda = 0$. In this case, we are solving the differential equation $y'' = 0$. This equation has a general solution given by

$$y(x) = C_1 + C_2 x \qquad \Rightarrow \qquad y'(x) = C_2 .$$

Applying the boundary conditions, we obtain

$$y(0) - y'(0) = C_1 - C_2 = 0 \qquad \text{and} \qquad y(\pi) = C_1 + C_2 \pi = 0.$$

Solving these equations simultaneously yields $C_1 = C_2 = 0$. Thus, we again find only the trivial solution.

Case 3: $\lambda > 0$. In this case, the roots of the associated auxiliary equation are $r = \pm\sqrt{\lambda}i$. Therefore, a general solution is given by

$$y(x) = C_1 \cos\left(\sqrt{\lambda}x\right) + C_2 \sin\left(\sqrt{\lambda}x\right)$$
$$\Rightarrow \qquad y'(x) = -\sqrt{\lambda}C_1 \sin\left(\sqrt{\lambda}x\right) + \sqrt{\lambda}C_2 \cos\left(\sqrt{\lambda}x\right).$$

Applying the boundary conditions, we obtain

$$y(0) - y'(0) = C_1 - \sqrt{\lambda}C_2 = 0 \qquad \Rightarrow \qquad C_1 = \sqrt{\lambda}C_2 ,$$

and

$$y(\pi) = C_1 \cos\left(\sqrt{\lambda}\pi\right) + C_2 \sin\left(\sqrt{\lambda}\pi\right) = 0.$$

By combining these results, we obtain

$$C_2 \left[\sqrt{\lambda}\cos\left(\sqrt{\lambda}\pi\right) + \sin\left(\sqrt{\lambda}\pi\right)\right] = 0.$$

Therefore, in order to obtain a solution other than the trivial one, we solve the equation

$$\sqrt{\lambda}\cos\left(\sqrt{\lambda}\pi\right) + \sin\left(\sqrt{\lambda}\pi\right) = 0.$$

Simplifying yields

$$\tan\left(\sqrt{\lambda}\pi\right) = -\sqrt{\lambda}.$$

Copyright © 2012 Pearson Education, Inc. Publishing as Addison-Wesley.

To see that there exist values of $\lambda > 0$, which satisfy this equation, we examine the graphs of the functions $y = -x$ and $y = \tan(\pi x)$. For any value of $x > 0$, where these two graphs intersect, we set $\lambda = x^2$. These values of λ are the eigenvalues that we are looking for. From the graph in Fig. **10–A** on page **10–A**, we see that there are (countably) infinitely many such eigenvalues. These eigenvalues λ_n satisfy

$$\tan\left(\sqrt{\lambda_n}\pi\right) + \sqrt{\lambda_n} = 0.$$

For n large, we can conclude from the graph that these eigenvalues are close to the square of odd multiples of $1/2$. That is,

$$\lambda_n \approx \left(\frac{2n-1}{2}\right)^2.$$

Corresponding to the eigenvalue λ_n, we obtain solutions

$$y_n(x) = C_{1,n}\cos\left(\sqrt{\lambda_n}x\right) + C_{2,n}\sin\left(\sqrt{\lambda_n}x\right) = \sqrt{\lambda_n}C_{2,n}\cos\left(\sqrt{\lambda_n}x\right) + C_{2,n}\sin\left(\sqrt{\lambda_n}x\right)$$

(since $C_{1,n} = \sqrt{\lambda_n}C_{2,n}$). Thus,

$$y_n(x) = c_n\left[\sqrt{\lambda_n}\cos\left(\sqrt{\lambda_n}x\right) + \sin\left(\sqrt{\lambda_n}x\right)\right],$$

where $c_n := C_{2,n}$ is arbitrary.

17. We are solving a boundary value problem

$$\frac{\partial u(x,t)}{\partial t} = 3\frac{\partial^2 u(x,t)}{\partial x^2}, \qquad 0 < x < \pi, \quad t > 0,$$
$$u(0,t) = u(\pi,t) = 0, \qquad t > 0,$$
$$u(x,0) = \sin x - 7\sin 3x + \sin 5x.$$

A solution to this partial differential equation, satisfying the first boundary condition, is given in the equation (15) in the text. By letting $\beta = 3$ and $L = \pi$ in this equation, we obtain the series

$$u(x,t) = \sum_{n=1}^{\infty} c_n e^{-3n^2 t}\sin nx. \tag{10.3}$$

To satisfy the initial condition, we let $t = 0$ and obtain

$$u(x,0) = \sum_{n=1}^{\infty} c_n \sin nx = \sin x - 7\sin 3x + \sin 5x.$$

 Copyright © 2012 Pearson Education, Inc. Publishing as Addison-Wesley.

Equating the coefficients, we see that $c_1 = 1$, $c_3 = -7$, $c_5 = 1$, and all other c_n's equal zero. Plugging in these values into (10.3) yields the solution

$$
\begin{aligned}
u(x,t) &= e^{-3(1)^2 t} \sin x - 7 e^{-3(3)^2 t} \sin 3x + e^{-3(5)^2 t} \sin 5x \\
&= e^{-3t} \sin x - 7 e^{-27t} \sin 3x + e^{-75t} \sin 5x \,.
\end{aligned}
$$

21. By letting $\alpha = 3$ and $L = \pi$ in the formula (24) in the text, we see that the solution we seek has the form

$$
u(x,t) = \sum_{n=1}^{\infty} [a_n \cos 3nt + b_n \sin 3nt] \sin nx \,. \tag{10.4}
$$

Therefore,

$$
\frac{\partial u}{\partial t} = \sum_{n=1}^{\infty} [-3na_n \sin 3nt + 3nb_n \cos 3nt] \sin nx \,.
$$

In order to satisfy the initial conditions, we have to find a_n and b_n such that

$$
u(x,0) = \sum_{n=1}^{\infty} a_n \sin nx = 6 \sin 2x + 2 \sin 6x,
$$

and

$$
\frac{\partial u(x,0)}{\partial t} = \sum_{n=1}^{\infty} 3nb_n \sin nx = 11 \sin 9x - 14 \sin 15x.
$$

From the first condition, we see that $a_2 = 6$ and $a_6 = 2$, and all of the other a_n's must be zero. By comparing coefficients in the second condition, we find that

$$
3(9)b_9 = 11 \quad \text{or} \quad b_9 = \frac{11}{27} \quad \text{and} \quad 3(15)b_{15} = -14 \quad \text{or} \quad b_{15} = -\frac{14}{45}\,.
$$

The values of all other for b_n's must be zero. Therefore, by substituting these coefficients into (10.4), we obtain the solution of the vibrating string problem with $\alpha = 3$, $L = \pi$, and $f(x)$ and $g(x)$ as given. This solution is

$$
u(x,t) = 6\cos(3\cdot2\cdot t)\sin 2x + 2\cos(3\cdot6\cdot t)\sin 6x + \frac{11}{27}\sin(3\cdot9\cdot t)\sin 9x - \frac{14}{45}\sin(3\cdot15\cdot t)\sin 15x.
$$

Simplifying yields

$$
u(x,t) = 6\cos 6t \sin 2x + 2\cos 18t \sin 6x + \frac{11}{27}\sin 27t \sin 9x - \frac{14}{45}\sin 45t \sin 15x.
$$

23. We know from the equation (15) in the text that a formal solution to the heat flow problem is given by

$$
u(x,t) = \sum_{n=1}^{\infty} c_n e^{-2(n\pi)^2 t} \sin n\pi x = \sum_{n=1}^{\infty} c_n e^{-2n^2\pi^2 t} \sin n\pi x \,, \tag{10.5}
$$

Copyright © 2012 Pearson Education, Inc. Publishing as Addison-Wesley.

where we have made the substitution $\beta = 2$, $L = 1$. For this function to be a solution to the given problem, the initial condition, $u(x, 0) = f(x)$, $0 < x < 1$, should be satisfied. Therefore, we let $t = 0$ in the equation (10.5) and obtain

$$u(x, 0) = \sum_{n=1}^{\infty} c_n \sin n\pi x = f(x) = \sum_{n=1}^{\infty} \frac{1}{n^2} \sin n\pi x.$$

By equating coefficients, we see that $c_n = n^{-2}$. Substituting these c_n into the equation (10.5) yields

$$u(x, t) = \sum_{n=1}^{\infty} n^{-2} e^{-2n^2\pi^2 t} \sin n\pi x.$$

EXERCISES 10.3: Fourier Series

1. Since

$$f(-x) = (-x)^3 + \sin[2(-x)] = -x^3 - \sin 2x = -\left(x^3 + \sin 2x\right) = -f(x)$$

for any x, this function is odd.

3. We have

$$f(-x) = \left[1 - (-x)^2\right]^{-1/2} = \left(1 - x^2\right)^{-1/2} = f(x).$$

Thus, the given function is even.

5. Note that $f(-x) = e^x \cos(-3x) = e^x \cos 3x$. Since

$$f(-x) = e^x \cos 3x \neq \pm e^{-x} \cos 3x = \pm f(x),$$

unless $x = 0$ or $\cos 3x = 0$, we see that this function is neither even nor odd.

7. **(a)** Since $f(-x) = f(x)$ and $g(-x) = g(x)$, for any x in their common domain we have

$$(fg)(-x) := f(-x)g(-x) = f(x)g(x) = (fg)(x).$$

Thus, fg is an even function.

(b) We have $f(-x) = -f(x)$ and $g(-x) = -g(x)$. Thus, for any x in the common domain of f and g,

$$(fg)(-x) := f(-x)g(-x) = [-f(x)][-g(x)] = f(x)g(x) = (fg)(x).$$

Thus, fg is an even function.

     Copyright © 2012 Pearson Education, Inc. Publishing as Addison-Wesley.

(c) This time $f(-x) = f(x)$ and $g(-x) = -g(x)$. Therefore,

$$(fg)(-x) := f(-x)g(-x) = f(x)[-g(x)] = -f(x)g(x) = -(fg)(x),$$

which means that fg is an odd function.

9. Since $f(x) = x$ is odd and, for any n, $\cos nx$ is even, $x \cos nx$ is odd (see Problem 7). Similarly, $x \sin nx$ is an even function. Thus, by Theorem 1, the Fourier coefficients

$$a_n = \frac{1}{\pi} \int_{-\pi}^{\pi} x \cos nx \, dx = 0, \quad b_n = \frac{1}{\pi} \int_{-\pi}^{\pi} x \cos nx \, dx = \frac{2}{\pi} \int_0^{\pi} x \sin nx \, dx.$$

Integrating by parts (or using the integral table on the inside front cover of the text) yields

$$b_n = \frac{2}{\pi} \left(-\frac{1}{n} \right) \left(x \cos nx \Big|_0^{\pi} - \int_0^{\pi} \cos nx \, dx \right)$$

$$= -\frac{2}{\pi n} \left(\pi \cos n\pi - \frac{1}{n} \sin nx \Big|_0^{\pi} \right) = \frac{2(-1)^{n+1}}{n},$$

where we have used that $\cos n\pi = (-1)^n$. Thus,

$$x \sim \sum_{n=1}^{\infty} \frac{2(-1)^{n+1}}{n} \sin nx.$$

11. Here, $L = 2$. The given function is a piecewise defined function. Using formulas (9) and (10), splitting the integrals, and integrating by parts yields

$$a_n = \frac{1}{2} \int_{-2}^{2} f(x) \cos \frac{n\pi x}{2} \, dx = \frac{1}{2} \left(\int_{-2}^{0} \cos \frac{n\pi x}{2} \, dx + \int_0^{2} x \cos \frac{n\pi x}{2} \, dx \right) = \frac{2\left((-1)^n - 1\right)}{\pi^2 n^2},$$

$$b_n = \frac{1}{2} \int_{-2}^{2} f(x) \sin \frac{n\pi x}{2} \, dx = \frac{1}{2} \left(\int_{-2}^{0} \sin \frac{n\pi x}{2} \, dx + \int_0^{2} x \sin \frac{n\pi x}{2} \, dx \right) = -\frac{(-1)^n + 1}{\pi n},$$

$n = 1, 2, \ldots$. Also,

$$a_0 = \frac{1}{2} \int_{-2}^{2} f(x) \, dx = \frac{1}{2} \left(\int_{-2}^{0} dx + \int_0^{2} x \, dx \right) = 2.$$

Therefore,

$$f(x) \sim 1 + \sum_{n=1}^{\infty} \left[\frac{2\left((-1)^n - 1\right)}{\pi^2 n^2} \cos \frac{n\pi x}{2} - \frac{(-1)^n + 1}{\pi n} \sin \frac{n\pi x}{2} \right],$$

which is equivalent to the answer in the text.

Copyright © 2012 Pearson Education, Inc. Publishing as Addison-Wesley.

Chapter 10

13. In this problem, $L = 1$. Thus, by Definition 1 in the text, the Fourier series for this function is given by

$$\frac{a_0}{2} + \sum_{n=1}^{\infty} (a_n \cos n\pi x + b_n \sin n\pi x). \tag{10.6}$$

To compute a_0, we use the equation (9). Thus, we have

$$a_0 = \int_{-1}^{1} x^2 \cos(0 \cdot \pi x)\, dx = \frac{x^3}{3}\Big|_{-1}^{1} = \frac{1}{3} - \frac{-1}{3} = \frac{2}{3}.$$

To find a_n, $n = 1, 2, 3, \ldots$, we again use the equation (9). This yields

$$a_n = \int_{-1}^{1} x^2 \cos n\pi x\, dx = 2\int_{0}^{1} x^2 \cos n\pi x\, dx,$$

where we have used the fact that $x^2 \cos n\pi x$ is an even function. Thus, using integration by parts twice, we obtain

$$
\begin{aligned}
a_n &= 2\int_{0}^{1} x^2 \cos n\pi x\, dx = 2\left[x^2 \frac{\sin n\pi x}{n\pi}\Big|_0^1 - \frac{2}{n\pi}\int_0^1 x\sin n\pi x\, dx \right] \\
&= 2\left[\left(\frac{\sin n\pi}{n\pi} - 0\right) - \frac{2}{n\pi}\left(-x\frac{\cos n\pi x}{n\pi}\Big|_0^1 + \frac{1}{n\pi}\int_0^1 \cos n\pi x\, dx \right) \right] \\
&= 2\left[0 + \frac{2}{n^2\pi^2}(\cos n\pi - 0) - \frac{2}{n^2\pi^2}\left(\frac{1}{n\pi}\sin n\pi x\Big|_0^1 \right) \right] \\
&= \frac{4}{n^2\pi^2}(-1)^n - \frac{4}{n^3\pi^3}(\sin n\pi - 0) = \frac{4}{n^2\pi^2}(-1)^n.
\end{aligned}
$$

To calculate b_n's, we note that $x^2 \sin n\pi x$ is an odd function (see Problem 7). Since this function is also continuous, by Theorem 1, we have

$$b_n = \int_{-1}^{1} x^2 \sin n\pi x\, dx = 0.$$

Plugging in the coefficients into (10.6), we see that the Fourier series, associated with x^2, is given by

$$\frac{1}{3} + \sum_{n=1}^{\infty} \frac{4(-1)^n}{n^2\pi^2} \cos n\pi x.$$

Copyright © 2012 Pearson Education, Inc. Publishing as Addison-Wesley.

15. Using the integration formulas, given on the inside front cover of the text (or integration by parts), yields

$$a_n = \frac{1}{\pi} \int_{-\pi}^{\pi} e^x \cos nx \, dx = \frac{(-1)^n \left(e^{2\pi} - 1\right) e^{-\pi}}{\pi(1 + n^2)} = \frac{2 \sinh(\pi)}{\pi} \frac{(-1)^n}{1 + n^2},$$

$$b_n = \frac{1}{\pi} \int_{-\pi}^{\pi} e^x \sin nx \, dx = -\frac{(-1)^n n \left(e^{2\pi} - 1\right) e^{-\pi}}{\pi(1 + n^2)} = \frac{2 \sinh(\pi)}{\pi} \frac{(-1)^{n+1} n}{1 + n^2},$$

$n = 1, 2, \ldots$. We also compute

$$a_0 = \frac{1}{\pi} \int_{-\pi}^{\pi} e^x \, dx = \frac{e^\pi - e^{-\pi}}{\pi} = \frac{2 \sinh(\pi)}{\pi}.$$

Substituting these coefficients into the Fourier series and simplifying, we get

$$e^x \sim \frac{\sinh(\pi)}{\pi} \left[1 + 2 \sum_{n=1}^{\infty} \left(\frac{(-1)^n}{1 + n^2} \cos nx + \frac{(-1)^{n+1} n}{1 + n^2} \sin nx \right) \right].$$

17. Since $f(x) = x$, being a polynomial, is continuously differentiable, Theorem 2 applies. Thus, for any x in $(-\pi, \pi)$, the Fourier series, that we found in Problem 9, converges to $f(x)$, i.e.,

$$x = \sum_{n=1}^{\infty} \frac{2(-1)^{n+1}}{n} \sin nx, \quad -\pi < x < \pi.$$

Since

$$f(-\pi^+) = -\pi, \quad f(\pi^-) = \pi \quad \Rightarrow \quad \frac{1}{2} \left[f(-\pi^+) + f(\pi^-) \right] = 0,$$

the associated Fourier series converges to zero at $x = \pm\pi$. (One can also confirm this fact by substituting $x = \pm\pi$ into the above series.)

Since the Fourier series in Problem 9 is 2π-periodic, it converges to

$$g(x) = \begin{cases} x - 2k\pi, & (2k-1)\pi < x < (2k+1)\pi, \\ 0, & x = (2k-1)\pi, \end{cases} \quad k = 0, \pm 1, \pm 2, \ldots .$$

19. Since both, $f(x)$ and $f'(x)$, are piecewise continuous on $(-2, 2)$, we use Theorem 2 to conclude that the Fourier series in Problem 11 converges to $f(x)$ at any point x, where $f(x)$ is continuous, that is, on $(-2, 0)$ and $(0, 2)$.

At $x = 0$, it converges to

$$\frac{1}{2} \left[f(0^+) + f(0^-) \right] = \frac{0 + 1}{2} = \frac{1}{2}.$$

Copyright © 2012 Pearson Education, Inc. Publishing as Addison-Wesley.

At $x = \pm 2$, it converges to

$$\frac{1}{2}\left[f\left(-2^+\right) + f\left(2^-\right)\right] = \frac{1+2}{2} = \frac{3}{2}.$$

By periodicity, the Fourier series converges for all real x to 4-periodic extension of the function

$$g(x) = \begin{cases} 1, & -2 < x < 0, \\ x, & 0 < x < 2, \\ 1/2, & x = 0, \\ 3/2, & x = \pm 2. \end{cases}$$

21. Here, we use Theorem 2. Notice that $f(x) = x^2$ and $f'(x) = 2x$ are continuous on $[-1, 1]$. Thus, the Fourier series for $f(x)$ converges to this function for $-1 < x < 1$. Furthermore,

$$f\left(-1^+\right) = \lim_{x \to -1^+} x^2 = 1 \quad \text{and} \quad f\left(1^-\right) = \lim_{x \to 1^-} x^2 = 1.$$

Hence,

$$\frac{1}{2}\left[f\left(-1^+\right) + f\left(1^-\right)\right] = \frac{1+1}{2} = 1,$$

and so, by Theorem 2, the sum of the Fourier series equals 1 for $x = \pm 1$. Therefore, the Fourier series converges to

$$f(x) = x^2 \quad \text{for} \quad -1 \le x \le 1.$$

Since the sum function of this Fourier series is periodic with period 2, it gives the 2-periodic extension of $f(x)$, which we can write as

$$g(x) = (x - 2n)^2, \quad 2n - 1 \le x < 2n + 1, \quad n = 0, \pm 1, \pm 2, \ldots.$$

23. Since $f(x) = e^x$ is continuously differentiable on $(-\pi, \pi)$ (actually, on the whole real line), Theorem 2 guarantees the convergence of its Fourier series to $f(x)$ for any x in $(-\pi, \pi)$. At the endpoints, $x = \pm\pi$, the series converges to

$$\frac{1}{2}\left[f\left(-\pi^+\right) + f\left(\pi^-\right)\right] = \frac{e^{-\pi} + e^{\pi}}{2} = \cosh(\pi).$$

By periodicity, the series converges to 2π-periodic extension of the function

$$g(x) = \begin{cases} e^x, & -\pi < x < \pi, \\ \cosh(\pi), & x = \pm\pi. \end{cases}$$

 Copyright © 2012 Pearson Education, Inc. Publishing as Addison-Wesley.

29. To calculate the coefficients of this expansion, we use the formula (20) in the text. Thus, we have

$$a_0 = \frac{\int_{-1}^{1} f(x)\,dx}{\|P_0\|^2} = \frac{0}{\|P_0\|^2} = 0,$$

where we have used the fact that $f(x)$ is an odd function. (See Theorem 1.) To find a_1, we first calculate the denominator

$$\|P_1\|^2 = \int_{-1}^{1} P_1^2(x)\,dx = \int_{-1}^{1} x^2\,dx = \left.\frac{x^3}{3}\right|_{-1}^{1} = \frac{2}{3}.$$

Therefore,

$$a_1 = \frac{3}{2}\int_{-1}^{1} f(x)P_1(x)\,dx = 3\int_{0}^{1} x\,dx = \left.\frac{3x^2}{2}\right|_{0}^{1} = \frac{3}{2}.$$

Notice that in order to calculate the above integral, we used the fact that the product of two odd functions $f(x)$ and $P_1(x)$ is even (see Problem 7). To find a_2, we observe that, since $f(x)$ is odd and $P_2(x)$ is even, their product is odd, and so we have

$$\int_{-1}^{1} f(x)P_2(x)\,dx = 0$$

by Theorem 1. Hence,

$$a_2 = \frac{\int_{-1}^{1} f(x)P_2(x)\,dx}{\|P_2\|^2} = \frac{0}{\|P_2\|^2} = 0.$$

31. We have to show that

$$\int_{-\infty}^{\infty} H_m(x)H_n(x)e^{-x^2}\,dx = 0,$$

for $m \neq n$, where $m, n = 0, 1, 2$. Therefore, we need to calculate several integrals. Let us begin with $m = 0$, $n = 2$. Here, we see that

$$\int_{-\infty}^{\infty} H_0(x)H_2(x)e^{-x^2}\,dx = \int_{-\infty}^{\infty} \left(4x^2 - 2\right)e^{-x^2}\,dx$$

$$= \lim_{N\to\infty}\int_{0}^{N} \left(4x^2 - 2\right)e^{-x^2}\,dx + \lim_{M\to\infty}\int_{-M}^{0} \left(4x^2 - 2\right)e^{-x^2}\,dx. \quad (10.7)$$

Copyright © 2012 Pearson Education, Inc. Publishing as Addison-Wesley.

We compute the indefinite integral in (10.7) using the integration by parts with a substitution

$$u = x, \qquad dv = 2xe^{-x^2}\,dx$$
$$\Rightarrow \qquad du = dx, \quad v = -e^{-x^2}.$$

Thus, we find

$$\begin{aligned}\int \left(4x^2-2\right)e^{-x^2}\,dx &= 2\int 2x^2e^{-x^2}\,dx - 2\int e^{-x^2}\,dx\\ &= 2\left[-xe^{-x^2}+\int e^{-x^2}\,dx\right] - 2\int e^{-x^2}\,dx = -2xe^{-x^2},\end{aligned}$$

where we have chosen zero integration constant. Substituting this antiderivative into (10.7) and using L'Hôpital's rule to find the limits yields

$$\begin{aligned}\int_{-\infty}^{\infty} H_0(x)H_2(x)e^{-x^2}\,dx &= \lim_{N\to\infty}\left(-2xe^{-x^2}\Big|_0^N\right) + \lim_{M\to\infty}\left(-2xe^{-x^2}\Big|_{-M}^0\right)\\ &= \lim_{N\to\infty}\left(\frac{-2N}{e^{N^2}}+0\right) + \lim_{M\to\infty}\left(0-\frac{2M}{e^{M^2}}\right)\\ &= -\lim_{N\to\infty}\frac{2N}{e^{N^2}} - \lim_{M\to\infty}\frac{2M}{e^{M^2}} = -\lim_{N\to\infty}\frac{2}{2Ne^{N^2}} - \lim_{M\to\infty}\frac{2}{2Me^{M^2}} = 0.\end{aligned}$$

When $m = 0$, $n = 1$ and $m = 1$, $n = 2$, the integrals are, respectively,

$$\int_{-\infty}^{\infty} H_0(x)H_1(x)e^{-x^2}\,dx = \int_{-\infty}^{\infty} 2xe^{-x^2}\,dx \tag{10.8}$$

and

$$\int_{-\infty}^{\infty} H_1(x)H_2(x)e^{-x^2}\,dx = \int_{-\infty}^{\infty} 2x(4x^2-2)e^{-x^2}\,dx\,. \tag{10.9}$$

In each case, the integrand is an odd function, and hence the integral over a symmetric interval $(-N, N)$ is zero. Since it is easy to show that the improper integrals in (10.8) and (10.9) are convergent, we get

$$\int_{-\infty}^{\infty}\cdots = \lim_{N\to\infty}\int_{-N}^{N}\cdots = \lim_{N\to\infty} 0 = 0\,,$$

where "$\cdots$" means either of integrands in (10.8) or (10.9). Since we have shown that all three integrals above equal to zero, the first three Hermite polynomials are orthogonal.

 Copyright © 2012 Pearson Education, Inc. Publishing as Addison-Wesley.

33. Using the norm definition (16) and the orthogonality relation (14) in the text, we get

$$\|f_m + f_n\|^2 = \int_a^b \left[f_m(x) + f_n(x)\right]^2 w(x)\, dx = \int_a^b \left[f_m^2(x) + 2f_m(x)f_n(x) + f_n^2(x)\right] w(x)\, dx$$

$$= \int_a^b f_m^2(x) w(x)\, dx + 2\int_a^b f_m(x)f_n(x)w(x)\, dx + \int_a^b f_n^2(x)w(x)\, dx$$

$$= \int_a^b f_m^2(x)w(x)\, dx + 2(0) + \int_a^b f_n^2(x)w(x)\, dx = \|f_m\|^2 + \|f_n\|^2 \ .$$

EXERCISES 10.4: Fourier Cosine and Sine Series

3. **(a)** The π-periodic extension $\widetilde{f}(x)$ on the interval $(-\pi, \pi)$ is

$$\widetilde{f}(x) = \begin{cases} 0, & -\pi < x < -\pi/2, \\ 1, & -\pi/2 < x < 0, \\ 0, & 0 < x < \pi/2, \\ 1, & \pi/2 < x < \pi, \end{cases}$$

with $\widetilde{f}(x + 2\pi) = \widetilde{f}(x)$. The graph of this function is shown in Fig. **10–B** on page 620.

(b) Using the formula for the odd 2π-periodic extension f_o to $(-\pi, \pi)$ of a function f, defined originally on the interval $(0, \pi)$ (see the text), we get

$$f_o(x) = \begin{cases} f(x), & 0 < x < \pi, \\ -f(-x), & -\pi < x < 0 \end{cases} = \begin{cases} 0, & 0 < x < \pi/2, \\ 1, & \pi/2 < x < \pi, \\ -1, & -\pi < x < -\pi/2, \\ 0, & -\pi/2 < x < 0, \end{cases}$$

with $f_o(x + 2\pi) = f_o(x)$. The graph of $f_o(x)$ is depicted in Fig. **10–C** on page 620.

(c) Using the formula for the even 2π-periodic extension $f_e(x)$ of $f(x)$ to the interval $(-\pi, \pi)$, we get

$$f_e(x) = \begin{cases} f(x), & 0 < x < \pi, \\ f(-x), & -\pi < x < 0 \end{cases} = \begin{cases} 0, & 0 < x < \pi/2, \\ 1, & \pi/2 < x < \pi \\ 1, & -\pi < x < -\pi/2, \\ 0, & -\pi/2 < x < 0 \end{cases}$$

with $f_e(x + 2\pi) = f_e(x)$. The graph of $f_e(x)$ is given in Fig. **10–D** on page 620.

Copyright © 2012 Pearson Education, Inc. Publishing as Addison-Wesley.

5. In this problem, $L = 1$ and $f(x) = -1$ on $(0, L)$. Applying the formula (7) for the coefficients of the Fourier sine series (6) of $f(x)$ (see Definition 2) yields

$$b_n = 2 \int_0^1 (-1) \sin(n\pi x)\, dx = \frac{2}{n\pi} \cos(n\pi x) \Big|_0^1 = \frac{2}{n\pi} (\cos n\pi - \cos 0) = \frac{2\left[(-1)^n - 1\right]}{n\pi}.$$

Clearly,

$$b_{2k} = \frac{2\left[(-1)^{2k} - 1\right]}{2k\pi} = 0,$$

$$b_{2k-1} = \frac{2\left[(-1)^{2k-1} - 1\right]}{(2k-1)\pi} = -\frac{4}{(2k-1)\pi}, \qquad k = 1, 2, \dots.$$

Thus,

$$f(x) \sim \sum_{n=1}^{\infty} b_n \sin(n\pi x) = \sum_{k=1}^{\infty} b_{2k-1} \sin[(2k-1)\pi x] = -\frac{4}{\pi} \sum_{k=1}^{\infty} \frac{1}{2k-1} \sin[(2k-1)\pi x].$$

7. In this problem $L = \pi$ and $f(x) = x^2$. We use the equation (7) in Definition 2 to calculate the coefficients of the Fourier sine series of $f(x)$. Thus, we have

$$f(x) = \sum_{n=1}^{\infty} b_n \sin nx \qquad \text{with} \qquad b_n = \frac{2}{\pi} \int_0^{\pi} x^2 \sin nx\, dx.$$

To compute the coefficients, we use the integration by parts twice to obtain

$$\frac{\pi}{2} b_n = \int_0^{\pi} x^2 \sin nx\, dx = -x^2 \frac{\cos nx}{n} \Big|_0^{\pi} + \frac{2}{n} \int_0^{\pi} x \cos nx\, dx$$

$$= -\frac{\pi^2 \cos n\pi}{n} + 0 + \frac{2}{n} \left[x \frac{\sin nx}{n} \Big|_0^{\pi} - \frac{1}{n} \int_0^{\pi} \sin nx\, dx \right]$$

$$= -\frac{\pi^2 \cos n\pi}{n} + \frac{2}{n} \left[0 - \frac{1}{n} \left(-\frac{\cos nx}{n} \Big|_0^{\pi} \right) \right]$$

$$= -\frac{\pi^2 \cos n\pi}{n} + \frac{2}{n^3} (\cos n\pi - \cos 0), \qquad n = 1, 2, 3, \dots.$$

Since $\cos n\pi = 1$ if n is even and $\cos n\pi = -1$ if n is odd, we see that

$$\frac{\pi}{2} b_n = -\frac{\pi^2 (-1)^n}{n} + \frac{2[(-1)^n - 1]}{n^3}, \qquad n = 1, 2, 3, \dots.$$

Therefore, for $n = 1, 2, 3, \dots$, we have

$$b_n = \frac{2\pi(-1)^{n+1}}{n} + \frac{4[(-1)^n - 1]}{\pi n^3}.$$

 Copyright © 2012 Pearson Education, Inc. Publishing as Addison-Wesley.

Substituting these coefficients into the Fourier sine series for $f(x) = x^2$, yields

$$\sum_{n-1}^{\infty} \left\{ \frac{2\pi(-1)^{n+1}}{n} + \frac{4[(-1)^n - 1]}{\pi n^3} \right\} \sin nx.$$

Since $f(x) = x^2$ and $f'(x) = 2x$ are piecewise continuous on the interval $(0, \pi)$, Theorem 2 implies that this Fourier series converges pointwise to $f(x)$ on the interval $(0, \pi)$. Hence, we can write

$$x^2 = \sum_{n=1}^{\infty} \left\{ \frac{2\pi(-1)^{n+1}}{n} + \frac{4[(-1)^n - 1]}{\pi n^3} \right\} \sin nx, \quad 0 < x < \pi.$$

But, since the odd 2π-periodic extension of $f(x)$ is discontinuous at odd multiples of π, the Gibbs' phenomenon (see Problem 39 in Exercises 10.3) occurs near these points, and so the convergence of this Fourier sine series is not uniform on $(0, \pi)$.

13. Since $f(x) = e^x$ is piecewise continuous on the interval $(0, 1)$, we can use Definition 2 in the text to find its Fourier cosine series. Thus, we have

$$\frac{a_0}{2} + \sum_{n=1}^{\infty} a_n \cos n\pi x, \quad \text{where} \quad a_n = 2 \int_0^1 e^x \cos n\pi x \, dx.$$

Using the fact that $\cos 0 = 1$, we find the coefficient a_0 to be

$$a_0 = 2 \int_0^1 e^x \, dx = 2(e - 1).$$

We now use the integration by parts twice (or the table of integrals on the inside front cover of the text) to calculate the integrals for the remaining coefficients. Since

$$\int e^x \cos n\pi x \, dx = \frac{e^x (\cos n\pi x + n\pi \sin n\pi x)}{1 + n^2\pi^2},$$

where $n = 1, 2, 3, \ldots$, the remaining coefficients are given by

$$a_n = 2 \int_0^1 e^x \cos n\pi x \, dx = \frac{2e^x (\cos n\pi x + n\pi \sin n\pi x)}{1 + n^2\pi^2} \Bigg|_0^1$$

$$= \frac{2e(\cos n\pi)}{1 + n^2\pi^2} - \frac{2}{1 + n^2\pi^2} = \frac{2[(-1)^n e - 1]}{1 + n^2\pi^2}, \quad n = 1, 2, 3, \ldots.$$

(We have used the fact that $\cos n\pi = 1$ if n is even and $\cos n\pi = -1$ if n is odd.) By substituting these coefficients into the Fourier cosine series for f, we obtain

$$e^x = e - 1 + 2 \sum_{n=1}^{\infty} \frac{(-1)^n e - 1}{1 + n^2\pi^2} \cos n\pi x, \quad 0 < x < 1.$$

Copyright © 2012 Pearson Education, Inc. Publishing as Addison-Wesley.

Note that, for $0 < x < 1$, e^x equals its Fourier cosine series (by Theorem 2), because this series converges uniformly. To see this, first notice that the even 2π-periodic extension of $f(x) = e^x$, $0 < x < 1$, is given by

$$f_e(x) = \begin{cases} e^{-x}, & -1 < x < 0, \\ e^x, & 0 < x < 1, \end{cases}$$

with $f_e(x + 2\pi) = f_e(x)$. Since this extension is continuous on $(-\infty, \infty)$ and $f_e'(x)$ is piecewise continuous on $[-1, 1]$, Theorem 3 states that its Fourier series (which is the one we found above) converges uniformly to $f_e(x)$ on $[-1, 1]$, and so it converges uniformly to $f(x) = e^x$ on $(0, 1)$.

15. Here, $T = \pi$ and $f(x) = \sin x$. Applying the formula (5) in Definition 2 for the coefficients of the Fourier cosine series, we obtain

$$a_0 = \frac{2}{\pi} \int_0^\pi \sin x \, dx = -\frac{2 \cos x}{\pi}\bigg|_0^\pi = \frac{4}{\pi}$$

and, for $n = 1, 2, \ldots$,

$$\begin{aligned}
a_n &= \frac{2}{\pi} \int_0^\pi \sin x \cos nx \, dx = \frac{1}{\pi} \int_0^\pi [\sin(n+1)x - \sin(n-1)x] \, dx \\
&= \frac{1}{\pi} \left[\frac{\cos(n-1)x}{n-1} - \frac{\cos(n+1)x}{n+1} \right]\bigg|_0^\pi \\
&= \frac{1}{\pi} \left[\left(\frac{(-1)^{n-1}}{n-1} - \frac{(-1)^{n+1}}{n+1} \right) - \left(\frac{1}{n-1} - \frac{1}{n+1} \right) \right] = \frac{2}{\pi(n^2 - 1)} \left[(-1)^{n+1} - 1 \right].
\end{aligned}$$

Since

$$(-1)^{n+1} - 1 = \begin{cases} 0, & n = 2k - 1, \\ -2, & n = 2k, \end{cases} \qquad k = 1, 2, \ldots,$$

we have

$$a_{2k-1} = 0, \quad a_{2k} = -\frac{4}{\pi(4k^2 - 1)}, \quad k = 1, 2, \ldots .$$

Thus,

$$\sin x \sim \frac{a_0}{2} + \sum_{n=1}^\infty a_n \cos nx = \frac{2}{\pi} - \frac{4}{\pi} \sum_{k=1}^\infty \frac{1}{4k^2 - 1} \cos 2kx, \quad 0 < x < \pi,$$

which is equivalent to the answer in the text.

 Copyright © 2012 Pearson Education, Inc. Publishing as Addison-Wesley.

17. This problem is the same as the heat flow problem in the text with $f(x) = 1 - \cos 2x$, $\beta = 5$, and $L = \pi$. Therefore, a formal solution to this problem is given by the equation (15), Section 10.2 in the text, that is,

$$u(x,t) = \sum_{n=1}^{\infty} c_n e^{-5n^2 t} \sin nx \qquad 0 < x < \pi, \quad t > 0, \tag{10.10}$$

with

$$u(x,0) = f(x) = 1 - \cos 2x = \sum_{n=1}^{\infty} b_n \sin nx.$$

Therefore, we have to find the Fourier sine series for $1 - \cos 2x$. To this end, we use the equation (7) in Definition 2. Hence, the coefficients are given by

$$b_n = \frac{2}{\pi} \int_0^{\pi} (1 - \cos 2x) \sin nx \, dx$$

$$= \frac{2}{\pi} \int_0^{\pi} \sin nx \, dx - \frac{2}{\pi} \int_0^{\pi} \cos 2x \sin nx \, dx, \qquad n = 1, 2, 3, \ldots.$$

Calculating the first integral above yields

$$\frac{2}{\pi} \int_0^{\pi} \sin nx \, dx = -\frac{2}{n\pi}(\cos n\pi - 1) = \frac{2}{n\pi}[1 - (-1)^n],$$

where we have used the fact that $\cos n\pi = (-1)^n$. To calculate the second integral, we use the trigonometric identity $2\cos\alpha\sin\beta = \sin(\beta - \alpha) + \sin(\beta + \alpha)$ and obtain

$$-\frac{2}{\pi} \int_0^{\pi} \cos 2x \sin nx \, dx = -\frac{1}{\pi}\left\{ \int_0^{\pi} \sin[(n-2)x] \, dx + \int_0^{\pi} \sin[(n+2)x] \, dx \right\}$$

$$= \frac{1}{\pi(n-2)}\{\cos[(n-2)\pi] - 1\} + \frac{1}{\pi(n+2)}\{\cos[(n+2)\pi] - 1\}$$

$$= \frac{1}{\pi(n-2)}[(-1)^n - 1] + \frac{1}{\pi(n+2)}[(-1)^n - 1].$$

Combining these two integrals yields

$$b_n = \frac{2}{n\pi}[1 - (-1)^n] + \frac{1}{\pi(n-2)}[(-1)^n - 1] + \frac{1}{\pi(n+2)}[(-1)^n - 1]$$

$$= \begin{cases} 0, & \text{if } n \text{ is even,} \\ 4/(n\pi) - 2/[\pi(n-2)] - 2/[\pi(n+2)], & \text{if } n \text{ is odd,} \end{cases}$$

Copyright © 2012 Pearson Education, Inc. Publishing as Addison-Wesley.

$n = 1, 2, 3, \ldots$. Hence, we obtain a formal solution to this problem by substituting these coefficients into the equation (10.10) and setting $n = 2k - 1$. Therefore, we have

$$u(x, t) = \frac{2}{\pi} \sum_{k=1}^{\infty} \left[\frac{2}{2k-1} - \frac{1}{2k+1} - \frac{1}{2k-3} \right] e^{-5(2k-1)^2 t} \sin(2k-1)x.$$

EXERCISES 10.5:   The Heat Equation

3. If we let $\beta = 3$, $L = \pi$, and $f(x) = x$, we see that this problem has the same form as the one in Example 1 in the text. Therefore, we can find the formal solution to this problem by substituting these values of parameters into the equation (14). Hence, we have

$$u(x, t) = \frac{a_0}{2} + \sum_{n=1}^{\infty} a_n e^{-3n^2 t} \cos nx, \tag{10.11}$$

where a_n's are the Fourier cosine series coefficients for f. Thus, we have to find the Fourier cosine series coefficients for $f(x) = x$, $0 < x < \pi$. (Note that the even 2π-extension of $f(x) = x$, $0 < x < \pi$, given by

$$f_e(x) = |x|, \quad -\pi \le x \le \pi,$$

with $f_e(x + 2\pi) = f_e(x)$, is continuous and its derivative is piecewise continuous. Therefore, the Fourier series for this extension converges uniformly to $f_e(x)$. This means that the equality sign in (10.11) holds.) To find the required Fourier series coefficients, we use the equations (4) and (5) given in Definition 2, Section 10.4. Hence, we have

$$x = \frac{a_0}{2} + \sum_{n=1}^{\infty} a_n \cos nx.$$

We compute

$$a_0 = \frac{2}{\pi} \int_0^{\pi} x \, dx = \frac{2}{\pi} \left. \frac{x^2}{2} \right|_0^{\pi} = \pi \quad \text{and} \quad a_n = \frac{2}{\pi} \int_0^{\pi} x \cos nx \, dx$$

for $n = 1, 2, 3, \ldots$. We calculate the second integral using the integration by parts.

$$
\begin{aligned}
a_n &= \frac{2}{\pi} \int_0^{\pi} x \cos nx \, dx = \frac{2}{\pi} \left[\left. \frac{x}{n} \sin nx \right|_0^{\pi} - \frac{1}{n} \int_0^{\pi} \sin nx \, dx \right] \\
&= \frac{2}{\pi} \left[0 - \frac{1}{n} \left(\left. -\frac{\cos nx}{n} \right|_0^{\pi} \right) \right] = \frac{2}{\pi n^2} (\cos n\pi - 1) = \frac{2}{\pi n^2} \left[(-1)^n - 1 \right].
\end{aligned}
$$

 Copyright © 2012 Pearson Education, Inc. Publishing as Addison-Wesley.

Combining these results yields

$$a_n = \begin{cases} \pi, & \text{if } n = 0, \\ -4/(\pi n^2), & \text{if } n \text{ is odd,} \\ 0, & \text{if } n \neq 0 \text{ is even,} \end{cases} \quad n = 0, 1, 2, \ldots.$$

The formal solution to this problem can be found by substituting these coefficients into the equation given in (10.11). Thus, we have

$$\begin{aligned} u(x,t) &= \frac{\pi}{2} e^{-0} \cos 0 - \sum_{k=0}^{\infty} \frac{4}{\pi(2k+1)^2} e^{-3(2k+1)^2 t} \cos(2k+1)x \\ &= \frac{\pi}{2} - \sum_{k=0}^{\infty} \frac{4}{\pi(2k+1)^2} e^{-3(2k+1)^2 t} \cos(2k+1)x . \end{aligned}$$

7. This problem has nonhomogeneous boundary conditions and so has the same form as the one in Example 2 in the text. By comparing these two problems, we see that for this problem $\beta = 2$, $L = \pi$, $U_1 = 5$, $U_2 = 10$, and $f(x) = \sin 3x - \sin 5x$. Thus, we assume that the solution consists of a steady state solution $v(x)$ and a transient solution $w(x, t)$. The steady state solution is given in the equation (24) in the text.

$$v(x) = 5 + \frac{(10 - 5)x}{\pi} = 5 + \frac{5}{\pi} x .$$

The formal transient solution is given by equations (39) and (40). Using these equations and making appropriate substitutions, we obtain

$$w(x, t) = \sum_{n=1}^{\infty} c_n e^{-2n^2 t} \sin nx , \tag{10.12}$$

where the coefficients c_n's satisfy

$$f(x) - v(x) = \sin 3x - \sin 5x - 5 - \frac{5}{\pi} x = \sum_{n=1}^{\infty} c_n \sin nx , \qquad 0 < x < \pi.$$

Therefore, we have to find the Fourier sine series coefficients for the function $f(x) - v(x)$, $0 < x < \pi$. Since the function $f(x) = \sin 3x - \sin 5x$ is already in the form of a sine series, we only need to find the Fourier sine series for $-v(x) = -5 - (5x)/\pi$ and then add to $\sin 3x - \sin 5x$. The resulting coefficients are the ones that we need. (Note that the Fourier sine series for $-5 - (5x)/\pi$ converges pointwise, but not uniformly to $-5 - (5x)/\pi$

for $0 < x < \pi$.) We now use the equations (6) and (7) in Definition 2, Section 10.4. Thus, with appropriate substitutions, we have

$$-5 - \frac{5x}{\pi} = \sum_{n=1}^{\infty} b_n \sin nx, \qquad \text{where} \qquad b_n = \frac{2}{\pi} \int_0^{\pi} \left(-5 - \frac{5x}{\pi}\right) \sin nx \, dx.$$

Using the integration by parts, we get

$$
\begin{aligned}
b_n &= -\frac{10}{\pi} \int_0^{\pi} \sin nx \, dx - \frac{10}{\pi^2} \int_0^{\pi} x \sin nx \, dx \\
&= \frac{10}{n\pi}(\cos n\pi - 1) - \frac{10}{\pi^2}\left[-\frac{x}{n}\cos nx\Big|_0^{\pi} + \frac{1}{n}\int_0^{\pi}\cos nx\,dx\right] \\
&= \frac{10}{n\pi}(\cos n\pi - 1) - \frac{10}{\pi^2}\left[-\frac{\pi}{n}\cos n\pi + 0\right] \\
&= \frac{10}{n\pi}(2\cos n\pi - 1) = \frac{10}{n\pi}\left[2(-1)^n - 1\right], \qquad n = 1, 2, 3, \ldots.
\end{aligned}
$$

Thus, the Fourier sine series for $\sin 3x - \sin 5x - 5 - (5x)/\pi$ is given by

$$
\begin{aligned}
\sin 3x - \sin 5x - 5 - \frac{5x}{\pi} &= \sin 3x - \sin 5x + \sum_{n=1}^{\infty} \frac{10}{n\pi}\left[2(-1)^n - 1\right]\sin nx \\
&= \sin 3x - \sin 5x - \frac{30}{\pi}\sin x + \frac{10}{2\pi}\sin 2x - \frac{30}{3\pi}\sin 3x + \frac{10}{4\pi}\sin 4x \\
&\qquad\qquad\qquad\qquad - \frac{30}{5\pi}\sin 5x + \sum_{n=6}^{\infty}\frac{10}{n\pi}\left[2(-1)^n - 1\right]\sin nx \\
&= -\frac{30}{\pi}\sin x + \frac{5}{\pi}\sin 2x + \left(1 - \frac{10}{\pi}\right)\sin 3x + \frac{5}{2\pi}\sin 4x \\
&\qquad\qquad - \left(1 + \frac{6}{\pi}\right)\sin 5x + \sum_{n=6}^{\infty}\frac{10}{n\pi}\left[2(-1)^n - 1\right]\sin nx.
\end{aligned}
$$

We, therefore, obtain the formal transient solution by substituting c_n's into (10.12).

$$
\begin{aligned}
w(x,t) &= -\frac{30}{\pi}e^{-2(1)^2 t}\sin x + \frac{5}{\pi}e^{-2(2)^2 t}\sin 2x + \left(1 - \frac{10}{\pi}\right)e^{-2(3)^2 t}\sin 3x \\
&\quad + \frac{5}{2\pi}e^{-2(4)^2 t}\sin 4x - \left(1 + \frac{6}{\pi}\right)e^{-2(5)^2 t}\sin 5x + \sum_{n=6}^{\infty}\frac{10}{n\pi}\left[2(-1)^n - 1\right]e^{-2n^2 t}\sin nx,
\end{aligned}
$$

and so the formal solution to the original problem is given by

$$u(x,t) = v(x) + w(x,t)$$

 Copyright © 2012 Pearson Education, Inc. Publishing as Addison-Wesley.

$$= 5 + \frac{5}{\pi}x - \frac{30}{\pi}e^{-2t}\sin x + \frac{5}{\pi}e^{-8t}\sin 2x + \left(1 - \frac{10}{\pi}\right)e^{-18t}\sin 3x + \frac{5}{2\pi}e^{-32t}\sin 4x$$

$$- \left(1 + \frac{6}{\pi}\right)e^{-50t}\sin 5x + \sum_{n=6}^{\infty}\frac{10}{n\pi}\left[2(-1)^n - 1\right]e^{-2n^2t}\sin nx.$$

9. Notice that the given equation is a nonhomogeneous partial differential equation, and has the same form as that in Example 3 in the text. Comparing these two problems, we see that here $\beta = 1$, $P(x) = e^{-x}$, $L = \pi$, $U_1 = U_2 = 0$, and $f(x) = \sin 2x$. As in Example 3, we assume that the solution is the sum of a steady state solution $v(x)$ and a transient solution $w(x,t)$. The steady state solution is the solution to the boundary value problem

$$v''(x) = -e^{-x}, \qquad 0 < x < \pi, \qquad v(0) = v(\pi) = 0.$$

Thus, the steady state solution can be found either by solving this boundary value problem for an ordinary differential equation or by substituting appropriate values into the equation (35) in the text. By either method we find

$$v(x) = \frac{e^{-\pi} - 1}{\pi}x - e^{-x} + 1.$$

The formal transient solution is given by the equations (39) and (40). By making appropriate substitutions into these equations, we obtain

$$w(x,t) = \sum_{n=1}^{\infty} c_n e^{-n^2 t}\sin nx, \qquad (10.13)$$

where c_n's are determined by

$$f(x) - v(x) = \sin 2x - \frac{e^{-\pi} - 1}{\pi}x + e^{-x} - 1 = \sum_{n=1}^{\infty} c_n \sin nx.$$

Hence, we have to find the coefficients of the Fourier sine series for $f(x) - v(x)$. The first term, $f(x) = \sin 2x$, is already in the desired form. Hence, the Fourier sine series for $f(x) - v(x)$ is

$$\sin 2x + \sum_{n=1}^{\infty} b_n \sin nx = b_1 \sin x + (b_2 + 1)\sin 2x + \sum_{n=3}^{\infty} b_n \sin nx,$$

where b_n's are the Fourier sine coefficients for $-v(x)$. These coefficients are

$$b_n = \frac{2}{\pi}\int_0^\pi \left(-\frac{e^{-\pi} - 1}{\pi}x + e^{-x} - 1\right)\sin nx\, dx$$

Copyright © 2012 Pearson Education, Inc. Publishing as Addison-Wesley.

$$= \frac{2}{\pi}\left(-\frac{e^{-\pi}-1}{\pi}\right)\int_0^{\pi} x\sin nx\, dx + \frac{2}{\pi}\int_0^{\pi} e^{-x}\sin nx\, dx - \frac{2}{\pi}\int_0^{\pi}\sin nx\, dx\,.$$

We evaluate each integral separately. The first one can be found integrating by parts.

$$\frac{2}{\pi}\left(-\frac{e^{-\pi}-1}{\pi}\right)\int_0^{\pi} x\sin nx\, dx = \frac{-2(e^{-\pi}-1)}{\pi^2}\left(-\frac{x}{n}\cos nx\Big|_0^{\pi} + \frac{1}{n}\int_0^{\pi}\cos nx\, dx\right)$$

$$= \frac{-2(e^{-\pi}-1)}{\pi^2}\left(-\frac{\pi}{n}\cos n\pi + 0 + 0\right) = \frac{2(e^{-\pi}-1)}{n\pi}(-1)^n.$$

To find the second integral, we use the table of integrals on the inside front cover of the text (or use integration by parts twice) to obtain

$$\frac{2}{\pi}\int_0^{\pi} e^{-x}\sin nx\, dx = \frac{2}{\pi}\left(\frac{-e^{-\pi}n\cos n\pi + n}{1+n^2}\right) = \frac{2n}{(1+n^2)\pi}\left[e^{-\pi}(-1)^{n+1}+1\right].$$

Finally,

$$-\frac{2}{\pi}\int_0^{\pi}\sin nx\, dx = \frac{2}{n\pi}(\cos n\pi - 1) = \frac{2}{n\pi}\left[(-1)^n - 1\right].$$

Thus, the Fourier sine coefficients for $-v(x)$ are by

$$b_n = \frac{2(e^{-\pi}-1)}{n\pi}(-1)^n + \frac{2n}{\pi(1+n^2)}\left[(-1)^{n+1}e^{-\pi}+1\right] + \frac{2}{n\pi}\left[(-1)^n - 1\right].$$

Therefore, the coefficients for the formal transient solution are given by

$$c_n = \begin{cases} \dfrac{2(e^{-\pi}-1)}{n\pi}(-1)^n + \dfrac{2n}{\pi(1+n^2)}\left[(-1)^{n+1}e^{-\pi}+1\right] + \dfrac{2}{n\pi}\left[(-1)^n - 1\right], & n \neq 2, \\[3mm] \dfrac{e^{-\pi}-1}{\pi} + \dfrac{4}{5\pi}\left(1 - e^{-\pi}\right) + 1, & n = 2. \end{cases}$$

Since a formal solution to the partial differential equation, given in this problem, is the sum of its steady state solution and its transient solution, the answer is

$$u(x,t) = v(x) + w(x,t) = \frac{e^{-\pi}-1}{\pi}x - e^{-x} + 1 + \sum_{n=1}^{\infty} c_n e^{-n^2 t}\sin nx\,.$$

11. Let $u(x,t) = X(x)T(t)$. Substituting $u(x,t) = X(x)T(t)$ into the given equation yields

$$T'(t)X(x) = 4X''(x)T(t) \qquad \Rightarrow \qquad \frac{T'(t)}{4T(t)} = \frac{X''(x)}{X(x)} = -\lambda,$$

 Copyright © 2012 Pearson Education, Inc. Publishing as Addison-Wesley.

where λ is a constant. Substituting the solution $u(x,t) = X(x)T(t)$ into the boundary conditions, we obtain

$$X'(0)T(t) = 0, \qquad X(\pi)T(t) = 0, \qquad t > 0.$$

Thus, we conclude that $X'(0) = 0$ and $X(\pi) = 0$. Therefore, we have a boundary value problem

$$\begin{aligned} X''(x) + \lambda X(x) = 0, \quad 0 < x < \pi, \\ X'(0) = X(\pi) = 0, \end{aligned} \tag{10.14}$$

and

$$T'(t) + 4\lambda T(t) = 0, \qquad t > 0. \tag{10.15}$$

To solve the boundary value problem (10.14), we examine three cases.

Case 1: $\lambda = 0.$ The equation in (10.14) becomes $X'' = 0$. Therefore, a general solution is $X(x) = C_1 x + C_2$, where C_1 and C_2 are arbitrary constants. To find these constants, we use the boundary conditions. Thus, we have

$$X'(0) = C_1 = 0 \qquad \Rightarrow \qquad X(x) = C_2 \,,$$

and so $X(\pi) = C_2 = 0$. Therefore, in this case, we have only the trivial solution.

Case 2: $\lambda < 0.$ In this case, the auxiliary equation for (10.14) is $r^2 + \lambda = 0$. Solving yields $r = \pm\sqrt{-\lambda}$. Thus, a general solution is given by

$$X(x) = C_1 e^{\sqrt{-\lambda}\,x} + C_2 e^{-\sqrt{-\lambda}\,x},$$

where C_1 and C_2 are arbitrary constants. To find these constants, we use the boundary conditions in (10.14). First, we note that

$$X'(x) = C_1\sqrt{-\lambda}\,e^{\sqrt{-\lambda}\,x} - C_2\sqrt{-\lambda}\,e^{-\sqrt{-\lambda}\,x}.$$

Therefore,

$$X'(0) = C_1\sqrt{-\lambda} - C_2\sqrt{-\lambda} = 0 \quad \Rightarrow \quad C_1 = C_2 \quad \Rightarrow \quad X(x) = C_1\left(e^{\sqrt{-\lambda}\,x} + e^{-\sqrt{-\lambda}\,x}\right).$$

The other boundary condition implies that

$$X(\pi) = C_1\left(e^{\sqrt{-\lambda}\,\pi} + e^{-\sqrt{-\lambda}\,\pi}\right) = 0 \quad \Rightarrow \quad C_1\left(e^{2\sqrt{-\lambda}\,\pi} + 1\right) = 0.$$

Copyright © 2012 Pearson Education, Inc. Publishing as Addison-Wesley.

The only way for this equation to hold is to have $C_1 = 0$. Therefore, we again obtain only the trivial solution.

<u>Case 3: $\lambda > 0$.</u> In this case, the auxiliary equation for the boundary value problem (10.14) has the roots $r = \pm\sqrt{-\lambda} = \pm i\sqrt{\lambda}$. Therefore, a general solution is

$$X(x) = C_1 \sin\left(\sqrt{\lambda}x\right) + C_2 \cos\left(\sqrt{\lambda}x\right)$$

$$\Rightarrow \qquad X'(x) = C_1\sqrt{\lambda}\cos\left(\sqrt{\lambda}x\right) - C_2\sqrt{\lambda}\sin\left(\sqrt{\lambda}x\right).$$

Using the boundary condition $X'(0) = 0$, we obtain

$$0 = X'(0) = C_1\sqrt{\lambda}\cos 0 - C_2\sqrt{\lambda}\sin 0 = C_1\sqrt{\lambda} \qquad \Rightarrow \qquad C_1 = 0.$$

Hence, $X(x) = C_2 \cos\left(\sqrt{\lambda}x\right)$. Applying the second boundary condition yields

$$0 = X(\pi) = C_2 \cos\left(\sqrt{\lambda}\pi\right)$$

$$\Rightarrow \qquad \sqrt{\lambda}\pi = (2n+1)\frac{\pi}{2} \qquad \Rightarrow \qquad \lambda = C_n = \frac{(2n+1)^2}{4}, \qquad n = 0, 1, 2, \ldots.$$

Therefore, nontrivial solutions to the boundary value problem (10.14) are given by

$$X_n(x) = c_n \cos\left(\frac{2n+1}{2}x\right), \qquad n = 0, 1, 2, \ldots.$$

Substituting λ_n's into the equation (10.15), we obtain

$$T'(t) + (2n+1)^2 T(t) = 0, \qquad t > 0.$$

This is a separable differential equation, and we find

$$\frac{dT}{T} = -(2n+1)^2\, dt \qquad \Rightarrow \qquad \int \frac{dT}{T} = -(2n+1)^2 \int dt$$

$$\Rightarrow \quad \ln|T| = -(2n+1)^2 t + C \qquad \Rightarrow \qquad T = T_n = b_n e^{-(2n+1)^2 t}, \qquad n = 0, 1, 2, \ldots$$

(where $b_n = \pm e^C$). By the superposition principle, since

$$u_n(x,t) = X_n(x)T_n(t),$$

we see that the formal solution to the original partial differential equation is

$$u(x,t) = \sum_{n=0}^{\infty} b_n e^{-(2n+1)^2 t} c_n \cos\left(\frac{2n+1}{2}x\right)$$

 Copyright © 2012 Pearson Education, Inc. Publishing as Addison-Wesley.

$$= \sum_{n=0}^{\infty} a_n e^{-(2n+1)^2 t} \cos\left[\left(n + \frac{1}{2}\right)x\right], \tag{10.16}$$

where $a_n := b_n c_n$. To find the coefficients a_n, we use the initial condition to obtain

$$u(x,0) = f(x) = \sum_{n=0}^{\infty} a_n \cos\left[\left(n + \frac{1}{2}\right)x\right]. \tag{10.17}$$

Therefore, the formal solution to this problem is given by (10.16), where a_n's are determined by (10.17).

17. This problem is similar to the problem in Example 4 in the text with $\beta = 1$, $L = W = \pi$, and $f(x,y) = y$. The formal solution to this problem is given in the equation (52) with coefficients computed in (54) and (55). By making appropriate substitutions, we see that the formal solution to this problem is

$$u(x,y,t) = \sum_{m=0}^{\infty}\sum_{n=1}^{\infty} a_{mn} e^{-(m^2+n^2)t} \cos mx \sin ny. \tag{10.18}$$

We can find the coefficients a_{0n}, $n = 1, 2, 3, \ldots$, by using the equation (54) in the text with appropriate values of parameters. This yields

$$a_{0n} = \frac{2}{\pi^2} \int_0^\pi \int_0^\pi y \sin ny \, dx \, dy = \frac{2}{\pi^2} \int_0^\pi y \sin ny \left(\int_0^\pi dx\right) dy$$

$$= \frac{2}{\pi} \int_0^\pi y \sin ny \, dy = \frac{2}{\pi}\left(-\frac{y}{n}\cos ny\Big|_0^\pi + \frac{1}{n}\int_0^\pi \cos ny \, dy\right)$$

$$= \frac{2}{\pi}\left[-\frac{\pi}{n}\cos n\pi + \left(\frac{1}{n^2}\sin ny\Big|_0^\pi\right)\right] = \frac{2}{\pi}\left(-\frac{\pi}{n}\cos n\pi\right) = \frac{2}{n}(-1)^{n+1}.$$

We now use the equation (55) in the text to find the other coefficients. Thus, for $m \geq 1$ and $n \geq 1$, we have

$$a_{mn} = \frac{4}{\pi^2} \int_0^\pi \int_0^\pi y \cos mx \sin ny \, dx \, dy$$

$$= \frac{4}{\pi^2} \int_0^\pi y \sin ny \left(\int_0^\pi \cos mx \, dx\right) dy = \frac{4}{\pi^2} \int_0^\pi y \sin ny \cdot (0) \, dy = 0.$$

The formal solution to the given problem can be found by substituting these coefficients into (10.18). To this end, we note that the coefficients for terms with $m \neq 0$ equal zero.

Copyright © 2012 Pearson Education, Inc. Publishing as Addison-Wesley.

Hence, only terms having the index $m = 0$ will appear in the summation. Therefore, the formal solution is given by

$$u(x, y, t) = \sum_{n=1}^{\infty} \frac{2}{n} (-1)^{n+1} e^{-n^2 t} \sin ny = 2 \sum_{n=1}^{\infty} \frac{(-1)^{n+1}}{n} e^{-n^2 t} \sin ny \,.$$

EXERCISES 10.6: The Wave Equation

1. This problem has the form of that given in equations (1)–(4) in the text. Here, $\alpha = 1$, $L = 1$, $f(x) = x(1 - x)$, and $g(x) = \sin 7\pi x$. This problem is consistent because

$$f(0) = f(1) = 0 \quad \text{and} \quad g(0) = g(1) = 0.$$

A general solution to the wave equation was derived in Section 10.2 of the text. Making appropriate substitutions yields the formal solution

$$u(x, t) = \sum_{n=1}^{\infty} [a_n \cos n\pi t + b_n \sin n\pi t] \sin n\pi x \,. \tag{10.19}$$

To find the coefficients a_n, we note that they are the Fourier sine series coefficients of $x(1 - x)$, and so can be found by using the formula (7), Section 10.4. Thus, for $n = 1, 2, 3, \ldots$, we have

$$a_n = 2 \int_0^1 x(1 - x) \sin n\pi x \, dx = 2 \left(\int_0^1 x \sin n\pi x \, dx - \int_0^1 x^2 \sin n\pi x \, dx \right).$$

Integration by parts yields

$$\int_0^1 x \sin n\pi x \, dx = -\frac{1}{n\pi} \cos n\pi = -\frac{(-1)^n}{n\pi}$$

and

$$\int_0^1 x^2 \sin n\pi x \, dx = -\frac{1}{n\pi} \cos n\pi - \frac{2}{n^2\pi^2} \left(-\frac{1}{n\pi} \cos n\pi + \frac{1}{n\pi} \right) = -\frac{(-1)^n}{n\pi} + \frac{2\left[(-1)^n - 1\right]}{n^3\pi^3} \,.$$

Therefore, for $n = 1, 2, 3, \ldots$, we see that

$$a_n = 2 \left\{ -\frac{(-1)^n}{n\pi} + \frac{(-1)^n}{n\pi} - \frac{2\left[(-1)^n - 1\right]}{n^3\pi^3} \right\} = -\frac{4\left[(-1)^n - 1\right]}{n^3\pi^3} \,.$$

 Copyright © 2012 Pearson Education, Inc. Publishing as Addison-Wesley.

Note that

$$a_n = \begin{cases} 0, & \text{if } n \text{ is even,} \\ 8/(n^3\pi^3), & \text{if } n \text{ is odd.} \end{cases}$$

The b_n's can be found from the equation (7) in this section. We have

$$\sin 7\pi x = \sum_{n=1}^{\infty} n\pi b_n \sin n\pi x.$$

We see that, for $n = 7$,

$$7\pi b_7 = 1 \quad \Rightarrow \quad b_7 = \frac{1}{7\pi} \quad \text{and} \quad b_n = 0 \quad \text{for} \quad n \neq 7.$$

By substituting these coefficients into the formal solution (10.19) we obtain

$$u(x,t) = \frac{1}{7\pi} \sin 7\pi t \sin 7\pi x + \sum_{k=0}^{\infty} \frac{8}{[(2k+1)\pi]^3} \cos[(2k+1)\pi t] \sin[(2k+1)\pi x].$$

5. First we note that this problem is consistent because

$$g(0) = g(L) = 0 \quad \text{and} \quad f(0) = f(L) = 0.$$

The formal solution to this problem is given in the equation (5) in the text with the coefficients determined in (6) and (7). From the equation (7), we see that

$$g(x) = \sum_{n=1}^{\infty} b_n \frac{n\pi\alpha}{L} \sin\left(\frac{n\pi x}{L}\right) \equiv 0.$$

Thus, $b_n = 0$ for all n. Therefore, the formal solution, given in the equation (5), becomes

$$u(x,t) = \sum_{n=1}^{\infty} a_n \cos\left(\frac{n\pi\alpha t}{L}\right) \sin\left(\frac{n\pi x}{L}\right). \tag{10.20}$$

Since, by (6) in the text, a_n's are the Fourier sine series coefficients for $f(x)$, we use the equation (7), Section 10.4, to find that

$$a_n = \frac{2}{L} \int_0^L f(x) \sin\left(\frac{n\pi x}{L}\right) dx = \frac{2}{L} \left[\frac{h_0}{a} \int_0^a x \sin\left(\frac{n\pi x}{L}\right) dx + h_0 \int_a^L \frac{L-x}{L-a} \sin\left(\frac{n\pi x}{L}\right) dx \right]$$

$$= \frac{2h_0}{L} \left[\frac{1}{a} \int_0^a x \sin\left(\frac{n\pi x}{L}\right) dx + \frac{L}{L-a} \int_a^L \sin\left(\frac{n\pi x}{L}\right) dx - \frac{1}{L-a} \int_a^L x \sin\left(\frac{n\pi x}{L}\right) dx \right],$$

Copyright © 2012 Pearson Education, Inc. Publishing as Addison-Wesley.

Chapter 10

$n = 1, 2, 3, \ldots$. By using the integration by parts, we find

$$\int x \sin\left(\frac{n\pi x}{L}\right) dx = -\frac{xL}{n\pi}\cos\left(\frac{n\pi x}{L}\right) + \frac{L^2}{n^2\pi^2}\sin\left(\frac{n\pi x}{L}\right).$$

Therefore, for $n = 1, 2, 3, \ldots$, the coefficients are

$$a_n = \frac{2h_0}{L}\left\{\frac{1}{a}\left[-\frac{aL}{n\pi}\cos\left(\frac{n\pi a}{L}\right) + \frac{L^2}{n^2\pi^2}\sin\left(\frac{n\pi a}{L}\right)\right] - \frac{L^2}{n\pi(L-a)}\left[\cos n\pi - \cos\left(\frac{n\pi a}{L}\right)\right]\right.$$
$$\left. -\frac{1}{L-a}\left[-\frac{L^2}{n\pi}\cos n\pi + \frac{aL}{n\pi}\cos\left(\frac{n\pi a}{L}\right)\right] + \frac{L^2}{n^2\pi^2}\left[\sin n\pi - \sin\left(\frac{n\pi a}{L}\right)\right]\right\}.$$

Simplifying yields

$$a_n = \frac{2h_0 L^2}{n^2\pi^2 a(L-a)}\sin\left(\frac{n\pi a}{L}\right), \qquad n = 1, 2, 3, \ldots.$$

Substituting this result into the equation (10.20), we obtain the formal solution

$$u(x,t) = \frac{2h_0 L^2}{\pi^2 a(L-a)}\sum_{n=1}^{\infty}\frac{1}{n^2}\sin\left(\frac{n\pi a}{L}\right)\cos\left(\frac{n\pi\alpha t}{L}\right)\sin\left(\frac{n\pi x}{L}\right).$$

7. With $\alpha = 1$, $h(x,t) = tx$, $L = \pi$, $f(x) = \sin x$, and $g(x) = 5\sin 2x - 3\sin 5x$, this problem is of the form given in Example 1 in the text. The formal solution to this problem is, therefore, given in the equation (16). Thus, with appropriate substitutions, we get the formal solution

$$u(x,t) = \sum_{n=1}^{\infty}\left\{a_n\cos nt + b_n\sin nt + \frac{1}{n}\int_0^t h_n(s)\sin[n(t-s)]\,ds\right\}\sin nx. \qquad (10.21)$$

According to (14) in the text,

$$f(x) = \sin x = \sum_{n=1}^{\infty}a_n\sin nx.$$

Thus, the only nonzero term in this infinite series is the first one. Namely, $a_1 = 1$ and $a_n = 0$ for $n > 1$. The coefficients b_n are given in the equation (15) in the text.

$$g(x) = 5\sin 2x - 3\sin 5x = \sum_{n=1}^{\infty}nb_n\sin nx,$$

which implies that

$$2b_2 = 5 \quad \Rightarrow \quad b_2 = \frac{5}{2} \quad \text{and} \quad 5b_5 = -3 \quad \Rightarrow \quad b_5 = -\frac{3}{5},$$

 Copyright © 2012 Pearson Education, Inc. Publishing as Addison-Wesley.

and $b_n = 0$ for all other values of n. To calculate the integral, given in (10.21), we need the functions $h_n(t)$, $n = 1, 2, \ldots$. These functions (see Example 1) are the Fourier sine series coefficients of $h(x, t) = tx$, with t fixed. Thus, we have

$$
\begin{aligned}
h_n(t) &= \frac{2}{\pi} \int_0^\pi tx \sin nx \, dx = \frac{2t}{\pi} \int_0^\pi x \sin nx \, dx \\
&= \frac{2t}{\pi} \left(-\frac{\pi}{n} \cos n\pi + 0 + \frac{1}{n^2} \sin n\pi - \sin 0 \right) = -\frac{2t}{\pi} \cos n\pi = \frac{2t}{\pi} (-1)^{n+1},
\end{aligned}
$$

$n = 1, 2, 3, \ldots$. Substituting this result into the integral in (10.21) yields

$$
\begin{aligned}
\int_0^t h_n(s) \sin[n(t - s)] \, ds &= \int_0^t \frac{2s}{\pi} (-1)^{n+1} \sin[n(t - s)] \, ds \\
&= \frac{2(-1)^{n+1}}{n} \left(\frac{t}{n} - \frac{\sin nt}{n^2} \right) = \frac{2(-1)^{n+1}}{n^3} (nt - \sin nt),
\end{aligned}
$$

where $n = 1, 2, 3, \ldots$. By plugging in a_n's, b_n's, and the result we just found into the equation (10.21), we obtain the formal solution to this problem given by

$$
\begin{aligned}
u(x, t) &= \cos t \sin x + \frac{5}{2} \sin 2t \sin 2x - \frac{3}{5} \sin 5t \sin 5x \\
&\qquad + \sum_{n=1}^\infty \frac{1}{n} \left[\frac{2(-1)^{n+1}}{n^3} (nt - \sin nt) \right] \sin nx \\
&= \cos t \sin x + \frac{5}{2} \sin 2t \sin 2x - \frac{3}{5} \sin 5t \sin 5x \\
&\qquad + 2 \sum_{n=1}^\infty \frac{(-1)^{n+1}}{n^3} \left(t - \frac{\sin nt}{n} \right) \sin nx \, .
\end{aligned}
$$

11. We assume that a solution to this problem has the form $u(x, t) = X(x)T(t)$. Substituting this expression into the partial differential equation yields

$$
X(x)T''(t) + X(x)T'(t) + X(x)T(t) = \alpha^2 X''(x)T(t).
$$

Dividing this equation by $\alpha^2 X(x)T(t)$ yields

$$
\frac{T''(t) + T'(t) + T(t)}{\alpha^2 T(t)} = \frac{X''(x)}{X(x)} \, .
$$

Since these two expressions must be equal for all x in $(0, L)$ and all $t > 0$, they cannot vary. This gives us the following two ordinary differential equations

$$
\frac{T''(t) + T'(t) + T(t)}{\alpha^2 T(t)} = -\lambda \qquad \Rightarrow \qquad T''(t) + T'(t) + \left(1 + \alpha^2 \lambda\right) T(t) = 0 \qquad (10.22)
$$

Copyright © 2012 Pearson Education, Inc. Publishing as Addison-Wesley.

and

$$\frac{X''(x)}{X(x)} = -\lambda \qquad \Rightarrow \qquad X''(x) + \lambda X(x) = 0. \qquad (10.23)$$

Substituting $u(x,t) = X(x)T(t)$ into the boundary conditions, $u(0,t) = u(L,t) = 0$ for $t > 0$, we obtain

$$X(0)T(t) = X(L)T(t) = 0, \qquad t > 0.$$

Since we are looking for a nontrivial solution to this problem, we do not want $T(t) \equiv 0$. Therefore, $X(0) = X(L) = 0$. Combining this fact with the equation (10.23) yields a boundary value problem

$$X''(x) + \lambda X(x) = 0, \qquad X(0) = X(L) = 0.$$

This problem was solved in Section 10.2 of the text (see page 570). There we have found that for $\lambda_n = (n\pi/L)^2$, $n = 1, 2, 3, \ldots$, nontrivial solutions are

$$X_n(x) = A_n \sin\left(\frac{n\pi x}{L}\right), \qquad n = 1, 2, 3, \ldots . \qquad (10.24)$$

Plugging in these values of λ_n's into the equation (10.22) yields the family of linear ordinary differential equations with constant coefficients

$$T''(t) + T'(t) + \left(1 + \frac{\alpha^2 n^2 \pi^2}{L^2}\right)T(t) = 0, \qquad n = 1, 2, 3, \ldots . \qquad (10.25)$$

The auxiliary equations, associated with (10.25), are

$$r^2 + r + \left(1 + \frac{\alpha^2 n^2 \pi^2}{L^2}\right) = 0 .$$

By using the quadratic formula, we obtain

$$r = \frac{-1 \pm \sqrt{1 - 4\left(1 + \dfrac{\alpha^2 n^2 \pi^2}{L^2}\right)}}{2}$$

$$= -\frac{1}{2} \pm \frac{\sqrt{L^2 - 4L^2 - 4\alpha^2 n^2 \pi^2}}{2L} = -\frac{1}{2} \pm \frac{\sqrt{3L^2 + 4\alpha^2 n^2 \pi^2}}{2L} i ,$$

$n = 1, 2, 3, \ldots$. Hence, general solutions to the linear equations in (10.25) are

$$T_n(t) = e^{-t/2}\left[B_n \cos\left(\frac{\sqrt{3L^2 + 4\alpha^2 n^2 \pi^2}}{2L} t\right) + C_n \sin\left(\frac{\sqrt{3L^2 + 4\alpha^2 n^2 \pi^2}}{2L} t\right)\right],$$

 Copyright © 2012 Pearson Education, Inc. Publishing as Addison-Wesley.

$n = 1, 2, 3, \ldots$. By letting

$$\beta_n = \frac{\sqrt{3L^2 + 4\alpha^2 n^2 \pi^2}}{2L}, \quad n = 1, 2, 3, \ldots, \tag{10.26}$$

this family of solutions can be written as

$$T_n(t) = e^{-t/2} \left[B_n \cos \beta_n t + C_n \sin \beta_n t \right]. \tag{10.27}$$

Combining (10.24) and (10.27) yields solutions to the original partial differential equation

$$u_n(x, t) = X_n(x) T_n(t) = A_n e^{-t/2} \left[B_n \cos \beta_n t + C_n \sin \beta_n t \right] \sin \left(\frac{n\pi x}{L} \right), \quad n = 1, 2, 3, \ldots.$$

By the superposition principle, a general solution has the form

$$u(x, t) = \sum_{n=1}^{\infty} e^{-t/2} \left[a_n \cos \beta_n t + b_n \sin \beta_n t \right] \sin \left(\frac{n\pi x}{L} \right),$$

where β_n's are given in (10.26), $a_n = A_n B_n$, and $b_n = A_n C_n$. To find the coefficients a_n and b_n, we use the initial conditions $u(x, 0) = f(x)$ and $\partial u(x, 0)/\partial t = 0$. Therefore, we have

$$\frac{\partial u(x, 0)}{\partial t} = \sum_{n=1}^{\infty} \left\{ -\frac{a_n}{2} + b_n \beta_n \right\} \sin \left(\frac{n\pi x}{L} \right) = 0,$$

since

$$\frac{\partial u(x, t)}{\partial t} = \sum_{n=1}^{\infty} \left\{ (-1/2) \, e^{-t/2} \left[a_n \cos \beta_n t + b_n \sin \beta_n t \right] \right.$$

$$\left. + e^{-t/2} \left[-a_n \beta_n \sin \beta_n t + b_n \beta_n \cos \beta_n t \right] \right\} \sin \left(\frac{n\pi x}{L} \right),$$

Hence, each term in this infinite series must be zero, which implies that

$$-\frac{a_n}{2} + b_n \beta_n = 0 \quad \Rightarrow \quad b_n = \frac{a_n}{2\beta_n}, \quad n = 1, 2, 3, \ldots.$$

Thus, we can write

$$u(x, t) = \sum_{n=1}^{\infty} a_n e^{-t/2} \left[\cos \beta_n t + \frac{1}{2\beta_n} \sin \beta_n t \right] \sin \left(\frac{n\pi x}{L} \right). \tag{10.28}$$

To find a_n's, we use the remaining initial condition to obtain

$$u(x, 0) = \sum_{n=1}^{\infty} a_n \sin \left(\frac{n\pi x}{L} \right) = f(x).$$

Copyright © 2012 Pearson Education, Inc. Publishing as Addison-Wesley.

Therefore, a_n's are the Fourier sine series coefficients of $f(x)$, and so

$$a_n = \frac{2}{L} \int_0^L f(x) \sin\left(\frac{n\pi x}{L}\right) dx. \tag{10.29}$$

Thus, the formal solution to the telegraph problem is given by the equation (10.28), where β_n's and a_n's are given in (10.26) and (10.29), respectively.

15. This problem has the form of that solved in Example 2 in the text with $f(x) = g(x) = x$. There it was found that d'Alembert's formula (32) gives a solution to this problem. By making appropriate substitutions in this formula (and noting that $f(x + \alpha t) = x + \alpha t$ and $f(x - \alpha t) = x - \alpha t$), we obtain

$$
\begin{aligned}
u(x,t) &= \frac{1}{2}\left[(x+\alpha t) + (x - \alpha t)\right] + \frac{1}{2\alpha}\int_{x-\alpha t}^{x+\alpha t} s\,ds = x + \frac{1}{2\alpha}\left(\left.\frac{s^2}{2}\right|_{x-\alpha t}^{x+\alpha t}\right) \\
&= x + \frac{1}{4\alpha}\left[(x+\alpha t)^2 - (x - \alpha t)^2\right] = x + \frac{1}{4\alpha}4\alpha tx = x + tx.
\end{aligned}
$$

17. We apply d'Alembert's formula (32) with $f(x) = e^{-x^2}$ and $g(x) = \sin x$. Since

$$
\begin{aligned}
f(x + \alpha t) + f(x - \alpha t) &= e^{-(x+\alpha t)^2} + e^{-(x-\alpha t)^2} \\
&= e^{-(x-\alpha t)^2}\left[e^{(x-\alpha t)^2 - (x+\alpha t)^2} + 1\right] = e^{-(x-\alpha t)^2}\left(e^{-4\alpha xt} + 1\right)
\end{aligned}
$$

and

$$\int_{x-\alpha t}^{x+\alpha t} g(s)\,ds = \int_{x-\alpha t}^{x+\alpha t} \sin s\,ds = \cos(x - \alpha t) - \cos(x + \alpha t) = 2\sin x \sin \alpha t,$$

we get

$$u(x,t) = \frac{e^{-(x-\alpha t)^2}\left(e^{-4\alpha xt} + 1\right)}{2} + \frac{\sin x \sin \alpha t}{\alpha},$$

which is equivalent to the answer in the text.

EXERCISES 10.7:   Laplace's Equation

3. To solve this problem using separation of variables, we assume that a solution has the form $u(x, y) = X(x)Y(y)$. Making this substitution into the partial differential equation yields

$$X''(x)Y(y) + X(x)Y''(y) = 0.$$

Dividing by $X(x)Y(y)$, we obtain

$$\frac{X''(x)}{X(x)} + \frac{Y''(y)}{Y(y)} = 0.$$

Since this equation must be true for $0 < x < \pi$ and $0 < y < \pi$, there must be a constant λ such that

$$\frac{X''(x)}{X(x)} = -\frac{Y''(y)}{Y(y)} = -\lambda, \qquad 0 < x < \pi, \quad 0 < y < \pi.$$

This leads to ordinary differential equations

$$X''(x) + \lambda X(x) = 0, \tag{10.30}$$

$$Y''(y) - \lambda Y(y) = 0. \tag{10.31}$$

By making the substitution $u(x,y) = X(x)Y(y)$ into the first boundary condition, that is, $u(0,y) = u(\pi,y) = 0$, we obtain

$$X(0)Y(y) = X(\pi)Y(y) = 0, \quad 0 < y < \pi.$$

Since $Y(y) \equiv 0$ yields the trivial solution, we conclude that

$$X(0) = X(\pi) = 0.$$

Combining these boundary conditions with (10.30) yields the boundary value problem

$$X''(x) + \lambda X(x) = 0, \quad X(0) = X(\pi) = 0.$$

We have already solved this problem in Section 10.2 (see page 570) and we obtain the eigenvalues $\lambda_n = n^2, n = 1, 2, 3, \ldots$ with corresponding eigenfunctions

$$X_n(x) = B_n \sin nx.$$

To solve the differential equation, given in (10.31), we use these values of the λ's. This yields the family of ordinary linear differential equations

$$Y''(y) - n^2 Y(y) = 0, \qquad n = 1, 2, 3, \ldots .$$

The corresponding auxiliary equations, $r^2 - n^2 = 0$, have the real roots $r = \pm n$, $n = 1, 2, 3, \ldots$. Hence, general solutions to this family of differential equations are given by

$$Y_n(y) = C_n e^{ny} + D_n e^{-ny}, \qquad n = 1, 2, 3, \ldots .$$

Copyright © 2012 Pearson Education, Inc. Publishing as Addison-Wesley.

With substitutions $K_{1,n} = C_n + D_,n$ and $K_{2,n} = C_n - D_n$, so that

$$C_n = \frac{K_{1,n} + K_{2,n}}{2} \quad \text{and} \quad D_n = \frac{K_{1,n} - K_{2,n}}{2},$$

we see that these solutions can be written as

$$
\begin{aligned}
Y_n(y) &= \frac{K_{1,n} + K_{2,n}}{2} e^{ny} + \frac{K_{1,n} - K_{2,n}}{2} e^{-ny} \\
&= K_{1,n} \frac{e^{ny} + e^{-ny}}{2} + K_{2,n} \frac{e^{ny} - e^{-ny}}{2} = K_{1,n} \cosh ny + K_{2,n} \sinh ny \quad (10.32)
\end{aligned}
$$

(10.32) can in turn be expressed as

$$Y_n(y) = A_n \sinh(ny + E_n), \tag{10.33}$$

where $A_n = K_{2,n}^2 - K_{1,n}^2$ and $E_n = \tanh^{-1}(K_{1,n}/K_{2,n})$. (See Problem 18.)

The last boundary condition, $u(x, \pi) = X(x)Y(\pi) = 0$, implies that $Y(\pi) = 0$. Therefore, by substituting $y = \pi$ into (10.33), we obtain

$$Y_n(\pi) = A_n \sinh(n\pi + E_n).$$

Since we do not want $A_n = 0$, this implies that $\sinh(n\pi + E_n) = 0$ and so $n\pi + E_n = 0$ or, in other words, $E_n = -n\pi$. Substituting these expressions for E_n into the family of solutions (10.32) yields

$$Y_n(y) = A_n \sinh(ny - n\pi).$$

Hence, substituting $X(x)$ and $Y(y)$ into $u_n(x, y) = X_n(x)Y_n(y)$, we see that

$$u_n(x, y) = a_n \sin nx \sinh(ny - n\pi),$$

where $a_n = A_n B_n$. By the superposition principle, a formal solution to the original partial differential equation is given by

$$u(x, y) = \sum_{n=1}^{\infty} u_n(x, y) = \sum_{n=1}^{\infty} a_n \sin nx \sinh(ny - n\pi). \tag{10.34}$$

In order to find the coefficients a_n, we apply the remaining boundary condition, that is, $u(x, 0) = f(x)$. This leads to

$$u(x, 0) = f(x) = \sum_{n=1}^{\infty} a_n \sin nx \sinh(-n\pi),$$

  Copyright © 2012 Pearson Education, Inc. Publishing as Addison-Wesley.

which implies that $a_n \sinh(-n\pi)$ are the coefficients of the Fourier sine series of $f(x)$. Therefore, by the equation (7), Section 10.4, we see that (with $L = \pi$)

$$a_n \sinh(-n\pi) = \frac{2}{\pi} \int_0^\pi f(x) \sin nx \, dx \qquad \Rightarrow \qquad a_n = \frac{2}{\pi \sinh(-n\pi)} \int_0^\pi f(x) \sin nx \, dx \, .$$

Thus, the formal solution to this problem is given in the equation (10.34) with the coefficients a_n found above.

5. This problem has two nonhomogeneous boundary conditions and, therefore, we have to solve two initial-boundary value problems. These problems are

$$\frac{\partial^2 u}{\partial x^2} + \frac{\partial^2 u}{\partial y^2} = 0, \qquad 0 < x < \pi, \quad 0 < y < 1,$$

$$\frac{\partial u(0, y)}{\partial x} = \frac{\partial u(\pi, y)}{\partial x} = 0, \qquad 0 \le y \le 1,$$

$$u(x, 0) = \cos x - \cos 3x, \qquad u(x, 1) = 0, \qquad 0 \le x \le \pi,$$

and

$$\frac{\partial^2 u}{\partial x^2} + \frac{\partial^2 u}{\partial y^2} = 0, \qquad 0 < x < \pi, \quad 0 < y < 1,$$

$$\frac{\partial u(0, y)}{\partial x} = \frac{\partial u(\pi, y)}{\partial x} = 0, \qquad 0 \le y \le 1,$$

$$u(x, 0) = 0, \qquad u(x, 1) = \cos 2x, \qquad 0 \le x \le \pi.$$

If u_1 and u_2 are solutions to the first and the second problem, respectively, then $u = u_1 + u_2$ is a solution to the original problem. Indeed,

$$\frac{\partial^2 u}{\partial x^2} + \frac{\partial^2 u}{\partial y^2} = \left(\frac{\partial^2 u_1}{\partial x^2} + \frac{\partial^2 u_2}{\partial x^2} \right) + \left(\frac{\partial^2 u_1}{\partial y^2} + \frac{\partial^2 u_2}{\partial y^2} \right)$$

$$= \left(\frac{\partial^2 u_1}{\partial x^2} + \frac{\partial^2 u_1}{\partial y^2} \right) + \left(\frac{\partial^2 u_2}{\partial x^2} + \frac{\partial^2 u_2}{\partial y^2} \right) = 0 + 0 = 0,$$

$$\frac{\partial u(0, y)}{\partial x} = \frac{\partial u_1(0, y)}{\partial x} + \frac{\partial u_2(0, y)}{\partial x} = 0 + 0 = 0,$$

$$\frac{\partial u(\pi, y)}{\partial x} = \frac{\partial u_1(\pi, y)}{\partial x} + \frac{\partial u_2(\pi, y)}{\partial x} = 0 + 0 = 0,$$

$$u(x, 0) = u_1(x, 0) + u_2(x, 0) = \cos x - \cos 3x + 0 = \cos x - \cos 3x \, ,$$

$$u(x, 1) = u_1(x, 1) + u_2(x, 1) = 0 + \cos 2x = \cos 2x \, .$$

(This is an application of the superposition principle.)

Copyright © 2012 Pearson Education, Inc. Publishing as Addison-Wesley.

The first of these two problems has the form of that discussed in Example 1 with $a = \pi$, $b = 1$, and $f(x) = \cos x - \cos 3x$. A formal solution to this problem is given in the equation (10) in the text. Thus, by making appropriate substitutions, we find that a formal solution to the first problem is

$$u_1(x, y) = E_0(y - 1) + \sum_{n=1}^{\infty} E_n \cos nx \sinh(ny - n). \tag{10.35}$$

To find the coefficients E_n, we use the boundary condition

$$u_1(x, 0) = \cos x - \cos 3x.$$

Thus, we have

$$u_1(x, 0) = \cos x - \cos 3x = -E_0 + \sum_{n=1}^{\infty} E_n \cos nx \sinh(-n).$$

For $n = 1$,

$$E_1 \sinh(-1) = 1 \quad \Rightarrow \quad E_1 = \frac{1}{\sinh(-1)}$$

and, for $n = 3$,

$$E_3 \sinh(-3) = -1 \quad \Rightarrow \quad E_1 = \frac{-1}{\sinh(-3)}.$$

For all other values of n, $E_n = 0$. By substituting these coefficients into the solution (10.35), we obtain the formal solution to the first of our two problems, namely,

$$u_1(x, y) = \frac{\cos x \sinh(y - 1)}{\sinh(-1)} - \frac{\cos 3x \sinh(3y - 3)}{\sinh(-3)}. \tag{10.36}$$

To solve the second problem, we note that, except for the last two boundary conditions, it is similar to the problem solved in Example 1. Using the separation of variables technique, we find that the boundary value problem

$$X''(x) + \lambda X(x) = 0, \qquad X'(0) = X'(\pi) = 0,$$

has solutions $X_n(x) = a_n \cos nx$, when $\lambda_n = n^2$, $n = 1, 2, 3, \ldots$. Substituting these values for λ's into the equation

$$Y''(y) - \lambda Y(y) = 0,$$

we find that the family of solutions to this differential equation is given by

$$\begin{aligned} Y_0(y) &= A_0 + B_0 y, \\ Y_n(y) &= C_n \sinh\left[n\left(y + D_n\right)\right], \qquad n = 1, 2, 3 \ldots. \end{aligned} \tag{10.37}$$

Copyright © 2012 Pearson Education, Inc. Publishing as Addison-Wesley.

The problem, we are solving, differs from that in Example 1. The boundary condition $u_2(x,0) = X(x)Y(0) = 0$, $0 \le x \le \pi$, implies that $Y(0) = 0$ (since we do not want the trivial solution). Applying this boundary condition to each of the solutions in (10.37) yields

$$Y_0(0) = A_0 + 0 = 0 \qquad \Rightarrow \qquad A_0 = 0,$$
$$Y_n(0) = C_n \sinh(nD_n) = 0 \quad \Rightarrow \qquad D_n = 0,$$

where we have used the fact that $\sinh x = 0$ if and only if $x = 0$. By substituting these results into the solutions (10.37), we obtain

$$Y_0(y) = B_0 y,$$

$$Y_n(y) = C_n \sinh ny, \qquad n = 1, 2, 3 \dots.$$

Combining these solutions with $X_n(x) = a_n \cos nx$ yields

$$u_{2,0}(x,y) = X_0(x)Y_0(y) = a_0 B_0 y \cos 0 = E_0 y,$$
$$u_{2,n}(x,y) = X_n(x)Y_n(y) = a_n C_n \cos nx \sinh ny = E_n \cos nx \sinh ny,$$

where $E_0 = a_0 B_0$ and $E_n = a_n C_n$. Thus, by the superposition principle, we find that a formal solution to the second problem is given by

$$u_2(x,y) = E_0 y + \sum_{n=1}^{\infty} E_n \cos nx \sinh ny.$$

Applying the second boundary condition, namely $u_2(x,1) = \cos 2x$, to these solutions, we see that

$$u_2(x,1) = E_0 + \sum_{n=1}^{\infty} E_n \cos nx \sinh n = \cos 2x.$$

Therefore, when $n = 2$,

$$E_2 \sinh 2 = 1 \qquad \Rightarrow \qquad E_2 = \frac{1}{\sinh 2},$$

and $E_n = 0$ for all other values of n. By substituting these coefficients into the solution $u_2(x,y)$, we obtain the formal solution to the second problem, that is,

$$u_2(x,y) = \frac{\cos 2x \sinh 2y}{\sinh 2}.$$

By the superposition principle (as it was noticed earlier), the formal solution to the original partial differential equation is the sum of $u_1(x,y)$ and $u_2(x,y)$. Thus,

$$u(x,y) = \frac{\cos x \sinh(y-1)}{\sinh(-1)} - \frac{\cos 3x \sinh(3y-3)}{\sinh(-3)} + \frac{\cos 2x \sinh 2y}{\sinh 2}.$$

Copyright © 2012 Pearson Education, Inc. Publishing as Addison-Wesley.

7. The solution to the Dirichlet boundary value problem, in general form, is given in Example 2 in the text. In this problem, the radius of the disc is $a = 2$ and the boundary value function is $f(\theta) = |\theta|$. We use the formulas (23) and (24) to compute coefficients in the series solution (22).

$$a_0 = \frac{1}{\pi} \int_{-\pi}^{\pi} |\theta| \, d\theta = \frac{2}{\pi} \int_0^{\pi} \theta \, d\theta = \pi,$$

$$a_n = \frac{1}{\pi} \int_{-\pi}^{\pi} |\theta| \cos n\theta \, d\theta = \frac{2}{\pi} \int_0^{\pi} \theta \cos n\theta \, d\theta = \frac{2}{\pi} \left(\frac{\theta \sin n\theta}{n} + \frac{\cos n\theta}{n^2} \right) \Big|_0^{\pi} = \frac{2}{\pi} \frac{(-1)^n - 1}{n^2},$$

$$b_n = \frac{1}{\pi} \int_{-\pi}^{\pi} |\theta| \sin n\theta \, d\theta = 0, \quad n = 1, 2, \dots .$$

Here, computing integrals, we used the fact that $|\theta|$ is an even function. Further, we note that

$$a_{2k} = \frac{2}{\pi} \frac{(-1)^{2k} - 1}{(2k)^2} = 0, \quad a_{2k-1} = \frac{2}{\pi} \frac{(-1)^{2k+1} - 1}{(2k-1)^2} = -\frac{4}{\pi(2k-1)^2}, \quad k = 1, 2, \dots .$$

Thus, the formula (22) yields the solution

$$u(r, \theta) = \frac{\pi}{2} + \sum_{k=1}^{\infty} \left(\frac{r}{2} \right)^{2k-1} \left[-\frac{4}{\pi(2k-1)^2} \right] \cos(2k-1)\theta$$

$$= \frac{\pi}{2} - \frac{1}{\pi} \sum_{k=1}^{\infty} \frac{r^{2k-1}}{(2k-1)^2 2^{2k-3}} \cos(2k-1)\theta,$$

which is equivalent to the answer in the text.

11. In this problem, the technique of separation of variables leads to ordinary differential equations

$$r^2 R''(r) + r R'(r) - \lambda R(r) = 0 \quad \text{and} \quad T''(\theta) + \lambda T(\theta) = 0. \tag{10.38}$$

As in Example 2, we require the solution $u(r, \theta)$ to be continuous on its domain. Therefore, $T(\theta)$ must be periodic with period 2π. This implies that $T(-\pi) = T(\pi)$ and $T'(-\pi) = T'(\pi)$. Thus, the family of solutions for the second equation in (10.38), which satisfies these periodic boundary conditions, is

$$T_0(\theta) = B_0 \quad \text{and} \quad T_n(\theta) = A_n \cos n\theta + B_n \sin n\theta, \quad n = 1, 2, 3, \dots .$$

 Copyright © 2012 Pearson Education, Inc. Publishing as Addison-Wesley.

In solving this problem, it was found that $\lambda = \lambda_n = n^2$, $n = 0, 1, 2 \ldots$. Again, as in Example 2, substituting these values of λ_n into the first equation in (10.38) leads to the solutions

$$R_0(r) = C_0 + D_0 \ln r \quad \text{and} \quad R_n(r) = C_n r^n + D_n r^{-n}, \qquad n = 1, 2, 3, \ldots.$$

Here, however, we are not concerned with what happens when $r = 0$. By our assumption that $u(r, \theta) = R(r)T(\theta)$, we see that solutions to the partial differential equation, given in this problem, have the form

$$u_0(r, \theta) = B_0(C_0 + D_0 \ln r) \quad \text{and} \quad u_n(r, \theta) = \left(C_n r^n + D_n r^{-n}\right)\left(A_n \cos n\theta + B_n \sin n\theta\right),$$

where $n = 1, 2, 3, \ldots$. Thus, by the superposition principle, we see that the formal solution to this Dirichlet problem is given by

$$u(r, \theta) = B_0 C_0 + B_0 D_0 \ln r + \sum_{n=1}^{\infty} \left(C_n r^n + D_n r^{-n}\right)\left(A_n \cos n\theta + B_n \sin n\theta\right),$$

or

$$u(r, \theta) = a_0 + b_0 \ln r + \sum_{n=1}^{\infty} \left[\left(c_n r^n + e_n r^{-n}\right)\cos n\theta + \left(d_n r^n + f_n r^{-n}\right)\sin n\theta\right], \quad (10.39)$$

where $a_0 = B_0 C_0$, $b_0 = B_0 D_0$, $c_n = C_n A_n$, $e_n = D_n A_n$, $d_n = C_n B_n$, and $f_n = D_n B_n$. To find these coefficients, we apply the boundary conditions $u(1, \theta) = \sin 4\theta - \cos \theta$ and $u(2, \theta) = \sin \theta$, $-\pi \leq \theta \leq \pi$. From the first boundary condition, we see that

$$u(1, \theta) = a_0 + \sum_{n=1}^{\infty}\left[(c_n + e_n)\cos n\theta + (d_n + f_n)\sin n\theta\right] = \sin 4\theta - \cos \theta,$$

which implies that $a_0 = 0$, $d_4 + f_4 = 1$, $c_1 + e_1 = -1$ and, for all other values of n, $c_n + e_n = 0$ and $d_n + f_n = 0$. From the second boundary condition, we have

$$u(2, \theta) = b_0 \ln 2 + \sum_{n=1}^{\infty}\left[\left(c_n 2^n + e_n 2^{-n}\right)\cos n\theta + \left(d_n 2^n + f_n 2^{-n}\right)\sin n\theta\right] = \sin \theta,$$

which implies that $b_0 = 0$, $2d_1 + 2^{-1}f_1 = 1$ and, for all other values of n, $2^n c_1 + 2^{-n} e_1 = 0$ and $2^n d_1 + 2^{-n} f_1 = 0$. By combining these results, we obtain three systems of two equations in two unknowns,

$$
\begin{array}{lll}
d_4 + f_4 = 1, & c_1 + e_1 = -1, & d_1 + f_1 = 0, \\
2^4 d_4 + 2^{-4} f_4 = 0 & 2c_1 + 2^{-1}e_1 = 0 & 2d_1 + 2^{-1}f_1 = 1
\end{array}
$$

Copyright © 2012 Pearson Education, Inc. Publishing as Addison-Wesley.

(where the first equation in each system was derived from the first boundary condition and the second equation in each system was derived from the second boundary condition), and for all other values of n, $c_n = 0$, $e_n = 0$, $d_n = 0$, $f_n = 0$. Solving these systems yields

$$d_4 = -\frac{1}{255}, \qquad f_4 = \frac{256}{255}, \qquad c_1 = \frac{1}{3}, \qquad e_1 = -\frac{4}{3}, \qquad d_1 = \frac{2}{3}, \qquad f_1 = -\frac{2}{3}.$$

Substituting these values into (10.39), we find that the solution to this Dirichlet problem is given by

$$u(r, \theta) = \left(\frac{1}{3}r - \frac{4}{3}r^{-1}\right)\cos\theta + \left(\frac{2}{3}r - \frac{2}{3}r^{-1}\right)\sin\theta + \left(-\frac{1}{255}r^4 + \frac{256}{255}r^{-4}\right)\sin 4\theta.$$

15. Here, as in Example 2 in the text, the technique of separation of variables leads to

$$r^2 R''(r) + r R'(r) - \lambda R(r) = 0 \qquad \text{and} \qquad T''(\theta) + \lambda T(\theta) = 0.$$

Since we want the nontrivial solution, the boundary condition $u(r, 0) = R(r)T(0) = 0$ implies that $T(0) = 0$ and the boundary condition $u(r, \pi) = R(r)T(\pi) = 0$ implies that $T(\pi) = 0$. Therefore, we seek a nontrivial solution to an initial value problem

$$T''(\theta) + \lambda T(\theta) = 0 \qquad \text{with} \qquad T(0) = T(\pi) = 0. \tag{10.40}$$

We encountered this problem several times before (see page 570 of the text). The eigenvalues are $\lambda = \lambda_n = n^2$, $n = 1, 2, 3, \ldots$, with corresponding eigenfunctions

$$T_n(\theta) = B_n \sin n\theta.$$

Substituting the λ_n's into the differential equation

$$r^2 R''(r) + r R'(r) - \lambda R(r) = 0,$$

yields

$$r^2 R''(r) + r R'(r) - n^2 R(r) = 0, \qquad n = 1, 2, 3, \ldots.$$

This is the same Cauchy-Euler equation that was solved in Example 2 (see (19) in the text). There it was found that the solutions have the form

$$R_n(r) = C_n r^n + D_n r^{-n}, \qquad n = 1, 2, 3, \ldots.$$

Since we require that $u(r, \theta)$ is bounded on its domain, we see that $u_n(r, \theta) = R_n(r)T_n(\theta)$ must be bounded about $r = 0$. This implies that $R_n(\theta)$ must be bounded. Therefore, $D_n = 0$, and so the solutions to this Cauchy-Euler equation are given by

$$R_n(r) = C_n r^n, \qquad n = 1, 2, 3, \ldots .$$

Thus, we see that formal solutions to the original partial differential equation are

$$u_n(r, \theta) = B_n C_n r^n \sin n\theta = c_n r^n \sin n\theta,$$

where $c_n = B_n C_n$. Therefore, by the superposition principle, the formal solution is

$$u(r, \theta) = \sum_{n=1}^{\infty} c_n r^n \sin n\theta$$

to the given Dirichlet problem. The last boundary condition yields

$$u(1, \theta) = \sin 3\theta = \sum_{n=1}^{\infty} c_n \sin n\theta .$$

This implies that $c_3 = 1$ and, for all other values of n, $c_n = 0$. Substituting these coefficients into the formal solution found above, yields

$$u(r, \theta) = r^3 \sin 3\theta .$$

17. We solve this problem by separation of variables. In Example 2, it was found that we have to solve two ordinary differential equations,

$$r^2 R''(r) + r R'(r) - \lambda R(r) = 0 \tag{10.41}$$

$$T''(\theta) + \lambda T(\theta) = 0 \qquad \text{with} \qquad T(\pi) = T(n\pi), \quad T'(\pi) = T'(n\pi). \tag{10.42}$$

In Example 2, we also found that for $\lambda = n^2$, $n = 0, 1, 2, \ldots$, the linear differential equation in (10.42) has nontrivial solutions of the form

$$T_n(\theta) = A_n \cos n\theta + B_n \sin n\theta , \qquad n = 0, 1, 2, \ldots ,$$

and the equation (10.41) has solutions

$$R_0(r) = C + D \ln r \qquad \text{and} \qquad R_n(r) = C_n r^n + D_n r^{-n}, \qquad n = 1, 2, 3, \ldots .$$

Copyright © 2012 Pearson Education, Inc. Publishing as Addison-Wesley.

Thus, since we are assuming that $u(r,\theta) = R(r)T(\theta)$, we see that solutions to the original partial differential equation are given by

$$u_0(r,\theta) = A_0(C_0 + D_0 \ln r) = a_0 + b_0 \ln r,$$

$$u_n(r,\theta) = \left(C_n r^n + D_n r^{-n}\right)\left(A_n \cos n\theta + B_n \sin n\theta\right)$$

$$= \left(a_n r^n + b_n r^{-n}\right)\cos n\theta + \left(c_n r^n + d_n r^{-n}\right)\sin n\theta, \qquad n = 1, 2, 3, \ldots.$$

Thus, by the superposition principle, we see that the formal solution to the partial differential equation, given in this problem, has the form

$$u(r,\theta) = a_0 + b_0 \ln r + \sum_{n=1}^{\infty}\left[\left(a_n r^n + b_n r^{-n}\right)\cos n\theta + \left(c_n r^n + d_n r^{-n}\right)\sin n\theta\right]. \qquad (10.43)$$

By applying the first boundary condition, we obtain

$$u(1,\theta) = f(\theta) = a_0 + \sum_{n=1}^{\infty}\left[\left(a_n + b_n\right)\cos n\theta + \left(c_n + d_n\right)\sin n\theta\right],$$

where we have used the fact that $\ln 1 = 0$. So, we see that a_0, $(a_n + b_n)$, and $(c_n + d_n)$ are the Fourier coefficients of $f(\theta)$ (with $T = \pi$). Therefore,

$$a_0 = \frac{1}{2\pi}\int_{-\pi}^{\pi} f(\theta)\,d\theta,$$

$$a_n + b_n = \frac{1}{\pi}\int_{-\pi}^{\pi} f(\theta)\cos n\theta\,d\theta, \qquad (10.44)$$

$$c_n + d_n = \frac{1}{\pi}\int_{-\pi}^{\pi} f(\theta)\sin n\theta\,d\theta, \qquad n = 1, 2, 3\ldots.$$

To apply the last boundary condition, we find $\partial u/\partial r$.

$$\frac{\partial u(r,\theta)}{\partial r} = \frac{b_0}{r} + \sum_{n=1}^{\infty}\left[\left(a_n n r^{n-1} - b_n n r^{-n-1}\right)\cos n\theta + \left(c_n n r^{n-1} - d_n n r^{-n-1}\right)\sin n\theta\right].$$

Applying the last boundary condition yields

$$\frac{\partial u(3,\theta)}{\partial r} = \frac{b_0}{3} + \sum_{n=1}^{\infty}\left[\left(a_n n 3^{n-1} - b_n n 3^{-n-1}\right)\cos n\theta + \left(c_n n 3^{n-1} - d_n n 3^{-n-1}\right)\sin n\theta\right].$$

So,

$$\frac{b_0}{3}, \qquad n3^{n-1}a_n - n3^{-n-1}b_n, \qquad \text{and} \qquad n3^{n-1}c_n - n3^{-n-1}d_n$$

 Copyright © 2012 Pearson Education, Inc. Publishing as Addison-Wesley.

are the Fourier coefficients of $g(\theta)$ (with $L = \pi$). Thus, we have

$$b_0 = \frac{3}{2\pi} \int_{-\pi}^{\pi} g(\theta)\, d\theta \,,$$

$$n3^{n-1}a_n - n3^{-n-1}b_n = \frac{1}{\pi} \int_{-\pi}^{\pi} g(\theta) \cos n\theta\, d\theta \,, \tag{10.45}$$

$$n3^{n-1}c_n - n3^{-n-1}d_n = \frac{1}{\pi} \int_{-\pi}^{\pi} g(\theta) \sin n\theta\, d\theta \,, \qquad n = 1, 2, 3 \ldots .$$

Therefore, the formal solution to this boundary value problem is given by (10.43) with the coefficients defined in (10.44) and (10.45).

21. Since the given function $f(\theta, z) = \sin z$ is independent of θ, the only functions in the coefficient formulas (45) and (46), which depend on θ, are $\cos n\theta$ and $\sin n\theta$. Clearly,

$$\int_{-\pi}^{\pi} \cos n\theta\, d\theta = \int_{-\pi}^{\pi} \sin n\theta\, d\theta = 0 \,.$$

Therefore, $a_{mn} = b_{mn} = 0$ for all $n \geq 1$. With $a = b = \pi$, the integral in (44) becomes

$$\int_0^{\pi} \int_{-\pi}^{\pi} \sin z \sin mz\, d\theta dz = \int_0^{\pi} \sin z \sin mz\, dz \int_{-\pi}^{\pi} d\theta = 2\pi \int_0^{\pi} \sin z \sin mz\, dz \,.$$

For $m \geq 2$, we get

$$2\pi \int_0^{\pi} \sin z \sin mz\, dz = \pi \int_0^{\pi} [\cos(m-1)z - \cos(m+1)z]\, dz$$

$$= \pi \left[\frac{\sin(m-1)z}{m-1} - \frac{\sin(m+1)z}{m+1} \right]\Bigg|_0^{\pi} = 0 \,.$$

For $m = 1$,

$$2\pi \int_0^{\pi} \sin z \sin mz\, dz = \pi \int_0^{\pi} (1 - \cos 2z)\, dz = \pi \left(z - \frac{\sin 2z}{2} \right)\Bigg|_0^{\pi} = \pi^2 \,.$$

Therefore,

$$a_{1,0} = \frac{\pi^2}{\pi^2 I_0(\pi)} = \frac{1}{I_0(\pi)} \,,$$

and all other coefficients in (42) equal zero. Hence,

$$u(r, \theta, z) = a_{1,0} I_0(r) \sin z = \frac{I_0(r) \sin z}{I_0(\pi)} \,.$$

Copyright © 2012 Pearson Education, Inc. Publishing as Addison-Wesley.

FIGURES

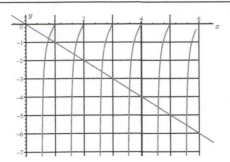

Figure 10–A: The intersection of the graphs $y = -x$ and $y = \tan(\pi x)$, $x > 0$.

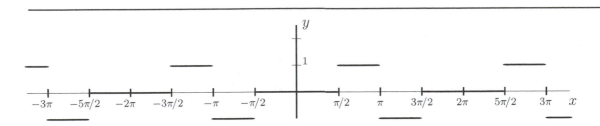

Figure 10–B: The graph of the π-periodic extension of f.

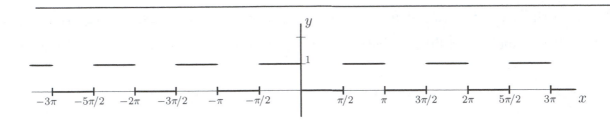

Figure 10–C: The graph of the odd 2π-periodic extension of f.

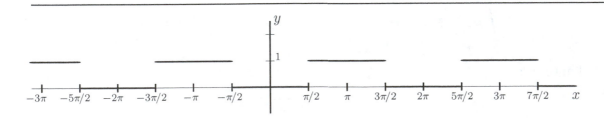

Figure 10–D: The graph of the even 2π-periodic extension of f.

 Copyright © 2012 Pearson Education, Inc. Publishing as Addison-Wesley.

CHAPTER 11: Eigenvalue Problems and Sturm-Liouville Equations

EXERCISES 11.2: Eigenvalues and Eigenfunctions

1. The auxiliary equation for this problem is $r^2 + 2r + 26 = 0$, which has roots $r = -1 \pm 5i$. Hence, a general solution to the differential equation $y'' + 2y' + 26y = 0$ is

$$y(x) = c_1 e^{-x} \cos 5x + c_2 e^{-x} \sin 5x.$$

We now try to determine constants c_1 and c_2 so that the boundary conditions are satisfied. Setting $x = 0$ and $x = \pi$, we find that

$$y(0) = c_1 = 1, \qquad y(\pi) = -c_1 e^{-\pi} = -e^{-\pi}.$$

Both boundary conditions give the same result, $c_1 = 1$. Hence, there is a one parameter family of solutions

$$y(x) = e^{-x} \cos 5x + c e^{-x} \sin 5x,$$

where $c := c_2$ is arbitrary.

3. The auxiliary equation in this problem is $r^2 - 4r + 13 = 0$, which has roots $r = 2 \pm 3i$. Hence, a general solution to the given differential equation is

$$y(x) = c_1 e^{2x} \cos 3x + c_2 e^{2x} \sin 3x.$$

We now determine constants c_1 and c_2 so that the boundary conditions are satisfied. Setting $x = 0$ and $x = \pi$, we find that

$$y(0) = c_1 = 0, \qquad y(\pi) = -c_1 e^{2\pi} = 0.$$

These boundary conditions are consistent, giving $c_1 = 0$. Hence, there is a one parameter family of solutions

$$y(x) = c e^{2x} \sin 3x,$$

where $c := c_2$ is any real number.

Copyright © 2012 Pearson Education, Inc. Publishing as Addison-Wesley.

5. The auxiliary equation for this problem is $r^2 + 1 = 0$, which has roots $r = \pm i$. Hence, a general solution to the corresponding homogeneous differential equation, $y'' + y = 0$, is

$$y_h(x) = c_1 \cos x + c_2 \sin x . \tag{11.1}$$

We now look for a particular solution to the given nonhomogeneous equation of the form (see the method of undetermined coefficients, Section 4.5)

$$y_p(x) = A \cos 2x + B \sin 2x$$
$$\Rightarrow \quad y_p'(x) = -2A \sin 2x + 2B \cos 2x \qquad \Rightarrow \qquad y_p''(x) = -4A \cos 2x - 4B \sin 2x .$$

Substituting into the given equation yields

$$-3A \cos 2x - 3B \sin 2x = \sin 2x \qquad \Rightarrow \qquad A = 0, \; B = -1/3 .$$

Therefore, a general solution is of the form

$$y(x) = y_h(x) + y_p(x) = c_1 \cos x + c_2 \sin x - (1/3) \sin 2x . \tag{11.2}$$

Since $\cos x$, $\sin x$ are periodic with period 2π and $\sin 2x$ is periodic with period π, the solution $y(x)$ in (11.2) is periodic with period 2π (so is $y'(x)$). Therefore, any solution given in (11.2) satisfies the boundary conditions, and so (11.2) gives the answer to this problem.

7. In Problem 5, we found a general solution (11.1) to the corresponding homogeneous equation. According to the method of undetermined coefficients (Section 4.5), a particular solution to the given equation has the form $y_p(x) = Ae^x$. Substituting yields

$$(Ae^x)'' + (Ae^x) = e^x \qquad \Rightarrow \qquad 2Ae^x = e^x \qquad \Rightarrow \qquad A = \frac{1}{2} .$$

Thus,

$$y(x) = y_h(x) + y_p(x) = c_1 \cos x + c_2 \sin x + \frac{e^x}{2}$$

is a general solution to the given nonhomogeneous equation.

The boundary conditions imply that

$$y(0) = c_1 + \frac{1}{2} = 0 \qquad \Rightarrow \qquad c_1 = -\frac{1}{2}$$

 Copyright © 2012 Pearson Education, Inc. Publishing as Addison-Wesley.

and

$$y(\pi) + y'(\pi) = \left[\left(c_1 \cos x + c_2 \sin x + \frac{e^x}{2} \right) + \left(-c_1 \sin x + c_2 \cos x + \frac{e^x}{2} \right) \right] \bigg|_{x=\pi}$$
$$= -c_1 - c_2 + e^\pi = 0.$$

Therefore, $c_2 = -c_1 + e^\pi = 1/2 + e^\pi$, and the answer is

$$y(x) = -(1/2) \cos x + (1/2 + e^\pi) \sin x + (1/2)e^x.$$

13. First, note that the auxiliary equation in this problem is $r^2 + \lambda = 0$. To find eigenvalues, which result nontrivial solutions, we consider the cases $\lambda < 0$, $\lambda = 0$, and $\lambda > 0$.

Case 1: $\lambda < 0$. In this case, the roots to the auxiliary equation are $\pm\sqrt{-\lambda}$. (Note that $-\lambda$ is positive.) Therefore, a general solution is given by

$$y(x) = c_1 e^{\sqrt{-\lambda}x} + c_2 e^{-\sqrt{-\lambda}x}.$$

By applying the first boundary condition, we obtain

$$y(0) = c_1 + c_2 = 0 \qquad \Rightarrow \qquad c_2 = -c_1.$$

Thus,

$$y(x) = c_1 \left(e^{\sqrt{-\lambda}x} - e^{-\sqrt{-\lambda}x} \right).$$

To apply the second boundary condition, we need $y'(x)$. We have

$$y'(x) = c_1 \sqrt{-\lambda} \left(e^{\sqrt{-\lambda}x} + e^{-\sqrt{-\lambda}x} \right),$$

and so

$$y'(1) = c_1 \sqrt{-\lambda} \left(e^{\sqrt{-\lambda}} + e^{-\sqrt{-\lambda}} \right) = 0. \tag{11.3}$$

Since $\sqrt{-\lambda} > 0$ implying that $e^{\sqrt{-\lambda}} + e^{-\sqrt{-\lambda}} \neq 0$, the only way that the equation (11.3) holds is $c_1 = 0$. So, in this case we have only the trivial solution. Thus, there are no eigenvalues $\lambda < 0$.

Case 2: $\lambda = 0$. In this case, we are solving the differential equation $y'' = 0$. This equation has a general solution

$$y(x) = c_1 + c_2 x \qquad \Rightarrow \qquad y'(x) = c_2.$$

Copyright © 2012 Pearson Education, Inc. Publishing as Addison-Wesley.

By applying the boundary conditions, we obtain

$$y(0) = c_1 = 0 \qquad \text{and} \qquad y'(1) = c_2 = 0.$$

Thus $c_1 = c_2 = 0$, $y(x) \equiv 0$, and $\lambda = 0$ is not an eigenvalue.

Case 3: $\lambda > 0$. In this case, the roots of the associated auxiliary equation are $r = \pm\sqrt{\lambda}i$. Therefore, a general solution is given by

$$y(x) = c_1 \cos\left(\sqrt{\lambda}x\right) + c_2 \sin\left(\sqrt{\lambda}x\right).$$

By applying the first boundary condition, we obtain

$$y(0) = c_1 = 0 \qquad \Rightarrow \qquad y(x) = c_2 \sin\left(\sqrt{\lambda}x\right).$$

In order to apply the second boundary condition, we find $y'(x)$. Thus, we have

$$y'(x) = c_2\sqrt{\lambda}\cos\left(\sqrt{\lambda}x\right),$$

and so

$$y'(1) = c_2\sqrt{\lambda}\cos\left(\sqrt{\lambda}\right) = 0.$$

Therefore, to obtain a solution other than the trivial one, we must have

$$\cos\left(\sqrt{\lambda}\right) = 0 \qquad \Rightarrow \qquad \sqrt{\lambda} = \left(n + \frac{1}{2}\right)\pi, \qquad n = 0, 1, 2, \ldots$$

$$\Rightarrow \qquad \lambda_n = \left(n + \frac{1}{2}\right)^2 \pi^2, \qquad n = 0, 1, 2, \ldots.$$

For these eigenvalues λ_n, the corresponding eigenfunctions are

$$y_n(x) = c_n \sin\left[\left(n + \frac{1}{2}\right)\pi x\right], \qquad n = 0, 1, 2, \ldots,$$

where c_n is an arbitrary nonzero constant.

19. The equation $(xy')' + \lambda x^{-1}y = 0$ can be rewritten as a Cauchy-Euler equation

$$x^2 y'' + xy' + \lambda y = 0, \qquad x > 0. \tag{11.4}$$

Substituting $y = x^r$ yields

$$x^2 r(r-1)x^{r-2} + xrx^{r-1} + \lambda x^r = 0 \qquad \Rightarrow \qquad x^r\left(r^2 + \lambda\right) = 0.$$

Therefore, $r^2 + \lambda = 0$ is the auxiliary equation for (11.4). Again, we consider the cases $\lambda < 0$, $\lambda = 0$, and $\lambda > 0$.

<u>Case 1: $\lambda < 0$.</u> Let $\lambda = -\mu^2$, $\mu > 0$. The roots of the auxiliary equation are $r = \pm\mu$, and so a general solution to (11.4) is given by

$$
\begin{aligned}
y(x) &= c_1 x^\mu + c_2 x^{-\mu} \\
\Rightarrow \quad y'(x) &= c_1 \mu x^{\mu-1} - c_2 \mu x^{-\mu-1} = \mu\left(c_1 x^{\mu-1} - c_2 x^{-\mu-1}\right).
\end{aligned}
$$

Substituting into the first boundary condition gives

$$
y'(1) = \mu\left(c_1 - c_2\right) = 0.
$$

Since $\mu > 0$, we conclude that

$$
c_1 - c_2 = 0 \quad \Rightarrow \quad c_1 = c_2 \quad \Rightarrow \quad y(x) = c_1\left(x^\mu + x^{-\mu}\right).
$$

Substituting this into the second boundary condition yields

$$
y(e^\pi) = c_1\left(e^{\mu\pi} + e^{-\mu\pi}\right) = 0. \tag{11.5}
$$

Since $e^{\mu\pi} + e^{-\mu\pi} \neq 0$, (11.5) is true only if $C_1 = 0$. So, in this case we, have the trivial solution only. Thus, there is no eigenvalue $\lambda < 0$.

<u>Case 2: $\lambda = 0$.</u> In this case, we are solving the differential equation $(xy')' = 0$. Integrating yields

$$
xy' = c_1 \quad \Rightarrow \quad y' = \frac{c_1}{x} \quad \Rightarrow \quad y(x) = c_2 + c_1 \ln x.
$$

Applying the boundary conditions, we obtain

$$
y'(1) = c_1 = 0 \quad \text{and} \quad y(e^\pi) = c_2 + c_1 \ln(e^\pi) = c_2 + c_1\pi = 0.
$$

Solving these equations simultaneously yields $c_1 = c_2 = 0$. Thus, we again get only the trivial solution. Therefore, $\lambda = 0$ is not an eigenvalue.

<u>Case 3: $\lambda > 0$.</u> Let $\lambda = \mu^2$, $\mu > 0$. The roots of the auxiliary equation are $r = \pm\mu i$, and so a general solution to (11.4) is

$$
y(x) = c_1 \cos\left(\mu \ln x\right) + c_2 \sin\left(\mu \ln x\right).
$$

Copyright © 2012 Pearson Education, Inc. Publishing as Addison-Wesley.

Next, we find $y'(x)$.

$$y'(x) = -c_1 \left(\frac{\mu}{x}\right) \sin\left(\mu \ln x\right) + c_2 \left(\frac{\mu}{x}\right) \cos\left(\mu \ln x\right).$$

By applying the first boundary condition, we obtain

$$y'(1) = c_2 \mu = 0 \qquad \Rightarrow \qquad c_2 = 0.$$

Applying the second boundary condition yields

$$y\left(e^\pi\right) = c_1 \cos\left(\mu \ln(e^\pi)\right) = c_1 \cos\left(\mu \pi\right) = 0.$$

Therefore, in order to get nontrivial solutions, we must have

$$\cos\left(\mu \pi\right) = 0 \qquad \Rightarrow \qquad \mu \pi = \left(n + \frac{1}{2}\right)\pi, \quad n = 0, 1, 2, \ldots$$

$$\Rightarrow \qquad \mu = n + \frac{1}{2} \qquad \Rightarrow \qquad \lambda_n = \left(n + \frac{1}{2}\right)^2, \quad n = 0, 1, 2, \ldots .$$

Corresponding to the eigenvalues λ_n's, we have the eigenfunctions

$$y_n(x) = c_n \cos\left[\left(n + \frac{1}{2}\right) \ln x\right], \qquad n = 0, 1, 2, \ldots ,$$

where c_n's are arbitrary nonzero constants.

25. The auxiliary equation in this problem is $r^2 + \lambda = 0$ (see Problem 13). To find eigenvalues, giving nontrivial solutions, we again consider three cases $\lambda < 0$, $\lambda = 0$, and $\lambda > 0$.

Case 1: $\lambda < 0$. The roots of the auxiliary equation are $r = \pm\sqrt{-\lambda}$, and so a general solution to the differential given equation is

$$y(x) = c_1 e^{\sqrt{-\lambda}x} + c_2 e^{-\sqrt{-\lambda}x}.$$

By applying the first boundary condition, we obtain

$$y(0) = c_1 + c_2 = 0 \qquad \Rightarrow \qquad c_2 = -c_1.$$

Thus,

$$y(x) = c_1 \left(e^{\sqrt{-\lambda}x} - e^{-\sqrt{-\lambda}x}\right).$$

Applying the second boundary condition yields

$$y\left(1 + \lambda^2\right) = c_1 \left[e^{\sqrt{-\lambda}(1+\lambda^2)} - e^{-\sqrt{-\lambda}(1+\lambda^2)}\right] = 0.$$

 Copyright © 2012 Pearson Education, Inc. Publishing as Addison-Wesley.

Multiplying by $e^{\sqrt{-\lambda}(1+\lambda^2)}$, we obtain

$$c_1 \left[e^{2\sqrt{-\lambda}(1+\lambda^2)} - 1 \right] = 0.$$

Since $\lambda < 0$, $e^{2\sqrt{-\lambda}(1+\lambda^2)} - 1 \neq 0$, and so $C_1 = 0$. Hence, there are no negative eigenvalues.

Case 2: $\lambda = 0$. In this case, we are solving the differential equation $y'' = 0$. This equation has a general solution given by

$$y(x) = c_1 + c_2 x.$$

By applying the boundary conditions, we obtain

$$y(0) = c_1 = 0 \qquad \text{and} \qquad y\left(1 + \lambda^2\right) = c_1 + c_2\left(1 + \lambda^2\right) = 0.$$

Solving these equations simultaneously yields $c_1 = c_2 = 0$. Thus, we find that $\lambda = 0$ is not an eigenvalue.

Case 3: $\lambda > 0$. The roots of the auxiliary equation are $r = \pm\sqrt{\lambda}i$, and so a general solution is

$$y(x) = c_1 \cos\left(\sqrt{\lambda}x\right) + c_2 \sin\left(\sqrt{\lambda}x\right).$$

Substituting into the first boundary condition yields

$$y(0) = c_1 \cos\left(\sqrt{\lambda} \cdot 0\right) + c_2 \sin\left(\sqrt{\lambda} \cdot 0\right) = c_1 = 0.$$

By applying the second boundary condition to $y(x) = c_2 \sin\left(\sqrt{\lambda}x\right)$, we obtain

$$y\left(1 + \lambda^2\right) = c_2 \sin\left[\sqrt{\lambda}(1 + \lambda^2)\right] = 0.$$

Therefore, in order to have nontrivial solutions, we must have

$$\sin\left[\sqrt{\lambda}(1 + \lambda^2)\right] = 0 \qquad \Rightarrow \qquad \sqrt{\lambda}(1 + \lambda^2) = n\pi, \qquad n = 1, 2, 3, \ldots.$$

Hence, the eigenvalues λ_n, $n = 1, 2, 3, \ldots$, satisfy $\sqrt{\lambda_n}(1 + \lambda_n^2) = n\pi$, and the corresponding eigenfunctions are

$$y_n(x) = c_n \sin\left(\sqrt{\lambda_n}x\right), \qquad n = 1, 2, 3, \ldots,$$

where c_n's are arbitrary nonzero constants.

Copyright © 2012 Pearson Education, Inc. Publishing as Addison-Wesley.

33. **(a)** We assume that $u(x,t) = X(x)T(t)$. Then

$$u_{tt} = X(x)T''(t), \qquad u_x = X'(x)T(t), \quad \text{and} \quad u_{xx} = X''(x)T(t).$$

Substituting these expressions into $u_{tt} = u_{xx} + 2u_x$, we obtain

$$X(x)T''(t) = X''(x)T(t) + 2X'(x)T(t).$$

Separating variables yields

$$\frac{T''(t)}{T(t)} = \frac{X''(x) + 2X'(x)}{X(x)} = -\lambda, \tag{11.6}$$

where λ is a constant. The second equation in (11.6) gives

$$X''(x) + 2X'(x) + \lambda X(x) = 0.$$

Let us now examine the boundary conditions. From $u(0,t) = 0$ and $u(\pi,t) = 0$, $t > 0$, we conclude that

$$X(0)T(t) = 0 \qquad \text{and} \qquad X(\pi)T(t) = 0, \qquad t > 0.$$

Hence, either $T(t) = 0$ for all $t > 0$, which implies $u(x,t) \equiv 0$, or $X(0) = X(\pi) = 0$. Ignoring the trivial solution, $u(x,t) \equiv 0$, we obtain the boundary value problem

$$X''(x) + 2X'(x) + \lambda X(x) = 0, \qquad X(0) = X(\pi) = 0.$$

(b) The auxiliary equation in this problem, $r^2 + 2r + \lambda = 0$, has roots $r = -1 \pm \sqrt{1-\lambda}$. To find eigenvalues, which give nontrivial solutions, we consider the following three cases: $1 - \lambda < 0$, $1 - \lambda = 0$, and $1 - \lambda > 0$.

Case 1: $1 - \lambda < 0 \ (\lambda > 1)$. Let $\mu = \sqrt{-(1-\lambda)} = \sqrt{\lambda - 1}$. In this case, the roots of the auxiliary equation are $r = -1 \pm \mu i$. Therefore, a general solution to the differential equation is given by

$$X(x) = c_1 e^{-x} \cos \mu x + c_2 e^{-x} \sin \mu x.$$

By applying the boundary conditions, we obtain

$$X(0) = c_1 = 0 \qquad \text{and} \qquad X(\pi) = e^{-\pi}(c_1 \cos \mu\pi + c_2 \sin \mu\pi) = 0.$$

Copyright © 2012 Pearson Education, Inc. Publishing as Addison-Wesley.

Solving these equations simultaneously yields $c_1 = 0$ and $c_2 \sin \mu\pi = 0$. Therefore, in order to obtain a solution other than the trivial one, we must have

$$\sin \mu\pi = 0 \quad \Rightarrow \quad \mu\pi = n\pi \quad \Rightarrow \quad \mu = n, \quad n = 1, 2, 3, \ldots .$$

Since $\mu = \sqrt{\lambda - 1}$,

$$\sqrt{\lambda - 1} = n \quad \Rightarrow \quad \lambda = n^2 + 1, \quad n = 1, 2, 3, \ldots .$$

Thus, the eigenvalues are given by

$$\lambda_n = n^2 + 1, \quad n = 1, 2, 3, \ldots .$$

Corresponding to the eigenvalue λ_n, we have eigenfunctions

$$X_n(x) = c_n e^{-x} \sin nx, \quad n = 1, 2, 3, \ldots ,$$

where $c_n \neq 0$ is arbitrary.

Case 2: $1 - \lambda = 0$ ($\lambda = 1$). In this case, the associated auxiliary equation has a double root $r = -1$. Therefore, a general solution is given by

$$X(x) = c_1 e^{-x} + c_2 x e^{-x}.$$

By applying the boundary conditions, we obtain

$$X(0) = c_1 = 0 \quad \text{and} \quad X(\pi) = e^{-\pi}(c_1 + c_2\pi) = 0.$$

Solving this system yields $c_1 = c_2 = 0$. So, in this case we have only the trivial solution. Thus, $\lambda = 1$ is not an eigenvalue.

Case 3: $1 - \lambda > 0$ ($\lambda < 1$). Let $\mu = \sqrt{1 - \lambda}$. In this case, the roots of the auxiliary equation are $r = -1 \pm \mu$. Therefore, a general solution to the given differential equation is

$$X(x) = c_1 e^{(-1-\mu)x} + c_2 e^{(-1+\mu)x}.$$

By applying the first boundary condition, we find that

$$X(0) = c_1 + c_2 = 0 \quad \Rightarrow \quad c_2 = -c_1.$$

So, we can express $X(x)$ as

$$X(x) = c_1 \left[e^{(-1-\mu)x} - e^{(-1+\mu)x} \right],$$

Copyright © 2012 Pearson Education, Inc. Publishing as Addison-Wesley.

and the second boundary condition gives us

$$X(\pi) = c_1 \left[e^{(-1-\mu)\pi} - e^{(-1+\mu)\pi} \right].$$

Since $e^{(-1-\mu)\pi} - e^{(-1+\mu)\pi} \neq 0$, we conclude that $c_1 = 0$ and, again, we have only the trivial solution. Thus, there are no eigenvalues for $\lambda < 1$.

Summarizing, the eigenvalues are $\lambda_n = n^2 + 1$, $n = 1, 2, 3, \ldots$, with the corresponding eigenfunctions $X_n(x) = c_n e^{-x} \sin nx$, $n = 1, 2, 3, \ldots$, $c_n \neq 0$ are arbitrary.

35. Assuming that $u(x, y) = X(x)Y(y)$ and substituting this expression into the given equation, we conclude that

$$X''(x)Y(y) + X(x)Y''(y) + \lambda X(x)Y(y) = 0$$
$$\Rightarrow \quad X''(x)Y(y) + X(x)\left[Y''(y) + \lambda Y(y) \right] = 0$$
$$\Rightarrow \quad \frac{X''(x)}{X(x)} = -\frac{Y''(y) + \lambda Y(y)}{Y(y)}.$$

Since the left-hand side is a function of x alone and the right-hand side depends only on y, we must have

$$\frac{X''(x)}{X(x)} = -\frac{Y''(y) + \lambda Y(y)}{Y(y)} = -\nu,$$

where ν is a constant. Thus, the Helmholtz equation results two ordinary linear homogeneous equations of the second order, namely

$$X''(x) + \nu X(x) = 0 \quad \text{and} \quad Y''(y) + (\lambda - \nu) Y(y) = 0.$$

The boundary conditions also split.

$$u(0, y) = u(a, y) = 0 \quad \Rightarrow \quad X(0)Y(y) = X(a)Y(y) = 0 \quad \Rightarrow \quad X(0) = X(a) = 0,$$
$$u(x, 0) = u(x, b) = 0 \quad \Rightarrow \quad X(x)Y(0) = X(x)Y(b) = 0 \quad \Rightarrow \quad Y(0) = Y(b) = 0.$$

Therefore, we have two boundary value problems,

$$X''(x) + \nu X(x) = 0, \quad X(0) = X(a) = 0; \tag{11.7}$$
$$Y''(y) + (\lambda - \nu) Y(y) = 0, \quad Y(0) = Y(b) = 0. \tag{11.8}$$

These are Dirichlet boundary value problems (see (17) in the text). Thus, the eigenvalues and eigenfunctions for (11.7) are

$$\nu_n = \left(\frac{n\pi}{a} \right)^2, \quad X_n(x) = a_n \sin\left(\frac{n\pi x}{a} \right), \quad n = 1, 2, \ldots.$$

  Copyright © 2012 Pearson Education, Inc. Publishing as Addison-Wesley.

Hence, (11.8) generates a sequence of Dirichlet boundary value problems

$$Y''(y) + (\lambda - \nu_n)Y(y) = 0, \quad Y(0) = Y(b) = 0, \quad n = 1, 2, \ldots.$$

For each n, we have a sequence of eigenvalues and eigenfunctions

$$\lambda - \nu_n = \left(\frac{m\pi}{b}\right)^2, \quad Y_m(y) = b_m \sin\left(\frac{m\pi y}{b}\right), \quad m = 1, 2, \ldots.$$

Therefore, the eigenvalues for the Helmholtz boundary value problem are

$$\lambda_{m,n} = \nu_n + \left(\frac{m\pi}{b}\right)^2 = \left(\frac{n\pi}{a}\right)^2 + \left(\frac{m\pi}{b}\right)^2$$

with the corresponding eigenfunctions

$$U_{m,n}(x, y) = X_n(x)Y_m(y) = c_{m,n} \sin\left(\frac{n\pi x}{a}\right) \sin\left(\frac{m\pi y}{b}\right), \quad m, n = 1, 2, \ldots,$$

where $c_{m,n} := a_n b_m$ are arbitrary constants.

EXERCISES 11.3:   Regular Sturm-Liouville Boundary Value Problems

1. Comparing the given equation with a general form (3) in the text, we see that $A_2(x) \equiv 1$ and $A_1(x) \equiv 6$. Using the formula (4), we find an integrating factor

$$\mu(x) = \frac{1}{A_2(x)} \exp\left[\int \frac{A_1(x)}{A_2(x)} \, dx\right] = \exp\left[\int (6) dx\right] = e^{6x},$$

where we have chosen zero integration constant. Multiplying the given differential equation by $\mu(x)$ yields

$$e^{6x}y'' + 6e^{6x}y' + \lambda e^{6x}y = 0 \quad \Rightarrow \quad \left(e^{6x}y'\right)' + \lambda e^{6x}y = 0.$$

3. Here $A_2 = x(1 - x)$ and $A_1 = -2x$. Using the formula (4) in the text, we find

$$\begin{aligned}
\mu(x) &= \frac{1}{x(1-x)} \exp\left[\int \frac{A_1(x)}{A_2(x)} \, dx\right] = \frac{1}{x(1-x)} \exp\left[-\int \frac{2x}{x(1-x)} \, dx\right] \\
&= \frac{1}{x(1-x)} \exp\left(2\int \frac{dx}{x-1}\right) = \frac{e^{2\ln|x-1|}}{x(1-x)} = \frac{(x-1)^2}{x(1-x)} = \frac{1-x}{x}.
\end{aligned}$$

Multiplying the original equation by $\mu(x) = (1-x)/x$, we get

$$(1-x)^2 y''(x) - 2(1-x)y'(x) + \lambda \frac{1-x}{x} y(x) = 0$$

$$\Rightarrow \quad \left[(1-x)^2 y'(x)\right]' + \lambda \frac{1-x}{x} y(x) = 0.$$

Copyright © 2012 Pearson Education, Inc. Publishing as Addison-Wesley.

Chapter 11

5. Here $A_2 = x^2$ and $A_1 = x$. Using the formula (4) in the text yields

$$\mu(x) = \frac{1}{x^2} \exp\left[\int \frac{x}{x^2}\,dx\right] = \frac{1}{x^2} \exp\left(\int \frac{dx}{x}\right) = \frac{1}{x^2} e^{\ln x} = \frac{1}{x}.$$

Multiplying the original equation by $\mu(x)$, we get

$$xy'' + y' + \lambda \frac{1}{x}\,y = 0 \qquad \Rightarrow \qquad (xy')' + \lambda x^{-1} y = 0.$$

7. The linear differential operator in this problem is $L = D^2 + 2D + 5$, where $D := d/dx$, with boundary conditions $y(0) = y(\pi) = 0$. We have to check if $(u, L[v]) = (L[u], v)$ for all twice continuously differentiable functions $u(x)$ and $v(x)$ satisfying the boundary conditions. Taking $u(x) = \sin x$, $v(x) = \sin 2x$ yields

$$L[u] = (\sin x)'' + 2(\sin x)' + 5(\sin x) = 4\sin x + 2\cos x\,,$$
$$L[v] = (\sin 2x)'' + 2(\sin 2x)' + 5(\sin 2x) = \sin 2x + 4\cos 2x\,.$$

Computing integrals, we get

$$(u, L[v]) = \int_0^\pi \sin x(\sin 2x + 4\cos 2x)\,dx = \frac{1}{2}\int_0^\pi (\cos x - \cos 3x + 4\sin 3x - 4\sin x)dx = -\frac{8}{3}$$

while

$$(L[u], v) = \int_0^\pi (4\sin x + 2\cos x)\sin 2x\,dx = \int_0^\pi (2\cos x - 2\cos 3x + \sin 3x + \sin x)dx = \frac{8}{3}\,.$$

Therefore, L is not selfadjoint.

9. Here, we consider the linear differential operator $L[y] := y'' + \lambda y$ with boundary conditions $y(0) = -y(\pi)$ and $y'(0) = -y'(\pi)$. We have to show that

$$(u, L[v]) = (L[u], v),$$

where $u(x)$ and $v(x)$ are arbitrary functions in the domain of L. Since

$$(u, L[v]) = \int_0^\pi u(x)\,[v''(x) + \lambda v(x)]\,dx = \int_0^\pi u(x)v''(x)\,dx + \lambda\int_0^\pi u(x)v(x)\,dx$$

and

$$(L[u], v) = \int_0^\pi [u''(x) + \lambda u(x)]\,v(x)\,dx = \int_0^\pi u''(x)v(x)\,dx + \lambda\int_0^\pi u(x)v(x)dx\,,$$

 Copyright © 2012 Pearson Education, Inc. Publishing as Addison-Wesley.

it is sufficient to show that

$$\int_0^\pi u(x)v''(x)\,dx = \int_0^\pi u''(x)v(x)\,dx.\tag{11.9}$$

Integrating the right-hand side of (11.9) by parts twice, we get

$$\int_0^\pi u''(x)v(x)\,dx = u'(x)v(x)\,\big|_0^\pi - u(x)v'(x)\,\big|_0^\pi + \int_0^\pi u(x)v''(x)\,dx.$$

Hence, to establish (11.9), we need to show that

$$u'(x)v(x)\,\big|_0^\pi - u(x)v'(x)\,\big|_0^\pi = 0.$$

Expanding yields

$$u'(x)v(x)\,\big|_0^\pi - u(x)v'(x)\,\big|_0^\pi = u'(\pi)v(\pi) - u'(0)v(0) - u(\pi)v'(\pi) + u(0)v'(0).$$

Since u is in the domain of L, we have $u(0) = -u(\pi)$ and $u'(0) = -u'(\pi)$. Hence,

$$u'(x)v(x)\,\big|_0^\pi - u(x)v'(x)\,\big|_0^\pi = u'(\pi)\left[v(\pi) + v(0)\right] - u(\pi)\left[v'(\pi) + v'(0)\right].$$

But v is also in the domain of L and, hence, $v(0) = -v(\pi)$ and $v'(0) = -v'(\pi)$. This results zero expressions in the brackets, and so

$$u'(x)v(x)\,\big|_0^\pi - u(x)v'(x)\,\big|_0^\pi = 0.$$

Therefore, L is selfadjoint.

17. In Problem 13, Section 11.2, we found the eigenvalues

$$\lambda_n = \left(n + \frac{1}{2}\right)^2 \pi^2, \qquad n = 0, 1, 2, \dots,$$

for this boundary value problem with the corresponding eigenfunctions

$$y_n(x) = c_n \sin\left[\left(n + \frac{1}{2}\right)\pi x\right], \qquad n = 0, 1, 2, \dots,$$

where c_n's are arbitrary nonzero constants.

(a) We have to find c_n's such that

$$\int_0^1 y_n(x)^2\,dx = \int_0^1 c_n^2 \sin^2\left[\left(n + \frac{1}{2}\right)\pi x\right] dx = 1.$$

Copyright © 2012 Pearson Education, Inc. Publishing as Addison-Wesley.

Computing yields

$$\int_0^1 c_n^2 \sin^2\left[\left(n+\frac{1}{2}\right)\pi x\right] dx = \frac{c_n^2}{2}\int_0^1 (1-\cos\left[(2n+1)\pi x\right]) \, dx$$

$$= \frac{c_n^2}{2}\left(x-\frac{1}{(2n+1)\pi}\sin\left[(2n+1)\pi x\right]\right)\Big|_0^1 = \frac{c_n^2}{2}.$$

Hence, $c_n = \sqrt{2}$, which gives

$$\phi_n(x) = \sqrt{2}\sin\left[\left(n+\frac{1}{2}\right)\pi x\right], \quad n=0,1,2\dots,$$

as an orthonormal system of eigenfunctions.

(b) To obtain the eigenfunction expansion for $f(x)=x$, we use the formula (25) in the text. Thus,

$$c_n = \int_0^1 x\sqrt{2}\sin\left[\left(n+\frac{1}{2}\right)\pi x\right] dx = \sqrt{2}\int_0^1 x\sin\left[\left(n+\frac{1}{2}\right)\pi x\right] dx.$$

Using the integration by parts with $u=x$ and $dv=\sin\left[(n+1/2)\pi x\right] dx$, we find that

$$c_n = \frac{-\sqrt{2}x\cos[(n+1/2)\pi x]}{(n+1/2)\pi}\Big|_0^1 + \int_0^1 \frac{\sqrt{2}x\cos[(n+1/2)\pi x] \, dx}{(n+1/2)\pi}$$

$$= \frac{-\sqrt{2}\cos[(n+1/2)\pi]}{(n+1/2)\pi} + \frac{\sqrt{2}\sin[(n+1/2)\pi x]}{(n+1/2)^2\pi^2}\Big|_0^1$$

$$= 0 + \frac{\sqrt{2}\sin[(n+1/2)\pi]}{(n+1/2)^2\pi^2} = \frac{(-1)^n\sqrt{2}}{(n+1/2)^2\pi^2}.$$

Therefore,

$$x = \sum_{n=0}^{\infty} c_n\sqrt{2}\sin\left[\left(n+\frac{1}{2}\right)\pi x\right] = \sum_{n=0}^{\infty}\frac{2(-1)^n}{(n+1/2)^2\pi^2}\sin\left[\left(n+\frac{1}{2}\right)\pi x\right]$$

$$= \frac{8}{\pi^2}\sum_{n=0}^{\infty}\frac{(-1)^n}{(2n+1)^2}\sin\left[\left(n+\frac{1}{2}\right)\pi x\right].$$

23. In Problem 19 of Section 11.2, we found the eigenvalues

$$\lambda_n = \left(n+\frac{1}{2}\right)^2, \quad n=0,1,2,\dots,$$

 Copyright © 2012 Pearson Education, Inc. Publishing as Addison-Wesley.

with the corresponding eigenfunctions

$$y_n(x) = c_n \cos\left[\left(n + \frac{1}{2}\right) \ln x\right], \qquad n = 0, 1, 2, \dots,$$

where c_n's are arbitrary nonzero constants.

(a) We need to choose c_n's such that

$$\int_1^{e^\pi} y_n(x)^2 dx = \int_1^{e^\pi} c_n^2 \cos^2\left[\left(n + \frac{1}{2}\right) \ln x\right] \frac{dx}{x} = 1.$$

To compute the integral, we let $u = \ln x$ (so that $du = dx/x$). Substituting yields

$$\int_1^{e^\pi} c_n^2 \cos^2\left[\left(n + \frac{1}{2}\right) \ln x\right] \frac{dx}{x} = c_n^2 \int_0^\pi \cos^2\left[\left(n + \frac{1}{2}\right) u\right] du$$

$$= \frac{c_n^2}{2} \int_0^\pi \{1 + \cos\left[(2n + 1)u\right]\} \, du$$

$$= \frac{c_n^2}{2} \left\{u + \frac{1}{2n + 1} \sin\left[(2n + 1)u\right]\right\}\Big|_0^\pi = \frac{\pi c_n^2}{2}.$$

Hence, we take $c_n = \sqrt{2/\pi}$, which gives us

$$\phi_n(x) = \sqrt{\frac{2}{\pi}} \cos\left[\left(n + \frac{1}{2}\right) \ln x\right], \qquad n = 0, 1, 2, \dots,$$

as an orthonormal system of eigenfunctions.

(b) To obtain the eigenfunction expansion for $f(x) = x$, we use the formula (25) in the text. Thus, with $w(x) = x^{-1}$, we have

$$c_n = \int_1^{e^\pi} x \sqrt{\frac{2}{\pi}} \cos\left[\left(n + \frac{1}{2}\right) \ln x\right] x^{-1} dx.$$

Let $u = \ln x$. Then $du = dx/x$, and we have

$$c_n = \sqrt{\frac{2}{\pi}} \int_0^\pi e^u \cos\left[\left(n + \frac{1}{2}\right) u\right] du$$

$$= \sqrt{\frac{2}{\pi}} \frac{e^u \cos[(n + 1/2)u] + e^u(n + 1/2)\sin[(n + 1/2)u]}{1 + (n + 1/2)^2}\Big|_0^\pi$$

$$= \sqrt{\frac{2}{\pi}} \frac{e^\pi(n + 1/2)\sin[(n + 1/2)\pi] - 1}{1 + (n + 1/2)^2} = \sqrt{\frac{2}{\pi}} \frac{(-1)^n e^\pi(n + 1/2) - 1}{1 + (n + 1/2)^2}.$$

Copyright © 2012 Pearson Education, Inc. Publishing as Addison-Wesley.

Therefore,

$$x = \sum_{n=0}^{\infty} c_n \sqrt{\frac{2}{\pi}} \cos\left[\left(n + \frac{1}{2}\right) \ln x\right]$$

$$= \frac{2}{\pi} \sum_{n=0}^{\infty} \frac{(-1)^n e^{\pi}(n + 1/2) - 1}{1 + (n + 1/2)^2} \cos\left[\left(n + \frac{1}{2}\right) \ln x\right].$$

EXERCISES 11.4: Nonhomogeneous Boundary Value Problems and the Fredholm Alternative

3. Here, the differential operator is given by $L[y] = (1 + x^2)\, y'' + 2xy' + y$. Substituting into the formula (3) in the text, we obtain

$$\begin{aligned} L^+[y] &= \left[(1 + x^2)y\right]'' - (2xy)' + y = \left[2xy + (1 + x^2)y'\right]' - 2y - 2xy' + y \\ &= 2y + 2xy' + 2xy' + (1 + x^2)y'' - 2y - 2xy' + y = (1 + x^2)y'' + 2xy' + y. \end{aligned}$$

7. In this problem, the differential operator and the boundary conditions are given by

$$L[y] = y'' - 2y' + 10y; \qquad y(0) = y(\pi) = 0.$$

Hence,

$$L^+[v] = v'' + 2v' + 10v.$$

To find $D\left(L^+\right)$, we need

$$P(u, v)(x)\, \big|_0^{\pi} = 0 \tag{11.10}$$

for all u in $D(L)$ and v in $D\left(L^+\right)$. Using the formula (9) in the text for $P(u, v)$ with $A_1 = -2$ and $A_2 = 1$, we find that

$$P(u, v) = -2uv - uv' + u'v.$$

Evaluating $P(u, v)$ at $x = \pi$ and $x = 0$ in the condition (11.10) yields

$$[-2u(\pi)v(\pi) - u(\pi)v'(\pi) + u'(\pi)v(\pi)] - [-2u(0)v(0) - u(0)v'(0) + u'(0)v(0)] = 0.$$

Since u is in $D(L)$, we know that $u(0) = u(\pi) = 0$. Thus, the above equation becomes

$$u'(\pi)v(\pi) - u'(0)v(0) = 0.$$

$u'(\pi)$ and $u'(0)$ can take on any values. Thus, we must have $v(0) = v(\pi) = 0$ for this equation to hold for all u in $D(L)$. Hence, $D\left(L^+\right)$ consists of all functions v, having continuous second derivatives on $[0, \pi]$ and satisfying the boundary conditions $v(0) = v(\pi) = 0$.

 Copyright © 2012 Pearson Education, Inc. Publishing as Addison-Wesley.

11. Here, the differential operator is given by

$$L[y] = y'' + 6y' + 10y; \qquad y'(0) = y'(\pi) = 0.$$

Hence,

$$L^+[v] = v'' - 6v' + 10v.$$

To find $D(L^+)$, we use

$$P(u, v)(x)\,\big|_0^\pi = 0 \qquad\qquad (11.11)$$

valid for all u in $D(L)$ and v in $D(L^+)$. Again, using the formula (9) in the text for $P(u, v)$ with $A_1 = 6$ and $A_2 = 1$, we find that

$$P(u, v) = 6uv - uv' + u'v.$$

Evaluating $P(u, v)$ at $x = \pi$ and $x = 0$, from (11.11) we conclude that

$$6u(\pi)v(\pi) - u(\pi)v'(\pi) + u'(\pi)v(\pi) - 6u(0)v(0) + u(0)v'(0) - u'(0)v(0) = 0.$$

Applying the boundary conditions $u'(0) = u'(\pi) = 0$ to this equation yields

$$6u(\pi)v(\pi) - u(\pi)v'(\pi) - 6u(0)v(0) + u(0)v'(0) = 0$$

$$\Rightarrow \qquad u(\pi)\left[6v(\pi) - v'(\pi)\right] - u(0)\left[6v(0) - v'(0)\right] = 0.$$

Since $u(\pi)$ and $u(0)$ can be arbitrary, we must have

$$\begin{cases} 6v(\pi) - v'(\pi) = 0 \\ 6v(0) - v'(0) = 0 \end{cases}$$

in order for the equation to hold for all u in $D(L)$. Therefore, the adjoint boundary value problem is

$$L^+[v] = v'' - 6v' + 10v; \qquad 6v(\pi) = v'(\pi) \quad \text{and} \quad 6v(0) = v'(0).$$

17. In Problem 7, we found that the adjoint boundary value problem is

$$L^+[v] = v'' + 2v' + 10v; \qquad v(0) = v(\pi) = 0. \qquad\qquad (11.12)$$

The auxiliary equation for (11.12), which is

$$r^2 + 2r + 10 = 0,$$

Copyright © 2012 Pearson Education, Inc. Publishing as Addison-Wesley.

has roots $r = -1 \pm 3i$. Hence, a general solution to the differential equation in (11.12) is given by

$$y(x) = c_1 e^{-x} \cos 3x + c_2 e^{-x} \sin 3x.$$

Using the boundary conditions in (11.12) to determine c_1 and c_2, we find that

$$y(0) = c_1 = 0 \qquad \text{and} \qquad y(\pi) = -c_1 e^{-\pi} = 0.$$

Thus $c_1 = 0$ and c_2 is arbitrary. Therefore, every solution to the adjoint boundary value problem (11.12) has the form

$$y(x) = c_2 e^{-x} \sin 3x.$$

It follows from the Fredholm alternative that, if h is continuous, then the nonhomogeneous problem has a solution if and only if

$$\int_0^\pi h(x) e^{-x} \sin 3x \, dx = 0.$$

21. In Problem 11, we found the adjoint boundary value problem to the given problem,

$$L^+[v] = v'' - 6v' + 10v; \qquad 6v(\pi) = v'(\pi) \quad \text{and} \quad 6v(0) = v'(0). \qquad (11.13)$$

The auxiliary equation for (11.13) is $r^2 - 6r + 10 = 0$, which has roots $r = 3 \pm i$. Hence a general solution to the differential equation in (11.13) is given by

$$y(x) = c_1 e^{3x} \cos x + c_2 e^{3x} \sin x.$$

To apply the boundary conditions in (11.13), we compute $y'(x)$.

$$y'(x) = 3c_1 e^{3x} \cos x - c_1 e^{3x} \sin x + 3c_2 e^{3x} \sin x + c_2 e^{3x} \cos x.$$

Applying the first condition yields

$$-6c_1 e^{3\pi} = -3c_1 e^{3\pi} - c_2 e^{3\pi} \qquad \Rightarrow \qquad 3c_1 = c_2.$$

Applying the second condition, we get

$$6c_1 = 3c_1 + c_2 \qquad \Rightarrow \qquad 3c_1 = c_2.$$

Thus, $c_2 = 3c_1$, where c_1 is arbitrary. Therefore, every solution to the adjoint problem (11.13) has the form

$$y(x) = c_1 e^{3x} (\cos x + 3 \sin x).$$

 Copyright © 2012 Pearson Education, Inc. Publishing as Addison-Wesley.

It follows from the Fredholm alternative that, for a continuous function h, the nonhomogeneous problem has a solution if and only if

$$\int_0^\pi h(x)e^{3x}(\cos x + 3\sin x)\,dx = 0.$$

EXERCISES 11.5: Solution by Eigenfunction Expansion

3. In Example 1 in the text we noticed that the boundary value problem

$$y'' + \lambda y = 0; \qquad y(0) = 0, \quad y(\pi) = 0,$$

has eigenvalues $\lambda_n = n^2$, $n = 1, 2, 3, \ldots$, with corresponding eigenfunctions

$$\phi_n(x) = \sin nx, \qquad n = 1, 2, 3, \ldots .$$

Here, $r(x) \equiv 1$, so we need to determine coefficients γ_n satisfying

$$f(x) = \frac{f(x)}{r(x)} = \sum_{n=1}^\infty \gamma_n \sin nx = \sin 2x + \sin 8x.$$

Clearly $\gamma_2 = \gamma_8 = 1$ and the remaining γ_n's equal zero. Since $\mu = 4 = \lambda_2$ and $\gamma_2 = 1 \neq 0$, there is no solution to this problem.

5. The Neumann boundary value problem (18) of Section 11.2, that is,

$$y'' + \lambda y = 0; \qquad y'(0) = 0, \quad y'(\pi) = 0,$$

has eigenvalues $\lambda_n = n^2$, $n = 0, 1, 2, \ldots$, with corresponding eigenfunctions

$$\phi_n(x) = \cos nx, \qquad n = 0, 1, 2, \ldots .$$

Here, $r(x) \equiv 1$, so we need to determine coefficients γ_n such that

$$f(x) = \frac{f(x)}{r(x)} = \sum_{n=0}^\infty \gamma_n \cos nx = \cos 4x + \cos 7x.$$

Clearly, $\gamma_4 = \gamma_7 = 1$ and the remaining γ_n's are zero. Since $\mu = 1 = \lambda_1$ and $\gamma_1 = 0$,

$$(\mu - \lambda_1)c_1 - \gamma_1 = 0$$

Copyright © 2012 Pearson Education, Inc. Publishing as Addison-Wesley.

is satisfied for any value of c_1. Calculating c_4 and c_7, we get

$$c_4 = \frac{\gamma_4}{\mu - \lambda_4} = \frac{1}{1-16} = -\frac{1}{15}, \qquad c_7 = \frac{\gamma_7}{\mu - \lambda_7} = \frac{1}{1-49} = -\frac{1}{48}.$$

Hence, a one parameter family of solutions is

$$\phi(x) = \sum_{n=0}^{\infty} c_n \phi_n(x) = c_1 \cos x - \frac{1}{15}\cos 4x - \frac{1}{48}\cos 7x,$$

where c_1 is arbitrary.

9. We first find the eigenvalues and corresponding eigenfunctions for this problem. Note that the auxiliary equation for this problem is $r^2 + \lambda = 0$. To find eigenvalues which yields nontrivial solutions we will consider the three cases $\lambda < 0$, $\lambda = 0$, and $\lambda > 0$.

Case 1: $\lambda < 0$. Let $\mu = \sqrt{-\lambda}$. Then the roots to the auxiliary equation are $r = \pm\mu$, and a general solution to the differential equation is given by

$$y(x) = c_1 \sinh \mu x + c_2 \cosh \mu x.$$

Since

$$y'(x) = c_1\mu \cosh \mu x + c_2\mu \sinh \mu x,$$

applying the boundary conditions, we obtain

$$y'(0) = c_1\mu = 0 \qquad \text{and} \qquad y(\pi) = c_1 \sinh \mu\pi + c_2 \cosh \mu\pi = 0.$$

Hence, $c_1 = 0$ and $y(\pi) = c_2 \cosh \mu\pi = 0$, implying that $c_2 = 0$ as well. Thus, we have only the trivial solution.

Case 2: $\lambda = 0$. In this case, the differential equation becomes $y'' = 0$. This equation has a general solution given by

$$y(x) = c_1 + c_2 x.$$

Since $y'(x) = c_2$, applying the boundary conditions, we obtain

$$y'(0) = c_2 = 0 \qquad \text{and} \qquad y(\pi) = c_1 + c_2\pi = 0.$$

Thus, $c_1 = c_2 = 0$, and we again find only the trivial solution.

Case 3: $\lambda > 0$. Let $\lambda = \mu^2$, $\mu > 0$. The roots of the auxiliary equation are $r = \pm\mu i$. So,

$$y(x) = c_1 \cos \mu x + c_2 \sin \mu x$$

Copyright © 2012 Pearson Education, Inc. Publishing as Addison-Wesley.

is a general solution. Since

$$y'(x) = -c_1\mu \sin \mu x + c_2\mu \cos \mu x\,,$$

using the first boundary condition, we find

$$y'(0) = -c_1\mu \sin(\mu \cdot 0) + c_2\mu \cos(\mu \cdot 0) = 0 \qquad \Rightarrow \qquad c_2\mu = 0 \qquad \Rightarrow \qquad c_2 = 0.$$

Thus, substituting into the second boundary condition yields

$$y(\pi) = c_1 \cos \mu\pi = 0.$$

Therefore, in order to obtain a solution other than the trivial one, we must have

$$\cos \mu\pi = 0 \qquad \Rightarrow \qquad \mu = n + \frac{1}{2}\,, \qquad n = 0, 1, 2, \dots.$$

Hence, $\lambda_n = (n + 1/2)^2$, $n = 0, 1, 2, \dots$, and

$$y_n(x) = c_n \cos\left[\left(n + \frac{1}{2}\right) x\right],$$

where the c_n's are arbitrary nonzero constants.

Next, we have to choose c_n's so that

$$\int_0^\pi c_n^2 \cos^2\left[\left(n + \frac{1}{2}\right) x\right] dx = 1.$$

Computing the integral, we find that

$$\int_0^\pi c_n^2 \cos^2\left[\left(n + \frac{1}{2}\right) x\right] dx = \frac{c_n^2}{2} \int_0^\pi \left\{1 + \cos[(2n + 1)x]\right\} dx$$

$$= \frac{c_n^2}{2} \left\{x + \frac{1}{2n + 1} \sin[(2n + 1)x]\right\}\Bigg|_0^\pi = \frac{\pi c_n^2}{2}\,.$$

We obtain an orthonormal system of eigenfunctions if we take $c_n = \sqrt{2/\pi}$, i.e.,

$$\phi_n(x) = \sqrt{\frac{2}{\pi}} \cos\left[\left(n + \frac{1}{2}\right) x\right], \qquad n = 0, 1, 2, \dots.$$

Now, $f(x)$ has the eigenfunction expansion

$$f(x) = \sum_{n=0}^{\infty} \gamma_n \sqrt{\frac{2}{\pi}} \cos\left[\left(n + \frac{1}{2}\right) x\right],$$

Copyright © 2012 Pearson Education, Inc. Publishing as Addison-Wesley.

Chapter 11

where

$$\gamma_n = \sqrt{\frac{2}{\pi}} \int_0^\pi f(x) \cos\left[\left(n + \frac{1}{2}\right)x\right] dx.$$

Therefore, with γ_n given above, the solution to this boundary value problem has a formal expansion

$$\phi(x) \sim \sum_{n=0}^\infty \frac{\gamma_n}{1 - \lambda_n} \sqrt{\frac{2}{\pi}} \cos\left[\left(n + \frac{1}{2}\right)x\right] = \sum_{n=0}^\infty \frac{\gamma_n}{1 - (n + 1/2)^2} \sqrt{\frac{2}{\pi}} \cos\left[\left(n + \frac{1}{2}\right)x\right].$$

EXERCISES 11.6: Green's Functions

1. A general solution to the corresponding homogeneous equation, namely, $y'' = 0$, is $y_h(x) = Ax + B$. Thus, we seek for particular solutions $z_1(x)$ and $z_2(x)$ of this form satisfying

$$\begin{aligned} z_1(0) &= 0, \\ z_2'(\pi) &= 0. \end{aligned} \tag{11.14}$$

The first equation yields

$$z_1(0) = B = 0.$$

Since A is arbitrary, we choose $A = 1$, and so $z_1(x) = x$. Next, from the second equation in (11.14) we get

$$z_2'(\pi) = A = 0.$$

Taking $B = 1$, we obtain $z_2(x) = 1$.

With $p(x) \equiv 1$, we now compute

$$C = p(x)W\left[z_1, z_2\right](x) = (1)[(x)(0) - (1)(1)] = -1.$$

Thus, the Green's function is

$$G(x, s) = \begin{cases} -z_1(s)z_2(x)/C, & 0 \le s \le x, \\ -z_1(x)z_2(s)/C, & x \le s \le \pi \end{cases} = \begin{cases} s, & 0 \le s \le x, \\ x, & x \le s \le \pi. \end{cases}$$

3. A general solution to the homogeneous problem, $y'' = 0$, is $y_h(x) = Ax + B$, so $z_1(x)$ and $z_2(x)$ must be of this form. To get $z_1(x)$, we have to choose A and B so that

$$z_1(0) = B = 0.$$

  Copyright © 2012 Pearson Education, Inc. Publishing as Addison-Wesley.

Since A is arbitrary, we set $A = 1$, and so $z_1(x) = x$. Next, to get $z_2(x)$, we need A and B so that

$$z_2(\pi) + z_2'(\pi) = A\pi + B + A = 0.$$

Thus, $B = -(1 + \pi)A$. Taking $A = 1$, we obtain $z_2(x) = x - 1 - \pi$.

We now compute

$$C = p(x)W[z_1, z_2](x) = (1)[(x)(1) - (1)(x - 1 - \pi)] = 1 + \pi.$$

Therefore, the Green's function is given by

$$G(x, s) = \begin{cases} \dfrac{-z_1(s)z_2(x)}{C}, & 0 \leq s \leq x, \\[2ex] \dfrac{-z_1(x)z_2(s)}{C}, & x \leq s \leq \pi \end{cases} = \begin{cases} -\dfrac{s(x - 1 - \pi)}{1 + \pi}, & 0 \leq s \leq x, \\[2ex] -\dfrac{x(s - 1 - \pi)}{1 + \pi}, & x \leq s \leq \pi. \end{cases}$$

5. The corresponding homogeneous differential equation, $y'' + 4y = 0$, has the characteristic equation $r^2 + 4 = 0$, whose roots are $r = \pm 2i$. Hence, a general solution to the homogeneous problem is given by

$$y_h(x) = c_1 \cos 2x + c_2 \sin 2x.$$

A solution $z_1(x)$ must satisfy the first boundary condition, $z_1(0) = 0$. Substitution yields

$$z_1(0) = c_1 \cos(2 \cdot 0) + c_2 \sin(2 \cdot 0) = 0 \quad \Rightarrow \quad c_1 = 0.$$

Setting $c_2 = 1$, we get $z_1(x) = \sin 2x$. For $z_2(x)$, we have to find constants c_1 and c_2 such that the second boundary condition is satisfied. Since

$$y_h'(x) = -2c_1 \sin 2x + 2c_2 \cos 2x,$$

we have

$$z_2'(\pi) = -2c_1 \sin(2\pi) + 2c_2 \cos(2\pi) = 2c_2 = 0 \quad \Rightarrow \quad c_2 = 0.$$

With $c_1 = 1$, $z_2(x) = \cos 2x$.

Next we find

$$C = p(x)W[z_1, z_2](x) = (1)[(\sin 2x)(-2 \sin 2x) - (\cos 2x)(2 \cos 2x)] = -2.$$

Thus, the Green's function in this problem is given by

$$G(x, s) = \begin{cases} -z_1(s)z_2(x)/C, & 0 \leq s \leq x, \\ -z_1(x)z_2(s)/C, & x \leq s \leq \pi \end{cases} = \begin{cases} (\sin 2s \cos 2x)/2, & 0 \leq s \leq x, \\ (\sin 2x \cos 2s)/2, & x \leq s \leq \pi. \end{cases}$$

Copyright © 2012 Pearson Education, Inc. Publishing as Addison-Wesley.

Chapter 11

13. In Problem 3, we found the Green's function

$$
\begin{cases}
-\dfrac{s(x-1-\pi)}{1+\pi}, & 0 \le s \le x, \\[2mm]
-\dfrac{x(s-1-\pi)}{1+\pi}, & x \le s \le \pi,
\end{cases}
$$

corresponding to this boundary value problem. For $f(x) = x$, the solution is given by the equation (16) in the text. Substituting $f(x)$ and $G(x,s)$ yields

$$
y(x) = \int_a^b G(x,s) f(s)\, ds = \int_0^\pi G(x,s) s\, ds
$$

$$
= \int_0^x \frac{-s^2(x-1-\pi)}{1+\pi}\, ds + \int_x^\pi \frac{-xs(s-1-\pi)}{1+\pi}\, ds.
$$

Computing

$$
\int_0^x \frac{-s^2(x-1-\pi)}{1+\pi}\, ds = -\frac{(x-1-\pi)}{1+\pi}\left(\frac{s^3}{3}\right)\Bigg|_{s=0}^x
$$

$$
= -\frac{(x-1-\pi)}{1+\pi}\left(\frac{x^3}{3}\right) = -\frac{x^4}{3(1+\pi)} + \frac{x^3}{3}
$$

and

$$
\int_x^\pi \frac{-xs(s-1-\pi)}{1+\pi}\, ds = -\frac{x}{1+\pi}\left[\frac{s^3}{3} - \frac{(1+\pi)s^2}{2}\right]\Bigg|_{s=x}^\pi = -\frac{x}{1+\pi}\left[\frac{\pi^3}{3} - \frac{(1+\pi)\pi^2}{2}\right]
$$

$$
+ \frac{x}{1+\pi}\left[\frac{x^3}{3} - \frac{(1+\pi)x^2}{2}\right] = -\frac{\pi^3 x}{3(1+\pi)} + \frac{\pi^2 x}{2} + \frac{x^4}{3(1+\pi)} - \frac{x^3}{2},
$$

we finally get

$$
y(x) = \left[-\frac{x^4}{3(1+\pi)} + \frac{x^3}{3}\right] + \left[-\frac{\pi^3 x}{3(1+\pi)} + \frac{\pi^2 x}{2} + \frac{x^4}{3(1+\pi)} - \frac{x^3}{2}\right]
$$

$$
= -\frac{x^3}{6} + \left[\frac{\pi^2}{2} - \frac{\pi^3}{3(1+\pi)}\right] x = -\frac{x^3}{6} + \frac{(3\pi^2 + \pi^3)x}{6 + 6\pi}.
$$

17. A general solution to the corresponding homogeneous problem, $y'' - y = 0$, is

$$
y_h(x) = c_1 e^x + c_2 e^{-x}.
$$

So, $z_1(x)$ and $z_2(x)$ must be of this form. For $z_1(x)$, we choose constants c_1, c_2 so that

$$
z_1(0) = c_1 e^0 + c_2 e^{-0} = c_1 + c_2 = 0.
$$

   Copyright © 2012 Pearson Education, Inc. Publishing as Addison-Wesley.

Let $c_1 = 1$. Then $c_2 = -1$, and so $z_1(x) = e^x - e^{-x}$. Likewise, to find $z_2(x)$, we need c_1 and c_2 such that

$$z_2(1) = c_1 e^1 + c_2 e^{-1} = 0 \quad \Rightarrow \quad c_2 = -c_1 e^2 .$$

We let $c_1 = 1$, $c_2 = -e^2$, and $z_2(x) = e^x - e^2 e^{-x} = e^x - e^{2-x}$. We now compute

$$C = p(x)W[z_1, z_2](x) = (1)\left[\left(e^x - e^{-x}\right)\left(e^x + e^{2-x}\right) - \left(e^x + e^{-x}\right)\left(e^x - e^{2-x}\right)\right] = 2e^2 - 2.$$

Thus, the Green's function in this problem is

$$
G(x, s) = \begin{cases} -z_1(s)z_2(x)/C, & 0 \le s \le x, \\ -z_1(x)z_2(s)/C, & x \le s \le 1 \end{cases}
$$

$$
= \begin{cases} \left(e^s - e^{-s}\right)\left(e^x - e^{2-x}\right)/\left(2 - 2e^2\right), & 0 \le s \le x, \\ \left(e^x - e^{-x}\right)\left(e^s - e^{2-s}\right)/\left(2 - 2e^2\right), & x \le s \le 1. \end{cases}
$$

Since $f(x) = -x$, using the Green's function, we obtain

$$
y(x) = \int_a^b G(x, s)f(s)\, ds
$$

$$
= \int_0^x \frac{(e^s - e^{-s})(e^x - e^{2-x})(-s)}{2 - 2e^2}\, ds + \int_x^1 \frac{(e^x - e^{-x})(e^s - e^{2-s})(-s)}{2 - 2e^2}\, ds.
$$

Computing integrals yields

$$
\int_0^x \frac{(e^s - e^{-s})(e^x - e^{2-x})(-s)}{2 - 2e^2}\, ds = -\frac{e^x - e^{2-x}}{2 - 2e^2} \int_0^x \left(se^s - se^{-s}\right) ds
$$

$$
= -\frac{e^x - e^{2-x}}{2 - 2e^2} \left. \left(se^s - e^s + se^{-s} + e^{-s}\right) \right|_0^x
$$

$$
= -\frac{e^x - e^{2-x}}{2 - 2e^2} \left(xe^x - e^x + xe^{-x} + e^{-x}\right)
$$

and

$$
\int_x^1 \frac{(e^x - e^{-x})(e^s - e^{2-s})(-s)}{2 - 2e^2}\, ds = -\frac{e^x - e^{-x}}{2 - 2e^2} \int_x^1 \left(se^s - se^{2-s}\right) ds
$$

$$
= -\frac{e^x - e^{-x}}{2 - 2e^2} \left. \left(se^s - e^s + se^{2-s} + e^{2-s}\right) \right|_x^1
$$

$$
= -\frac{e^x - e^{-x}}{2 - 2e^2} \left[2e - \left(xe^x - e^x + xe^{2-x} + e^{2-x}\right)\right].
$$

Copyright © 2012 Pearson Education, Inc. Publishing as Addison-Wesley.

Thus,

$$y(x) = -\frac{e^x - e^{2-x}}{2 - 2e^2}\left(xe^x - e^x + xe^{-x} + e^{-x}\right) - \frac{e^x - e^{-x}}{2 - 2e^2}\left[2e - \left(xe^x - e^x + xe^{2-x} + e^{2-x}\right)\right]$$

$$= \frac{-2x + 2xe^2 - 2e^{1+x} + 2e^{1-x}}{2 - 2e^2} = -x + \frac{e^{1+x} - e^{1-x}}{e^2 - 1}.$$

25. **(a)** Substituting $y = x^r$ into the corresponding homogeneous Cauchy-Euler equation

$$x^2 y'' - 2xy' + 2y = 0,$$

we obtain the auxiliary equation

$$r(r - 1) - 2r + 2 = r^2 - 3r + 2 = (r - 1)(r - 2) = 0.$$

Hence, a general solution to the homogeneous equation is

$$y_h(x) = c_1 x + c_2 x^2.$$

To get $z_1(x)$, we choose constants c_1 and c_2 so that

$$z_1(1) = c_1 + c_2 = 0 \qquad \Rightarrow \qquad c_2 = -c_1.$$

Let $c_1 = 1$, $c_2 = -1$, and so $z_1(x) = x - x^2$. Next, we find $z_2(x)$ satisfying

$$z_2(2) = 2c_1 + 4c_2 = 0 \qquad \Rightarrow \qquad c_1 = -2c_2.$$

Hence, taking $c_2 = -1$, we get $c_1 = 2$ and $z_2(x) = 2x - x^2$. We now compute (see the formula for $K(x, s)$ in Problem 22)

$$C(s) = A_2(s)W[z_1, z_2](s) = (s^2)\left[(s - s^2)(2 - 2s) - (1 - 2s)(2s - s^2)\right]$$

$$= (s^2)\left(2s - 4s^2 + 2s^3 - 2s + 5s^2 - 2s^3\right) = s^4,$$

and

$$K(x, s) = \begin{cases} -z_1(s)z_2(x)/C(s), & 1 \le s \le x, \\ -z_1(x)z_2(s)/C(s), & x \le s \le 2 \end{cases}$$

$$= \begin{cases} -(s - s^2)(2x - x^2)s^{-4}, & 1 \le s \le x, \\ -(x - x^2)(2s - s^2)s^{-4}, & x \le s \le 2. \end{cases}$$

Simplifying yields

$$K(x, s) = \begin{cases} -x(2 - x)(s^{-3} - s^{-2}), & 1 \le s \le x, \\ -x(1 - x)(2s^{-3} - s^{-2}), & x \le s \le 2. \end{cases}$$

 Copyright © 2012 Pearson Education, Inc. Publishing as Addison-Wesley.

(b) The solution to the boundary value problem with $f(x) = -x$ is

$$
\begin{aligned}
y(x) &= \int_a^b K(x,s)f(s)\,ds = \int_1^x K(x,s)f(s)\,ds + \int_x^2 K(x,s)f(s)\,ds \\
&= \int_1^x [-x(2-x)(s^{-3}-s^{-2})](-s)\,ds + \int_x^2 [-x(1-x)(2s^{-3}-s^{-2})](-s)\,ds \\
&= (2x-x^2)\int_1^x \left(s^{-2}-s^{-1}\right)ds + (x-x^2)\int_x^2 \left(2s^{-2}-s^{-1}\right)ds \\
&= (2x-x^2)\left(-s^{-1}-\ln s\right)\Big|_1^x + (x-x^2)\left(-2s^{-1}-\ln s\right)\Big|_x^2 \\
&= x^2\ln 2 - x\ln x - x\ln 2\,.
\end{aligned}
$$

29. Let $f(x) = \delta(x-s)$, and let $H(x,s)$ be the solution to

$$
\frac{\partial^4 H(x,s)}{\partial x^4} = -\delta(x-s) \tag{11.15}
$$

that satisfies the given boundary conditions, the jump condition

$$
\lim_{x\to s^+}\frac{\partial^3 H(x,s)}{\partial x^3} - \lim_{x\to s^-}\frac{\partial^3 H(x,s)}{\partial x^3} = -1,
$$

and such that $H, \partial H/\partial x, \partial^2 H/\partial x^2$ are continuous on the square $[0,\pi]\times[0,\pi]$. Integrating (11.15) yields

$$
\frac{\partial^3 H(x,s)}{\partial x^3} = -u(x-s) + C_1,
$$

where u is the unit step function and C_1 is a constant. (Recall that $u'(t-a) = \delta(t-a)$. See Section 7.8.) $\partial^3 H/\partial x^3$ is not continuous along the line $x = s$, but it does satisfy the jump condition

$$
\begin{aligned}
\lim_{x\to s^+}\frac{\partial^3 H(x,s)}{\partial x^3} - \lim_{x\to s^-}\frac{\partial^3 H(x,s)}{\partial x^3} &= \lim_{x\to s^+}[-u(x-s)+C_1] - \lim_{x\to s^-}[-u(x-s)+C_1] \\
&= (-1+C_1) - C_1 = -1.
\end{aligned}
$$

We want $H(x,s)$ to satisfy the boundary condition $y'''(\pi) = 0$. Thus, we solve

$$
\frac{\partial^3 H(\pi,s)}{\partial x^3} = -u(\pi-s) + C_1 = -1 + C_1 = 0
$$

and obtain $C_1 = 1$. Therefore,

$$
\frac{\partial^3 H(x,s)}{\partial x^3} = -u(x-s) + 1. \tag{11.16}
$$

Copyright © 2012 Pearson Education, Inc. Publishing as Addison-Wesley.

We now integrate (11.16) with respect to x and conclude that

$$\frac{\partial^2 H(x, s)}{\partial x^2} = x - u(x - s)(x - s) + C_2.$$

(Verify this equation by differentiating it with respect to x.) We have chosen this particular form of the antiderivative because we want $\partial^2 H/\partial x^2$ to be continuous on $[0, \pi] \times [0, \pi]$. (The factor $(x - s)$ takes care of the jump of $u(x - s)$ at $x = s$.) Since

$$\lim_{x \to s} \frac{\partial^2 H(x, s)}{\partial x^2} = s + C_2,$$

we define

$$\frac{\partial^2 H(s, s)}{\partial x^2} = s + C_2$$

to have a continuous function. Next, we want that $y''(\pi) = 0$. Solving yields

$$0 = \frac{\partial^2 H(\pi, s)}{\partial x^2} = \pi - u(\pi - s)(\pi - s) + C_2 = \pi - (\pi - s) + C_2 = s + C_2.$$

Thus, we see that $C_2 = -s$ and

$$\frac{\partial^2 H(x, s)}{\partial x^2} = (x - s) - u(x - s)(x - s).$$

We integrate this equation with respect to x once again to get

$$\frac{\partial H(x, s)}{\partial x} = \frac{x^2}{2} - sx - u(x - s)\frac{(x - s)^2}{2} + C_3,$$

which is continuous on $[0, \pi] \times [0, \pi]$. The boundary condition, $y'(0) = 0$, satisfied if

$$0 = \frac{\partial H}{\partial x}(0, s) = -u(0 - s)\frac{s^2}{2} + C_3 = C_3.$$

Hence,

$$\frac{\partial H(x, s)}{\partial x} = \frac{x^2}{2} - sx - u(x - s)\frac{(x - s)^2}{2}.$$

Integrating one more time with respect to x, we get

$$H(x, s) = \frac{x^3}{6} - \frac{sx^2}{2} - u(x - s)\frac{(x - s)^3}{6} + C_4.$$

$H(x, s)$ is continuous on $[0, \pi] \times [0, \pi]$, and we want that $H(x, s)$ satisfies the boundary condition $y(0) = 0$. Solving for C_4 yields

$$0 = H(0, s) = -u(0 - s)\frac{(0 - s)^3}{6} + C_4 = C_4.$$

    Copyright © 2012 Pearson Education, Inc. Publishing as Addison-Wesley.

Hence,

$$H(x, s) = \frac{x^3}{6} - \frac{sx^2}{2} - u(x - s)\frac{(x - s)^3}{6},$$

which can be written in the form

$$H(x, s) = \begin{cases} s^2(s - 3x)/6, & 0 \le s \le x, \\ x^2(x - 3s)/6, & x \le s \le \pi. \end{cases}$$

EXERCISES 11.7: Singular Sturm-Liouville Boundary Value Problems

1. This is a typical singular Sturm-Liouville boundary value problem. The condition (ii) of Lemma 1 holds because

$$\lim_{x \to 0^+} p(x) = \lim_{x \to 0^+} x = 0$$

and $y(x)$, $y'(x)$ remain bounded, as $x \to 0^+$. Since

$$\lim_{x \to 1^-} p(x) = p(1) = 1$$

and $y(1) = 0$, an analogue of the condition (i) of Lemma 1 holds at the right endpoint. Hence, L is selfadjoint.

The corresponding homogeneous equation is a Bessel's equation of order $\nu = \sqrt{4} = 2$. As we observed in the text, the eigenfunctions for this boundary value problem are given by

$$y_n(x) = c_n J_2\left(\alpha_{2n}x\right),$$

where $\alpha_{2n} = \sqrt{\mu_n}$ is the increasing sequence of real zeros of $J_2(x)$.

Now, to find the eigenfunction expansion for given nonhomogeneous equation, we compute the eigenfunction expansion for $f(x)/x$ (see Section 11.5).

$$\frac{f(x)}{x} \sim \sum_{n=1}^{\infty} a_n J_2\left(\alpha_{2n}x\right),$$

where

$$a_n = \frac{\int_0^1 f(x)J_2(\alpha_{2n}x)\,dx}{\int_0^1 xJ_2^2(\alpha_{2n}x)\,dx}, \qquad n = 1, 2, 3, \ldots.$$

Therefore,

$$y(x) = \sum_{n=1}^{\infty} \frac{a_n}{\mu - \alpha_{2n}^2} J_2\left(\alpha_{2n}x\right) = \sum_{n=1}^{\infty} b_n J_2\left(\alpha_{2n}x\right), \qquad b_n = \frac{\int_0^1 f(x)J_2(\alpha_{2n}x)\,dx}{(\mu - \alpha_{2n}^2)\int_0^1 xJ_2^2(\alpha_{2n}x)\,dx}.$$

Copyright © 2012 Pearson Education, Inc. Publishing as Addison-Wesley.

3. This is a typical singular Sturm-Liouville boundary value problem. L is selfadjoint since the condition (ii) of Lemma 1 in the text holds at the left endpoint and an analogue of this condition holds at the right endpoint.

The corresponding homogeneous equation is a Bessel's equation of order 0. As we found in the text, $J_0\left(\sqrt{\mu}x\right)$ satisfies the boundary conditions at the origin. At the right endpoint, we want $J_0'\left(\sqrt{\mu}\right) = 0$. Now, it follows from the equation (32) of Section 8.8 that the zeros of J_0' and J_1 are the same. So, if we let $\sqrt{\mu_n} = \alpha_{1n}$ denote the increasing sequence of zeros of J_1, then $J_0'(\alpha_{1n}) = 0$. Hence, the eigenfunctions are given by

$$ y_n(x) = c_n J_0\left(\alpha_{1n}x\right), \qquad n = 1, 2, 3, \ldots. $$

To find an eigenfunction expansion for the solution to the given nonhomogeneous problem, we expand $f(x)/x$ (see Section 11.5) in a series

$$ \frac{f(x)}{x} \sim \sum_{n=1}^{\infty} b_n J_0\left(\alpha_{1n}x\right), $$

where

$$ b_n = \frac{\int_0^1 f(x)J_0(\alpha_{1n}x)\,dx}{\int_0^1 x J_0^2(\alpha_{1n}x)\,dx}, \qquad n = 1, 2, 3, \ldots. $$

Therefore,

$$ y(x) = \sum_{n=1}^{\infty} \frac{b_n}{\mu - \alpha_{1n}^2} J_0\left(\alpha_{1n}x\right) = \sum_{n=1}^{\infty} a_n J_0\left(\alpha_{1n}x\right), \qquad a_n = \frac{\int_0^1 f(x)J_0(\alpha_{1n}x)\,dx}{(\mu - \alpha_{1n}^2)\int_0^1 x J_0^2(\alpha_{1n}x)\,dx}. $$

11. (a) Let $\phi(x)$ be an eigenfunction for

$$ \frac{d}{dx}\left[x\frac{dy}{dx}\right] - \frac{\nu^2}{x}y + \lambda x y = 0. $$

Then

$$ \frac{d}{dx}\left[x\phi'(x)\right] - \frac{\nu^2}{x}\phi(x) + \lambda x\phi(x) = 0 $$

$$ \Rightarrow \qquad \phi'(x) + x\phi''(x) - \frac{\nu^2}{x}\phi(x) + \lambda x\phi(x) = 0. $$

Multiplying both sides by $\phi(x)$ and integrating from $x = 0$ to $x = 1$, we obtain

$$ \int_0^1 \phi(x)\phi'(x)\,dx + \int_0^1 x\phi(x)\phi''(x)\,dx - \int_0^1 \frac{\nu^2}{x}\phi(x)^2\,dx + \int_0^1 \lambda x\phi(x)^2\,dx = 0. \quad (11.17) $$

Copyright © 2012 Pearson Education, Inc. Publishing as Addison-Wesley.

Integrating by parts with $u = \phi(x)\phi'(x)$ and $dv = dx$ yields

$$\int_0^1 \phi(x)\phi'(x)\,dx = x\phi(x)\phi'(x)\,\big|_0^1 - \int_0^1 x\left[\phi'(x)\phi'(x) + \phi(x)\phi''(x)\right]dx$$

$$= x\phi(x)\phi'(x)\,\big|_0^1 - \int_0^1 x\phi'(x)^2\,dx - \int_0^1 x\phi(x)\phi''(x)\,dx.$$

Since $\phi(1) = 0$, we have

$$x\phi(x)\phi'(x)\,\big|_0^1 = 0,$$

$$\int_0^1 \phi(x)\phi'(x)\,dx = -\int_0^1 x\phi'(x)^2\,dx - \int_0^1 x\phi(x)\phi''(x)\,dx.$$

Thus, the equation (11.17) reduces to

$$-\int_0^1 x\phi'(x)^2\,dx - \int_0^1 x\phi(x)\phi''(x)\,dx + \int_0^1 x\phi(x)\phi''(x)\,dx$$

$$-\nu^2 \int_0^1 x^{-1}\phi(x)^2\,dx + \lambda \int_0^1 x\phi(x)^2\,dx = 0.$$

Simplifying yields

$$-\int_0^1 x\phi'(x)^2\,dx - \nu^2 \int_0^1 x^{-1}\phi(x)^2\,dx + \lambda \int_0^1 x\phi(x)^2\,dx = 0. \qquad (11.18)$$

(b) First note that each integrand in (11.18) is nonnegative on the interval $(0, 1)$. Hence, each integral is nonnegative. Moreover, since $\phi(x)$ is an eigenfunction, it is a continuous function, not identically zero. Thus, the second and third integrals are strictly positive. If $\nu > 0$, then λ must be positive in order for the left-hand side of (11.18) to sum to zero.

(c) If $\nu = 0$, then only the first and third terms remain on the left-hand side of the equation (11.18). Since the first integral needs only be nonnegative, we require λ to be nonnegative in order for the equation (11.18) be satisfied.

To show that $\lambda = 0$ is not an eigenvalue, we solve the Bessel's equation with $\nu = 0$, that is, we solve

$$xy'' + y' = 0 \qquad \Rightarrow \qquad x^2 y'' + xy' = 0,$$

Copyright © 2012 Pearson Education, Inc. Publishing as Addison-Wesley.

which is a Cauchy-Euler equation. Solving this Cauchy-Euler equation, we find a general solution

$$y(x) = c_1 + c_2 \ln x.$$

Since $\lim_{x \to 0+} y(x) = -\infty$ if $c_2 \neq 0$, we take $c_2 = 0$. Now, $y(x) = c_1$ satisfies the boundary condition (17) in the text. The right endpoint boundary condition (18) is $y(1) = 0$. So, we want $y(1) = c_1 = 0$. Hence, the only solution to the Bessel's equation of order 0, satisfying the boundary conditions (17) and (18), is the trivial one. Hence, $\lambda = 0$ is not an eigenvalue.

EXERCISES 11.8:   Oscillation and Comparison Theory

1. Comparing the given equation with a general form (1) of the Sturm-Liouville equation, we find that $p(x) = e^x$ and $Q(x) = x$. Since $p(x)$, $p'(x)$, and $Q(x)$ are continuous on $(-\infty, \infty)$, and $p(x) > 0$, Theorem 13 applies. The given solution $\phi(x)$ has infinitely many zeros on, say, $[0, 1] \subset (-\infty, \infty)$. Therefore, $\phi(x) \equiv 0$.

3. The function $\phi_0(x) = x$ satisfies the given equation. Indeed,

$$\frac{d}{dx}\left[(x^2 + 1)\frac{d\phi_0}{dx}\right] - 2\phi_0 = \frac{dx^2 + 1}{dx} - 2x = 0.$$

We also note that $p(x) = x^2 + 1$, $p'(x) = 2x$, and $Q(x) \equiv -2$ are continuous on $(-\infty, \infty)$, and $p(x) > 0$. Therefore, Theorem 14 applies. Since $\phi_0(0) = 0$, any other solution $\phi(x)$, satisfying this initial condition, is a constant multiple of $\phi_0(x)$ (meaning that ϕ and ϕ_0 are linearly dependent). Thus,

$$\phi(x) = c\phi_0(x) = cx.$$

5. To apply the Sturm fundamental theorem to the equation

$$y'' + (1 - e^x)y = 0, \qquad 0 < x < \infty, \tag{11.19}$$

we must find functions $Q(x)$ and $\phi(x)$ such that $Q(x) \geq 1 - e^x$, $0 < x < \infty$, and $\phi(x)$ is a solution to

$$y'' + Q(x)y = 0, \qquad 0 < x < \infty. \tag{11.20}$$

Since, for $x > 0$, $1 - e^x < 0$, we choose $Q(x) \equiv 0$. Hence, the equation (11.20) becomes $y'' = 0$. The function $\phi(x) = x + 4$ is a nontrivial solution to this differential equation,

 Copyright © 2012 Pearson Education, Inc. Publishing as Addison-Wesley.

which does not vanish for $x > 0$. Therefore, any nontrivial solution to (11.19) can have at most one zero in $0 < x < \infty$. To use the Sturm fundamental theorem in showing that any nontrivial solution to

$$y'' + (1 - e^x)\, y = 0\,, \qquad -\infty < x < 0\,, \tag{11.21}$$

has infinitely many zeros, we must find $Q(x)$ and $\phi(x)$ such that $q(x) \le 1 - e^x$, $x < 0$, and $\phi(x)$ is a solution to

$$y'' + q(x)y = 0\,, \qquad -\infty < x < 0\,.$$

Since $1 - e^{-1} \approx 0.632$, we choose $Q(x) \equiv 1/4$ and consider the interval $(-\infty, -1)$ only. Thus, we obtain

$$y'' + \frac{1}{4}\, y = 0,$$

which has a nontrivial solution $\phi(x) = \sin(x/2)$. Now, the function $\phi(x)$ has infinitely many zeros in $(-\infty, -1)$ and between any two consecutive zeros of $\phi(x)$ any nontrivial solution to (11.21) must have a zero. Hence, any nontrivial solution to (11.21) has infinitely many zeros in $(-\infty, -1)$.

7. Legendre polynomials P_{m-1} and P_m satisfy the equations

$$\frac{d}{dx}\left[\left(1 - x^2\right)\frac{dy}{dx}\right] + (m - 1)my = 0\,,$$

$$\frac{d}{dx}\left[\left(1 - x^2\right)\frac{dy}{dx}\right] + m(m + 1)y = 0\,,$$

respectively. Using notations of the Picone comparison theorem, we have

$$p(x) := p_1(x) = p_2(x) = 1 - x^2\,, \quad Q_1(x) \equiv (m - 1)m\,, \quad Q_2(x) \equiv m(m + 1)\,.$$

These functions and $p'(x) = -2x$ are continuous on $(-1, 1)$, and $p(x) > 0$ there. Therefore, the Picone theorem applies and, since $Q_1(x) < Q_2(x)$, we get the required.

9. First, we express the equation

$$y'' + x^{-2}y' + \left(4 - e^{-x}\right)y = 0$$

in Strum-Liouville form by multiplying it by an integrating factor $e^{-1/x}$.

$$e^{-1/x}y'' + e^{-1/x}x^{-2}y' + e^{-1/x}\left(4 - e^{-x}\right)y = 0 \quad \Rightarrow \quad \left(e^{-1/x}y'\right)' + e^{-1/x}\left(4 - e^{-x}\right)y = 0.$$

Copyright © 2012 Pearson Education, Inc. Publishing as Addison-Wesley.

Chapter 11

Now, when x is getting large, we have

$$\sqrt{\frac{p}{q}} \approx \sqrt{\frac{e^{-1/\text{large}}}{e^{-1/\text{large}}\left(4 - e^{-\text{large}}\right)}} \approx \sqrt{\frac{1}{(1)(4 - \text{small})}} \approx \sqrt{\frac{1}{4}} = \frac{1}{2}.$$

Hence, the distance between consecutive zeros is approximately $\pi/2$.

11. We apply Corollary 5 with $p(x) = 1+x$, $q(x) = e^{-x}$, and $r(x) \equiv 1$ to a nontrivial solution on the interval $[0, 5]$. On this interval, we have $p_M = 6$, $p_m = 1$, $q_M = 1$, $q_m = e^{-5}$, and $r_M = r_m = 1$. Therefore, for

$$\lambda > \max\left\{\frac{-q_M}{r_M}, \frac{-q_m}{r_m}, 0\right\} = 0,$$

the distance between two consecutive zeros of a nontrivial solution $\phi(x)$ to the given equation is bounded by

$$\pi\sqrt{\frac{p_m}{\lambda r_M + q_M}} = \pi\sqrt{\frac{1}{\lambda + 1}} \qquad \text{and} \qquad \pi\sqrt{\frac{p_M}{\lambda r_m + q_m}} = \pi\sqrt{\frac{6}{\lambda + e^{-5}}}.$$

13. Using the product rule, we get

$$y' = \left(ux^{-1/2}\right)' = u'x^{-1/2} - (1/2)ux^{-3/2},$$

$$y'' = \left[u'x^{-1/2} - (1/2)ux^{-3/2}\right]' = u''x^{-1/2} - u'x^{-3/2} + (3/4)ux^{-5/2}.$$

Substituting these expressions into (11) yields

$$
\begin{aligned}
x^2 y'' + xy' + \left(x^2 - \nu^2\right)y &= x^2\left[u''x^{-1/2} - u'x^{-3/2} + (3/4)ux^{-5/2}\right] \\
&\quad + x\left[u'x^{-1/2} - (1/2)ux^{-3/2}\right] + \left(x^2 - \nu^2\right)\left[ux^{-1/2}\right] \\
&= u''x^{3/2} + \left[x^2 - \nu^2 + (1/4)\right]ux^{-1/2} = 0.
\end{aligned}
$$

Dividing this equation by $x^{3/2}$, we obtain

$$u'' + \left[x^2 - \nu^2 + (1/4)\right]ux^{-2} = 0 \qquad \Rightarrow \qquad u'' + \left[1 + \frac{(1/4) - \nu^2}{x^2}\right]u = 0,$$

which is equivalent to (12).

       Copyright © 2012 Pearson Education, Inc. Publishing as Addison-Wesley.

CHAPTER 12: Stability of Autonomous Systems

EXERCISES 12.2: Linear Systems in the Plane

3. The characteristic equation for this system is $r^2 + 2r + 10 = 0$, which has roots $r = -1 \pm 3i$. Since the real part of these roots is negative, the trajectories approach the origin, and so the origin is an asymptotically stable spiral point.

7. The critical points are solutions to the system

$$-4x + 2y + 8 = 0,$$
$$x - 2y + 1 = 0.$$

Solving this system, we obtain a unique critical point $(3, 2)$. Now, we use the change of variables

$$x = u + 3, \qquad y = v + 2,$$

to translate the critical point $(3, 2)$ to $(0, 0)$. Substituting into the giving system and simplifying yields a system of differential equations in u and v, that is,

$$
\begin{aligned}
u' = x' = -4(u + 3) + 2(v + 2) + 8 &= -4u + 2v, \\
v' = y' = (u + 3) - 2(v + 2) + 1 &= u - 2v.
\end{aligned}
\qquad (12.1)
$$

The characteristic equation for this system is $r^2 + 6r + 6 = 0$ with roots $r = -3 \pm \sqrt{3}$. Since both roots are distinct and negative, the origin is an asymptotically stable improper node for the system (12.1). Therefore, the critical point $(3, 2)$ is an asymptotically stable improper node for the original system.

9. The critical point is the solution to the system

$$2x + y + 9 = 0,$$
$$-5x - 2y - 22 = 0.$$

Solving this system, we obtain a critical point $(-4, -1)$. We now use the change of variables

$$x = u - 4, \qquad y = v - 1,$$

to translate the critical point $(-4, -1)$ to the origin. Substituting into the given system and simplifying yields a system of differential equations in u and v, namely,

$$\frac{du}{dt} = \frac{dx}{dt} = 2(u-4) + (v-1) + 9 = 2u + v,$$

$$\frac{dv}{dt} = \frac{dy}{dt} = -5(u-4) - 2(v-1) - 22 = -5u - 2v.$$

The characteristic equation for this system is $r^2 + 1 = 0$, which has roots $r = \pm i$. Since the roots are pure imaginary, the origin is a stable center for the (u, v)-system. Therefore, the critical point $(-4, -1)$ is a stable center for the original system.

15. The characteristic equation for this system is $r^2 + r - 12 = 0$, which has roots $r = -4$ and $r = 3$. Since the roots are real and have opposite signs, the origin is an unstable saddle point. To sketch the phase plane diagram, we first determine two lines passing through the origin that correspond to the transformed axes. Thus, we make a substitution $y = mx$ into

$$\frac{dy}{dx} = \frac{dy/dt}{dx/dt} = \frac{5x - 2y}{x + 2y}$$

and conclude that

$$m = \frac{5x - 2mx}{x + 2mx}.$$

Solving for m yields

$$m(x + 2mx) = 5x - 2mx \quad \Rightarrow \quad 2m^2 + 3m - 5 = 0 \quad \Rightarrow \quad (2m+5)(m-1) = 0.$$

So $m = -5/2$ or $m = 1$. Hence, the transformed axes are $y = -(5/2)x$ and $y = x$. On the line $y = x$, we have

$$\frac{dx}{dt} = 3x,$$

so that the trajectories flow away from the origin. On the line $y = -5x/2$, one finds that

$$\frac{dy}{dt} = -4y,$$

and the trajectories flow towards the origin. A phase plane diagram is depicted in Fig. B.56 in the answers of the text.

19. The characteristic equation for this system is $(r+2)^2 = 0$, which has a double root $r = -2$. Since the roots are real, equal, and negative, the origin is an asymptotically stable improper node. To sketch the phase plane diagram, we determine slopes of two

 Copyright © 2012 Pearson Education, Inc. Publishing as Addison-Wesley.

lines, passing through the origin, that correspond to the transformed axes. Substituting $y = mx$ into

$$\frac{dy}{dx} = \frac{dy/dt}{dx/dt} = \frac{-2y}{-2x + y},$$

we obtain

$$m = \frac{-2mx}{-2x + mx}.$$

Solving for m yields

$$m(-2x + mx) = -2mx \qquad \Rightarrow \qquad m^2 = 0 \qquad \Rightarrow \qquad m = 0.$$

Since there is only one line, $y = 0$, through the origin that is a trajectory, the origin is an improper node. A phase plane diagram is shown in Fig. B.58 in the answers of the text.

EXERCISES 12.3:   Almost Linear Systems

5. This system is almost linear since $ad - bc = (1)(-1) - (5)(-1) \neq 0$ and, in addition, the functions $F(x, y) = G(x, y) = -y^2$ involve only a higher order term in y. The characteristic equation for this system is $r^2 + 4 = 0$, which has pure imaginary roots $r = \pm 2i$. Thus, the origin is either a center or a spiral point, and the stability is indeterminate.

7. To see that this system is almost linear, we expand e^{x+y}, $\cos x$, and $\cos y$ into their respective Maclaurin series. This gives

$$\frac{dx}{dt} = \left[1 + (x + y) + \frac{(x + y)^2}{2!} + \cdots\right] - \left[1 - \frac{x^2}{2!} + \cdots\right]$$
$$= x + y + (\text{higher order terms}) = x + y + F(x, y),$$
$$\frac{dy}{dt} = \left[1 - \frac{y^2}{2!} + \cdots\right] + x - 1 = x + (\text{higher order terms}) = x + G(x, y).$$

This system is almost linear since $ad - bc = (1)(0) - (1)(1) \neq 0$ and $F(x, y)$, $G(x, y)$ involve only higher order terms in x and y. The characteristic equation for this system is $r^2 - r - 1 = 0$, which has roots $r = (1 \pm \sqrt{5})/2$. Since these roots are real and have different signs, the origin is an unstable saddle point.

9. The critical points for this system are the solutions to the system

$$16 - xy = 0,$$
$$x - y^3 = 0.$$

Copyright © 2012 Pearson Education, Inc. Publishing as Addison-Wesley.

Solving the second equation for x in terms of y and substituting the result into the first equation yields

$$16 - y^4 = 0,$$

which has real solutions $y = \pm 2$. Hence, there are two critical points $(2^3, 2) = (8, 2)$ and $((-2)^3, -2) = (-8, -2)$.

Consider the critical point $(8, 2)$. Using the change of variables

$$x = u + 8, \quad y = v + 2,$$

we obtain the system

$$\frac{du}{dt} = 16 - (u + 8)(v + 2),$$

$$\frac{dv}{dt} = (u + 8) - (v + 2)^3,$$

which simplifies to the almost linear system

$$\frac{du}{dt} = -2u - 8v - uv,$$

$$\frac{dv}{dt} = u - 12v - 6v^2 - v^3.$$

The characteristic equation for this system is $r^2 + 14r + 32 = 0$, which has distinct negative roots $r = -7 \pm \sqrt{17}$. Hence, $(8, 2)$ is an improper node, which is asymptotically stable.

Next, we consider the critical point $(-8, -2)$. Using the change of variables

$$x = u - 8, \quad y = v - 2,$$

we get the system

$$\frac{du}{dt} = 16 - (u - 8)(v - 2),$$

$$\frac{dv}{dt} = (u - 8) - (v - 2)^3,$$

which simplifies to the almost linear system

$$\frac{du}{dt} = 2u + 8v - uv,$$

$$\frac{dv}{dt} = u - 12v + 6v^2 - v^3.$$

 Copyright © 2012 Pearson Education, Inc. Publishing as Addison-Wesley.

The characteristic equation for this system is $r^2 + 10r - 32 = 0$ with distinct real roots $r = -5 \pm \sqrt{57}$. Since these roots are of different signs, $(-8, -2)$ is an unstable saddle point.

13. The critical points for this system are solutions to the pair of equations (system)

$$1 - xy = 0,$$
$$x - y^3 = 0.$$

Solving the second equation for x in terms of y and substituting the result into the first equation, we obtain

$$1 - y^4 = 0$$

which has (real) solutions $y = \pm 1$. Hence, the critical points are $(1, 1)$ and $(-1, -1)$.

First, consider the critical point $(1, 1)$. Using substitutions $x = u + 1$ and $y = v + 1$, we obtain the almost linear system

$$\frac{du}{dt} = 1 - (u + 1)(v + 1) = -u - v - uv,$$
$$\frac{dv}{dt} = (u + 1) - (v + 1)^3 = u - 3v - 3v^2 - v^3.$$

The characteristic equation for this system is $r^2 + 4r + 4 = 0$, which has a double negative root $r = -2$. Hence, $(1, 1)$ is either an improper or proper node, or an asymptotically stable spiral point.

Next, we consider the critical point $(-1, -1)$. Using the change of variables $x = u - 1$ and $y = v - 1$, we obtain the almost linear system

$$\frac{du}{dt} = 1 - (u - 1)(v - 1) = u + v - uv,$$
$$\frac{dv}{dt} = (u - 1) - (v - 1)^3 = u - 3v + 3v^2 - v^3.$$

The characteristic equation for this system is $r^2 + 2r - 4 = 0$, which has roots $r = -1 \pm \sqrt{5}$. Since these roots are real and are of opposite signs, $(-1, -1)$ is an unstable saddle point. A phase plane diagram is given in Fig. B.59 in the answers of the text.

21. <u>Case 1: $h = 0$.</u> The critical points for this system are solutions to the system of equations

$$x(1 - 4x - y) = 0,$$
$$y(1 - 2y - 5x) = 0.$$

Copyright © 2012 Pearson Education, Inc. Publishing as Addison-Wesley.

To solve this system, we use the first equation and assume first that $x = 0$. Then the second equation yields $y(1 - 2y) = 0$. So, $y = 0$ or $y = 1/2$. Thus, $(0, 0)$ and $(0, 1/2)$ are critical points.

Assuming next that $1 - 4x - y = 0$, which gives $x = (1 - y)/4$, from the second equation we conclude that $y[1 - 2y - 5(1 - y)/4] = 0$. Solving yields $y = 0$ or $y = -1/3$. Therefore, $(1/4, 0)$ and $(1/3, -1/3)$ are two other critical points.

At the critical point $(0, 0)$, the characteristic equation is $r^2 - 2r + 1 = 0$, which has a double positive root $r = 1$. Hence $(0, 0)$ is an improper or proper node, or an unstable spiral point. From Fig. B.61 in the text, we see that $(0, 0)$ is an improper node.

Next, we consider the critical point $(0, 1/2)$. Using the change of variables (moving this critical point to the origin) $x = u$ and $y = v + 1/2$, we obtain an almost linear system

$$\frac{du}{dt} = u\left(1 - 4u - v - \frac{1}{2}\right) = \frac{1}{2}u - 4u^2 - uv,$$

$$\frac{dv}{dt} = \left(v + \frac{1}{2}\right)(1 - 2v - 1 - 5u) = -\frac{5}{2}u - v - 2v^2 - 5uv.$$

The characteristic equation for this system is $r^2 + (1/2)r - (1/2) = 0$, which has roots $r = 1/2$ and $r = -1$. Since these roots are real and have different signs, $(0, 1/2)$ is an unstable saddle point.

Now, consider the critical point $(1/4, 0)$. Using the change of variables $x = u + 1/4$ and $y = v$, we obtain an almost linear system

$$\frac{du}{dt} = \left(u + \frac{1}{4}\right)(1 - 4u - 1 - v) = -u - \frac{1}{4}v - 4u^2 - uv,$$

$$\frac{dv}{dt} = v\left(1 - 2v - 5u - \frac{5}{2}\right) = -\frac{1}{4}v - 2v^2 - 5uv.$$

The characteristic equation for this system is $r^2 + (5/4)r + (1/4) = 0$, which has roots $r = -1/4$ and $r = -1$. Since these roots are distinct and negative, $(1/4, 0)$ is an improper node, which is asymptotically stable.

At the critical point $(1/3, -1/3)$, we use the change of variables $x = u + 1/3$ and $y = v - 1/3$ to obtain an almost linear system

$$\frac{du}{dt} = \left(u + \frac{1}{3}\right)\left(1 - 4u - \frac{4}{3} - v + \frac{1}{3}\right) = -\frac{4}{3}u - \frac{1}{3}v - 4u^2 - uv,$$

$$\frac{dv}{dt} = \left(v - \frac{1}{3}\right)\left(1 - 2v + \frac{2}{3} - 5u - \frac{5}{3}\right) = \frac{5}{3}u + \frac{2}{3}v - 2v^2 - 5uv.$$

    Copyright © 2012 Pearson Education, Inc. Publishing as Addison-Wesley.

The characteristic equation for this system is $r^2 + (2/3)r - (1/3) = 0$ with roots $r = 1/3$ and $r = -1$. Again, since these roots are real and of different signs, $(1/3, -1/3)$ is an unstable saddle point, but not of interest since $y < 0$. Species x survives while species y dies off. A phase plane diagram is given in Fig. B.61 in the answers of the text.

Case 2: $h = 1/32$. The critical points for this system are solutions to the pair of equations

$$x(1 - 4x - y) - \frac{1}{32} = 0,$$
$$y(1 - 2y - 5x) = 0.$$

To solve this system, we first set $y = 0$ and solve $x(1 - 4x) - 1/32 = 0$, which has solutions $x = (2 \pm \sqrt{2})/16$.

If $y \neq 0$, we have $1 - 2y - 5x = 0$. So $y = (1/2) - (5/2)x$. Substituting, into the first equation, we obtain

$$x\left[1 - 4x - \left(\frac{1}{2} - \frac{5}{2}x\right)\right] - \frac{1}{32} = 0.$$

Simplifying yields

$$-\frac{3}{2}x^2 + \frac{1}{2}x - \frac{1}{32} = 0,$$

which has the solutions $x = 1/4$ and $x = 1/12$. When $x = 1/4$, we have

$$y = \frac{1}{2} - \frac{5}{2}\left(\frac{1}{4}\right) = -\frac{1}{8},$$

and $x = 1/12$ gives

$$y = \frac{1}{2} - \frac{5}{2}\left(\frac{1}{12}\right) = \frac{7}{24}.$$

Hence, the critical points are

$$\left(\frac{2 - \sqrt{2}}{16}, 0\right), \quad \left(\frac{2 + \sqrt{2}}{16}, 0\right), \quad \left(\frac{1}{4}, -\frac{1}{8}\right), \quad \text{and} \quad \left(\frac{1}{12}, \frac{7}{24}\right).$$

At the critical point $\left(\dfrac{2 - \sqrt{2}}{16}, 0\right)$, we use the change of variables $x = u + \dfrac{2 - \sqrt{2}}{16}$ and $y = v$ to obtain an almost linear system

$$\frac{du}{dt} = \left(u + \frac{2 - \sqrt{2}}{16}\right)\left(1 - 4u - \frac{2 - \sqrt{2}}{4} - v\right) - \frac{1}{32} = \frac{\sqrt{2}}{2}u - \frac{2 - \sqrt{2}}{16}v - 4u^2 - uv,$$

$$\frac{dv}{dt} = v\left(1 - 2v - 5u - \frac{10 - 5\sqrt{2}}{16}\right) = \frac{6 + 5\sqrt{2}}{16}v - 2v^2 - 5uv.$$

Copyright © 2012 Pearson Education, Inc. Publishing as Addison-Wesley.

The characteristic equation for this system is

$$\left(r - \frac{\sqrt{2}}{2}\right)\left(r - \frac{6 + 5\sqrt{2}}{16}\right) = 0,$$

which has distinct positive roots. Hence, $\left(\dfrac{2 - \sqrt{2}}{16}, 0\right)$ is an unstable improper node.

Now, consider the critical point $\left(\dfrac{2 + \sqrt{2}}{16}, 0\right)$, where we use the change of variables

$y = v$ and $x = u + \dfrac{2 + \sqrt{2}}{16}$ to obtain an almost linear system

$$\frac{du}{dt} = \left(u + \frac{2 + \sqrt{2}}{16}\right)\left(1 - 4u - \frac{2 + \sqrt{2}}{4} - v\right) - \frac{1}{32} = -\frac{\sqrt{2}}{2}u - \frac{2 + \sqrt{2}}{16}v - 4u^2 - uv,$$

$$\frac{dv}{dt} = v\left(1 - 2v - 5u - \frac{10 + 5\sqrt{2}}{16}\right) = \frac{6 - 5\sqrt{2}}{16}v - 2v^2 - 5uv.$$

The characteristic equation for this system is

$$\left(r + \frac{\sqrt{2}}{2}\right)\left(r - \frac{6 - 5\sqrt{2}}{16}\right) = 0,$$

which has distinct negative roots. Hence, $\left(\dfrac{2 + \sqrt{2}}{16}, 0\right)$ is an asymptotically stable improper node.

For the critical point $(1/12, 7/24)$, the change of variables $x = u + 1/12$ and $y = v + 7/24$ leads to an almost linear system

$$\frac{du}{dt} = \left(u + \frac{1}{12}\right)\left(1 - 4u - \frac{1}{3} - v - \frac{7}{24}\right) - \frac{1}{32} = \frac{1}{24}u - \frac{1}{12}v - 4u^2 - uv,$$

$$\frac{dv}{dt} = \left(v + \frac{7}{24}\right)\left(1 - 2v - \frac{7}{12} - 5u - \frac{5}{12}\right) = -\frac{35}{24}u - \frac{7}{12}v - 2v^2 - 5uv.$$

The characteristic equation for this system is $r^2 + (13/24)r - (7/48) = 0$, which has roots $r = (-13 \pm \sqrt{505})/48$. Since these roots have opposite signs, $(1/12, 7/24)$ is an unstable saddle point.

Finally, for the critical point is $(1/4, -1/8)$, the change of variables $x = u + 1/4$ and $y = v - 1/8$ leads to an almost linear system

$$\frac{du}{dt} = \left(u + \frac{1}{4}\right)\left(1 - 4u - 1 - v + \frac{1}{8}\right) - \frac{1}{32} = -\frac{7}{8}u - \frac{1}{4}v - 4u^2 - uv,$$

 Copyright © 2012 Pearson Education, Inc. Publishing as Addison-Wesley.

$$\frac{dv}{dt} = \left(v - \frac{1}{8}\right)\left(1 - 2v + \frac{1}{4} - 5u - \frac{5}{4}\right) = \frac{5}{8}u + \frac{1}{4}v - 2v^2 - 5uv.$$

The characteristic equation for this system is $r^2 + (5/8)r - (1/16) = 0$, which has roots $r = (-5 \pm \sqrt{41})/16$. Since these roots have opposite signs, $(1/4, -1/8)$ is an unstable saddle point. But, since $y < 0$, this point is not of interest.

Hence, this is a competitive exclusion: one species survives while the other dies off. A phase plane diagram is given in Fig. B.62 in the answers of the text.

Case 3: $h = 5/32$. The critical points for this system are solutions to

$$x(1 - 4x - y) - \frac{5}{32} = 0,$$
$$y(1 - 2y - 5x) = 0.$$

To solve this system, we first set $y = 0$ and solve

$$x(1 - 4x) - \frac{5}{32} = 0,$$

which has complex solutions. If $y \neq 0$, then we have

$$1 - 2y - 5x = 0 \qquad \Rightarrow \qquad y = \frac{1}{2} - \frac{5}{2}x.$$

Substituting this expression into the first equation, we obtain

$$x\left[1 - 4x - \left(\frac{1}{2} - \frac{5}{2}x\right)\right] - \frac{5}{32} = 0.$$

Simplifying yields

$$-\frac{3}{2}x^2 + \frac{1}{2}x - \frac{5}{32} = 0,$$

which also has only complex solutions. Hence, there are no critical points. The phase plane diagram shows that species y survives while species x dies off. A phase plane diagram is given in Fig. B.63 in the answers of the text.

EXERCISES 12.4:   Energy Methods

3. Here $g(x) = x^2/(x - 1) = x + 1 + 1/(x - 1)$. By integrating $g(x)$, we obtain the potential function

$$G(x) = \frac{x^2}{2} + x + \ln|x - 1| + C,$$

Copyright © 2012 Pearson Education, Inc. Publishing as Addison-Wesley.

and so

$$E(x, v) = \frac{v^2}{2} + \frac{x^2}{2} + x + \ln|x - 1| + C.$$

Since $E(0,0) = 0$ implies that $C = 0$, we have

$$E(x, v) = \frac{v^2}{2} + \frac{x^2}{2} + x + \ln|x - 1|.$$

Now, we are interested in $E(x, v)$ near the origin, so we can assume that $|x - 1| = 1 - x$ (for x near 0, $x - 1 < 0$). Therefore,

$$E(x, v) = \frac{v^2}{2} + \frac{x^2}{2} + x + \ln(1 - x).$$

9. Here, we have $g(x) = 2x^2 + x - 1$ and, hence, the potential function is

$$G(x) = \frac{2x^3}{3} + \frac{x^2}{2} - x.$$

The local maxima and minima of $G(x)$ occur when $G'(x) = g(x) = 2x^2 + x - 1 = 0$. Thus, the phase plane diagram has critical points at $(-1, 0)$ and $(1/2, 0)$. Since $G(x)$ has a strict local minimum at $x = 1/2$, the critical point $(1/2, 0)$ is a center. Furthermore, since $x = -1$ is strict local maximum, the critical point $(-1, 0)$ is a saddle point. The potential plane and the phase plane diagram are depicted in Fig. B.65 in the answers of the text.

11. Here, we have $g(x) = x/(x - 2) = 1 + 2/(x - 2)$, so that the potential function is

$$G(x) = x + 2\ln|x - 2| = x + 2\ln(2 - x)$$

for x near zero. Local maxima and minima of $G(x)$ occur when

$$G'(x) = g(x) = \frac{x}{x - 2} = 0.$$

Thus, the phase plane diagram has a critical point $(0, 0)$. Further, we note that $x = 2$ is not in the domain of either $g(x)$ or $G(x)$. Now, $G(x)$ has a strict local maximum at $x = 0$. Hence, the critical point $(0, 0)$ is a saddle point. A sketch of the potential plane along with the phase plane diagram for $x < 2$ is given in Fig. B.66 in the answers of the text.

 Copyright © 2012 Pearson Education, Inc. Publishing as Addison-Wesley.

13. We first observe that $vh(x,v) = v^2 > 0$ for $v \neq 0$. Hence, the energy is continually decreasing along a trajectory. The level curves for the energy function, which is

$$E(x,v) = \frac{v^2}{2} + \frac{x^2}{2} - \frac{x^4}{4},$$

are just the integral curves in Example 2(a) shown in Fig. 12.22 in the text. The critical points for this damped system are the same as those in Example 2(a) and, moreover, they are of the same type. The resulting phase plane is given in Fig. B.67 in the answers of the text.

EXERCISES 12.5:   Lyapunov's Direct Method

3. We compute $\dot{V}(x,y)$ for $V(x,y) = x^2 + y^2$.

$$
\begin{aligned}
\dot{V}(x,y) &= V_x(x,y)f(x,y) + V_y(x,y)g(x,y) \\
&= 2x\left(y^2 + xy^2 - x^3\right) + 2y\left(-xy + x^2y - y^3\right) \\
&= 4x^2y^2 - 2x^4 - 2y^4 = -2\left(x^2 - y^2\right)^2.
\end{aligned}
$$

According to Theorem 3, since $\dot{V}$ is negative semidefinite, V is positive definite function, and $(0,0)$, the origin, is an isolated critical point of the system which is stable.

5. The origin is an isolated critical point for the given system. Using the hint, we compute $\dot{V}(x,y)$, where $V(x,y) = x^2 - y^2$. Differentiating, we obtain

$$
\begin{aligned}
\dot{V}(x,y) &= V_x(x,y)f(x,y) + V_y(x,y)g(x,y) \\
&= 2x\left(2x^3\right) - 2y\left(2x^2y - y^3\right) = 4x^4 - 4x^2y^2 + 2y^2 = 2x^4 + \left(x^2 - y^2\right)^2,
\end{aligned}
$$

which is positive definite. Now, $V(0,0) = 0$, and so, in every disk centered at the origin, V is positive at some point (namely, at those points where $|x| > |y|$). Therefore, by Theorem 4, the origin is unstable.

7. We compute $\dot{V}(x,y)$ for $V(x,y) = ax^4 + by^2$.

$$
\begin{aligned}
\dot{V}(x,y) &= V_x(x,y)f(x,y) + V_y(x,y)g(x,y) \\
&= 4ax^3\left(2y - x^3\right) + 2by\left(-x^3 - y^5\right) \\
&= 8ax^3y - 4ax^6 - 2bx^3y - 2by^6.
\end{aligned}
$$

Copyright © 2012 Pearson Education, Inc. Publishing as Addison-Wesley.

To eliminate the x^3y term, we let $a = 1$ and $b = 4$, then

$$\dot{V}(x, y) = -4x^6 - 8y^6,$$

and we conclude that $\dot{V}$ is negative definite. Since V is positive definite and the origin is an isolated critical point, according to Theorem 3, the origin is asymptotically stable.

11. Here, we set

$$y = \frac{dx}{dt} \qquad \Rightarrow \qquad \frac{dy}{dt} = \frac{d^2x}{dt^2}.$$

Thus, we obtain the system

$$\frac{dx}{dt} = y,$$

$$\frac{dy}{dt} = -\left(1 - y^2\right)y - x.$$

Clearly, this system has the trivial solution $x \equiv 0$, $y \equiv 0$. To apply Lyapunov's direct method, we try a positive definite function $V(x, y) = ax^2 + by^2$ and compute $\dot{V}$.

$$\begin{aligned}
\dot{V}(x, y) &= V_x(x, y)f(x, y) + V_y(x, y)g(x, y) \\
&= 2ax(y) + 2by\left[-\left(1 - y^2\right)y - x\right] = 2axy - 2by^2 + 2by^4 - 2bxy.
\end{aligned}$$

To eliminate the terms involving xy, we choose $a = b = 1$, so that

$$\dot{V}(x, y) = -2y^2 + 2y^4 = -2y^2\left(1 - y^2\right).$$

Hence, $\dot{V}$ is negative semidefinite for $|y| < 1$ and, by Theorem 3, the origin is stable.

EXERCISES 12.6:   Limit Cycles and Periodic Solutions

5. We compute $r(dr/dt)$.

$$\begin{aligned}
r\frac{dr}{dt} &= x\frac{dx}{dt} + y\frac{dy}{dt} \\
&= x\left[x - y + x\left(r^3 - 4r^2 + 5r - 3\right)\right] + y\left[x + y + y\left(r^3 - 4r^2 + 5r - 3\right)\right] \\
&= x^2 - xy + x^2\left(r^3 - 4r^2 + 5r - 3\right) + xy + y^2 + y^2\left(r^3 - 4r^2 + 5r - 3\right) \\
&= r^2 + r^2\left(r^3 - 4r^2 + 5r - 3\right) = r^2\left(r^3 - 4r^2 + 5r - 2\right).
\end{aligned}$$

Hence,

$$\frac{dr}{dt} = r\left(r^3 - 4r^2 + 5r - 2\right) = r(r - 1)^2(r - 2).$$

 Copyright © 2012 Pearson Education, Inc. Publishing as Addison-Wesley.

Now, $dr/dt = 0$ when $r = 0, 1, 2$. The critical point is represented by $r = 0$ and, when $r = 1$ or $r = 2$, we have a limit cycle of radius 1 and 2, respectively. When r lies in $(0, 1)$, we have $dr/dt < 0$, so a trajectory in this region spirals into the origin. Therefore, the origin is an asymptotically stable spiral point. Now, when r lies in $(1, 2)$, we again have $dr/dt < 0$, so that a trajectory in this region spirals into the limit cycle $r = 1$. This tells us that $r = 1$ is a semistable limit cycle. Finally, when $r > 2$, $dr/dt > 0$, so that a trajectory in this region spirals away from the limit cycle $r = 2$. Hence, $r = 2$ is an unstable limit cycle.

To find directions of the trajectories, we compute $r^2(d\theta/dt)$.

$$
\begin{aligned}
r^2 \frac{d\theta}{dt} &= x\frac{dy}{dt} - y\frac{dx}{dt} \\
&= x\left[x + y + y\left(r^3 - 4r^2 + 5r - 3\right)\right] - y\left[x - y + x\left(r^3 - 4r^2 + 5r - 3\right)\right] \\
&= x^2 + xy + xy\left(r^3 - 4r^2 + 5r - 3\right) - xy + y^2 - xy\left(r^3 - 4r^2 + 5r - 3\right) \\
&= x^2 + y^2 = r^2 .
\end{aligned}
$$

Hence, $d\theta/dt = 1$, which tells us that the trajectories revolve counterclockwise about the origin. A phase plane diagram is given in Fig. B.74 in the answers of the text.

11. We compute $r(dr/dt)$.

$$
\begin{aligned}
r\frac{dr}{dt} &= x\frac{dx}{dt} + y\frac{dy}{dt} = x\left[y + x\sin\left(\frac{1}{r}\right)\right] + y\left[-x + y\sin\left(\frac{1}{r}\right)\right] \\
&= xy + x^2\sin\left(\frac{1}{r}\right) - xy + y^2\sin\left(\frac{1}{r}\right) = r^2\sin\left(\frac{1}{r}\right) .
\end{aligned}
$$

Hence,

$$
\frac{dr}{dt} = r\sin\left(\frac{1}{r}\right),
$$

and so $dr/dt = 0$ when $r = 1/(n\pi)$, $n = 1, 2, \dots$. Consequently, the origin $(r = 0)$ is not an isolated critical point. Observe that

$$
\frac{dr}{dt} > 0 \qquad \text{for} \qquad \frac{1}{(2n+1)\pi} < r < \frac{1}{2n\pi}
$$

and

$$
\frac{dr}{dt} < 0 \qquad \text{for} \qquad \frac{1}{2n\pi} < r < \frac{1}{(2n-1)\pi} .
$$

Thus, trajectories spiral into the limit cycles $r = 1/(2n\pi)$ and spiral away from the limit cycles $r = 1/[(2n+1)\pi]$. Therefore, $r = 1/(2n\pi)$ are stable limit cycles and

Copyright © 2012 Pearson Education, Inc. Publishing as Addison-Wesley.

Chapter 12

$r = 1/[(2n + 1)\pi]$ are unstable ones. To determine the direction of the spiral, we compute $r^2(d\theta/dt)$.

$$
\begin{aligned}
r^2 \frac{d\theta}{dt} &= x\frac{dy}{dt} - y\frac{dx}{dt} = x\left[-x + y\sin\left(\frac{1}{r}\right)\right] - y\left[y + x\sin\left(\frac{1}{r}\right)\right] \\
&= -x^2 + xy\sin\left(\frac{1}{r}\right) - y^2 - xy\sin\left(\frac{1}{r}\right) = -r^2 .
\end{aligned}
$$

Hence, $d\theta/dt = -1$, which tells us that the trajectories revolve clockwise. A phase plane diagram is given in Fig. B.77 in the answers of the text.

15. We compute $f_x + g_y$ in order to apply Theorem 6. Thus,

$$
f_x(x, y) + g_y(x, y) = \left(-8 + 3x^2\right) + \left(-7 + 3y^2\right) = 3\left(x^2 + y^2 - 5\right),
$$

which is negative in the given domain. Hence, by Theorem 6, there are no nonconstant periodic solutions in the disk $x^2 + y^2 < 5$.

19. It is easily seen that $(0, 0)$ is a critical point. However, it is not that that it is *the only* critical point for this system. Using the Lyapunov function $V(x, y) = 2x^2 + y^2$, we compute $\dot{V}(x, y)$.

$$
\begin{aligned}
\dot{V}(x, y) &= V_x(x, y)\frac{dx}{dt} + V_y(x, y)\frac{dy}{dt} \\
&= 4x\left(2x - y - 2x^3 - 3xy^2\right) + 2y\left(2x + 4y - 4y^3 - 2x^2y\right) \\
&= 8x^2 - 8x^4 - 16x^2y^2 + 8y^2 - 8y^4 = 8\left(x^2 + y^2\right) - 8\left(x^2 + y^2\right)^2 .
\end{aligned}
$$

Therefore, $\dot{V}(x, y) < 0$ for $x^2 + y^2 > 1$ and $\dot{V}(x, y) > 0$ for $x^2 + y^2 < 1$. Let E_1 denote the ellipse $2x^2 + y^2 = 1/2$, which lies inside of $x^2 + y^2 = 1$, and let E_2 be the ellipse $2x^2 + y^2 = 3$, which lies outside of $x^2 + y^2 = 1$. Now, $\dot{V}(x, y) > 0$ on E_1 and $\dot{V}(x, y) < 0$ on E_2. Hence, we let R be the region bounded by E_1 and E_2. Any trajectory, that enters R, remains in R. By Theorem 7, the system has a nonconstant periodic solution in R.

25. To apply Theorem 8, we check if all five conditions hold. Here, we have $g(x) = x$ and $f(x) = x^2(x^2 - 1)$. Clearly, $f(x)$ is even, hence condition (a) holds. Now,

$$
F(x) = \int_0^x s^2\left(s^2 - 1\right) ds = \frac{x^5}{5} - \frac{x^3}{3} .
$$

Thus, $F(x) < 0$ for $0 < x < \sqrt{5/3}$ and $F(x) > 0$ for $x > \sqrt{5/3}$. Therefore, condition (b) holds. Furthermore, condition (c) holds since $F(x) \to +\infty$, as $x \to +\infty$, monotonically

     Copyright © 2012 Pearson Education, Inc. Publishing as Addison-Wesley.

for $x > \sqrt{5/3}$. Since $g(x) = x$ is an odd function with $g(x) > 0$ for $x > 0$, condition (d) holds. Finally,

$$G(x) = \int_0^x s\,ds = \frac{x^2}{2}$$

and, clearly, $G(x) \to +\infty$, as $x \to +\infty$. Hence, condition (e) holds as well. It follows from Theorem 8, that the Lienard equation has a unique nonconstant periodic solution.

EXERCISES 12.7: Stability of Higher-Dimensional Systems

5. From the characteristic equation, $-(r-1)(r^2+1) = 0$, we find that the eigenvalues are $r = 1$ and $r = \pm i$. Since at least one eigenvalue, namely, $r = 1$, has a positive real part, the trivial solution is unstable.

9. The characteristic equation in this problem is $(r^2+1)(r^2+1) = 0$, which has double eigenvalues $r = \pm i$. Next, we determine the eigenspace for the eigenvalue $r = i$. Computing yields

$$
\begin{vmatrix}
i & -1 & -1 & 0 \\
1 & i & 0 & -1 \\
0 & 0 & i & -1 \\
0 & 0 & 1 & i
\end{vmatrix}
\quad \Rightarrow \quad
\begin{vmatrix}
1 & i & 0 & 0 \\
0 & 0 & 1 & 0 \\
0 & 0 & 0 & 1 \\
0 & 0 & 0 & 0
\end{vmatrix}.
$$

Hence, the eigenspace degenerates and (see Problem 8(c)) the zero solution is unstable. (It can be shown that the eigenspace for the eigenvalue $r = -i$ degenerates as well.)

13. To find the fundamental matrix for this system, we first recall the Taylor series for e^x, $\sin x$, and $\cos x$. These are

$$e^x = 1 + x + \frac{x^2}{2!} + \frac{x^3}{3!} + \cdots, \quad \sin x = x - \frac{x^3}{3!} + \frac{x^5}{5!} - \cdots, \quad \cos x = 1 - \frac{x^2}{2!} + \frac{x^4}{4!} - \cdots.$$

Hence,

$$\frac{dx_1}{dt} = \left(1 - x_1 + \frac{x_1^2}{2!} - \cdots\right) + \left(1 - \frac{x_2^2}{2!} + \cdots\right) - 2$$

$$= -x_1 + \left(\frac{x_1^2}{2!} - \cdots\right) + \left(-\frac{x_2^2}{2!} + \cdots\right),$$

$$\frac{dx_2}{dt} = -x_2 + \left(x_3 - \frac{x_3^3}{3!} + \cdots\right) = -x_2 - x_3 + \left(-\frac{x_3^3}{3!} + \cdots\right),$$

Copyright © 2012 Pearson Education, Inc. Publishing as Addison-Wesley.

$$\frac{dx_3}{dt} = 1 - \left[1 + (x_2 + x_3) + \frac{(x_2 + x_3)^2}{2!} + \cdots \right] = -x_2 - x_3 - \left[\frac{(x_2 + x_3)^2}{2!} + \cdots \right].$$

Thus,

$$\mathbf{A} = \begin{bmatrix} -1 & 0 & 0 \\ 0 & -1 & 1 \\ 0 & -1 & -1 \end{bmatrix}.$$

Calculating the eigenvalues yields

$$|\mathbf{A} - r\mathbf{I}| = \begin{vmatrix} -1 - r & 0 & 0 \\ 0 & -1 - r & 1 \\ 0 & -1 & -1 - r \end{vmatrix} = 0.$$

Hence, the characteristic equation is $-(r + 1)(r^2 + 2r + 2) = 0$, and so the eigenvalues are $r = -1$, $-1 \pm i$. Since real parts of eigenvalues are negative, the zero solution is asymptotically stable.

15. For critical points, we have a system

$$-x_1 + 1 = 0,$$
$$-2x_1 - x_2 + 2x_3 - 4 = 0,$$
$$-3x_1 - 2x_2 - x_3 + 1 = 0.$$

Solving, we find that the only solution is $(1, -2, 2)$. We now use the change of variables

$$x_1 = u + 1, \qquad x_2 = v - 2, \qquad x_3 = w + 2$$

to translate the critical point to the origin. Substituting, we obtain the following system

$$\frac{du}{dt} = -u,$$
$$\frac{dv}{dt} = -2u - v + 2w,$$
$$\frac{dw}{dt} = -3u - 2v - w.$$

Here, the matrix $\mathbf{A}$ is given by

$$\mathbf{A} = \begin{bmatrix} -1 & 0 & 0 \\ -2 & -1 & 2 \\ -3 & -2 & -1 \end{bmatrix}.$$

Writing the characteristic equation, $-(r + 1)(r^2 + 2r + 5) = 0$, we conclude that the eigenvalues are $r = -1$, $-1 \pm 2i$. Since each eigenvalue has a negative real part, the critical point $(1, -2, 2)$ is asymptotically stable.

 Copyright © 2012 Pearson Education, Inc. Publishing as Addison-Wesley.

CHAPTER 13: Existence and Uniqueness Theory

EXERCISES 13.1: Introduction: Successive Approximations

1. In this problem, $x_0 = 1$, $y_0 = y(x_0) = 4$, and $f(x, y) = x^2 - y$. Thus, applying the formula (3) in the text yields

$$y(x) = y_0 + \int_{x_0}^{x} f(t, y(t))\, dt = 4 + \int_{1}^{x} \left[t^2 - y(t) \right] dt. \tag{13.1}$$

Since

$$\int_{1}^{x} t^2\, dt = \left. \frac{t^3}{3} \right|_{1}^{x} = \frac{x^3}{3} - \frac{1}{3},$$

the answer (13.1) becomes

$$y(x) = 4 + \int_{1}^{x} t^2\, dt - \int_{1}^{x} y(t)\, dt = 4 + \frac{x^3}{3} - \frac{1}{3} - \int_{1}^{x} y(t)\, dt = \frac{11}{3} + \frac{x^3}{3} - \int_{1}^{x} y(t)\, dt.$$

3. The initial conditions are $x_0 = 1$ and $y_0 = -3$, and $f(x, y) = (y - x)^2$. Therefore,

$$y(x) = y_0 + \int_{x_0}^{x} f(t, y(t))\, dt = -3 + \int_{1}^{x} \left[y^2(t) - t \right]^2 dt.$$

Using the linear property of integrals, we find that

$$\int_{1}^{x} \left[y^2(t) - t \right]^2 dt = \int_{1}^{x} \left[y^2(t) - 2ty(t) + t^2 \right] dt$$

$$= \int_{1}^{x} y^2(t)\, dt - 2\int_{1}^{x} ty(t)\, dt + \int_{1}^{x} t^2\, dt = \int_{1}^{x} \left[y^2(t) - 2ty(t) \right] dt + \frac{x^3}{3} - \frac{1}{3}.$$

Hence, we have

$$y(x) = \frac{x^3}{3} - \frac{10}{3} + \int_{1}^{x} y^2(t)\, dt - 2\int_{1}^{x} ty(t)\, dt. \tag{13.2}$$

Copyright © 2012 Pearson Education, Inc. Publishing as Addison-Wesley.

Note that we can rewrite the last integral using integration by parts in terms of integrals of the function $y(x)$ alone. Namely,

$$\int_1^x ty(t)\,dt = t\int_1^t y(s)\,ds\Big|_{t=1}^{t=x} - \int_1^x\int_1^t y(s)\,ds\,dt = x\int_1^x y(s)\,ds - \int_1^x\int_1^t y(s)\,ds\,dt$$

Thus, another form of the answer (13.2) is

$$y(x) = \frac{x^3}{3} - \frac{10}{3} + \int_1^x y^2(t)\,dt + x\int_1^x y(t)\,dt - \int_1^x\int_1^t y(s)\,ds\,dt.$$

5. In this problem, we have

$$g(x) = \frac{1}{2}\left(x + \frac{3}{x}\right).$$

Thus, the recurrence formula (7) in the text yields

$$x_{n+1} = g(x_n) = \frac{1}{2}\left(x_n + \frac{3}{x_n}\right), \qquad n = 0, 1, \dots.$$

With $x_0 = 3$ as an initial approximation, we compute

$$x_1 = \frac{1}{2}\left(x_0 + \frac{3}{x_0}\right) = \frac{1}{2}\left(3 + \frac{3}{3}\right) = 2.0, \qquad x_2 = \frac{1}{2}\left(x_1 + \frac{3}{x_1}\right) = \frac{1}{2}\left(2 + \frac{3}{2}\right) = 1.75,$$

and so on. The results of these computations are given in Table **13–A** on page 687.

Since $x_4 - x_5 < 10^{-8}$, we can take $x \approx 1.7320508$.

7. Since $g(x) = 1/(x^2 + 4)$, we have the recurrence formula

$$x_{n+1} = g(x_n) = \frac{1}{x_n^2 + 4}, \qquad n = 0, 1, \dots$$

with an initial approximation $x_0 = 0.5$. Hence,

$$x_1 = \frac{1}{x_0^2 + 4} = \frac{1}{(0.5)^2 + 4} = \frac{4}{17} \approx 0.23529412,$$

$$x_2 = \frac{1}{x_1^2 + 4} \approx \frac{1}{(0.23529412)^2 + 4} \approx 0.24658703, \quad \text{etc.}$$

Further approximations are listed in Table **13–B** on page 687.

Since the error $x_6 - x_7 < 10^{-8}$, we take $x \approx 0.2462662$.

 Copyright © 2012 Pearson Education, Inc. Publishing as Addison-Wesley.

9. To start the method of successive substitutions, we observe that

$$g(x) = \left(\frac{5-x}{3}\right)^{1/4}.$$

Therefore, according to the equation (7) in the text, we can find the next approximation from the previous one by using the recurrence relation

$$x_{n+1} = g(x_n) = \left(\frac{5-x_n}{3}\right)^{1/4}.$$

We start the procedure at $x_0 = 1$. Thus, we obtain

$$x_1 = \left(\frac{5-x_0}{3}\right)^{1/4} = \left(\frac{5-1}{3}\right)^{1/4} = \left(\frac{4}{3}\right)^{1/4} \approx 1.07456993,$$

$$x_2 = \left(\frac{5-x_1}{3}\right)^{1/4} \approx \left(\frac{5-1.07456993}{3}\right)^{1/4} \approx 1.06952637.$$

By continuing this process, we fill in Table **13–C** on page 687.

Noticing that $x_7 - x_6 \approx 10^{-8}$, we stop after seven steps and take $x \approx 1.0698479$.

11. First, we derive an integral equation corresponding to the given initial value problem. We have $f(x, y) = -y$, $x_0 = 0$, $y_0 = y(0) = 2$, and so the formula (3) yields

$$y(x) = 2 + \int_0^x [-y(t)]\, dt = 2 - \int_0^x y(t)\, dt.$$

Thus, Picard's recurrence formula (15) becomes

$$y_{n+1}(x) = 2 - \int_0^x y_n(t)\, dt, \qquad n = 0, 1, \ldots .$$

Starting with $y_0(x) \equiv y_0 = 2$, we compute

$$y_1(x) = 2 - \int_0^x y_0(t)\, dt = 2 - \int_0^x 2\, dt = 2 - 2t\, \Big|_{t=0}^{t=x} = 2 - 2x,$$

$$y_2(x) = 2 - \int_0^x y_1(t)\, dt = 2 - \int_0^x (2 - 2t)\, dt = 2 + (t-1)^2 \Big|_{t=0}^{t=x} = 2 - 2x + x^2.$$

13. In this problem, $f(x, y) = 3x^2$, $x_0 = 1$, $y_0 = y(1) = 2$, and so Picard's iterations to the solution of the given initial value problem are given by

$$y_{n+1}(x) = 2 + \int_1^x \left(3t^2\right) dt = 2 + t^3\, \Big|_{t=1}^{t=x} = x^3 + 1.$$

Copyright © 2012 Pearson Education, Inc. Publishing as Addison-Wesley.

Since the right-hand side does not depend on n, the sequence of iterations $y_n(x)$, where $n = 1, 2, \ldots,$ is a constant sequence. That is,

$$y_n(x) = x^3 + 1 \qquad \text{for any } n \geq 1.$$

In particular, $y_1(x) = y_2(x) = x^3 + 1$.

(In this connection, note the following. If it happens that one of the iterations, say, $y_k(x)$, obtained via (15) matches the exact solution to the integral equation (3), then all the subsequent iterations will give the same function $y_k(x)$. In other words, the sequence of iterations will become a constant sequence starting from its kth term. In the given problem, the first application of (15) gives the *exact* solution, $y = x^3 + 1$, to the original initial value problem and, hence, to the corresponding integral equation (3).)

15. We first write this differential equation as an integral equation. Integrating both sides from $x_0 = 0$ to x and using the fact that $y(0) = 0$, we obtain

$$y(x) - y(0) = \int_0^x \left[y(t) - e^t \right] dt \qquad \Rightarrow \qquad y(x) = \int_0^x \left[y(t) - e^t \right] dt .$$

Hence, by the equation (15) in the text, the Picard iterations are given by

$$y_{n+1}(x) = \int_0^x \left[y_n(t) - e^t \right] dt .$$

Thus, starting with $y_0(x) \equiv y_0 = 0$, we calculate

$$y_1(x) = \int_0^x \left[y_0(t) - e^t \right] dt = -\int_0^x e^t dt = 1 - e^x ,$$

$$y_2(x) = \int_0^x \left[y_1(t) - e^t \right] dt = \int_0^x \left(1 - 2e^t \right) dt = 1 - e^x = 2 + x - 2e^x .$$

17. First of all, remark that the function $f(x, y(x))$ in the integral equation (3), that is,

$$y(x) = y_0 + \int_{x_0}^x f(t, y(t)) \, dt ,$$

is a continuous function as the composition of $f(x, y)$ and $y(x)$, which are both continuous by our assumption. Next, if $y(x)$ satisfies (3), then

$$y(x_0) = y_0 + \int_{x_0}^{x_0} f(t, y(t)) \, dt = y_0 ,$$

 Copyright © 2012 Pearson Education, Inc. Publishing as Addison-Wesley.

because the integral term is zero as a definite integral of a continuous function with equal limits of integration. Therefore, $y(x)$ satisfies the initial condition in (1).

Differentiating $y(x)$ yields

$$y'(x) = \left(y_0 + \int_{x_0}^{x} f(t, y(t))\, dt \right)' = f(t, y(t))\, \big|_{t=x} = f(x, y(x)),$$

where we have used the fundamental theorem of calculus to find the derivative of the integral with variable upper bound. So, $y(x)$ satisfies the differential equation in (1).

19. The graphs of the functions $y = (x^2 + 1)/2$ and $y = x$ on the same coordinate plane are sketched in Fig. **13–A**, page 688. By examining this figure, we see that these two graphs intersect only at $(1, 1)$. We can find this point by solving the equation $x = (x^2 + 1)/2$ for x. Thus, we have

$$2x = x^2 + 1 \quad \Rightarrow \quad x^2 - 2x + 1 = 0 \quad \Rightarrow \quad (x - 1)^2 = 0 \quad \Rightarrow \quad x = 1.$$

Since $y = x$, the only point of intersection of the curves is $(1, 1)$.

To approximate the solution to the equation $x = (x^2+1)/2$ using the method of successive substitutions, we use the recurrence relation $x_{n+1} = (x_n^2 + 1)/2$.

  (i) Starting this method at $x_0 = 0$, we obtain the approximations given in Table **13–D**, page 687. These approximations do appear to be approaching the solution $x = 1$.

  (ii) However, if we start the process at the point $x = 2$, we obtain the approximations given in Table **13–E**, page 688. We observe that these approximations are getting larger and so do not seem to converge.

These conclusions can be confirmed by examining the pictorial representation for the method of successive substitutions given in Fig. **13–A**. Substituting $x_0 = 0$ into the function $(x^2 + 1)/2$, we find that $P_0 = (0, 0.5)$. Then, by moving horizontally from the point P_0 toward the line $y = x$, we find the point $Q_0 = (0.5, 0.5)$. Next, by moving vertically from the point Q_0 toward the curve $y = (x^2 + 1)/2$, we find that P_1 has coordinates $(0.5, 0.625)$. Continuing this process leads us slowly, in a step fashion, to the point $(1, 1)$. However, starting this process at $x_0 = 2$, we observe that this method moves us through larger and larger steps away from the point of intersection $(1, 1)$.

Copyright © 2012 Pearson Education, Inc. Publishing as Addison-Wesley.

Note that, for this equation, the direction of motion in the method of successive substitutions is to the right. This is because the term $(x_n^2 + 1)/2$ in the recurrence relation is increasing for $x > 0$. Thus, *the sequence of approximations $\{x_n\}$ is an increasing sequence*. Starting at any nonnegative point $x_0 < 1$, we will approach the fixed point at $x = 1$, but starting at any point $x_0 > 1$ yields ever increasing values for approximations and, therefore, moves us away from the fixed point.

EXERCISES 13.2:   Picard's Existence and Uniqueness Theorem

1. In order to determine whether this sequence of functions converges uniformly, we compute $\|y_n - y\|$. Since

$$y_n(x) - y(x) = \left(1 - \frac{x}{n}\right) - 1 = -\frac{x}{n},$$

we have

$$\|y_n - y\| = \max_{x \in [-1,1]} |y_n(x) - y(x)| = \max_{x \in [-1,1]} \frac{|x|}{n} = \frac{1}{n}.$$

Thus,

$$\lim_{n \to \infty} \|y_n - y\| = \lim_{n \to \infty} \frac{1}{n} = 0$$

and so $\{y_n(x)\}$ converges to $y(x)$ uniformly on $[-1, 1]$.

3. In order to determine whether or not this sequence of functions converges uniformly, we need to compute $\|y_n - y\|$.

$$\|y_n - y\| = \|y_n\| = \max_{x \in [0,1]} |y_n| = \max_{x \in [0,1]} \left|\frac{nx}{1 + n^2 x^2}\right| = \max_{x \in [0,1]} \frac{nx}{1 + n^2 x^2}.$$

We now use calculus methods to find this maximum value. Thus, we differentiate the function $y_n(x)$ and obtain

$$y_n'(x) = \frac{n(1 - n^2 x^2)}{(1 + n^2 x^2)^2}.$$

Setting $y_n'(x) = 0$ and solving yields

$$n\left(1 - n^2 x^2\right) = 0 \qquad \Rightarrow \qquad n^2 x^2 = 1 \qquad \Rightarrow \qquad x = \pm \frac{1}{n}.$$

Since we are interested in the values of x on the interval $[0, 1]$, we examine the critical point $x = 1/n$ only. By the first derivative test, we conclude that the function $y_n(x)$ has a local maximum value at the point $x = 1/n$, which is

$$y_n\left(\frac{1}{n}\right) = \frac{n\left(n^{-1}\right)}{1 + n^2 \left(n^{-1}\right)^2} = \frac{1}{2}.$$

 Copyright © 2012 Pearson Education, Inc. Publishing as Addison-Wesley.

Computing

$$y_n(0) = \frac{n(0)}{1 + n^2(0)^2} = 0,$$

$$y_n(1) = \frac{n(1)}{1 + n^2(1)^2} = \frac{n}{1 + n^2} < \frac{1}{n} \le \frac{1}{2} \quad \text{for} \quad n \ge 2,$$

we conclude that

$$\max_{x \in [0,1]} y_n(x) = \frac{1}{2}, \qquad n \ge 2.$$

Therefore,

$$\lim_{n \to \infty} \|y_n - y\| = \lim_{n \to \infty} \frac{1}{2} = \frac{1}{2} \ne 0,$$

and so the given sequence of functions does *not* converge uniformly to the function $y(x) \equiv 0$ on the interval $[0, 1]$.

However, this sequence of functions converges pointwise to the function $y(x) \equiv 0$ on the interval $[0, 1]$. Indeed, for any fixed $x \in (0, 1]$, we have

$$\lim_{n \to \infty} [y_n(x) - y(x)] = \lim_{n \to \infty} \frac{nx}{1 + n^2 x^2} = \lim_{n \to \infty} \frac{n}{2n^2 x} = \lim_{n \to \infty} \frac{1}{2nx} = 0,$$

where we have used the L'Hospital's rule. For $x = 0$, we observe that

$$\lim_{n \to \infty} [y_n(0) - y(0)] = \lim_{n \to \infty} \frac{n(0)}{1 + n^2(0)} = \lim_{n \to \infty} 0 = 0.$$

Thus, we have pointwise, but *not* uniform, convergence. See Fig. **13–B**, page 689 for the graphs of functions $y_1(x)$, $y_{10}(x)$, $y_{30}(x)$, and $y_{00}(x)$.

5. We know (see the equation (3), Section 8.2) that for all x, satisfying $|x| < 1$, the geometric series $\sum_{k=0}^{\infty} x^k$ converges to the function $f(x) = 1/(1 - x)$. Thus, for all x in $[0, 1/2]$,

$$\frac{1}{1 - x} = 1 + x + x^2 + \cdots + x^k + \cdots = \sum_{k=0}^{\infty} x^k.$$

We see that

$$\|y_n - y\| = \max_{x \in [0,1/2]} |y_n(x) - y(x)|$$

$$= \max_{x \in [0,1/2]} \left| \sum_{k=0}^{n} x^k - \sum_{k=0}^{\infty} x^k \right| = \max_{x \in [0,1/2]} \sum_{k=n+1}^{\infty} x^k = \sum_{k=n+1}^{\infty} \left(\frac{1}{2} \right)^k.$$

Since

$$\sum_{k=n+1}^{\infty} \left(\frac{1}{2} \right)^k = \left(\frac{1}{2} \right)^{n+1} \sum_{m=0}^{\infty} \left(\frac{1}{2} \right)^m = \left(\frac{1}{2} \right)^{n+1} \frac{1}{1 - (1/2)} = \frac{1}{2^n},$$

Copyright © 2012 Pearson Education, Inc. Publishing as Addison-Wesley.

we find that

$$\lim_{n\to\infty} \|y_n - y\| = \lim_{n\to\infty} \frac{1}{2^n} = 0.$$

Thus, the given sequence of functions converges uniformly to $y(x) = 1/(1-x)$ on the interval $[0, 1/2]$.

7. Let $x \in [0, 1]$ be fixed.

If $x = 0$, then $y_n(0) = n^2(0) = 0$ for any n, and so $\lim_{n\to\infty} y_n(0) = \lim_{n\to\infty} 0 = 0$.

For $x > 0$, we let $N_x := [2/x] + 1$ with $[\cdot]$ denoting the integer part of a number. Then, for $n \geq N_x$, one has

$$n \geq \left[\frac{2}{x}\right] + 1 > \frac{2}{x} \qquad \Rightarrow \qquad x > \frac{2}{n}.$$

Hence, in evaluating $y_n(x)$, the third line in its definition must be used. This yields $y_n(x) = 0$ for all $n \geq N_x$, which implies that $\lim_{n\to\infty} y_n(x) = \lim_{n\to\infty} 0 = 0$.

Therefore, for any fixed $x \in [0, 1]$,

$$\lim_{n\to\infty} y_n(x) = 0 = y(x).$$

On the other hand, for any n, the function $y_n(x)$ is a continuous piecewise linear function, which is increasing on $[0, 1/n]$, decreasing on $(1/n, 2/n)$, and zero on $[1/n, 1]$. Thus, it attains its maximum value at $x = 1/n$, which is

$$y_n\left(\frac{1}{n}\right) = n^2\left(\frac{1}{n}\right) = n.$$

Therefore,

$$\lim_{n\to\infty} \|y_n - y\| = \lim_{n\to\infty} \|y_n\| = \lim_{n\to\infty} n = \infty,$$

and the sequence does not converge uniformly on $[0, 1]$. See Fig. **13–C** on page 689 for the graphs of $y_5(x)$, $y_{10}(x)$, $y_{15}(x)$, and $y_{20}(x)$.

9. We have to find $h > 0$ such that $h < \min\left(h_1, \alpha_1/M, 1/L\right)$. Since

$$R_1 = \{(x, y) : |x - 1| \leq 1, \ |y| \leq 1\} = \{(x, y) : 0 \leq x \leq 2, -1 \leq y \leq 1\},$$

we have $h_1 = 1$ and $\alpha_1 = 1$.

For M, we require that it satisfies the condition

$$|f(x, y)| = |y^2 - x| \leq M$$

 Copyright © 2012 Pearson Education, Inc. Publishing as Addison-Wesley.

for all (x, y) in R_1. To find such an upper bound for $|f(x, y)|$, we need the maximum and the minimum values of $f(x, y)$ in R_1. (Since $f(x, y)$ is a continuous function in the closed and bounded region R_1, it has absolute maximum and minimum in this region.) We involve calculus methods to find these extremal values. The first partial derivatives of $f(x, y)$ are given by

$$f_x(x, y) = -1 \qquad \text{and} \qquad f_y(x, y) = 2y\,.$$

Since $f_x(x, y)$ is never zero, the maximum and minimum must occur on the boundary of R_1. Notice that R_1 is bounded on the left by the line $x = 0$, on the right by the line $x = 2$, on the top by the line $y = 1$, and on the bottom by the line $y = -1$. Therefore, we examine the behavior of $f(x, y)$ on each edge of the rectangle R_1.

<u>Case 1:</u> If $x = 0$, then the function $f(x, y)$ becomes a function in the single variable y,

$$f(0, y) = F_1(y) = y^2 - 0 = y^2, \qquad y \in [-1, 1].$$

This function attains its maximum at $y = \pm 1$ and minimum at $y = 0$. Thus, we see that, on the left side of R_1, f has the maximum value of $f(0, \pm 1) = 1$ and the minimum value of $f(0, 0) = 0$.

<u>Case 2:</u> Let $x = 2$. The function $f(x, y)$ becomes a function in y only,

$$f(2, y) = F_2(y) = y^2 - 2, \qquad y \in [-1, 1].$$

This function also attains its maximum at $y = \pm 1$ and a minimum at $y = 0$. Thus, on the right side of R_1, the function $f(x, y)$ has the maximum value of $f(2, \pm 1) = -1$ and the minimum value of $f(2, 0) = -2$.

<u>Case 3:</u> On the top and bottom sides of R_1, where $y = \pm 1$, respectively, the function $f(x, y)$ becomes a function in x given by

$$f(x, \pm 1) = F_3(x) = (\pm 1)^2 - x = 1 - x, \qquad x \in [0, 2].$$

This function has the maximum at $x = 0$ and the minimum at $x = 2$. Thus, on the top and bottom of the region R_1 the function $f(x, y)$ has the maximum value of $f(0, \pm 1) = 1$ and the minimum value of $f(2, \pm 1) = -1$.

Copyright © 2012 Pearson Education, Inc. Publishing as Addison-Wesley.

Summarizing, we see that, on the boundary of R_1, the maximum value of $f(x, y)$ is 1 and the minimum value is -2. Thus, $|f(x, y)| \le 2$ for all (x, y) in R_1, and so we can choose $M = 2$.

To find L, we observe that L is an upper bound for

$$\left| \frac{\partial f}{\partial y} \right| = |2y| = 2|y|, \qquad (x, y) \in R_1 .$$

Since $|y| \le 1$, we see that

$$\left| \frac{\partial f}{\partial y} \right| \le 2, \qquad (x, y) \in R_1 .$$

Therefore, we can take $L = 2$.

Theorem 3 guarantees that the given initial value problem has a unique solution on the interval $[1 - h, 1 + h]$ for any $h > 0$ satisfying

$$h < \min \left(h_1, \frac{\alpha_1}{M}, \frac{1}{L} \right) = \min \left(1, \frac{1}{2}, \frac{1}{2} \right) = \frac{1}{2} .$$

11. We are given that the recurrence relation for these approximations is $y_{n+1} = T[y_n]$. Using the definition of $T[y]$ yields

$$y_{n+1} = x^3 - x + 1 \int_0^x (u - x)y_n(u)\, du.$$

Thus, starting approximations with

$$y_0(x) = x^3 - x + 1 ,$$

we obtain

$$
\begin{aligned}
y_1(x) &= x^3 - x + 1 + \int_0^x (u - x)y_0(u)\, du = x^3 - x + 1 + \int_0^x (u - x)\left(u^3 - u + 1\right) du \\
&= x^3 - x + 1 + \int_0^x \left(u^4 - u^2 + u - xu^3 + xu - x\right) du \\
&= x^3 - x + 1 + \left(\frac{x^5}{5} - \frac{x^3}{3} + \frac{x^2}{2} - \frac{x^5}{4} + \frac{x^3}{2} - x^2\right) .
\end{aligned}
$$

Simplifying gives

$$y_1(x) = -\frac{1}{20}x^5 + \frac{7}{6}x^3 - \frac{1}{2}x^2 - x + 1.$$

 Copyright © 2012 Pearson Education, Inc. Publishing as Addison-Wesley.

Substituting this result into the recurrence relation yields

$$y_2(x) = x^3 - x + 1 \int_0^x (u-x) y_1(u) \, du$$

$$= x^3 - x + 1 \int_0^x (u-x) \left(-\frac{1}{20} u^5 + \frac{7}{6} u^3 - \frac{1}{2} u^2 - u + 1 \right) du$$

$$= x^3 - x + 1 + \left(-\frac{1}{140} x^7 + \frac{7}{30} x^5 - \frac{1}{8} x^4 - \frac{1}{3} x^3 + \frac{1}{2} x^2 \right.$$

$$\left. + \frac{1}{120} x^7 - \frac{7}{24} x^5 + \frac{1}{6} x^4 + \frac{1}{2} x^3 - x^2 \right),$$

which reduces to

$$y_2(x) = \frac{1}{840} x^7 - \frac{7}{120} x^5 + \frac{1}{24} x^4 + \frac{7}{6} x^3 - \frac{1}{2} x^2 - x + 1.$$

13. Using properties of limits and the linearity of integrals, we can rewrite the statement

$$\lim_{n \to \infty} \int_a^b y_n(x) \, dx = \int_a^b y(x) \, dx$$

in an equivalent form, that is,

$$\lim_{n \to \infty} \left[\int_a^b y_n(x) - \int_a^b y(x) \, dx \right] = 0 \qquad \Leftrightarrow \qquad \lim_{n \to \infty} \int_a^b [y_n(x) - y(x)] \, dx = 0. \qquad (13.3)$$

The sequence $\{y_n\}$ converges uniformly to y on $[a, b]$, which means, by the definition of uniform convergence, that

$$\|y_n - y\| := \max_{x \in [a,b]} |y_n(x) - y(x)| \to 0 \quad \text{as} \quad n \to \infty.$$

Since

$$\left| \int_a^b [y_n(x) - y(x)] \, dx \right| \leq \int_a^b |y_n(x) - y(x)| \, dx \leq (b-a) \|y_n - y\|,$$

we conclude that

$$0 \leq \lim_{n \to \infty} \left| \int_a^b [y_n(x) - y(x)] \, dx \right| \leq (b-a) \lim_{n \to \infty} \|y_n - y\| = 0,$$

and (13.3) follows. (Recall that $\lim_{n \to \infty} a_n = 0$ if and only if $\lim_{n \to \infty} |a_n| = 0$.)

Copyright © 2012 Pearson Education, Inc. Publishing as Addison-Wesley.

Chapter 13

15. (a) In the given initial value problem,

$$
\begin{aligned}
x'(t) &= -y^2(t), & x(0) &= 0; \\
y'(t) &= z(t), & y(0) &= 1; \\
z'(t) &= x(t)y(t), & z(0) &= 0.
\end{aligned}
\tag{13.4}
$$

Replacing t by s, integrating the differential equations from $s = 0$ to $s = t$, and using the fundamental theorem of calculus, we obtain

$$
\begin{aligned}
\int_0^t x'(s)\,ds &= -\int_0^t y^2(s)\,ds \\
\int_0^t y'(s)\,ds &= \int_0^t z(s)\,ds \\
\int_0^t z'(s)\,ds &= \int_0^t x(s)y(s)\,ds
\end{aligned}
\qquad\Rightarrow\qquad
\begin{aligned}
x(t) - x(0) &= -\int_0^t y^2(s)\,ds, \\
y(t) - y(0) &= \int_0^t z(s)\,ds, \\
z(t) - z(0) &= \int_0^t x(s)y(s)\,ds.
\end{aligned}
$$

Substituting the initial conditions from (13.4), i.e., $x(0) = 0$, $y(0) = z(0) = 1$, into this system yields

$$
\begin{aligned}
x(t) &= -\int_0^t y^2(s)\,ds, \\
y(t) - 1 &= \int_0^t z(s)\,ds, \\
z(t) - 1 &= \int_0^t x(s)y(s)\,ds,
\end{aligned}
\tag{13.5}
$$

which is equivalent to the given system of integral equations.

Conversely, differentiating the equations in (13.5) and using the fundamental theorem of calculus, we conclude that solutions $x(t)$, $y(t)$, and $z(t)$ to (13.5) also satisfy differential equations in (13.4). Clearly,

$$
\begin{aligned}
x(0) &= -\int_0^0 y^2(s)\,ds = 0, \\
y(0) - 1 &= \int_0^0 z(s)\,ds = 0, \\
z(t) - 1 &= \int_0^0 x(s)y(s)\,ds = 0,
\end{aligned}
$$

and so the initial conditions in (13.4) are satisfied. Therefore, (13.5) implies (13.4).

(b) With initial values $x_0(t) \equiv x(0) = 0$, $y_0(t) \equiv y(0) = 1$, and $z_0(t) \equiv z(0) = 1$, we

  Copyright © 2012 Pearson Education, Inc. Publishing as Addison-Wesley.

compute the first Picard iterations $x_1(t)$, $y_1(t)$, and $z_1(t)$.

$$x_1(t) = -\int_0^t y_0^2(s)\,ds = -\int_0^t (1)^2\,ds = -t,$$

$$y_1(t) = 1 + \int_0^t z_0(s)\,ds = 1 + \int_0^t (1)\,ds = 1 + t,$$

$$z_1(t) = 1 + \int_0^t x_0(s)y_0(s)\,ds = 1 + \int_0^t (0)\,ds = 1.$$

Applying given recurrence formulas again yields

$$x_2(t) = -\int_0^t y_1^2(s)\,ds = -\int_0^t (1+s)^2\,ds = -(1+s)^3/3 \,\big|_0^t = -t - t^2 - \frac{t^3}{3}\,,$$

$$y_2(t) = 1 + \int_0^t z_1(s)\,ds = 1 + \int_0^t 1\,ds = 1 + t\,,$$

$$z_2(t) = 1 + \int_0^t x_1(s)y_1(s)\,ds$$

$$= 1 - \int_0^t s(1+s)\,ds = 1 - (s^2/2 + s^3/3) \,\big|_0^t = 1 - \frac{t^2}{2} - \frac{t^3}{3}\,.$$

EXERCISES 13.3:   Existence of Solutions of Linear Equations

1. In this problem,

$$\mathbf{A}(t) = \begin{bmatrix} \cos t & \sqrt{t} \\ t^3 & -1 \end{bmatrix}, \qquad \mathbf{f}(t) = \begin{bmatrix} \tan t \\ e^t \end{bmatrix}.$$

In $\mathbf{A}(t)$, functions $\cos t$, t^3, and -1 are continuous on $(-\infty, \infty)$ while $\sqrt{t}$ is continuous on $[0, \infty)$. Therefore, $\mathbf{A}(t)$ is continuous on $[0, \infty)$. In $\mathbf{f}(t)$, the exponential function is continuous everywhere, but $\tan t$ has infinite discontinuities at $t = (k + 1/2)\pi$, where $k = 0, \pm 1, \pm 2, \ldots$. The largest interval, containing the initial point $t_0 = 2$, where $\tan t$ and, therefore, $\mathbf{f}(t)$, is continuous is $(\pi/2, 3\pi/2)$. Since $\mathbf{A}(t)$ is also continuous on $(\pi/2, 3\pi/2)$, by Theorem 6, the given initial value problem has a unique solution on this interval.

3. By comparing this problem with the problem, given in Theorem 7 of the text, we see that

$$p_1(t) = -\ln t, \qquad p_2(t) \equiv 0, \qquad p_3(t) = \tan t, \quad \text{and} \quad g(t) = e^{2t}.$$

We also observe that $t_0 = 1$. Thus, we have to find an interval, containing $t_0 = 1$, on which the functions $p_1(t)$, $p_2(t)$, $p_3(t)$, and $g(t)$ are simultaneously continuous. Since $p_2(t)$ and $g(t)$ are continuous everywhere, $p_1(t)$ is continuous on the interval $(0, \infty)$, and

Copyright © 2012 Pearson Education, Inc. Publishing as Addison-Wesley.

the largest interval, containing t_0, on which $p_3(t)$ is continuous, is $(-\pi/2, \pi/2)$, these four functions are simultaneously continuous on $(0, \pi/2)$. Therefore, Theorem 7 guarantees that we have a unique solution to this initial value problem on the whole interval $(0, \pi/2)$.

5. In this problem, we use Theorem 5. Since

$$\mathbf{f}(t, \mathbf{x}) = \begin{bmatrix} \sin x_2 \\ 3x_1 \end{bmatrix},$$

we have

$$\frac{\partial \mathbf{f}}{\partial x_1}(t, \mathbf{x}) = \begin{bmatrix} 0 \\ 3 \end{bmatrix}, \qquad \frac{\partial \mathbf{f}}{\partial x_2}(t, \mathbf{x}) = \begin{bmatrix} \cos x_2 \\ 0 \end{bmatrix}.$$

Vectors $\mathbf{f}$, $\partial \mathbf{f}/\partial x_1$, and $\partial \mathbf{f}/\partial x_2$ are continuous on

$$R = \{-\infty < t < \infty, -\infty < x_1 < \infty, -\infty < x_2 < \infty\}$$

(which is the whole space $\mathbb{R}^3$) since their components are. Moreover,

$$\left|\frac{\partial \mathbf{f}}{\partial x_1}(t, \mathbf{x})\right| = 3, \qquad \left|\frac{\partial \mathbf{f}}{\partial x_2}(t, \mathbf{x})\right| = |\cos x_2| \leq 1$$

for any $(t, \mathbf{x})$, and so the condition (3) of Theorem 5 is satisfied with $L = 3$. Hence, the given initial value problem has a unique solution on the whole real axis $-\infty < t < \infty$.

7. The equation

$$y'''(t) - (\sin t)y'(t) + e^{-t}y(t) = 0$$

is a linear homogeneous equation and, hence, has a trivial solution $y(t) \equiv 0$. Clearly, this solution satisfies the initial conditions, $y(0) = y'(0) = y''(0) = 0$. All that remains is to note that the coefficients, $-\sin t$ and e^{-t}, are continuous on $(-\infty, \infty)$. By Theorem 7, the solution $y \equiv 0$ is unique.

EXERCISES 13.4: Continuous Dependence of Solutions

3. To apply Theorem 9, we first determine the constant L for $f(x, y) = e^{\cos y} + x^2$. We observe that

$$\frac{\partial f(x, y)}{\partial y} = -e^{\cos y} \sin y.$$

Now, on any rectangle R_0, we have

$$\left|\frac{\partial f(x, y)}{\partial y}\right| = |-e^{\cos y} \sin y| = e^{\cos y} |\sin y| \leq e.$$

 Copyright © 2012 Pearson Education, Inc. Publishing as Addison-Wesley.

(More detailed analysis shows that this function attains its maximum value $1.4585\ldots$ at $y = (\sqrt{5} - 1)/2$.) Thus, since $h = 1$, by Theorem 9 we have

$$|\phi(x, y_0) - \phi(x, \widetilde{y}_0)| \le |y_0 - \widetilde{y}_0| \, e^e \, . \tag{13.6}$$

Since we are given that $|y_0 - \widetilde{y}_0| \le 10^{-2}$, (13.6) yields

$$|\phi(x, y_0) - \phi(x, \widetilde{y}_0)| \le 10^{-2} e^e \approx 0.151543 \, .$$

9. We can use the inequality (18) of Theorem 10 to obtain the bound, but first we must determine the constants L and ε. Here, $f(x, y) = \sin x + (1 + y^2)^{-1}$, $F(x, y) = x + 1 - y^2$. We compute

$$\left| \frac{\partial f(x, y)}{\partial y} \right| = \left| \frac{2y}{(1 + y^2)^2} \right| = \frac{2|y|}{(1 + y^2)^2}$$

and

$$\left| \frac{\partial F(x, y)}{\partial y} \right| = |-2y| \le 2 \quad \text{since} \quad -1 \le y \le 1.$$

To find an upper bound for $|\partial f / \partial y|$ on R_0, we maximize $y/(1 + y^2)^2$ on $[0, 1]$. To this end, we compute

$$\left[\frac{y}{(1 + y^2)^2} \right]' = \frac{(1 + y^2)^2 - y \cdot 2(1 + y^2)2y}{(1 + y^2)^4} = \frac{(1 + y^2) - 4y^2}{(1 + y^2)^3} = \frac{1 - 3y^2}{(1 + y^2)^3} \tag{13.7}$$

and solve

$$\frac{1 - 3y^2}{(1 + y^2)^3} = 0 \quad \Rightarrow \quad 1 - 3y^2 = 0 \quad \Rightarrow \quad y = \frac{1}{\sqrt{3}} \quad (y \ge 0) \, .$$

Since at $y = 1/\sqrt{3}$ the derivative (13.7) changes its sign from plus to minus, the function $y/(1 + y^2)^2$ has the absolute maximum at this point on $[0, 1]$. Thus,

$$\left| \frac{\partial f(x, y)}{\partial y} \right| \le \frac{2(1/\sqrt{3})}{\left[1 + (1/\sqrt{3})^2 \right]^2} = \frac{3\sqrt{3}}{8} \, ,$$

and so we can take $L = 3\sqrt{3}/8$. To obtain ε, we need an upper bound for

$$|f(x, y) - F(x, y)| = \left| \sin x + \frac{1}{1 + y^2} - x - 1 + y^2 \right|$$

$$\le |\sin x - x| + \left| \frac{1}{1 + y^2} - 1 + y^2 \right| \, .$$

Using Taylor's theorem with remainder, we have

$$\sin x = x - \frac{x^3 \cos \xi}{3!} \, ,$$

Copyright © 2012 Pearson Education, Inc. Publishing as Addison-Wesley.

where ξ is between 0 and x. Thus, for $-1 \le x \le 1$, we obtain

$$|\sin x - x| = \left| x - \frac{x^3 \cos \xi}{3!} - x \right| = \frac{|x|^3 \cos \xi}{3!} \le \frac{1}{6}.$$

Denoting $g(y) := (1 + y^2)^{-1} - 1 + y^2$, we compute

$$g'(y) = -2y(1 + y^2)^{-2} + 2y,$$
$$g''(y) = -2(1 + y^2)^{-2} + 2(1 + y^2)^{-3}(2y)^2 + 2,$$
$$g'''(y) = 4(1 + y^2)^{-3}(2y) - 6(1 + y^2)^{-4}(2y)^3 + 2(1 + y^2)^{-3}(8y)$$

and apply Taylor's theorem with remainder to $g(y)$. Since $g(0) = g'(0) = g''(0) = 0$, we have

$$(1 + y^2)^{-1} - 1 + y^2 = \frac{g'''(\xi)}{3!},$$

where ξ is between 0 and y. Hence,

$$
\begin{aligned}
\left| (1 + y^2)^{-1} - 1 + y^2 \right| &= \left| \frac{g'''(\xi)}{3!} \right| \\
&\le \frac{1}{6} \left[8(1 + y^2)^{-3}|y| + 48(1 + y^2)^{-4}|y|^3 + 16(1 + y^2)^{-3}|y| \right] \\
&\le \frac{1}{6}(8 + 48 + 16) = 12.
\end{aligned}
$$

Thus,

$$|f(x, y) - F(x, y)| \le \frac{1}{6} + 12 = \frac{73}{6}.$$

It now follows from the inequality (18) of Theorem 10 that

$$|\phi(x) - \psi(x)| \le \frac{73}{6} e^{3\sqrt{3}/8} \approx 23.294541, \qquad x \in [-1, 1].$$

 Copyright © 2012 Pearson Education, Inc. Publishing as Addison-Wesley.

TABLES

Table 13–A: Approximations for a solution of $x = \dfrac{1}{2}\left(x + \dfrac{3}{x}\right)$.

$x_0 = 3.00$ $x_3 \approx 1.73214286$
$x_1 = 2.00$ $x_4 \approx 1.73205081$
$x_2 = 1.75$ $x_5 \approx 1.73205081$

Table 13–B: Approximations for the solution of $x = \dfrac{1}{x^2 + 4}$.

$x_0 = 0.5$ $x_4 \approx 0.24626646$
$x_1 \approx 0.23529412$ $x_5 \approx 0.24626616$
$x_2 \approx 0.24658703$ $x_6 \approx 0.24626617$
$x_3 \approx 0.24625658$ $x_7 \approx 0.24626617$

Table 13–C: Approximations for the solution of $x = \left(\dfrac{5 - x}{3}\right)^{1/4}$.

$x_0 = 1.0$ $x_4 \approx 1.06984638$
$x_1 \approx 1.07456993$ $x_5 \approx 1.06984797$
$x_2 \approx 1.06952637$ $x_6 \approx 1.06984786$
$x_3 \approx 1.06986975$ $x_7 \approx 1.06984787$

Table 13–D: Approximations for the solution of $x = \dfrac{x^2 + 1}{2}$ starting at $x_0 = 2$.

$x_1 = 2.5$ $x_4 \approx 25.4946594$
$x_2 = 3.625$ $x_5 \approx 325.488829$
$x_3 = 7.0703125$ $x_6 \approx 52971.9891$

Copyright © 2012 Pearson Education, Inc. Publishing as Addison-Wesley.

Table 13–E: Approximations for the solution of $x = \dfrac{x^2+1}{2}$ starting at $x_0 = 0$.

$x_1 = 0.5$	$x_{15} \approx 0.8985984$
$x_2 = 0.625$	$x_{20} \approx 0.9198875$
$x_3 = 0.6953125$	$x_{30} \approx 0.9433716$
$x_4 \approx 0.7417297$	$x_{40} \approx 0.9561175$
$x_5 \approx 0.7750815$	$x_{50} \approx 0.9641451$
$x_{10} \approx 0.8610982$	$x_{99} \approx 0.9810285$

FIGURES

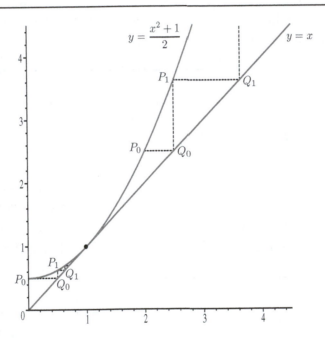

Figure 13–A: The method of successive substitution for the equation $x = \dfrac{x^2+1}{2}$.

 Copyright © 2012 Pearson Education, Inc. Publishing as Addison-Wesley.

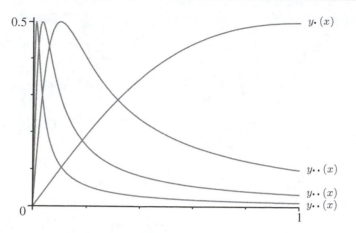

Figure 13–B: Graphs of functions in Problem 3.

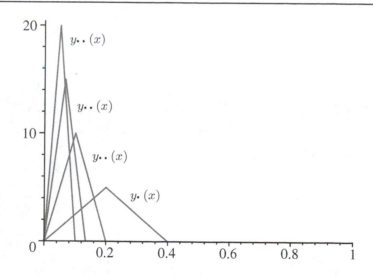

Figure 13–C: Graphs of functions in Problem 7.

Copyright © 2012 Pearson Education, Inc. Publishing as Addison-Wesley.

Appendix A: Review of Integration Techniques

1. Substitution $u = 1 + x^2$, $du = 2x\,dx$ yields

$$\int x^3 \left(1 + x^2\right)^{3/2} dx = \int x^2 (1 + x^2)^{3/2} x\,dx = \frac{1}{2} \int (u - 1) u^{3/2} du$$

$$= \frac{1}{2} \int \left(u^{5/2} - u^{3/2}\right) du = \frac{1}{7} \left(1 + x^2\right)^{7/2} - \frac{1}{5} \left(1 + x^2\right)^{5/2} + C$$

$$= (1 + x^2)^{5/2} \left[\frac{1}{7}(1 + x^2) - \frac{1}{5}\right] + C = (1 + x^2)^{5/2} \frac{(5x^2 - 2)}{35} + C.$$

3. We substitute $u = 6t^5$, $4t^4 dt = (2/15)du$ and obtain

$$\int 4t^4 e^{6t^5} dt = \frac{2}{15} \int e^u du = \frac{2}{15} e^{6t^5} + C.$$

5. Using double angle formula for cosine function, we get

$$\int \cos^2(\pi\theta) d\theta = \int \frac{1 + \cos(2\pi\theta)}{2} d\theta = \frac{1}{2} \left[\theta + \frac{1}{2\pi} \sin(2\pi\theta)\right] + C.$$

7. We split the integral and use substitution $u = t^2 + 4$ in the first one.

$$\int \frac{t}{t^2 + 4} dt + \int \frac{1}{t^2 + 4} dt = \frac{1}{2} \int \frac{du}{u} + \frac{1}{2} \arctan\left(\frac{t}{2}\right) = \frac{1}{2} \left[\ln\left(t^2 + 4\right) + \arctan\left(\frac{t}{2}\right)\right] + C.$$

9. Substituting $u = \sqrt{2}x$, we obtain

$$\int \frac{dx}{\sqrt{1 - 2x^2}} = \frac{1}{\sqrt{2}} \int \frac{du}{\sqrt{1 - u^2}} = \frac{1}{\sqrt{2}} \arcsin u + C = \frac{1}{\sqrt{2}} \arcsin\left(\sqrt{2}x\right) + C.$$

11. We use partial fraction decomposition.

$$\frac{4}{(x - 1)(x - 2)(x - 3)} = \frac{A}{x - 1} + \frac{B}{x - 2} + \frac{C}{x - 3}.$$

Multiplying by $x - 1$ and then substituting $x = 1$ yields

$$\frac{4}{(x - 2)(x - 3)} = A + (x - 1)\left(\frac{B}{x - 2} + \frac{C}{x - 3}\right) \quad \Rightarrow \quad A = 2.$$

Similarly, we find that $B = -4$, $C = 2$. Therefore, the given integral equals

$$2 \int \frac{dx}{x - 1} - 4 \int \frac{dx}{x - 2} + 2 \int \frac{dx}{x - 3} = 2 \ln|x - 1| - 4 \ln|x - 2| + 2 \ln|x - 3| + C.$$

 Copyright © 2012 Pearson Education, Inc. Publishing as Addison-Wesley.

13. We use trigonometric substitution $3x = \sec t$ and the identity $\sec^2 t - 1 = \tan^2 t$.

$$\int \frac{\sqrt{9x^2 - 1}}{x} \, dx = \int \frac{\tan t}{\sec t} \sec t \tan t \, dt = \int \tan^2 t \, dt = \int \left(\sec^2 t - 1 \right) dt = \tan t - t + C.$$

All that remains is to make back substitution to conclude that the given integral equals

$$\sqrt{9x^2 - 1} - \arccos \left(\frac{1}{3x} \right) + C.$$

15. We apply partial fractions.

$$\frac{x^3 + 2x^2 + 8}{x^2 \left(x^2 + 4 \right)} = \frac{A}{x} + \frac{B}{x^2} + \frac{Cx + D}{x^2 + 4}.$$

Multiplying by the common denominator, we get

$$(Ax + B) \left(x^2 + 4 \right) + x^2 (Cx + D) = x^3 + 2x^2 + 8.$$

For $x = 0$, $4B = 8$ so that $B = 2$. Comparing coefficients or x^3, x^2, and x, we obtain three equations for A, C, and D: $A + C = 1$, $D + 2 = 2$, and $4A = 0$. So, $A = D = 0$, $C = 1$, and

$$\int \frac{x^3 + 2x^2 + 8}{x^2 \left(x^2 + 4 \right)} \, dx = \int \left(\frac{2}{x^2} + \frac{x}{x^2 + 4} \right) dx = -\frac{2}{x} + \frac{1}{2} \ln \left(x^2 + 4 \right) + C.$$

17. With $u = \ln x$ we have $du = dx/x$ and so

$$\int \frac{dx}{x \ln x} = \int \frac{du}{u} = \ln |u| + C = \ln |\ln x| + C.$$

19. We use integration by parts.

$$\int y \sinh y \, dy = \int y d(\cosh y) = y \cosh y - \int \cosh y \, dy = y \cosh y - \sinh y + C.$$

21. We use trigonometric substitution $x = 2 \tan t$, $dx = 2 \sec^2 t \, dt$, and the trigonometric identity $1 + \tan^2 t = \sec^2 t$.

$$\int \frac{dx}{\left(x^2 + 4 \right)^{3/2}} = \int \frac{2 \sec^2 t \, dt}{\left(4 \sec^2 t \right)^{3/2}} = \frac{1}{4} \int \frac{dt}{\sec t} = \frac{1}{4} \int \cos t \, dt = \frac{\sin t}{4} + C = \frac{x}{4\sqrt{x^2 + 4}} + C.$$

23. We apply integration by part with $u = \ln(t + 3)$ and $dv = t \, dt$.

$$\int t \ln(t + 3) dt = \frac{t^2}{2} \ln(t + 3) - \frac{1}{2} \int \frac{t^2}{t + 3} \, dt.$$

Copyright © 2012 Pearson Education, Inc. Publishing as Addison-Wesley.

Appendix A

For the integral term on the right-hand side we use long division and get

$$\int \frac{t^2}{t+3} \, dt = \int \left(t - 3 + \frac{9}{t+3} \right) dt = \frac{t^2}{2} - 3t + 9 \ln(t+3) + C.$$

Thus, the answer is

$$\frac{1}{2} \left(t^2 - 9 \right) \ln(t+3) - \frac{t^2}{4} + \frac{3t}{2} + C.$$

25. We substitute $u = \cos t$ to get

$$\int \frac{\sin t \, dt}{1 + \cos^2 t} = -\int \frac{du}{1 + u^2} = -\arctan(\cos t) + C.$$

27. We use $u = \sin x$ and the identity $\sin^2 x + \cos^2 x = 1$ to obtain

$$\int \sin^{2/5} x \cos^3 x \, dx = \int u^{2/5} \left(1 - u^2 \right) du$$

$$= \frac{5u^{7/5}}{7} - \frac{5u^{17/5}}{17} + C = \frac{5 \sin^{7/5} x}{7} - \frac{5 \sin^{17/5} x}{17} + C.$$

29. Product-to-sum identity for cosine function yields

$$\int \cos(3x) \cos(7x) dx = \frac{1}{2} \int \left[\cos(10x) + \cos(4x) \right] dx = \frac{\sin(10x)}{20} + \frac{\sin(4x)}{8} + C.$$

31. We make the substitution $u = \tan \theta$, $du = \sec^2 \theta \, d\theta$, and use $\sec^4 \theta = \left(1 + u^2 \right)^2$ to get

$$\int \tan^4 \theta \sec^6 \theta \, d\theta = \int u^4 \left(1 + u^2 \right)^2 du$$

$$= \int \left(u^4 + 2u^6 + u^8 \right) du = \frac{\tan^5 \theta}{5} + \frac{2 \tan^7 \theta}{7} + \frac{\tan^9 \theta}{9} + C.$$

33. We use a double angle formula to find that the given integral equals

$$\frac{1}{4} \int \sin^2(6x) dx = \frac{1}{8} \int [1 - \cos(12x)] dx = \frac{1}{8} \left[x - \frac{\sin(12x)}{12} \right] + C.$$

 Copyright © 2012 Pearson Education, Inc. Publishing as Addison-Wesley.